Handbook of Humidity Measurement

Methods, Materials and Technologies

Vol. 1: Spectroscopic Methods of Humidity Measurement

Handbook of Humidity Measurement: Methods, Materials and Technologies, 3 Volume Set
(ISBN: 978-1-138-29787-6)

Handbook of Humidity Measurement,
Volume 1: Spectroscopic Methods of Humidity Measurement
(ISBN: 978-1-138-30021-7)

Handbook of Humidity Measurement,
Volume 2: Electronic and Electrical Humidity Sensors
(ISBN: 978-1-138-30022-4)

Handbook of Humidity Measurement,
Volume 3: Sensing Materials and Technologies
(ISBN: 978-1-138-48287-6)

Handbook of Humidity Measurement

Methods, Materials and Technologies

Vol. 1: Spectroscopic Methods of Humidity Measurement

Ghenadii Korotcenkov

CRC Press is an imprint of the
Taylor & Francis Group, an **informa** business

CRC Press
Taylor & Francis Group
6000 Broken Sound Parkway NW, Suite 300
Boca Raton, FL 33487-2742

First issued in paperback 2020

ISBN-13: 978-0-367-57188-7 (pbk)
ISBN-13: 978-1-138-30021-7 (hbk)

Visit the Taylor & Francis Web site at
http://www.taylorandfrancis.com

and the CRC Press Web site at
http://www.crcpress.com

Contents

SECTION II Humidity Monitoring Using Absorption of Electromagnetic Radiation

Preface

On account of unique water properties, humidity greatly affects living organisms, including humans and materials. The amount of water vapor in the air can affect the human comfort, and the efficiency and the safety of many manufacturing processes, including drying of products such as paint, paper, matches, fur, and leather; packaging and storage of different products such as tea, cereal, milk, and bakery items; and manufacturing of food products such as plywood, gum, abrasives, pharmaceutical powder, ceramics, printing materials, and tablets. Moreover, industries discussed above are only a small part of the industries where the humidity should be controlled. In agriculture, the measurement of humidity is important for the plantation protection (dew prevention), the soil moisture monitoring, and so on. In the medical field, a humidity control should be used in respiratory equipment, sterilizers, incubators, pharmaceutical processing, and biological products. Humidity measurements at the Earth's surface are also required for meteorological analysis and forecasting, for climate studies, and for many special applications in hydrology, aeronautical services, and environmental studies, because water vapor is the key agent in both weather and climate.

This means that the determination of humidity is of great importance. Therefore, humidity control becomes imperative in all fields of our activity, from production management to creating a comfortable environment for our living, and for understanding the nature of the changes happening to the climate. Humidity can change in a wide range. This means that there are necessary devices and sensors, capable of carrying out the measurement of humidity in the entire range of possible changes in humidity. It is clear that these sensors and measurement systems must be able to work in a variety of climatic conditions, ensuring the functioning of the control and surveillance systems for a long time.

In the past decade, it was done a lot for the development of new methods for measuring humidity, improvements, and optimization of manufacturing technology of already developed humidity sensors, and for the development of different measuring systems with an increased efficiency. As a result, the field of humidity sensors has broadened and expanded greatly. At present, humidity sensors are being used in medicine, agriculture, industry, environmental control, and other fields. Humidity sensors are being widely used for the continuous monitoring of humidity in diverse applications such as the baking and drying of food, cigar storage, civil engineering to detect water ingress in soils or in the concrete in civil structures, medical applications, and many other fields. However, the process of developing new humidity sensors and improving older types of devices used for humidity measurement is still ongoing. New technologies and the toughening of ecological standards require more sensitive instruments with faster response times, better selectivity, and improved stability. It is therefore time to resume the developments carried out during this time and identify ways for further development in this area of research. This is important, as currently too many approaches and types of devices that can be used for measuring humidity are proposed. They use different measuring principles, different humidity-sensitive materials, and various configurations of devices, making it difficult to conduct an objective analysis of these devices' capabilities. We hope that data presented in this book with detailed information on various types of humidity sensors that are developed by different teams, accompanied by an analysis of their strengths and weaknesses, will allow to make the cross-comparison and the selection of suitable sensing method for specific applications. As a result, all the conditions will be created for the development of devices to ensure accurate, reliable, economically viable, and efficient humidity measurements.

This issue is organized as follows. Taking into account the current trends in the development of instruments for measuring humidity, this publication is divided into three parts: The first volume, *Spectroscopic Methods of Humidity Measurement* focuses on the review of devices based on optical principles of measurement such as optical UV and fluorescence hygrometers and optical and fiber-optic sensors of various types. As it was indicated before, atmospheric water plays a key role in the climate. Therefore, various methods for monitoring the atmosphere have been developed in recent years, on the basis of measuring the absorption of electromagnetic field in different spectral ranges. All these methods, covering the optical (FTIR and Lidar techniques), and a microwave and THz ranges are discussed in this volume, and their analysis of strengths and weaknesses is given. The role of humidity-sensitive materials in optical and fiber-optic sensors is also detailed. This volume also describes the reasons that cause us to control the humidity, features of water and water vapors, and units used for humidity measurement. I am sure that this information will certainly be cognitive and interesting for readers.

The second volume, *Electronic and Electrical Humidity Sensors*, as the title implies, is entirely devoted to the consideration of different types of solid-state devices, the operating principles of which are based on other physical principles. There will be given a detailed information, including advantages and disadvantages about the capacitive, resistive, gravimetric, hygrometric, field ionization, microwave, solid-state electrochemical, and thermal conductivity-based humidity sensors, followed by a relevant analysis of the properties of humidity-sensitive materials, used for the development of such devices. Humidity sensors based on thin film and field-effect transistors, heterojunctions, flexible substrates, and integrated humidity sensors are also discussed in this volume. Today is an age of automation and control. Therefore, in addition to interest in the properties of sensors, such as accuracy and long-term drift, there is also interest to durability in different environments, component size, digitization, simple and quick system integration, and last but not least, the price. This means that modern humidity sensors should be able to integrate all these demands into one sensor. The experiment showed that these capabilities can be fully realized in electric and electronic sensors manufactured using semiconductor solid-state technology. Such humidity sensors can be fabricated with low cost and are used for moisture control more conveniently. Great attention in this volume is also paid to the consideration of conventional devices, which were used for the measurement of humidity for several centuries. It is important to note that many of these methods are widely used so far.

The title of the third volume, *Sensing Materials and Technologies* indicates that the main focus of this volume will be focused on considering the properties of various materials suitable for the development of humidity sensors. Polymers, metal oxides, porous semiconductors (Si, SiC), carbon-based materials (black carbon, carbon nanotubes, and graphene), zeolites, silica, and some others, are included in the list of these materials. Features of fabrication of humidity sensors and related materials used in their manufacturing are also considered. Market forces naturally lead to ever more specialized and innovative products; sensors should be smaller in size, cheaper, more robust, and accurate in measurement, they should have better sensitivity and stability. This challenge requires new technological solutions, some of which, such as the integration and miniaturization, are considered in this volume. Specificity of the humidity sensor calibration and the market of humidity sensors are also an object of analysis.

The content of these books shows that materials play a key role in the humidity sensor functioning, and that the range of materials that can be used in the development of humidity sensors is very broad. Each material has its advantages and disadvantages, and therefore the selection of optimal sensing material for humidity sensors is a complicated and multivariate task. However, the number of published books or reviews describing the analysis of all possible materials through their application in the field of humidity sensors is very limited. Therefore, it is very difficult to conduct a comparative analysis of various materials and to choose a humidity-sensing material optimal for particular applications. The content of the present book contributes to the solution of this problem. In these three volumes, the readers, including scientists, can find a comparative analysis of all materials acceptable for the humidity sensor design and can estimate their real advantages and shortcomings. Moreover, throughout these books, numerous strategies for the fabrication and characterization of humidity-sensitive materials and sensing structures, employed in sensor applications, are described. This means that one can consider the present books as a selective guide to materials for the humidity sensor manufacture.

Thus, these books provide an up-to-date account of the present status of humidity sensors, from understanding the concepts behind humidity sensors to the practical knowledge, necessary to develop and manufacture such sensors. In addition, these books contain a large number of tables with information necessary for the humidity sensor design. The tables alone make these books very helpful and comfortable for the user. Therefore, this issue can be utilized as a reference book for researchers and engineers as well as graduate students who are either entering the field of humidity measurement for the first time, or who are already conducting research in these areas but are willing to extend their knowledge in the field of humidity sensors. In this case, these books will act as an introduction to the world of humidity sensors, which may encourage further study, and estimate the role that humidity sensors may play in the future. I hope that these books will also be useful to university students, postdocs, and professors. The structure of these books offers the basis for courses in the field of material sciences, chemical sensors, sensor technologies, chemical engineering, semiconductor devices, electronics, medicine, and environmental monitoring. We believe that practicing engineers, measurement experts, laboratory technicians, and project managers in industries and national laboratories who are interested in looking for solutions to humidity measurement tasks in industrial and environmental applications, but do not know how to design or to select optimal devices for these applications, will also find useful information that help them to make the right choices concerning technical solutions and investments.

Ghenadii Korotcenkov

Acknowledgments

My sincere gratitude goes to CRC Press for the opportunity to write this book. I also like to give acknowledgment to Gwangju Institute of Science and Technology, Gwangju, South Korea and State University of Moldova, Chisinau, the Republic of Moldova for inviting me and supporting my research in various programs and projects. I would like to thank my wife and the love of my life, Irina Korotcenkova for always being there for me, inspiring and supporting all my endeavors. She gives me purpose, motivates me to continue my work and makes my life so much more exciting. Also I am grateful to my daughters Anya and Anastasia for being a part of my life and for encouraging my work. My special thanks go to my friends, colleagues, and my coauthors for their support and collaboration over the years. Great thanks to all of you, this would not be possible without you being by my side.

Ghenadii Korotcenkov

Author

Ghenadii Korotcenkov earned his PhD in material sciences from the Technical University of Moldova, Chisinau, Moldova in 1976 and his doctor of science degree in physics from the Academy of Science of Moldova in 1990 (Highest Qualification Committee of the USSR, Moscow). He has more than 40 years of experience as a scientific researcher. For a long time, he was the leader of the gas sensor group and manager of various national and international scientific and engineering projects carried out in the Laboratory of Micro- and Optoelectronics, Technical University of Moldova. His research had financial support from international foundations and programs such as the CRDF, the MRDA, the ICTP, the INTAS, the INCO-COPERNICUS, the COST, and the NATO. From 2007 to 2008, he was an invited scientist in the Korea Institute of Energy Research, Daejeon, South Korea. After which, until the end of 2017 G. Korotcenkov was a research professor at the School of Materials Science and Engineering at the Gwangju Institute of Science and Technology, Gwangju, South Korea. Currently G. Korotcenkov is the chief scientific researcher at the Department of Physics and Engineering at the State University of Moldova, Chisinau, Rep. of Moldova.

Specialists from the former Soviet Union know G. Korotcenkov's research results in the field of study of Schottky barriers, MOS structures, native oxides, and photoreceivers on the basis of III–Vs compounds such as InP, GaP, AlGaAs, and InGaAs. His present scientific interests starting from 1995 include material sciences, focusing on the metal oxide film deposition and characterization, surface science, and the design of thin film gas sensors and thermoelectric convertors. These studies were carried out in cooperation with scientific teams from Ioffe Institute (St. Petersburg, Russia), University of Michigan (Ann Arbor, USA), Kiev State University (Kiev, Ukraine), Charles University (Prague, Czech Republic), St. Petersburg State University (St. Petersburg, Russia), Illinois Institute of Technology (Chicago, USA), University of Barcelona (Barcelona, Spain), Moscow State University (Moscow, Russia), University of Brescia (Brescia, Italy), Belarus State University (Minsk, Belarus), South-Ukrainian University (Odessa, Ukraine).

G. Korotcenkov is the author or editor of 35 books, including the 11-volume *Chemical Sensors* series published by the Momentum Press (United States), 15-volume *Chemical Sensors* series published by Harbin Institute of Technology Press (China), 3-volume *Porous Silicon: From Formation to Application* published by CRC Press (United States), and 2-volume *Handbook of Gas Sensor Materials* published by Springer (United States). At present, G. Korotcenkov is a series' editor of *Metal Oxides* series, which is published by Elsevier.

G. Korotcenkov is author and coauthor of more than 550 scientific publications, including 20 review papers, 35 book chapters, and more than 250 articles published in peer-reviewed scientific journals (h-factor = 38 [Scopus] and h-factor = 44 [Google Scholar citation]). He is a holder of 17 patents. He has presented more than 200 reports at national and international conferences, including 15 invited talks. G. Korotcenkov was co-organizer of several international conferences. His name and activities have been listed by many biographical publications, including Who's Who. His research activities are honored by an Award of the Supreme Council of Science and Advanced Technology of the Republic of Moldova (2004); Prize of the Presidents of the Ukrainian, Belarus, and Moldovan Academies of Sciences (2003); and National Youth Prize of the Republic of Moldova in the field of science and technology (1980), among others. G. Korotcenkov also received a fellowship from the International Research Exchange Board (IREX, United States, 1998), Brain Korea 21 Program (2008–2012), and Brainpool Program (Korea, 2015–2017).

Section I

Introduction to Humidity Measurement

1 Water, Water Vapors, and Humidity

1.1 WATER AND WATER VAPORS

Water is a special substance (Cracolice and Edward 2006; Petrucci et al. 2017). Water makes the Earth unique. Water covers 71% of the Earth's surface. It is vital for all known forms of life. Life, the climate, and the weather all exist as they do because gaseous, liquid, and solid forms of water can coexist on the planet. Water is a transparent and nearly colorless chemical substance that is the main constituent of Earth's streams, lakes, and oceans, and the fluids of most living organisms. Its chemical formula is H_2O, meaning that its molecule contains one oxygen and two hydrogen atoms, that are connected by covalent bonds. Figure 1.1 shows the schematic of water molecule, whereas Table 1.1 shows some of its properties. As the water molecule is not linear and the oxygen atom has a higher electronegativity than hydrogen atoms, it is a polar molecule, with an electrical dipole moment: the oxygen atom carries a slight negative charge, whereas the hydrogen atoms are slightly positive. On account of its polarity, a molecule of water in the liquid or solid state can form up to four hydrogen bonds with neighboring molecules. This attraction explains many of the properties of water, such as solvent action (Campbell and Farrell 2007).

Although hydrogen bonding is a relatively weak attraction compared to the covalent bonds within the water molecule itself, it is responsible for a number of water's physical properties (Cracolice and Edward 2006; Petrucci et al. 2017). These properties include its relatively high melting and boiling point temperatures: more energy is required to break the hydrogen bonds between water molecules. In contrast, hydrogen sulfide (H_2S) has much weaker hydrogen bonding because of sulfur's lower electronegativity. H_2S is a gas at room temperature, in spite of H_2S having nearly twice the molar mass of water. Water is one of the few known substances whose solid form is less dense than the liquid. The extra bonding between water molecules also gives liquid water a large specific heat capacity. This high heat capacity makes water a good heat storage medium (coolant) and heat shield. On account of their polarity, water molecules are strongly attracted to one another, which gives water a high surface tension. The molecules at the surface of the water *stick together* to form a type of *skin* on the water, strong enough to support very light objects. Insects that walk on water are taking advantage of this surface tension. Surface tension causes water to clump in drops rather than spreading out in a thin layer. It also allows water to move through plant roots and stems and the smallest blood vessels in your body—as one molecule moves up the tree root or through the capillary, it *pulls* the others with it. Water also has an exceptionally high heat of vaporization. Water's heat of vaporization is 41 kJ/mol.

Pure water containing no exogenous ions is an excellent insulator, but not even *deionized* water is completely free of ions. Water undergoes autoionization in the liquid state, when two water molecules form one hydroxide anion (OH^-) and one hydronium cation (H_3O^+). As water is such a good solvent, it almost always has some solute dissolved in it, often a salt. If water has even a tiny amount of such an impurity, then it can conduct electricity far more readily.

It is important that the air around us always has some content of water vapor and this content can be varied in wide range (Wexler 1965). For example, atmospheric humidity ranges from nearly 0% to more than 4% of the mass of air, making it the third most common atmospheric constituent. Other gases presented in atmosphere are listed in Table 1.2. However, it is necessary to take into account that since air is seldom totally without water vapor, the above-mentioned percentages change depending on the amount of water vapor that is mixed with other gases.

Like other gases, water vapor can be considered to behave as an ideal gas, except its behavior near saturation. However, in average environmental conditions, water can also be presented in the liquid and solid phases. Therefore, usually we speak of water vapor instead of water gas.

Whether one likes it or not, water and water vapor can be found everywhere (Visscher 1999). On account of the asymmetrical distribution of their electric charge, water molecules are easily adsorbed on almost any surface, where they are present as a mono- or multimolecular

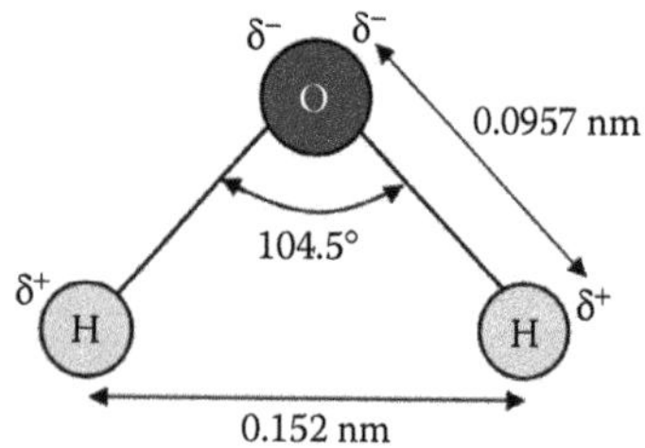

FIGURE 1.1 Schematic of the H_2O molecule. (Reprinted from *Sens. Actuators A*, 233, Sikarwar, S. and Yadav, B.C., Opto-electronic humidity sensor: A review, 54–70, Copyright 2015, with permission from Elsevier.)

TABLE 1.1
Some Properties of H_2O Molecule

Property	Value
H–O–H angle	104.5°
OH–bond length	0.957 Å
Molar weight	18.01528
Molecule size	~2.8 Å
OH–dissociation energy	498 kJ/mol (5.18 eV)
Dipole moment	6.17×10^{-30} C·m (1.85 D)

TABLE 1.2
Gases Present in Standard Amounts in a Dry Atmosphere

Constituent	Volume Ratio (%)	Parts per Million
Nitrogen, N_2	78.084	780,840
Oxygen, O_2	20.946	209,460
Argon, Ar	0.934	9340
Carbon dioxide, CO_2	0.0314	314.0
Neon, Ne	0.001818	18.18
Helium, He	0.000524	5.24
Methane, CH_4	0.00016	1.6
Krypton, Kr	0.000114	1.14
Hydrogen	0.00005	0.5
Nitrous oxide, N_2O[a]	0.00005	0.5

Source: Smith, F.G. (Ed.), *Atmospheric Propagation of Radiation*. Volume 2 of *The Infrared & Electro-Optical Systems Handbook*, J.S. Accetta, D.L. Shumaker (Eds.), Infrared Information Analysis Center, Ann Arbor, MI; SPIE Optical Engineering Press, Bellingham, Washington, DC, 1993.

[a] Has varying concentration in polluted air.

layer of molecules. Therefore, the American Heritage Dictionary defines humidity as: dampness, especially of air. Usually, water vapor in the air or any other gas is called humidity, in liquids and solids; it is generally designated as moisture.

Water and water vapor in the environment are constantly involved in various processes such as evaporation (Figure 1.2a), sublimation, condensation, and deposition. Whenever a water molecule leaves a surface and diffuses into a surrounding gas, it is said to have evaporated. Each individual water molecule, which transitions between a more associated (liquid) and a less associated (vapor/gas) state does so through the absorption or release of kinetic energy. The aggregate measurement of this kinetic energy transfer is defined as thermal energy and occurs only when there is differential in the temperature of the water molecules. Liquid water that becomes water vapor takes a parcel of heat with it, in a process called evaporative cooling. The amount of water vapor in the air determines how fast each molecule will return to the surface. When a net evaporation occurs, the body of water will undergo a net cooling directly related to the loss of water. Sublimation is another form of evaporation; water molecules become gaseous directly, leaving the surface of ice without first becoming liquid water. Sublimation accounts for the slow mid-winter disappearance of ice and snow at temperatures too low to cause melting. Condensation is the reverse process of evaporation. Water vapor will only condense onto another surface when that surface is cooler than the dew point temperature (Section 1.2), or when the water vapor equilibrium in air has been exceeded. When water vapor condenses onto a surface, a net warming occurs on that surface. The water molecule brings heat energy with it. In turn, the temperature of the atmosphere drops slightly. In the atmosphere, condensation produces clouds, fog, and precipitation (Figure 1.2b). Fog and clouds form through condensation around cloud condensation nuclei. In the absence of nuclei, condensation will only occur at much lower temperatures. Deposition is a phase transition separate from condensation, which leads to the direct formation of ice from water vapor. Frost and snow are examples of deposition. Water vapor is totally invisible. If you see a cloud, fog, or mist, these are all liquid water, not water vapor.

The number of variables that interact and affect the humidity levels in the environment are myriad. Changes in the water vapor content in the atmosphere not only varies with the time of day and latitude but also is affected by seasonal changes. The amount of water vapor in air depends also on the temperature of

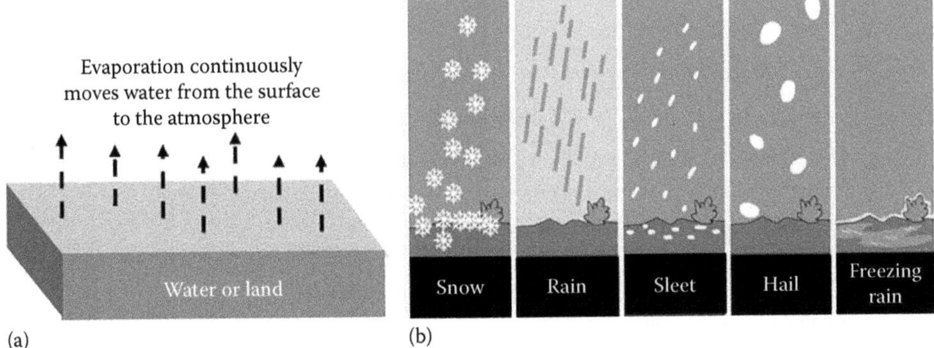

FIGURE 1.2 (a) Water evaporation and (b) forms of precipitation. Precipitation can be in the form of rain, snow, sleet, and hail. When water vapor is frozen directly into a solid without first forming a liquid, it forms tiny ice crystals called snow. Sleet is a frozen rain that forms when rain droplets encounter a cold air and freezes into ice before falling from the sky. Hail is the rounded lumps of ice that falls from the sky, whereas rain consists of droplets of liquid water that falls from the sky.

the air. At lower temperatures, very little water vapor can be held in air. At higher temperatures, the amount of water vapor in air increases. At the boiling point of water, the air has reached the maximum water vapor that it can hold, and the excess is condensed into water and seen as steam. This means that air humidity is not a constant value and can significantly change in a short time.

As it will be shown in Chapter 2, water vapor has a significant effect on the properties of the materials. The water molecules presented in air change the length of organic materials, the conductivity and weight of hygroscopic materials and chemical absorbents, and, in general, the impedance of almost any materials. Water absorbs infrared (IR) and ultraviolet radiation. It changes the color of chemicals, the refractive index (RI) of air and liquids, the velocity of sound in air or electromagnetic radiation in solids, and the thermal conductivity of gases and that of liquids and solids. It is probably difficult to find a material that is inert to water molecules and with which it would be impossible, with some physical method to measure the presence of water (Visscher 1999).

In addition, water vapor is key agent in both weather and climate. As it is known, water vapor is the most abundant of all greenhouse gases. Water vapor, like a green lens that allows green light to pass through it but absorbs red light, is a *selective absorber*. Along with other greenhouse gases, water vapor is transparent to most solar energy, as you can literally see. But it absorbs the IR energy emitted (radiated) upward by the Earth's surface, which is the reason that humid areas experience very little nocturnal cooling but dry desert regions cool considerably at night.

As shown by observations and research (Chapter 2), air humidity affects our security, our living conditions, our quality of life, and efficiency of industrial production. A huge variety of manufacturing, storage, and testing processes are humidity-critical. All this means that the measurement of humidity and constant monitoring of its change when using various sensors becomes mandatory for meteorological observations and industrial control (Nitta 1981; Fleming 1981a, b; Spomer and Tibbitts 1997; Wiederhold 1997; Rittersma 2002; Srivastava 2012). It should be noted that people began to realize the significance of this process many years ago. Even in the fifteenth century, Leonardo da Vinci built the first device for the air humidity measurement. According to other data, the first hydrometer was invented in the same fifteenth century by German Nicolaus de Cusa. It was a wool gravimetric hygrometer. There is also evidence that the accurate measurement of relative humidity (RH) was a very important activity in ancient China, and that there were dire consequences to those who got their weather predictions wrong. The death penalty was reportedly very easily imposed if it rained during a major ceremonial event! In Japan, the Beard Plant was used in the Edo era to give an indication of RH conditions to maintain product quality in the silkworm industry (Hattingh 2001).

1.2 THE MAIN GAS LAWS OF PHYSICS

Before starting discussions concerning humidity, the main ideal gas laws, helping to understand how a humidity level shifts, depending on the environment, will be presented (Table 1.3). The ideal gas law is the equation of state of a hypothetical ideal gas. It is a good approximation of the behavior of many gases under many conditions.

In Table 1.3, P is the pressure of the gas; V is the volume of the gas; n is the amount of substance of

TABLE 1.3
The Gas Laws

Name	Definition	Law
Boyle–Mariotte's law	"The absolute pressure exerted by a given mass of an ideal gas is inversely proportional to the volume it occupies if the temperature and amount of gas remain unchanged within a closed system" or "At constant temperature, the product of the volume and pressure of a given amount of gas is a constant"	$P=\frac{k}{V}$ or $P\bullet V=k=\text{const.}$
Charles's law	"When the pressure on a sample of a dry gas is held constant, the Kelvin temperature and the volume will be directly related" or "At constant pressure, the volume of a given quantity of gas is proportional to absolute temperature (°K). Or at constant volume, the pressure of a given quantity of gas is proportional to absolute temperature"	$V=q\bullet T$ or $P=j\bullet T$
Dalton's law of partial pressures	"The total pressure of a mixture of gases is equal to the sum of the pressures that each gas would exert if it were present alone"	$P_t=P_1+P_2+P_3+\ldots$
Avogadro's law	"Equal volumes of gases at the same temperature and pressure contain equal numbers of molecules" *Example:* one liter of any ideal gas at a temperature of 0°C and a pressure of 101.3 kPa, contains 2.688 × 10^{22} molecules	
Volume of a mole of gas at STP	As one liter of gas at STP contains 2.688 × 10^{22} molecules (or atoms in the case of a mono atomic gas), it follows that a mole of gas (6.022 × 10^{23} molecules) occupies a volume of 22.4 l at STP	
Ideal gas law	The product of volume and pressure of a given amount of gas is proportional to absolute temperature	$P\bullet V=nRT$
Van der Waals equation	The van der Waals equation (or van der Waals equation of state) is an equation relating the density of gases to the pressure (p), volume (V), and temperature (T) conditions (*i.e.*, it is a thermodynamic equation of state). It can be viewed as an adjustment to the ideal gas law that takes into account the nonzero volume of gas molecules, which are subject to an interparticle attraction	$\left(p+\frac{n^2a}{V^2}\right)(V-nb)=nRT$

gas (also known as number of moles); R is the ideal, or universal, gas constant, equal to the product of the Boltzmann constant and the Avogadro constant; T is the absolute temperature of the gas; STP is standard conditions of temperature and pressure; a is a measure of the average attraction between particles; and b is the volume excluded by a mole of particles.

While summarizing the data presented in Table 1.3, it turns out that Boyle's law states that the pressure–volume product is constant, Charles's law shows that the volume is proportional to the absolute temperature, Gay-Lussac's law says that the pressure is proportional to the absolute temperature, and Dalton's law states that all gases make contribution to a total pressure in proportion to their content. Knowledge of these laws is important because the water vapor is one of several gases that makes up air, and in most cases it behaves as an ideal gas. This means that the water vapors such as nitrogen, oxygen, and other trace gases contribute to the total pressure of the air. The portion that water vapors contribute to the total pressure of the air is called the partial pressure of water vapor. Further it will be shown that the partial pressure of water vapor is a key metric, found as a component in the formulas that define all other humidity parameters. It is important to note that the change in the total pressure of a gas mixture, at constant composition, results in the same change in the partial pressure of each component.

At the end of this section, some important physical thermodynamic constants will be presented. These constants are displayed in Table 1.4.

TABLE 1.4
Some Physical Constants in Thermodynamics

Physical Value	Symbol	Unit	Numerical Value
Absolute temperature (IPTS-68)	T_{68}	K	$273.15+T_{68}$
Physical standard temperature	T_0	K	273.15
Triple point of water	T_{tp}	K	273.16
Physical standard pressure	p_0	Pa	101325
Molar standard volume of ideal gas	$V_m = R{\cdot}T_0/p_0$	m³/mol	0.02241383
Universal molar or ideal gas constant	R	J/K·mol	8.3144598
Gas constant of water vapor	$R_w = R/M_v$	J/Kg	0.461520
Molar molecular mass of water	M_v	g/mol	18.01528
Gas constant of dry air	$R_a = R/M_a$	J/Kg	0.287055
Molar molecular mass of dry air	M_a	g/mol	28.9645
Relation of molecular mass	$\gamma = M_v/M_a$		0.62198
Mass of water molecules	$m_{H_2O} = M_v = N_A$	g	$2.991555{\cdot}10^{-23}$
Avogadro constant	N_A	mol⁻¹	$6.022045{\cdot}10^{23}$
Loschmidt constant	$N_L=N_A/V_m$	m⁻³	$2.686754{\cdot}10^{25}$
Saturated vapor pressure by triple point	$P_{SW}(T_{tp})$	Pa	611.657

Source: Wernecke, R. and Wernecke, J., *Industrial Moisture and Humidity Measurement: A Practical Guide*, Wiley-VCH, Weinheim, Germany, 2014.

1.3 UNITS FOR HUMIDITY MEASUREMENT

Humidity measurement determines the amount of water vapor presents in a gas that can be a mixture, such as air, or a pure gas, such as nitrogen or argon. There are many different ways to express humidity. On the basis of measurement techniques, the most commonly used units for humidity measurement are RH, dew/frost point (D/F PT), and parts per million (ppm) (Spomer and Tibbitts 1997; Wiederhold 1997; Chen and Lu 2005). Other parameters that can be used to indicate moisture levels are tabulated in Table 1.5.

It should be noted that the humidity can be fully described with a single parameter (any of them) if the gas pressure and temperature are known, but many national

TABLE 1.5
Terms Relevant to Moisture

Term	Definition	Unit
Absolute humidity	Ratio of vapor mass to the volume occupied by the air	g/m³
Volumetric concentration	Ratio of vapor volume × 10⁶ to the volume of the dry gas	ppmv
	Ratio of vapor weight × 10⁶ to the weight of the dry gas	ppmw
Saturated water vapor pressure (above water and ice)		Pa, kPa
Partial water pressure		kPa
Mixing ratio or mass ratio	Ratio of water vapor mass to the mass of dry gas	%, g/kg
Relative humidity	Ratio of vapor mass to the mass of saturated vapor or ratio of actual vapor pressure to saturation vapor pressure	% RH
Specific humidity	Ratio of vapor mass to total mass of the gas	%, g/kg
Dew point	Temperature (above 0°C) at which the water vapor in a gas condenses to liquid water	°C
Frost point	Temperature (below 0°C) at which the water vapor in a gas condenses to ice	°C
Volume ratio	Ratio of vapor partial pressure to partial pressure of dry gas	% by volume

Note: ppm—parts per million.

humidity laboratories have chosen the dew point temperature as the primary humidity parameter for practical reasons: the realization of a dew point temperature scale provides a simple source of traceability in a wide range of humidity at a good uncertainty level.

D/F PT is the temperature (above 0°C) at which the water vapor in a gas condenses to liquid water. At this temperature, water is either evaporated to a gas, or is condensed from gas to water. Frost point is the temperature (below 0°C) at which the vapor condenses to ice. D/F PT is a function of the pressure of the gas but is independent of temperature and is therefore defined as absolute humidity measurement. An increase in pressure will also increase the dew point. Harrison (1965) defined the dew point temperature of moist air as follows: the thermodynamic dew point temperature t_d of moist air at pressure P and with mixing ratio r is the temperature at which moist air, saturated with respect to water at the given pressure, has a saturation mixing ratio r_w equal to the mixing ratio r. Here, $r_w = r_w(P, t_d)$ is the mixing ratio of moist air saturated with respect to a plane surface of clean liquid water when the system consisting of the water and moist air is at a uniform temperature t_d, and the system is at a pressure P equal to that which exists in the given sample of moist air having the mixing ratio r. In addition, any plane surface that are below the dew or frost point temperature will acquire a dew or frost layer.

The vapor concentration or absolute humidity of a mixture of water vapor and dry air is defined as the ratio of the mass of water vapor M_w to the volume V occupied by the mixture:

$$D_V = \frac{M_V}{V}, \text{ expressed in g/m}^3, \tag{1.1}$$

where:

$M_w = n_w{\cdot}m_w$, n_w is a number of moles of water vapor present in the volume V
m_w is the molecular mass of water

As an example, Table 1.6 shows the approximate mass of water (in grams) contained in a cubic meter (m^3) of saturated air at a total pressure of 101325 Pa (1013.25 mbar). As shown in Table 1.6, 1 m^3 of air at 20°C (100% saturated) contains about 17 g water. If it is warmed to 37°C, the absolute humidity remains the same but the RH is only 39% because at 37°C, 1 m^3 of air contains 44 g water vapor when fully saturated and the ratio of 17–44 gives a value of 39%.

TABLE 1.6
Mass of Water Vapor per Cubic Meter of Saturated Air

Temperature, °C	0	5	10	15	20	25	30	35
Water vapor, g/m^3	4.9	6.8	9.4	12.9	17.4	23.1	30.5	39.8

Source: Bell, S., A beginner's guide to humidity measurement, National Physical Laboratory, Teddington, Middlesex, UK, TW11 OLW, http://www.npl.co.uk; http://www.rotronic.com/, 2012.

Specific humidity is the ratio of the mass M_w of water vapor to the mass ($M_w + M_a$) of moist air:

$$Q = \frac{M_w}{(M_w + M_a)}, \text{ expressed in \% or g/kg} \tag{1.2}$$

The mixing ratio r of moist air is the ratio of the mass M_w of water vapor to the mass M_a of dry air with which the water vapor is associated:

$$r = \frac{M_w}{M_a} \tag{1.3}$$

One should note that except in solid physics, volumetric units are rarely used.

PPM represents water vapor content by volume fraction (ppmv) or, if multiplied by the ratio of the molecular weight of water to that of air, as ppmw. PPM is also an absolute measurement. Although this measurement unit is more difficult to conceive, it has extensive applications in industry especially for trace moisture measurement. The vapor pressure of water is the pressure at which water vapor is in thermodynamic equilibrium with its condensed state. At higher pressures water would condense.

RH is the ratio of the partial pressure of water vapor (p_w) presented in a gas to the saturation vapor pressure (p_{ws}) in the gas at a given temperature. The value of RH is very sensitive to temperature, and thus it is a relative measurement. The RH measurement is expressed as a percentage. It is important to know that the RH in the closed chamber of Figure 1.3 is exactly 100% RH when equilibrium is achieved. In this situation, air contains a well-defined maximum quantity of water vapor. When this saturation vapor pressure is reached, any further addition of water vapor results in condensation.

$$\text{RH (in \%)} = \frac{p_w}{p_{ws}} \cdot 100 \tag{1.4}$$

The expression equilibrium relative humidity (ERH) refers to a condition where there is no net exchange of water vapor between a moisture-containing material (paper, medicines, foodstuffs, tobacco, seeds, etc.) and its environment. It is equivalent for water activity, a_w, used in the field of biology or food technology, generally

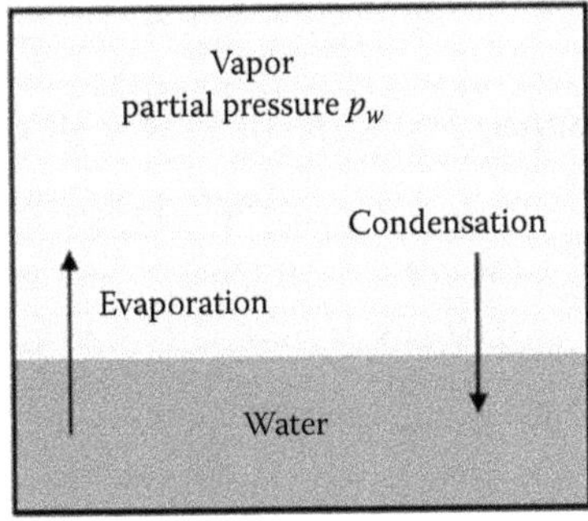

FIGURE 1.3 Closed chamber with liquid water and gaseous water particles (vapor). In equilibrium, the number of water particles leaving the surface of the liquid water is equal to the number rejoining it. The gaseous water particles exert a pressure p_w to the wall.

expressed as a ratio rather than a percentage (i.e., 0.6 instead of 60%).

In some cases, for the humidity characterization it is being used the heat index (HI), which indicates how the human body feels temperature. Therefore the result is also known as the *felt air temperature* or *apparent temperature*. For example, when the temperature is 32°C with 70% RH, the HI is 41°C. If RH is low, human body cools itself by perspiration, dissipating heat from the body. At higher RH, the evaporation rate from the human skin is lower. In that case, the body cannot dissipate heat as easily as it is the case in dry air. The HI is based on subjective measurements and is only meaningful above 25°C and 40% RH. The most popular definition of the HI is the one of the National Weather Service and Weather Forecast Office of the National Oceanic and Atmospheric Administration (NOAA). The HI in °C is given by the Equation 1.5 with the coefficients listed in Table 1.7.

TABLE 1.7
Coefficients for Heat Index Formula

Coefficient	Value
c_{00}	−8.7847
C_{10}	1.6114
c_{01}	2.3385
C_{11}	−0.1461
C_{20}	−0.0123
c_{02}	−0.0164
C_{21}	$2.2117 \cdot 10^{-3}$
C_{12}	$7.2546 \cdot 10^{-4}$
C_{22}	$-3.5820 \cdot 10^{-6}$

Source: Data extracted from http://www.sensirion.com.

$$\begin{aligned} \mathrm{HI} = {} & c_{00} + c_{10}T + c_{01}U_w + c_{11}TU_w + c_{20}T^2 \\ & + c_{02}U_w^2 + c_{21}T^2U_w + c_{12}TU_w^2 + c_{22}T^2U_w^2 \end{aligned} \tag{1.5}$$

where:
U_w is RH
T is temperature in °C

Figure 1.4 and Table 1.8 show the correlation among RH, ppmv, and the D/F PT. At present there are different approximations, which can be used for description of these correlations. However, many of them are complicated. As a result, conversions between the different parameters used to be cumbersome. However, they are now becoming standard operation since the introduction of the microprocessors.

As a rule, the conversions include the intermediate step of calculating the actual vapor pressure of water and the saturated vapor pressure of water at the temperature of interest. For example, absolute humidity, which is defined by the mass of water vapor m_{H_2O} per humid air volume V and can be expressed as $d_V = m_{H_2O}/V$, specific humidity (Q), mixing ratio (r), and ppmv can be calculated as follows:

$$d_V = 2.167 \cdot \left(\frac{p_w}{273.15 + T} \right), \ [\mathrm{g/m^3}] \tag{1.6}$$

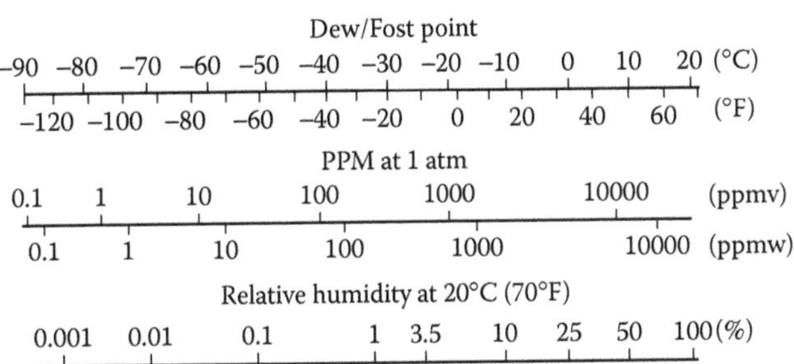

FIGURE 1.4 Correlation among humidity units: relative humidity (RH), dew/frost point (D/F PT), and parts per million by volume fraction (ppmv). (From Chen, Z., and Lu, C., *Sensor Lett.*, 3, 274–295, 2005.)

TABLE 1.8
Relations between Psychometric Parameters

Dew Point, °C	Relative Humidity at 20°C, %	Water Content in Air, ppmv
−75	0.002	~1
−45	0.3	70
−20	4.5	10^3
0	25	$6.1 \cdot 10^3$
10	50	$1.3 \cdot 10^3$
20	100	$2.3 \cdot 10^4$

$$Q = \frac{1000p_w}{(1.6078P_{total} - 0.6078p_w)}, \text{[g/kg]} \quad (1.7)$$

$$r = 621.97 \cdot \frac{p_w}{P_{total} - p_w}, \text{[g/kg]} \quad (1.8)$$

$$\text{PPMV} = 10^6 \cdot \frac{p_w}{P_{total} - p_w} \quad (1.9)$$

where:

P_{total} is total or barometric pressure in Pa
p_w is water vapor partial pressure in Pa
T is in °C

As it is known (Dalton's law), total pressure can be expressed as sum of partial pressures of gases presented in air:

$$P_{total} = p_{N_2} + p_{O_2} + p_{H_2O} + p_{other} \quad (1.10)$$

A relatively simple equation for the calculation of the saturation vapor pressure $p_{ws}(T)$ in the pure phase with respect to water is the Magnus formula (WMO 2008):

$$p_{ws}(T) = 6.112 \cdot \exp\left(\frac{17.62 \cdot T}{243.12 + T}\right) \quad (1.11)$$

Saturation vapor pressure over ice can be calculated using the equation (WMO 2008):

$$p_{IS}(T) = 6.112 \cdot \exp\left(\frac{22.46 \cdot T}{272.62 + T}\right) \quad (1.12)$$

In these equations, $p_w(T)$ and $p_{IS}(T)$ are kPa and T in °C. Uncertainty of these calculations does not exceed 0.6%–1% of value. Saturation water vapor pressures calculated in the temperature range from 0°C to 100°C are listed in Table 1.9. The change of the water vapor pressure over the wider temperature range from −100°C to +400°C is shown in Figure 1.5. The whole of this range is used in industry. Vapor pressures to be measured thus span about 10 orders of magnitude.

When water freezes, the molecules assume a structure that permits the maximum number of hydrogen-bonding interactions between molecules. As hydrogen-bonding is stronger in ice than in liquid water, it follows that intermolecular attraction forces are the strongest in ice. For that reason, vapor pressure above ice is less than the vapor pressure above liquid water (Table 1.10).

TABLE 1.9
Saturation Vapor Pressure of Water (0°C–100°C)

Temperature		Vapor Pressure			
T, °C	*T*, °F	*P*, kPa	*P*, mbar	*P*, torr	*P*, atm
0	32	0.6113	6.113	4.5851	0.0060
5	41	0.8726	8.726	6.5450	0.0086
10	50	1.2281	12.281	9.2115	0.0121
15	59	1.7056	17.056	12.7931	0.0168
20	68	2.3388	23.388	17.5424	0.0231
25	77	3.1690	31.690	23.7695	0.0313
30	86	4.2455	42.455	31.8439	0.0419
35	95	5.6267	56.267	42.2037	0.0555
40	104	7.3814	73.814	55.3651	0.0728
45	113	9.5898	95.898	71.9294	0.0946
50	122	12.3440	123.440	92.5876	0.1218
55	131	15.7520	157.520	118.1497	0.1555
60	140	19.9320	199.320	149.5023	0.1967
65	149	25.0220	250.220	187.6804	0.2469
70	158	31.1760	311.760	233.8392	0.3077
75	167	38.5630	385.630	289.2463	0.3806
80	176	47.3730	473.730	355.3267	0.4675
85	185	57.8150	578.150	433.6482	0.5706
90	194	70.1170	701.170	525.9208	0.6920
95	203	84.5290	845.290	634.0196	0.8342
100	212	101.3200	1013.200	759.9625	1.0000

Source: Lide, D.R. (Ed.), *CRC Handbook of Chemistry and Physics*, CRC Press, Boca Raton, FL, 2005.

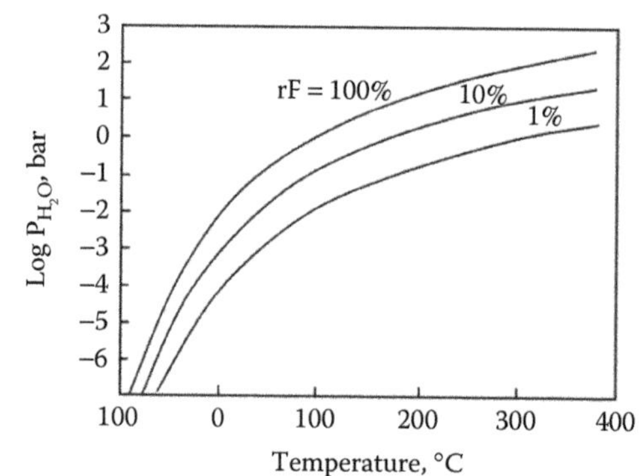

FIGURE 1.5 Water vapor pressure versus temperature rF = 100% is the saturation/dewpoint curve, 10% and 1% show the pressures at the respective relative humidity. (Reprinted from *Sens. Actuators A*, 12, Heber, K.V., Humidity measurement at high temperatures, 145–157, Copyright 1987, with permission from Elsevier.)

TABLE 1.10
Comparison of Vapor Pressure Under Water and Ice

Temperature (°C)	Vapor Pressure Liquid (kPa)	Vapor Pressure Ice (kPa)	Ratio Ice/Liquid
0	0.611	0.611	1.00
−5	0.422	0.402	0.95
−10	0.287	0.260	0.91
−15	0.191	0.165	0.86
−20	0.126	0.103	0.82
−25	0.081	0.064	0.78
−30	0.049	0.037	0.75

It is important to know that the saturation vapor pressure depends only on temperature. There is no effect of total pressure and there is no difference between the situation in an open space and that in a closed container. From the above it follows that

1. In an open space, at constant moisture level and temperature, % RH is directly proportional to the total pressure. However, the value of % RH is limited to 100% as P_w cannot be greater than P_{sw}.
2. In an open space, at constant moisture level and pressure, % RH decreases strongly as temperature increases.
3. In a closed container of fixed volume, % RH decreases as temperature increases, however not quite as strongly as in the situation of the open space.

As regards the relationship of the water vapor pressure with a dew point temperature (T_d), at standard pressure p_N, and a gas temperature T from −3°C to 70°C, the T_d can be determined by

$$T_d = 241.2°\text{C} \cdot \left[17.5043 \cdot \left(ln \frac{e_w}{6.11213hPa} \right)^{-1} - 1 \right]^{-1} \quad (1.13)$$

where hPa = mbar (e_w–p_w). In deriving this formula it was taken into account that the T_d of a gas is the temperature at complete saturation with water vapor. This means that in this point a water vapor pressure p_w is equal to the saturated water vapor pressure p_{sw}. Consequently, with the further addition of water vapor, dewing or wetting will occur. An equivalent statement is that a gas must be cooled to the T_d to achieve dewing or wetting. The results of these calculations for different humidity levels make it possible to establish a correlation between the air temperature, dew point, and RH. This correlation is shown in Figure 1.6.

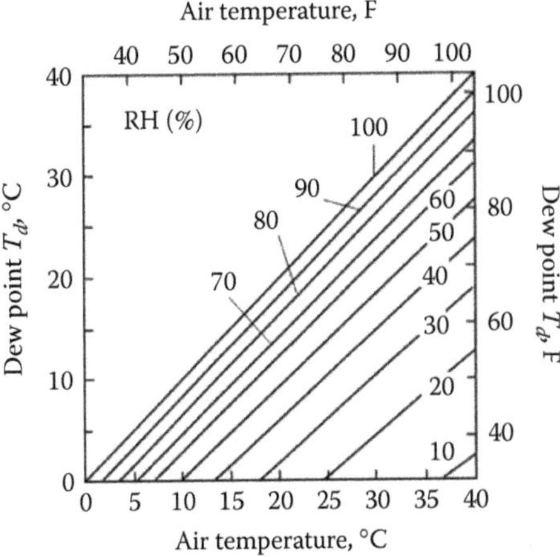

FIGURE 1.6 The chart illustrating correlation between the air temperature, dew point, and RH. (Data extracted from http://www.asge-online.com.)

There is also a very simple approximation that allows conversion between the dew point, temperature, and RH. This approach is accurate to within about ±1°C as long as the RH is above 50%:

$$T_{dp} \cong T - \frac{100 - \text{RH}}{5} \quad (1.14)$$

and

$$\text{RH} \cong 100 - 5(T - T_{dp}) \quad (1.15)$$

This can be expressed as a simple rule of thumb: for every 1°C difference in the dew point and dry bulb temperatures (T_{DB}), the RH decreases by 5%, starting with RH = 100% when the dew point equals the T_{DB}. The derivation of this approach, a discussion of its accuracy, comparisons to other approximations, and more information on the history and applications of the dew point are given in the Bulletin of the American Meteorological Society (Lawrence 2005). The relationship between the dew point temperatures and the RH in the digital form is given in Table 1.11.

Absolute humidity in relation to the water vapor pressure can be presented as following:

$$a = \frac{e_w M_w}{R_w T} \quad (1.16)$$

where R_w is gas constant of water vapor. Due to the strong temperature dependence of the absolute humidity, a normalization is usually carried out in technical

TABLE 1.11
Values of Relative Humidity at Selected Temperatures and Dew Points

Dew Point Temperature, °C	Air Temperature, °C 0	5	10	15	20	25	30	35	40	50
	Relative Humidity, % RH									
0	100	70	50	36	26	19	14	11	8	5
5	–	100	71	51	37	28	21	16	12	7
10	–		100	72	53	39	29	23	17	10
15	–	–		100	73	54	40	30	23	14
20	–	–	–		100	74	55	42	32	19
25	–	–	–	–		100	75	56	43	26
30	–	–	–	–	–		100	75	58	34
35	–	–	–	–	–	–		100	76	46
40	–	–	–	–	–	–	–		100	60
50	–	–	–	–	–	–	–	–	–	100

Source: Bell, S., A beginner's guide to humidity measurement, National Physical Laboratory, Teddington, Middlesex, UK, TW11 OLW, http://www.npl.co.uk; http://www.rotronic.com/, 2012.

applications. The normalization factor is related to the atmospheric pressure and a gas temperature of $T = 0°C$.

Graphically, the relation between temperature, absolute humidity, and RH can be represented as a psychometric chart shown in Figure 1.7.

This method is a graphical representation of properties of various mixtures of air and water vapor was invented early in the twentieth century by a German engineer named Richard Mollier. At first, the chart can be rather daunting, because it displays so much information in a small space. However, once the basic information elements are understood, the chart becomes an essential reference tool when designing temperature and humidity control systems (Bindon 1965). T_{DB} on the chart is the temperature that we measure with a standard thermometer that has no water on its surface. This chart was calculated for standard atmospheric pressure and temperatures of 0°C–40°C, which are adequate for most greenhouse or livestock housing applications. When people refer to the temperature of the air, they are commonly referring to its T_{DB}. Psychrometric charts are printed mostly for a sea level atmospheric pressure.

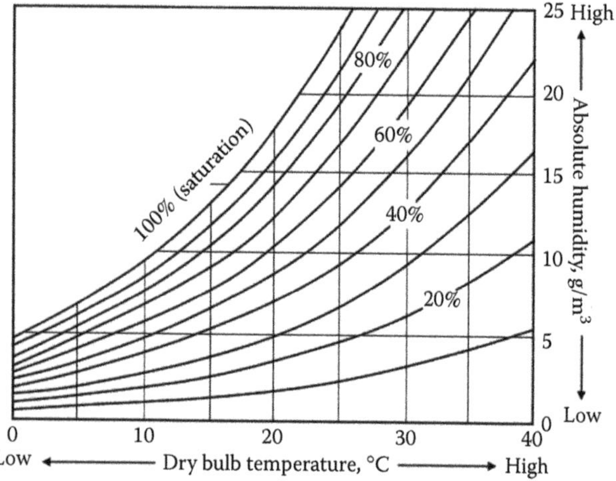

FIGURE 1.7 Simplest psychrometric chart at 1 atm total pressure. (Data extracted from http://www.sensirion.com.)

A psychrometric chart contains a lot of information packed into an odd-shaped graph. A detailed description of using a psychrometric charts for obtaining necessary information on the properties of the moist air one can find in the book by Harriman (2002). Harriman (2002) in his book describes each of the properties of moist air in turn, and then shows how these can be found quickly by using a psychrometric chart. One can also use the Internet sites of the companies, developing psychrometers. The chart is useful both for the information it contains, and the relationships it shows between air at different conditions. It not only shows the *trees* in the psychrometric jungle but also shows the whole *forest* as well, allowing an engineer to gain a sense of how easy or difficult it might be to change the air from one condition to another. For example, it provides an invaluable aid in illustrating and diagnosing environmental problems such as why heated air can hold more moisture, and conversely, how allowing moist air to cool will result in condensation.

For example, follow any horizontal line (representing a specific amount of moisture in the air) from left to right on the chart (i.e., from lower to higher temperature levels): note that RH decreases as temperature increases, so long as the quantity of moisture in the air does not change. For example, the water load in air at 25°C at 80% RH equates to about 25,000 ppm. At 55°C, this same 25,000 ppm is less than 21% RH. Again, follow any horizontal line, but this time from right to left (i.e., from higher to lower temperatures): note that RH levels increase, although the amount of moisture in the air remains constant. Finally, 100% RH is reached at the left edge of the chart, when the temperature drops to the dew point. The air is now saturated and will have to give up water (e.g., as condensation) at any lower temperature. Note also that moving from left to right on the chart along the upward curving RH lines correspond to increasing amounts of moisture in the air. This shows that maintaining a constant level of RH, as the temperature rises, requires adding moisture to the air. Conversely, if the temperature falls, the downward sloping

RH lines indicate that water has to be removed from the air to maintain RH at a constant level. More detailed description of a psychrometric chart one can find on many sites related to psychrometrics and psychrometers.

1.3.1 Other Absolute Moisture Scales

In addition to the previously mentioned values such as absolute humidity, D/F PT, and volume ratio, which are absolute scales and do not change with temperature, there are other absolute scales as well:

- % Moisture by volume, $\%M_V$
- Grams per cubic meter, g/m^3
- Humidity ratio (LB water/LB dry air or g water/g dry Air), W
- Specific humidity (LB water/LB mixture or g water/g mixture), q

Comparison of indicated values is shown in Table 1.12.

The $\%M_V$ can be defined in a least two ways:

$$\%M_V = \frac{\text{Number of } H_2O \text{ molecules per unit volume}}{\text{Total number of molecules per unit volume}} \quad (1.17)$$

$$\%M_V = \frac{P_W}{P_T} \quad (1.18)$$

where:

P_W is the partial pressure due to water vapor
P_T is the total pressure (usually atmospheric pressure)

In some countries a popular moisture measurement scale is *grams of water vapor per cubic meter.* This scale is based on the density of water vapor at STP. There are several different STP values that are given by various standard organizations. Grams of water vapor per cubic meter calculated for different air humidity are listed in the last column in Table 1.12. In our case, we use the STP of the International Union of Pure and Applied Chemistry (IUPAC). The STP of the IUPAC is 0°C and 100 kPa. The density of air at STP was given as 1275.4 g/m^3. If 100% water vapor could exist at STP, it would have a density of ~800.00 g/m^3. However, one should take into account that pure water vapor (100% H_2O by volume) cannot exist at STP because 0.6% water vapor by volume has a dew point of 0°C at this pressure. The last column on Table 1.12 is calculated using a value of 7.9317 g/m^3 for each 1% of water vapor by volume (MAC 1999).

TABLE 1.12
Moisture/Humidity Scales

% Moisture by Volume	Humidity Ratio, W	Specific Humidity, q	Dew Point Temperature, t_d		Grams per Cubic Meter
$\%M_V$	$\frac{LBH_2O}{\text{LB Dry air}}$	$\frac{LBH_2O}{\text{LB Mixture}}$	°F	°C	g/m^3
0.0	0	0	−460	−273.3	0.00
0.5	0.00313	0.00312	28	−2.2	3.81
1.0	0.00628	0.00624	45	7.2	7.62
2.0	0.0127	0.0125	64	17.8	15.24
5.0	0.0327	0.0317	92	33.3	38.10
10.0	0.0691	0.0646	115	46.1	76.19
20.0	0.155	0.135	141	60.6	152.38
30.0	0.267	0.210	157	69.4	228.58
40.0	0.415	0.293	169	76.1	304.77
50.0	0.622	0.383	179	81.7	380.96
60.0	0.933	0.483	187	86.1	457.15
70.0	1.45	0.592	194	90.0	533.35
80.0	2.49	0.713	201	93.9	609.54
90.0	5.60	0.848	207	97.2	685.73
95.0	11.80	0.922	209	98.3	723.83
100.0	∞	1	212	100.0	761.92

Source: MAC, *The Humidity/Moisture Handbook*, Machine Applications Corporation, Sandusky, OH, http://www.macinstruments.com/pdf/handbook.pdf, 1999.

The humidity ratio W is sometimes referred to as moisture content or the mixing ratio. It is the mass of water vapor per unit mass of dry air. The humidity ratio (W) can be calculated if the % moisture by volume ($\%M_V$) is known.

$$\text{Humidity ratio} = W = 0.622 \cdot \frac{\%M_V}{(100 - \%M_V)} \quad (1.19)$$

This equation is valid only for the normal mixture of gases in the atmosphere. When a different mixture of gases is present as is found inside a boiler flue, the factor 0.622 must change. This factor is the ratio of the molecular weight of water vapor (18.015) to the average molecular weight of the other gases (28.965 in the case of air):

$$\frac{18.015}{28.965} = 0.622 \quad (1.20)$$

Note that the $\%M_V$ scale is totally independent of the molecular weights of the other gases in the mixture, as in a boiler or direct fired oven.

Specific humidity is the ratio of the mass of water vapor to the total mass of the mixture of water vapor and dry air. The specific humidity (q) can be calculated if the $\%M_V$ is known.

$$\text{Specific humidity} = q = 0.622 \cdot \frac{\%M_V}{(100 - \%M_V) + 0.622 \cdot \%M_V} \quad (1.21)$$

The factor 0.622 is for normal air only. It must be corrected if the average molecular weight of the gases is different than air.

One should note that the above-mentioned moisture scales are preferred by people that work in specific disciplines (MAC 1999). When dealing with human comfort at normal ambient temperatures *RH* is the preferred scale. Weather forecasters (meteorologists) and heating, ventilating, and air-conditioning engineers (HVAC) use RH regularly.

$\%M_V$ is the most intuitive of the absolute scales. People who work in the areas of combustion and pollution control engineering routinely measure flue gas constituents in $\%M_V$. On account of the linear nature of this scale it is easy to display and easy to regulate using normal set point proportional–integral–derivative (P.I.D.) controllers. For these reasons, $\%M_V$ is also used in the areas of food processing, product drying, and product humidifying.

Humidity Ratio (W) is preferred by people who work in the product drying process because it can be directly used in energy calculations. This scale is also commonly used as the vertical axis on most psychometric charts. As this scale is very nonlinear and goes through many orders of magnitude, it is a difficult scale to display or use in a control mode. Since it is simple to convert between scales, $\%M_V$ is used for display and control of the moisture level, and then converted to humidity ratio for calculations.

Dew Point Temperature (T_d) is widely used by people who are concerned with the possibility of water condensing in pipes carrying compressed air or other gases. Dew point is also used by those working with sampling systems for the same reason. Condensation in lines can be avoided by maintaining the working fluid at a temperature well above its dew point or by drying a fluid to a dew point well below the lowest temperature to which it will be exposed.

REFERENCES

Bell S. (2012) A beginner's guide to humidity measurement. National Physical Laboratory, Teddington, Middlesex, UK, TW11 OLW (http://www.npl.co.uk; https://www.rotronic.com/).

Bindon H.H. (1965) A critical review of tables and charts used in psychrometry. In: A. Wexler (Ed.), *Humidity and Moisture*, Vol. 1. Reinhold, New York, pp. 3–15.

Campbell M.K., Farrell S.O. (2007) *Biochemistry* (6th ed.), Tomson Leaning, Belmont, CA, pp. 37–38.

Chen Z., Lu C. (2005) Humidity sensors: A review of materials and mechanisms. *Sensor Lett.* 3(4), 274–295.

Cracolice M.S., Edward P.I. (2006) *Basics of Introductory Chemistry.* Thompson, Brooks/Cole Publishing Company, Belmont, CA.

Fleming W.J. (1981a) A physical understanding of solid state humidity sensors. *Soc. Automot. Eng. Trans.* Section 2, 90, 1656–1667.

Fleming W.J. (1981b) A physical understanding of solid state humidity sensors. SAE Technical Paper 810432.

Harriman III L.G. (Ed.) (2002) *The Dehumidification Handbook*, Munters Corporation, Amesbury, MA.

Harrison L.P. (1965) Fundamental concepts and definitions relating to humidity. In: Wexler A. (Ed.), *Humidity and Moisture*, Vol. III, Reinhold Publishing Corporation, New York, pp. 3–70.

Hattingh J. (2001) The importance of relative humidity measurements in the improvement of product quality. In: *Proceedings of 2001 NCSL International Workshop & Symposium* (http://www.ncsli.org/i/c/TransactionLib/C01_R7.pdf).

Heber K.V. (1987) Humidity measurement at high temperatures. *Sens. Actuators.* 12, 145–157.

Lawrence M.G. (2005) The relationship between relative humidity and the dew point temperature in moist air: A simple conversion and applications, *Bull. Am. Meteorol. Soc.* 86, 225–233.

Lide D.R. (Ed.) (2005) *CRC Handbook of Chemistry and Physics*. CRC Press, Boca Raton, FL.

MAC (1999) *The Humidity/Moisture Handbook*, Machine Applications Corporation, Sandusky, OH (http://www.macinstruments.com/pdf/handbook.pdf).

Nitta T. (1981) Ceramic humidity sensor. *Ind. Eng. Chem. Prod. Res. Dev.* 20, 669–674.

Petrucci R.H., Herring F.G., Madura J.D., Bissonnette C. (2017) *General Chemistry: Principles & Modern Applications: AIE (Hardcover)*. Pearson/Prentice Hall, Upper Saddle River, NJ.

Rittersma Z.M. (2002) Recent achievements in miniaturised humidity sensors–A review of transduction techniques. *Sens. Actuators A.* 96, 196–210.

Sikarwar S., Yadav B.C. (2015) Opto-electronic humidity sensor: A review, *Sens. Actuators A.* 233, 54–70.

Smith F.G. (Ed.) (1993) *Atmospheric Propagation of Radiation*. Volume 2 of *The Infrared & Electro-Optical Systems Handbook*, J.S. Accetta, D.L. Shumaker (Eds.), Infrared Information Analysis Center, Ann Arbor, MI; SPIE Optical Engineering Press, Bellingham, Washington, DC.

Spomer L.A., Tibbitts T.W. (1997) Humidity. In: Langhans R.W. and Tibbitts T.W. (Eds.), *Plant Growth Chamber Handbook*, Iowa State University, Ames, IA, pp. 43–64.

Srivastava R. (2012) Humidity sensor: An overview. *Intern. J. Green Nanotechnol.* 4, 302–309.

Visscher G.J.W. (1999) Humidity and moisture measurement. In: Webster J.G. (Ed.), *The Measurement, Instrumentation, and Sensors: Handbook*, CRC Press, Boca Raton, FL, Chapter 72.

Wernecke R., Wernecke J. (2014) *Industrial Moisture and Humidity Measurement: A Practical Guide*, Wiley-VCH, Weinheim, Germany.

Wexler A. (Ed.) (1965) *Humidity and Moisture*, Volumes. 1–3, Reinhold Publishing Corporation, New York.

Wiederhold P.R. (1997) *Water Vapor Measurement: Methods and Instrumentation*, CRC Press, New York.

WMO (2008) World Meteorological Organization. Guide to Meteorological Instruments and Methods of Observation, Appendix 4B, WMO–No. 8 (CIMO Guide), Geneva, Switzerland.

2 Why Do We Need to Control Humidity?

On account of unique water properties, humidity influences greatly both living organisms, including human and materials (Lancaster 1990; Fitzpatrick et al. 2002; Amin et al. 2004; Puthoff et al. 2010; Prowse et al. 2011; Srivastava 2012). Let's consider the features of this influence.

2.1 CLIMATE COMFORT

Although human tolerance to humidity variations is much greater than tolerance to temperature variations, the humidity control is also important. Studies have shown that the air humidity have a strong impact on the perception of air quality (Fang et al. 1998, 1999; Toftum and Jørgensen 1998; Bradshaw 2006). The human body is sensitive to the climate, as it is most often exposed to the surrounding atmosphere. Typical human perception of the air humidity is shown in Table 2.1. When the climate begins to change, you'll notice more evidence of your body's sensitivity. The water content of your body is a convenient source. When the air temperature is high, the human body uses the evaporation of perspiration to cool down, with the cooling effect directly related to how fast the perspiration evaporates. The rate at which perspiration can evaporate depends on how much moisture is in the air and how much moisture the air can hold. As the air becomes drier, the body begins to lose moisture. This loss of water pulled from your skin is not immediately noticeable, but if not then restore the water content in the body, there may be serious complications related to dehydration. The rate at which perspiration can evaporate depends on how much moisture is in the air and how much moisture the air can hold. Therefore, particularly dangerous is the combination of high humidity with high temperatures. This combination reduces the rate of evaporative cooling of the body (air already laden with moisture cannot absorb much more from the skin) and can cause a considerable discomfort or lead to a heat stroke, exhaustion, and possibly death. High humidity can also be accompanied by condensation of water vapor. Condensation causes damage to interior paintwork, the inside surface of wall linings, floor coverings, curtains, and furnishings, which creates additional inconvenience to a human being. It also results in increased heating costs (as additional energy is required to convert condensation back into vapor, which is taken up by the air as the temperature rises).

Moisture exiting the body can also cause increased sensitive to cold temperatures. You may feel like it is colder than it really is. If the air in your home, office, or other indoor space was considerably dry, you may also feel discomfort from the low humidity level. Relative humidity (RH) levels below 25% are associated with drying of the mucous membranes and skin, which can lead to chapping and irritation. This situation often occurs in winter, when the dew point is low (below around −30°C). Low RH (10%–15%) also increases static electricity, which causes discomfort and can hinder the operation of computers and paper processing equipment. Even without noticeable symptoms, drying houseplants and increased presence of static electricity. The appearance of cracks and gaps in your hardwood floors and furniture are also common indicators of low humidity. Remember, low humidity can occur at any time of the year, not just during the winter.

One should note that there is no one temperature and humidity condition at which everyone is comfortable. People are comfortable at a range of temperature and humidity. Comfort zones determined by different institutes are shown in Figure 2.1. Notice that most people are comfortable at higher temperatures if there is a lower humidity. As the temperature drops, higher humidity levels are still within the comfort zone. Until recently, it was a general practice to design for 22°C–24°C and 35%–40% RH in winter, and 24°C–26°C and 50%–55% RH in summer. At air humidity above 60% RH and temperature exceeding 25°C the skin feels sweaty. Occupational Safety & Health Administration (OSHA) recommends that indoor air be maintained at 20°C–24.5°C with a 20%–60% RH. Nevertheless, as it will be shown below, some types of industrial applications, such as textile manufacturing, optical lens grinding, and food storage, maintain an RH above 60% because of equipment, manufacturing processes, or product storage requirements. At the other extreme, certain pharmaceutical products, plywood cold pressing, and some other processes require an RH below 20%.

It should be noted that in different countries, depending on geographic location, comfort zones can be different. For example, there is some acclimation to higher dew points by those who inhabit tropical and subtropical

TABLE 2.1
Influence of Temperature and Humidity on Human Perception of Air

RH at 32°C (90°F)	Dew Point		Human Perception
≥73%	Over 26°C	Over 80°F	Severely high. Even deadly for asthma-related illnesses
62%–72%	24°C–26°C	75°F–80°F	Extremely uncomfortable, fairly oppressive
52%–61%	21°C–24°C	70°F–74°F	Very humid, quite uncomfortable
44%–51%	18°C–21°C	65°F–69°F	Somewhat uncomfortable for most people at upper edge
37%–43%	16°C–18°C	60°F–64°F	OK for most, but all perceive the humidity at upper edge
31%–36%	13°C–16°C	55°F–59°F	Comfortable
26%–30%	10°C–12°C	50°F–54°F	Very comfortable
≤25%	Under 10°C	Under 50°F	A bit dry for some

climates. Thus, a resident of Darwin or Miami, for example, might have a higher threshold for discomfort than a resident of a temperate climate such as London or Chicago. Those accustomed to temperature climates often begin to feel uncomfortable when the dew point reaches between 15°C and 20°C, whereas others might find dew points below 18°C comfortable. Most inhabitants of these areas will consider dew points above 21°C oppressive and tropical-like.

One should note that the position of any upper humidity limit has great economic significance, particularly in hot and arid parts of the country. For example, in the West, it affects the need for billions of dollars of new peak electrical generating capacity. It also directly affects a substantial fraction of the cooling load in hot, humid climates. Under such economic imperatives, it is desirable to carefully examine the position of any upper humidity limit. Ideally, one would be able to assess the health risks against the economic benefits for any given humidity limit. At present, there is not enough information on this subject to even begin such an analysis. There is little conclusive evidence to show that either high or low humidity are of themselves detrimental to the health of normal people. It seems logical to assume that, in general, extremes of humidity are undesirable and that it is

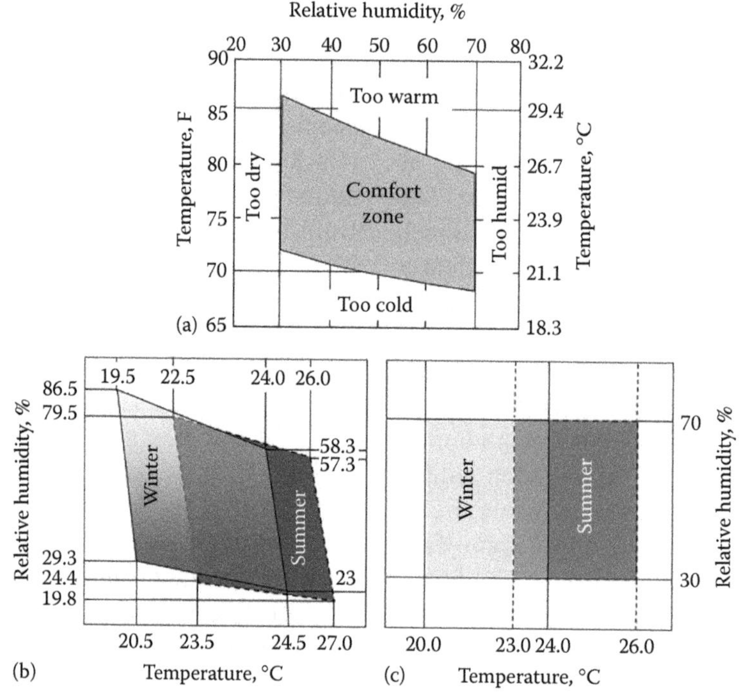

FIGURE 2.1 Chart illustrating zone for comfort living. (Data from (a) http://www.osha.gov, (b) ASHRAE, BSR/ASHRAE Standard 55-1992R: Thermal environmental conditions for human occupancy, American Society of Heating, Refrigerating and Air-conditioning Engineering, Atlanta, GA, http://www.archive.org/details/ASHRAE551992, 2001. With permission, and (c) ISO, ISO 7730:2005: Ergonomics of the thermal environment: Analytical determination and interpretation of thermal comfort using calculation of the PMV and PPD indices and local thermal comfort criteria, http://www.iso.org, 2005. With permission.)

desirable to keep RH at values within a broad range of from 30% to 70%. There is, however, no firm basis for establishing such limits so far as the health and comfort of most people are concerned. However, in most countries, for example, in many Canadian cities, ideal indoor RH levels are 35% in the winter and 50% in the summer.

It should be noted that it is needed to maintain a comfortable climate not only for humans but also for the many instruments used by man, including musical instruments, such as guitar, piano, and organ. If the moisture content becomes too high, the wood starts to swell; if it's too low, it shrinks. These actions may easily destroy the glued junctions and in extreme cases even lead to the ripping of the wood itself. If the wood swells, the organ is no longer easy to play. The reaction time of the various components is not correct anymore; the organ does not sound good. Obviously, this effect results in a loss of value, and it seems rather logical, that many alterations and reconstructions of old instruments have been done for that reason. So, the goal is to prevent the instruments from being damaged or even destroyed. Obviously, it is much easier to prevent than to repair. The humidity should always be in a range of 50%–60% RH.

To create a comfortable climate typically there are used a variety of devices, which, depending on the tasks provide drying humid air, which is more often called dehumidification, or moistening of air (humidification). Very often these devices are built in the air-conditioning system or in the air ventilation.

2.2 INJURIOUS ACTIVITIES OF MICROORGANISMS

An excessive moisture content can indirectly contribute to the destruction of organic material by increased microbial activity. Mold, mildew, and fungi are all different types of microorganisms participated in this process. The spores lie dormant until suitable conditions of temperature and humidity are achieved. The actual temperature conditions for germination may vary widely between different types of molds. When a RH exceeds 70%–80%, it creates favorable conditions for the rapid growth of mold spores, which are present everywhere. Outdoor air is well endowed with these microorganisms, which are small enough to be carried indoors and will settle on materials. As a result, in a warm and moist environment, mold appears rapidly. For example, observations have shown that the intense growth of the mold on the surface of pine and spruce begins when the RH is 95% and the temperature is between +20°C and +40°C. It was established that at 100% RH, 80°F (27°C), mold starts within two days. However, the growth speed was clearly slowed down when the RH dropped below 90%. If the temperature exceeded 30°C, molding began already at 80% RH in six weeks. According to research Image Permanence Institute's research (IPI's), mold will not grow below 65% RH, below 30°F (−1°C) or above 110°F (+43°C). In general, the spores will not germinate below 60% RH.

At moisture condensation, it is being observed as a considerable increase in the intensity of destructive processes and the acceleration in the development of various types of fungi and microorganisms, many of which are active biodestructors. The condensed moisture on materials acts as a medium conducive to the growth of microorganisms. This microorganism growth is injurious to materials, as it not only results in decomposition but also mechanical weakening of the products.

With increasing humidity biogenic and chemical corrosion are being enhanced, and most of the materials are being destroyed: the metal corrodes, caked friable materials get caked, wood, and fabric are being covered with mold and start rotting, cardboard boxes get wet and deformed, and electronic equipment has functional problems. At high temperature, the activity of the microorganisms increase but a certain amount of activity occurs even at very low temperatures.

Experiment has shown that the moisture content in the product is also important for the growth of microorganisms. Table 2.2 lists the water activity limits for the growth of microorganisms significant to public health and examples of foods in those ranges. In this Table 2.2, a_w is a parameter, which corresponds to the food moisture content, and is called *water activity* (a_w = % ERH/100, where ERH = equilibrium RH) (Scott 1957; Fontana 1998; McMinn and Magee 1999). It should be noted that water activity, not water content, determines the lower limit of available water for microbial growth. Water activity influences a food's shelf life as well as its odor, color, flavor, and texture. Research data indicate that the lowest a_w at which the vast majority of food spoilage bacteria will grow is about 0.90. *Staphylococcus aureus* under anaerobic conditions is inhibited at a_w of 0.91, but aerobically the a_w level is 0.86. The a_w for mold and yeast growth is about 0.61 with the lower limit for growth of mycotoxigenic molds at 0.78 a_w (Beuchat 1981; McMinn and Magee 1999).

From the previous definition, it is easy to understand how the water activity is useful in predicting a food safety and stability with respect to microbial growth, chemical/biochemical reaction rates, and physical properties. By measuring and controlling the water activity of foodstuffs, it is possible to predict which microorganisms will be potential sources of spoilage and infection;

TABLE 2.2
The Range of Water Activity Values for Various Common Food Types and Groups

Range of a_w	Microorganisms Generally Inhibited by Lowest a_w in This Range	Foods Generally within This Range
1.00–0.95	*Pseudomonas*, *Escherichia*, *Proteus*, *Shigella*, *Bacillus*, *Clostridium perfringens*, some yeasts	Highly perishable (fresh) foods and canned fruits, vegetables, meat, fish, and milk
0.95–0.91	*Salmonella*, *Vibrio parahaemolyticus*, *C. botulinum*, *Serratia*, *Lactobacillus*, *Pediococcus*, some molds, yeasts (*Rhodotorula*, *Pichia*)	Some cheeses (Cheddar, Swiss, Muenster, Provolone), cured meat (ham)
0.91–0.87	Many yeasts (*Candida*, *Torulopsis*, *Hansenula*), *Micrococcus*	Fermented sausage (salami), sponge cakes, dry cheeses, and margarine
0.87–0.80	Most molds (mycotoxigenic *Penicillium*), *Staphylococcus aureus*, most *Saccharomyces* (bailii) spp., *Debaryomyces*	Fruit juice concentrates, sweetened condensed milk, syrups
0.80–0.75	Most halophilic bacteria, mycotoxigenic *aspergillus*	Jam, marmalade
0.75–0.65	Xerophilic molds (*Aspergillus chevalieri*, *A. candidus*, *Wallemia sebi*), *Saccharomyces bisporus*	Jelly, molasses, raw cane sugar, some dried fruits, nuts
0.65–0.60	Osmophilic yeasts (*Saccharomyces rouxii*), few molds (*Aspergillus echinatus*, *Monascus bisporus*)	Dried fruits containing 15%–20% moisture, some toffees and caramels, honey
<0.60	No microbial proliferation	
0.50		Some pasta
0.40		Whole egg powder containing 5% moisture
0.30		Cookies, crackers, bread crusts
0.20		Fruit cake

Source: Fontana, A.J., Water activity: Why it is important for food safety. In: *Proceedings of the First NSF International Conference on Food Safety*, Albuquerque, NM, pp. 177–185, November 16–18, 1998.

maintain the chemical stability of foods; minimize nonenzymatic browning reactions and spontaneous autocatalytic lipid oxidation reactions; prolong the activity of enzymes and vitamins in food; and optimize the physical properties of foods, such as texture and shelf life.

As it was indicated before, at a relatively low humidity (<70%–80% RH) and low water activity, the activity of biodestructors, destroying materials and constructions, significantly slows down or stops. In most cases, bacterial growth can be arrested if RH is maintained below 45%. This means that RH of 50% is enough for the storage of the most materials, subjected to the influence of microorganisms. If the RH is comparatively low, the temperature plays only a minor role.

2.3 HUMAN HEALTH AND THE AIR HUMIDITY

Human health is not affected by high levels of humidity. The exact moisture-induced agents that cause health effects are yet unknown. However, there seems to be a significant association. Bornehag et al. (2001) reviewed 61 studies that concern moisture-related health effects concluding that there is strong evidence for a true association between dampness and health effects. In addition, Peat and Dickerson (1998) reviewed papers accessible via MEDLINE that investigate respiratory health outcomes in relation to housing characteristics or the presence of damp or mold in the home. Approximately half of the reviewed studies showed a significant association between respiratory symptoms, especially cough and wheeze, and the presence of damp. These studies conclude that health effects related to high humidity are primarily caused by the growth and spread of biotic agents under elevated humidity. The review of the literature identifies a number of health-related agents that are affected by indoor humidity (Baughman and Arens 1996a, b). Biological pollutants include pathogens such as bacteria (e.g., *Streptococcus*, *Legionella*), viruses (e.g., common cold, flu), and fungi (e.g., *Aspergillus fumigatus*). Moisture damages can be associated with several different health effects and symptoms in different studies. Commonly reported mold or moisture related health effects are for example (Haverinen 2002):

1. Irritative and general symptoms such as rhinitis, sore throat, hoarseness, cough, phlegm, shortness of breath, eye irritation, eczema, tiredness, headache, nausea, difficulties in concentration, and fever
2. Infections such as common cold, otitis, maxillary sinusitis, and bronchitis
3. Allergic diseases such as allergy, asthma, and alveolitis

The irritative and general symptoms in the first group do not cause permanent health hazards. The symptoms typically disappear within a few weeks after the end of the exposure. The same holds for repeated infections, but possibly not until after several months. However, a prolonged moisture damage may also lead to allergy or hyperergia. One should note that about 10% of the population is estimated to suffer from allergies. Allergic reactions (e.g., asthma, rhinitis) in many cases are connected with the presence of the dust mites (dried body parts and fecal excreta) and fungi. Several fungi such as *Aspergillus* and bacteria are also a reason of nonallergic immunologic reactions (e.g., hypersensitivity pneumonitis) (Arundel et al. 1986). Fungi are also a reason of the appearance of mycotoxicosis.

Most infectious diseases spread when pathogens are transmitted through their inhalation from the air: usually through human-to-human contact when droplet nuclei form as a result of sneezing or coughing and are subsequently inhaled by a human receptor, although some of them have effects through the skin. Biological agents require appropriate conditions in the building for their germination, growth, release to the air, and transport to the human host (Section 2.5.9). For example, fungal contamination occurs primarily as a result of condensation on surfaces and/or water damage. Field and laboratory studies suggested that fungal growth does not become an issue below 70% or even 80% RH unless there are other factors influencing their growth on building surfaces.

Studies aimed at establishing the correlation between the human health and air humidity have also shown that an environment with RH lower than 50% RH will increase the spreading rate of influenza virus (Hemmers et al. 1960). More recently, the U.S. National Institute for Occupational Safety and Health has also demonstrated that RH can be a factor in controlling the spread of flu (NIOSH 2013). At low humidity and low temperatures working together, you may subject your health to issues such as dry skin, itchy eyes, bloody nose, and irritation of the throat and sinus. Cracked skin is more than a cosmetic concern, without a firm layer of skin protecting the body's internal system; you run greater risk of getting sick and transmitting the illness to others. Extremely low (below 20%) relative humidities may also cause eye irritation. Dry air evaporates tears stored in your tear ducts, interrupting the natural balance of moisture in the eyes. In the same time maintaining moderate to high levels of humidity can contribute to reducing severity of asthma. RH may also directly affect the mucous membranes of individuals with bronchial constriction, rhinitis, or cold and influenza-related symptoms. Several reports, apparently based on the experience of physicians with patients who complained of dryness of the nose and throat during low RH, have also argued that indoor RH should be kept above 30%–40% to prevent drying of the mucous membranes and to maintain adequate nasal mucus transport and ciliary activity (Arundel et al. 1986). Moreover, studies have shown that an increase in humidity can reduce a nose tissue inflammation (Hashiguchi et al. 2008). A study of the effect of RH on nasal mucus has shown that the viscidity reduces by half when RH drops from 100% to 60% RH. It was also found that at a temperature of 21°C, influenza survival in the air is lowest at a midrange of 40%–60% RH. Allergic reactions (e.g., asthma, rhinitis) and dust mites are all affected by the amount of humidity in the air. Humidity values less than 50% RH are fatal to the dust mite (Baughman and Arens 1996a, b). This means that keeping your RH below 50% will minimize the presence of these pollutants.

Airborne levels of nonbiological pollutants, such as formaldehyde and ozone, may also be affected by humidity through influences on off gassing and surface reaction rates. Formaldehyde generation is exacerbated in some materials by higher humidity. A strong impact of temperature and humidity on the emission of formaldehyde from chipboard was observed by Andersen et al. (1975). They found that within the temperature range 14°C–35°C, the chemical emission rate of formaldehyde was doubled for each 7°C temperature rise. In addition, the emission rate was doubled when the relative air humidity increased from 30% to 70% at 22°C. The emission of volatile organic compounds (VOCs) from building materials is also humidity sensitive (Fang et al. 1999). For example, Wolkoff (1998) established that increasing the RH by 50% was seen to increase the emission of three VOCs emitted from carpet, sealant, and in particular, wall paint. Finally, the occupants' susceptibility to these agents may also be a function of humidity, although this appears to be a problem primarily at low humidity, when respiratory ailments result from dry mucous membranes (Baughman and Arens 1996b).

2.4 THE HUMIDITY INFLUENCE ON THE MATERIAL PROPERTIES

Plants and animals contain a high proportion of water, and it is therefore not surprising that their products—organic materials, such as wood, paper, cotton, linen, wool, silk, parchment, leather, fur, feathers, ivory, bone, and horn—also retain moisture. These materials are hygroscopic. They can and will absorb or give off moisture until they reach a state of equilibrium with the air that surrounds them. When the surrounding air is very dry, organic materials will give off some of their moisture: they become brittle and may shrink, warp, split, or crack. When the surrounding air is damp, the materials will absorb some of the moisture from the air: they may swell, cockle, warp, change shape, and/or lose strength. Dampness can also cause mold and fungal growth on organic materials.

Inorganic materials (glass, ceramics, metals, and minerals) are also affected by high or low humidity. Materials that have a natural salt content may suffer from efflorescence when the air is dry. The salts in deteriorated glass, porous ceramics, and some geological material are carried to the surface by moisture (which may have entered the pores during a period of higher humidity). The moisture evaporates and the salts crystallize on the surface. Other inorganic materials are affected by high humidity: metals (particularly iron and copper alloys) corrode; dyes and pigments fade more readily; and geological material can suffer from pyrite decay.

Thus, a humidity change can affect many parameters of materials, including the following:

- *Weight and/or volume (density change)*: For example, plastics usually absorb water 0%–0.5% of their dry weight (epoxies 0.08%–0.15%). Plastics based on pulp and polyamides (nylons) are an exception. Nylon PA6 can retain a maximum of 10% of water when immersed. Wooden materials, cardboards, and paper can absorb water 10%–30% of their weight if the RH of the air exceeds 70%. Two-component epoxy resin absorbs water about 0.1% of its weight at a temperature of 23°C and about 0.2% at a 100°C after immersion of 8 h. Cardboard materials absorb water about 15%–20% of their weight if the RH of the air is 100% near the room temperature. The quantity of water is almost independent of temperature in range 5°C–40°C. At 50% RH water content is about 8% and at humidity 90% about 16% of the weight of cardboard (Hienonen and Lahtinen 2007). Conversely, the loss of moisture can cause the shrinkage of materials.

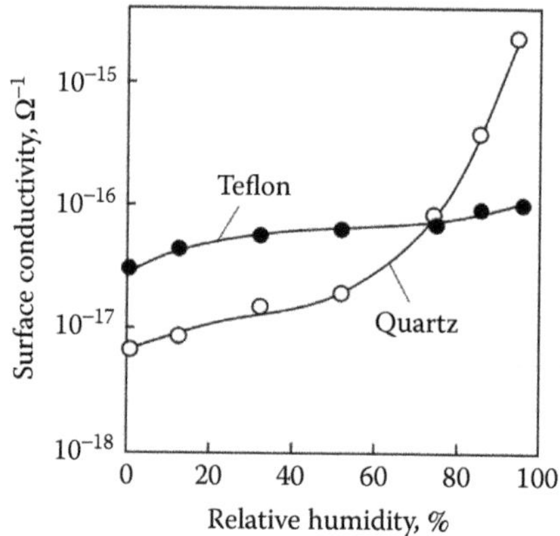

FIGURE 2.2 Surface conductivity of Teflon and quartz as a function of relative humidity at 23°C. (Reprinted with permission from Awakunit, Y. and Calderwood, J.H., Water vapor adsorption and surface conductivity in solids, *J. Phys. D: Appl. Phys.*, 5, 1038–1045, 1972. Copyright 1972, Institute of Physics.)

- *Electric conductivity, including surface conductivity* (Awakunit and Calderwood 1972): Normal variations in RH can alter the value of surface resistivity by several orders of magnitude (Figure 2.2), enormously affecting a material's ability to dissipate charge. For example, some static dissipative plastics (sometimes referred to as *antistatic*) use additives that rely on atmospheric water to decrease their surface resistivities. If these are used in low RH conditions, they may be just as insulating as the untreated plastic. If they have been specially selected to avoid an ignition hazard due to static, low RH could even mean the unwitting introduction of the ignition source (McCartney 2012).
- *Thermal conductivity and fluid viscosity*: As a result, humidity affects the heat transfer (Still et al. 1998; Lakatos 2017). For example, the moisture induces the degradation of the thermal properties of silica aerogel blankets (Lakatos 2017). Experimental studies show that the thermal conductivity of the aerogel blankets can be increased significantly (approximately 20%–40%) after wetting them (Figure 2.3).
- *The speed of chemical reactions*: As a result, the change in the activity of catalysts takes place (Elías et al. 2014), as well as in the gas sensors characteristics (Korotchenkov et al. 1999) and in the effectiveness of chemical reactions (Primavera et al. 1998; Chughtai et al. 2003). For example, it was established that increased

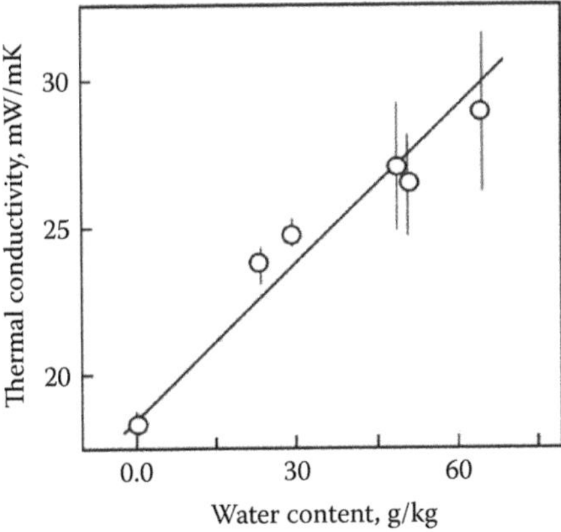

FIGURE 2.3 The change in the thermal conductivity of silica aerogel in the function of moisture content. (Reprinted from *Energ. Buildings*, 139, Lakatos, A., Investigation of the moisture induced degradation of the thermal properties of aerogel blankets: Measurements, calculations, simulations, 506–516, Copyright 2017, with permission from Elsevier.)

humidity decreases the degradation rates of trichloroethylene (TCE) during photocatalysis. As it is shown in Figure 2.4, the increase of RH higher than 75% made the reaction of photocatalytic degradation proceed slowly compared to that below 50% (Kim and Lee 2001). At that the humidity influence is specific for each reaction. For example, the rate of hydrogen sulfide oxidation (Figure 2.5), depending on humidity, has pronounced maximum (Primavera et al. 1998). Water also participates in oxygen reduction reactions, and interdiffusion of metals in ceramics may be enhanced by moisture due to surface OH (Hardy et al. 2015).

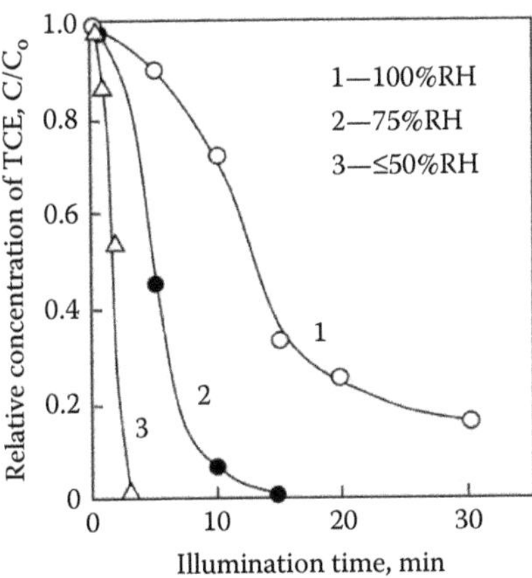

FIGURE 2.4 Effect of water vapor on the gas-phase photocatalysis of TCE over the nontreated TiO_2–GP. (With kind permission from Springer Science+Business Media: *Korean J. Chem. Eng.*, Effect of humidity on the photocatalytic degradation of trichloroethylene in gas phase over TiO_2 thin films treated by different conditions, 18, 2001, 935–940, Kim, J.S. and Lee, T.K.)

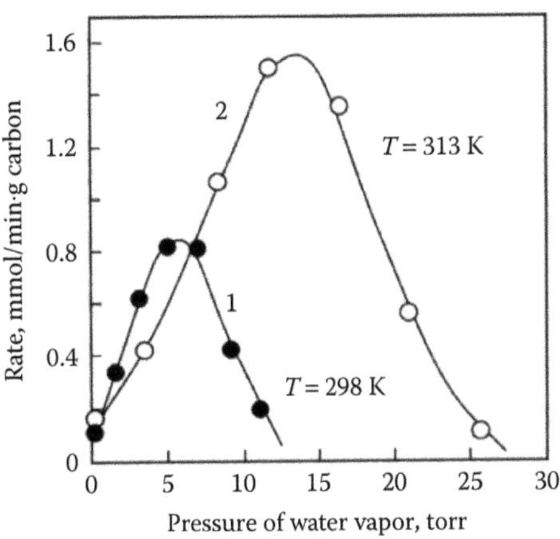

FIGURE 2.5 Rate of hydrogen sulfide oxidation at different H_2O pressures. Reaction conditions: 1-T = 298 K, 2-T = 313 K; Reactant gas flow rate/reactor volume = 134200 h^{-1}, 30 mg catalyst, $P(H_2S)$ = 26.1 torr, $P(O_2)$ = 14.5 torr; balance CO_2. (Reprinted from *Appl. Catal. A: General*, 173, Primavera, A. et al., The effect of water in the low-temperature catalytic oxidation of hydrogen sulfide to sulfur over activated carbon, 185–192, Copyright 1998, with permission from Elsevier.)

- *Product quality*: Excessive hygroscopic moisture of products contributes to the accelerated bacteriological growth, chemical composition change in reactions with water vapor, caking and clumping of powder and friable products, and so on (Guy 2001).
- *Mechanical properties, including ultimate strength, friction, surface adhesion, elasticity, plasticity, and so on* (Lancaster 1990; Walter et al. 2010; Muñoz and García-Manrique 2015; Gor et al. 2017): For example, it was found that compared with dry sliding, water usually reduces friction of materials (Lancaster 1990). It was also established that numerous assemblies with polymers degrade when exposed to elevated temperature and moisture (Walter et al. 2010). All these alter the mechanical properties of the materials at the interfaces. This then may lead to adhesion loss, interface delamination, and finally to crack propagation. Changes in moisture content cause paper to become thicker or thinner, flatter or curlier, harder or softer, larger or smaller, and limp or brittle. Moisture penetration into polymers reduces the mechanical performance by its plasticization and degradation effects, as well as by decreasing the elastic modules, fracture toughness, and yield strength.

TABLE 2.3
Moisture Dependent Parameters of Epoxy Resins

Material	Conditions	Density, g/cm³	T_g, K	Moisture Uptake, %	WLF C_1	WLF C_2
Epoxy 2/1–2	Dry	1.16	373		129	1912
	85°C/85% RH		349	4.78	25	121
Epoxy 2/1–4	Dry	1.14	373		33	174
	85°C/85% RH		337	4.86	13	71
Epoxy 2/1–6	Dry	1.12	363		34	157
	85°C/85% RH		333	3.97	12	66

Source: Walter H. et al., Influence of moisture on humidity sensitive material parameters of microelectronics relevant polymers. In *Proceedings of NSTI-Nanotech 2010*, Vol. 2, pp. 178–181, http://www.nsti.org, 2010.

Note: T_g is a glass-transition temperature, C_1 and C_2 are coefficients of Williams–Landel–Ferry (WLF)-function describing viscoelastic properties.

For example, experiments with thermosetting polymers, for example, epoxy resins, widely used in electronic applications, has shown that the humidity diffusion does not only effect a shift in the glass-transition temperature and reduction in the modulus, but that it also influences the viscoelastic properties significantly (Table 2.3). In the case of low humidity, the removal of moisture from hydroscopic materials can change the mechanical structure and cause embrittlement. Krautgasser et al. (2015) have shown that air humidity provides the decrease of the strength of low temperature cofired ceramics (LTCC). LTCC consist of ceramic grains (i.e., alumina) embedded in a silicate glass matrix. It was established that with rising absolute humidity more water molecules are present in the environment leading to an enhanced reaction rate and thus an accelerated subcritical crack growth rate. Just this effect causes a decrease of strength. Tanner et al. (1999) have shown that humidity is a strong factor in the wear of rubbing surfaces in polysilicon micromachines. It was demonstrated that very low humidity can lead to very high wear. As the humidity decreases, the wear debris generated increases. For the higher humidity levels, the formation of surface hydroxides may act as a lubricant.

At high humidity, the quality of the resistance of isolating materials, including the air itself as the electrical insulator, is getting worse. This leads to uncontrolled failures that may lead to the big accidents and disasters, mainly due to short circuits.

Metals, having almost no hygroscopicity, in air become a subject of corrosion (Hienonen and Lahtinen 2007). Corrosion is defined as destruction of a metal or alloy by chemical or electrochemical reaction with its environment. In most instances, the reaction of corrosion is electrochemical in nature. The main factor affecting climatic corrosion is perhaps temperature through humidity, because it affects the prevailing humidity conditions. Temperature as such has an accelerating effect on chemical reactions according to the Arrhenius equation. Corrosion caused by humidity can progress quickly even at 0°C–30°C temperatures. When the temperature changes the RH of the air also changes. If the temperature drops, the RH will at some point exceed 100% RH and a layer of water is formed on the surface. After the water layer is formed, the speed of corrosion also accelerates to several thousand times faster than at the starting point. The molecular layer of water eventually permits ionic conduction, which accelerates the rate of corrosion. The ferrous ions or ions of other metals may react with hydroxyl ions in water to form metal hydroxides. In normal conditions (RH < 50%) a layer of water 1–3 molecules thick can be adsorbed onto the surface without starting the corrosion reactions. If the layer grows to 20–50 molecules thick, the corrosion reactions start to prevail. This means that higher humidity may lead to higher condensation of water on the metal surfaces and thus stimulate corrosion. When the temperature rises again, the water evaporates and corrosion nearly stops. Thus, low humidity guarantees a low rate of corrosion. Practically, there is no corrosion of iron at a RH of 40%–45%. A small iron corrosion begins with an increase in RH from 40%–45% to 60%–70% (the so-called *critical* value of humidity). Above this value, the rate of corrosion of iron and other metals increases sharply (according to a logarithmic dependence), and a rapid destruction of the metal takes place. The critical humidity level, which is at 45%, is approximately the same for clear and polluted air; however, the rate of corrosion is faster where the surfaces are exposed to polluted air in combination with high RH. For example, at presence even small amounts of gaseous industrial pollutants such as SO_2, SO_3, NO_x, and others, «critical» humidity for iron and many other metals is being sufficiently decreased.

It is important that even metal oxide layer can degrade in humid atmosphere. For example, it is reported that capacitive humidity sensors containing Al_2O_3 are degrading at high humidity and that ALD–Al_2O_3 films, which are stored in water or humid environment, show degradation or dissolve almost completely (Ruckerl et al. 2017).

It was established that humidity can influence even the structure of material, formed during synthesis process. Elgh et al. (2014) studied mesoporous titanium films prepared at low temperature from a micellar reaction mixture via the evaporation-induced self-assembly (EISA) methodology and found that the RH employed during aging of the prepared films showed profound effects on mesoscopic order, porosity, the pore size, crystallinity, and photocatalytic activity of the films. An increased porosity, as well as an increase in the pore diameter from 2 to 4 nm was observed as the RH was increased from 3.8% to 75%. Ordered mesopores were only observed in scanning electron microscopy (SEM) and transmission electron microscopy (TEM) images for the film aged at 75% RH. For films aged at a RH of 75%, anatase crystallites were readily found incorporated into the pore walls, whereas for films aged at 3.8% few or no crystallites could be found. This shows the critical importance of RH of the environment around the EISA films for the formation of crystalline pore walls at room temperature.

Cai and Gevelber (2017) have shown that even diameter of electrospun polymer-based fibers depended on air humidity. The fiber diameters obtained from poly (ethylene oxide) (PEO)/water and polyvinylpyrrolidone (PVP)/alcohol solutions decreased as RH increased. It was established that RH directly affects evaporation rate, which in turn affects both jet length (stretching time) and the force balance in the bending region.

If the humidity of the air changes frequently, hygroscopic materials will swell and shrink repeatedly. This causes internal stress and damage, and can particularly be a problem in composite objects where the different materials have different rates of shrinkage. The expansion of one material may force changes in the dimensions of another, causing considerable tension, crack propagation, and eventually damage of composite materials or items made on the basis of different materials. This means that a RH can dramatically affect each property of all materials and the measurement of the environmental humidity should, therefore, always be included in any material testing or research program concerned with study of any their physical and chemical properties. The most common consequences for materials and devices, caused by the humidity influence, are reflected in Table 2.4.

Basing of the example of the humidity influence on the solid oxide fuel cell cathodes, one can judge how strong and overall could be the humidity influence on the material properties and the device parameters.

TABLE 2.4
Humidity Effect on the Material Parameters and Devices

Humidity	Principal Effects	Typical Failure
High relative humidity	Moisture absorption or adsorption Swelling Loss of mechanical strength Corrosion and electrolysis Increased conductivity of insulators Increase of dielectric constant	Physical breakdown Insulation failure Mechanical failure Increase of dielectric losses
Low relative humidity	Desiccation Embrittlement Loss of mechanical strength Shrinkage Abrasion of moving contacts	Mechanical failure Cracking

For example, Hardy et al. (2015) established that for $La_{0.8}Sr_{0.2}MnO_3$ (LSM), $La_{0.6}Sr_{0.4}Co_{0.2}Fe_{0.8}O_3$ (LSCF), and Yttria-stabilized zirconia (YSZ)-based cathodes the following effects can be observed:

- Humidity in the cathode air can cause an immediate increase in the polarization resistance of LSM/YSZ cathodes; this effect is reversible when the humidity is removed.
- Humidity in the cathode air can produce an increase in the LSM/YSZ cathode performance degradation rate, with the effect being more pronounced at higher current densities.
- Chemical effects of humidity on LSM/YSZ cathodes can include Mn interdiffusion, Mn oxide formation at the LSM-YSZ interfaces, Mn enrichment at the YSZ grain boundaries, and expansion of LSM lattice parameters.
- Humidity in the cathode air can cause an increase in the polarization resistance of LSCF cathodes; the effect is more pronounced at lower temperatures.
- Humidity in the cathode air can cause an increased degradation rate in LSCF at temperatures below 700°C–750°C, but can result in a decreased degradation rate at higher temperatures.

- Humidity can lead to increased Sr or Co segregation in the LSCF cathodes as well as an increased rate of compositional evolution of minor iron cobalt spinel phases.
- Infiltration is a possible mean of mitigating the effects of humidity on the LSM/YSZ and LSCF cathodes.

The humidity control is also required at exploitation of polymer electrolyte membrane fuel cell (PEMFC). The RH of the air is important, because it affects the hydration of the membrane. Water is the product of the chemical reaction at the cathode; water is also used for proton transport within the membrane from the anode to the cathode. If the membrane is poorly hydrated it would result in a poor ion transfer, and a drying of the membrane. Therefore, the reactants when supplied to the membrane especially air, are prehumidified to achieve a better hydration of the membrane. The best conditions are being achieved when humidity is about 100%, not lesser than 80% and not more than 100%. If the hydration of the membrane is less than 80% it results in the drying of the fuel cell membrane, and if it is more than 100% it results in the overflooding of the membrane.

2.5 HUMIDITY EFFECT ON INDUSTRY

The amount of water vapor in the air can affect human comfort, as well as efficiency of many manufacturing processes, including drying of products such as paint, paper, matches, fur, and leather, packaging, and storage of different products such as tea, cereal, milk, and bakery items, manufacturing of food products such as plywood, gum, abrasives, pharmaceutical powder, ceramics, printing materials, and tablets. The moisture present in the air along with temperature has a long-term and devastating effect on machine and material. The storage, manufacture, and transportation of material often takes place in a humid environment, which is not suited to the moisture sensitivity of the material, leading to deterioration of stored material, machinery, equipment, and reduced product appeal. The damage, which can be caused by excessive RH, is mainly related to the occurrence of the three processes, discussed previously: (1) corrosion of steel and metals, (2) deteriorated characteristics of hygroscopic material, and (3) increased harmful activity of microorganisms. Therefore, humidity measurement and control with purpose to improve the product quality is vital to many industries from the textiles, food processing, paper and car production, agriculture and from medicine to semiconductor industry, and petrochemical ones (Jefferies 1993; Traversa 1995; Wiederhold 1997; Patissier 1999; Hattingh 2001; Chen and Lu 2005; Fraden 2010; Liu et al. 2011; Khanna 2012; Alwis et al. 2013). Every industry of the mechanized world is affected by humidity both in terms of material and money. Let us consider briefly the role of humidity control in various industries.

2.5.1 Automobile Industry

Automobile industry which includes sophisticated production processes and uses different materials also requires constant monitoring of humidity. Therefore, correct humidity levels will improve productivity in several areas within an automotive manufacturing facility. For example, in automobile industry, humidity or moisture levels needs to be properly controlled and monitored in the motor assembly lines, ventilation and air-conditioning systems, and in the storage of the components and car painting. The fabrication of safety glass in car also requires humidity control. The thin, transparent plastic film, which serves as the adhesive between layers of safety glass is quite hygroscopic. If allowed absorbing moisture, the film will boil it off in processing, creating steam bubbles which get trapped in laminated glass.

When depositing coatings on the car, humidity also plays important role because moisture-laden air cannot hold as much solvent as dry air. Therefore, high RH can retard the rate of solvent evaporation presented in paint. This is especially observed in the use of water-based paints. For this reason, the maximum RH at which coatings can be applied and dried within a reasonable time is generally set at 60%. The humidity will make the dry time a lot longer, and can cause the paint to look hazy when finished, because of moisture trapped between paint layers. Some coatings, however, require moisture to cure. At the same time the optimum humidity for the transfer of paint from the spray nozzle to bodywork is 72% RH (http://www.humidity.com). At this level evaporation of the paint is reduced allowing it to reach the bodywork as the manufacturer intended. Reducing moisture evaporation from the paint also reduces the amount of paint dust that is introduced to the spray booth. These benefits combine to improve the finish quality of the bodywork, reduce sanding requirements, and lower paint costs by avoiding atmospheric losses. Application at incorrect humidity and temperature can cause defects such as blistering, pinholing, cratering, dry spray, and mud cracking. Therefore, for determining optimal conditions for painting it is important to check the specifications of the coating and the paint used for their forming.

As for the polishing of the car body, then in this case there is an optimum humidity ~55% RH (http://www.condair.kr). By maintaining 55% RH in sanding decks, dust is suppressed and static build-up reduced. This prevents airborne paint dust from being attracted to the surface of the bodywork, greatly decreasing sanding time and improving the quality of the finish.

One should note that the using of the humidity control is not limited by use just in the production of vehicles. Humidity should also be controlled in engine testing cells, because exhaust emission testing must be carried out within certain temperature and humidity ranges. The humidity control also allows improving the effectiveness of engine and creation a necessary cabin comfort (http://www.sensirion.com). The basic principles of climate control discussed earlier in Section 2.1 can be also applied to vehicles. In addition, there may be safety considerations. For instance, high humidity inside a vehicle can lead to problems of condensation, such as misting of windshields and shorting of electrical components. In sealed vehicles and pressure vessels such as pressurized airliners, submersibles, and spacecraft, these considerations may be critical to a safety, and the complex environmental control systems, including equipment to maintain pressure needed. The humidity sensors can detect the presence of fog on the inside of the windshield, and defogger or glass heating elements can be activated. Development of alternative fuel vehicles also requires the use of the humidity sensors.

2.5.2 Semiconductor (Electronic) Industry

The electronics industry is an industry where the humidity is being controlled very carefully (Hienonen and Lahtinen 2007). A high percentage of output, improved device quality, and long-term operation are possible only in case when the manufacture of integrated circuits (ICs), testing, and packaging of electrical and electronic devices will be conducted in strictly specified climatic conditions. Therefore, as semiconductor devices become smaller and more complex, the industry is driven to control and monitor the contaminants in the process gases at extremely low levels. Until recently, the gas suppliers guaranteed the moisture in gases at <100 ppb level, today 10 ppb specifications are common, whereas gases with humidity levels below 1 ppb are already required for some future developments.

Changing the properties of materials due to the humidity change can be accompanied by the deterioration of the accuracy of alignment in the photolithography process, and then change the parameters of semiconductor devices. Due to the presence of moisture, some chemically inactive coatings may have inferior adhesion qualities (silicon and some halogenized coatings,

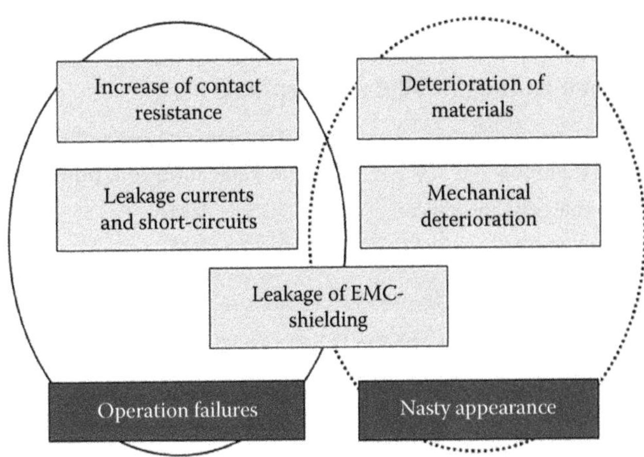

FIGURE 2.6 Basic problems caused by corrosion in electronics. (Idea and data from Hienonen R. and Lahtinen R., Corrosion and climatic effects in electronics, VTT Technical Research Centre of Finland, Otakaari, Finland, 2007. With permission.)

e.g., poly(vinyl chloride) [PVC]). Humidity influences the insulation resistance of the printed board. In optoelectronics, especially operating in the infrared band, the materials used can degrade in a humid atmosphere. This means that the performance of expensive equipment operating in the infrared band will be worsening. The problems that arise in the electronics due to various types of corrosion are reflected in Figure 2.6.

Experience has demonstrated that many printed circuit boards assembled by a number of different Commercial Companies have experienced blow holes or voids in the solder joints from the outgassing from the resins. One of the main causes of outgassing during the soldering process is a moisture: moisture absorbed by the resin during long-term storage prior to assembly in area with uncontrolled humidity, is released in gaseous form during the extreme temperature during soldering, resulting in poor solder joints, pinholes, blow holes, or the rupturing of plated through holes. The rupturing of a plated through hole in multilayer boards can also create such problems as the disconnecting of internal tracks and the resultant high rework costs.

Experience has shown that in the insides of the device the cases and electronics should be maintained as dry as possible in all situations, because the presence of water always increases the risk for corrosion considerably and weakens the isolation of the surfaces. In addition, the water absorbed into the materials increases a material loss and changes conductivity and the dielectric constant ε_r. All these changes in the elements of IC at high humidity can alter timing circuits, change the frequency of oscillator circuits, change the current level in a constant

TABLE 2.5
Device Failure in Wet Atmosphere

Failure Modes	Failure-Free Upper Limits of Water Vapor Concentration, ppmv
Aluminum disappearance	1000
Gold migration	1000
Nichrome disappearance	500
MOS inversion	200

Source: Kovac, M.G. et al., *Solid State Technol.*, 21, 35–39, 1978.

current source, result in loss of sensitivity or reduce the input impedance on high impedance amplifiers. For example, failure analysis of IC packages identified moisture trapped within the hermetically sealed enclosure as a major reliability problem throughout the semiconductor industry (Kovac et al. 1978). Water vapor can cause nichrome disappearance, gold migration, and the appearance of inversion layer in metal-oxide-semiconductor (MOS) structures inversion. Table 2.5 shows the failure-free upper limit of moisture level for device reliability. It is seen that the failure-free upper limits are less than 1000 ppmv. Therefore, it would be important to stop the dripping of water resulting from condensation onto the component boards or the connectors in indoor and outdoor conditions. This means that the semiconductor sealing process should eliminate or minimize the presence of moisture in the case. It is important when the products are packaged that the RH of the packaging space is as low as possible.

On the other hand, too low humidity is also undesirable at certain stages of the production of electronic devices, as in a dry environment the static electricity and static discharge increase dramatically. For example, voltages as high as 20 kV can be generated by a person walking across a carpet when the humidity levels are below 30% RH. Under high humidity conditions the same walker may only generate about 1.5 kV. If this person picks up a device without following the correct antistatic procedures, the damage of semiconductor devices can occur. It must be pointed out that the damage can be caused by a static discharge as low as 30 V in some metal–oxide–semiconductor field-effect transistor (MOSFET) technologies, where the gate oxide is very thin. Usually such a situation arises during the final stages of manufacture such devices as testing ones. In addition, it is necessary to take into account that a static charge attracts dust: but a build-up of dust in an unsealed unit can result in malfunction of the device.

2.5.3 The Pharmaceutical Industry

The pharmaceutical industry, like any health-related industry, makes the highest demands to the climate and air quality. The variety of requirements is determined by the variety of technological processes. For example, dehumidification is extremely important in such processes as milling, mixing, granulation, drying, tableting, coating, production of *effervescent* tablets, gelatin capsules, suppositories, ophthalmic dosage forms, and storage of finished pharmaceutical products. Each of these processes prevents uncontrolled air humidity and deserves special consideration. In particular, in the pharmaceutical industry, chemists have common problems of decomposition and difficulty in compression of tablets leading to breaking of tablets; lumping and caking of dry powders; improper adhesion under pressure of tablets; and improper drying of gelatin capsules. The presence of high humidity in the air is the cause of all these problems. Moreover, some chemicals and compounds on absorbing moisture lose their medicinal value. Therefore, in the production and storage of certain drugs maximum permitted parameters of the air are limited by the range of +12°C and 20% RH.

2.5.4 Production of Lithium Batteries

High technology often entails working under precisely controlled environmental conditions. Lithium ion batteries with increased energy capacity are a classic example of a product, the production of which is impossible without effective deep dehumidification. Lithium is a highly hygroscopic alkaline metal and therefore Li-based electrodes are extremely sensitive to the moisture. For example, Zhang et al. (2015) reported that while $LiFeSO_4F$ remained intact for several months when the RH is below 50%, there was full and rapid decomposition of $LiFeSO_4F$ into $FeSO_4 \cdot nH_2O$ (n = 1, 4, 7) and LiF in a highly humid environment (>62% RH at 25°C). Thus, the Li sensitivity to water and instability of Li-based electrodes in humid air makes the production of lithium batteries impossible without a reliable humidity control. The interaction of lithium and Li-based electrodes with even a small amount of water vapor can significantly reduce the shelf life of the product during storage and use. Moreover, lithium, plutonium, and other high-energy metals are hazardous because they ignite when atmospheric water vapor makes them corrode. Therefore, the humidity level in the air in the production zones of lithium batteries should be extremely low, up to 0.25 g/m^3. Dehumidifiers make it possible to work with such metals quite safely in open air.

2.5.5 The Food Industry

Food Industry is area where the humidity is very important. Higher temperatures and humidity are virtually always harmful across the spectrum of food products. For example, potato chips, dry breakfast cereals, and soda crackers exhibit an affinity for water when exposed to high humid conditions and will become soggy and unappetizing. Powdered foods tend to agglomerate or lump together. The result is that their movement through the manufacturing or packaging process is greatly inhibited. In addition, humidity may interfere with their processing and packaging as well. But it is often more than just a matter of lowering temperatures or reducing humidity; the relationship between these two conditions is interdependent and must be carefully controlled. For example, if a product, say a chocolate-covered pretzel, is kept in 90% RH, then it will become stale. By the same token, if it is stored at 95°F (35°C), then the chocolate will undoubtedly melt. When the two conditions work together, usually in the presence of light, the result oxidation can complicate the situation even more.

Strict control of humidity is needed in the production process of other foods such as fish drying, production of biscuits and cookies, cheese-making, the production of sausages and meat products, and many others. For example, in the production of raw sausage, a transition from the cold smoking process (at a temperature of 20°C–24°C and 90%–95% RH) to maturing and drying (to 10°C–15°C and 70%–75% RH), should be gradual and incremental to prevent the formation of condensation on the surface.

The ripening of cheese should also occur under certain temperature and humidity conditions: the temperature should be of 12°C–16°C (10°C–12°C), at RH of 80%–85%. Dry yeast is also a complex process that requires a dry cool air, because the high-temperature drying destroys microorganisms. The moisture content of the air for drying the yeast should be maintained at a level of 1.4–2.0 g/kg, which corresponds to a dew point of −13°C–−8°C.

In the production and storage of sugar its hygroscopic properties can lead to undesirable results such as sintering, the formation of lumps, solidification, and caramelization. RH in the storage areas of sugar should be maintained at 20% at a temperature of 24°C.

It is extremely important to provide tight microclimatic conditions during production and storage of gelatin, which is a raw material not only for the food industry but also for the manufacture of cosmetic products, films, capsules (in the pharmaceutical industry), and others. At breakdown in technological process, gelatin, which is inherently sensitive to moisture and temperature, can become brittle or, conversely, can melt. In addition, high humidity also leads to microbial contamination, which is extremely undesirable, taking into account the use of gelatin in the food and pharmaceutical industries. In many sections of the production of gelatin, the temperature is being maintained within ±1°C and a RH should not exceed 20%.

2.5.6 Agriculture

It should be noted that a comfortable climate is important not only for human but also for the plant, which means that the plants also need a comfortable moisture level for their development (Cockshull 1988). RH directly influences the water relations of plant and indirectly affects leaf growth, photosynthesis, pollination, occurrence of diseases, and finally economic yield. Humidity can affect the plant turgor pressure, which is an indicator of the amount of water in the plant cells. When humidity is low, moisture evaporates from the plants very fast. When this happens, plants can wilt rapidly if too much water was pulled out of the plant cells through transpiration. Conversely, when the humidity and temperature are both high, plants can get overheated, because transpiration is reduced, thus restricting evaporative cooling.

Humidity influences pollination. At high RH, pollen may not be dispersed from the anthers. Seed germination is also dependent on the humidity. Moderately low air humidity is favorable for seed set in many crops, provided soil moisture supply is adequate. For example, seed set in wheat was high at 60% RH compared to 80% when water availability in the soil was not limiting.

Humid air directly contributes to the problems such as foliar and root diseases, slow drying of the growing medium, plant stress, loss of quality, loss in yields, and so on. For example, the blight diseases of potato and tea spread more rapidly under humid conditions (Cockshull 1988). As it was shown before fungi and mold grow and spread rapidly when humidity is high. Several insects such as aphids and jassids also thrive better under moist conditions. Therefore, more pesticides are needed for disease control and plants tend to have weak, stretched growth making the plant less desirable.

As a result, we observe strong correlation between humidity and grain yield. Very high or very low RH is not conducive for high grain yield. Under high humidity, RH is negatively correlated with grain yield of maize. For example, Tnau Agritech Portal (http://www.agritech.tnau.ac.in/) informed that the yield reduction in India was 144 kg/ha with an increase in 1% of mean monthly RH. Similarly, wheat grain yield is reduced in high RH. It can be attributed to adverse effect of RH on pollination and high incidence of pests. On the contrary, increase in

RH during panicle initiation to maturity increased grain yield of sorghum under low humidity conditions due to favorable influence of RH on water relations of plants and photosynthesis. With similar amount of solar radiation, crops that are grown with irrigation gives less yield compared to those grown with equal amount of *water as rainfall*. This is because the dry atmosphere, which is little affected by irrigation, independently suppresses the growth of crops.

Humidity can also affect the fruit set of some plant species. An example is the bean (*Phaseolus* spp.), which has been shown to respond negatively to low humidity during fruit set. If the humidity was too low, the plant growth is often compromised as crops take much longer to obtain the saleable size. In addition, lower leaves often drop off, the growth slows down, and the overall quality is not very good. Whether the humidity is too high or too low, the loss of quality reduces the selling price of crops and increases the production costs, which lead to drop in profit.

Humidity also has a huge impact on the animals. Moreover, it was found that the humidity in the incubator effect even on chicken embryonic development (Noiva et al. 2014).

2.5.7 Paper and Printing Industry

Humidity is a very important parameter in the paper and printing industry. Every production superintendent in the paper industry is familiar with the excessive scrap losses and customer complaints that can result from the following wintertime headaches such as curling of paper, loss of package and container strength, production delays when sheets fail to go through machines smoothly due to static electricity, and gluing failures. All of the above-mentioned wintertime problems have a common cause—dry paper caused by low indoor RH. This behavior of paper and cardboard is a consequence of the strong dependence of their properties on the moisture content. It was established that moisture content with the RH in storage and work areas, responsible for any change that may occur in the product. These changes affect the change in a great number of physical properties of paper. The weight per surface unit, thickness, and volume increase with the moisture content. When the paper humidity is not correct, the paper tends to curl, and hence is unacceptable to customers. Paper curling, generally caused by the expansion and contraction of an unprotected sheet of paper, takes place when too dry an atmosphere draws moisture from the exposed surface, which shrinks and curls. The curl will be with the grain of the sheet. This trouble is most pronounced with very lightweight stocks or with cover stocks and coated-one-side papers. Changes in moisture content result in swelling or shrinking of paper fibers, whereas humidity cycles result impermanent dimensional changes as internal tensions in paper are released. Similarly, mechanical properties of paper, as well as printing properties, are influenced by the moisture content. This means that humidity influences the characteristics of a printing process as well. Moisture content is therefore an important production parameter, which should be monitored and controlled.

For example, from time to time, screen printers will experience difficulties due to static electricity phenomena, such as paper sheets sticking together. This happens mostly when the air is too dry and the paper is too dry. Testing has shown that a paper moisture content range of 5%–7% is essential to maintain satisfactory strength and workability of paper. This requires an indoor RH of about 40%–50%, depending on the composition of the paper. It has been found that when both the paper and air are in the 40%–50% range of RH the problem with screen printers seldom occurs. Moisture contents of different types of papers will vary slightly but will follow an identical pattern. Decrease in the moisture content induces paper to become thinner. Paper in stacks or rolls also shows deformation if too much moisture is exchanged with the surrounding air through the edges of the stack or roll. This is due to the uneven distribution of this moisture as it is exchanged with the ambient air during storage or transport.

It is well known that dry paper, more than moist, tends to generate dust. Severe dust problems in offset printing machines are due to the fact that the brittle fibers at the surface of the dry pager are easily detached. The result is the appearance of a dust layer on intermediary rolls causing poor printing. The swelling of the paper due to air humidity change also creates some problems for printing process. The variation in ERH from 50% to 10% results in a change of typically 0.1%–0.2% in the length of the paper. This means that such a humidity difference gives a dimensional variation of 1–2 mm on a 1 × 1 meter paper and therefore probabilities of poor, inaccurate printing due to positioning problems are high.

2.5.8 Textile Industry

One should note that there are not many occasions where humidity is more of an issue than during the processing of textiles. Get it wrong and it can stop production, damage machinery, and harm staff. Get it right and you can maximize product weights, improve quality, and increase machine speeds.

All textiles are hygroscopic, that is, they absorb or release moisture depending on the RH of the surrounding air. If the atmosphere is drier than the textile's ERH, then the textile will give up its moisture to the air. If the air is very humid then the textile's moisture content will increase. This moisture loss and gain occurs at every stage from the initial processing of the fibers through to final garment manufacturing, distribution, and use by the consumer. This change in moisture content has a direct impact on the properties of textiles, such as tensile strength, elasticity, fiber diameter, and friction. A drop in the ERH of a textile may cause it to be weaker, thinner, less elastic, and therefore more brittle. It will also have more imperfections. By maintaining the air humidity while processing the fibers, this loss in moisture to the atmosphere is minimized.

Moisture loss during processing cannot be totally eliminated as the act of processing will increase the temperature of the material, which will cause it to become drier. However, by increasing the humidity of the air surrounding the textile directly after processing, the material experiences *regain*. Moisture is reabsorbed by the textile, thus improving the quality and performance of the fabric. This regain also has a direct impact on the weight of the textile. As textile yarns are sold by weight, if a drop in humidity leads to a 4% reduction in weight, this will require 4% more fiber to be included in the sale product. For a mill manufacturing 100 tons of textile per day, this can lead to a loss of 4000 kg of product per day due to incorrect humidity control.

It should be borne in mind that in a dry atmosphere static electricity occurs. Static electrical build-up will cause materials to stick together and be less manageable. This in turn will slow machinery, directly effecting production schedules. By maintaining humidity at around 50% RH, static build-up is eliminated and all these associated problems are avoided. Another advantage of maintaining the correct humidity in the processing facilities is that it reduces airborne particles. A higher humidity encourages airborne lint, dust, and fly to precipitate out of the atmosphere. Thus humidity control provides creation of a healthier, less polluted, more pleasant atmosphere for workers, and a more productive workforce.

Experiment has shown that production of different type of textile requires different optimal humidity of the air. Cotton and linen have to be processed at very high levels, around 70%–80% RH, because they are very brittle. By humidifying each process, from the combing of the raw material, through carding, twisting, spinning, and weaving, the manufacturer can ensure that the product remains flexible and is prevented from breaking. This is important because the longer the fiber, the finer the thread that can be spun from it. Wool is similarly susceptible to dry air, although a little more forgiving, requiring humidity levels of around 65% RH. Man-made fibers also require the correct, albeit lower, level of humidity because below 45% RH they are prone to a build-up of static electricity. Silk should be processed at between 65% and 70% RH, although artificial silk spinning requires a higher level of 85% RH.

2.5.9 Humidity Effect on Building Structure

Buildings contain a large number of hygroscopic materials and, therefore, excess moisture and its variation in buildings have several effects both on the durability of structures and on the indoor air of building (Voutilainen 2005). These effects may occur once the moisture stress toward a structure is larger than the structure can endure, that is, moisture damage occurs (Nilsson 1980). For example, materials sensitive to moisture may experience changes in their physical dimensions and mechanical properties when moisture is present (Oliver 1997). Such changes include freezing of a moist porous material, swelling due to moisture absorption, and decreased strength. As a consequence, the materials, in particular concrete products, may be damaged. The mechanism of frost-induced deterioration is a hydraulic pressure originating from the sudden expansion of the water in the pores of the material when it freezes (Nilsson 1980). For this to happen, the moisture content of the material has to be near saturation simultaneously as the temperature drops below 0°C. Swelling due to moisture may lead to cracking, skewness, and arching of structures. At extreme consequences, parquet may blister, wooden floors may move partition walls, and ceiling panels may break. With many materials, most of the moisture-induced swelling occurs at the upper hygroscopic range, say above 75% RH. A form of physical damage concerning especially wood products as their moisture content increases is the loss of strength and increased elastic and plastic deformations.

Corrosion of metals also contributes to reducing the strength of structures used in buildings. For example, polished steel, typically used as reinforcement in concrete, has a critical moisture condition of ca. 80% RH both in free air and when embedded in concrete.

Deterioration of a polymer-based floor adhesive that is used to attach, for example, a PVC carpet to a concrete surface, is also influenced by moisture. In addition, moisture causes the carpet to swell, thus causing a strain on the adhesive. If the adhesive is too deteriorated, damage will occur (Nilsson 1980).

Material emissions from construction materials also depend on the humidity (Voutilainen 2005). For example, significant material emissions may occur if concrete structures are covered when they are still wet. Different covering materials have different critical moisture conditions varying between the RH of 80% and 97%. Especially chemical reactions in the combination of concrete, mortar, adhesive, and the covering material may produce harmful compounds into the indoor air. The adhesive used to attach the covering material may experience decay reactions that may develop emissions hazardous to health. The emissions have been noticed to increase significantly when RH exceeds 90%.

Different microbes, appearing in a humid atmosphere, cause biological damages to especially wooden structures. Typical microbes involved with moist wood include stain fungi, molds, and rots (Samson et al. 1994). Table 2.6 shows the list of indicator microbes, which can be observed in buildings. The microbes have been grouped by the minimum RH that they need in the material for growth in the typical building environment. *Hydrophilic microbes* may grow only in very moist conditions, whereas *xerophile microbes* may grow in drier conditions. As a consequence, at an early stage of moisture damage, the damaged material may be occupied by xerophile microbes. The growth of xerophile microbes may begin when the RH of a material is 65%–70%. If the damage is prolonged, they are gradually superseded by microbes that require moister conditions until only hydrophilic microbes remain. This phenomenon is known as *microsuccession* (Dini-Andreote et al. 2015).

TABLE 2.6
Indicator Microbes Observed in Buildings

Relative Humidity	Microbe
High (RH > 90%)	*Aspergillus fumigatus*
hydrophilic microbes	*Trichoderma*
	Exophiala
	Stachybotrys
	Phialophora
	Fusarium
	Ulocladium
	Yeasts, such as *Rhodotorula*
	Actinomycetes
	Gram-negative bacteria
Moderately high (85% < RH < 90%)	*Aspergillus versicolor*
Lower (RH < 85%)	*Aspergillus versicolor*
xerophile microbes	*Eurotium*
	Wallemia
	Penicillia

Source: Samson, R.A. (Eds.) et al., *Health Implications of Fungi in Indoor Environments*, Elsevier Science, Amsterdam, the Netherlands, pp. 529–538, 1994.

Damage effect depends on the nature of fungi. *Stain fungi* grow on and in timber without causing weakening or decay in the structure. Instead, they cause discoloration of wood, typically as blue or gray stains. Thus, the damage is mostly esthetic. The significance of stain fungi is mostly as an indicator of high moisture levels that could support the growth of more hazardous fungi. The growth of *mold* in structures is mostly superficial and thus the possibly arising problems are typically health-related or merely esthetic. Mold can usually be seen as green and white blotches on the surface of the structure. At the same time, *rot fungus* breaks down the wood cells, which weakens the durability and strength of the wooden structures. The decay process is initiated by the fungal hyphae that grow in the wood.

As it was shown in Section 2.2, the growth of stain fungi, mold, and rot fungus on a material is affected by several factors (Pasanen et al. 1994): the ambient RH and the moisture content of the material, the prevailing temperature, the time of exposure, the type of material and its nutritive status, and the fungal species involved. For example, the critical conditions for the initiation of decay caused by *Coniophora puteana* (cellar fungus) in pine and spruce sapwood at 20°C has been reported to be approximately one year in a RH of 93%–94% or one month in a RH near 100% (Viitanen 1997). Thus, the moisture conditions are significantly higher than those of mold fungi. Dry rot, on the other hand, can cause decay at a moisture quotient as low as 90% RH (Oliver 1997).

In addition to damaging structures, microbe growth has significant health effects that are discussed in Section 2.3. Microbe growth in buildings may manifest itself in several different ways (Pekkanen and Lampi 2015). Even if the growth is not visible, it may be recognized from a moldy or cellar-like smell or from the symptoms of the people using the building. Microbe-related health effects may be caused by several factors, including microbial volatile organic compounds (MVOCs), mycotoxins, allergens, and airborne microbe spores and fungal particles. MVOCs are chemical compounds that are released when some microbes grow. In fact, they are typically the same compounds as the VOCs of chemical origin. MVOCs also cause the typical smell of mold. Mycotoxins are toxic compounds produced by some microbes. Among the indicator list of Table 2.6, especially *Stachybotruys*, *Fusarium*, and *Aspergillus versicolor* are toxigenic (Samson et al. 1994). In addition, some microbes include proteins that are allergens, that is, compounds that have the ability to cause allergy. Microbe spores and fungal particles may both cause symptoms themselves and transport toxins in the indoor air (Pekkanen and Lampi 2015).

2.5.10 Other Fields

Industries discussed previously are only a small part of the industries where the humidity should be controlled (Table 2.7). In agriculture, measurement of humidity is important for the plantation protection (dew prevention), soil moisture monitoring, and so on. In the medical field, a humidity control should be used in respiratory equipment, sterilizers, incubators, pharmaceutical processing, and biological products. It is recommended that the temperature in the storage area of sterile materials must be maintained between 18°C and 22°C and RH between 35% and 50%, arguing that the maintenance of optimal conditions minimizes the potential for contamination of sterile products (De Moraes Bruna and Graziano 2012). In the fertilizer industry, dry fertilizers may agglomerate in the presence of high humidity. In leather processing, RH, which is maintained uniformly in the 40%–60% range (in some production areas the humidity can be higher), reduces cracking, minimizes loss of pliability, helps to maintain quality and appearance, and reduces the dust problem in the plant. The life cycle of machines is being increased when there is less humidity in the air. As a result, servicing intervals are reduced because of the lower levels of corrosion. The humidity control can be also applied in the flood-warning systems and for actuating early warning systems, or adjusting the amount of water in agricultural irrigation systems.

The measuring the moisture and oxygen contents of the flue gases promotes the increase in effectiveness of the fuel combustion for heat and power production. The monitor and control of the key parameters, for example, vapor concentration, temperature, humidity, and oxygen content, are also necessary methods to prevent or recede a potential explosion accident during the gasoline storage and transportation process (Qi et al. 2017).

There are many domestic applications, such as intelligent control of the living environment in buildings, storage of various goods, cooking control for microwave

TABLE 2.7
Impact of Humidity Measurement and Control on Selected Industries and Technologies

Industry or Technology	Process, Operation, or Phenomenon	Nature of Impact or Reason for Measurement or Control
Aerospace	Controlling capsule environment, life support systems, sealed cabins, and pressure suits	Removing and recovering water vapor from expired breath. Ensuring aviator and astronaut survival and comfort
	Simulating high altitude and space conditions in environmental chambers	At high altitude and space temperatures extreme dryness is essential to prevent icing and freezing
	Studying planetary atmospheres from orbiters and landers	Moisture is an indicator of possible life
	Measuring upper air humidity	Humidity affects aircraft icing, condensation trails, carburettor icing, visibility, fog, clouds, precipitation, UV transmission, ozone depletion, propagation of electromagnetic energy, and ballistic trajectories
	Dehydrating contaminated fuel tanks of jet aircraft and detecting moisture in jet fuel	Moisture can cause freeze-up of fuel lines
Nuclear	Treasuring moisture in high-pressure reactor cooling gases	Presence of moisture indicates leakage of water into reactor
	Operating high-temperature nuclear reactors	Moisture in helium causes graphite corrosion leading to unsafe operation
Electrical	Assembling lamp bulbs	Moisture in inert gas used in filling bulbs decomposes allowing oxygen to oxidize filaments and shorten life
	Insulating transformers with dielectric liquids	Moisture causes breakdown, arcing, and corrosion
	Sorption of moisture by electrical insulating materials	Causes reduction in resistivity and voltage breakdown
	Electrolytic corrosion	Causes deterioration
	Assembling motors, coils, and transformers	Low humidity required to prevent moisture sorption
Marine	Mothballing naval vessels	High humidity causes rust and oxidation of metal structures instruments components
	Storing shipboard cargo	High humidity can result in condensation on cargo
	Drying high-pressure air on naval ships, submarines for missile launching, and so on	Icing can cause locking and blocking of valves, orifices, and filters

(Continued)

TABLE 2.7 (*Continued*)
Impact of Humidity Measurement and Control on Selected Industries and Technologies

Industry or Technology	Process, Operation, or Phenomenon	Nature of Impact or Reason for Measurement or Control
Metallurgy	Welding titanium, stainless steel, and other alloys	High water content causes oxidation of weld
	Heat treating, carburizing, nitriding, dry cyaniding; polishing, brazing, sintering, annealing of metals, and alloys	Water vapor in gases or furnace atmospheres can decompose yielding oxygen, which causes oxidation of metal surfaces; causes decarburization of surface; and affects carbon potential
	Storage of ferrous metals	High humidity causes rust
	Measuring moisture in sinter mix, coke and iron ore for blast furnaces	Moisture content affects the quality and product efficiency in manufacturing
Missiles, munitions	Monitoring missile tank gas	Too much moisture causes freeze-up bleed valves
	Measuring atmospheric humidity	Required for computing trajectories
	Monitoring solid propellent mixing rooms	Dry atmosphere required for quality control and to prevent corrosion
	Monitoring missile sites	
	Controlling humidity around powder type fuses	Prevention of condensation on components
	Storing liquid rocket propellants	RH is held to 1% so that moisture content is controlled to 0.1% by weight; more moisture gives slower timing
		Small quantities (few tenths of 1%) can cause corrosion rates high enough to cause structural or component failure
Petro-chemical	Transporting, transmitting, and storing natural gas and liquefied natural gas	High moisture content will cause freeze-ups below 0°C and form hydrates, which freeze above 0°C
	Monitoring moisture level of hydrogen and butane feed lines in butane isomerization process	Moisture impairs catalyst life and activity
	Monitoring water vapor in recycle hydrogen streams of catalytic reforming processes for producing high-octane gasoline	Provides operational guidance during start-up, normal operation, test run, and abnormal conditions. Moisture poisons catalyst
	Purging and blanketing operations with dry gas	Reduces explosive hazards; prevents reaction of chemicals with moisture
	Liquid-phase drying of such chemicals as benzene, toluene, xylene, butane, propane, trichloroethylene, refrigerants, methyl chloride	Moisture inhibits drying

Source: Wexler, A., A study of the National humidity and moisture system, NBSIR 75-933, National Bureau of Standards, Washington, DC, 1975.

ovens, and intelligent control of laundry, and so on. RH maintained at 30%–40% stops splitting, checking, shrinkage, and glue joint failure in paneling and furnishings, adds life to carpeting and draperies. Electronic office equipment such as computers, xerographic copiers, and phone systems require a constant RH of 40%–50% to guard against harmful electrical transients.

2.6 STORAGE

2.6.1 The Storage of Library and Archival Materials

Books, papers, and other items in the library and archival collections are made up of a variety of components. The useful life of these materials is determined by the inherent characteristics of these components and by the environment in which they are housed. Strict environmental control is required to slow the rate of deterioration because the useful life of documentary materials is significantly affected by the levels of temperature, RH, light, and air pollution in which they are stored (http://www.lyrasis.org). For example, paper manufactured since the middle of the nineteenth century, is highly acidic and thus is subject to rapid deterioration in humid air. High humidity, similar to high temperature, accelerates the rate of chemical reactions and increases the rate of deterioration of library and archival materials. One should note that the selection of optimal air humidity is not a trivial task (Lull and Banks 1995; Ogden 1999). The recommended level of RH is a compromise among several requirements such as (1) a level of moisture, high

TABLE 2.8
Related Humidity Recommended for the Storage of Library and Archival Materials

Media	Relative Humidity	Allowable Fluctuations
Books and paper	40%–50% (21°C)	±3%
Papers on parchment and leather	60%	±3%
Photographic materials (black and white prints)	30%–50% (21°C)	±3%
Motion picture films	20%–50% (21°C)	±3%
Books, papers, and photos	40%–50%	±3%
Magnetic media on polyester base	30% (17°C)	±3%
Optical discs media	20%–50% (<23°C)	±10%

Source: Data extracted from CCA, Basic conservation of archival materials: Revised edition, Chapter 3: Environment, Canadian Council of Archives, http://www.cdncouncilarchives.ca/rbch3_en.pdf, 2003, http://www.lyrasis.org.

enough to maintain flexibility, (2) a level low enough to slow deterioration of materials and to control insects and mold, and (3) a level which will not cause a structural damage to buildings due to condensation in cold weather.

Optimal humidity values for the storage of various items, set by experiment, are presented in Table 2.8. A certain level of RH is recommended for paper to retain its flexibility. Film and other photographic media require a lower level of RH for optimum storage conditions. Most collection materials are composite structures made of various components. For example, books can be constructed of paper, board, string, adhesive, cloth, and leather. The components of books and photographic materials stretch or shrink at differing rates in response to changes in their moisture content, which is directly related to the level of RH in the surrounding air. Thus, the book components tend to fall apart and photographic materials tend to flake or peel away from their paper supports when the RH fluctuates.

Minimizing fluctuations in the temperature and RH is an attainable goal, and it retards chemical deterioration (PD 5454 2012). Environmental conditions for documentary materials stored separately from the areas used by patrons can and should be maintained at more stringent levels than for materials stored in the areas used by people.

2.6.2 Museum Collections

It should be noted that the preservation of the museum collections requires a careful choice of storage conditions, because all items collected in the museums are a big variety of materials, differing both in composition and in their combinations. In the museum storages one can find an object of organic origin (e.g., bones), as well as paper articles, wood, metal, cloth, plastic, and stone. All this naturally requires a specific approach to their preservation, especially if the museum exhibits represent the composition of different materials (Thomson 1986; Michalski 1993; Alten 1999; NPS 2016). At that, stable environmental conditions are among the most important factors in the preservation of museum collections.

For example, over centuries, wood has been the material of choice for panel supports for tempera and oil paintings, furniture, structural systems in the buildings, and a thousand other things. It is still used today and is a material that is very hygroscopic. The dimensions of wood vary with the humidity and the magnitude of these changes depends on the changes in RH (Michalski 1994). It is important that other groups of materials in contact with wood, are also affected by a moisture, but the speed of dimensional responses of these materials are different. As a result, cracks may appear on the panel supports, separation and cracking of the paint is also possible, which contributes to the destruction of the art exhibit, which is particularly intensive at very high levels of RH, when wood loses its strength and becomes flexible.

Hide glue (gelatin) is also a RH-responsive material and it is present in nearly all cultural collections. It is found as the adhesive for bonding parts and veneers in wood furniture, it is used as the size in traditional canvas paintings and some watercolor papers, and it is used to make gesso. When refined into gelatin, it is used as the image emulsion in photographic materials. At low RH levels the material is still ductile and not brittle. At high humidity levels, hide glue loses strength and this is a critical factor in the moisture-related damage to paintings. Thus, fluctuations in the air humidity can cause damage to museum items and should therefore be avoided or minimized.

Analysis of the exhibits preservation showed that deterioration can occur when RH is too high, variable, or too low. When RH is high, chemical reactions may increase, just as when temperature is elevated. Many chemical reactions require water; if there is a lot of it available, then chemical deterioration can proceed more quickly. Examples include metal corrosion or fading of dyes. High RH levels cause swelling and warping of wood and ivory. High RH can make adhesives or sizing softer or sticky. Paper may cockle or buckle; stretched canvas paintings may become too slack. High humidity also supports biological activity. Mold growth is more likely as RH rises above 65%. The insect activity may increase. At the same time very low RH levels cause shrinkage, warping, and cracking of wood and ivory; shrinkage, stiffening, cracking, and flaking of photographic emulsions and leather; desiccation of paper and adhesives; and dessication of basketry fibers. Below 30% of RH level some objects may become stiff and brittle. In the case of variable conditions, changes in the surrounding, RH can affect the water content of objects, which can result in dimensional changes in hygroscopic materials. They swell or contract, constantly adjusting to the environment until the rate or magnitude of change is too great and deterioration occurs. Deterioration may occur in imperceptible increments, and therefore go unnoticed for a long time (e.g., cracking paint layers). The damage may also occur suddenly (e.g., cracking of wood). Materials particularly at high risk due to fluctuations are laminate and composite materials such as photographs, magnetic media, veneered furniture, paintings, and other similar objects.

With regard to the recommended storage conditions, the maximum level for RH is determined by the point at which the mold and fungal growth starts (Michalski 1993, 1994, 1998). Mold and other fungi need a humidity of at least 70% RH; therefore, the recommended maximum level for RH is 65%–70%. Below 40% humidity-sensitive items can become unacceptably dry and brittle; therefore, the recommended minimum level for RH is 40%. Within these outer limits, some materials require more specifically controlled levels of RH. Ideally, fluctuations should not exceed ±5% from a set point, each month. RH optimum ranges for various materials housed in the museum collection are listed in Table 2.9.

TABLE 2.9
Recommended Conditions for the Storage of the Museum Exhibits

Museum Piece	Recommended Humidity
Archeological Materials	
• Negligible climate-sensitive materials	30%–65%
• Climate-sensitive materials	30%–55%
• Significantly climate-sensitive materials	30%–40%
• Metals	<35%
Natural History Materials	
• Biological specimens	40%–60%
• Bone and teeth	45%–60%
• Paleontological specimens	45%–55%
• Pyrite specimens	<30%
Painting	40%–65%
Paper	45%–55%
Photographs/Films/Negatives	30%–40%
Other organics (wood, textiles, ivory)	45%–60%
Ceramics, glass, stone	40%–60%

Source: Thomson, G., *The Museum Environment*, 2nd ed, Butterworths, London, UK, 1986; Michalski, S., Relative humidity in museums, galleries, and archives: Specifications and control. In *Bugs, Mold & Rot II: a Workshop on Control of Humidity for Health, Artifacts, and Buildings*, National Institute of Building Sciences, Washington, DC, pp. 51–62, 1993; Alten, H., How temperature and relative humidity affect collection deterioration rates. In *Collections Caretaker*, Vol. 2(2) Temperature and Relative Humidity, Northern States Conservation Center, http://www.collectioncare.org/pubs/v2n2p1.html, 1999; NPS Museum Handbook, Part I: Museum Collection, National Park Service, Chapter 4, U.S. Department of the Interior, http://www.nps.gov/museum/publications/MHI/mushbkI.html, 2016.

As follows from the analysis, the conditions of the museum collection storage aimed at the elimination of all three possible decay processes: (1) chemical, (2) biological, and (3) mechanical. Unfortunately, to date, there is no model to compare the amount of damage from chemical, biological, and mechanical decay on a particular collection. Chemical deterioration, sometimes called *natural aging*, is when a chemical reaction occurs, causing damage to an object. Chemical deterioration includes metal corrosion, increased fading, and glass decomposition. Plastics and organic materials have inherent and spontaneous deterioration reactions whose rate is determined by temperature and RH. It is considered that for chemical decay, usually the temperature has the largest influence on deterioration. Chemical reaction rates really increase with higher temperature, and cooler temperatures provide substantial increases in the material life. However, water is a reactant in many decay processes connected with chemical reactions. Increased RH increases the concentration of water participating in these processes. For biological and mechanical decay, we have another situation. As a rule, RH plays the larger role in these processes. Exactly RH determines whether various microorganisms flourish or exist at all (Section 2.2). Moreover, all biological decay is deterred by lower RH. Therefore, museums should strive to keep low, nonfluctuating temperatures, and a nonfluctuating RH below the level established for every group of museum piece.

2.6.3 Drying and Storage of Grain and Other Agricultural Products

Drainage and humidity control at the storage of grain, which is the main product of agriculture, is of paramount importance. For example, according to the Food and Agricultural Organization (FAO) at the United Nations, the annual losses of cereals account for more than 10% of total production, and in some less developed countries there are 30%–50%. It should be added that most of these losses occur at high humidity and high temperature of grain. Grain is the raw material, having the storage stability under appropriate conditions, if it was precleaned from impurities and the excess moisture was removed in a timely manner.

Optimal results are obtained by drying the grain by warm (45°C) and, that is very important, by dry air with a RH of not more than 2%. When drying the grain, hygroscopic moisture of the grain should be reduced to 18% or more, but not less than 14%.

Quite different storage requirements imposed on storage of fruits and vegetables. In this case too dry atmosphere is unacceptable. The loss of moisture due to transportation and storage in the dry atmosphere often limits the shelf life of fruits and vegetables. Fruits such as bell peppers (capsicums), for example, are mostly placed in cardboard boxes that are stored at a RH below 90% (Dijkink et al. 2004). This results mainly in precocious desiccation of the peel lead to a superficial shriveling. Other products that dry out too much during storage include pears, currants, and avocado.

2.6.4 Food and Industrial Products Storage

Experience shows that the preservation of manufactured products at the moment is no less important than the process of these products production (Guy 2001; Perez Flores et al. 2017). Globalization of the economy and the growth of widespread distribution channels make the manufacturers and distributors be cared about ensuring the product stability throughout extended shelf lives. As it was shown before, the temperature and humidity are the key factors that accelerate physical and chemical changes in foods and thus bring about the end of shelf life. It was shown that moisture gain above critical values lead to severe undesirable changes in the appearance and texture of the expanded product, which result in a product nonsuitable for consumption (Perez Flores et al. 2017). This means that the storage and transport of many goods, including foodstuffs, must be carried out under a strict temperature and humidity conditions. In this case, we will be able to avoid damage of the product caused by the processes described earlier, including chemical, biological, and mechanical decay processes.

TABLE 2.10
Humidity Recommended for the Storage of the Products and for Some Industrial Facilities

Object	RH, %	Object	RH, %
Storage of sugar	20–35	Unpackaged drugs	20–35
Cloth storage	60–75	Medical syrups	30–40
Storage of ground coffee	30–40	Storage of capsules	30–45
Storage of dried milk	20–35	Storage of powders	30–45
Seed storage	35–45	Laboratory electronica	45–60
Archives	40–55	Explosives[a]	35–50
Paper storage	35–45	Musical instruments	45–55
Leather products	60–75	Metal storage	<55
		([b]rust prevention)	<40

[a] Explosives should not be in conditions with relative humidity less than 30%, because static electricity can cause sparks and an explosion.

[b] Up to 55%—creeping corrosion, up to 40%—zero corrosion.

Table 2.10 shows by way of example some of recommendation data on a RH in the rooms aimed for storage, which in specific real cases may differ somewhat from the given data.

2.7 SAFETY

Humidity directly impacts the safety of many industrial and manufacturing processes. Water content in the air must be monitored, as either too much or too little humidity can be accompanied by the appearance of critical situations. At that, either very dry or very wet atmosphere is a source of danger. For example, without proper humidity control, low humidity and static charge present an increased explosion hazard in the areas where fine dust or powder is present. This situation is the most dangerous in the manufacture of explosives. But it is possible in the paper and textile industries as well. As the fibers lose moisture, they increase their electrical resistance. This means they can no longer easily dissipate the electrical charge which is generated by the frictional contact with the machinery. In a textile production facility with a low humidity, static discharges can jump up to 4–5 inches and, although they have a low current, can build-up to several hundred thousand volts. This presents a danger to staff working with the machines as it is not only very uncomfortable if they are shocked, but it can cause a person to jump and fall, which presents extreme risks when working near to textile machinery. The static discharge can also present a direct health risk to people with weak hearts or pace makers fitted. In addition, static charges can cause damage of the equipment responsible for security. As most machines are now microprocessor controlled, an uncontrolled electrical discharge in the wrong place can damage the electronics of the unit resulting in expensive repair bills and significant downtime.

On the other hand, when the humidity is very high due to condensation of water vapor, short-circuit can occur; and it may be accompanied by the failure of electrical or electronic equipment and fires. Such a situation is possible in the storehouses of agricultural products, in cultivation of mushrooms, beer production, and so on.

TABLE 2.11
An Inventory of the Hydrosphere

Component	Percentage of Mass of Hydrosphere
Oceans	97
Ice	2.4
Fresh water (underground)	0.6
Fresh water in lakes, rivers, and so on	0.02
Atmosphere	0.001

Source: Lamb, H.H., *Climate: Present, Pat and Future*, Methuen Publishing, London, UK, p. 482, 1972.
Total mass = $1.36 \cdot 10^{21}$ kg = $2.66 \cdot 10^{6}$ kg/m^2 over surface of earth.

2.8 METEOROLOGICAL MEASUREMENTS

Humidity measurements at the Earth's surface are also required for meteorological analysis and forecasting, for climate studies, and for many special applications in hydrology, aeronautical services, and environmental studies, because water vapor is key agent in both weather and climate (Peixoto and Oort 1996). The climatic hydrological cycle at global scale is shown in Figure 2.7.

It is known that the oceans play the central role in feedback loop via heat storage and transport around the globe. As it is seen in Table 2.11, oceans contain 97% of all water in the Earth. Water can absorb and store significant amounts of heat without raising its temperature. This high heat capacity allows water storing heat, which is transported by the ocean currents to increase the temperature of colder parts of the planet. As the water vapor rises in the atmosphere, it cools by releasing heat to the outer space. This cooling triggers the formation of clouds. These clouds continue to release heat and eventually form precipitation as rain, snow, or hail. Since the oceans cover about 71% of the Earth's surface, it is easy to comprehend how changes in the oceans could influence water vapor/precipitation and thus contribute to the extreme changes in weather and the appearance of extreme events such as drought and floods.

In addition, atmospheric water acts as the most important greenhouse gas. For example, models and

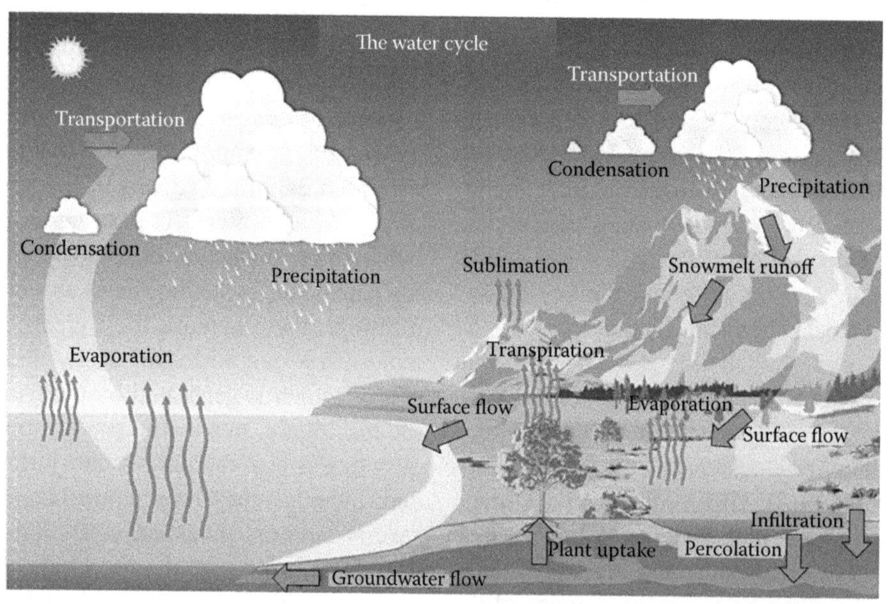

FIGURE 2.7 The climatic hydrological cycle at global scale. (Adapted from NOAA National Weather Service, http://www.pmm.nasa.gov.)

observations show that the water vapor increases as the climate warms, which in turn tends to further warm the atmosphere. As the air is warmer, the RH can be higher (in essence, the air is able to *hold* more water when it's warmer), leading to more water *vapor* in the atmosphere. As a greenhouse gas, the higher concentration of water vapor is then able to absorb more thermal IR energy radiated from the Earth, thus further warming the atmosphere. The warmer atmosphere can then hold more water vapor and so on. This is referred to as a *positive feedback loop*. Thus, water acts together with other external forces that affect the climate system, such as increases in atmospheric carbon dioxide (CO_2). However, huge scientific uncertainty exists in defining the extent and importance of this feedback loop, because at the same time water vapors influence solar radiation. Through the formation of clouds, water vapor leads to the reflection of sunlight back into space; thus allowing less energy to reach the Earth's surface and heat it up. However, the question of feedback between the water vapor mixing ratio in the upper troposphere (UT) and surface temperature is still unanswered. It is unclear whether the warming produces more water due to further evaporation (positive feedback) or if the attendant increased upwelling causes a drying (negative feedback). For example, the results of calculations by Manabe and Wetherald (1967) from a radiative-convective model with constant humidity suggested that the exponential increase of absolute humidity due to the sea surface temperature rise would exert a strong positive feedback. Other analyses from the complex general circulation models are generally consistent with Manabe's and Weatherald's conclusions showing similarly large positive feedback. Ellsaesser (1984) on the other hand, argued that an increase in the strength of convection in the tropics would cause an increase in the Hadley Cell circulation. An increase in the strength of the circulation will lead to drying or negative feedback, rather than a moistening of the UT (Ellsaesser 1984; Lindzen 1990; Sun and Lindzen 1993). In short, as far as the greenhouse effect and climatic change are concerned, it is still not clear whether thermodynamics or dynamics controls tropospheric water vapor. And over geological time-scales, the waxing and waning of ice sheets changes the reflection of the sun's light back into space, and largely determines the sea level.

So, now there is no adequate model explaining in detail the processes occurring in the atmosphere and giving quantitative relationship between water vapor, clouds, and heat exchange near the Earth's surface (Sun and Oort 1995; Lindzen et al. 2001; Kukkonen et al. 2012). The lack of knowledge on water vapor also leads to a large uncertainty in the prediction of climate change. If a model could not replicate the observed water vapor distribution, it nonetheless can reproduce the correct distribution of clouds and predict climate. Therefore, it is important to understand climate sensitivity and the factors governing the climate change.

To develop an adequate model, we need more observations of the state of the atmosphere, including the measurement of pressure, temperature, and humidity at all points of the Earth, because limited observations restrict understanding. This means that the future monitoring of atmospheric processes involving water vapor will be critical to fully understand the feedbacks in the climate system leading to global climate change. It is important that monitoring of the atmosphere and the control of the water vapor concentration should be conducted at all heights from the UT to lower stratosphere (LS) up to mesosphere (Figure 2.8 and Table 2.12).

According to Weinstock et al. (2009), an accurate and continuous global water vapor measurement record in the UT and LS and the tropical tropopause layer (TTL) is pivotal for (1) unraveling the relative

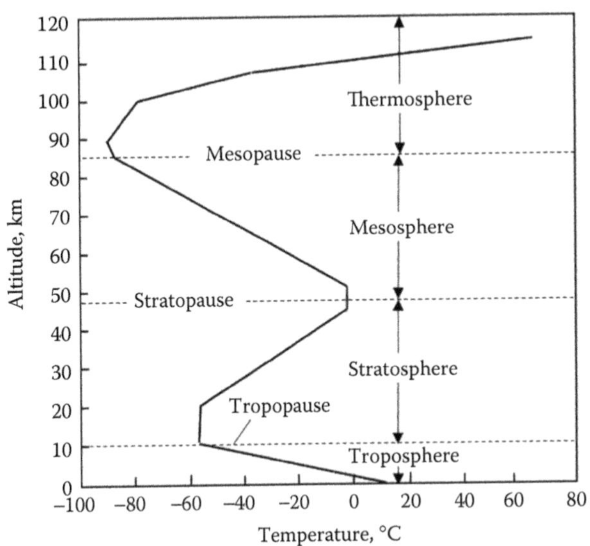

FIGURE 2.8 Structure of the atmosphere. The changes in the air pressure in millibars and temperature are also shown. The reasons for the temperature inversion in the stratosphere are due to the ozone absorption of ultraviolet solar energy. Although maximum ozone concentration occurs at 25 km, the lower air density at 50 km allows solar energy to heat up temperature there at a much greater degree. Also, much solar energy is absorbed in the upper stratosphere and cannot reach the level of ozone maximum. (Idea and data from Aguado, E. et al., *Understanding Weather and Climate*, Prentice Hall, Upper Saddle River, NJ, 2003. With permission.)

TABLE 2.12
Main Characteristics of the Atmosphere

Part of the Atmosphere	Features
Troposphere *(the prefix "tropo" comes from the Greek word for "turning over")*	• The lowest atmospheric layer • Average height is 10–11 km; 15–16 km at the equator; 5–6 km at the poles • Contains 75%–80% of the mass • Surface heated by solar radiation *f* • Temperature generally decreases with height; Average lapse rate is 6.5°/km. The coldest tropopause temperatures are over the tropics, where it can be −70°C or colder, whereas over the poles tropopause temperatures of −40°C are found. The tropopause is generally higher in summer than in winter, and is expected to rise in warmer climates as well • Strong vertical motion *f* • Referred to as the *weather layer*
Stratosphere *f*	• From 11 to 45–50 km • Contains ~20%–25% of total mass • Lid to the troposphere and important to aircraft • Weak vertical motions • The air in the stratosphere is very dry; Clouds are very rare here • Dominated by radiative processes *f* • Temperature is constant with height initially, then it increases with height, known as an inversion (temperature increases w/z) • Heated by ozone absorption of solar ultraviolet (UV) radiation
Mesosphere *(from the Greek "meso" meaning middle)*	• From 45 to 92 km • Less than 1% of the total mass • Sharp decline in temperature −93°C at top • Heated by solar radiation at the base *f* • Heat dispersed upward by vertical motion
Thermosphere *(from the Greek "therme" meaning heat) f*	• Increase in temperatures (extremely high) due to absorption of radiation • Very little mass; very few molecules resulting in very low pressure • Topped by exosphere • Aurora Borealis and Aurora Australis

Source: Hartmann, D.L., *Global Physical Climatology*, 2nd ed., Elsevier Science, New York, 1994; Ahrens, C.D., *Meteorology Today: An Introduction to Weather, Climate, and the Environment*, 9th ed., Cengage Learning, Belmont, CA, 2008.

importance of the dehydration mechanisms proposed to control the water vapor budget of the stratosphere; (2) determining the distribution of relative humidities within and in the vicinity of thin cirrus clouds; (3) quantifying the heterogeneous loss of stratospheric ozone in the Arctic and Antarctic both from water vapor's direct impact on the heterogeneous removal of ozone as well as its potential impact on vortex temperature; and (4) understanding the radiative properties of the TTL and stratosphere, especially at a time when the relationship between global climate change and surface and atmospheric temperatures must be clearly established.

2.9 SUMMARY

As follows from our consideration, humidity plays an important role in our life, determining our health and the way we feel, the Earth's climate and the effectiveness of the majority of the processes used in the industry. In many industrial processes, the measurement of moisture and humidity is important for the maintenance of optimal conditions in manufacturing. This means that the determination of humidity and moisture, as in prediction of floods, fog, conditions for the appearance of plant diseases, and so on, is of great economic importance. Therefore, humidity control becomes imperative in all fields of our activity, from production management to creating a comfortable environment for our living, as well for understanding the nature of the changes happening in the climate.

The tabulations of environmental requirements in several industries and to residence are listed in Table 2.13. To sum up some data presented in Table 2.13, an optimal temperature and RH conditions for applications in industry and at home may be presented as a chart shown in Figure 2.9.

It is seen that the optimal values of humidity, depending on technological processes, may vary widely. It is interesting to note that there were different environmental conditions defined for the various production stages of the same product. This means that (1) it is necessary to control the humidity at all stages of production and (2) to maintain moisture in the desired range required, there are necessary sensors, capable of carrying out the measurement of humidity in the entire range of possible changes in humidity. It is clear that these sensors must be able to work in a variety of climatic conditions, ensuring the functioning of the control and surveillance systems for a long time.

TABLE 2.13
Environmental Condition Requirements in Major Industries Activities

Market Segment	Manufacturing	*T*, °C	RH, %
Residential rooms and offices	Indoor comfort	22–24	30–70
Home appliances	Clothes dryer	80	0–40
	Microwave oven	5–100	2–100
	Air-conditioners	5–40	40–70
Textile (cotton)	Carding	23–25	50–55
	Combing	23–25	60–65
	First twisting	23–25	50–60
	Spinning	15–25	50–70
	Weaving	18–23	85
Textile	Spinning	20	85
(rayon)	Twisting	20	60
	Weaving	23–27	60–75
Agriculture	Soil humidity	5–30	0–90
	Greenhouse	5–40	0–100
	Cereal stocking	15–20	0–45
	Protection of plantations	−10 to +60	50–100
Baking (confectionery)	Frosting	20	50
	Mixing	22	65
	Leaven room	25	76–80
	Alimentary paste preservation	0–5	76–85
	Moulding cooling	21	60–70
	Preparation room	22–25	55–70
	Wrapping in paraffin paper	25	55
	Security boxes	30–32	80–90
	Flour storage	17–22	55–65
	Fermentation stage	0–7	60–75
Food processing and storage	Dairy	15	60
	Cooling dairy products	5	60
	Cereals preparation	15–20	38
	Alimentary paste preparation	20–25	38
	Brewing industry	–	35–45
	Food dehydration	50–100	0–50
	Meats tender	5	80
	Slicing of salami	15	45
	Apples storage	−1 to +1	75–85
	Lemons storage	0	80
	Grapes storage	−1	80
	Frozen meat storage	−17 to +15	85
	Defrozen meat storage	2	85
	Sugar storage	25	45
Printing	Binding	20	45
	Folding	24	65
	Printing room	23	60–80
	Printing lithographic room	23–35	50–60
	Impression cylinders' storage	20–30	50–55
	Web printing	–	40–55

(Continued)

TABLE 2.13 (*Continued*)
Environmental Condition Requirements in Major Industries Activities

Market Segment	Manufacturing	*T*, °C	RH, %
Pharmacy	Deliquescent powders	22	35
	Effervescent pills	25	40
	Powder liver extracts	20	20–30
	Powder and pills	20–25	30–35
	Tableting	20–25	40
	Packing	25	40
Chemical	Chemical laboratory	21–24	30–45
	Spray painting	20–80	30–50
	Plastics industry	–	5–30
Electronics, semiconductors, and optics	Semiconductor production	21–24	30–50
	Clean room	21	36–39
	EDS control	22	30–70
	Lithium batteries	–	up to 2
	Computer peripherals		50–50
	Production of hard disks		40–50
	Photo and film optics		40–55
Automobile	Car-window demisters	−20 to +80	50–100
	Motor assembly line	17–25	40–55
Materials	Wood drying	45–62	25–35
	Ceramics drying	5–100	0–50
Medical/hospitals	Operating room	20–25	50–60
	Infant incubator	10–30	50–80
	Pharmaceuticals	20–25	20–40
Medical equipment	Respiratory	20–30	80–100
	Sterilizers	>100	0–100
	Incubators	10–30	25–30
Meteorology	Weather radiosonde	–50 to +40	0–100

Source: Carr-Brion, K.G. et al., *Humidity and Moisture Measurement Course Notes*, Sira Test & Certification, Hawarden, UK, 1994; Fenner, R. and Zdankiewicz, E., *IEEE Sensors J.* 4, 309–317, 2001.

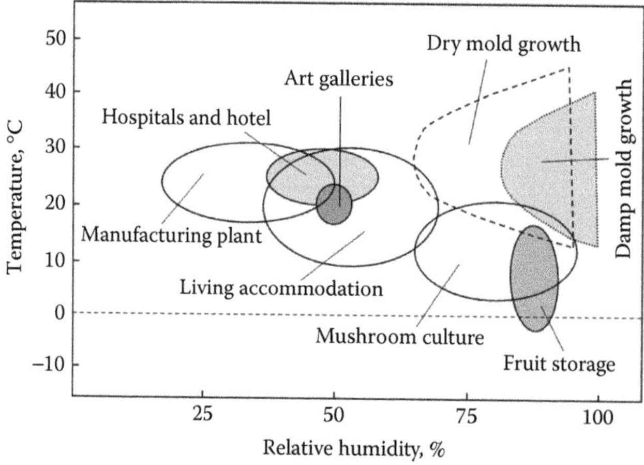

FIGURE 2.9 Various optimal temperature and relative humidity conditions for everyday applications in industry and at home. (Idea and data from Carr-Brion, K.G. et al., *Humidity and Moisture Measurement Course Notes*, Sira Test & Certification, Hawarden, UK, 1994. With permission.)

REFERENCES

Aguado E., Burt J.E., Burt J. (2003) *Understanding Weather and Climate*, Prentice Hall, Upper Saddle River, NJ.

Ahrens C.D. (2008) *Meteorology Today: An Introduction to Weather, Climate, and the Environment*, 9th ed. Cengage Learning, Belmont, CA.

Alten H. (1999) How temperature and relative humidity affect collection deterioration rates. In: *Collections Caretaker*, Vol. 2(2) Temperature and Relative Humidity, Northern States Conservation Center (http://www.collectioncare.org/pubs/v2n2p1.html)

Alwis L., Sun T., Grattan K.T.V. (2013) Optical fibre–based sensor technology for humidity and moisture measurement: Review of recent progress. *Measurement* 46, 4052–4074.

Amin M.N., Hossain M.A., Roy K.C. (2004) Effects of moisture content on some physical properties of lentil seeds. *J. Food Eng.* 65 (1), 83–87.

Andersen I., Lundquist G.R., Mølhave L. (1975) Indoor air pollution due to chipboard used as a construction material. *Atmospheric Environ.* 9, 1121–1127.

Arundel A.V., Sterling E.M., Biggin J.H., Sterling T.D. (1986) Indirect health effects of relative humidity in indoor environments. *Environ. Health Perspect.* 65, 351–361.

ASHRAE (2001) *BSR/ASHRAE Standard 55-1992R: Thermal Environmental Conditions for human Occupancy*. American Society of Heating, Refrigerating and Air-conditioning Engineering, Atlanta, GA (https://www.archive.org/details/ASHRAE551992).

Awakunit Y., Calderwood J.H. (1972) Water vapour adsorption and surface conductivity in solids. *J. Phys. D: Appl. Phys.* 5, 1038–1045.

Baughman A., Arens E. (1996a) Indoor humidity and human health. Part 1: Literature review of health effects of humidity–influenced indoor pollutants. *ASHRAE Trans.* 102 (1), 193–211.

Baughman A., Arens E. (1996b) Indoor humidity and human health: Part II: Buildings and their systems. *ASHRAE Trans.* 102 (1), 212–221.

Beuchat L.R. (1981) Microbial stability as affected by water activity. *C. F. W.* 26(7), 345–349.

Bornehag C.-G., Blomquist G., Gyntelberg F., Jarvholm B., Malmberg P., Nordvall L., Nielsen A., Pershagen G., Sundell J. (2001) Dampness in buildings and health. *Indoor Air* 11(2), 72–86.

Bradshaw V. (2006) Human comfort and health requirements. In: *The Building Environment: Active and Passive Control Systems*, 3rd ed. Wiley, Hoboken, NJ, pp. 3–37.

Cai Y., Gevelber M. (2017) Analysis of bending region physics in determining electrospun fiber diameter: Effect of relative humidity on evaporation and force balance. *J. Mater. Sci.* 52, 2605–2627.

Carr-Brion K.G., Forton A.G., Simpson R.J., Slight A. (1994) *Humidity and moisture measurement course notes*, Sira Test & Certification, Hawarden, UK.

CCA (2003) Basic conservation of archival materials: Revised edition, Chapter 3: Environment. Canadian Council of Archives (http://www.cdncouncilarchives.ca/rbch3_en.pdf).

Chen Z., Lu C. (2005) Humidity sensors: A review of materials and mechanisms. *Sensor Lett.* 3 (4), 274–295.

Chughtai A.R., Kim J.M., Smith D.M. (2003) The effect of temperature and humidity on the reaction of ozone with combustion soot: Implications for reactivity near the Tropopause, *J. Atmosp. Chem.* 45, 231–243.

Cockshull K.E. (Ed.) (1988) The effects of high humidity on plant growth in energy-saving greenhouses. In: *Proc. workshop held at Littlehampton*, UK, Commission of the European Communities, Brussels, Luxembourg. March 25 and 26, 1987.

De Moraes Bruna C.Q., Graziano K.U. (2012) Temperature and humidity in the storage area of sterile materials: A literature review. *Rev. Esc. Enferm. USP* 46 (5), 1212–1217.

Dijkink B.H., Tomassen M.M., Willemsen J.H.A., Doorn W.G. (2004) Humidity control during bell pepper storage, using a hollow fiber membrane contactor system. *Postharvest Biol. Technol.* 32, 311–320.

Dini-Andreote F., Stegen J., Dirk van Elsas J., Falcão Salles J. (2015) Disentangling mechanisms that mediate the balance between stochastic and deterministic processes in microbial succession. *PNAS* 112 (11), E1326–E1332.

Elgh B., Yuan N., Cho H.S. et al. (2014) Controlling morphology, mesoporosity, crystallinity, and photocatalytic activity of ordered mesoporous TiO_2 films prepared at low temperature. *APL Mater.* 2, 113313.

Elías V.R., Sabre E.V., Winkler E.L. et al. (2014) Influence of the hydration by the environmental humidity on the metallic speciation and the photocatalytic activity of Cr/MCM-41. *J. Solid State Chem.* 213, 229–234.

Ellsaesser H.W. (1984) The climatic effect of CO_2: A different view. *Atmos. Env.* 18, 431–434.

Fang L., Clausen G., Fanger P.O. (1998) Impact of temperature and humidity on perception of indoor air quality. *Indoor Air* 8, 80–90.

Fang L., Clausen G., Fanger P.O. (1999) Impact of temperature and humidity on chemical and sensory emissions from building materials. *Indoor Air* 9, 193–201.

Fenner R., Zdankiewicz E. (2001) Micro–machined water vapor sensors: A review of sensing technologies. *IEEE Sensors J.* 4 (1), 309–317.

Fitzpatrick S., McCabe J.F., Petts C.R., Booth S.W. (2002) Effect of moisture on polyvinylpyrrolidone in accelerated stability testing. *Int. J. Pharmaceut.* 246 (1–2), 143–151.

Fontana A.J. (1998) Water activity: Why it is important for food safety. In: *Proceedings of the First NSF International Conference on Food Safety*, Albuquerque, NM, pp. 177–185, November 16–18, 1998.

Fraden J. (2010) *Handbook of Modern Sensors: Physics, Designs, and Applications*. Springer-Verlag, New York.

Gor G.Y., Huber P., Bernstein N. (2017) Adsorption-induced deformation of nanoporous materials—A review. *Appl. Phys. Rev.* 4, 011303.

Guy R. (2001) Snack foods. In: Guy, R. (Ed.), *Extrusion Cooking: Technologies and Applications*. Woodhead Publishing, Cambridge, UK, pp. 161–181.

Hardy J., Stevenson J., Singh P., Mahapatra M., Wachsman E., Liu M., Gerdes K. (2015) Effects of humidity on solid oxide fuel cell cathodes, U.S. Department of Energy, Report PNNL-24115.

Hartmann D.L. (1994) *Global Physical Climatology*, 2nd ed. Elsevier Science, New York.

Hashiguchi N., Hirakawa M., Tocihara Y., Kaji Y., Karaki C. (2008). Effects of humidifiers on thermal conditions and subjective responses of patients and staff in a hospital during winter. *Appl. Ergon.* 39 (2), 158–165.

Hattingh J. (2001) The importance of relative humidity measurements in the improvement of product quality. In: *Proceedings of 2001 NCSL International Workshop & Symposium* (http://www.ncsli.org/i/c/TransactionLib/C01_R7.pdf)

Haverinen U. (2002) Modeling moisture damage observations and their association with health symptoms. PhD Thesis, National Public Health Institute, Kuopio, Finland.

Hemmers J.H., Winkler K.C., Kool S.M. (1960) Virus survival as a seasonal factor in influenza & poliomyelities. *Nature* 188 (4748), 430–431.

Hienonen R., Lahtinen R. (2007) Corrosion and Climatic Effects in Electronics. VTT Technical Research Centre of Finland, Otakaari, Finland.

HIOKI (1994), HIOKI Brochure, F8005E- 4ZM-05K, December 15, 1994.

ISO (2005) ISO 7730:2005: Ergonomics of the thermal environment: Analytical determination and interpretation of thermal comfort using calculation of the PMV and PPD indices and local thermal comfort criteria (https://www.iso.org).

Jefferies J. (1993) Product quality improvement with correct moisture measurement in thermal processes using electrolytic hygrometers. Industrial Heating (http://meeco.com).

Khanna V.K. (2012) Detection mechanisms and physic–chemical models of solid–state humidity sensors. In: Korotcenkov G. (Ed.) *Chemical Sensors: Simulation and Modelling*, Vol. 3, Solid State Devices, Momentum Press, New York.

Kim J.S., Lee T.K. (2001) Effect of humidity on the photocatalytic degradation of trichloroethylene in gas phase over TiO_2 thin films treated by different conditions. *Korean J. Chem. Eng.* 18 (6), 935–940.

Korotchenkov G., Brynzari V., Dmitriev S. (1999) Electrical behavior of SnO_2 thin films in humid atmosphere. *Sens. Actuators B* 54, 197–201.

Kovac M.G., Chleck D., Goodman P. (1978) A new moisture sensor for in-situ monitoring of sealed packages. *Solid State Technol.* 21, 35–39.

Krautgasser C., Danzer R., Supancic P., Bermejo R. (2015) Influence of temperature and humidity on the strength of low temperature co-fired ceramics. *J. Eur. Cer. Soc.* 35, 1823–1830.

Kukkonen J., Olsson T., Schultz D.M. et al. (2012) A review of operational, regional-scale, chemical weather forecasting models in Europe. *Atmos. Chem. Phys.* 12, 1–87.

Lakatos A. (2017) Investigation of the moisture induced degradation of the thermal properties of aerogel blankets: Measurements, calculations, simulations. *Energ Buildings* 139, 506–516.

Lamb H.H. (1972) *Climate: Present, Pat and Future*. Methuen Publishing, London, UK, p. 482.

Lancaster J.K. (1990) A review of the influence of environmental humidity and water on friction, lubrication and wear. *Tribology Int.* 23 (6), 371–389.

Lindzen R.S. (1990) Some coolness concerning global warming. *Bull. Am. Meteorol. Soc.* 71, 288–289.

Lindzen R.S., Chou M.D., Hou A.Y. (2001) Does the earth have an adaptive infrared iris? *Bull. Amer. Meteor. Soc.* 82 (3), 417–432.

Liu Y.J., Shi J., Zhang F., Liang H., Xu J., Lakhtakia A., Fonash S.J., Huang T.J. (2011) High–speed optical humidity sensors based on chiral sculptured thin films. *Sens. Actuators B* 156, 593–598.

Lull W.P., Banks P.N. (1995) *Conservation Environment Guidelines for Libraries and Archives*. Canadian Council of Archives, Ottawa, Canada.

Manabe S., Wetherald R.T. (1967) Thermal equilibrium of the atmosphere with a given distribution of relative humidity. *J. Atmos. Sci.* 24, 241–259.

McCartney A. (2012) Static electricity and relative humidity. In: *Fire Protection Practice Specialty*, Fire line 2012 (http://www.asse.org)

McMinn W.A.M., Magee T.R.A. (1999) Principles, methods and applications of the convection drying of foodstuffs. *Food Bioprod. Proc.* 77 (3), 175–193.

Michalski S. (1993) Relative humidity in museums, galleries, and archives: Specifications and control. In: *Bugs, Mold & Rot II: A Workshop on Control of Humidity for Health, Artifacts, and Buildings*, National Institute of Building Sciences, Washington, DC, pp. 51–62.

Michalski S. (1994) *Humidity Response Times of Wooden Objects*. Canadian Conservation Institute, Ottawa, Canada, pp. 1–4.

Michalski S. (1998) Climate control priorities and solutions for collections in historic buildings. *Forum* 12 (4), 8–14.

Muñoz E., García-Manrique J.A. (2015) Water absorption behaviour and its effect on the mechanical properties of flax fibre reinforced bioepoxy composites. *Int. J. Polymer Sci.* 2015, 390275.

Nilsson L.-O. (1980) Hygroscopic moisture in concrete: Drying, measurements & related material properties. PhD Thesis, Division of Building Materials, Lund Institute of Technology, Lund, Sweden.

NIOSH. (2013) *High Humidity Leads to Loss of Infectious Influenza Virus from Simulated Coughs*. National Institute for Occupational Safety and Health, Washington, DC.

Noiva R.M., Menezes A.C., Peleteiro M.C. (2014) Influence of temperature and humidity manipulation on chicken embryonic development. *BMC Veterinary Res.* 10, 234.

NPS Museum Handbook (2016) Part I: Museum Collection, National Park Service. Chapter 4, U.S. Department of the Interior (https://www.nps.gov/museum/publications/MHI/mushbkI.html).

Ogden S. (Ed.) (1999) *Guidelines for Preservation of Library & Archival Materials: A Manual*, 3rd ed. Northeast Document Conservation Center, Andover, MA, Chapter 2.1.

Oliver A. (1997) *Dampness in Buildings*, 2nd ed. revised by James Douglas and J. Stewart Stirling. Blackwell Science, Oxford, UK.

Qi S., Du Y., Zhang P., Li G., Zhou Y., Wang B. (2017) Effects of concentration, temperature, humidity, and nitrogen inertdilution on the gasoline vapor explosion. *J. Hazardous Mater.* 323, 593–601.

Pasanen A.L., Kalliokoski P., Jantunen M. (1994) Recent studies of fungal growth on buildings materials. In: Samson R.A., Flannigan B., Flannigan M.E., Verhoeff A.P., Adan O.C.G., Hoekstra E.S. (Eds.) *Health Implications of Fungi in Indoor Environments*, Elsevier, Amsterdam, the Netherlands, pp. 485–493.

Patissier B. (1999) Humidity sensors for automotive, appliances and consumer applications. *Sens. Actuators B* 59, 231–234.

PD 5454 (2012) Guide for the storage and exhibition of archival materials. British Standards Institution, London, UK.

Peat J.K., Dickerson J., Li J. (1998) Effects of damp and mold in the home on respiratory health: A review of the literature. *Allergy* 53(2), 120–128.

Peixoto J.P., Oort A.H. (1996) The climatology of relative humidity in the atmosphere. *J. Climate* 9, 3443–3463.

Pekkanen J., Lampi J. (2015) Moisture and mold damages of buildings in relation to health. *Duodecim.* 131(19), 1749–1755.

Perez Flores J.G., Ordaz J.J., Morga J.A., Gonzalez Olivares L.G., Ovando A.C., Contreras Lopez E. (2017) Influence of water sorption phenomena on the shelf life of third generation snacks. *J. Food Process Eng.* 40, e12328.

Primavera A., Trovarelli A., Andreussi P., Dolcetti G. (1998) The effect of water in the low-temperature catalytic oxidation of hydrogen sulfide to sulfur over activated carbon. *Appl. Catal. A: General* 173, 185–192.

Prowse M.S., Wilkinson M., Puthoff J.B., Mayer G., Autumn K. (2011) Effects of humidity on the mechanical properties of gecko setae. *Acta Biomater.* 7 (2), 733–738.

Puthoff J.B., Prowse M.S., Wilkinson M., Autumn K. (2010) Changes in materials properties explain the effects of humidity on gecko adhesion. *J. Exp. Biol.* 213, 3699–3704.

Ruckerl A., Zeisel R., Mandl M., Costina I, Schroeder T., Zoellner M.H. (2017) Characterization and prevention of humidity related degradation of atomic layer deposited Al_2O_3. *J. Appl. Phys.* 121, 025306.

Samson R.A., Flannigan B., Flannigan M.E., Verhoeff A.P., Adan O.C.G., Hoekstra E.S. (Eds.) (1994) *Health Implications of Fungi in Indoor Environments*, Elsevier Science, Amsterdam, the Netherlands, pp. 529–538.

Scott W.J. (1957) Water relations of food spoilage microorganisms. *Adv. Food Res.* 7, 83–127.

Srivastava R. (2012) Humidity sensor: An overview. *Intern. J. Green Nanotechnol.* 4, 302–309.

Still M., Venzke H., Durst F., Melling A. (1998) Influence of humidity on the convective heat transfer from small cylinders. *Exp. Fluids* 24, 141–150.

Sun D.Z., Lindzen R.S. (1993) Distribution of tropical tropospheric water vapour. *J. Atmos. Sci.* 50, 1643–1660.

Sun D.-Z., Oort A.H. (1995) Humidity–temperature relationships in the tropical troposphere. *J. Climate* 8, 1974–1987.

Tanner D.M., Walraven J.A., Irwin L.W., Dugger M.T., Smith N.F., Eaton W.P., Miller W.M., Miller S.L. (1999) The effect of humidity on the reliability of a surface micromachined microengine. In: *Proceedings of IEEE International Reliability Physics Symposium*, San Diego, CA, pp. 189–197, March 21–25, 1999.

Thomson G. (1986) *The Museum Environment*, 2nd ed. Butterworths, London, UK.

Toftum J., Jørgensen A.S. (1998) Effect of humidity and temperature of inspired air on perceived comfort. *Energ. Build.* 28, 15–23.

Traversa E. (1995) Ceramic sensors for humidity detection: the state–of–the–art and future developments. *Sens. Actuators B* 23, 1335–1356.

Viitanen H. (1997) Critical time of different humidity and temperature conditions for the development of brown rot decay in pine and spruce. *Holzforschung* 51(2), 99–1067.

Voutilainen J. (2005) Methods and instrumentation for measuring moisture in building structures. PhD Thesis, Helsinki University of Technology, Espoo, Finland.

Walter H., Dermitzaki E., Wunderle B., Michel B. (2010) Influence of moisture on humidity sensitive material parameters of microelectronics relevant polymers. In: *Proceedings of NSTI–Nanotech 2010*, Vol. 2, pp. 178–181 (http://www.nsti.org).

Weinstock E.M., Smith J.B., Sayres D.S. et al. (2009) Validation of the Harvard Lyman-α in situ water vapor instrument: Implications for the mechanisms that control stratospheric water vapor. *J. Geophys. Res.* 114, D23301.

Wexler A. (1975) A study of the National humidity and moisture system. NBSIR 75-933. National Bureau of Standards, Washington, DC.

Wiederhold P.R. (1997) *Water Vapor Measurement: Methods and Instrumentation*, CRC Press, New York.

Wolkoff P. (1998) Impact of air velocity, temperature, humidity, and air on long–term VOC emissions from building products. *Atmos. Environ.* 32, 2659–2668.

Zhang L., Tarascon J.-M., Sougrati T.M., Rousse G., Chen G. (2015) Influence of relative humidity on the structure and electrochemical performance of sustainable $LiFeSO_4F$ electrodes for Li–ion batteries. *J. Mater. Chem. A* 3, 16988–16997.

Section II

Humidity Monitoring Using Absorption of Electromagnetic Radiation

3 Optical Hygrometers

3.1 INTRODUCTION

As it was shown in Chapter 2, the determination of humidity is of great importance. Therefore, humidity control becomes imperative in all fields of our activity, from production management to creating a comfortable environment for our living, as well for understanding the nature of the changes happening in the climate. In the second volume, *Electronic and Electrical Humidity Sensors*, it will be shown that there are a large number of various devices that allow controlling humidity. Along with conventional devices such as mechanical (hair) hygrometers, psychrometers, chilled mirror hygrometers, the Dunmore cells, coulometric hygrometers, and so on, humidity can be controlled using capacitive-, resistive-, surface acoustic wave (SAW)-, quartz crystal microbalance (QCM)-, cantilever-, field effect transistor (FET)-, Schottky barrier-, field emission-based humidity sensors, and many others. In the past decades, it was done a lot for the development of these devices and for the development of different measuring systems with an increased efficiency. However, the process of developing new humidity sensors as well as improving older types of devices used for humidity measurement is still ongoing. New technologies and the toughening of ecological standards require more sensitive instruments with faster response times, better selectivity, and improved stability. In this regard, optical humidity sensors, which are considered in this book, are of particular interest.

Optical hygrometry is a classic method of humidity measurement. However, unlike above-mentioned devices, the work of optical hygrometers is based on other principles. Detection and determination of the water vapor concentration take place through the measurements of optical properties. The first known measurements of the water vapor absorption in the infrared (IR) were undertaken by Fowle (1912) for the 1.13 and 1.37 μm bands using spectroscopic techniques. However, optical methods had the greatest development only in the last 30 years, when there appeared the low-cost optoelectronic components such as light-emitting diodes (LEDs), laser diodes (LDs), photodetectors (PDs), and a fast data acquisition, and the data processing technique.

Optical hygrometers form a sufficiently large class of devices, which can be divided into subgroups on the basis of (1) the waveband used for the detection of water vapor (visible, infrared [IR], and ultraviolet [UV]) (Table 3.1), (2) the fraction of light, which is detected after interaction with the sample (i.e., the transmitted, reflected, or absorbed fraction), and (3) spectral width of light (e.g., monochromatic, polychromatic, and reference light with continuous spectrum).

The most studied and the most popular are optical monochromatic hydrometers of absorption type, working in IR (1–14 μm) and UV (~100 nm) spectral ranges (Cerni 1994; Wiederhold 1997), where there is a large number of absorption bands, attributed to different gases (Figure 3.1). The areas of the electromagnetic (EM) spectrum that are absorbed by atmospheric gases such as oxygen, carbon dioxide, ozone, and other gases are known as absorption bands. Absorption bands corresponding to various gases in the IR и UV spectral regions are presented in Table 3.2. The main absorption bands, peculiar to the liquid water and the water vapor, are in the same band (Table 3.2 and Figure 3.1). It is important that the water molecule absorbs electromagnetic radiation both in a range of wavebands, and on discrete wavelengths.

It was established that absorption of electromagnetic radiation by water depends on the state of the water; and absorption in different spectral ranges has different nature. The water molecule, in the gaseous state, has three types of transitions that can give rise to absorption of electromagnetic radiation:

- Rotational transitions, in which the molecule gains a quantum of rotational energy.
- Vibrational transitions in which a molecule gains a quantum of vibrational energy.
- Electronic transitions in which a molecule is promoted to an excited electronic state.

It is important to know that the electronic levels are associated with the energy of the electron subsystem. The transition from one electronic level to another can be considered as the transition of one of the electrons from one orbital to another. The vibration levels are related with the vibration motions of the molecules. The transitions between them have almost no effect on the electron

TABLE 3.1
Spectral Regions Important for Optical Sensors

Range	$h\nu$, eV	λ, nm	ν, cm^{-1}
Infrared (IR)	0.025–0.4	3000–50000	200–3300
Near-infrared (NIR)	0.4–1.6	780–3000	3100–13000
Visual (Vis)	1.6–3.3	380–780	13000–26000
Ultraviolet (UV)	2.3–6.2	200–380	26000–50000

Source: Korotcenkov, G. (Ed.), *Electrochemical and Optical Sensors*, Momentum Press, New York, pp. 311–476.
λ is wavelength; $\nu = \lambda^{-1}$ is wave number; $h\nu$ is energy of photons.

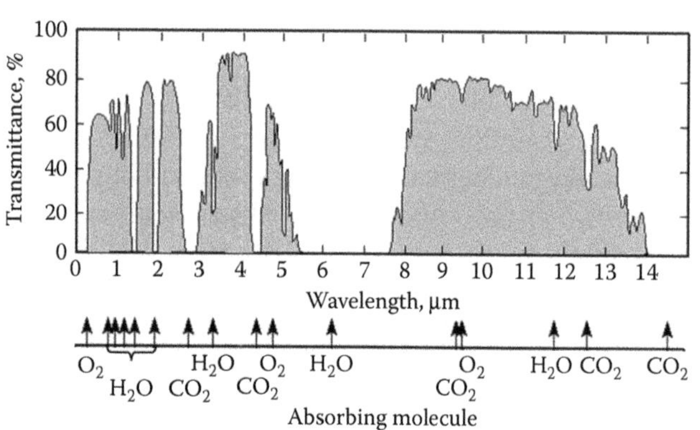

FIGURE 3.1 Transmission spectra of atmosphere. It is seen that the atmospheric transparence is controlled mainly by absorption by atmospheric gases and vapors such CO_2, O_2 and H_2O. (Data from http://www.web.archive.org; http://www.vision-systems.com.)

subsystem. The rotational levels arise from rotations of molecules as a whole.

The energies of the transitions increase in order: rotational<vibrational<electronic (Tkachenko 2006). This is the order of decreasing mass of the considered subsystem: molecule<atom>electron. A smaller mass results in a higher frequency of oscillators ($\nu \sim 1/\sqrt{m}$ for harmonic oscillators). Typical ranges of the transition are summarized in Table 3.3 and presented in Figure 3.2.

As it is seen in Table 3.4, energy of transition can be presented in different units such as energy (ev), wavelength, and frequency. One should note that the wavelength, frequency, and energy are equivalent measures in spectroscopy. This takes place because the energy of a photon determines frequency (and wavelength) of the electromagnetic wave (by famous Planck formula $E = h\nu$ or $E = hc/\lambda$ in vacuum). Appropriate functions which can be used for conversion of these units are presented in Table 3.4. One practical inconvenience of this is that numerous units used to characterize one and the same parameter-transition energy. Usually the wavelengths and energy are being used in UV spectroscopy. IR spectroscopy usually uses the wavelength and the wavenumber. The wavenumber is used in IR spectroscopy, because the wavenumber is directly proportional to the frequency and energy ($\upsilon = \nu/c = E/hc$), and therefore it is convenient to use this unit when energy or frequency dependence is presented. The microwave and the terahertz (THz) spectroscope usually use the frequency.

TABLE 3.2
Main Visible and Near-IR Absorption Bands of Atmospheric Gases

Gas	Center		Band Interval
	ν (cm^{-1})	λ (µm)	(cm^{-1})
H_2O	3703	2.7	2500–4500
	5348	1.87	4800–6200
	7246	1.38	6400–7600
	9090	1.1	8200–9400
	10638	0.94	10100–11300
	12195	0.82	11700–12700
	13888	0.72	13400–14600
	visible		15000–22600
CO_2	2526	4.3	2000–2400
	3703	2.7	3400–3850
	5000	2.0	4700–5200
	6250	1.6	6100–6450
	7143	1.4	6850–7000
O_3	2110	4.74	2000–2300
	3030	3.3	3000–3100
	visible		10600–22600
O_2	6329	1.58	6300–6350
	7874	1.27	7700–8050
	9433	1.06	9350–9400
	13158	0.76	12850–13200
	14493	0.69	14300–14600
	15873	0.63	14750–15900
N_2O	2222	4.5	2100–2300
	2463	4.06	2100–2800
	3484	2.87	3300–3500
CH_4	3030	3.3	2500–3200
	4420	2.20	4000–4600
	6005	1.66	5850–6100
CO	2141	4.67	2000–2300
	4273	2.34	4150–4350

Source: Korotcenkov, G. (Ed.), *Electrochemical and Optical Sensors*, Momentum Press, New York, pp. 311–476. With permission.

TABLE 3.3
Energies and Spectral Ranges of Different Types of Transitions

Type	Energy, J	Frequency, Hz	Wavelength, μm
Electronic	$(2–10)\cdot 10^{-19}$	$(3–15)\cdot 10^{14}$	0.2–1.0
Vibrational	$(2–20)\cdot 10^{-20}$	$(3–30)\cdot 10^{13}$	1.0–10
Rotational	$(2–20)\cdot 10^{-21}$	$(3–30)\cdot 10^{12}$	10–100

Source: Tkachenko, N.V., *Optical Spectroscopy. Methods and Instrumentations*, Elsevier, Amsterdam, the Netherlands, 2006.

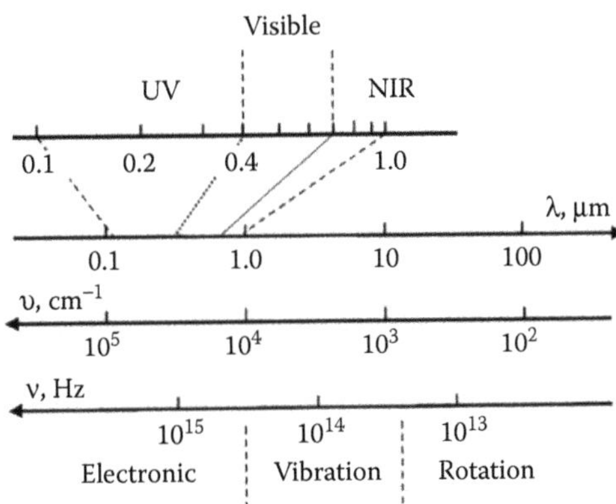

FIGURE 3.2 Relations between different scales used in spectroscopy: Wavelength, λ, wavenumber, υ, and frequency, ν. (Data extracted from Tkachenko, N.V., *Optical Spectroscopy. Methods and Instrumentations*, Elsevier, Amsterdam, the Netherlands, 2006.)

TABLE 3.4
Parameters and Units Usually Used in Spectroscopy

Quantity	Relationship	Usually Used Units
Wavelength	λ	μm, nm
Wavenumber	$\upsilon = 1/\lambda$ (cm)	cm^{-1}
Frequency	$\nu = c/\lambda$ (m) ($c = 3 \times 10^8\ m\cdot s^{-1}$)	Hz
Energy	$h\nu = 1.24/\lambda$ (μm)	eV
Velocity of light	$C = \nu\ \lambda$ (m)	$m\ s^{-1}$

Note: h is Planck's constant ($6.62606957 \times 10^{-34}$ J·s; c is the speed of light in vacuum (299792458 m/s).

As it is seen in Figure 3.2, rotational transitions are responsible for absorption in the microwave and far-infrared (FIR), vibrational transitions in the mid-infrared (MIR) and near-infrared (NIR). In the liquid water, the rotational transitions are effectively quenched. In reality, vibrations of molecules in the gaseous state are accompanied by rotational transitions, giving rise to a vibration–rotation spectrum. Furthermore, vibrational overtones and combination bands occur in the NIR region. Therefore, vibrational bands have rotational fine structure. In crystalline ice the vibrational spectrum is also affected by the hydrogen bonding and there are lattice vibrations causing absorption in the FIR. Electronic transitions occur in the UV regions. Electronic transitions of gaseous molecules will show both vibrational and rotational fine structure.

UV and IR absorption optical hygrometers offer several advantages in various fields of application (Wernecke and Wernecke 2014). These devices are capable of observing a sample in its dynamic environment, no matter how distant, difficult to reach, or hostile this environment is. These devices are intrinsically safe, involving a low optical power, and are nonelectrical at the sensing point. These hygrometers are electrically passive and immune to electromagnetic disturbances, are geometrically flexible and corrosion-resistant, and are compatible with telemetry. In principle, these devices are capable of continuous use. These features impart to the optical hygrometers immense potential importance in biomedical, process, and environmental monitoring applications.

It should be noted that there are also hygrometers, which operate in the THz and microwave regions of the spectrum. These devices differ significantly from the optical hygrometers, because basically forced to use other radiants and radiation detectors. However, the base of these sensors operation is also the absorption of radiation by molecules of water vapor. Therefore, the techniques, which are often classified as optical hygrometry, more correctly to classify as electromagnetic radiation (EMR) absorption hygrometry (WMO 1992, 2011).

It is important that in contrast to the absorption bands, there are areas of the EM spectrum, where the atmosphere is transparent (little or no absorption of energy) to specific wavelengths (Figure 3.3). These wavelength bands are known as atmospheric *windows* because they allow the energy to easily pass through the atmosphere to the Earth's surface. As it is seen in Figures 3.1 and 3.3, a clear electromagnetic spectral transmission *window* can be seen between 8 and 14 μm. A fragmented part of the *window* spectrum (one might say a louvered part of the *window*) can also be seen in the visible to mid-wavelength IR between 0.2 and 5.5 μm. You can see that there is plenty of atmospheric transmission of radiation at 0.5, 2.5, and 3.5 μm, but in contrast there is

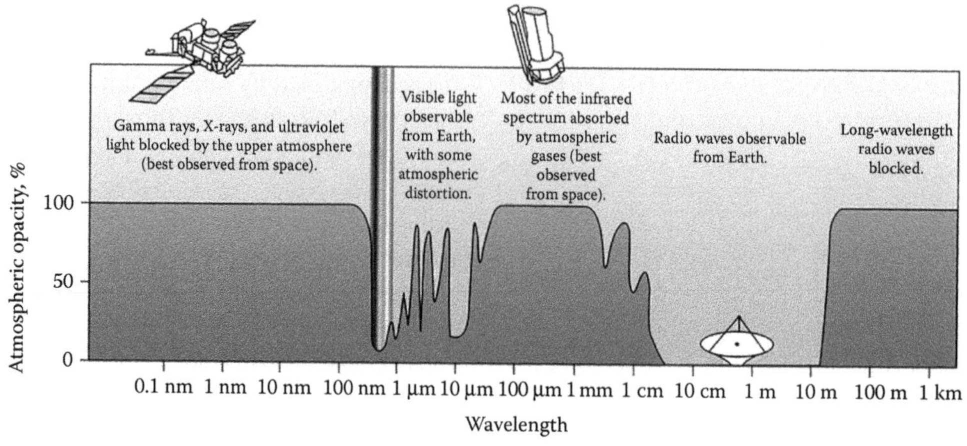

FIGURE 3.3 Rough plot of the Earth atmospheric transmittance (or opacity) to various wavelengths of electromagnetic radiation, including visible light. (From http://www.nasa.gov. This file is in the public domain in the United States because it was solely created by NASA. NASA copyright policy states that "NASA material is not protected by copyright unless noted.")

a great deal of atmospheric absorption at 2.0, 3.0, and about 7.0 μm. Atmospheric *windows* are also observed in spectral range from 4 cm to 20 m. It is understandable, that both passive and active remote-sensing technologies do best if they operate within the atmospheric windows. For example, the range from 4 cm to 20 m is used for operation of the radio telescopes, studying the Universe.

Among hygrometers operated using the absorption of electromagnetic radiation, it is also necessary to allocate the hygrometers, which are based on the detection of the fluorescence intensity of water molecules induced by the UV irradiation, and the hygrometers, using for the water vapors detection a remote sounding principles such as light detection and ranging (LIDAR).

3.2 PRINCIPLES OF OPERATION OF ELECTROMAGNETIC RESONANCE HYGROMETERS BASED ON ABSORPTION

The principle of the method, based on the absorption of electromagnetic radiation, is to determine the attenuation of radiation in a waveband that is specific to water–vapor absorption, along the path between a source of the radiation and a receiving device. Thus, the transmitted radiation carries the sought for information on the analyte. In particular, its intensity (I) is related to the analyte concentration (C), and this relation is expressed by the Beer–Lambert law, which is a combination of the Bouguer–Lambert law and the Beer law. The former gives an expression for the light intensity attenuation dI due to absorption and scattering along an optical path of dl. The latter describes the dependency of the intensity attenuation dI from the concentration c of sample. The differential Beer–Lambert law Equation 3.1 states that a variation in intensity dI along a differential path length dl is related to an absorption coefficient $\alpha(\lambda)$ and to the concentration C of the sample:

$$\frac{dI}{dl} = -\alpha(\lambda)\cdot C \cdot I \tag{3.1}$$

Integration of both sides of Equation 3.1 yields

$$I(l) = I_0 e^{-\alpha(\lambda)Cl} \text{ or} \tag{3.2}$$

$$ln\left(\frac{I_0}{I}\right) = \alpha(\lambda) lC = A \tag{3.3}$$

where:

I_0 is the intensity of the incident radiation
$\alpha(\lambda)$ is molar absorptivity
l is the length of the optical path in the absorbing phase

The molar absorptivity is characteristic of a species at a given wavelength. Thus, the analyte concentration is a linear function of the quantity log(I_0/I), which is defined as absorbance (A).

In our case, the Beer–Lambert principle states that when light energy at certain wavelengths travel through gas, a certain amount of the energy is absorbed by the water within the path. The amount of light energy lost is related to the concentration of water.

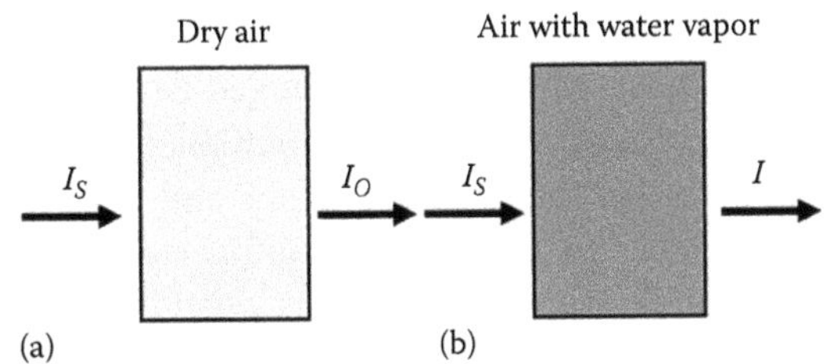

FIGURE 3.4 (a,b) Two transmittance measurements necessary for determining the concentration of water vapors in gas.

One should note that this equation is valid only under the assumptions of

- Parallel and monochromatic incident light
- Homogeneous absorption in the medium
- No other interaction processes within the medium
- Negligible scattering and reflection at entrance and exit windows

Figure 3.4 shows the two transmittance measurements that are necessary to use absorption to determine the concentration of water vapors in the gas. The diagram in Figure 3.4a is for gas only, that is, dry gas, and the diagram in Figure 3.4b is for an absorbing sample in the same gas, that is, for air or gas contenting the water vapors. In this example, I_S is the source light power that is incident on a sample; I is the measured light power after passing through the wet gas or air, and sample holder; and I_0 is the measured light power after passing through only the dry gas and sample holder. The measured transmittance in this case is attributed to only the analyte, that is, water vapors.

With regard to the limitations to Beer's law, then one can select the following: at first, Beer's law has fundamental limitations. This law is valid only for low concentrations of analyte. At higher concentrations, the individual particles of analyte no longer behave independently of each other. The resulting interaction between particles of analyte may change the analyte's absorptivity. In addition, the analyte's concentration via the sample's refractive index (RI) can influence the values of $\alpha(\lambda)$. Only for sufficiently low concentrations of analyte, the RI is essentially constant and the calibration curve is linear. At second, Beer's law has chemical limitations. A chemical deviation from Beer's law may occur if the analyte is involved in an equilibrium reaction. At third, Beer's law has instrumental limitations. There are two principal instrumental limitations to Beer's law. The first limitation is that Beer's law assumes that the radiation reaching the sample is of a single wavelength—that is, that the radiation is purely monochromatic. However, even the best wavelength selector passes radiation with a small, but finite effective bandwidth. Polychromatic radiation always gives a deviation from Beer's law, but the effect can be reduced to make absorbance measurements at the top of a broad absorption peak. Stray radiation is the second contribution to instrumental deviations from Beer's law. Stray radiation arises from imperfections in the wavelength selector that allow light to enter the instrument and reach the detector without passing through the sample.

3.3 INSTRUMENTATION AND CONFIGURATION OF OPTICAL HYGROMETERS—GENERAL VIEW

In principle optical devices, which can be used as optical hygrometers, can be classified depending on the way how the specific wavelength of interest is extracted from the light source. On one hand, a dispersive spectrometer includes an optical device such as a prism or a diffraction grating to spread the light spectrum and isolate a specific wavelength band. It uses their absorption characteristics to measure ingredients and quantity of a sample. On the other hand, nondispersive spectrometer isolates the specific wavelength band of the light that corresponds to the absorption spectrum of the target gas using narrowband transmission filters. Fourier transform spectroscopy represents another gas detection technique, especially in the IR spectra Fourier transform infrared (FTIR) spectroscopy, which is widely used for multipurpose applications. It is based on Michelson interferometer and a later Fourier transform to obtain the spectrum. Of course, nondispersive technique as the easiest to use, does not require the use of expensive and bulky spectrometers and enables the development of compact devices. It is the most suitable for the development of optical hygrometers.

The instrumentation associated with optical hygrometer is similar to that associated with conventional spectrophotometric techniques (Smith 2007). Typical optical hygrometers consist of an IR or UV light source (L), one or more optical filters (F), entrance and exit windows, a chopper for light pulsing (C), and an IR or UV detectors (D) with associated electronics (signal converter and amplifier, and digital and analog processing units). The components are assembled together into a sampling chamber (S) that acts as a sealed optical bench, while allowing humidified gas to flow through (Figure 3.5). The utilized wavelength has to be fitted to the specific measurement situation, taking into account technical capabilities (the presence of appropriate source and detector of radiation), and the presence of other components that may also

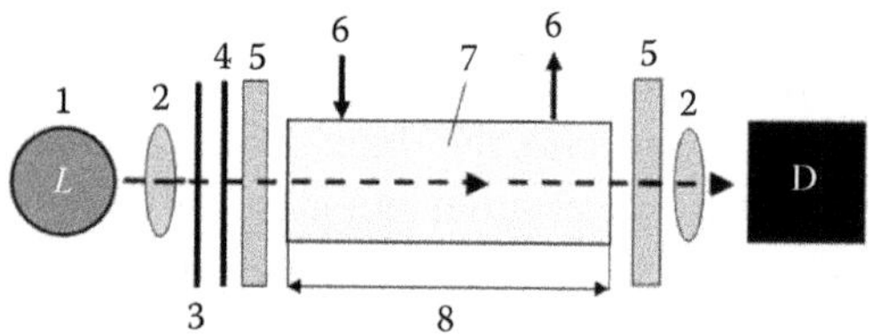

FIGURE 3.5 Scheme of principal setups for optical humidity measurement in gas: 1-light source; 2-focusing elements; 3-modulator; 4-chopping; 5-optical filters; 6-sample (air, gas); 7-measuring cell; 8-measurement path; 9-detector.

interact with the radiation in selected range, and expected percent of water vapor in the sample.

The most commonly used source of radiation in optical hygrometers is hydrogen lamps, halogen lamps, tungsten light bulbs, LEDs or LDs. Until recently, the most widely used source in IR range hygrometers as well as in UV range hygrometers were tungsten lamps, filtered to isolate a pair of wavelengths in required region. They are small and rugged. In addition, the emission spectrum of tungsten lamp can be modified by variation of lamp pressure and addition of other gas components. Such additives increase the number of available spectral lines. However, in the past decade the usage of LEDs and LDs has received widespread use. Recent developments in a diode laser technology, including the relatively recent availability of novel quantum cascade laser (QCL) sources, have led to the development of highly sensitive and selective *diode laser sensing systems.* One should note that while IR spectroscopy such as FTIR is well suited to multicomponent analysis of gases and other chemical species for a range of processes, the laser spectroscopy is the method of choice for the trace gas analysis because of its higher sensitivity and specificity, which arises when a spectrally narrow laser source probes a narrow absorption feature of the analyte. In addition, the application of (monochromatic) laser sources provides the capability of multicomponent detection. Comprehensive reviews of diode laser–based gas sensor systems have been provided by Werle (1996, 1998, 2004), Werle et al. (2002) and Allen (1998). The reasons of such wide application of lasers in optical spectroscopy are their small dimensions, their low energy demands, and their manufacturing technology, which is compatible with production of IC and solid-state sensors. LDs are made from the same materials as LEDs, but their structure is somewhat more complex, and they are processed differently. At present, LEDs and LDs, sources with relatively intense radiation of narrow bandwidth, are available over a range of wavelengths. The description of these devices can be found in the book by Bass and Van Stryland (2002).

Photomultiplier tubes, photodiodes, photodiode arrays, phototransistors, and the charge-coupled devices (CCDs) are used as optical detectors. Photomultiplier tubes are the most sensitive photodetection systems and are preferred for a low-level light detection. Phototransistors and photodiodes (which are also known as quantum detectors) offer compactness and miniaturization of analytical systems, and are useful for measurements in the UV, visible, and NIR spectral regions. Photodiode arrays and CCDs are multichannel detectors that can perform simultaneous detection of the dispersed optical radiation.

The optical filters are among typical wavelength selectors employed in optical chemical hygrometer. The optical filters are selected to optimize detection only in that part of the spectrum around the water absorption lines.

At present, a multitude of configurations have been adopted in optical hygrometers. However, they are mostly attributed to two principal methods used for determination of the degree of attenuation of the radiation, such as (WMO 1992, 2011):

1. Transmission of narrow-band radiation at a fixed intensity to a calibrated receiver. This approach is illustrated in Figure 3.5.
2. Transmission of radiation at two wavelengths, one of which is strongly absorbed by water vapor and the other being either not absorbed or only very weakly absorbed. If a single source is used to generate the radiation at both wavelengths, the ratio of their emitted intensities may be accurately known, so that the attenuation at the absorbed wavelength can be determined by measuring the ratio of their intensities at the receiver.

The simplest embodiment of the measurement method is shown in Figure 3.6. The two beams, each with a different

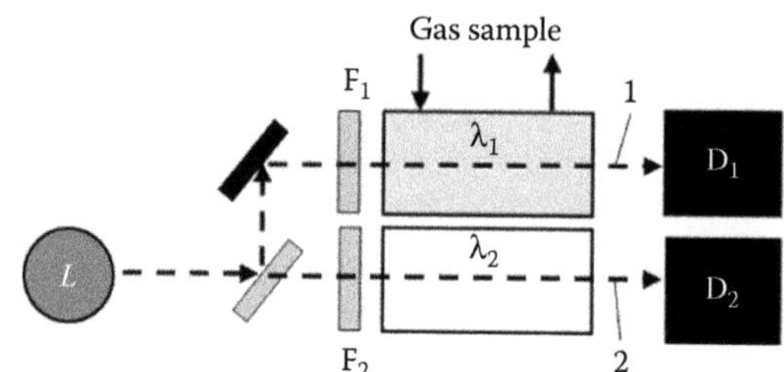

FIGURE 3.6 Setup of a two-beam measurement device: 1-measurement path, 2-reference path. The two-channel transmission cell combines reference and signal pathways. One cell is filled with reference gas, which does not interact with infrared radiation. The gas to be measured can be inserted into other cell through the gas inlet. Two detectors D_1 and D_2 act as receivers to record the transmitted intensity through both pathways.

wavelength, are created by a radiation source with a broad spectrum (e.g., lamp of LED) and an electronic or mechanical chopper that switches between two optical filters F_1 and F_2, which are transparent for different wavelengths. This results in a wavelength modulation between λ_1 and λ_2 (or more). This approach can also be realized using two lasers with different spectrum of radiation.

As follows from the previous discussions, optical measurements of the transmittance of the gaseous medium serve as the physical basis for instrumentally determining the volume concentration of a specific gaseous component of the atmosphere from its absorption bands. However, experiments have shown that optimal measurements took place only for transmittance changes in the range from 15% to 85%; outside this range, the errors caused by errors in the photometry and the zero setting of the device became substantial. Unfortunately, in practice, we usually encounter other situations, and during direct absorption measurements in the atmosphere with low concentration of water vapor we have to resolve very small changes in a large signal; essentially, this is the equivalent of finding a needle in a haystack. Thus, measuring low concentrations of the sample material is limited by the noise present in the measurement of the background. It was found that a good way to solve the above-mentioned problem is to measure not the transmittance, but its derivative (differential). That is why optical humidity analyzers mainly present a differential optical absorption spectroscopy (DOAS) (Sigrist 1994; Kosterev and Tittel 2002). In a DOAS, to reduce errors due to fluctuations of the source, temperature, and so on, a second channel with cell is utilized for reference. The measurement signal and reference signal can then be combined to give a measurement value. A simplified optical scheme for such a humidity hygrometer is shown in Figure 3.7. The optical resolution is determined by the total length and precision of mirror displacement.

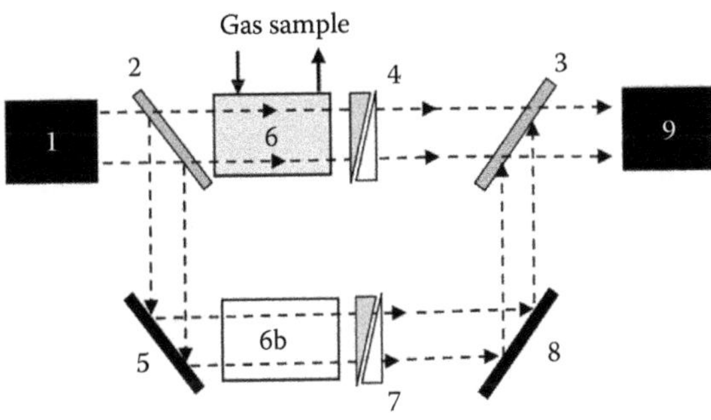

FIGURE 3.7 Optical scheme for moisture measurement based on differential spectroscopy: 1, light source; 2 and 3, semitransparent mirrors; 4 and 7, attenuators; 5 and 8, mirrors; 6, sample cell; 6b, reference cell; 9, analyzer.

A structural two-channel layout can be implemented in which the signal is formed as the difference between the working and reference fluxes, using a null method of comparison, with optical, electrical, or gaseous compensation. This method assumes that the signals from the two channels are equal when there is no absorbing gas in the measurement cell. So, in comparison with direct spectroscopy, DOAS produces a signal which is directly proportional to the concentration of the species. The dynamic range of gas concentrations that can be measured in the best devices with zero compensation is limited by errors in setting and maintaining in time the equality of the channels, as well as by the photometric accuracy, which is usually maintained to within 1%–3% (Mirumyants and Maksimyuk 2002). The incorporation of various automatic compensation systems makes it possible to use simple methods to compensate for the errors associated with ambient temperature variation, the dustiness of the optical elements of the system, and so on.

However, it should be recognized that the most effective approach to the development of optical hygrometers is an approach based on the use of tunable diode lasers (TDLs). The hygrometers developed based on this principle are called a tunable diode laser absorption spectrometers (TDLAS), or TDLAS hygrometers. Advantages of such devices are listed in Table 3.5. Examples of the TDLAS hygrometers realization will be considered in Section 3.4.2. As is clear from the

TABLE 3.5
Advantages and Disadvantages of TDLAS Hygrometers

Advantages	Disadvantages
• Very fast response in both directions: dry to wet or wet to dry • Noncontact measurements • No sensing surface to degrade due to exposure during exploitation in various atmospheres • High long-term stability • No zero or span gases needed • Based on fundamental measurement • Immune to many contaminants present in the air or natural gas	• Relatively expensive • Must be calibrated using a test gas with the same basic major components of the measured gas • Measurement is made a close to atmospheric pressure • Strong pressure and temperature control • Skilled support is required because troubleshooting and repair can be very difficult

title, the basis of these devices constitutes the TDLs appeared in the 1960s. The main advantage of TDLs is the ability to change the wavelength of the radiation. In principle, it is possible to change the frequency of the emitted light by changing the temperature. But in TDLs, the changing of its wavelength is reached by ramping the injection current. Emission bandwidths for the TDLs are on the order of 10^{-4}–10^{-5} cm^{-1}. During the measurements, the laser temperature is kept constant by using of a thermoelectric (Peletier) heat pump array. The laser is also modulated at high frequency. Consequently, by passing light at the water absorption frequency through a sample chamber containing air or natural gas of a certain moisture content, it is possible to precisely establish the water content by measuring the amount of loss in the absorption spectrum obtained by adjustment of the laser radiation. One should note that TDLAS has the ability to rapidly tune the lasers, so techniques such as wavelength modulation spectroscopy (WMS), which yields dramatic sensitivity enhancements over a direct absorption approach, are easily implemented (Linnerud et al. 1998). In WMS, the wavelength of the optical source is modulated, so that a small region of the absorption spectrum of an analyte, containing a distinct absorption peak, is repetitively scanned. As the wavelength of the source scans through the spectral feature of interest the relative attenuation of the source power is varied. Thus, by modulating the wavelength of the source, a time-varying signal (i.e., the optical power) is produced at the detector. A phase-sensitive detection scheme is used to detect the presence of, and quantify, the time-varying signal produced by the presence of the analyte species. Since the line widths of the lasers are much narrower than the individual absorption lines observed for the gas-phase molecules, the TDLAS technique obtains an extremely high spectral resolution, which results in the ability to isolate a single rotation–vibrational transition line of the analyte species (Figure 3.8), thereby reducing (close to eliminating) the background interference encountered by conventional IR–NIR spectrometers. Such ability of TDLAS provides both extremely high specificity for the analyte, even in a complex sample matrix, and detection limit that is several orders below a conventional absorption measurement.

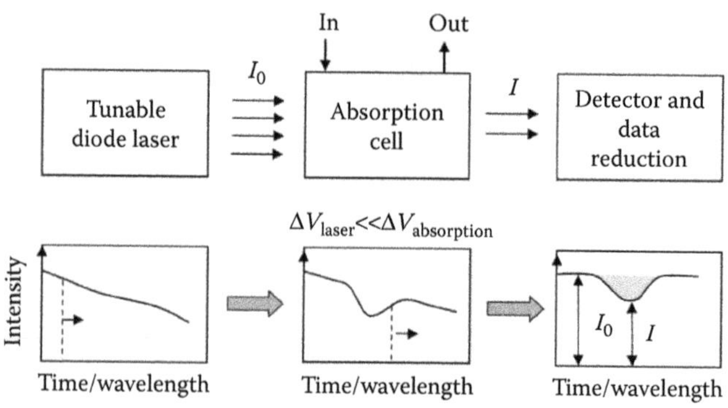

FIGURE 3.8 Basic setup of tunable laser absorption spectroscopy (TLAS). (Idea from Bange, J. et al., *Airborne Measurements for Environmental Research*, Wiley-VCH, Weinheim, Germany, 2013.)

Phase-sensitive detection, normally employed in WMS, permits the measurement of different multiples of the drive frequency used to modulate the output of the tunable diode laser. Most implementations of TDLAS technology have made use of the second harmonic signal (known as 2F) for two main reasons. First, if the 1F spectrum is a very close approximation to the derivative of the absorbance peak, whereas the second harmonic signal approximates the second derivative of the absorption spectrum. Such operation removes sloping backgrounds and offsets that are the result of the less than ideal output characteristics of the TDLs. This means that the 2F signal produces a zero baseline signal. And second, the second harmonic spectra display a peak, which coincides with the peak in the absorbance spectrum. Moreover, the 2F peak height is directly proportional to the partial pressure of water in the absorption cell. By simultaneously measuring the cell total pressure the concentration of water vapor can be determined. The transmitted light intensity can be related to the concentration of the absorbing water vapor by the Lambert–Bouguer law. The simultaneous measurement of the gas temperature and total pressure and other humidity parameters as absolute humidity, a dew point can be determined with a high degree of precision using psychrometric equations.

One should note that TDLAS hygrometers can be designed using different optical schemes such as single path, multipath beam, single modulation, double modulation, open path, and closed path (Buchholz et al. 2013). Some of them will be described in Section 3.4. The optical path length necessary for TDLAS depends on the required S/N ratio, the dynamic range, the line strength of the transition, the temporal response, but also on detector properties, choice of electronics, optical noise by fibers and other optical elements, and many other factors, making the TDLAS system design a nontrivial optimization problem (Buchholz et al. 2013). Path length folding, such as in White (White 1942; Lübken et al. 1999), Herriott (Herriot et al. 1964; May 1998), and astigmatic Herriott (McManus et al. 1995) arrangements, is a frequently applied principle, but difficult to optimize in applications.

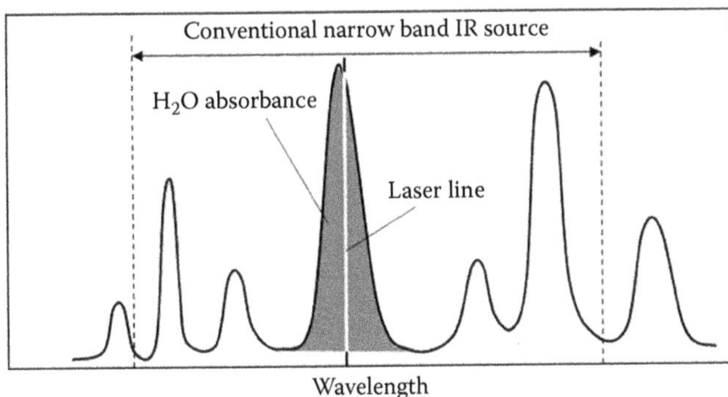

FIGURE 3.9 Illustration showing TDLAS narrow emission bandwidth versus conventional IR techniques.

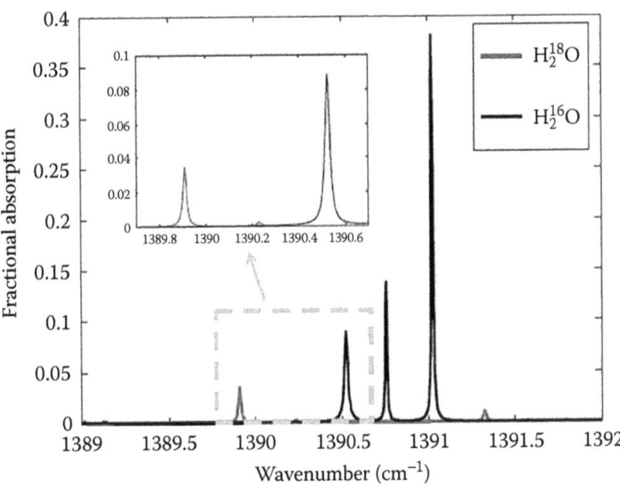

FIGURE 3.10 The selected lines for $H_2^{16}O$ and $H_2^{18}O$ are: transition $5_{14} \leftarrow 5_{41}$ at 1390.52 cm^{-1} and transition $4_{32} \leftarrow 5_{41}$ at 1389.91 cm^{-1}, respectively, in the ν_2 rotational-vibrational band of H_2O. (Reprinted with permission from Wang, W.E. et al., *Rev. Sci. Instrum.*, 85, 093103, 2014. Copyright 2014 by the American Institute of Physics.)

Additional information related to TDLAS, one can find in numerous reviews and original articles (Allen 1998; Werle 2004; Buchholz et al. 2013; White and Cataluna 2015).

It is noted that conventional IR and NIR spectrometers do not have such high wavelength resolution (Figure 3.9), and therefore such hygrometers one can use only the spectral ranges, where there is not much overlap in the spectra of the absorbing species.

TDLAS hygrometers are capable of fast response. The optical response is about 2 s. However, it takes time to purge the absorption cell and sampling system. Therefore, typical system response times are in the range of 3–10 min for a 90% step change. The typical accuracy of a TDLAS hygrometer is 2% of reading in terms or the mole fraction or ppmv. As a rule, modern TDLAS hygrometers use a self-calibrating data evaluation strategy based on the first principles approach and known parameters such as the absorption line strength, pressure, gas temperature, and absorption path length. The best TDLAS hygrometers elaborated recently have the detection limit ~50 ppb (0.05 ppm). This is achieved using a multipass absorption cell to achieve a longer optical path length and operating the cell at a vacuum pressure.

One should note that application of lasers with such cell allows developing very compact, robust, and lightweight instruments. Advantages, indicated previously and listed in Figure 3.9, as well as performances of TDLAS make it very suitable for a number of various applications. For example, using TDL spectrometry, one can conduct the measurement of water vapor $H_2^{16}O$ and $H_2^{18}O$ isotopes (Wang et al. 2014). For these purposes, Wang et al. (2014), proposed to use a field-deployable QCL-based spectrometer operated in the 7.12 μm region. The target lines shown in Figure 3.10 have been selected based on four criteria: (1) both lines should provide approximately equal peak absorption to allow probing both isotopes with significantly different natural abundance using the same optical path, (2) the lower state energies of both transitions should be within ±6.8 cm^{-1} to relax the requirement for the sample gas temperature stability to only ±1 K (assuming the targeted precision of the delta value is $\Delta\delta = 0.1‰$), (3) minimal interference from other species present in the air, and (4) center frequencies of the selected lines should be within ~1 cm^{-1} from each other to enable simultaneous access to both lines within a single current scan of a typical single-frequency distributed feedback (DFB) QCL. The system optical layout and functional block diagram of the water vapor isotope analyzer designed by Wang et al. (2014) are detailed in Figure 3.11. Experiment has shown that the achieved 1 s detection limits for $H_2^{16}O$, $H_2^{18}O$, and $\delta^{18}O$ measurements using this instrument were 2.2 ppm, 7.0 ppb, and 0.25%, respectively. The Allan deviation analysis indicated that after 160 s averaging the detection limits achievable for $H_2^{16}O$, $H_2^{18}O$, and $\delta^{18}O$ measurements can be 0.6 ppm, 1.7 ppb, and 0.05%, respectively.

One should note the measurement of water vapor isotopes is important for atmosphere monitoring. Measurements of water vapor isotopes ($H_2^{16}O$ and $H_2^{18}O$) enable the fingerprinting of water fluxes in the urban environment, including the partitioning of latent heat flux between evaporation (bare soil and impervious surfaces) and the plant transpiration. Thus, stable isotope measurements can provide transformative observational capabilities for environmental monitoring and climatological assessments at regional and global scales.

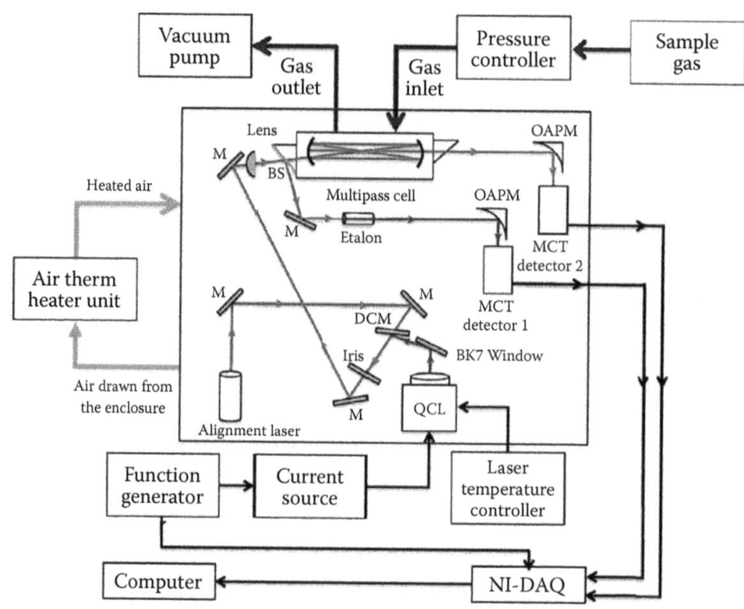

FIGURE 3.11 System schematic and optical layout. M, mirror; BS, beam splitter; QCL, quantum cascade laser; DCM, dichroic mirror; MCT detector, mercury cadmium telluride detector; OAPM, off-axis parabolic mirror; NI-DAQ, National Instruments data acquisition board. (Reprinted with permission from Wang, W.E. et al., *Rev. Sci. Instrum.*, 85, 093103, 2014. Copyright 2014 by the American Institute of Physics.)

Furthermore, water vapor isotopes can be used to assess the role of spatial heterogeneities of the land surface composition on turbulent fluxes of water vapor and sensible heat. These capabilities are particularly important for examination of the water cycle in urban environments. Isotopic measurements are critical for hydrological cycle studies, the gas exchange and the transport processes between vegetation and the atmosphere, and understanding spatial and temporal variations (Gat 1996; Yakir and Sternberg 2000; Wen et al. 2008, 2012)

With regard to the disadvantages of these devices, they are listed in Table 3.4. In addition, the measurement is not flow dependant, but care must be taken to control the sample temperature and pressure. Changes in the sample temperature and pressure affect the line shape of the rotational–vibrational transitions and will cause changes in the instrument readings. Drawback to this technology is also the temperature sensitivity of the TDL. The junction temperature of the laser diode is critical to the measurement; any small changes in temperature of the diode will shift the center wavelength of the emission and can result in erroneous measurements or alarm conditions in the analyzer. While most manufacturers use conventional thermoelectric coolers and thermocouples to maintain a stable temperature of the laser diode, the use of onboard moisture reference cells has been used to provide the TDLAS system with *line-lock* capability and provide feedback to the temperature control loop of the TDL to increase reliability and confidence in the measurement. Part of the laser beam is passed through the reference cell assembly where the spectra of the analyte sample in the reference cell are monitored and any shift in the observed peak is used as a feedback signal for the temperature control of the tunable laser diode. A final disadvantage is the range and accuracy capability of the TDLAS device. Although it is acceptable to use these instruments in the traditional pipeline natural gas applications, their use for low-level detection of water vapor is limited due to the detection capability in a methane-based background. Specialized techniques such as differential spectroscopy add maintenance, cost, and potential errors to the measurement systems.

3.4 IR OPTICAL HYGROMETERS

3.4.1 General View

IR light is defined as electromagnetic radiation with a wavelength longer that in the range of visible light, that is $\lambda > 800$ nm. The IR range can be further divided into

Near-infrared (NIR)	$\lambda = 780$–3000 nm (0.78–3.0 μm)
Middle-infrared (MIR)	$\lambda = 3000$–8000 nm (3.0–8.0 μm)
Far-infrared (FIR)	$\lambda > 8000$ nm (>8.0 μm)

Range definitions vary throughout the literature and should be understood as a general guideline.

Studies have shown that in this region of the spectrum there are multiple absorption bands due to the interaction of radiation with different gases present in the atmosphere (Table 3.2), which makes this spectral range, particularly NIR and MIR, attractive for various gas analyzers, including optical hygrometers (Smith 1993). In the more distant area there is an absorption band of H_2O at 1594.8 cm^{-1} (6.3 μm). Important advantage of IR hygrometers is a low cross-sensitivity to the temperature fluctuations and variations of gas density and particle contribution (e.g., dust and particles).

It should be noted that little difference in the position of the absorption bands can be observed in various sources. As is known, pressure, temperature, and variance in the composition of the sample gas can affect the position of these bands. For example, the absolute signal for water at the same volume ratio at two different pressures will be different due to the pressure *broadening.*

It should be understood that in addition to discrete absorption bands of water vapor, there is also a so-called continuous absorption spectrum, for example, in the 200–1200 cm^{-1}, which is also attributed to the water vapor.

Continuum absorption by water vapor is defined as any observed absorption by water vapor, not attributable to the Lorentz line contribution within 25 cm^{-1} of each line. It has been suggested that it results from the accumulated absorption of the distant wings of lines in the FIR. This absorption is caused by collision broadening between H_2O molecules (called *self-broadening*) and between H_2O and nonabsorbing molecules (N_2) (called *foreign broadening*). The most recent work suggested that the large portion of the continuum might be due to collision-induced transitions and does not relate to the line wings.

The presence of such amount of the strips forces to approach carefully when selecting the spectral range in which the hygrometer can operate. Of course, this choice should take into account the technical capabilities such as the presence of appropriate source and the detector of radiation, the presence of other components that may also interact with the radiation in selected range, the presence of free water, and expected percent of water vapor in the sample. In this case, the selected wavelength must satisfy such requirements as follows:

- High selectivity with regard to water in reflection, transmission, or the absorption properties
- High contrast in optical properties between water and the other components of the measured material

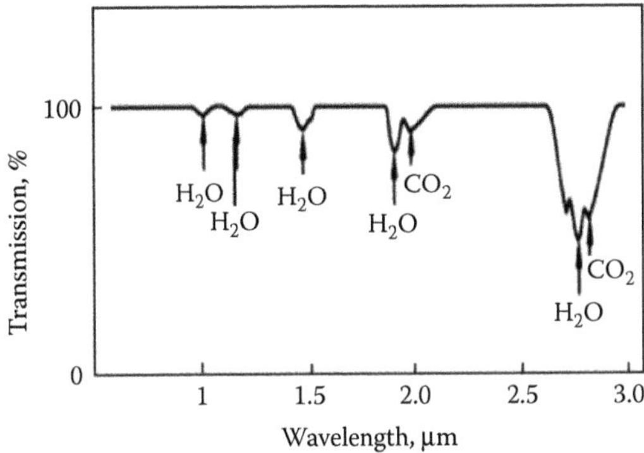

FIGURE 3.12 Infrared absorption spectra of the air in near IR range. (Data extracted from Wernecke, R. and Wernecke, J., *Industrial Moisture and Humidity measurement: A Practical Guide*, Wiley-VCH, Weinheim, Germany, 2014.)

Comparison of all these factors indicates that the preferred spectral range for IR optical hygrometers is around the strong absorption peaks of water at $\lambda = 1.47$ and 1.94 μm (Figure 3.12). These wavelengths are also well separated from the absorption peaks of CO_2 and CO, which are often present in a gas mixture. For the humidity measurement, one can also use the absorption bands near 2.7 μm and in the region of 5–7 μm (Figure 3.13) (Silver and Stanton 1987). In a short interval from 5.7 to 6.6 μm water-vapor absorption is very intense. Of course, the wavelengths in real hygrometers may differ slightly from the specified values indicated in Table 3.6 that is associated sometimes with a limited selection of lasers with the required wavelength.

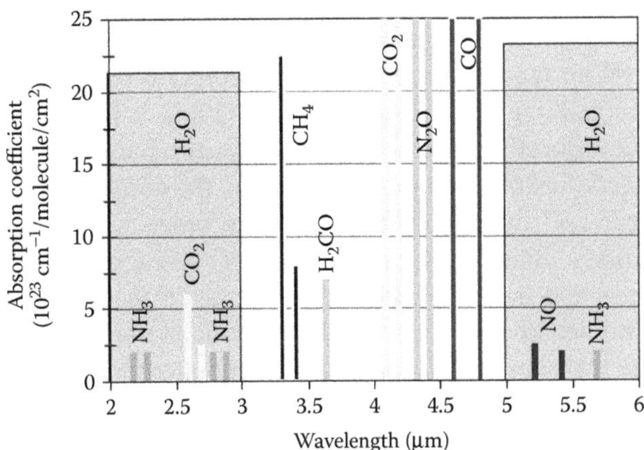

FIGURE 3.13 Infrared absorption spectra of a number of relevant gas species in the area between 2 and 6 μm. This is a simplified graphic presentation to show the importance of the individual absorption lines, as well as the shadowing effect of water vapor. (With kind permission from Springer Science+Business Media: *Solid State Gas Sensor,* 2009, Springer, New York, Comini, E. [Eds.] et al.)

TABLE 3.6
The Wavelengths of Monochromatic Radiation Used in Developing IR Hygrometers

Wavelength	References
6.5 μm	Silver and Stanton (1987)
2.59–2.7 μm	Cerni (1994), Wunderle et al. (2015), Ebert and Buchholz (2016)
1.854–1.90 μm	Chen and Mitsuta (1967), Green and Kohsiek (1995), Amerov et al. (2007)
1.32–1.4 μm	Fosket et al. (1953), Silver and Hovde (1994), Wilson et al. (1995), Edwards et al. (2000), Seidel et al. (2012), Buchholz and Ebert (2013)

But we must understand that the use of wavelengths, not coinciding with the absorption band of water vapor will be accompanied by a reduction of hygrometer sensitivity. As a result, when developing the hygrometers, it is commonly used radiation in the near IR spectral range between 1.3 and 2.7 μm. It is also important to note that at present, for given spectral range they were developed the diode lasers (DL) with unique properties such as very high spectral resolution and spectral power density, continuous tunability in combination with excellent technical properties, that is, rather low cost, very low size, weight and power consumption, long laser life time, and good beam quality. That is why over the last years IR hygrometers operated in this spectral range had huge influence on the atmospheric hygrometry. There is also a large set of highly sensitive receivers for this spectral range, capable of working without cooling, which greatly simplifies the work of hygrometers.

With regard to the preferred absorption band in this spectral region, Buchholz et al. (2013) believed that the absorption line near 1.37 μm is the most optimal. According to Buchholz et al. (2013), the absorption line near 1.37 μm should be selected in order to minimize cross-sensitivity to other gases or neighboring H_2O lines and the temperature dependence of the line area. Another advantage of this line is excellent availability of the high-quality telecom-grade fiber optical components and the fiber-coupled laser modules, which minimizes size, weight, and cost of the instrument and allows effective suppression of parasitic water absorption in the air absorption paths outside the intended measurement cell.

Earlier optical IR hygrometers as radiation sources mainly used halogen lamp, tungsten light bulb, and the heat radiation sources, filtered to isolate a pair of wavelengths in required region. Thermal radiation sources such as Globar, Nernst pin (lamp), and incandescent light bulb,

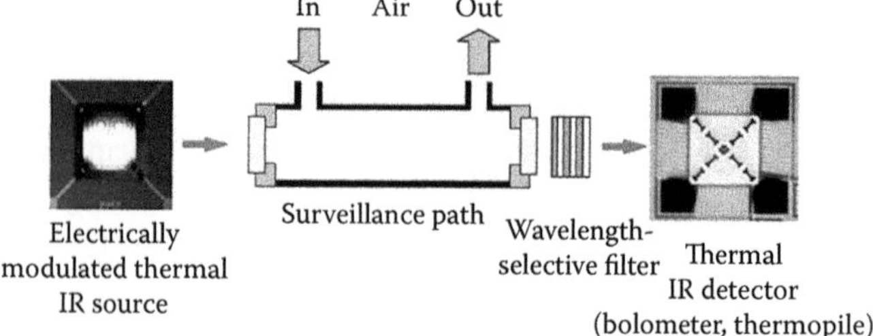

FIGURE 3.14 Principle architecture of a NDIR gas-sensing microsystem. (With kind permission from Springer Science+Business Media: *Solid State Gas Sensor*, 2009, Springer, New York, Comini, E. [Eds.] et al.)

giving a continuous spectrum of radiation, were mainly used for measurements in the FIR region. However, in the last few decades as the radiation sources are used primarily LEDs, and especially LDs, which advantages for spectroscopic applications have been considered previously. In addition, the main absorption bands of water vapor that can be used for the measurements are in NDIR (nondispersive infrared) and MIR spectral ranges, which coincide with the spectral range of the radiation of LEDs and lasers offered on the market. Of course, there are now developments using thermal radiation sources (Bauer et al. 1996). But these thermal sources are being made using a modern technology, which permits to realize their miniaturization (Spannhake et al. 2005) (Figure 3.14). In addition, they permit to reach high modulation frequencies. Small time constants are important parameters to directly modulate the IR radiation. Nowadays, thermal emitters permit direct modulation of radiation within the frequency range up to several Hz. Other IR radiation sources which can be used in hygrometers are listed in Table 3.7.

But it should be recognized that the thermal emitters (blackbody radiation sources) are not optimal for use in hygrometers and their parameters are significantly inferior to LEDs, and especially LDs (Wilson et al. 1995). First, thermal emitters need a high operating temperature (up to 2000K) to obtain large radiation power in the MIR range, and this can cause unwanted heating in a compact instrument. But the additional emitted radiation can be accompanied by degradation of system performance; it warms up detectors and electronics elements and generates parasitic components of the detector signal (Cozzani et al. 2007). Second, their emitting spectrum covers from visible to FIR. Therefore, they emit a too wide spectrum of light and they are not efficient from a power consumption point of view. Third, the extremely broadband radiation produced by a blackbody requires optical filtering for selection required spectral range, and this often removes most of the optical power.

TABLE 3.7
Types of Infrared Radiation Sources

Type	Method	Material	Radiation Source Example	Wavelength (μm)	Remark
Thermal radiation	Resistor heating by current flow	Tungsten	Infrared bulb	1–2.5	Long wavelength region is cut off by external bulb (glass). Secondary radiation is emitted trough the tube
		Nichrome Kanthal	Electric heater	2–5	
		Silicon carbide (siliconate)	Globar	1–50	Constant voltage, large current
		Ceramic	Nernst glower	1–50	Preheating is needed
	Secondary heating by other power source	Metal (stainless steel, etc.)	Sheath heater	4–10	
		Ceramic	IRS type lamp	4–25	
			Radiant burner	1–20	Heating by gas burning
	Heating by discharge	Carbon	Carbon arc lamp	2–25	Causes some environmental problems such as soot
Cold radiation	Gas discharge	Mercury	Mercury lamp	0.8–2.5	Long wavelength region is cut off by external bulb (glass). Secondary radiation is emitted trough the tube
		Cesium	Xenon lamp		
		Xenon			
Stimulated emission	Laser reaction	Carbon dioxide	CO_2 laser	9–11	
		Gallium arsenic compounds	InGaAs laser	1.1–1.5	
		Lead compounds	PbSnTe laser	1.2–6 to 7	

Source: http://www.hamamatsu.com.

Also, narrow-bandpass filters sometimes transmit light at wavelengths outside the bandpass region that are still detected. Fourth, the emitting area of a blackbody source is usually many times larger than that of a semiconductor light source, so that obtaining a uniform and collimated light beam can be difficult. Finally, blackbody sources tend to have a slower speed of response than semiconductor devices, and this makes them less suitable for applications where high-speed intensity or frequency modulation is required (Wilson et al. 1995).

If you look at Figure 3.15, it is seen that the modern lasers that can be used for the development of optical hygrometers that span the NIR and MIR spectral regions, with wavelengths ranging from 0.6 to >2 μm for conventional DL and from 2 to >4 μm for recently developed QCLs (Hvozdara et al. 2002; Kosterev and Tittel 2002). QCLs are semiconductor lasers based on

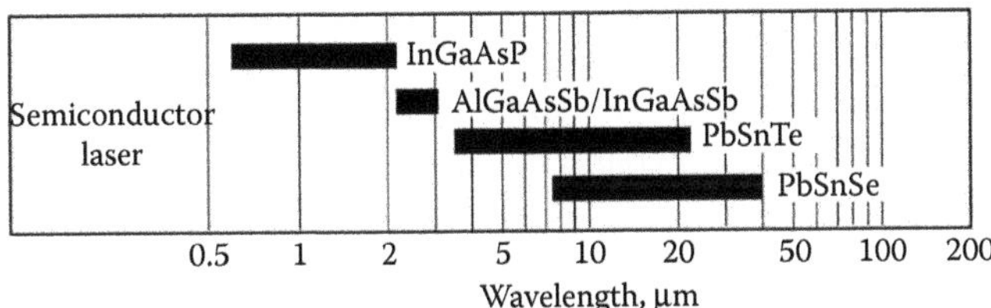

FIGURE 3.15 Approximate spectral windows of commercially available room-temperature diode laser sources.

transitions in a multiple-quantum-well heterostructure (Faist et al. 1994). The emission wavelength depends mainly on the quantum well thickness rather than on the size of the band gap, as is the case with conventional DL. This has allowed, using the same base semiconductors (InGaAs and AlInAs grown on InP), the manufacture of lasers with an emission wavelengths ranging from 3.5 to 17 μm (Werle et al. 2002). Thus, QCLs can operate at wavelengths in the MIR range, which match well with the fundamental vibrational absorption bands of water vapor and other chemical species—in comparison to conventional diode sources, where the laser emission generally matches the weaker overtone bands (Grouiez et al. 2008). In addition, due to specific mechanism of recombination, QCLs can operate at near room temperature and produce more optical power (milliwatts) of radiation within a broad range of frequencies. For comparison, lead-salt DL (PbSnTe and PbSnSe) designed for the 5–8 μm wavelength region can operate only at liquid nitrogen temperature. This greatly limits their application in the development of hygrometers, as the hygrometer in this case should have a cryogenic refrigeration diode laser system. But in this case, hygrometer is bulky and cannot be used for stand-alone operation.

Of course, LEDs and LDs also have disadvantages. They exhibit strong dependency of the power and

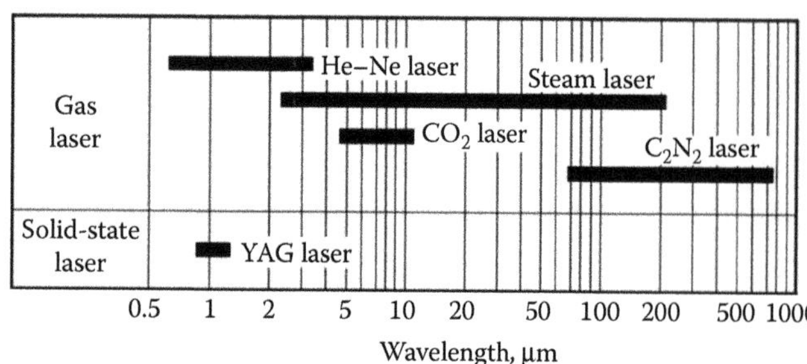

FIGURE 3.16 Spectral windows of commercially available gas and solid-state lasers. (Data extracted from http://www.hamamatsu.com.)

spectrum characteristics on the temperature and their price significantly exceeds a thermal sources costs. But their usage considerably improves the performance of hygrometers and therefore, these disadvantages do not limit the application of LEDs and LDs in the hygrometers. The gas and solid-state lasers (Figure 3.16) can be used in the same region of the spectrum. But the size and the cost of these lasers are inferior to semiconductor lasers, which considerably limit their application in the development of portable devices.

Signal can be measured using heterodyne or direct detection systems. According to Grund et al. (1995), these systems have the following differences:

- Diffraction-limited optics are usually used in heterodyne LIDARs, and the aperture and focusing of these optics should be at least approximately matched to the strength and range of the return; the limited field of view reduces background light, but the direct detection systems can make use of larger optics.
- The heterodyne signal-to-noise ratio (SNR) is approximately constant out to long range and can be optimized at long range by focusing, whereas the direct detection SNR falls rapidly with range; consequently, very good differential absorption LIDAR (DIAL) results can be obtained using direct detection at short ranges, but the long range results are impaired.

Thus somewhat smaller optics can be used for the heterodyne systems and the results from the smaller heterodyne system are superior to those from the direct detection LIDAR at long range but inferior at short range (Grund et al. 1995).

With regard to PDs, suitable for use in IR optical hygrometer, at present time there is a large variety of semiconductor detectors and thermal detectors, which

TABLE 3.8
Thermal Detectors

Detector Types	Method of Operation
Bolometer	Change in electrical conductivity
Metal	
Semiconductor	
Superconductor	
Ferroelectric	
Hot electron	
Thermocouple/Thermopile	Voltage generation, caused by change in temperature of the junction of two dissimilar materials
Pyroelectric	Changes in spontaneous electrical polarization
Golay cell/Gas microphone	Thermal expansion of a gas

are able to cover the entire spectral range, suitable for the development of optical hygrometers (Tables 3.8 and 3.9 and Figure 3.17). Figure 3.17 shows the normalized detectivity characteristics for different types of detectors. Quantum-type semiconductor detectors show excellent detectivity in the IR range and fast response, but they are strongly wavelength dependent (Figure 3.17) and, in addition, they generally need to be cooled for accurate measurement. In thermal detectors such as pyroelectric sensors, bolometers, and thermopiles the absorbed photon results in a temperature rise of the detector, which entails an alteration of its electrical properties.

Despite of its smaller detectivity compared to quantum detectors (Figure 3.17), thermal detectors are widely used due to its uncooled operation, their small dependency on the wavelength, and the flat response. At that thermopiles show better sensitivity than pyroelectric and bolometer detectors (Schilz 2000). Thermopile detectors fabricated by PerkinElmer are shown in Figure 3.18. Modern thermopiles fabricated in micromachining technology are extremely sensitive, with a fast response time due to its small size (Graf et al. 2007; Fonollosa et al. 2009). One should note that nowadays, silicon-based thermopiles are commonly used for IR detection in NDIR spectrometers.

Filters are also important elements of the optical hygrometer for improving the selectivity of the specific optical narrowband filters. They are usually placed directly upon the IR detector set. A bandpass optical filter usually consists of a number of dielectric layers on a substrate. The thickness, the number and the material of deposited layers on the substrate determine the transmission characteristics of the filter (Rancourt 1996).

TABLE 3.9
Types of Infrared Detectors and Their Characteristics

Type			Detector	Spectral Response (μm)	Operating Temperature (K)	D* (cm·Hz$^{1/2}$/W)
Thermal type	Thermocouple (Thermopile)		Golay cell,	Dependens on	300	D* (λ, 10, 1) = 6·10^8
	Bolometer		condenser-	window	300	D* (λ, 10, 1) = 1·10^8
	Pneumatic cell		microphone	material	300	D* (λ, 10, 1) = 1·10^9
	Pyroelectric detector		PZT, TGS, $LiTaO_3$		300	D* (λ, 10, 1) = 2·10^9
Quantum type	Intrinsic type	Photoconductivity type	PbS	1–3.6	300	D* (500,600,1) = 1·10^9
			PbSe	1.5–5.8	300	D* (500,600,1) = 1·10^8
			InSb	2–6	213	D* (500,1200,1) = 2·10^9
			HgCdTe	2–16	77	D* (500,1000,1) = 2·10^{10}
		Photovoltaic type	Ge	0.8–1.8	300	D* (λp) = 1·10^{11}
			InGaAs	0.7–1.7	300	D* (λp) = 5·10^{12}
			Ex. InGaAs	1.2–2.55	253	D* (λp) = 2·10^{11}
			InAs	1–3.1	77	D* (500,1200,1) = 2·10^9
			InSb	1–5.5	77	D* (500,1200,1) = 2·10^{10}
			HgCdTe	2–16	77	D* (500,1000,1) = 1·10^{10}
	Extrinsic type		Ge:Au	1–10	77	D* (500,900,1) = 1·10^{11}
			Ge:Hg	2–14	4.2	D* (500,900,1) = 8·10^9
			Ge:Cu	2–30	4.2	D* (500,900,1) = 5·10^9
			Ge:Zn	2–40	4.2	D* (500,900,1) = 5·10^9
			Si:Ga	1–17	4.2	D* (500,900,1) = 5·10^9
			Si:As	1–23	4.2	D* (500,900,1) = 5·10^9

Source: http://www.hamamatsu.com.

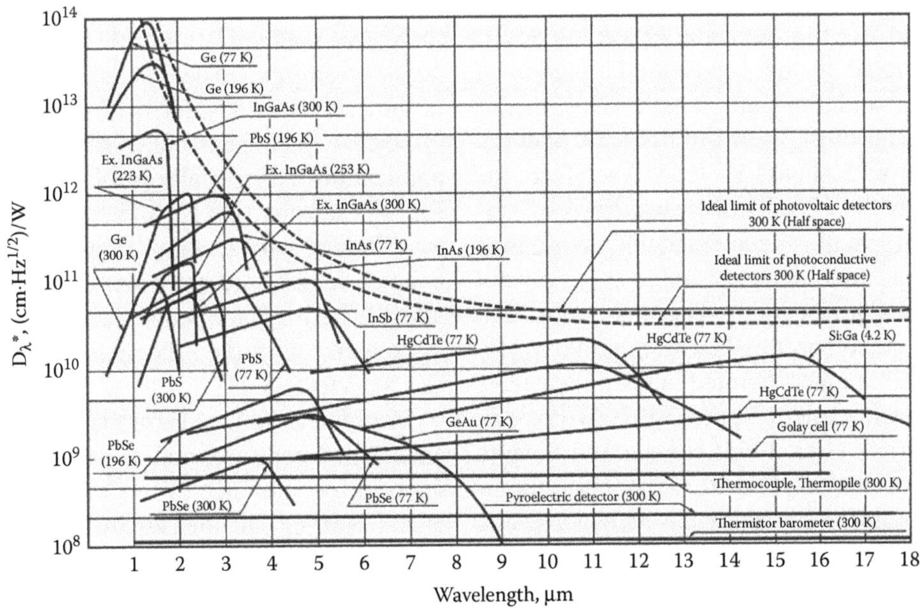

FIGURE 3.17 Normalized detectivity characteristics for different type of detectors. (From http://www.hamamatsu.com.)

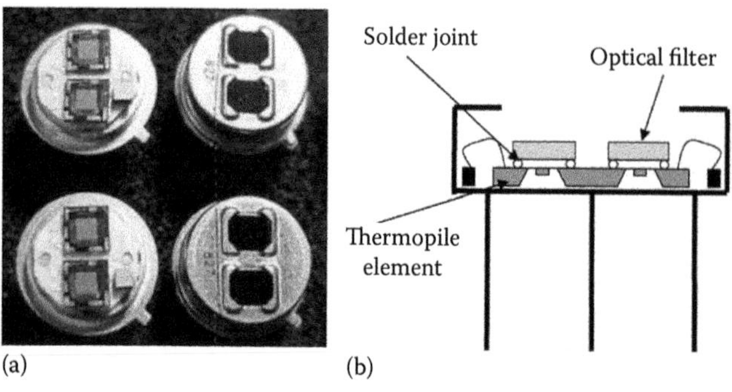

FIGURE 3.18 (a) Photo of PerkinElmer dual-thermopile detectors. The left side shows open detectors. Clearly the two thermopiles can be seen. The gray-colored squares in the middle of each sensor chip are the absorber areas, which collect the IR light to be measured. The small cube near to the lower chip is the thermistor, which senses the ambient temperature. The right side shows the detectors covered by a cap holding the two different IR-filters. (From http://www.perkinelmer.com.) (b) Schematics of the detection unit in a TO8 package, with thermopile elements fabricated using micromachining technology, optical filters, and solder joints. (Reprinted from *Sens. Actuators A* 149, Fonollosa, J. et al., Limits to the integration of filters and lenses on thermoelectric IR detectors by flip–chip techniques, 65–73, Copyright 2009, with permission from Elsevier.)

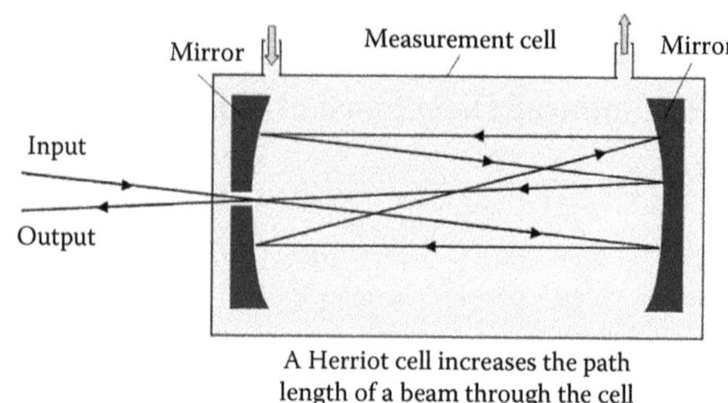

FIGURE 3.19 Diagram showing the principle of Herriot cell operation.

Usually, silicon is used as a substrate to take advantage of its related technologies. Systems with narrowband but tunable optical filters are also designed (Santander et al. 2005).

As it is known, the sensitivity of optical instruments is dependent on the absorption path length. For example, according to the Lambert–Beer law, the transmission of the IR radiation through the absorbing gas is inversely proportional to the exponential function of gas concentration and the path length. This means that for achievement of high sensitivity in IR spectral range, the optical path should be long. Therefore, the measuring path in the IR optical hygrometers is normally greater than 1 m. It is understandable that so long optical path lengths are often undesirable or simply not possible in industrial applications. Therefore, instead of single transmission, compact cells with multiple reflections inside the cell are incorporated in the device (e.g., Herriot measurement cells) (Figure 3.19). At present, Herriot cells (Herriot et al. 1964; Herriot and Schulte 1965) have been widely applied in atmospheric monitoring and are commercially available with astigmatic optics providing more than 100 m of nonoverlapping optical path length in a physical cell less than 1 m in length. Liu (2012) in his experiments, has used a multipass absorption cell with physical length about 35 cm, which provided an optical absorption path length exceeded 30 m. Thus, using this approach, the optical path length remains constant, while the cell dimensions can be significantly reduced. For comparison, the measuring path in the UV optical hygrometer is typically a few centimeters in length.

Of course, one can use a simplified version shown in Figure 3.20, when open-path is used and the dual emission pass is achieved by using a separated reflector (Seidel et al. 2012). The same approach was used in developing the hygrometer by Wilson et al. (1995). However, such an approach can only be used for stationary devices. An additional difficulty of this approach is that to achieve the required accuracy it is necessary to ensure that any target (reflector) movement such as rotation, translation, or tilt of the target should have a negligible influence on the accuracy and, if possible, on the precision of the spectrometer. In addition, the performance of the spectrometer should not be disturbed if there is a small deviation from the precise perpendicular target positioning. There should not be any dependence of the exact laser position on the target, which is not only important for adjustment, but also for future scanning applications for spatially resolved measurements. Also, for high sensitivity it is still necessary to use sufficiently long optical path lengths. However, the more the optical path lengths are, the more difficult is to perform the above listed requirements. Seidel et al. (2012) in their hygrometer used the reflection tape being in a distance of 75 cm to 1 m.

In order to reduce errors due to fluctuations of the source, temperature, and so on, a second cell is often utilized for reference (Silver and Hovde 1994; Amerov et al. 2007). The measurement signal and reference signal can then be combined to give a measurement value. As for errors and anomalies observed at measuring in the MIR region, they are discussed in detail by Chalmers (2002).

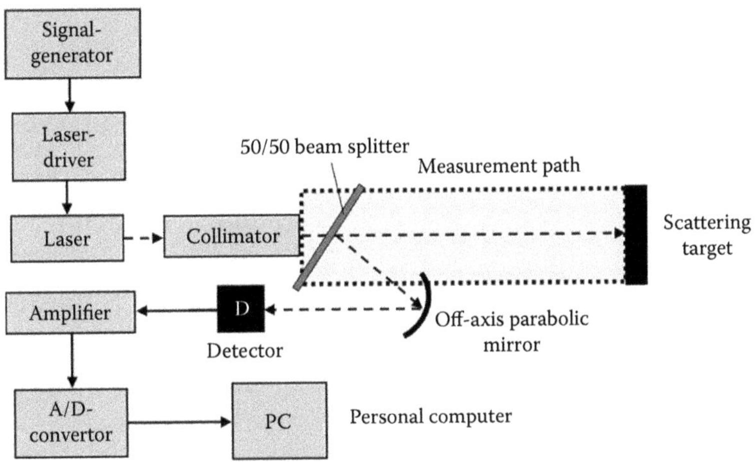

FIGURE 3.20 Optical setup of the spectrometer with sending and detecting side (left) and separate scattering target (right). Here the light of a fiber-coupled DFB-laser at 1370 nm was collimated and directed through a 50/50 dielectric beam splitter onto the target, where it was partially reflected/scattered. The beam-splitter serves for directing the part of the reflected light onto a photodiode placed outside the incoming beam-path. A micro prismatic reflection tape was used as reflector (scattering target). (With kind permission from Springer Science+Business Media: *Appl. Phys. B*, TDLAS–based open–path laser hygrometer using simple reflective foils as scattering targets, 109, 2012, 497–504, Seidel, A. et al.)

3.4.2 Realization

Usually modern optical IR hygrometers are being developed using two approaches: (1) transmission of radiation at two wavelengths; one wavelength corresponds to the water absorption line and second wavelength is used as a reference (Foskett et al. 1953; Chen and Mitsuta 1967), and (2) TDLAS (Amerov et al. 2007; Soleyn 2009; Buchholz and Ebert 2013; Nwaboh et al. 2017). The advantages of these approaches were considered earlier in Section 3.3. Importantly, that multichannel approach can be also used in TDLAS hygrometers. For example, to achieve an unprecedented dynamic range of 1–40,000 ppm, Ebert and Buchholz (2016) have used in his hygrometer simultaneously two lasers at 1.4 and 2.6 μm for high/low H_2O concentrations.

Successful representative of optical hydrometers, developed when using the first approach, is LI-7500A Open Path CO_2/H_2O Gas Analyzer, fabricated by LI-COR Inc. (http://www.licor.com) and shown in Figure 3.21. The LI-7500A analyzer is a high speed, high precision, NDIR gas analyzer that accurately measures densities of carbon dioxide and water vapor in the turbulent air structures. The LI-7500A sensor head has a 12.5 cm open optical path, with single pass optics and a large 8 mm diameter optical beam. Optical filters centered at 3.95 μm provide a reference signal for CO_2 and water vapor. Absorption at wavelengths centered at 4.26 and 2.59 μm provide for measurements of CO_2 (0–3000 ppm) and water vapour (0–60,000 ppm), respectively with accuracy ±1–2%. This design minimizes sensitivity to dirt and dust, which can be accumulated during normal operation. One should note that LI-COR Inc. designed also CO_2/H_2O Gas Analyzer, LI-7200, with enclosed optical cell (Figure 3.21b). This analyzer has the optical scheme similar to the optical scheme used in LI-750A.

As regards the second approach, here we have a large number of successful projects (Amerov et al. 2007; Soleyn 2009; Buchholz and Ebert 2013; Nwaboh et al. 2017). For example, Nwaboh et al. (2017) suggested TDLAL hygrometer for absolute measurements of H_2O in methane, ethane, propane, and a low CO_2 natural gas. The sensor was operated with a 2.7 μm DFB laser, equipped with a high pressure single pass gas cell, and used to measure H_2O amount of the substance fractions in the range of 0.31–25,000 μmol/mol. The operating total gas pressures were up to 5000 hPa. The relative reproducibility of H_2O amount of substance fraction measurements at 87 μmol/mol was 0.26% (0.23 μmol/mol). The maximum precision of the sensor was determined using a H_2O in methane mixture, and found to be 40 nmol/mol for a time resolution of 100 s. This corresponds to a normalized detection limit of 330 nmol $mol^{-1}{\cdot}m\ Hz^{-1/2}$. The relative combined uncertainty of the H_2O amount fraction measurements delivered by the sensor was 1.2%.

In the same spectral range the Pico-SDLA H_2O (hereafter Pico-SDLA) hydrometer is working (Durry et al. 2008; Ghysels et al. 2016). Pico-SDLA hydrometer is a lightweight spectrometer which measures the water vapor using laser absorption spectroscopy. Hygrometer was equipped with the probe diode laser emitted at a

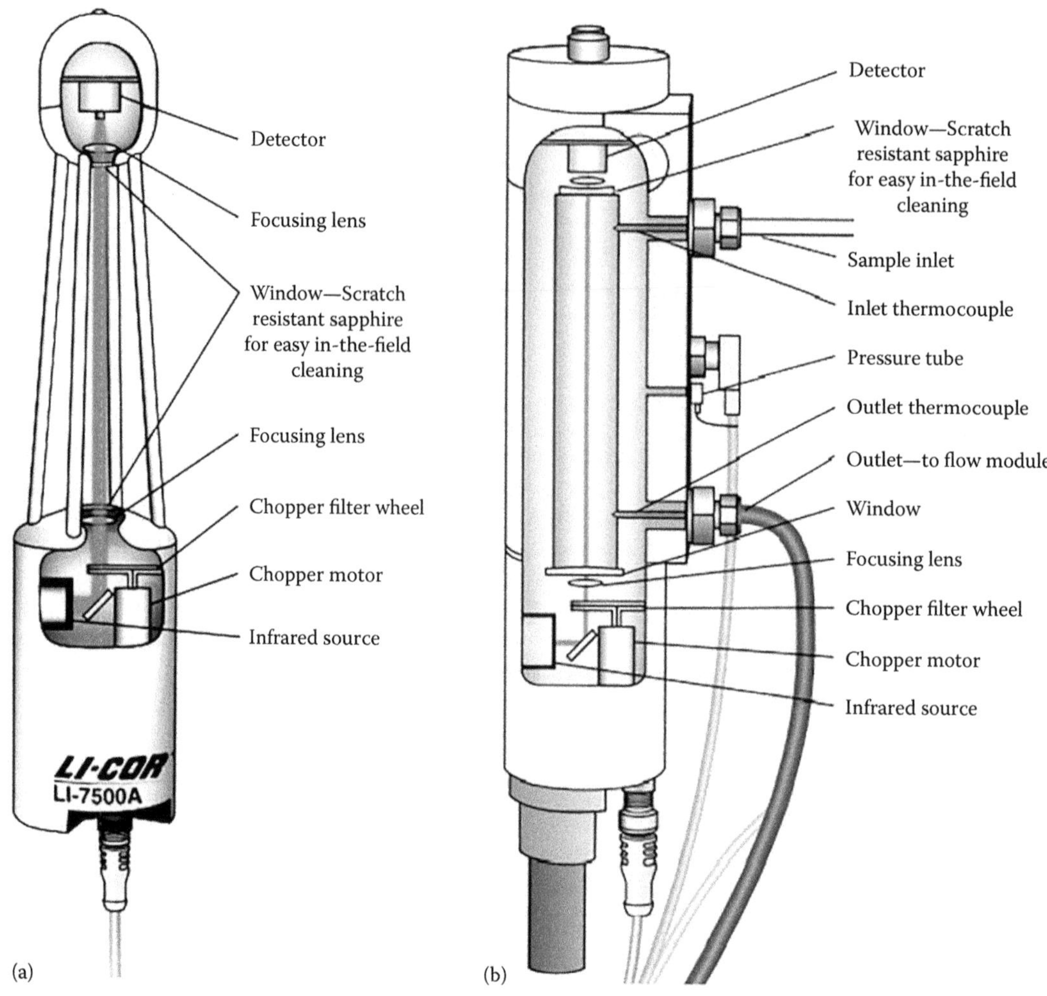

FIGURE 3.21 CO_2/H_2O Gas Analyzers designed by LI-COR Inc.: (a) LI-7500A analyzer with open path and (b) LI-7200 with optical cell. (From http://www.licor.com.)

wavelength of 2.63 μm and had a 1 m path length through ambient air. The water vapor absorption line was scanned by tuning the laser current and fixing the TEC temperature. As it was indicated before, the current modulation of the laser is the preferred method to scan the water vapor absorption line since the response time is much faster than for temperature modulation. After passing through the ambient-air sample, the laser beam was focused onto an indium arsenide detector using a sapphire lens. A feature of the hygrometer was that two different rotation–vibration absorption transitions of water vapor were probed during measurements. This transition was needed because of the large variation in mixing ratio occurring between the troposphere and the stratosphere. For measurements from the ground to around 200 hPa pressure level, the $4_{13} \leftarrow 4_{14}H^{16}{}_2O$ line at 3802.96561 cm^{-1} was used. While above 200 hPa pressure level, we use the $2_{02} \leftarrow 1_{01}H^{16}{}_2O$ line at 3801.41863 cm^{-1}. During in-flight measurements, the switch from one line to the other is automatically driven. Both sets of line parameters were obtained from HITRAN 2012 database (Rothman et al. 2013). The mass of the Pico-SDLA was less than 9 kg, making it suitable as a payload for small stratospheric balloons (500 and 1500 m^3). The hygrometer was able to measure the water vapor from the ground to altitudes of 35 km for concentrations ranging from 15,000 to less than 1 ppmv (Ghysels et al. 2016). The mixing ratio was extracted from the measured spectra using a nonlinear least squares fitting algorithm applied to the measured line shape. The authors of Ghysels et al. (2016) used the Beer–Lambert law to model the spectrum and used a Voigt profile (VP) to describe the molecular line shape.

Amerov et al. (2007) and AMETEK (http://www.ametekpi.com) have chosen the water vapor line at 1854 nm for the measurements. They proposed to use of an all-digital protocol for the modulation of the laser drive signal and the demodulation of the detector response. A reference cell used in parallel with the

FIGURE 3.22 The OptiPEAK TDL600 Tuneable Diode Laser Analyzer. The next generation TDLAS Analyzer for automatic online measurement of the moisture in variable compositions of natural gas and biomethane desihned by Michel Instruments. Range of 1–1000 ppmv. Limit detection ~1 ppmv. (From http://www.michell.com/uk.)

sample cell allowed to continuously validating the performance of the system. It was shown that the TDLAS hygrometer designed by Amerov et al. (2007) had a limit of detection of 5 ppmv for both the nitrogen and natural gas samples. Hygrometer can operate in the range of 5–2500 ppmv. The accuracy of the reading of hygrometer fabricated by AMETEK is <1 ppb by volume. Michel Instruments (Figure 3.22) reported that their hygrometers had limit detection ~1 ppmv. Liu (2012) has shown that by reducing the sample gas pressure, the spectral interference with background gas could be greatly reduced and effectively removed in real time, and therefore, enhanced specificity, improved accuracy, lower detection limit, faster response, and lower cost of ownership all become achievable. Liu (2012) demonstrated that while using this technique a detection limit approaching 10 ppbv, with accuracy around 5 ppbv, and the response time in minutes could be achieved. However, it should be noted that Liu (2012) in his experiments used a multipass absorption cell, large enough; its physical length was about 35 cm only but absorption path length was extended to approximately 30 m. The IR TDL absorption hygrometer, designed by Edwards et al. (2000), had the same accuracy, ~5 ppbv. For achievement such accuracy they used two-tone frequency modulation spectroscopy (TTFMS), a multipass absorption cell and a computer controlled diode laser operated at $\lambda = 1.393$ µm. Without using a multipass absorption cell, Hydrometers, as a rule, cannot provide such high sensitivity and accuracy. For example, 1.4 µm-TDLAS hygrometer with an internal optical cell with 1.5 m optical path length, elaborated by Buchholz and Ebert (2013) for airbone applications, had a precision of 33 ppbv.

Electrical and optical scheme of hygrometer designed by Amerov et al. (2007) is shown in Figure 3.23. A DFB laser, which was used as the source of light, produced an optical power of approximately 3 mW, when operated at the target wavelength for the water–vapor measurement. Output from the laser was coupled into a single-mode fiber, which was connected to a fiber-optic splitter, used to divide the optical power in a 70/30 ratio and simultaneously connect the DFB laser to the sample and reference cells. Signals from the InGaAs 0.5 mm^2 photodiode detectors were input to separate channels of the electronics unit. It was now possible to make simultaneous measurement of unknown samples and a known moisture reference, which was used to lock the output wavelength of the laser to the 1854 nm absorption line of water vapor. Further, the digital signal process methods employed in this system can successfully remove minor background interferences, caused by other species in the sample.

Figure 3.24 is a schematic that illustrates the components of a TDLAS hygrometer designed by Soleyn (2009). This hygrometer was adapted for its use in natural gas. The TDLAS hygrometer has very good correlation to the reference standard. The accuracy of the hygrometer exceeds ±4 ppmv in the range of 5–200 ppmv, and ±2% of reading in the range of 200–5000 ppmv. For use in natural gas the components of hygrometer, which can be wetted, are constructed of stainless steel with the exception of the optical window that consists of proprietary glass and the mirror that consists of proprietary polished metal alloy. Those components are selected for their resistance to corrosion and *optical purity*. It is known that the path length is related to a lower detection limit. Therefore, the path length in the hygrometer was optimized for a volume ratio of approximately 5–5000 ppmv. The photodiode and reference photodiode are housed in a hermetically sealed and dry enclosure. A platinum-resistance temperature detector (PRTD) measures the gas temperature and a silicon micromachined strain gauge pressure sensor measures the sample pressure. The temperature sensor is encased

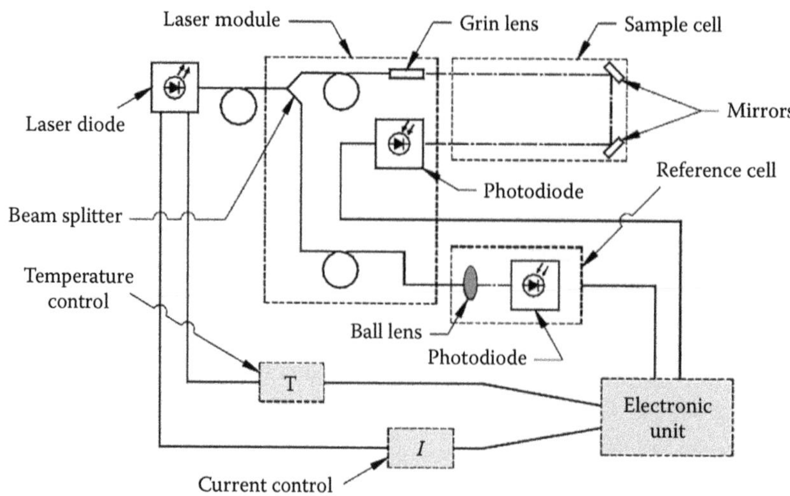

FIGURE 3.23 Electrical and optical scheme of TDLAS hygrometer operated at 1.854 µm. (Data extracted from Soleyn, K., *Proceedings of the 5th International Gas Analysis Symposium, GAS 2009*, Rotterdam, the Netherlands, February 11–13, 2009.)

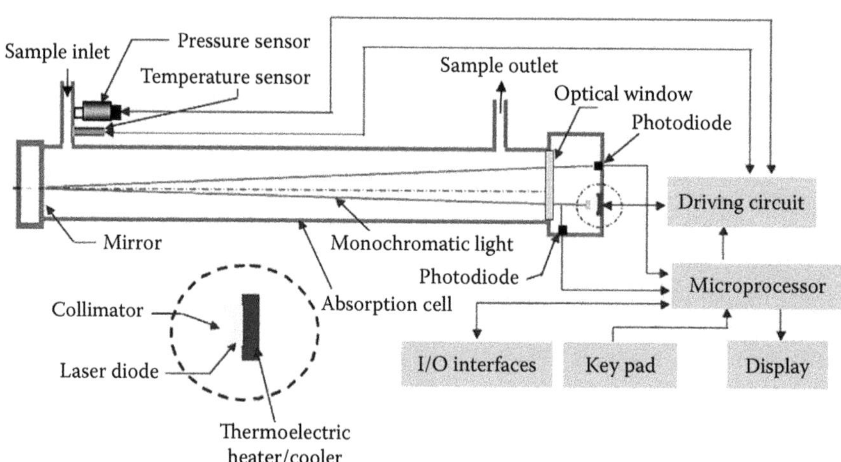

FIGURE 3.24 Diagram illustrating TDLAS hygrometer designed by Soleyn K. (Data extracted from Soleyn, K., *Proceedings of the 5th International Gas Analysis Symposium, GAS 2009*, Rotterdam, the Netherlands, February 11–13, 2009.)

in a stainless steel sheath and the pressure sensor is also constructed of stainless steel with a hastelloy–wetted diaphragm. For field installation, a pipeline-insertion membrane filter and the pressure regulator will separate liquids (hydrocarbons, liquid water, and glycol carry over from the dehydration process) and drop them back into the pipeline and also reduce the pressure. A second pressure regulator will decrease the pressure close to atmospheric. The sampling system and absorption cell are installed in a stainless steel enclosure that is heated by a thermostatically controlled electrical resistance heater.

At present, many companies fabricate IR hygrometers and these hygrometers present on the market. For example, General Electric Company developed Aurora Moisture Analyzer to measure moisture in natural gas. Sampling system of this tool is shown in Figure 3.25. Aurora is equipped with a two-stage turnkey sampling system. An optional first stage consists of a membrane filter/regulator installed directly in the pipeline. It prevents any liquid (hydrocarbon, glycol, or water in liquid phase) from entering the sample line. The pipeline pressure is reduced by means of a regulator. As the gas enters the second stage it flows through a coalescing filter, and a pressure regulator further reduces the pressure. The flow rate is adjusted with a needle valve. Only clean low-pressure gas enters the absorption cell. An optional heater may be installed in the enclosure for application in cold climates. The heater also serves to

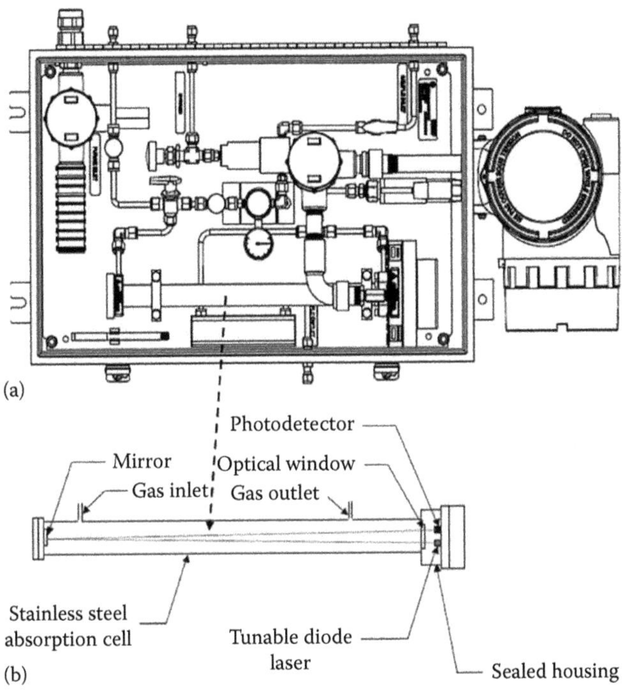

FIGURE 3.25 Sampling system (a) and measurement cell (b) of Aurora Moisture Analyzer. (From http://www.gesensinginspection.com)(Data extracted from Soleyn, K., *Proceedings of the 5th International Gas Analysis Symposium, GAS 2009*, Rotterdam, the Netherlands, February 11–13, 2009.)

keep the sample in the gas phase. The laser-based measurement system provides very fast response. The optical response is <2 s. The system is designed to operate continuously for many years with unsurpassed reliability. Factory service or calibration is recommended on a five-year interval.

Examples of other hygrometers are listed in Table 3.10. Most of these hygrometers use the TDLAS technology. As it is seen in the table, generally, these devices besides the water vapor control the content of CO_2 in the atmosphere. For these purposes, it is commonly used a dual laser TDLAS setup which consists of two TDLs, tuned on the characteristic absorption lines of the analytes in the sample. Optical scheme of the 5100 HD analyzer fabricated by AMETEK Process Instruments and intended for detecting H_2O and CO_2 is shown in Figure 3.26.

Optical hygrometers can operate at short-wave spectral region (1.0–0.7 μm). Studies have shown that in this

TABLE 3.10
Examples of Commercial IR Hygrometers

Sensor Type	Tested Gases	Produces	Country
LI 6262 (closed-path)	H_2O, CO_2	LI-COR Inc.	USA
LI 7500 (open-path)	H_2O, CO_2	LI-COR Inc.	USA
IR-3000 moisture transmitter	H_2O	MoistTech. Corp.	USA
OP-2 (open-path)	H_2O, CO_2	ATC Bioscientific Ltd.	UK
DF-700 moisture analyzer	H_2O	Servomex	UK
Ophir IR-2000 (open-path, very large)	H_2O	Ophir Corporation	USA
Gas Analyzer E-009 (open-path, very large)	H_2O, CO_2	Kysei Maschin. Trading Co. Ltd.	Japan
5100/5100 HD analyzer	H_2O, CO_2	AMETEK Process Instruments	USA
OptiPEAK TDL600 analyser	H_2O, gases	Michel Instruments	UK
GE's Aurora analyzer	H_2O	General Electric Company	USA
WVSS-II	H_2O	SpectraSensors Inc.	USA

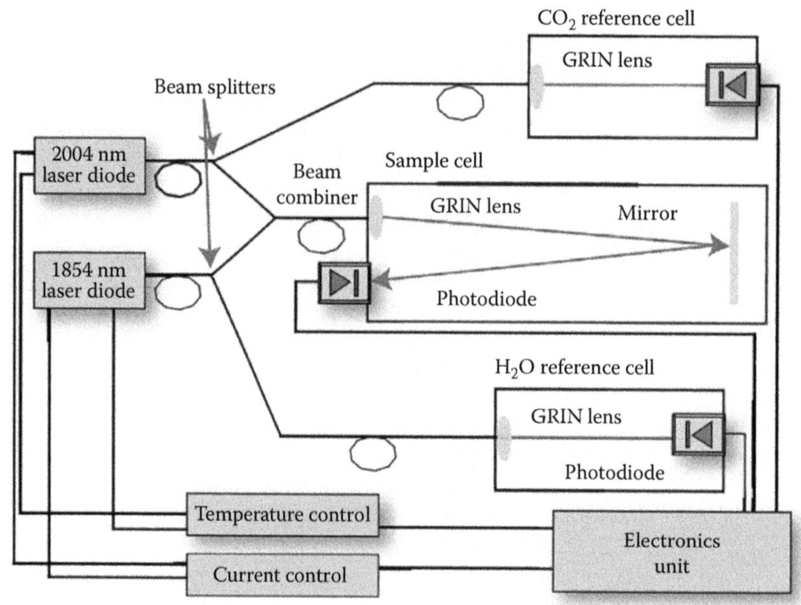

FIGURE 3.26 Optical scheme of the 5100 HD analyzer fabricated by AMETEK Process Instruments. The Model 5100 HD can be configured with single/dual lasers and single/dual gas cells. In a two-laser two-cell configuration, two different gas streams can be simultaneously monitored. One specific application is the measurement of wet and dry natural gas streams for H_2O levels. This dual cell configuration is superior to stream switching with a single gas cell which would require long equilibration periods when switching from wet to dry streams. Dual lasers systems can also be used for the monitoring of two gas species at the same time. (From http://www.ametek.com.)

spectral region there were also observed an absorption bands associated with the presence of water vapor. One of the examples of hygrometers working in this spectral region is a hydrometer, developed by Kebabian et al. (2002). Optical scheme of this hygrometer is shown in Figure 3.27. The hydrometer uses an argon emission line at 935.4 nm that can be Zeeman-split into two components. When the emission line is Zeeman-split by a longitudinal magnetic field of the proper strength, it is divided into one component that is only weakly absorbed, λ_1, and a second component that is strongly absorbed by the water vapor line, λ_2. This fact forms the basis for a differential absorption measurement. The major advantages of the gas discharge lamp, compared to the use of alternative light sources (e.g., NIR DL) are the stability of the spectral properties of the light source, which results in a relatively simple and inexpensive instrument. Moreover, the light source is reproducible at any time, based on its

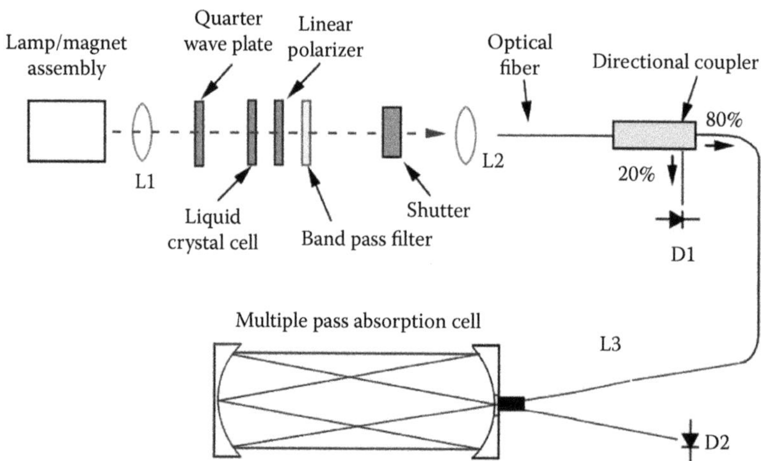

FIGURE 3.27 Schematic of optical water vapor hygrometer operated at 935.4 nm with all its optical elements. (From Kebabian, P. et al., *J. Geophys. Res. Atmosph.*, 107, 4670, 2002. With permission of American Geophysical Union.)

key physical dimensions (capillary diameter and length), gas fill pressure, and magnetic field. It also offers the advantages in its combination of time response and the ability to function properly in the presence of liquid water that cannot ordinarily be achieved by other types of sensors including the chilled mirror hygrometers and capacitance-based sensors. Kebabian et al. (2002) shows that this hygrometer is acceptable for *in situ* measurements of fine structure (over tens of meters distance) in the water vapor concentration profile both in free air and within and in the vicinity of clouds that could not be obtained using the commonly utilized technologies.

An improved model of optical hygrometer, working in this range, has been developed by Matthey et al. (2006). They used the same spectral range, 935 nm. However, as a source of radiation, they have used low-power continuous-wave DL. In addition, they used four single-mode frequency-stabilized optical signals. Three lasers were locked to three water–vapor absorption lines of different strengths (strong [935.4283 nm], medium [9935.3049 nm], and weak [935.6501 nm]), whereas the fourth lied outside any absorption line (935.600 nm). On-line stabilization was performed by wavelength-modulation spectroscopy using a compact water-vapor reference cells. An offset-locking technique implemented around an electrical filter was applied for the stabilization of the off-line slave laser to an online master laser. Matthey et al. (2006) believe that such configuration is promising for the in situ water vapor monitoring in the atmosphere at various altitudes. Strongly absorbing water vapor lines can be used for higher altitudes, whereas weakly absorbing lines can be served for lower altitudes, which show high water-vapor concentrations. The simultaneous use of four different wavelengths (three on-line and one off-line) enables measurement of H_2O profiles across the entire altitude ranging from the lower stratosphere to the boundary layer. The specific wavelengths must be selected depending on the atmospheric process under study, the desired altitude resolution and coverage and the required measurement precision. In particular, such approach can be promising for development of LIDAR systems discussed in Chapter 8 (Wulfmeyer and Walther 2001).

One should note that besides ground-based instruments there are TDL spectrometers, developed for aircraft application (Table 3.11).

TABLE 3.11
Tuneable Diode Laser (TDL) Spectrometers Designed for Aircraft Application

Device	Разработчик	References
JPL open-path tuneable diode laser spectrometers	Jet Propulsion Laboratory	Webster et al. (1988), May (1998)
The LaRC/ARC diode laser hygrometer (DLH)	NASAs Langley and Ames Research Centers (LaRC and ARC)	Vay et al. (1998, 2000), Cho et al. (2000)

3.5 UV OPTICAL HYGROMETERS

UV spectral range is also of interest for optical hygrometers development. In this spectral range there are several emission lines, which are of particular significance in the humidity measurement due to their high selectivity to water. These are, for example, the Lyman–alpha line and the line of krypton (123.6 nm). Lyman–alpha radiation $\lambda(L_\alpha)$ is emitted by hydrogen atoms $H(^2P–^2S)$ at a narrow line in the far UV portion of the spectrum at $\lambda = 121.56$ nm (Figure 3.28). It is produced by an electrical discharge in hydrogen. The wavelength of the Lyman–alpha line coincides with a deep absorption minimum in transmission spectra of air due to the presence of water vapor.

In principle, any wavelength satisfying the requirement, that there be large difference between the absorption cross section of the water vapor and the other constituents, can be used for humidity control. The advantage of using hydrogen Lyman–alpha line is that at this wavelength the difference in the cross sections amounts is almost three orders of magnitude. For example, the

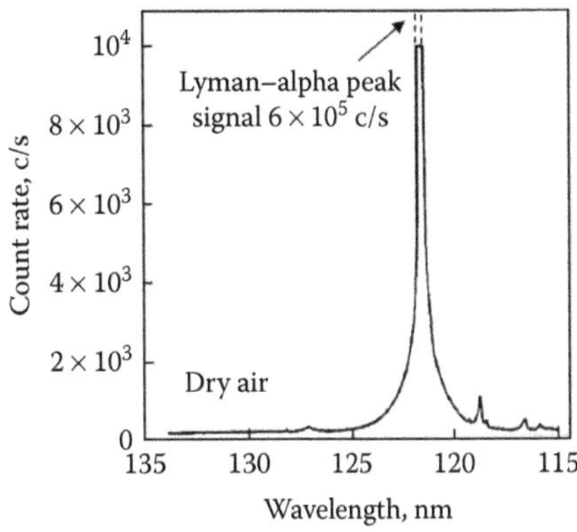

FIGURE 3.28 Output spectrum of the light source between 115 and 135 nm, which corresponds to the response interval of the NO ionization chamber. The spectrum was recorded with a PMT Hamamatsu 1259 (CsI photocathode and MgF_2 window). The resolution is 0.15 nm. Measurements were carried out in a 4.8-cm long absorption cell with dry air. At Lyman–alpha the peak signal is 6×10^5 counts/s. (From A. Zuber and G. Witt, *Appl. Opt.*, 26, 3083–3089, 1987. With permission of Optical Society of America.)

dominant absorber of radiation in the vacuum UV region (below 185 nm) is molecular oxygen with a pressure-dependent absorption cross section at $\lambda = 121.6$ nm of ~2.3×10^{-20} at STP (Kley 1984), whereas the absorption cross section for water vapor at the same wavelength is ~$1.57–1.59 \times 10^{-17}$ cm^2 (Kley 1984). These data show that while for Lyman–alpha light water vapor the absorption is very strong, the oxygen absorption is uniquely low, and most other common gases are relatively transparent, for example, nitrogen. In addition, oxygen effects can be removed using measurements of temperature and pressure to calculate the fractional oxygen density. Ozone absorption does represent a true interferent but only becomes significant at extremely high altitudes (stratosphere) where the natural ozone concentrations routinely reach significant levels. Thus, even the trace amounts of water vapor in a sample of the air dominate the absorption of this radiation. Therefore, the most commonly used source of UV radiation in hygrometers is the emission of hydrogen gas which includes the Lyman–alpha line at 121.6 nm (Buck 1976a, 1985; Beaton and Spowart 2012). In addition, the Lyman–alpha light sources have high output and spectral purity (Zuber and Witt 1987).

Lyman–alpha lamp is a direct current (DC)-powered hydrogen discharge tube with a temperature-controlled hydrogen reservoir made from a mixture of uranium and uranium hydride (Dieke and Cunningham 1952). Powdered uranium and a certain quantity of uranium hydride are contained in a side arm. The side arm can be heated and maintained at a constant temperature. Thus, the uranium hydride serves as a source of H_2, whereas the uranium simultaneously acts as an effective chemical absorber of gaseous contaminants that arise during the discharge. Such configuration of the lamp permits a low operating H_2 pressure for good spectral purity of the Lyman–alpha emission line, whereas the hydride restores the H_2 lost to the electrodes so as to give an acceptable lifetime. One should note that the Lyman–alpha hygrometer was developed at the beginning of the 1970s almost in parallel in the United States, the Soviet Union, and the former GDR (Buck 1973; Kretschmer and Karpovitsch 1973; Martini et al. 1973); and the American instrument was commercially produced by AIR Inc., Boulder CO. Other examples are listed in Table 3.12. Some of devices indicated in this table are no longer manufactured.

The embodiment of the UV hygrometer, using absorption at $\lambda(L_\alpha) = 121.6$ nm is shown in Figure 3.29. As in the IR hygrometers, the measurement with UV radiation is based on the detection of the transmitted or absorbed fraction of light after propagation through the sample. UV hygrometer shown in Figure 3.30 was developed by Beaton and Spowart (2012) for the airborne measurement of atmospheric water vapor. It is established that a significant fraction of radiation is absorbed over a few millimeters path length under normal conditions, which helps to minimize the adsorption surface and increase the exchange rate. On account of the extremely wide variation of the water vapor concentration between the lower and upper troposphere, Beaton and Spowart (2012) decided that it was necessary to adjust the sample path length to optimize the signal dynamic range for the humidity and the air density of primary interest. It is known that if the path length is too short, the mixing ratio noise will be excessive at low humidity. If the

TABLE 3.12
Commercial UV Hygrometers

Sensor Type	Produces	Country
L-5V	Buck Research	USA
Lyman–alpha hygrometer	MUERIJ METEO	NL
Lyman–alpha hygrometer	Wittich and Visser	NL
AIR-LA-1 Lyman–alpha hygrometer	AIR Now: Vaisala	USA Finland
KH2O–Krypton hygrometer	Campbell Sci.	USA

Source: Foken, T. *Micro–Meteorology*. Springer, Berlin, Germany, 2008.

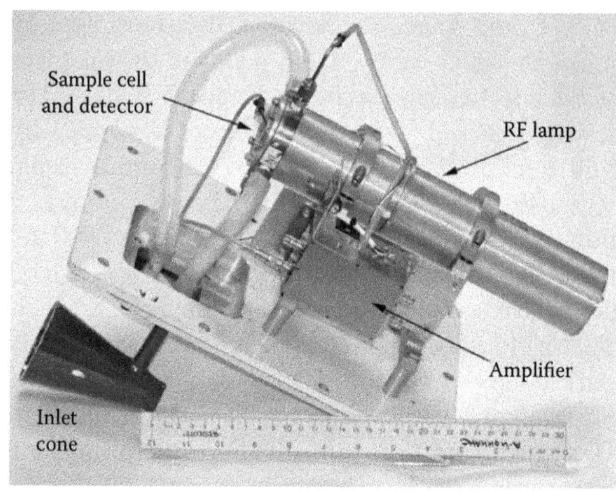

FIGURE 3.29 The UV hygrometer mounted to a C-130 aperture plate designed for aircraft applications. A radio-frequency (RF)-excited Lyman–alpha lamp (HHeLM-L) from Resonance, Ltd., which contains a temperature-controlled hydrogen sponge for improved spectral purity along with long life was used. The power supply box (15 cm × 15 cm × 5 cm) is not shown. The ruler is 30 cm long. The total cost was approximately U.S.$15,000 dollars (USD). (From Beaton, S.P. and Spowart, M., *J. Atmos. Oceanic. Technol.* 29, 1295–1303, 2012. With permission of American Meteorological Society.)

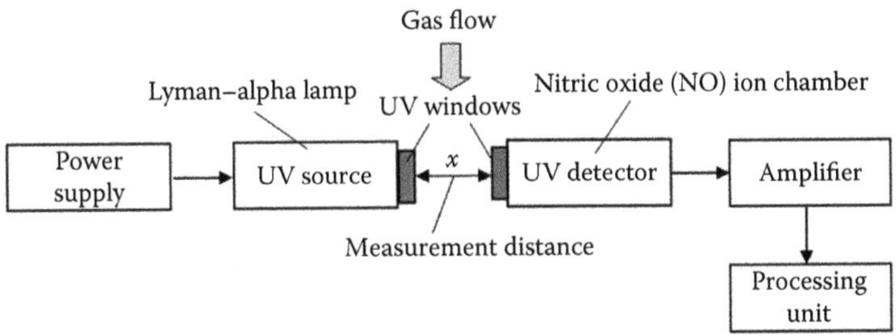

FIGURE 3.30 Components of a Lyman–alpha hygrometer.

path length is too long, there will be reduced sensitivity at high humidity resulting from non-Lyman–alpha light dominating in the signal. Therefore, the construction used in this hygrometer allowed varying the sample path length from 2.4 mm up to a useful maximum of 10 mm by placing spacers between the sample housing and the detector housing. The sample volume is a short cylinder 19 mm in diameter, giving a volume of 0.7 cm^3 when the path length is set at its minimum 2.4 mm.

In the first UV Lyman–alpha light hygrometers usually a nitric oxide ionization cell served as the detector. A photoionization chamber is a simple device containing an ionizing gas and two electrodes. Incoming radiation ionizes the gas (nitric oxide in this application), and an electric field maintained between the electrodes induces the electron and ion drift. The resultant current is proportional to the incident light intensity. Radiation enters the detector through another magnesium fluoride window aligned with the source across the sampling volume. The combination of the magnesium fluoride windows and nitric oxide as the ionization medium limit the response of the detector to incident wave lengths between 132 and 115 nm. This effectively filters almost everything except the Lyman–alpha line. Beaton and Spowart (2012) have shown that a solar-blind diamond photocathode vacuum photodiode (R6800U-26, Hamamatsu) with a spectral response of 115–220 nm could be also used. The light level is high enough that photomultiplier tubes are not needed, and the photodiodes are more compact and have lower microphonics and magnetic field sensitivity than photomultiplier tubes. In this case a highly stable high-voltage power supply is not needed, and in fact the photodiodes are rather insensitive to power supply voltage once it exceeds 10 V. In addition, the vacuum photodiodes have an unlimited shelf life and are mechanically and electrically robust. Their drawbacks compared to the NO ionization cell are lower quantum efficiency, ~10% versus ~60% for the NO cell (Carver and Mitchell 1964), and a smaller active area. There are also devices where a photomultiplier connected to a low-noise electrometer is used as detector of Lyman–alpha radiation (Mestayer et al. 1986, 1987).

Simulations have shown that the Lyman–alpha absorption hygrometer can offer a very fast response (~5 ms). It is great advantage of UV hygrometers: the fast response makes it a suitable instrument for ultrafast hygrometry (sample rates of 10–100 Hz), the water vapor flux measurement, or micrometeorological measurements inside and outside clouds. Of course, the time of response indicated before is theoretical limit. However, even real instruments have response time smaller than 1 s. It was also found that this device can measure the water vapor densities of 0.1–25 g/m^3 with a relative precision of 0.2% and an accuracy of 5%.

It is important for applications that in contrast to IR, the optical pathway required for measurement in UV range is much shorter that required for IR light. Therefore, the UV optical hygrometers can be directly embedded in the pipes where the humidity control is needed. For example, Figure 3.31 depicts a measurement device integrated into a gas flow. The source and detector are aligned perpendicularly to the direction of gas flow. The distance between the source and the detector L is variable and should be suited to the expected water content of the gas to be measured. L can be calculated

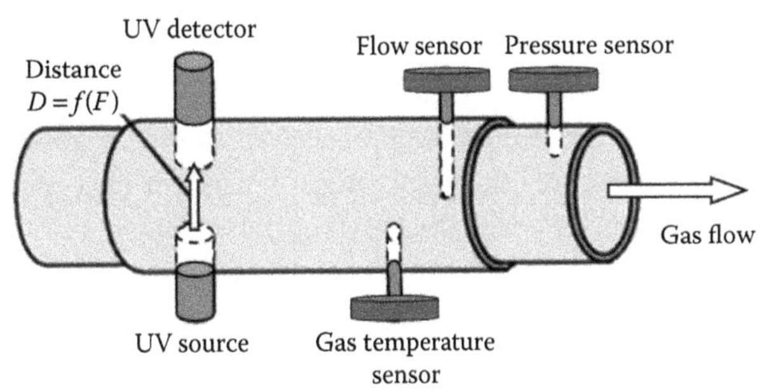

FIGURE 3.31 UV hygrometer incorporated in the gas pipeline. (Idea from Wernecke, R. and Wernecke, J., *Industrial Moisture and Humidity measurement: A Practical Guide*, Wiley-VCH, Weinheim, Germany, 2014. With permission.)

using the Lambert–Beer law, and is a function of the absolute water content and the effective absorption cross section. In real devices an optical path length can vary from several millimeters to 20 mm or more. The incident beam enters the sample chamber through the entrance window and penetrates the gas to be measured. The typical window materials are the quartz glass and magnesium fluoride.

However, the Lyman–alpha absorption hygrometer is a secondary measurement device and therefore it is not possible to use the single-beam Lyman–alpha absorption hygrometer alone to measure the absolute water vapor density (Schanot 1987). This means that this device must be regularly calibrated, and for this purpose an additional absolute water vapor sensor, such as a chilled mirror hygrometer, is needed. Therefore, data received using such devices are typically corrected by the means of data from a second, slow response hygrometer, and application of a suitable algorithm. Various correction schemes have been developed to make the data useful. But, of course, this situation is not comfortable for users. The Lyman–alpha light source aging, temperature drift, and optical window contamination are other factors that prevent a stable predictable calibration. One should note that the devices working in the UV range are affected more than those in the IR range. This means that the calibration characteristics of UV devices are subjected to large changes during the application time. Therefore, the Lyman–alpha absorption hygrometer usually is used together with more stable absolute hygrometer such as a dew point hygrometer. When a reference hygrometer is not available, a variable path length self-calibration technique in the case of a constant absolute humidity can be used. A first Lyman–alpha hygrometer with variable path length was proposed by Buck (1976a) and an updated version was proposed by Foken et al. (1995). In addition, for quantitative analysis of the concentration C of the absorbing species in single beam devices both I and I_0 must be measured simultaneously. Temporal changes in the output from the lamp seriously affect the accuracy of the measurement at small optical depths as well (Zuber and Witt 1987). Other complications are the possible presence of alternate absorbers with pressure- or temperature-dependent absorption cross sections as well as the requirement of a parallel beam which cannot be met without collimating optics. Zuber and Witt (1987) and Weinheimer and Schwiesow (1992) have shown that many indicated above problems could be resolved using a two-beam optical scheme (Figure 3.32). In this hygrometer, the absorption is determined along two separate paths containing the sample and the dry reference air, respectively, hence eliminating undesired absorption by O_2 and other gases presented in the air. At the same time, the differential absorption eliminates problems related to the intensity variations in the light source output. The dynamic range of the instrument can be tuned by diluting the sample with reference air. The sample and the reference air are admitted to their respective chambers with the aid of a gas handling system. Weinheimer and Schwiesow (1992) also proposed to use the second wavelength that is not strongly absorbed by water vapor to control the degradation of windows. Weinheimer and Schwiesow (1992) believe that solids or liquids in the path have broadband absorption features and are essentially opaque at UV wavelengths, so absorption at both primary and secondary wavelengths should be approximately the same for closely spaced cells and wavelengths.

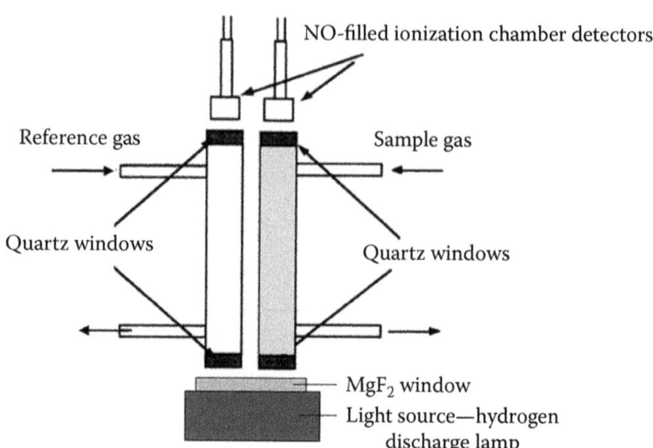

FIGURE 3.32 Schematic drawing of the optical scheme of two-beam UV hygrometer, which include light source (the hydrogen discharge lamp emitting the Lyman–alpha [121.6 nm]), absorption paths for sample and reference air, and no filled ionization chamber detectors. The output from the two detectors is passed to electronics units. (Idea from Zuber, A. and Witt, G., *Appl. Opt.*, 26, 3083–3089, 1987; Weinheimer, A.J. and Schwiesow, R.L., *J. Atmos. Oceanic Technol.*, 9, 407–419, 1992.)

Further critical disadvantages of a Lyman–alpha absorption hygrometer are a long-term instability of the UV hydrogen lamp, long-term intensity fluctuations, and degradation of the MgF_2 windows via solarization, which requires a frequent calibration. It is in many cases hampered the use of UV hygrometers and forced many producers to abandon a production of Lyman–alpha absorption hygrometers. In addition, these lamps were mainly handmade. Therefore, there is the requirement for careful selection of the light source before installing it in the gauge, since for the lamp there is a considerable scatter of the illumination parameters in relation to their careful preparation. Correspondingly, these problems affect the cost of a gauge and the hygrometer that is markedly higher than the cost of routine humidity meters. Beaton and Spowarti (2012) noted that a nitric oxide ionization cells were also unstable and they could fail in use or even while in the storage.

Experience has shown that krypton lamps are more stable, long-lived and inexpensive sources and their production is much easier. Therefore, during the past decades the UV krypton hygrometers have also been developed. First UV hygrometer using a krypton lamp was developed in 1985 (Campbell and Tanner 1985). The benefits of this instrument were a longer lifetime and easier production. But the absorption band in krypton hygrometers is not directly located in the Lyman–alpha band. Emission from the krypton tube exhibits a major band at 123.58 nm and a minor band at 116.49 nm.

Modern krypton hygrometers fabricated by Campbell Scientific Inc. is shown in Figure 3.33. This device (KH20 krypton hygrometer) is designed for measurement of rapid fluctuations in atmospheric water vapor (Foken and Falke 2012). High frequency response is suitable for eddy-covariance applications (up to 100 Hz). In the KH20, the source is a low-pressure krypton glow tube. As it was indicated before, emission from the krypton tube exhibits a major band at 123.58 nm and a minor band at 116.49 nm (Budovich et al. 2013). Radiation at 123.58 nm is strongly attenuated by the water vapor whereas the absorption by other gases in the optical path is relatively weak at this wavelength. However, at 123.6 nm the water vapor absorption strength is down by a factor of almost 2 in comparison with absorption at Lyman–alpha line, whereas the oxygen absorption is stronger by a factor of about 40 (Tillman 1965; Yoshino et al. 1996), making the oxygen interference much more severe and variable with air density, a particular problem for measurements from aircraft. Radiation at the shorter wavelength (116.49 nm) is also attenuated by the water vapor and oxygen molecules. This means that a much larger correction for oxygen is required than for Lyman–alpha device. In addition, a variable path length self-calibration technique cannot be applied in the krypton hygrometer, because this device works with two absorption lines, each with two absorbers. All this reduces the accuracy of measurement by krypton hygrometers. However, Campbell and Tanner (1985) determined that for their Kr lamp, this line (116.49 nm) contributes only 3% to the total lamp output. Since it is very near the nominal 115 nm cutoff for the magnesium fluoride windows, they concluded that it is greatly attenuated. Thus, the usage of MgF_2 windows fitted to the source and detector tubes should help to improve the accuracy of measurements. But it is necessary to note that magnesium fluoride is hygroscopic material. This material changes its transfer characteristics in humid environments by interaction of atmospheric constituents with UV photons. Windows had to be cleaned manually more or less frequent. Another disadvantage is the wide range, nonlinear output signal of both Lyman–alpha and krypton-based systems. This either requires a special analog signal processing or the use of wide range A/D-converters for accurate calculations. In addition, we should keep in mind that due to simplicity of configuration and its signal offset drift these hygrometers cannot be used for determination of absolute concentrations. A characteristic of the KH20 is that the source tube window experiences scaling when operating, especially in humid environments. This will arise through the disassociation of atmospheric constituents by UV photons and can be removed simply by wiping the windows.

No doubts that other wavelengths in the UV range can be also used for hygrometer design. For example, the emission line ($\lambda = 184.9499$ nm) of the hydrogen low

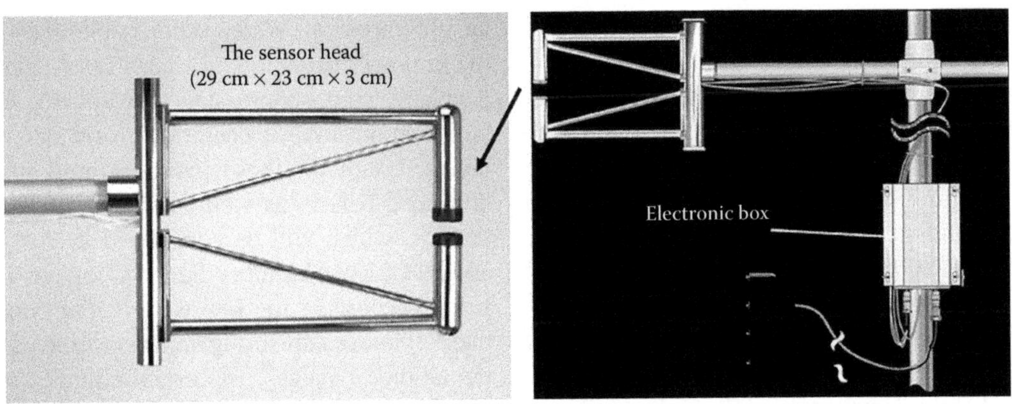

FIGURE 3.33 KH20 krypton hygrometers developed by Campbell Scientific: calibration range—1.7–19.5 g/m^3 (nominal), frequency response—100 Hz; operating temperature range—30 to +50°C; weight—6.8 kg. (From http://www.campbellsci.eu.)

pressure lamp can be used for these purposes (Wernecke and Wernecke 2014). However, the sensitivity and the accuracy of these devices will be much worse than in the devices considered previously. Gersh and Matthew (1988) for their research have used the radiation at wavelengths of 177 and 205 nm. Xenon filled, fused silica window flashlamp with filters was used as the source of UV radiation, while the CsTe photocathodes, which are *solar blind*, were used as detectors. They used this hygrometer to monitor and control the operation of the industrial drying chamber. The basic concept of this UV hygrometer was to measure the differential absorption by water vapor of UV radiation in two spectral bands and to relate this measurement to the absolute humidity of the air in the measurement volume. In the first absorption spectral band ($\lambda = 177$ nm), the UV radiation is strongly attenuated by the water vapor, whereas in the second ($\lambda = 205$ nm) reference band, the UV radiation is not affected by the presence of water vapor. Then, the ratio of the absorption to the reference band intensities is a reflection of the mass of water vapor per unit volume in the measurement region (which is the definition of the absolute humidity). No doubts this approach is correct, but the selected spectral range does not provide the necessary sensitivity. A complication arises from the fact that atmospheric oxygen also absorbs the UV radiation in the same spectral region as water vapor.

With regard to other sources of light in the desired spectral range, that unfortunately in this particular spectral range the set of sources is very limited (Heering 2004). LEDs can be manufactured to emit light in the UV range, although practical LED arrays are very limited below 365 nm. LED efficiency at 365 nm is about 5%–8%, whereas the efficiency at 395 nm is closer to 20%, and the power outputs at these longer UV wavelengths are also better. Only Taoyuan Electron Lmt. (HK) suggests the UVC Flashlight LED operated at 280 nm (http://www.ledwv.com). But this source has small power ~25 mW, stringent requirements for the operation and short service life. The UVLEDs should be used within three months. A similar situation is observed for UV DL. The Ce:LiSAF and Ce:LiCAF DL (Lawrence Livermore National University) can generate radiation in the range from 280 to 316 nm. Lasermate Group, Inc. provides a diode-pumped solid-state lasers (DPSS) working in the range from 261 to 2200 nm. Laser 2000 produces QL262 laser with wavelength 262 nm (http://www.laser2000.co.uk/). Pulsed laser systems *passively Q-switched laser*, emitting wavelengths of 213 nm was designed by CryLaS GmbH (http://www.crylas.de/). Only gas lasers can generate light with wavelength smaller than 200 nm: the UV argon-fluoride (ArF) excimer lasers operate at 193 nm, and Ar_2* excimer laser can operate at 126 nm. However, even in this case, the emission spectrum is not optimal for the hydrometer construction (Parkinson and Yoshino 2003). Furthermore, the use of gas lasers does not contribute to lower prices and sizes of developed hygrometers.

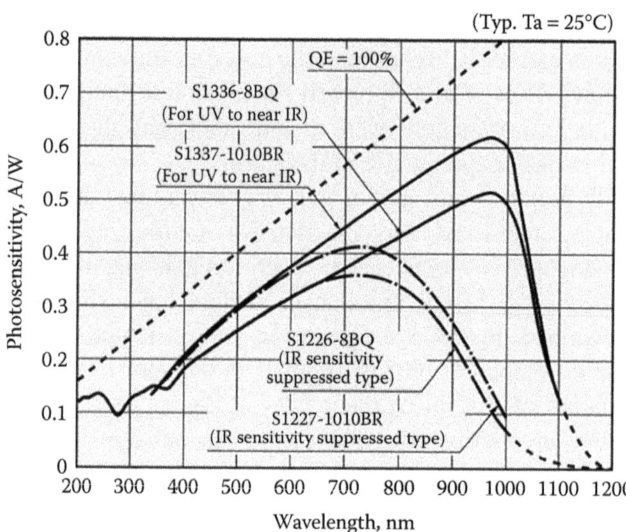

FIGURE 3.34 Examples of Si semiconductor photodetectors optimized for operation in the UV spectral range. (From http://www.hamamatsu.com.)

The use of semiconductor radiation detectors in the UV hygrometers is also limited because of low sensitivity in the current spectral range (Seib and Aukerman 1973; Razeghi and Rogalski 1996). Typical spectral characteristics of such photodetectors are shown in Figure 3.34.

3.6 FLUORESCENT-BASED HYGROMETERS

As was showed earlier, the water vapor plays a unique role in atmospheric processes as a key chemical and radiative component. Water vapor is present in all layers of the atmosphere including upper layers, such as the troposphere, stratosphere, and mesosphere, so it is important to control the concentration of water vapor in these layers. Using for these purposes most of the methods described before, as well as most of the solid humidity sensors, which will be considered in Vol. 2 "Electronic and Electrical Humidity Sensors" of our issue, is hampered because of the low water vapor concentrations at these altitudes and strong variations in temperature when the altitude changes. At certain altitudes, a strong temperature jumps are possible. Currently, a limited number of techniques can be used for *in situ* measurements of the upper troposphere and stratosphere humidity. Among them, the fluorescent technique, offering high accuracy and fast response, has proved reliable (Meyer et al. 2015). Devices using this technique worked well when used on the board balloons, high-altitude aircrafts and rockets.

3.6.1 Principles of Fluorescent-Based Technique

Fluorescence is a radiation that is emitted after a chemical species absorb a radiation of another wavelength. For excitation of fluorescence, the UV radiation is being used. It results from the instantaneous deactivation that takes place after the species go to a higher energy state upon the absorption of incident radiation. The excited species has a lifetime of about 1–100 ps, and returns to its ground (unexcited) state through the emission of radiation:

$$A^* \rightarrow A + h\nu \quad (3.4)$$

where:

A^* is an excited state of a substance A
h is Planck's constant
ν is the frequency of the photon

The emission occurs in all directions, and is of a longer wavelength than the incident radiation. The wavelength of the emitted radiation is influenced by the chemical structure of the fluorescent species. The fluorescence quantum yield gives the efficiency of the fluorescence process. It is defined as the ratio of the number of photons emitted to the number of photons absorbed:

$$\Phi_{\text{quantum yield}} = \frac{\text{number of photons reemitted}}{\text{number of photons absorbed}} \quad (3.5)$$

The radiant intensity of the emission conveys information on the concentration of the fluorescent species. For a weakly absorbing system (i.e., $A < 0.05$), the fluorescence intensity (I_F) is linearly related to the concentration (C) of the fluorescent species, as expressed in the following equation:

$$I_F = k_F I_0 C \quad (3.6)$$

where:

I_0 is the intensity of the incident radiation
k_F is a constant that is dependent on the absorption characteristics of the fluorescent species, the quantum efficiency of the fluorescence, and the configuration of the instrumentation system

In relation to water vapor, the theory of fluorescent-based technique has been developed by Kley and Stone (1978) and Bertaux and Delannoy (1978). Later refinements were made by Keramitsoglou et al. (2002). The fluorescent method of water vapor detection is based on the photodissociation of H_2O at wavelengths below 137 nm and the subsequent fluorescent relaxation of the exited OH* radical produced (Yushkov et al. 1995; Kley et al. 2000). For Layman–alpha dissociation of water vapor, the process can be expressed as

$$H_2O + h\nu(121.6\,\text{nm}) \rightarrow OH^*(A^2\Sigma^+) + H(^2S) \quad (3.7)$$

with a quantum yield far less than 1. About 10% of the absorbed photons result in excitation of the OH fragment to the $A^2\Sigma^+$ electronic state. The electronically excited OH* radical either fluoresces at 310 nm

$$OH^*(A^2\Sigma^+) \rightarrow OH(X^2\Pi) + h\nu(305\text{–}325\,\text{nm}) \quad (3.8)$$

or is quenching by collisions with air molecules

$$OH^*(A^2\Sigma^+) + N_2, O_2 \rightarrow \text{products} \rightarrow OH(X^2\Pi) + M \quad (3.9)$$

Fluorescence is seen in both (0→0) and (1→0) bands but the strongest fluorescence is from highly excited rotational levels (N = 20–22) of the (0→0) band.

A photomultiplier with an interference filter measures the intensity of fluorescence which is signature of the parent molecules of H_2O. The intensity of fluorescence is obtained as

$$J = |OH^*| \cdot A = \frac{|H_2O| \cdot F_\lambda \cdot \sigma_{H_2O} \cdot \varphi \cdot A}{A + k_q \cdot |air|} \cdot \exp\{-\sigma_{O_2} \cdot |O_2| \cdot L\} \quad (3.10)$$

where:

L is the length of the absorption between the lamp and interaction region
[OH*], [H_2O], [air], [O_2] are the number densities of OH*($A^2\Sigma^+$), H_2O, air and oxygen respectively
A is Einstein transition probability
F_λ is photon flux of the light source
σ_{H_2O} and σ_{O_2} are cross sections of water vapor and oxygen for Layman–alpha, respectively
φ is the quantum yield of photo dissociation
k_q quenching coefficient

Thus, by measuring the fluorescence radiation the H_2O abundance can be determined.

Of course, for correct measurement of the photon flux, the absorption by oxygen has to be taken into account.

However, at air pressure below 10^{-1} hPa the quenching by the air and oxygen absorption are negligible (Kley and Stone 1978) and the fluorescence intensity becomes

$$J = |H_2O| \cdot F_\lambda \cdot \sigma_{H_2O} \cdot \varphi \quad (3.11)$$

In the other limiting case $P_{air} > 10$ hPa, that is, in the atmosphere up to 20–35 km, k_q*[air]>> A and hence

$$J = \frac{|H_2O|}{|air|} \cdot \frac{F_\lambda \cdot \sigma_{H_2O} \cdot \varphi \cdot A}{k_q} \cdot \exp\{-\sigma_{O_2} \cdot |O_2| \cdot L\} = C \cdot \frac{|H_2O|}{|air|} \quad (3.12)$$

The factor C in Equation 3.12 summarizes molecular coefficients, known from the literature, as well as instrument specific quantities. For example, for C calculations one can use the values of the following parameters: Einstein coefficient for reaction Equation 3.8, $A = 1.26 \cdot 10^6\ s^{-1}$ (Crosley and Lenge 1975); and $k_p = 2.3 \cdot 10^{-11}\ cm^{-3}\ s^{-1}$ (Kley and Stone 1978). If C is a constant, the number of detected fluorescence photons is proportional to the H_2O mixing ratio [H_2O]/[air] for measurements in the troposphere and lower stratosphere. Then the water vapor mixing ratio (μ) may be expressed (following Kley and Stone 1978) as

$$\mu = \frac{J}{C \cdot F_\lambda} \quad (3.13)$$

For measurements at higher altitudes, Equation 3.11 has to be used to obtain correct water vapor mixing ratios. Thus, under condition with negligible oxygen absorption, the fluorescence gives a direct measurement of the atmospheric water vapor mixing ratio in this case (Yushkov et al. 1995). Of course, we have to keep in mind that in reality, C is a function of the photo dissociation rate of the reaction Equation 3.7, and thus depends on the photon flux in the fluorescence volume, which in turn depends on the variations of the lamp intensity (Zoger et al. 1999).

3.6.2 Examples of Fluorescent-Based Hygrometers Realization

The first fluorescent-based hygrometers suitable for research of atmosphere have been developed by Kley and Stone in 1978. However, the actual flight instrument for use in the stratosphere was described and characterized in 1979 (Kley et al. 1979). Later on fluorescent-based hygrometers, developed in other laboratories, have appeared (Goutail and Pommereau 1987; Weinstock et al. 1990, 1994; Kretova et al. 1991; Yushkov et al. 1995; Zoger et al. 1999). The most advanced Lyman–alpha fluorescence hygrometers have been developed in the laboratories at National Oceanic and Atmospheric Administration (NOAA) in Boulder, at Harvard and in Julich (Germany).

All existing fluorescent-based hygrometers are built on a similar principle, but constructively they may differ essentially. Their versions are shown in Figure 3.35.

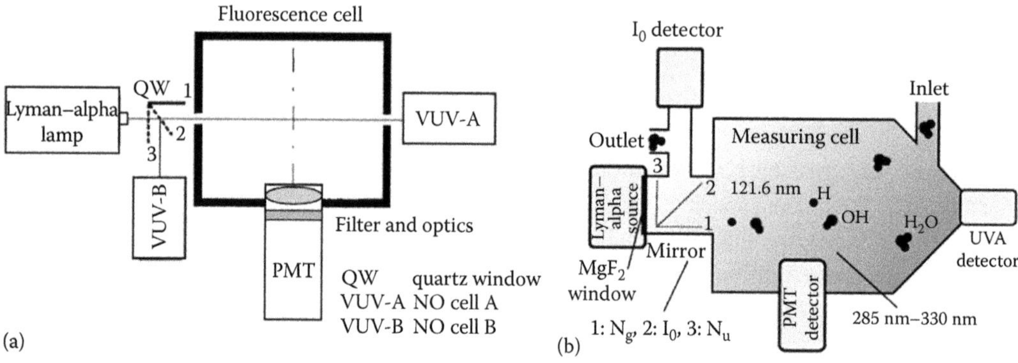

FIGURE 3.35 (a) Sketch of the fluorescence cell and the major components as Lyman–alpha radiation source, photomultiplier tube, vacuum UV (VUV) detectors and the mirror drive in the hygrometer designed by Zoger et al. (1999). (b) Conceptual diagram illustrating the principal mechanical and optical components of the airborne Fast In-situ Stratospheric Hygrometer (FISH). The size of the cell is 0.3 L in total. As the lamp is not monochromatic, the number of lamp background counts also has to be taken into account. Therefore a swiveling mirror is implemented between the lamp and the measuring cell. During one measurement cycle the mirror is placed in three different positions to determine the total fluorescence rate N_g (mirror position 1), the background rate N_u (mirror position 3) and the lamp intensity I_0 (mirror position 2). Thus, the mirror drive allows monitoring the intensity of the Lyman–alpha radiation (I_0) and a determination of the background and dark count rates. N_g is the number of detected fluorescence photons proportional to the water vapor mixing ratio. Thus, calibration factors can be determined which are nearly independent of changes of the output of the radiation source. The measurements using this hygrometer can be accomplished with a precision <0.2 ppmv at 1 s integration time. ((a) From Zoger, M. et al., *J. Geophys. Res.*, 104, 1807–1816, 1999. With permission. (b) from Meyer, J. et al., *Atmos. Chem. Phys.* 15, 8521–8538, 2015. Published by the European Geosciences Union as open access.)

Typically, the difference in design is associated with different designations of developed devices and different approach to calibrating fluorescent-based hygrometers. For example, the devices designed for aircraft applications, in contrast to the devices intended for balloon measurements (Yushkov et al. 1995), are not limited in size and weight, which allows entering additional elements, improving their parameters (Goutail and Pommereau 1987; Zoger et al. 1999). Typically, such devices have a measuring chambers and additional receivers for controlling the intensity of the UV lamp radiation (Figure 3.35). The presence of closed measuring chamber essentially extends the opportunity to use fluorescent-based hygrometers, as it allows using them during the day.

Hygrometer shown in Figure 3.36 does not have such a possibility, because the measurements take place in the open space. This means that the readings are sensitive to the ambient solar radiation, which usually exceeds the fluorescence radiation to be detected.

In this regard, present device can only be used at night time. However, the devices with an open structure have another significant advantage. They avoid contamination by desorption of H_2O from the walls, which is very important when measuring in the atmosphere with a very low concentration of water vapor.

For excitation of fluorescence, one can use the hydrogen or krypton vacuum UV lamps described earlier (Varier 1967; Buck 1976b; Zoger et al. 1999). As it was mentioned previously, the radiation of the hydrogen lamp contains a Layman line (121.6 nm). An oxygen spectral window at this line provides the most effective generation of excited hydroxyl radicals within the volume where fluorescence is being registered. The radiation of krypton vacuum UV lamp does not possess this property. Hydrogen vacuum UV lamps may have different design, which is associated with the use of various sources of hydrogen. The most common sources of hydrogen are the uranium hydride and a mixture of hydrogen and helium.

According to Varier (1967) and Buck (1976b), the intensity of a 121.6 nm line in glow-discharge lamps using a mixture of 2%–25% hydrogen and helium or argon grows through decreasing recombination of hydrogen atoms on the walls of the bulb. Such lamps are easier to manufacture and are more ecologically safe compared to those containing uranium hydride and uranium (Hutcheson 1972; Keramitsoglou et al. 2002). It was established that the uranium hydride could be replaced by the hydride of ZrCo, which possesses the same thermodynamic properties (Lykov et al. 2012). This replacement allows eliminating the thermostatic control of the lamp, which helps to reduce the size and weight of the hygrometer.

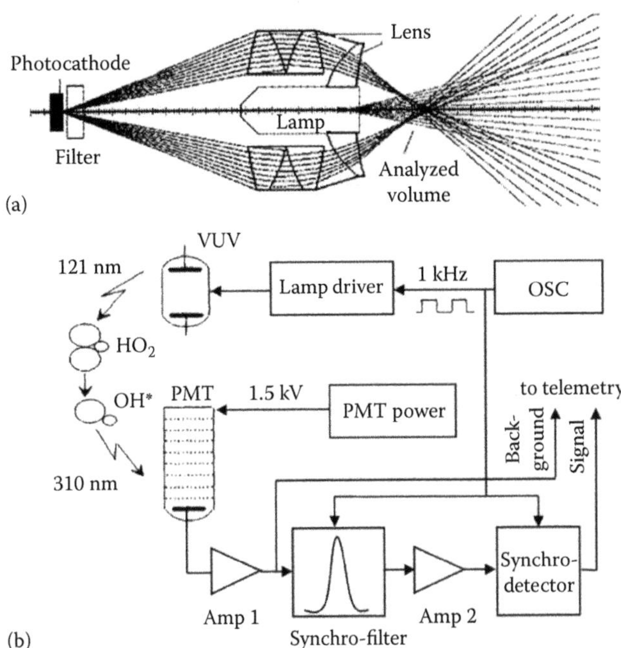

FIGURE 3.36 (a) Optical scheme and (b) electronics block diagram of fluorescent-based compact hygrometer designed by Yushkov et al. (1995) to carry out the high-resolution nighttime upper tropospheric and stratospheric water vapor balloon measurements. To increase the detection efficiency of the useful signal and to reduce the size of the device, the radiation source and the optical detection system are positioned coaxially. The lenses are made from U-Viol glassUS-49. The front lens has a diameter 50 mm and is sealed to the lamp body. The focal length is 10 mm which leads to a distance of 24 mm between the windows of vacuum ultraviolet (VUV) lamp and the analyzed volume. Hygrometer did not use special measurement cell. The measurements are carried out in the open space in the immediate vicinity of the objective. Therefore, the device does not need a suction system to sample the air to be analyzed. The total uncertainty of the measurement is less than 10% at the stratospheric mixing ratios greater than 3 ppmv increasing to about 20% at mixing ratios less than 3 ppmv. A description of improved version of hydrometer one can find in (Lykov et al. 2011, 2012). Compact and light hygrometer has a mass ~0.5 kg. The absence of the receiver, controlling the intensity of the lamp radiation, the authors explain by the high stability of the radiation intensity of lamps used in hygrometer. (Idea from Yushkov, V. et al., *SPIE Proc.*, 2506, 783–794, 1995.)

But, it was established that in the spectra of hydrogen UV lamps within 200–360 nm range an intensive hydrogen continuum is present, which overrides the spectrum of hydroxyl fluorescence and may cause a considerable noise during laboratory calibration of hygrometers in a closed chamber. Therefore, in optical fluorescence hygrometers, which use a hydrogen glow-discharge lamp, a 270–320 nm radiation should be rejected with a special window-filter. For such filter, MgF_2 is usually

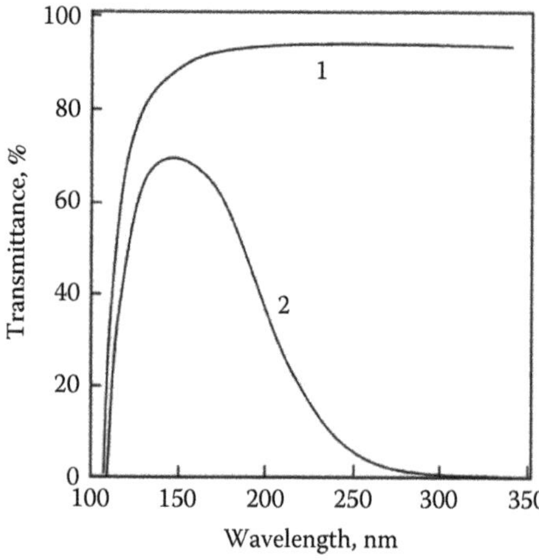

FIGURE 3.37 Typical spectra of the transmittance of monocrystalline magnesium fluoride window of hydrogen lamp without filter (1) and with filter (2) for selective absorption at 300 nm. (Data extracted from Yushkov, V. et al., *SPIE Proc.*, 2506, 783–794, 1995.)

TABLE 3.13
Examples of Lyman–Alpha Photofragment Fluorescence Hygrometers Designed and Fabricated for Atmosphere Monitoring Using Balloons and Aircrafts

Hygrometers	References
NOAA Aeronomy instrument (U.S.)	Kley and Stone (1978), Kelly et al. (1989)
Harvard Water Vapor (HWV) instrument (U.S.)	Weinstock et al. (1990, 1994), Hintsa et al. (1999)
Central Aerological Observatory (CAO) instrument (Russia)	Yushkov et al. (1995)
FISH (Fast in situ Stratospheric hygrometer) instrument (Germany)	Zoger et al. (1999)
UK Met Office instrument (UK)	Keramitsoglou et al. (2002)

used, spectral characteristics of which are shown in Figure 3.37.

To improve sensitivity, an interference filter centered at 318 nm, is also desirable. It allows selecting the spectral region, coincident with the emission from the upper rotational levels of the (0→0) band of the *A*→*X* system of OH. With regard to the sensitivity measurement of the photofragment fluorescence, it is usually carried out by sun-blind photomultiplier (PM).

Usually for this purposes a HAMAMATSU R647-P photomultiplier is being used. In some devices, in addition to indicated above elements, one (Goutail and Pommereau 1987) or two (Zoger et al. 1999) additional detectors may also be present. They are applied to control the intensity of radiation, used for fluorescence excitation. Typically, such detectors are the NO filled ionization chamber detectors. The energy required to ionize NO corresponds to a threshold wavelength of 134 nm. Therefore, the NO filled ionization cells are a highly selective radiation sensor for the vacuum UV (VUV) spectral region (Samson 1967). An iodine ionization cell that is sensitive from 115–135 nm can also be used (Kelly et al. 1989). The electronic system used in hygrometer designed for the lamp modulation and synchronous demodulation of the signal received. This technique improves noise-to-signal ratio in more than 100 times.

Studies have shown that the fluorescent-based hygrometers have high sensitivity. Besides, they are compact enough for the balloon and aircraft measurements (Table 3.13). Rocket-borne measurements in the mesosphere using this technique have also been reported (Khaplanov et al. 1996). In addition, the Lyman–alpha fluorescence technique can achieve a large dynamic range for measurements from the middle and upper troposphere at about 1000 ppmv into the dry stratosphere with only 2–5 ppmv, where changes on the order of 0.1 ppmv can be detected with a relative uncertainly of ±5%. At that high speed of response, when integration time is on the order of 1 s, enables the measurement of small-scale features in the atmosphere. However, the H_2O measurement range of these devices is limited to pressures lower than 300 hPa because of strong Lyman–alpha absorption in the lower troposphere. Other disadvantages of these hygrometers are the instability of emission hydrogen UV lamp, a relatively short service life, and the lack of commercially produced samples with reproducible parameters.

3.7 ADVANTAGES AND DISADVANTAGES OF OPTICAL HYGROMETERS

Numerous studies have confirmed that the optical hygrometers have several important advantages, which are not found in other hygrometers. At first, the measurement of water vapor using this optical (or spectroscopic) technique has received very favorable response from the industry because these devices employ a noncontact sensor technology for the detection of water vapor in various environments including natural gas: the sample never comes into direct contact with the sensor element. This *noncontact* approach significantly reduces maintenance requirements of the instruments and reduces the overall costs of maintaining the equipment.

Moreover, in principle, optical methods for humidity measurement can be used in extreme environmental conditions (high pressure, high temperature, aggressive gases, etc.), and the lack of delay of the measurement. The speed of response is limited only by the speed of the indicator or recorder. The time-constant of an optical hygrometer is typically just a few milliseconds. Therefore, these instruments are very useful for tracking fast humidity fluctuations, since the method does not require the detector to achieve vapor-pressure equilibrium with the sample. This makes the method ideal for use in measuring humidity from an airplane or under any circumstance where rapid changes of humidity can occur.

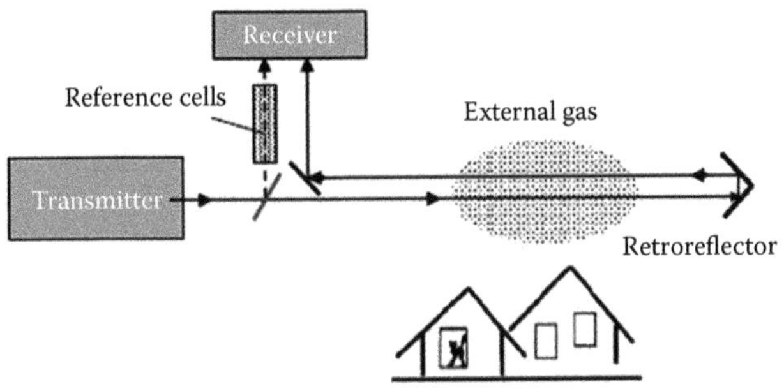

FIGURE 3.38 Experimental setup for remote atmosphere monitoring with retroreflector using open-path measurement. (From Korotcenkov, G. (Ed.), *Electrochemical and Optical Sensors*, Momentum Press, New York, pp. 311–476. With permission.)

In addition, optical method has specific sampling; it is possible to sample instantaneously any desired path length of atmosphere from a few millimeters to thousands of meters (Figure 3.38), and in so doing arrive at an integrated value of the absolute humidity in the path in question. This is co-called as open-path measurement. If the path length is appreciable, the value obtained is more representative than that obtained by spot sampling (Foskett et al. 1953). Of course, extractive or cell-based IR measurements, commonly used in the previously considered optical hygrometers, have the obvious advantage over the open-path setup that the physical conditions (concentration, pressure, temperature, absorbing path length, etc.) of the measurement can be readily controlled or changed at will (Table 3.14). This provides better conditions for prolonged measurements when high precision of determination is needed, for example, for fundamental studies in the gas spectroscopy or analytical method development. Open-path measurements, on the contrary, are more prone to technical difficulties connected to changing atmospheric (meteorological) conditions leading to uncertainties, but the *in situ* nature of such studies offer distinctive advantages as well.

The advantages of optical hygrometer can also include their ability to maintain high sensitivity even at low concentrations of water vapor, where the psychrometer, hair hygrometer, electrical hygrometer, and even the dew point apparatus all become less sensitive. Moreover, optical hygrometer will operate as well below

TABLE 3.14
Comparison of Extractive and Open-Path Techniques

	Extractive (Closed Cell)	Open-Path
Advantages	• Path-length adjustable, independent of plume size • Simplicity, repeatability • Lower detection limits • Stable conditions, longer measurement times • Temperature and pressure control • Quantitative reference and background spectra collected more easily	• In situ measurement, non-invasive sampling • Detection of very polar, reactive, and labile compounds • Continuous monitoring • Line-integrated (averaged) concentrations: area sources of pollution can be examined • Measurement at dangerous sites
Disadvantages	• Time delay between sampling and measurement • Very polar, reactive compounds difficult to sample (may lead to errors in the analysis) • Memory effects, losses on the cell wall	• Path-length limited by the size of plume • Field adjustment needed at every site • Weather dependence • No temperature and pressure control • Decreasing S/N for longer paths • Calibration spectra over the same path as the sample cannot be collected • Experiments are hardly repeatable

Source: Meyers, R.A. (Ed.), *Encyclopedia of Analytical Chemistry*, Wiley, Chichester, UK, 2000.

the freezing point and above it. In addition, this instrument has negligible drift and can generally operate over a wide humidity range. Optical hygrometers can be used for the measurement of high humidity, as well as for very low concentrations in the range of ppmv. Finally, the method in no way alters the sample concentration by either adding or subtracting water or changing the state of any part of the sample as occurs in the psychrometric or dew point methods.

We need to note that all mentioned above devices have advantages and also shortcomings. Optical techniques often require a high effort in equipment and analysis.

These are not diminutive. The size of measurement cell with required optical path length is a limiting factor in the development of IR hygrometers. Attempts to reduce the size of IR hygrometers are accompanied by a restriction of the measurement range and the sensitivity decrease. Miniaturization of UV hygrometers is impossible due to the size of existing UV light sources and the detectors of radiation in this area. Besides, optical hygrometers are expensive in comparison with other devices. They may also suffer from contamination. In addition, regular calibration is required for many traditional instruments, which is both inconvenient and expensive. Usually for this purposes they are used other hygrometers such as dew or frost point hygrometers. Therefore, according to WMO reports (1992, 2011), optical hygrometers are more suitable for measuring changes in the vapor concentration rather than absolute levels.

Who has the desire to learn more about optical sensors and optical spectroscopy can refer to reviews and books published in this field (Wolfbeis 1991, 1992, 2006, 2008; Ligler and Rowe Taitt 2002; Narayanaswamy and Wolfbeis 2004; Orellana and Moreno-Bondi 2005; Orellana 2006; Tkachenko 2006; McDonagh et al. 2008; Korotcenkov et al. 2011, and many other).

REFERENCES

Allen M.G. (1998) Diode laser absorption sensors for gas–dynamic and combustion flows. *Meas. Sci. Technol.* 9, 545–562.

Amerov A., Maskas M., Meyer W., Fiore R., Tran K. (2007) New process gas analyzer for the measurement of water vapor concentration. In: *Proceedings of ISA 52nd Symposium of Instrumentation, Systems, and Automation Society*, Houston, TX, pp. 63–74, April 15–19, 2007.

Amerov A., Fiore R., Langridge S. (2009) Process gas analyzer for the measurement of water and carbon dioxide concentrations. In: *Proceedings of the Annual ISA Analysis Division Symposium*, Vol. 478, pp. 184–197.

Bange J., Espositi M., Lenschow D.H., Brown P.R.A., Dreiling V., Giez A. et al. (2013) Measurement of aircraft state and thermodynamic and dynamic variables. In: Wendisch M., Brenguier J.-L. (Eds.), *Airborne Measurements for Environmental Research.* Methods and Instruments, Wiley-VCH, Weinheim, Germany, Chapter 2.

Bass M., van Stryland E.M. (Eds.) (2002) *Fiber Optics Handbook*, McGraw-Hill, New York.

Bauer D., Heeger M., Gebhard M., Benecke W. (1996) Design and fabrication of a thermal infrared emitter. *Sens. Actuators A* 55 (1), 57–63.

Beaton S.P., Spowart M. (2012) UV absorption hygrometer for fast–response airborne water vapor measurements. *J. Atmos. Oceanic. Technol.* 29, 1295–1303.

Bertaux J.-L., Delannoy A. (1978) Vertical distribution of H_2O in the stratosphere as determined by UV fluorescence in-situ measurements. *Geophys. Res. Lett.* 5, 1017–1020.

Buchholz B., Ebert V. (2013) Field tests of a new, extractive, airborne 1.4 µm–TDLAS hygrometer (SEALDH–I) on a Learjet 35A. *Geophys. Res.* Abstracts 15, EGU2013–4569–1.

Buchholz B., Kühnreich B., Smit H.G.J., Ebert V. (2013) Validation of an extractive, airborne, compact TDL spectrometer for atmospheric humidity sensing by blind intercomparison. *Appl. Phys. B* 110, 249–262.

Buck A.L. (1973) Development of an improved Lyman–alpha hygrometer. *Atmos. Technol.* 2, 213–240.

Buck A.L. (1976a) The variable–path Lyman–alpha hygrometer and its operating characteristics. *Bull. Amer. Meteor. Soc.* 57, 1113–1118.

Buck A.L. (1976b) Lyman-α radiation source with high spectral purity. *Appl. Opt.* 16, 2634–2638.

Buck A.L. (1985) The Lyman-alpha absorption hygrometer. In: *Proceedings, Moisture and Humidity Symposium*, Washington, D.C., Instrument Society of America, Research Triangle Park, NC, p. 411.

Budovich V.L., Polotnyuk E.B., Gerasimov N., Krylov B.E. (2013) Characteristics of a glow–discharge krypton lamp in the vacuum ultraviolet. *J. Opt. Technol.* 80 (11), 691–694.

Campbell G.S., Tanner B.D. (1985) A krypton hygrometer for measurement of atmospheric water vapor concentration. In: *Moisture and Humidity: Measurement and Control in Science and Industry.* Instrument Society of America, Research Triangle Park, Proceedings of the 1985 International Symposium on Moisture and Humidity, Instrument Society of America, pp. 609–614.

Carver J.H., Mitchell P. (1964) Ionization chambers for the vacuum ultra–violet. *J. Sci. Instrum.* 41, 555–557.

Cerni T.A. (1994) An infrared hygrometer for atmospheric research and routine monitoring. *J. Atmos. Oceanic Technol.* 11, 445–462.

Chalmers J.M. (2002) Mid-infrared spectroscopy: anomalies, artifacts and common errors. In: Chalmers J.M., Griffiths P.R. (Eds.), *Handbook of Vibrational Spectroscopy*, John Wiley & Sons, Chichester, UK, pp. 2327–2347.

Chen H.-S., Mitsuta Y. (1967) An infrared absorption hyhrometer and its application to the study of the water vapor flux near the ground. *Special, Contribution. Geophysical Institute, Kyoto University*, No. 7, pp. 83–94.

Cho J., Newell R.E., Sachse G.W. (2000) Anomolous scaling of mesoscale tropospheric humidity fluctuations. *Geophys. Res. Lett.* 27, 377–380.

Comini E., Faglia G., Sberveglieri G. (Eds.) (2009) *Solid State Gas Sensor.* Springer, New York.

Cozzani E., Summonte C., Belsito L., Cardinali G.C., Roncaglia A. (2007) Design study of micromachined thermal emitters for NDIR gas sensing in the 9–12 μm wavelength range. *IEEE Sens.* 1–3, 181–184.

Crosley D.R., Lengel R.K. (1975) Relative transition probabilities and the electronic transition moment in the A-X system of OH. *J. Quant. Spectrosc. Radiat. Tran.* 15, 579–591.

Dieke G.H., Cunningham S.P. (1952) A new type of hydrogen discharge tube. *J. Opt. Soc. Amer.* 42, 187–189.

Durry G., Amarouche N., Joly L., Liu X., Parvitte B., Zeninari, V. (2008) Laser diode spectroscopy of H_2O at 2.63 μm for atmospheric applications. *Appl. Phys. B* 90, 573–580.

Ebert V., Buchholz B. (2016) Absolute high–speed laser hygrometry: From aircraft applications to primary validation. In: *Proceedings of the Conference on Lasers and Electro–Optics (CLEO:2016): Science and Innovations*, OSA, San Jose, CA, STh4J.1, June 5–10, 2016.

Edwards C.S., Barwood G.P., Gill P., Stevens M., Schirmer B., Benyon R., Mackrod P. (2000) An IR tunable diode laser absorption spectrometer for trace humidity measurements at atmospheric pressure. In: *Proceedings of 2000 IEEE Sensor Conference*, Aventura, FL, pp. II3–II4, July 24–28, 2000.

Faist J., Capasso F., Sivco D.L., Sirtori C., Hutchinson A.L., Cho A.Y. (1994) Quantum cascade laser. *Science* 264, 553–555.

Foken T. (2008) *Micro–Meteorology.* Springer, Berlin, Germany.

Foken T., Falke H. (2012) Technical Note: Calibration instrument for the krypton hygrometer KH20. *Atmos. Meas. Tech. Discuss.* 5, 1695–1715.

Foken T., Dlugi R., Kramm G. (1995) On the determination of dry deposition and emission of gaseous compounds at the biosphere–atmosphere interface. *Meteorol. Z.* 4, 91–118.

Fonollosa J., Carmona M., Santander J., Fonseca L., Moreno M., Marco S. (2009) Limits to the integration of filters and lenses on thermoelectric IR detectors by flip–chip techniques. *Sens. Actuators A* 149, 65–73.

Foskett L.W., Foster N.B., Thickstun W.R., Wood R.C. (1953) Infrared absorption hygrometer. *Monthly Weather Rev.* 81 (9), 267–277.

Fowle E.E. (1912) The spectroscopic determination of aqueous vapor. *Astrophys. J.* 35, 149–162.

Gat J.R. (1996) Oxygen and hydrogen isotopes in the hydrological cycle. *Ann. Rev. Earth Planet. Sci.* 24, 225–262.

Gersh M.E. Matthew M.W. (1988) VaporSense contamination-resistant high temperature UV hygrometer. In: *SPIE* Vol. 961 Industrial Optical Sensing, pp. 52–67.

Ghysels M., Riviere E.D., Khaykin S., Stoeffler C., Amarouche N., Pommereau J.-P., Held G., Durry G. (2016) Intercomparison of in situ water vapor balloon-borne measurements from Pico-SDLA H_2O and FLASH-B in the tropical UTLS. *Atmos. Meas. Tech.* 9, 1207–1219.

Goutail F., Pommereau J.-P. (1987) Stratospheric water vapor in situ measurements from infra-red montgolfier. *Adv. Space Res.* 7 (7), 111–114.

Graf A., Arndt M., Sauer M., Gerlach G. (2007) Review of micromachined thermopiles for infrared detection. *Meas. Sci. Technol.* 18, R59–R75.

Green A.E., Kohsiek W. (1995) A fast response, open path, infrared hygrometer, using semiconductor source. *Bound. Layer Meteor.* 74, 353–370.

Grouiez B., Parvitte B., Joly I., Courtois D., Zeninari V. (2008) Comparison of a quantum cascade laser used in both CW and pulsed modes: Application to the study of SO_2 lines around 9 μm. *Appl. Phys. B* 90, 177–186.

Grund C.J., Hardesty R.M., Rye B.J. (1995) Feasibility of tropospheric water vapor profiling using infrared heterodyne differential absorption Lidar. In: *Proceedings of 5th Atmospheric Radiation Measurement (ARM) Science Team Meeting*, San Diego, CA, March 19–23. http://www.arm.gov/publications/proceedings/conf05/

Heering W. (2004) UV sources: Basics, properties and applications. *IUVA News* 6 (4), 7–13.

Herriot D., Kogelnik H., Kompfner R. (1964) Off-axis paths in spherical mirror interferometers. *Appl. Opt.* 3, 523–526.

Herriot D.R., Schulte H.J. (1965) Folded optical delay lines. *Appl. Opt.* 4, 883–889.

Hintsa E.J., Weinstock E.M., Anderson J.G., May R.D., Hurst D. (1999) On the accuracy of *in situ* water vapor measurements in the troposphere and lower stratosphere with the Harvard Lyman-α hygrometer. *J. Geophys. Res.* 104, 8183–8189.

Hvozdara L., Pennington N., Kraft M., Karlowatz M., and Miziakoff B. (2002) Quantum cascade lasers for midinfrared spectroscopy. *Vib. Spectrosc.* 30, 53–58.

Hutcheson E. (1972) Monitoring the thickness of thin MgF and LiF films on alluminium by reflectant measurements using 1216 Å line of hydrogen. *Appl. Opt.* 11 (7), 1590–1595.

Kebabian P., Kolb C.E., Freedman A. (2002) Spectroscopic water vapor sensor for rapid response measurements of humidity in the troposphere. *J. Geophys. Res. Atmosph.* 107 (23), 4670.

Kelly K.K., Tuck A.F., Murphy D.M. et al. (1989) Dehydration in the lower Antarctic stratosphere during late winter and early spring 1987. *J. Geophys. Res* 94 (D9), 11317–11357.

Keramitsoglou I., Harries J.E., Colling D.J., Barker R.A., Foot J.S. (2002) A study of the theory and operation of a resonance fluorescence water vapour sensor for upper tropospheric humidity measurements. *Meteorol. Appl.* 9, 443–453.

Khaplanov M., Gumbel J., Wilhelm N., Witt G. (1996) Hygrosonde-A direct measurement of water vapor in the stratosphere and mesosphere. *Geophys. Res. Lett.* 23, 1645–1648.

Kley D., Stone E. (1978) Measurements of water vapor in the stratosphere by photo dissociation with Ly-α (1216 Å) light. *Rev. Sci. Instrum.* 49 (6), 691–697.

Kley D., Stone E.J., Henderson W.R., Drummond J.W., Harrop W.J., Schmeltekopf A.L., Thompson T.L., Winkler R.H. (1979) *In situ* measurements of the mixing ratio of water vapor in the stratosphere. *J. Atmos. Sci.* 36, 2513–2524.

Kley D. (1984) Ly(α) absorption cross–sections of H_2O and O_2. *J. Atmos. Chem.* 2 (2), 203–210.

Kley D., Russell III J.M., Phillips C. (Eds.) (2000) SPARC assessment of upper tropospheric and stratospheric water vapour. World Climate Research Programme, SPARC Report No.2.

Korotcenkov G., Cho B.K., Sevilla III F., Narayanaswamy, R. (2011) Optical and fiber optic chemical sensors. In: *Chemical Sensors: Comprehensive Sensor Technologies.* Vol. 5: *Electrochemical and Optical Sensors*, Korotcenkov G. (Ed.), Momentum Press, New York, pp. 311–476.

Kosterev A.A., Tittel F.K. (2002) Chemical sensors based on quantum cascade lasers. *IEEE J. Quantum Electron.* 38, 582–590.

Kretova M., Khaplanov M., Yushkov V. (1991) Resonance-fluorescence hygrometer for humidity measurement in the stratosphere. *Prib. Tekh. Eksp.* 3, 195–198 (in Russian).

Kretschmer S.I., Karpovitsch J.V. (1973) Sensitive ultraviolet hygrometer. *Izv. AN SSSR, Fiz. Atm. i Okeana* 9, 642–645 (in Russian).

Ligler F.S., Rowe-Taitt C.A. (2002) *Optical Biosensors—Present and Future.* Elsevier, New York.

Linnerud I., Kaspersen P., Jaeger T. (1998) Gas monitoring in the process industry using diode laser spectrosocopy. *Appl. Phys. B* 67, 297–305.

Liu F. (2012) A new method for measuring trace moisture in natural gas. In: *Proceedings of ISA 57th Analysis Division Symposium*, Anaheim, CA. Sessions 7–2.

Lübken F.–J., Dingler F., von Lucke H., Anders J., Riedel W.J., Wolf H. (1999) MASERATI: A rocketborne tunable diode laser absorption spectrometer. *J. Anders, Appl. Opt.* 38, 5338–5349.

Lykov A., Yushkov V., Khaykin S., Astakhov V., Budovich V. (2011) New version of balloon hygrometer for in situ water vapor measurements in the upper troposphere and lower stratosphere (flash-BM). In: *Proceedings of 20th ESA Symposium on European Rocket and Balloon Programmes and Related Research*, Hyère, France, ESA SP-700, May 22–26, 2011.

Lykov A.D., Astakhov V.I., Korshunov L.I., Yushkov V.A., Budovich V.L., Budovich D.V., Dubakin A.D. (2012) Using vacuum ultraviolet sources in the design of a fluorescence hygrometer. *J. Opt. Technol.* 79 (8), 515–520.

Martini L., Stark B., Hunsalz G. (1973) Elektronisches Lyman–Alpha–feuchtigkeitsmessgerat. *Meteorol. Z.* 23, 313–322.

Matthey R., Schilt S., Werner D., Affolderbach C., Thevenaz L., Mileti G. (2006) Diode laser frequency stabilisation for water-vapour differential absorption sensing. *Appl. Phys. B* 85 (2), 477–485.

May R.D. (1998) Open–path, near–infrared tunable diode laser spectrometer for atmospheric measurements of H_2O. *J. Geophys. Res.* 103 (D15), 19161–19172.

McDonagh C., Burke C.S., MacCraith B.D. (2008) Optical chemical sensors. *Chem. Rev.* 108, 400–422.

McManus J., Kebabian P., Zahniser M. (1995) Astigmatic mirror multiple pass absorption cells for long pathlength spectroscopy. *Appl. Opt.* 34, 3336–3348.

Meyers R.A. (Ed.) (2000) *Encyclopedia of Analytical Chemistry.* Wiley, Chichester, UK.

Meyer J., Rolf C., Schiller C. et al. (2015) Two decades of water vapor measurements with the FISH fluorescence hygrometer: A review. *Atmos. Chem. Phys.* 15, 8521–8538.

Mestayer P., Rebattet C., Goutail F. (1986) Improved Lyman-alpha hygrometer for small-scale atmospheric turbulence measurements. Part I: Miniaturizing the sampling volume. *Rev. Sci. Instrum.* 57, 20–25.

Mestayer P., Rebattet C., Goutail F. (1987) Improved Lyman-alpha hygrometer for small-scale atmospheric turbulence measurements. Part II: The O_2 filtering technique. *Rev. Sci. Instrum.* 58, 2165.

Mirumyants S.O., Maksimyuk V.S. (2002) A portable optical gas analyzer for remote measurement of the methane concentration in closed spaces and rooms. *J. Opt. Technol.* 69 (12), 900–903.

Narayanaswamy R. and Wolfbeis O.S. (Eds.) (2004) *Optical Sensors—Industrial, Environmental and Diagnostic Applications*, Springer Series on Chemical Sensors and Biosensors, Vol. 1. Springer-Verlag, Berlin, Germany.

Nwaboh J.A., Pratzler S., Werhahn O., Ebert V. (2017) Tunable diode laser absorption spectroscopy sensor for calibration free humidity measurements in pure methane and low CO_2 natural gas. *Appl. Spectrosc.* 71 (5), 888–900.

Orellana G. (2006) Fluorescence-based sensors. In: Baldini F., Chester A.N., Homola J., Martellucci S. (Eds.), *Optical Chemical Sensors.* Springer-Verlag, Dordrecht, the Netherlands, pp. 99–116.

Orellana G., Moreno-Bondi M.C. (Eds.) (2005) *Frontiers in Chemical Sensors: Novel Principles and Techniques*, Springer Series on Chemical Sensors and Biosensors, Vol. 3. Springer-Verlag, Berlin, Germany.

Parkinson W.H., Yoshino K. (2003) Absorption cross–section measurements of water vapor in the wavelength region 181–199 nm. *Chem. Phys.* 294, 31–35.

Rancourt R.D. (1996) *Optical Thin Films: User Handbook*, SPIE Press, Bellingham, Washington, DC.

Razeghi M., Rogalski A. (1996) Semiconductor ultraviolet detectors. *J. Appl. Phys.* 79 (10), 7433–7473.

Rothman L.S., Gordon I.E., Babikov Y. et al. (2013) The HITRAN 2012 molecular spectroscopic database. *J. Quant. Spectrosc. Rad. Trans.* 130, 4–50.

Samson J.A.R. (1967) *Techniques of Vacuum Ultraviolet Spectroscopy.* John Wiley & Sons, New York.

Santander J., Sabate N., Rubio R., Calaza C., Fonseca L., Gracia I., Cane C., Moreno M., Marco S. (2005) Mirror electrostatic actuation of a medium infrared tuneable Fabry–Perot interferometer based on a surface micromachining process. *Sens. Actuators A* 123–124, 584–589.

Schanot A.J. (1987) An evaluation of the uses and limitations of Lyman-alpha hygrometer as an operational airborne humidity sensor. In: *Proceedings of 6th Symposium Meteorological Observations and Instrumentation*, American Meteorological Society, New Orleans, LA, pp. 257–260.

Schilz J. (2000) Thermophysica minima: Applications of Thermoelectric Infrared Sensors (Thermopiles): Gas Detection by Infrared absorption; NDIR, PerkinElmer Optoelectronics.

Seib D.H., Aukerman L.W. (1973) Photodetectors for the 0.1 to 1.0 μm spectral region. In: Morton L. (Ed.) *Advances in Electronics and Electron Physics*, Vol. 34, Academic Press, New York, pp. 95–221.

Seidel A., Wagner S., Ebert V. (2012) TDLAS–based open–path laser hygrometer using simple reflective foils as scattering targets. *Appl. Phys. B* 109, 497–504.

Sigrist M.W. (Ed.) (1994) *Air Monitoring by Spectrometric Techniques*. John Wiley & Sons, New York.

Silver J.A., Stanton A.C. (1987) Airborne measurements of humidity using a single–mode Pb–salt diode laser. *Appl. Opt.* 26, 2558–2566.

Silver J.A., Hovde D.C. (1994) Near–infrared diode laser airborne hygrometer. *Rev. Sci. Instrum.* 65 (6), 1691–1694.

Smith F.G. (Ed.) (1993) *Atmospheric Propagation of Radiation*, volume 2 of The Infrared & Electro–Optical Systems Handbook (J.S. Accetta, D.L. Shumaker Eds.), Infrared Information Analysis Center, Ann Arbor, MI & SPIE Optical Engineering Press, Bellingham, Washington, DC.

Smith W.J. (2007) *Modern Optical Engineering*, 4th ed., McGraw-Hill, New York.

Soleyn K. (2009) Development of a tunable diode laser absorption spectroscopy moisture analyzer for natural gas. In: *Proceedings of the 5th International Gas Analysis Symposium, GAS 2009*, Rotterdam, the Netherlands, February 11–13.

Spannhake J., Schulz O., Doll T. (2005) Design, development and operational concept of an advanced MEMS IR source for miniaturized gas sensor systems. In: *Proceedings of 2005 IEEE Sensor Conference*, Irvine, CA, pp. 762–765, October 30–November 3, 2005.

Tillman J.E. (1965) Water vapor density measurements utilizing the absorption of vacuum ultraviolet and infrared radiation. In: Ruskin R.E. (Ed.) *Humidity and Moisture*, Vol. 1, Principles and Methods of Measuring Humidity in Gases, Reinhold, New York, pp. 428– 443.

Tkachenko N.V. (2006) *Optical Spectroscopy. Methods and Instrumentations*. Elsevier, Amsterdam, the Netherlands.

Varier G. (1967) Experiment on the interaction of Ly-α radiation and atomic hydrogen. *Appl. Opt.* 6 (1), 167–171.

Vay S., Anderson B.E., Sachse G.W. et al. (1998) DC-8-based observations of aircraft CO, CH_4, N_2O, and $H_2O(g)$ emission indices during SUCCESS, *Geophys. Res. Lett.* 25, 1717–1720.

Vay S.A., Anderson B.E., Jensen E.J. et al. (2000) Tropospheric water vapor measurements over North Atlantic during subsonic assessment ozone and nitrogen oxide experiment (SONEX). *J. Geophys. Res.* 105, 3745–3756.

Wang W.E., Michel A.P.M., Wang L., Tsai T., Baeck M.L., S mith J.A., Wysocki G. (2014) A quantum cascade laser-based water vapor isotope analyzer for environmental monitoring. *Rev. Sci. Instrum.* 85, 093103.

Webster C.R., Menzies R.T., Hinkley E.D. (1988) Infrared laser absorption: Theory and applications. In: Measures R.M. (Ed.), *Laser Remote Chemical Analysis*, John Wiley & Sons, New York, Chapter 3.

Weinheimer A.J., Schwiesow R.L. (1992) A two–path, two–wavelength ultraviolet hygrometer. *J. Atmos. Oceanic Technol.* 9, 407–419.

Weinstock E., Schwab J., Nee J.B., Schwab M., Anderdson J. (1990) A cryogenically cooled photofragment fluorescence instrument for measuring stratospheric water vapor. *Rev. Sci. Instrum.* 61 (5), 1413–1432.

Weinstock E.M., Hintsa E.J., Dessler A.E., Oliver J.F., Hazen N.L. (1994) New fast response photofragment fluorescence hygrometer for use on the NASA ER-2 and the Perseus remotely piloted aircraft. *Rev. Sci. Instrum.* 65, 3544–3554.

Wen X.-F., Sun X.-M., Zhang S.-C., Yu G.-R., Sargent S.D., Lee X. (2008) Continuous measurement of water vapor D/H and 18O/16O isotope ratios in the atmosphere. *J. Hydrol.* 349, 489–500.

Wen X.-F., Lee X., Sun X.-M., Wang J.-L., Tang Y.-K., Li S.-G., Yu G.-R. (2012) Intercomparison of four commercial analyzers for water vapor isotope measurement. *J. Atmos. Oceanic Technol.* 29, 235–247.

Werle P. (1996) Spectroscopic trace gas analysis using semiconductor diode lasers. *Spectrochim. Acta A* 52 (8), 805–822.

Werle P. (1998) A review of recent advances in semiconductor laser based gas monitors. *Spectrochim. Acta A* 54 (2), 197–236.

Werle P., Slemr F., Maurer K., Kormann R., Mucke R., and Janker B. (2002) Near- and mid-infrared laser-optical sensors for gas analysis. *Opt. Lasers Eng.* 37, 101–114.

Werle P.W. (2004) Diode–laser sensors for in-situ gas analysis. In: Hering P., Lay J.P., Stry S. (Eds.) *Laser in Environmental and Life Sciences*, Springer-Verlag, Berlin, Germany, pp. 223–243.

Wernecke R., Wernecke J. (2014) *Industrial Moisture and Humidity measurement: A Practical Guide*. Wiley-VCH, Weinheim, Germany.

White J.U. (1942) Long optical paths of large aperture. *J. Opt. Soc. Am.* 32, 285–288.

White S.E., Cataluna M.A. (2015) Unlocking Spectral Versatility from Broadly–Tunable Quantum–Dot Lasers. *Photonics* 2, 719–744.

Wiederhold P.R. (1997) *Water Vapor Measurement, Methods and Instrumentation*. Marcel Dekker, New York.

Wilson A.C., Barnes T.H., Seakins P.J., Rolfe T.G., Meyer E.J. (1995) A low–cost, high–speed, near–infrared hygrometer. *Rev. Sci. Instrum.* 66 (12), 5618–5624.

WMO (1992) World Meteorological Organization: Measurement of Temperature and Humidity (Wylie R.G. and Lalas T. Eds.). (Technical Note No. 194), (WMO–No. 75), Geneva, Switzerland.

WMO (2011) World Meteorological Organization: Technical Regulations. Volume I – General Meteorological Standards and recommended Practices, (WMO–No. 49), Geneva, Switzerland.

Wolfbeis O.S. (1991) *Fiber Optic Chemical Sensors and Biosensors,* Vol. 1. CRC Press, Boca Raton, FL.

Wolfbeis O.S. (1992) *Fiber Optic Chemical Sensors and Biosensors,* Vol. 2. CRC Press, Boca Raton, FL.

Wolfbeis O.S. (2006) Fiber-optic chemical sensors and biosensors. *Anal. Chem.* 78, 3859–3874.

Wolfbeis O.S. (2008) Fiber-optic chemical sensors and biosensors. *Anal. Chem.* 80, 4269–4283.

Wulfmeyer V., Walther C. (2001) Future performance of ground-based and airborne water-vapor differential absorption lidar. I. Overview and theory. *Appl. Opt.* 40, 5304–5320.

Wunderle K., Rascher U., Pieruschka R., Schurr U., Ebert V. (2015) A new spatially scanning 2.7 μm laser hygrometer and new small–scale wind tunnel for direct analysis of the H_2O boundary layer structure at single plant leaves. *Appl. Phys. B* 118, 11–21.

Yakir D., Sternberg L.D.S.L. (2000) The use of stable isotopes to study ecosystem gas exchange. *Oecologia* 123, 297–311.

Yoshino K., Esmond J.R., Parkinson W.H., Ito K., Matsui T. (1996) Absorption cross section measurements of water vapor in the wavelength region 120 to 188 nm. *Chem. Phys.* 211, 387–391.

Yushkov V., Lukjanov A., Merkulov S., Khaplanov M., Shyshatzkaya L., Gumbel J. (1995) Optical fluorescent hygrometer for water vapor low concentration measurements. *SPIE Proc.* 2506, 783–794.

Zoger M., Afchine A., Eicke N. et al. (1999) Fast in situ stratospheric hygrometers: A new family of balloon-borne and airborne Lyman photofragment fluorescence hygrometers. *J. Geophys. Res.* 104 (D1), 1807–1816.

Zuber A., Witt G. (1987) Optical hygrometer using differential absorption of hydrogen Lyman–alpha radiation. *Appl. Opt.* 26 (15) 3083–3089.

4 Atmosphere Monitoring Using Methods of Absorption of Electromagnetic Radiation—Fourier Transform Infrared Spectroscopy

4.1 INTRODUCTION

As mentioned in the introduction, our life is in the *ocean of air*, and the state of the *Ocean* substantially affects many aspects of our lives and in some cases the existence of life itself. Well-known examples of these considerations are catastrophic weather events (typhoons, tornadoes, floods, etc.), leading to a huge number of human victims and causing a significant economic damage. In recent years, the problem of climate change and the destruction of the Earth's ozone layer of our planet become very urgent (Solomon et al. 2007; USGCRP 2008). The consequences of these processes are still difficult to predict with sufficient reliability, but they may lead to undesirable environmental, economic, and social consequences, such as changes in the temperature and rainfall, raising the level of the oceans, the increase in the ultraviolet (UV) light exposure of the Earth's surface, and so on. It is important to note that the state of the atmosphere and the lack of catastrophic weather events are mainly determined by the efficiency of the various production sectors such as transport, agriculture, forestry, water industry, and so on. It is important to note that water plays a major role in the processes taking place in the atmosphere (Spencer and Braswell 1997). Water vapor is the main responsible atmospheric gas that regulates the weather and climate and contributes with about 90% of the Earth's natural greenhouse effect. The continuous cycle of evaporation, vapor transport, cloud formation, and precipitation distributes water and energy around the globe. In this connection, it is a natural desire to know the laws of the phenomena, occurring in the atmosphere and determining the conditions of our lives. Unfortunately, current knowledges about the atmosphere are not sufficient. The object of the research is too spreading.

Features of measurements in the atmosphere are conditioned by the complexity of the studied system and by the specificity of the tasks. This is, above all, the general circulation of the atmosphere, the weather forecasts for various periods, the theory of climate and ozonosphere and so on. When studying the atmosphere and solving mentioned tasks we have to deal with an extremely extensive list of physical phenomena and processes, and their complex interaction.

In addition, when studying the system *atmosphere-surface*, we have to deal with a huge range of spatial-scaled processes, from the atomic level to the size of the globe. It is also important to emphasize that currently under the atmosphere should be understood not only the troposphere, but also higher layers such as stratosphere, mesosphere, thermosphere, exosphere and ionosphere. Weather at the surface of the Earth depends on the state of the atmosphere at all levels. This is because the state of the atmosphere substantially depends on the environmental factors such as a solar radiation, cosmic rays, and so on. Thus, we are interested in the atmosphere at heights from the surface of the Earth up to 100 km or more. This is because the state of the atmosphere substantially depends on the environmental factors such as a solar radiation, cosmic rays, and so on.

Taking into account that this system is open, spatial dimensions in a number of tasks have to be widening significantly. Another feature of the monitoring of the atmosphere, including a control of water vapor concentration, is also a very wide range of timing parameters studied, since they influence the processes in the atmosphere. It also determines the specificity of atmospheric parameters measurements. The quick changes in the atmospheric water vapor concentrations with time, their large horizontal gradients, and their decrease of several orders of magnitude with height make their accurate detection a challenging task for any measurement technique. This means that a permanent long-term monitoring of the condition of the surface-atmosphere system is required.

In connection with these factors, one of the important directions in the study of the atmosphere is the formation of the global observing system for the state of the land, oceans, and atmosphere. At the same time, there is no single system or program that could meet all the requirements for the observations needed to understand the Earth's system, and to control and forecast its condition.

To fulfill all the requirements, a contribution of many different observing systems is necessary. Optimal integration of these systems requires careful planning at the international and national levels in order to maximize the impact of existing and planned observing systems (IGOS 1998). The World Meteorological Organization (WMO), which includes 189 countries and territories, as well as other international and national agencies, is engaged in this planning. For instance, for these purposes the Global Atmosphere Watch (GAW) program has been established by the WMO.

Due to the diversity of the processes, taking place in the atmosphere, a set of requirements for measuring equipment used in the atmospheric monitoring, greatly expands. For example, it is believed that the air monitoring techniques need to be simple, unobtrusive, acquire real-time data, have increased sensitivity, and the ability to analyze many compounds simultaneously.

As we showed earlier, there are a large number of optical methods for in situ measurements of atmospheric parameters, including humidity. With the help of these methods, a large number of measurements of different parameters of the atmosphere and the surface is carried out (e.g., the measurement of temperature and humidity using regularly launched radiosondes). Contemporary radiosonde instruments measure a temperature and relative humidity with accuracies of ~0.2°C and ~3.5%, respectively, with diminishing performance in cold, dry regions. Although the advantages of in situ measurements that provide good vertical resolution are clear, radiosonde measurements have some serious disadvantages. Radiosondes are expendable, and the cost of these devices restricts the number of launches to twice daily at a limited number of stations. On account of these restrictions, radiosonde measurements inadequately resolve the temporal and spatial variability of water vapor, which occurs at scales much finer than the spatial and temporal variability of temperature or wind. In fact, limitations in the analysis of water vapor are the major source of error in short-term (0–24 h) forecasts of precipitation. This also means that to obtain detailed information about the state of the Earth's atmosphere on a regional and especially on a global scale is difficult and very expensive by the methods mentioned above, but in reality it is impossible.

Therefore, new techniques have been developed in recent decades, connected with atmosphere monitoring such as Fourier transform infrared (FTIR) spectroscopy, Microwave radiometry, Terahertz time-domain spectroscopy, and Lidar (Kampfer 2013). All these techniques are based on the methods of absorption of electromagnetic radiation. The main difference between these methods is that they allow realizing a remote control that is very important to improve atmospheric processes estimations (O'Brien and Rayner 2002).

4.2 FTIR SPECTROSCOPY

No doubts that high-precision FTIR spectrometer is one of the most suitable instruments to measure the concentration of greenhouse gases (GHG), including water vapors, from remote-sensing techniques. As a result, a number of infrared (IR) spectrometers from different laboratories have been used for atmosphere monitoring. In fact, it is the technique used by the ground-based the Total Carbon Column Observing Network (TCCON) and the network for the Detection of Atmospheric Composition Change (NDACC) networks as well as by several space-based remote sensors such as IASI on the board of the polar meteorological satellites (METOP) and TANSO on the board of the Greenhouse Gases Observing Satellite (GOSAT). IASI is the Infrared Atmospheric Sounding Interferometer and TANSO is the Thermal And Near-infrared Sensor for carbon Observation.

4.2.1 Principles of FTIR Spectroscopy

FTIR spectrometers are often simply called FTIRs. But for the purists, an FTIR is a method of obtaining IR spectra by first collecting an interferogram of a sample signal using an interferometer, and then performing a Fourier transform (FT) on the interferogram to obtain the spectrum. An FTIR spectrometer collects and digitizes the interferogram, performs the FT function, and displays the spectrum.

An FTIR spectrometer is an instrument which acquires broadband NIR (near infrared) to FTIR spectra. Unlike a dispersive instrument, that is, grating monochromator or spectrograph, FTIR spectrometer collects all wavelengths simultaneously. Thus, an interferometer provides an alternative approach for wavelength selection. Instead of filtering or dispersing the electromagnetic radiation, an interferometer allows source radiation of all wavelengths to reach the detector simultaneously. Among different types of interferometers, the double-beam interferometer designed by Michelson in 1891 is the most widely used in commercial FTIR spectrometers. An example of the Michelson Interferometer Experimental Setup is shown in Figure 4.1.

The interferometer consists of a beam splitter, a fixed mirror, and a mirror that translates back and forth, very

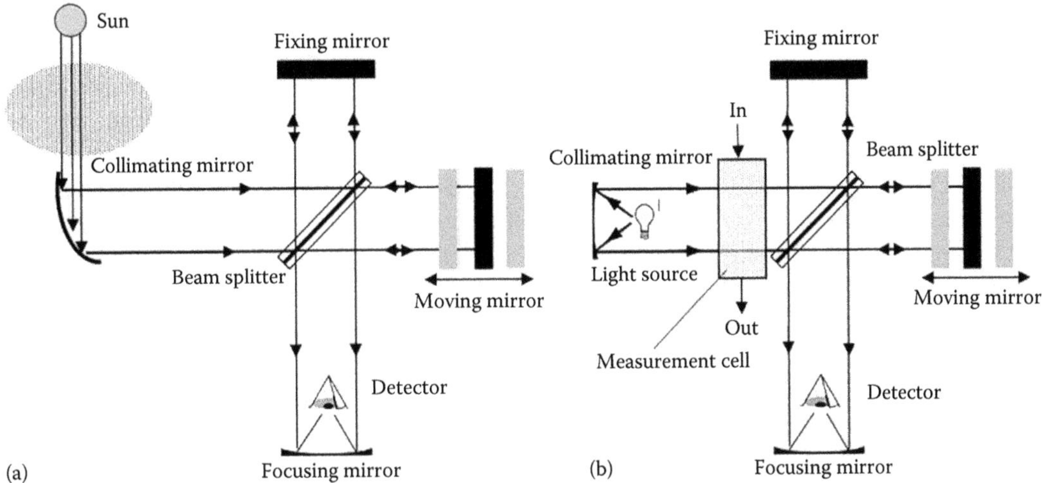

FIGURE 4.1 A schematic of a generic Michelson interferometers (a,b) used for atmosphere monitoring. The position of the sample (a) between a natural light source (sun) and a collimating mirror, or (b) between a light source and a beam splitter depends on the field of application.

precisely. The beam splitter is made of a special material that transmits half of the radiation striking it and reflects the other half. Radiation from the source strikes the beam splitter and separates into two beams. One beam is transmitted through the beam splitter to the fixed mirror and the second is reflected off the beam splitter to the moving mirror. The fixed and moving mirrors reflect the radiation back to the beam splitter. Again, half of this reflected radiation is transmitted and half is reflected at the beam splitter, resulting in one beam passing to the detector and the second back to the source. The radiation recombines at the beam splitter, where constructive and destructive interference determines, for each wavelength, the intensity of light reaching the detector. As the moving mirror changes position, the wavelengths of light experiencing maximum constructive interference and maximum destructive interference also changes. The path difference, also known as optical path difference (OPD), is measured with a monochromatic laser (normally a helium neon laser). The signal at the detector shows intensity as a function of the moving mirror's position, expressed in units of distance or time. The result is called an interferogram, or a time domain spectrum (Figure 4.2). After the recombined beam has passed through the sample, the detector will record the FT of the IR spectrum of the sample. A very detailed description of FT spectrometry can be found in the textbook of Davis et al. (2001).

On account of the qualitative and quantitative analysis, we need a frequency spectrum (a plot of the intensity at each individual frequency, that is, standard an IR spectrum, showing intensity as a function of the radiation's energy), the interferogram signal measured by FTIR spectrometer cannot be interpreted directly. A means of *decoding* the individual frequencies is required. This can be accomplished via a well-known mathematical technique called the FT (Smith 2011; Blum and John 2012). Of course, this transformation is made by a computer using special programs. This means that a computer is mandatory component of FTIR spectrometers.

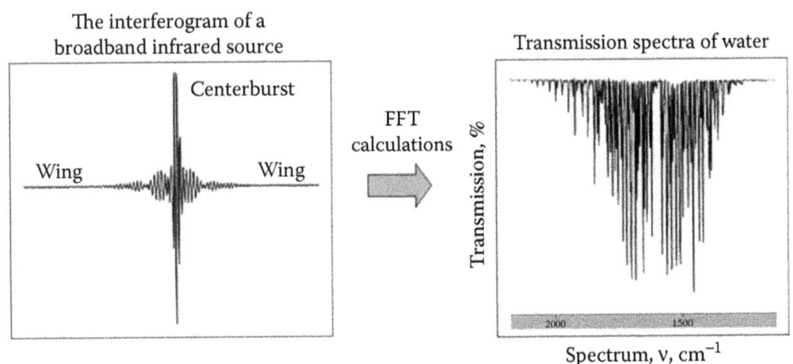

FIGURE 4.2 Transfer of the interferogram to the transmission spectrum.

In comparison to a monochromator, an interferometer has two significant advantages. The first advantage, which is termed Jacquinot's advantage, is the higher throughput of source radiation. As an interferometer does not use slits and has fewer optical components from which radiation can be scattered and lost, the throughput of radiation reaching the detector is 80–200× greater than that for a monochromator. The result is less noise. The second advantage, which is called Fellgett's advantage, is a savings in the time needed to obtain a spectrum. As the detector monitors all frequencies simultaneously, an entire spectrum takes approximately 1 s to record, as compared to 10–15 min with a scanning monochromator. In addition, FTIR instrument is mechanically simple with only one moving part. Thus, there is very little possibility of mechanical breakdown.

It should be noted that in addition to the traditional transmission FTIR (T-FTIR) methods, modern reflectance techniques are widely used today in agricultural, pharmaceuticals, and food studies. These modern techniques are attenuated total reflection FTIR (ATR-FTIR), and diffuse reflectance infrared Fourier transform spectroscopy (DRIFTS). However, for atmosphere monitoring a transmission FTIR method is the most appropriate.

The basic principle of transmission FTIR spectroscopy used in atmosphere monitoring is that most atmospheric molecules interact with electromagnetic radiation in the IR spectral region. This means that every gas has its own *fingerprint* or absorption spectrum in IR range. Therefore, if you have the opportunity to read the different fingerprints of the gases present in the air sample, then you have the ability to control the composition of the atmosphere. The potential value of FTIR spectroscopy to a wide range of environmental applications has been demonstrated by numerous research studies, results of which were summarized in the reviews made by Griffith (1996), McKelvy et al. (1998), Workman Jr. (1999), Griffith and Jamie (2000), Visser (2000), Russwurm and Childers (2002), Spellicy and Webb (2002), Bacsik et al. (2004, 2005), Simonescu (2012), Bezanila et al. (2014). A number of guideline documents and recommendations describing methods and procedures for FTIR monitoring of atmospheric gases have also been prepared by the U.S. Environmental Protection Agency (US EPA 1999a, b), Verein Deutscher Ingenieure (VDI 2000), and by the American Society for Testing and Materials (ASTM 2002).

In the case of air pollution, the FTIR instrument is used successfully for measuring the gas pollutants and air humidity due to its many advantages such as: multiple gas pollutants can be monitored in real time, the IR spectra of the sample can be analyzed and preserved for a long time, it can be used to detect and measure directly both criteria and toxic pollutants in ambient air, it also measures organic and inorganic compounds, it can be also used to characterize and analyze microorganisms and monitor biotechnological processes, it is generally installed at one location, but can be also portable and operated using battery for short-term survey, it presents sensitivity from very low parts per million to a high percent levels, it can be applied to the analysis of solids, liquids, and gases, no reagent is needed, and the data acquisition is faster than with other physicochemical techniques (Santos et al. 2010).

Generally, for a given concentration (partial pressure) of the target species the sensitivity of this detection technique increases with the

- Optical throughput of the sensor
- Transmittance and modulation efficiency of the FTIR spectrometer
- Sensitivity of the IR-detector element
- Optical path length
- Absorption cross section of the target molecule, that is, absorption coefficient
- Number of coadded spectra, that is, measurement time

For environmental monitoring, one can use open- and closed-path measurement techniques. However, open-path measurement techniques are more promising, due to the ability to conduct a remote control. Due to the high sensitivity and mobility of modern open-path FTIR spectrometers, there are number of FTIR applications in the field of atmosphere monitoring. At that, open-path measurement techniques can be divided into passive and active measurement techniques. The active open-path measurement technique is quite similar to the classical setup of a laboratory FTIR spectrometer. In an active remote-sensing measurement, the IR light is emitted by a hot IR source, and is received by the spectrometer optics. Possible configurations are shown in Figure 4.3. In the bistatic arrangement, the IR radiation source and the detector are mounted on separate platforms positioned at the opposite ends of the optical path. The monostatic system has both the IR source and the detector at the same end of the path. There are currently two techniques in use on monostatic systems for returning the beam along the optical path: one of them has an arrangement of mirrors that translate the beam slightly for its return path (Figure 4.3c); the other possibility is to place a retroreflector array consisting of corner-cube mirrors at the end of the path (Figure 4.3d). Thus, the emitted IR light propagates in direction of an open optical path, which is defined by the distance and position of the IR-source and the

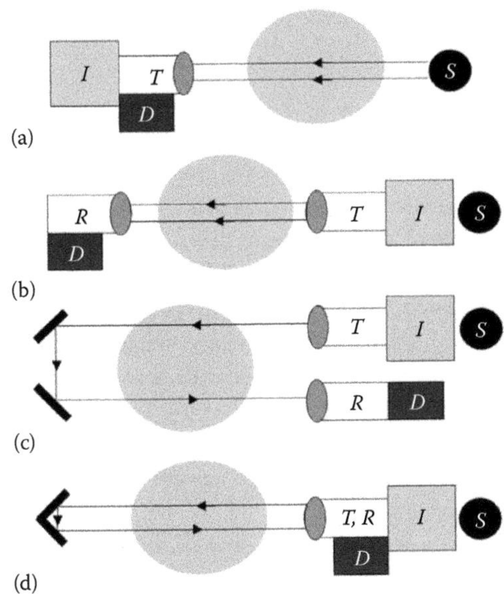

FIGURE 4.3 Schematic representation of different active OP configurations. (a) Bistatic configuration with separate IR source and receiving telescope attached to the interferometer; (b) bistatic configuration with separate receiving detector optics and detector; (c) monostatic configuration with translating retroreflector mirrors and separate receiving optics and detector; (d) monostatic configuration with a nontranslating retroreflector mirror. Abbreviations: *I*, interferometer; *S*, *IR* light source; *T*, transmitting optics; *R*, receiving optics; *D*, detector. (With kind permission from Taylor & Francis: *Appl. Spectroscopy Rev.*, FTIR spectroscopy of the atmosphere. I. Principles and methods, 39, 2004, 295–363, Bacsik, Z. et al.)

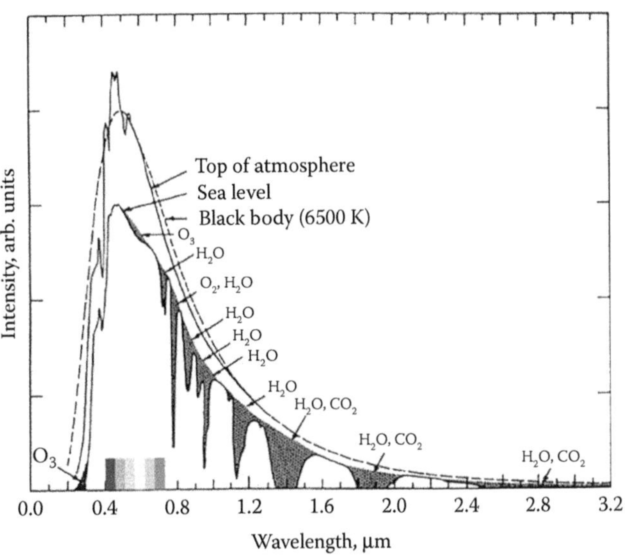

FIGURE 4.4 Solar spectrum+atmospheric absorption. (Data extracted from Iqbal, M., *An Introduction to Solar Radiation*, Academic Press, New York, 1983.)

spectrometer. Of course, both the bistatic and monostatic systems are equipped with telescope optics to produce and receive a parallel beam of light traversing the long optical path, and contain a computer for data collection, data storage, and analysis (Bacsik et al. 2004).

Passive remote-sensing works similar to the active setup described above. The only difference is that the ambient IR radiation is detected instead of the light of the artificial IR source (Beil et al. 1998). Thus, a passive FTIR measurement is one in which you use an emission spectrometer to collect naturally occurring IR radiance within the field-of-view (FOV) of the instrument. The FOV is typically restricted with a telescope so that you can look at a specific target. Your measured signal contains radiance from whatever sources are present. This could be direct emission from the atmosphere or any objects that are present. In particular, direct solar radiation is often used in FTIR spectroscopy used for atmosphere monitoring. Spectrum of solar radiation is shown in Figure 4.4. The usage of passive FTIR measurements results in the advantage of mobile and fast operation (man-held system), remote-sensing distance up to several kilometers, and easy handling. A more detailed comparison of active and passive methods of measurements one can find in Table 4.1. One should note that the radiation reflected from the moon can also be used for passive remote atmosphere monitoring at night time (Notholt et al. 1993; Notholt and Lehmann 2003; Wood et al. 2004). However, the first night time FTIR water vapor total column amount measurements (determined from lunar absorption measurements) were reported by Palm et al. in 2010. But, the Lunar-FTIR exhibited a low signal-to-noise ratio (SNR) and very sparse measurements necessitating long coincidence times to enable radiosonde comparisons. Therefore, the use of solar radiation is dominant when using FTIR spectroscopy.

The basic equation for analyzing IR solar absorption spectra is the Lambert Beer's law:

$$I(\lambda) = I_{\text{sun}}(\lambda) \cdot \exp\left(-\int_{\text{TOA}}^{\text{Obs.}} \sigma_x\big(\lambda, s(T,p)\big) \cdot x(s)ds \right) \quad (4.1)$$

where:

$I(\lambda)$ is the measured intensity at wavelength λ
I_{sun} is the extraterrestrial solar intensity
$\sigma_x(\lambda, s)$ is the absorption cross section
$x(s)$ the concentration of an absorber x at location s

The integration is performed along the path of the direct sunlight (between the Observer, Obs., and the Top of the Atmosphere, TOA). The cross section σ_x depends on temperature and pressure.

TABLE 4.1
Advantages and Disadvantages of Passive FTIR Remote Monitoring

	Advantages	Disadvantages (Complications)
Passive FTIR remote measurements	• No IR source is necessary • Mobile and fast operation • It can be used at remote-sensing distances up to several kilometers • Laboratory calibration is possible with the use of heated gas cells • The absorption and self emission of gases is extremely temperature sensitive; based on this, it is possible to judge about the temperature of the gas • Toxic cloud imaging with an FTIR spectrometer equipped with a detector array is also possible	• Reducing spectral data to quantitative information is significantly more complicated in field analysis utilizing emission measurements than in those based on absorption phenomena alone • The method suffers from all disadvantages connected with any field measurements (weather-related problems: wind, temperature, cloud conditions, and so on, and their fast changes) • A general problem of remote-sensing FTIR spectroscopy is to obtain suitable reference or background spectra • Passive FTIR spectroscopy is less sensitive than the active configuration

Source: Bacsik, Z. et al., *Appl. Spectroscopy Rev.* 39, 295–363, 2004.

Solar absorption spectra contain information about the absorbing gases present in the atmosphere (line positions), the amounts of each gas present (line depths/areas) and some information about the altitude distribution of each gas (line shapes). The characteristic absorption features are caused by molecules absorbing radiation at frequencies that correspond to the allowed transitions between different vibrational and rotational states. As a consequence of Heisenberg's uncertainty principle, the absorption lines are never infinitely narrow. The shorter the lifetime, the larger the uncertainty in a state's energy and the broader the absorption or emission line (as the energy uncertainty manifests itself as an uncertainty in the frequency of the line). There are mainly three effects responsible for the broadening of spectral lines:

- *Natural broadening* refers to the broadening due to the Heisenberg uncertainty in the energy levels of the gas molecules due to the finite natural lifetime of a molecule in an excited state. For vibrational-rotational states, this lifetime is usually very small ($<10^{-6}$ cm^{-1}). This effect therefore can be ignored in most practical situations, especially in the IR when compared to Lorentz or Doppler broadening in the atmosphere.
- *Collision broadening*, also known as Lorentz or pressure broadening, occurs when the collisions of atoms, ions, or gas molecules shorten the lifetime of states. In gases, it is proportional to pressure. This means that absorption lines from spectra taken through the whole atmosphere will have different shapes depending upon the vertical distribution of the absorbing gas in the atmosphere. Pressure broadening leads to a Lorentzian line shape contribution at a given wavenumber ν.
- *Doppler broadening* occurs because the molecules travelling with different velocities with respect to the light source absorb at different wavelengths (Doppler effect). Doppler broadening produces a Gaussian line shape due to the Gaussian distribution of molecular velocities.

Pressure broadening dominates in the troposphere, but its effects drop off rapidly with altitude as the pressure drops. Doppler broadening is temperature dependent, but its variation through the atmosphere is much smaller than pressure broadening. Stratospheric gas lines are primarily Doppler broadened. The two types of broadening become equally significant at around 30 km (for $\nu \sim 1000$ cm^{-1}). The convolution of Lorentzian and Gaussian line shapes produces a Voigt line shape. This variation of the shape and width of the absorbing gas lines with respect to the pressure means that spectra of atmospheric gases contain information about the altitude of the absorbing gas as well as the total number of absorbing molecules in the path. Therefore, the high-resolution FTIR spectra disclose not only the total column amount of the absorber, but also contain some information about its vertical distribution.

The height information can be derived by considering the temperature and pressure broadening of the spectral absorption lines. In the case of limb sounding, the profiles of trace gases are determined from several measurements with different elevation angles, thus sensing the various atmospheric layers with different weighting. Inversion methods used are, for example, the onion peeling or the global fit technique (e.g., Goldmann and Saunders 1979). In the first case, the atmosphere is peeled like an onion starting with the highest layer and working through the atmosphere to the lowest layer. Performing a global fit involves using all the measured spectra in one nonlinear least-square fit in order to generate the vertical profile of the trace gas under investigation. If, in addition, a priori knowledge of the atmosphere is taken into account, the retrieval procedure is called optimal estimation technique. In other words, when synthesizing spectra, the atmosphere is usually divided in many layers stacked in altitude and the concentration, pressure, and temperature broadening in each layer is taken into account to synthetically calculate how solar spectra in the IR region would be observed by the instrument after it has passed through the atmosphere. The gas concentration in each layer is calculated with an optimum estimation method (Rodgers 2000). To calculate how a gas in the atmosphere absorbs light, the HIgh Resolution TRANsmission (HITRAN) database (Rothman et al. 1998, 2003, 2005, 2009, 2013) can be used. The HITRAN database is a research-standard database containing parameters for each absorption line for the 63 most common gases in the atmosphere, including reference temperature and pressure line strengths ($S(\nu_0)$), pressure broadening (γ_{air}, γ_{self}), and shift parameters (δ_{air}), along with their temperature dependence exponent n_{air}, for the vibrational–rotational transitions of the most common atmospheric molecules. The database also contains lower state energy (E″) information necessary for the line strength calculation and several other quantum mechanical parameters. In addition to line-by-line parameters, cross section data and aerosol indices of refraction are also included for some species in some pressure, temperature, and wavenumber ranges. Using this information, it is possible to calculate how a specific gas at a specific temperature and pressure absorbs light.

$$\sigma_\nu = S(T) \cdot g\left(\nu - \nu_{\eta\eta'} - \delta\right) \tag{4.2}$$

where g is a function describing the broadening of the line. Figure 4.5 describes the different parameters in this equation. $\nu_{\eta\eta'}$ is the center frequency for the absorption line. $S_{\eta\eta'}$ is the line strength, that is, the total shaded area in the figure. δ is a shift in frequency caused by air-broadened pressure shift. The line strength $S(T)$ is dependent on the temperature since the temperature determines how strongly the upper and lower energy levels are populated. The effects that contribute to the line broadening are represented in Equation 4.2 by function g. The nature of these effects was discussed earlier.

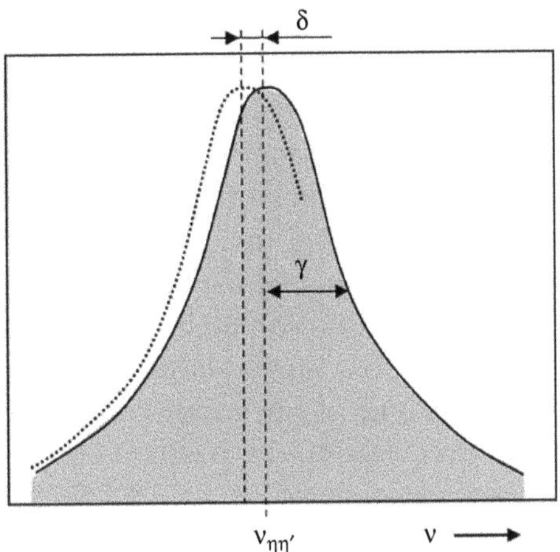

FIGURE 4.5 The figure shows fundamental spectroscopic parameters of a line transition in HITRAN. The dotted line refers to a perturbed transition (with a negative δ).

Profile measurements above the location of the instrument are possible in principle with spectrally highly resolving instruments. The higher the spectral resolution is, the larger is the amount of information about the absorbers vertical distribution (Paton-Walsh 2011). But the upper level of the altitude range depends on the wave number (for 1000 cm^{-1} at about 30 km and for 100 cm^{-1} at about 48 km) (Kley et al. 2000). This is caused by the fact that the shape of the absorption lines is dominated by the Doppler effect above this height level. In the case of H_2O, it is even more difficult to sound the stratosphere from the ground because there is only a small amount of the H_2O column above the tropopause. In order to achieve profiles with good vertical resolution in the middle atmosphere, the limb sounding technique, that is, viewing the atmosphere along tangential slant paths, has to be applied from balloon gondolas or high flying aircraft (Kley et al. 2000). Vertical profiles can in addition be derived from uplooking measurements during ascent or descent of the balloon or the aircraft.

One should note that the first water vapor profiles measured by a ground-based FTIR experiment were reported by the IMK–ASF FTIR group (Hase et al. 2004) (IMK–ASF: Institute for Meteorology and Climate

Research—Trace Constituents in the Stratosphere and Tropopause Region, Karlsruhe). At IMK–ASF, the water vapor algorithm has been continuously improved since 2005 (for a review, refer Schneider 2009). These efforts made it possible to monitor tropospheric H_2O profiles (including upper tropospheric amounts) and HDO/H_2O ratio profiles by ground-based FTIR experiments (Schneider et al. 2006, 2010a, b). Similar approach was also used by the ground-based FTIR group of IUP Bremen (Institute of Environmental Physics of the University of Bremen) (Palm et al. 2010). However, they used the water vapor signatures at 3268.6–3273.0 cm^{-1} and 3299.6–3305.0 cm^{-1}, corresponding to 3.02–3.05 μm. The IMK–IFU FTIR group used absorptions signatures at 11.7–11.9 μm.

Thus, summing up this review, we can say that remote-sensing FTIR (RS-FTIR) really is an important technology for the atmosphere monitoring in open environments. Many have noted that compared to the traditional point sampling detection methods, a remote-sensing by FTIR spectroscopy has several advantages:

- Capability of detecting several compounds simultaneously
- Fast analysis of multicomponent mixtures
- High resolution and high selectivity
- Real-time monitoring of atmosphere over the distance
- Large sampling area, no sample preparation and handling
- No sensor contamination during measurement
- Simple operation and little maintenance
- Possibility of automatic monitoring

As regards general errors related to FTIR measurements, the systematic errors (and random noise) in the interferogram can be grouped into three broad categories: additive errors, intensity errors, and phase errors. The reasons for their appearance and approaches used to reduce them considered in detail in Mertz (1965, 1967), Forman et al. (1966), Griffiths and de Haseth (1986), Remedios (1990), and Davis et al. (2001) and taken into account in the software of FTIR spectrometers. Therefore, a number of steps are involved in calculating the frequency domain spectrum. Instrumental imperfections and basic scan limitations need to be accommodated by performing a phase correction and apodization steps. These electronic and optical imperfections can cause erroneous readings due to a different time or phase delays of various spectral components. Apodization is used to correct spectral leakage, artificial creation of spectral features due to the

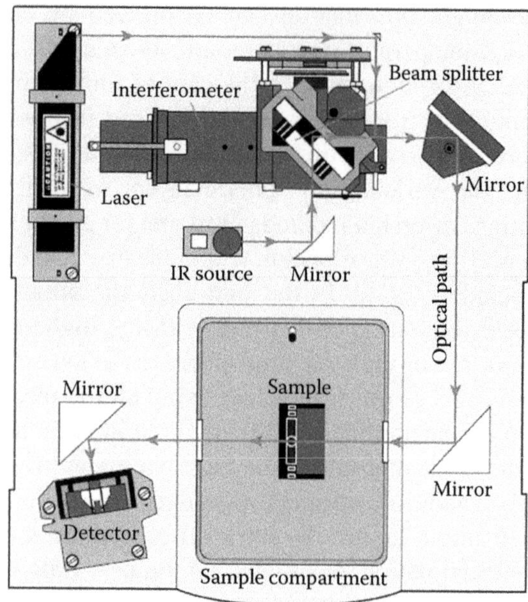

FIGURE 4.6 A simple layout of the spectrometer designed by Thermo Nicolet. These instruments employ a He–Ne laser as an internal wavelength calibration standard. These instruments are self-calibrating and never need to be calibrated by the user. A background spectrum needed for calibration is measured with no sample in the beam. This can be compared to the measurement with the sample in the beam to determine the *percent transmittance*. (From http://www.thermonicolet.com.)

truncation of the scan at its limits (a FT of sudden transition will have a very broad spectral content).

One should note that the FTIR spectrometers aimed for active closed-path measurements have been also developed. The traditional area of integration of such spectrometers is shown in Figure 4.6. At that these very devices are the most widely used, including atmospheric monitoring (Griffith et al. 2012). In this case, the measurement accuracy is higher, and the processing of the results is a lot easier than in the case of passive open pass measurements. However, this *in situ* monitoring may be carried out only in the surface layer. To build a profile of distribution of the water vapor concentration adjustment when using these devices, they need to be installed in the radiosonde or aircraft.

One should note that the He–Ne laser whose light is red and its wavelength is known precisely ($\lambda = 632.8$ nm) is a very important component in every modern FTIR spectrometer (Figure 4.6). First, all IR wavenumbers are measured relative to the He–Ne laser light line, which is determined precisely at 15798.637 cm^{-1} or 633 nm. Second, the He–Ne is being used for measurements of OPD properly, by tracking the position of the moving mirror.

As is shown in Figure 4.6, the beam splitter is another important part of the FTIR spectrometer. The beam splitter splits half of the radiation that impinges it at 45° and transmits the other half. Experiment has shown that unfortunately very limited amount of materials can be used for these purposes (Colthup et al. 1990). Potassium bromide (KBr) itself does not split the IR light because it is transparent to IR radiation. Therefore, the beam splitter on the base of this material consists of a thin coating of germanium sandwiched between two pieces of KBr. The function of the Ge coating is to split the IR beam. This beam splitter is usable in the range of 4000–400 cm^{-1}, covering the mid-IR very well. Such type of beam splitter is almost universally used FTIR. However, KBr is hygroscopic (it absorbs water from the atmosphere and fogs over), which is a drawback. Cesium iodide (CsI), transmitting from 4000–200 cm^{-1}, is another material which can be used in the mid-IR beam splitters. But, CsI is soft and very hygroscopic, which means that it should be replaced more frequently. The beam splitters fabricated from quartz, calcium fluoride (CaF_2), zinc sulfide (ZnS), and zinc selenide (ZnSe) are also able to handle IR light. The selection of material depends on the special spectral region to be investigated.

The detectors used in FTIR instruments must be sensitive to IR radiation and are therefore made usually of HgCdTe, InSb, InAs, InGaAs, PbS, or PbSe. The first three versions must be cooled by liquid nitrogen to work and the last three work at room temperature but have lower sensitivity.

Most FTIR instruments designed for atmosphere monitoring, use Off-Axis Replicated Parabolic Mirrors for collimating and focusing light external to the interferometer. Parabolic mirrors are devices ideally suited for collimating light from small sources and conversely for tightly focusing collimated beams of radiation. Light from a point source placed in a focus of a parabola will be transformed after reflection into an ideally parallel beam. Accordingly, a parallel beam will be focused into a tiny focal spot. This is true for any section of the parabola. So, an off-axis section of the parabolic mirror can be cut out for convenience (Figure 4.7). These gold coated mirrors are very broadband, from 0.7 to 10 μm they reflect more than 98%, and it stays in this range up to 25 μm (bear in mind that for wavelengths shorter than 0.6 μm, gold is a bad reflector; its reflectivity drops abruptly to less than 40%). An important feature of mirrors in general also is that they do not have any dispersion; there is no chromatic aberration so the focal spot stays at the same place for any wavelength. They do have monochromatic aberrations.

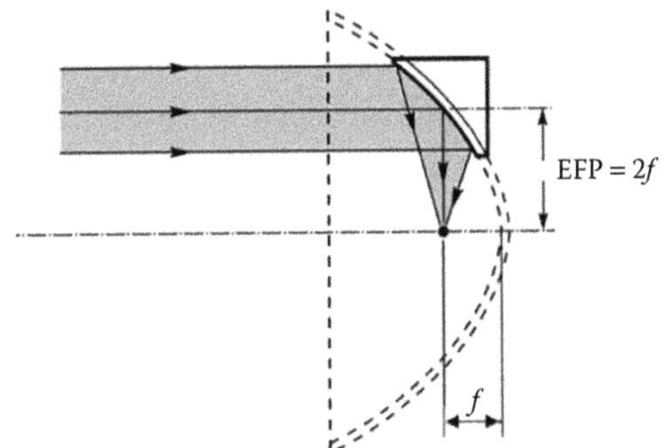

FIGURE 4.7 Section of off-axis parabolic mirror.

Despite the universality and wide usage of off-axis parabolic mirrors in the FTIR spectroscopy, they have certain disadvantages. They are pretty difficult to align; each reflection turns the beam through 90°, and this may make the system bulky. At low *f*-ratio, that is, large fields of view (high étendue), they suffer from significant aberrations. Therefore, in many applications, especially in the Near IR, lenses could be a good choice. For example, designers of Newport Corporation (http://www.newport.com/n/introduction-to-ftir-spectroscopy) recommend the use of CaF_2 lenses in the whole range where the CaF_2 beam splitter is applicable. In the very Near IR, up to 3 μm, fused silica lenses are fine, though the water absorption bands can cause some loss with lenses that are not *IR grade*. They are somewhat cheaper than CaF_2 lenses. A wide variety of materials are available for the mid-IR. You usually have a choice among performance, expense, durability, birefringence, and so on. The hygroscopic nature of some materials can be a big problem. The NaCl windows and KBr are two such popular materials. Some materials are transparent in the visible and others not; this can be a plus if you are trying to align in the visible, or a negative when you would prefer the material to act as a filter. A popular rugged and transparent material which is used for manufacturing lenses is ZnSe. It has, however, a very high index of refraction that pushes reflectance losses to relatively high levels: up to 30%. Antireflection coatings can help, but at further expense, and reduction of the spectral range. A second issue is a dispersion of the lens material. Lenses are definitely good for limited wavelength range applications. For example, the sensitivity range of an InGaAs detector is from 800 to 1700 nm. Using a lens should not pose a major problem, though we do see some dispersion in our labs with fused silica lenses over this range, that is, you can axially move the lens to optimize the long wavelength or short wavelength signal. For a wider wavelength range you should position the detector at the shortest focal length position, in other words, in the position of minimum spot size for the shortest wavelength. However, in this case the system efficiency is usually the lowest. These examples show that the auxiliary optics for an interferometer must be carefully chosen and arranged. Poor choices of components will lead to a lack of resolution or unnecessary system throughput limitations.

4.2.2 Examples of FTIR Spectrometers Designed for Water Vapor Monitoring

FTIR spectrometers, which were developed and used for atmosphere monitoring during last decades, are listed in Table 4.2. One should note that since the FTIR technique allows the determination of vertical distribution of many atmospheric trace constituents, usually these experiments were not designed primarily for the H_2O retrieval. A transmission spectrum of atmosphere in spectral region acceptable for FTIR for atmospheric monitoring is shown in Figure 4.8. It is seen that there are many strong water absorption bands, which can be used for water vapor measurements using FTIR instrument.

Exterior view of the station using a FTIR spectrometer for monitoring the atmosphere is shown in Figure 4.9. As a rule, the ground-based FTIR experiment consists in a high precise solar tracker that captures the direct solar light beam and couples it into a high resolution FT spectrometer. The high precision solar tracker is controlled by a combination of astronomical calculations and a solar quadrant sensor, or more recently by a digital camera (Gisi et al. 2011), for active tracker control.

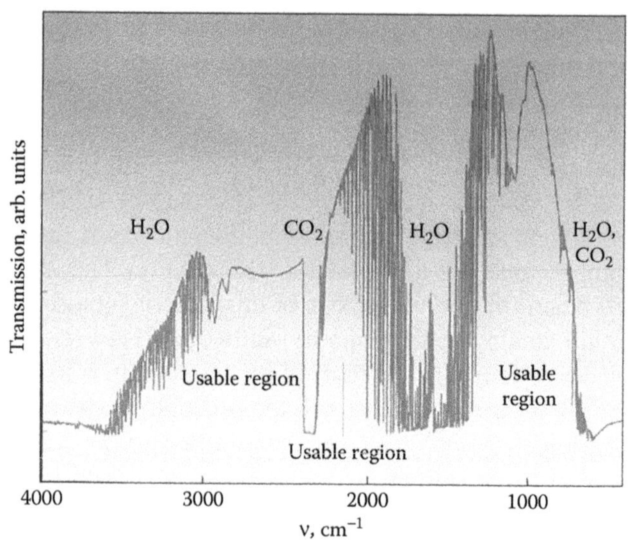

FIGURE 4.8 Transmission spectrum of atmosphere. (Reprinted with permission from Marshall, T.L. et al., *Environ. Sci. Technol.*, 28, 224–232, 1994. Copyright 1994 American Chemical Society.)

Let us consider in more detail some of the FTIR spectrometers indicated in Table 4.2. Information on other IR spectrometers from which H_2O profiles have been derived is found in Murcray et al. (1990) and Bacsik et al. (2004, 2005) and references therein.

Several FTIR spectrometers, known as Michelson interferometer for passive atmospheric sounding (MIPAS), using a special configuration to provide the

TABLE 4.2
Examples of FTIR Spectrometers Used for Remote Atmosphere Monitoring During Last Decades

FTIR Spectrometers	Developer	References
JPL MkIV interferometer	Jet Propulsion Laboratory, NASA, USA	Toon (1991)
Michelson Interferometer for Passive Atmospheric Sounding (MIPAS)	Institut für Meteorologie und Klimaforschung in Karlsruhe, Germany	Fischer (1993), Friedl-Vallon et al. (1995), Oelhaf et al. (1996)
SAO Far Infrared Spectrometer (FIRS-2)	Smithsonian Astrophysical Observatory, Harvard, USA	Johnson et al. (1995)
Bruker 120 HR FTIR spectrometer at Ny Alesund station, Spitsbergen	Institute of Environmental Physics, Universitat Bremen, Germany	Notholt et al. (1993), Palm et al. (2010)
DA8 Research Grade Fourier Transform Spectrometer (FTS) at the University of Toronto Atmospheric Observatory (TAO)	ABB Bomem Inc. (Quebec, Canada) University of Toronto, Canada	http://www.abb.ca/cawp, Wiacek (2006)
Solar Fourier Transform Infrared (FTIR) at Darwin TCCON station	University of Wollongong, Wollongong, Australia	Griffith et al. (2012)
Bruker IFS 120/5 HR instrument at the Izaña Atmospheric Research Center (IARC), Canary Island of Tenerife, Spain	Bruker Corporation; IMK–ASF: Institute for Meteorology and Climate Research—Trace Constituents in the Stratosphere and Tropopause Region, Karlsruhe, Germany	http://www.bruker.com, Schneider et al. (2006, 2010a, b)

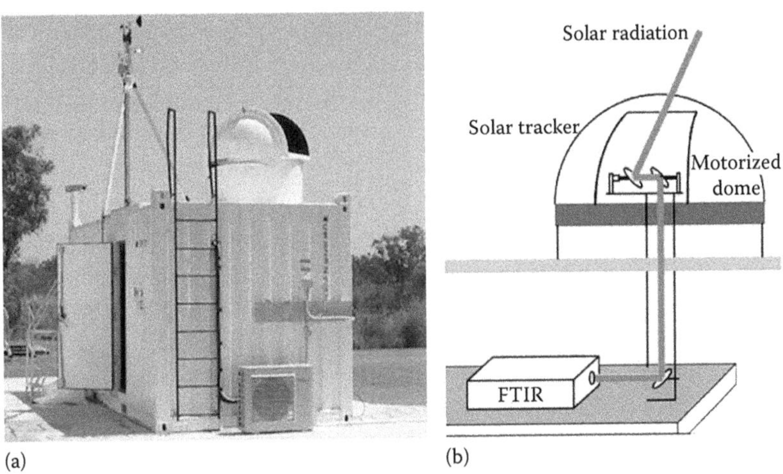

FIGURE 4.9 (a) Solar Fourier Transform Infrared (FTIR) at Darwin TCCON station. Ground based solar infrared remote sensing is able to detect and quantify a wide range of atmospheric trace gas species by analysis of the absorption of solar radiation in the infrared and near-infrared regions (From http://www.smah.uow.edu.au); (b) Schematic drawing of the solar absorption infrared experiment. (Idea from Bezanila, A. et al., *Atmósfera*, 27, 173–183, 2014.)

OPD, have been developed for operation on ground, aircraft, and stratospheric balloon gondolas by the Institut für Meteorologie und Klimaforschung in Karlsruhe, Germany (Fischer 1993). The last version of the interferometer provided two-sided interferograms with a maximum OPD of 15 cm, resulting in an unapodized spectral resolution of 0.033 cm^{-1} (Friedl-Vallon et al. 1995; Oelhaf et al. 1996). The four-channel detector system with Si:As detectors allowed the simultaneous coverage of the most important absorption bands of ozone- and climate-relevant trace gases between 5 and 14 μm. The instrument was equipped with suitable subsystems to precisely allow limb emission sounding of the stratosphere independent of external radiation sources. As such it had possibility to operate day and night, not being restricted to any solar zenith angle. Typical SNRs of single H_2O lines were between 50 and 200. Radiometric calibration was performed on the basis of blackbody and deep-space spectra that are recorded several times during a flight. The calibrated spectra were analyzed using multiparameter nonlinear least-square curve fitting in combination with the onion-peeling technique (von Clarmann et al. 1995; Wetzel et al. 1997). The estimation of the error budget was carried out using an elaborate scheme that taken into account random noise, mutual influence of fitted parameters, temperature and pointing uncertainties, onion-peeling error propagation, and errors in spectroscopic data. MIPAS was suitable to simultaneously obtain vertical profiles of ozone and a considerable number of species such as O_3, NO, NO_2, HNO_3, HO_2NO_2, N_2O_5, $ClONO_2$, COF_2, CH_4, N_2O, H_2O, HDO, CF_2Cl_2, CCl_3F, CHF_2Cl, CCl_4, CF_4, C_2H_6, and SF_6 with an altitude resolution of 2–3 km. The precision of the retrieved H_2O mixing ratio was typically near 5% with temperature errors being the dominant source of errors, and the absolute accuracy of H_2O profiles derived from MIPAS spectra was between 6% and 11% assuming a 5% uncertainty of the spectroscopic data (Kley et al. 2000).

The Izaña Atmospheric Research Center (IARC), located in the Canary Island of Tenerife, Spain, uses a Bruker IFS 120/5 HR instrument (http://www.bruker.com) that records direct solar spectra in the mid- and near-IR spectral region, using a set of different apertures, filters, and detectors. The mid-IR measurements are made between 740 and 4250 cm^{-1}, corresponding to 13.5–2.4 μm, with a spectral resolution of 0.005 cm^{-1}, defined as 0.9 = OPD_{max}, being the OPD_{max} the maximum OPD (180 cm). The near-IR measurements are made between 3800 and 15,500 cm^{-1}, corresponding to 2.6–0.65 μm, with a resolution of 0.02 cm^{-1} (OPD_{max} of 45 cm). In the mid-IR spectral region, a potassium chloride (KCl) beam splitter and either of two liquid-nitrogen-cooled semiconductor detectors are used for the record of the interferogram. The photovoltaic indium antimonide (InSb) detector and the photoconductive mercury–cadmiun–telluride (Hg–Cd–Te), MCT detector are sensitive to radiation from 1850–9600 cm^{-1} (5.4–1.1 μm) and 680–6000 cm^{-1} (14.7–1.7 μm), respectively, and are typically used in conjunction with narrow-bandpass filters and different apertures. In the near-IR spectral region, a CaF_2 beam splitter and an Indium Gallium Arsenide (InGaAs) (3800–12,000 cm^{-1}) or Silicon (Si) (9500–30,000 cm^{-1}) photodiode detectors are used at room temperature for the record of the interferograms. Recording of one spectrum using this instrument requires between some seconds to a few of tens minutes, depending on the spectral resolution and quality needed. For instance, one scan can be performed in 30 s, but normally several scans are coadded in order to increase the SNR. Therefore, the acquisition of one spectrum can last several minutes. The state of the atmosphere and the possibility of using the FTRI spectrometer allowed controlling the atmosphere up to 13–16 km.

Studies carried out on IARC have shown that the optimal estimation of atmospheric water vapor amounts from ground-based FTIR spectra is far from being a typical atmospheric inversion problem and, due to its large vertical gradient and variability, standard retrieval methods are not appropriate (Schneider et al. 2006, 2010a). Therefore, they applied the inversion strategy as explained in (Schneider et al. 2006, 2010a) and references therein, using the retrieval code PROFFIT 1. It was also proposed to analyze the spectral H_2O signatures from six spectral microwindows between 4564 cm^{-1} and 4702 cm^{-1} (Meier et al. 2004). These spectral windows contain weak absorption signatures of CO_2, N_2, and CH_4 (Rothman et al. 2009). The H_2O intercomparison with data obtained by Vaisala RS92 radionsondes has shown that the estimated precision was about 5% for precipitable water vapor and for the lower and middle tropospheric volume mixing ratio (VMR) (Vömel et al. 2007; Miloshevich et al. 2009; Schneider et al. 2010b). In the upper troposphere and for very dry conditions, it is poorer (about 10%–20%). However, Schneider et al. (2006) state that the ground-based FTIR technique used in IARC allows measuring a column amounts and volume mixing ratio (VMR) profiles of many different atmospheric gases often with an unprecedented precision. For instance, total column amounts of water vapor (H_2O) can be determined with a precision of 0.5%–1%.

MkIV interferometer developed in the Jet Propulsion Laboratory is a high resolution FTIR spectrometer, which was designed to remotely sense atmospheric composition by ground-based and aircraft observations (Toon 1991; Kley et al. 2000). The interferometer operates in a solar absorption mode, meaning that direct sunlight is spectrally analyzed and the amounts of various gases at different heights in the Earth's atmosphere are derived from the depths of their absorption lines. In the mid-IR spectral region, the Sun is so bright that emission from the atmosphere or the instrument is negligible. This provides spectra of high accuracy without the need for in-flight radiometric calibration. The entire 700–5700 cm^{-1} spectral region was observed simultaneously at 0.01 cm^{-1} resolution. Over this wide interval more than 30 different gases have spectral signatures that can be identified, including H_2O, CO_2, O_3, N_2O, CO, CH_4, N_2, O_2, NO, NO_2, HNO_3, HNO_4, N_2O_5, $ClNO_3$, HOCl, HCl, HF, SF_6, COF_2, CF_4, CH_3Cl, $CHFCl_2$, $CFCl_3$, CF_2Cl_2, CCl_4, OCS, HCN, C_2H_2, C_2H_6, and many isotopic variants. Profiles of these gases can be retrieved from cloud-top to balloon altitude. The vertical resolution of 2–3 km is limited mainly by the 1–3 km separation of the tangent altitudes, and by the 3.6 mrad diameter circular field-of-view of the instrument, which subtends 1.8 km at a tangent point 500 km distant.

The MkIV H_2O retrievals are performed by fitting 14 different spectral intervals between 1400 and 5500 cm^{-1}. Since there are a large range of strengths encompassed by these lines, the H_2O abundance from <1 to >20,000 ppmv is easily measured. For H_2O, a retrieval precision of 5% is estimated and limited mainly by uncertainty in the pointing and by the presence of horizontal gradients. The overall accuracy is estimated to be 7% in the upper stratosphere, limited mainly by uncertainties in the spectral line intensities. In the lower most stratosphere and troposphere, the accuracy progressively worsens to 12% due to inadequacies in the pressure-dependent molecular spectroscopic parameters (e.g., half-width, shifts) and due to the presence in the spectrum of poorly fitted interfering absorption.

The primary instrument at Toronto atmospheric observatory (TAO) is a high-resolution DA8 Research Grade Fourier transform spectrometer (FTS) manufactured by ABB Bomem Inc. (Quebec, Canada). The FTS is complemented by a commercially available weather station (Vantage Pro Plus manufactured by Davis Instruments Corp.) that records local meteorological variables, UV radiation and solar irradiance, and (on a campaign basis) by a differential optical absorption spectroscopy (DOAS), and a SunPhotoSpectrometer (SPS). FTIR solar absorption spectra are recorded under clear sky conditions, allowing for approximately 80 observation days per year in the first three years of operation. The measurements are semiautomated, involving an operator at start up and shut down to engage the suntracker, cool the detectors with liquid nitrogen, and initiate an automatic measurement sequence. The DA8 FTS is a modified Michelson interferometer with a maximum OPD of 250 cm, providing a maximum apodized resolution of 0.004 cm^{-1}, equivalent to an unapodized resolution of 0.0026 cm^{-1}. The FTS is currently equipped with KBr and CaF_2 beam splitters, and Indium Antimonide (InSb) and Mercury Cadmium Telluride (HgCdTe or MCT) detectors, for nominal coverage of the spectral range from 750 to 8500 cm^{-1}. For water monitoring the spectral range 2900–3500 cm^{-1} or 2.6–3.3 μm is used. Importantly, that FTS has been installed on two types of satellites

(Soucy et al. 2006). The CrIS2 (Cross-Track Infrared Sounder) provided over one thousand three hundred spectral channels of information and was able to measure temperature profiles with a vertical resolution of 1 km to an accuracy approaching 1°C.

The Smithsonian Astrophysical Observatory Far-infrared Spectrometer (FIRS) is a high resolution FTS that measures the thermal emission of the atmosphere from balloon and aircraft platforms, and from the ground (Johnson et al. 1995; Kley et al. 2000). The spectrometer produces mostly one-sided interferograms with an unapodized resolution of 0.004 cm^{-1}. In present configuration gallium-doped and copper-doped germanium photoconductors for the far- (80–330 cm^{-1}) and mid- (330–1250 cm^{-1}) IR bands, respectively, are used as detectors. With this wide spectral coverage, it is possible to make simultaneous measurements of many radical and reservoir species, including CO_2, O_2, O_3, OH, HO_2, H_2O_2, HCl, HOCl, $ClNO_3$, HBr, NO_2, HNO_3, HNO_4, N_2O_5, N_2O, CO, OCS, HCN, C_2H_6, C_3H_6O, HF, SF_6, CFC-11, CFC-12, HCFC-22, CCl_4, CH_3Cl, CH_4, and H_2O, as well as temperature and pressure (derived from CO_2). The instrument field-of-view is 0.22 and 0.16 in the far- and mid-IR, respectively, corresponding to 1.3 and 0.65 km at the limb, respectively. Typically, the limb is scanned at intervals of 3–4 km in tangent height, retrieving slant columns in an onion-peeling fashion. Then, a set of linear equations is solved for the smoothest mixing ratio profile which reproduces the observed slant columns (Johnson et al. 1996), producing profiles with a 1 km step size and a vertical resolution equal to the scan interval in tangent height.

The FTIR spectrometer at Ny Alesund station is operated in a solar or lunar absorption mode whenever weather conditions permit (Palm et al. 2010). Due to the measurement principle a clear sight of the illuminating object (Sun or Moon) is necessary during the measurement which takes up to 30 min. Changes in the optical depth of the atmosphere due to very thin clouds or phenomena like Arctic haze change the SNR of the measurement. The spectroscopic signatures are much broader (e.g., Ritter et al. 2005) than the signature of gases and do not interfere with the measurement of integrated water vapor (IWV). For water vapor monitoring two microwindows, 3268–3273 cm^{-1} and 3299.5–3305 cm^{-1} have been used to derive the IWV. The spectra were recorded using an Indium Antimonide (InSb) detector cooled with liquid nitrogen and a CaF_2 beam splitter. An SNR of better than 1400 can regularly be obtained for measurement times of about 10 min in a solar absorption spectroscopy. An SNR of about 100 for measurement times of about 30 min was obtained during winter in a lunar absorption mode. The HITRAN database (Rothman et al. 2003, version2k + updates from 2006) has been used to analyze the spectra.

It was reported (Kley et al. 2000; Schneider et al. 2006, 2010a, b; Barthlott et al. 2016) that the measurements of H_2O profiles can also include profiles for $H_2{}^{16}O$, $H_2{}^{17}O$, $H_2{}^{18}O$, and HDO. The precision (accuracy) of a typical profile was 4%, 60‰, 40‰, and 20‰ for $H_2{}^{16}O$, $\delta^{17}O$, $\delta^{18}O$, and δD, respectively (Kley et al. 2000).

It is important to note that for achieving a correct spectrum and high accuracy of measurements a constant solar input is required. If the intensity of the incoming solar radiation varies during the acquisition of an interferogram, which occurs when there are clouds in the path between the FTIR instrument and the Sun, the resulting spectrum will be distorted. This is due to the fact that the continuum level and the higher-resolution spectral structure will have a different gain signals. Although this distortion may be subtle, it can significantly alter the retrievals. Therefore, observations are only performed under homogeneous sky conditions (generally clear sky conditions). However, in the near-IR spectral region the detectors are supported by electronics that allow a solar intensity variation during data acquisition. And therefore, these variations can be corrected using the method described by Keppel-Aleks et al. (2007), resulting in higher quality spectra and less data loss during partly cloudy conditions. This correction does not work under thick clouds.

To achieve the required accuracy of measurement, FTIR spetrometers must also meet certain requirements. For example, in the case of determination of gas concentrations from stratosphere, the FTIR spetrometers have to be designed with a fine resolution (0.01 cm^{-1}) due to the lower atmospheric pressure, and with a lower resolution between 0.05 and 2 cm^{-1} for tropospheric gases determination. This is due to pressure broadening effects that result in broadened absorption lines. The strong interference of water vapor in troposphere can be overcome by detecting chemical substances in the narrow bands of the IR spectrum where water absorption is very weak. If interferences are observed (e.g., the analyte absorption features overlap with atmospheric absorption lines), more refined algorithms are needed, but univariate analysis can also be appropriate after some spectral preprocessing. In general, using various statistical data evaluation methods, the accuracy of concentration prediction and the detection limits can be improved (Thomas and Haaland 1990; Griffith 1996; Ingling and Isenhour 1999; Hart et al. 1999, 2000a, b; Griffith and Jamie 2000; Haaland and Melgaard 2001; Bacsik et al. 2004).

TABLE 4.3
The Advantages and Disadvantages of the Instruments with Low and High Resolution

Parameter	Low Resolution	High Resolution
V (field of view)	Large (+)	Small (−)
S/N	High (+)	Low (−)
Frequency precision	Low (−)	High (+)
Integrated intensity	High (+)	Low (−)
Line separation	Poor (−)	Good (+)
Long-term intensity stability	Poor (−)	Good (+)
Accuracy in the optical path difference	Good enough (+)	May be a problem (−)
Instruments	Cheaper (+)	More expensive (−)
Concentration range	Wider (+)	Narrower (−)
Spectral linearity	(+, −)	(+, −)
Temperature sensitivity	Poor (−)	Good (+)
Measurement time	Shorter (+)	Longer (−)

Source: Data extracted from Kauppinen, J.K., *Proceedings of 12th International Conference on Fourier Transform Spectroscopy*, Tokyo, Japan, pp. 47–49, August 22–27, 1999, and from Bacsik, Z. et al., *Appl. Spectroscopy Rev.* 39, 295–363, 2004.

However, the use of a tool with high resolution may create difficulties in conducting long-term measurements in a highly changing environment (Table 4.3). Generally, a spectroscopist would argue for high resolution to resolve all spectral bands and their rotational fine structure as much as possible, while an analytical chemist would argue for the lowest acceptable resolution to maximize S/N and quantitative precision (Griffith and Jamie 2000). The general conclusion from the long history of the *resolution battle* can be that the choice of optimal resolution depends on the actual analytical task and circumstances, so it is always application specific. The user has to settle the question by considering the advantages and disadvantages of low and high resolutions in relation to the given application (Bacsik et al. 2004).

It should also be borne in mind that the FTIR spectroscopy can be used not only for monitoring the atmosphere, but for controlling technological processes (Bacsik et al. 2005). For instance, Stallard et al. (1995) suggested using the FTIR spectroscopy for determination of trace levels of water vapor in N_2 and corrosive gases such as HCl and HBr used in semiconductor industry. Measurements were carried out in the mid-IR spectral region of 1600 and 3800 cm^{-1} where the strongest water absorption bands were observed. It has been shown that the proposed approach gives possibility to have a detection limit on water vapors of approximately 10 ppb in indicated before gases such as N2, HCl, HBr and etc. However, in this case as a rule it is usually used active measurement techniques with an additional light source. Active open-path technique is also widely used in airborne FTIR instruments. In particular, Kira et al. (2016) used such instrument for detection and quantification of water-based aerosols.

4.3 SPECTROSCOPY IN VISIBLE RANGE

The IR range is surely the most suitable spectral region for analyzing the water vapor in the atmosphere. However, the research has shown that the measurements in the visible range are also possible and may be effective. For example, Buscher and Lemke (1980) developed an optical analyzer, working in the range of 757–940 nm. As in the previous section 4.2, the solar radiation was used as a light source. A photovoltaic silicon PIN diode was chosen as the detector. Approximately the same spectral region, where $\lambda \sim 700$ nm, was used to develop "The Air Mass Corrected Differential Optical Absorption Spectroscopy" (AMC–DOAS). This method allowed retrieving a total water vapor column amounts from spectral measurements in the visible wavelength region around 700 nm (Noël et al. 1999, 2002). As it is seen in Figure 4.10, in this spectral range we have several lines, attributed to the absorption of water vapor.

As data in the visible spectral range were analyzed, the AMC–DOAS method was only applicable to measurements on the dayside and to (almost) cloud-free ground scenes. It is, of course, a significant disadvantage of this method. At the same time, the AMC-DOAS method has a significant advantage: the derived water vapor columns do not depend on additional external information, like a calibration using radiosonde data which is often used in the microwave spectral region. The AMC–DOAS water vapor columns therefore provide a completely independent data set. A comparison with in situ radiosonde data has shown that in general a good agreement between the different data sets, but AMC–DOAS water vapor columns were typically slightly (~0.2 g/cm^2) lower than, for example, Special Sensor Microwave Imager (SSM/I) and the European Centre for Medium-Range Weather Forecast (ECMWF) data. At the same time, intercomparisons between AMC–DOAS results based on

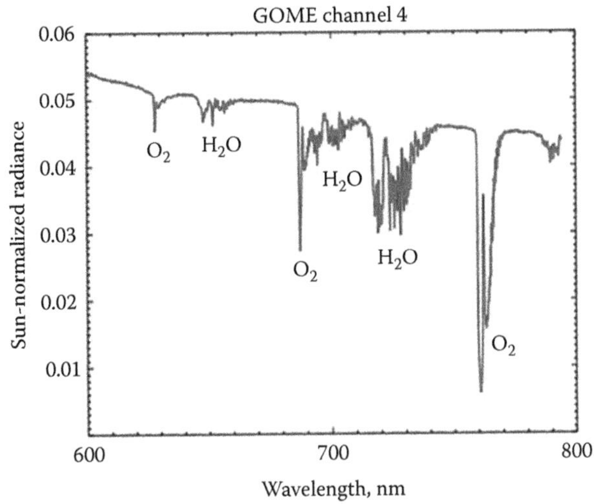

FIGURE 4.10 The example of a spectrum measured by the Global Ozone Monitoring Experiment GOME instrument. The GOME instrument is a space-based grating spectrometer measuring both the extraterrestrial solar irradiance and the Earthshine radiance in the spectral range between 240 and 800 nm. The algorithm to derive water vapor total column amounts from GOME data is based on the Differential Optical Absorption Spectroscopy (DOAS) approach. (Data extracted from Noël, S. et al., *Geophys. Res. Lett.*, 26, 1841–1844, 1999.)

Global Ozone Monitoring Experiment (GOME) and SCIAMACHY measurements showed a good agreement (Noël et al. 2005). SCIAMACHY system was developed by a bilateral Dutch/German activity. The primary scientific objective of SCIAMACHY program is the global measurement of various trace gases in the troposphere and stratosphere, which are retrieved from the instrument by observation of transmitted, backscattered, and reflected radiation from the atmosphere in the wavelength range between 240 and 2400 nm. In the SCIAMACHY optical system, the light from the atmosphere is fed by the scanner unit (consisting of an azimuth and an elevation scanner) into the telescope, which directs it onto the entrance slit of the spectrometer. The spectrometer contains a predisperser, which separates the light into three spectral bands followed by a series of dichromatic mirrors, which further divide the light into a total of eight channels. A grating is located in each channel to diffract the light into a high-resolution spectrum which is then focused onto eight detectors. Good correlation with SCIAMACHY data makes it possible to generate a combined GOME/SCIAMACHY water vapor climatology which is useful for climatological trend analysis applications (Mieruch et al. 2008).

REFERENCES

ASTM (2002) Standard practice for open-path Fourier transform infrared (OP/FT-IR) monitoring of gases and vapors in air. In: Annual Book of ASTM Standard; ASTM, Vol. 03.06.

Bacsik Z., Mink J., Keresztury G. (2004) FTIR spectroscopy of the atmosphere. I. Principles and methods. *Appl. Spectrosc. Rev.* 39 (3), 295–363.

Bacsik Z., Mink J., Keresztury G. (2005) FTIR spectroscopy of the atmosphere. II. Applications. *Appl. Spectrosc. Rev.* 40 (4), 327–390.

Barthlott S., Schneider M., Hase F., Blumenstock T., Kiel M., Dubravica D. (2016) Tropospheric water vapour isotopologue data ($H_2^{16}O$, $H_2^{18}O$ and $HD^{16}O$) as obtained from NDACC/FTIR solar absorption spectra. *Earth Syst. Sci. Data Discuss.* doi:10.5194/essd-2016-9.

Beil A., Daum R., Matz G., Harig R. (1998) Remote sensing of atmospheric pollution by passive FTIR spectrometry. *Proc. SPIE* 3493, 32–43.

Bezanila A., Kruger A., Stremme W., Grutter M. (2014) Solar absorption infrared spectroscopic measurements over Mexico City: Methane enhancements. *Atmósfera* 27(2), 173–183.

Blum M.-M., John H. (2012) Historical perspective and modern applications of attenuated total reflectance–Fourier Transform Infrared Spectroscopy (ATR-FTIR). *Drug Test. Analysis* 4, 298–302.

Buscher E., Lemke D. (1980) An infrared hygrometer for astronomical site testing. *Infrared Phys.* 20, 321–325.

Colthup N.B., Daly L.H., Wiberley S.E. (1990) *Introduction to Infrared and Raman Spectroscopy*, 3rd ed. Academic Press, New York.

Davis S.P., Abrams M.C., Brault J.W. (2001) *Fourier Transform Spectrometry*, Academic Press, New York.

Fischer H. (1993) Remote sensing of atmospheric trace gases. *Interdiscip. Sci. Res.* 10 (3), 185–191.

Forman M.L., Steel W.H., Vanasse G.A. (1966) Correction of asymmetric interferograms obtained in Fourier spectroscopy. *J. Opt. Soc. Am.* 56, 59–63.

Friedl-Vallon F., Maucher G., Oelhaf H., Seefeldner M. (1995) The new balloon-borne MIPASB2 limb emission sounder. In: *Proceedings of the 1995 International Geoscience and Remote Sensing Symposium.* IEEE Press, Piscataway, NJ, pp. 242–244.

Gisi M., Hase F., Dohe S., Blumenstock T. (2011) Camtracker: a new camera controlled high precision solar tracker system for FTIR-spectrometers. *Atmos. Meas. Tech.* 4, 47–54.

Goldmann A., Saunders R.S. (1979) Analysis of atmospheric infrared spectra for altitude, distribution of atmospheric trace constituents-I: method of analysis. *J. Q. S. R. T.* 21, 155–162.

Griffith D.W.T. (1996) Synthetic calibration and quantitative analysis of gas-phase FT-IR spectra. *Appl. Spectrosc.* 50(1), 59–70.

Griffiths P.R., de Haseth J.A. (1986) *Fourier Transform Infrared Spectrometry.* John Wiley & Sons, Etobicoke, ON.

Griffith D.W.T., Jamie I.M. (2000) Fourier transform infrared spectrometry in atmospheric and trace gas analysis. In: Meyers R.A. (Ed.) *Encyclopedia of Analytical Chemistry*. Wiley, Chichester, UK, pp. 1979–2007.

Griffith D.W.T., Deutscher N.M., Caldow C.G.R., Kettlewell G., Riggenbach M., Hammer S. (2012) A Fourier transform infrared trace gas analyser for atmospheric applications. *Atmos. Meas. Tech.* 5, 2481–2498.

Haaland D.M., Melgaard D.K. (2001) New classical least-squares/partial least-squares hybrid algorithm for spectral analyses. *Appl. Spectrosc.* 55, 1–8.

Hart B.K., Berry R.J., Griffiths P.R. (1999) Effects of resolution, spectral window, and background on multivariate calibrations used for open-path Fourier-transform infrared spectroscopy. *Field Anal. Chem. Tech.* 3, 117–130.

Hart B.K., Griffiths P.R. (2000a) Effect of resolution on quantification in open path Fourier transform infrared spectrometry under conditions of low detector noise 1. Classical least squares regression. *Environ. Sci. Tech.* 34, 1337–1345.

Hart B.K., Berry R.J., Griffiths P.R. (2000b) Effect of resolution on quantification in open-path Fourier transform infrared spectrometry under conditions of low detector noise 2. Partial least squares regression. *Environ. Sci. Tech.* 34, 1346–1351.

Hase F., Hannigan J., Coffey M., Goldman A., Hopfner M., Jones N., Rinsland C., Wood S. (2004) Intercomparison of retrieval codes used for the analysis of high-resolution, ground-based FTIR measurements. *J. Quant. Spectros. Radiat. Transfer* 87, 24–52.

IGOS (1998) Integrated Global Observing Strategy. European Space Agency, Committee on Earth Observation Satellites. CEOS Yearbook, http://www.igospartners.org/.

Ingling L., Isenhour T.L. (1999) Spectral matching quantitative open-path Fourier-transform infrared spectroscopy. *Field Anal. Chem. Tech.* 3, 37–43.

Iqbal M. (1983) *An Introduction to Solar Radiation*. Academic Press, New York.

Johnson D.G., Jucks K.W., Traub W.A., Chance K.V. (1995) Smithsonian stratospheric farinfrared spectrometer and data reduction system. *J. Geophys. Res.* 100, 3091–3106.

Johnson D.G., Traub W.A., Jucks K.W. (1996) Phase determination from mostly one-sided interferograms. *J. Geophys. Res. Atmos.* 101 (4D), 9031–9043.

Kampfer N.K. (2013) *Monitoring Atmosphere Water Vapors: Ground-based Remote Sensing and In-situ Methods*, Vol. 10. Springer, New York.

Kauppinen J.K. (1999) The lowest and the highest resolution in FT-IR spectroscopy. In: *Proceedings of 12th International Conference on Fourier Transform Spectroscopy*, Tokyo, Japan, pp. 47–49, August 22–27, 1999.

Keppel-Aleks G., Toon G.C., Wennberg P.O., Deutscher N.M. (2007) Reducing the impact of source brightness fluctuations on spectra obtained by Fourier-transform spectrometry. *Appl. Opt.* 46, 4774–4779.

Kira O., Linker R., Dubowski Y. (2016) Detection and quantification of water-based aerosols using active open-path FTIR. Sci. Reports 6, 25110.

Kley D., Russell III J.M., Phillips C. (Eds.) (2000) SPARC Assessment of Upper Tropospheric and Stratospheric Water Vapour, World Climate Research Programme, SPARC Report No.2.

Marshall T.L., Chaffin C.T., Hammaker R.M., Fateley W.G. (1994) An introduction to open-path FT-IR atmosphere monitoring. *Environ. Sci. Technol.* 28 (5), 224–232.

McKelvy M.L., Britt T.R., Davis B.L., Gillie J.K., Graves F.B., Lentz L.A. (1998) Infrared spectroscopy. *Anal. Chem.* 70, 119R–177R.

Meier A., Toon G.C., Rinsland C.P., Goldman A., Hase F. (2004) Spectroscopic atlas of atmospheric microwindows in the middle infra-red. In: IRF Technical Report No.48, ISSN 0248-1738, Kiruna, Sweden.

Mertz L. (1965) *Transformations in Optics*. John Wiley & Sons, New York.

Mertz L. (1967) Auxiliary computation for Fourier spectroscopy. *Infrared Phys.* 7, 17–23.

Mieruch S., Noël S., Bovensmann H., Burrows J.P. (2008) Analysis of global water vapour trends from satellite measurements in the visible spectral range. *Atmos. Chem. Phys.* 8, 491–504.

Miloshevich L.M., Vömel H., Whilteman D.N., Leblanc T. (2009) Accuracy assessment and correction of Vaisala RS92 radiosonde water vapor measurements. *J. Geophys. Res.* 114(D11305).

Murcray D., Goldman A., Kosters J. et al. (1990) Intercomparison of stratospheric water vapor profiles obtained during the balloon intercomparison campaign. *J. Atmos. Chem.* 10, 159–179.

Noël S., Buchwitz M., Bovensmann H., Hoogen R., Burrows J.P. (1999) Atmospheric water vapor amounts retrieved from GOME satellite data. *Geophys. Res. Lett.* 26(13), 1841–1844.

Noël S., Buchwitz M., Bovensmann H., Burrows J.P. (2002) Retrieval of total water vapour column amounts from GOME/ERS-2 data. *Adv. Space Res.* 29 (11), 1697–1702.

Noël S., Buchwitz M., Bovensmann H., Burrows J.P. (2005) Validation of SCIAMACHY AMC–DOAS water vapour columns. *Atmos. Chem. Phys.* 5, 1835–1841.

Notholt J., Lehmann R. (2003) The moon as light source for atmospheric trace gas observations: Measurement technique and analysis method. *J. Quant. Spectrosc. Rad. Transfer* 76, 435–445.

Notholt J., Neuber R., Schrems O., Clarmann T.V. (1993) Stratospheric trace gas concentrations in the Arctic polar night derived by FTIR-spectroscopy with the moon as IR light source. *Geophys. Res. Lett.* 20, 2059–2062.

O'Brien D.M., Rayner P.J. (2002) Global observations of the carbon budget, 2. CO_2 column from differential absorption of reflected sunlight in the 1.61 μm band of CO_2. *J. Geophys. Res.* 107 (D18), 4354.

Oelhaf H., von Clarmann T., Fischer H. et al. (1996) Remote sensing of the Arctic stratosphere with the new balloon-borne MIPAS-B2 instrument. In: *Proceedings of Polar Stratospheric Ozone, Proceedings of the 3rd European Workshop*, European Communications, Brussels, Luxembourg, pp. 270–275, September 18–22, 1995.

Palm M., Melsheimer C., Noel S., Heise S., Notholt J., Burrows J., Schrems O. (2010) Integrated water vapor above Ny Alesund, Spitsbergen: a multi-sensor intercomparison. *Atmos. Chem. Phys.* 10, 1215–1226.

Paton-Walsh C. (2011) Remote sensing of atmospheric trace gases by ground-based solar Fourier transform infrared spectroscopy. In: Nikolic G. (Ed.) *Fourier Transforms: New Analytical Approaches and FTIR Strategies*, InTech, Rijeka, pp. 459–478.

Remedios J.J. (1990) Spectroscopy for remote sounding of the atmosphere. PhD Thesis, University of Oxford.

Ritter C., Notholt J., Fischer J., Rathke C. (2005) Direct thermal radiative forcing of tropospheric aerosol in the Arctic measured by ground based infrared spectrometry. *Geophys. Res. Lett.* 32, L23816.

Rodgers C.D. (2000) *Inverse Methods for Atmospheric Sounding: Theory and Practice*, 1st ed. World Scientific Publishing, Hackensack, NJ.

Rothman L.S., Rinsland C.P., Goldman A. et al. (1998) The HITRAN molecular spectroscopic database and HAWKS (HITRAN Atmospheric Workstation): 1996 edition. *J. Quant. Spectrosc. Radiat. Transfer* 60, 665–710.

Rothman L., Barbe A., Benner D.C. et al. (2003) The HITRAN molecular spectroscopic database: Edition of 2000 including updates through 2001. *J. Quant. Spectrosc. Rad. Transfer* 82, 5–44.

Rothman L.S., Jacquemart D., Barbe A. et al. (2005) The HITRAN 2004 molecular spectroscopic database. *J. Quant. Spectrosc. Radiat. Transfer* 96, 139–204.

Rothman L., Gordon I., Barbe A. (2009) The HITRAN 2008 molecular spectroscopic database. *J. Quant. Spectrosc. Rad. Transfer* 110(9), 533–572.

Rothman L.S., Gordon I.E., Babikov Y. et al. (2013) The HITRAN 2012 molecular spectroscopic database. *J. Quant. Spectrosc. Rad. Transfer* 130, 4–50.

Russwurm G.M., Childers J.W. (2002) Open-path Fourier transform infrared spectroscopy. In: Chalmers J.M., Griffiths P.R. (Eds.) *Handbook of Vibrational Spectroscopy*, Vol. 2. Wiley, New York, pp. 1750–1773.

Santos C., Fraga M.E., Kozakiewicz Z., Lima N. (2010) Fourier transform infrared as a powerful technique for the identification and characterization of filamentous fungi and yeast. *Res. Microbiol* 161, 168–175.

Schneider M., Hase F., Blumenstock T. (2006) Water vapour profiles by ground-based FTIR spectroscopy: Study for an optimized retrieval and its validation. *Atmos. Chem. Phys.* 6, 811–830.

Schneider, M., Romero, P.M., Hase, F., Blumenstock, T., Cuevas, E., Ramos, R. (2009) Quality assessment of Izana'supper-air water vapour measurement techniques: FTIR, Cimel, MFRSR, GPS, and Vaisala RS92. *Atmos. Meas. Tech. Discuss.* 2, 1625–1662.

Schneider M., Toon G.C., Blavier J.-F., Hase F., Leblanc T. (2010a) H_2O and D profiles remotely-sensed from ground in different spectral infrared regions. *Atmos. Meas. Tech.* 3, 1599–1613.

Schneider M., Romero P.M., Hase F., Blumenstock T., Cuevas E., Ramos R. (2010b) Continuous quality assessment of atmospheric water vapour measurement techniques: FTIR, Cimel, MFRSR, GPS, and Vaisala RS92. *Atmos. Meas. Tech.* 3, 323–338.

Simonescu C.M. (2012) Application of FTIR Spectroscopy in environmental studies. In: Farrukh M.A. (Ed.), *Advanced Aspects of Spectroscopy.* InTech, Rijeka, Croatia, pp. 49–84.

Smith B.C. (2011) *Fundamentals of Fourier Transform Infrared Spectroscopy*, 2nd ed. CRC Press, Boca Raton, FL.

Solomon S., Qin D., Manning M. et al. (Eds.) (2007) Climate change 2007: The physical science basis. Contribution of Working Group I to the Fourth Assessment Report of the Intergovernmental Panel on Climate Change. Cambridge University Press, Cambridge, UK and New York.

Soucy M.-A., Châteauneuf F., Giroux J., Roy C. (2006) Looking at the air from space: FTIR atmospheric sounding applications for remote sensing satellites. In: Special Report Instrumentation & Analytics, ABB Review, pp. 50–53.

Spellicy R.L., Webb J.D. (2002) Atmospheric monitoring using extractive techniques. In: Chalmers J.M., Griffiths P.R. (Eds.), *Handbook of Vibrational Spectroscopy*, Vol. 2. Wiley, Chichester, UK.

Spencer R.W., Braswell W.D. (1997) How dry is the tropical free troposphere? Implications for global warming theory. *Bull. Am. Meteorol. Soc.* 78, 1097–1106.

Stallard B.R., Espinoza L.H., Rowe R.K., Garcia M.J., Niemczyk T.M. (1995) Trace water vapor detection in nitrogen and corrosive gases by FTIR spectroscopy. *J. Electrochem. Soc.* 142(8), 2778–2782.

Thomas, E.V., Haaland, D.M. (1990) Comparison of multivariate calibration methods for quantitative spectral analysis. *Anal. Chem.* 62, 1091–1099.

Toon G.C. (1991) The JPL MkIV Interferometer. *Opt. Photon. News* 2, 19–21.

VDI (2000) Atmospheric measurements near ground with FTIR spectroscopy, measurement of gaseous emissions and immissions fundamentals. VDI-Guideline 4211. In VDI/DIN Handbuch Reinhaltung der Luft; Part 1. Remote Sensing; VDI: Berlin, Germany, 2000.

Visser T. (2000) Infrared spectroscopy in environmental analysis. In: Meyers R.A. (Ed.) *Encyclopedia of Analytical Chemistry*, John Wiley & Sons, Chichester, UK, pp. 1–21.

Vömel H., Selkirk H., Miloshevich L. et al. (2007) Radiation dry bias of the Vaisala RS92 humidity sensor. *J. Atmos. Oceanic Technol.* 24, 953–963.

Von Clarmann T., Linden A., Oelhaf H. et al. (1995) Determination of the stratospheric organic chlorine budget in the spring arctic vortex from MIPAS-B limb emission spectra and air sampling experiments. *J. Geophys. Res.* 100, 13979–13997.

US EPA (1999a) Long-path open path Fourier transform infrared monitoring of atmospheric gases.

Compendium of Methods for the Determination of Toxic Organic Compounds in Ambient Air, Compendium Method TO-16, 2nd ed. US EPA: Cincinnati, OH, 1999EPA/625/R-96/010b.

US EPA (1999b) Measurement of vapor phase organic and inorganic emissions by extractive Fourier transform infrared (FTIR) spectroscopy. *Fed. Regist.* 1999, 64, 31937–31962, US EPA Test Method 320.

USGCRP (2008) Scientific assessment of the effects of global change on the United States. A Report of the Committee on Environment and Natural Resources National Science and Technology Council, 271 pp.

Wetzel G., Oelhaf H., von Clarmann T. et al. (1997) Vertical profiles of N_2O_5, HO_2NO_2, and NO_2 inside the Arctic vortex, retrieved from nocturnal MIPAS-B2 limb emission measurements in February 1995. *J. Geophys. Res.* 102, 19,177–19,186.

Wiacek A. (2006) First trace gas measurements using Fourier transform infrared solar absorption spectroscopy at the University of Toronto atmospheric observatory. PhD Thesis University of Toronto.

Wood S.W., Batchelor R.L., Goldman A., Rinsland C.P., Connor B.J., Murcray F.J., Stephen T.M., Heuff D.N. (2004) Groundbased nitric acid measurements at Arrival Heights, Antarctica, using solar and lunar Fourier transform infrared observations. *J. Geophys. Res.* 109(D18307), 2004.

Workman Jr. J.J. (1999) Review of process and non-invasive near-infrared and infrared spectroscopy: 1993–1999. *Appl. Spectrosc. Rev.* 34 (1–2), 1–89.

5 Atmosphere Monitoring Using Methods of Absorption of Electromagnetic Radiation—Microwave Absorption

5.1 MICROWAVE RADIOMETRY

It is known that microwaves with wavelengths ranging from 1 m to 1 mm or with frequencies between 300 MHz (100 cm) and 300 GHz (0.1 cm) are absorbed by the water vapor in the material. But in contrast to the radiation from infrared (IR) spectral range, where absorption is limited by the surface, microwaves can penetrate into deeper thicknesses of material. This means that the tested materials can be transparent to microwaves and thus the intensity of transmitted radiation will carry information about the amount of moisture in the sample investigated. Therefore, currently, this method is widely used in determining the moisture content in various materials (Kraszewski 1973).

Two simple microwave moisture meters, utilizing the electromagnetic wave absorption in wet materials are shown in Figure 5.1. A microwave generator is mounted on one side of the material. On the opposite side, a receiver measures the change in microwave amplitude and the extent of the phase shift. The amplitude loss divided by the phase shift is proportional to the moisture in the material. Microwave and IR absorption techniques are equally accurate, but since IR instruments are generally less costly, microwaves are typically used in deep beds or dark materials where IR techniques have limitations. As high-power microwave sources could be used specialized vacuum tubes. These devices operate on different principles from low frequency vacuum tubes, using the ballistic motion of electrons in a vacuum under the influence of controlling electric or magnetic fields, and include the magnetron (used in a microwave ovens), klystron, traveling-wave tube (TWT), and gyrotron. These devices work in the density-modulated mode, rather than the current modulated mode. This means that they work on the basis of clumps of electrons flying ballistically through them, rather than using a continuous stream of electrons. Solid-state devices such as the field-effect transistor (at least at lower frequencies), tunnel diodes, Gunn diodes, and IMPact ionization Avalanche Transit-Time (IMPATT) diodes can be used as low-power microwave sources. Low-power sources are available as benchtop instruments, rackmount instruments, embeddable modules, and in card-level formats. A maser is a solid-state device, which amplifies microwaves using similar principles to the laser, which amplifies higher frequency light waves.

Various types of microwave antennas (Tsui 2008), tuned to the appropriate wavelength and coupled with the sensors of heat or microwave radiation such as bolometers, thermocouples, microwave diodes (Schottky diodes, tunnel diodes, etc.), Pyroelectric InfraRed (PIR) sensors, and superconductivity-based sensors can be used as receivers of microwave radiation. In most cases, microwave receivers are built on the superheterodyne scheme, because as usual, this scheme provides the highest sensitivity, and it is easier to implement it practically than a direct amplification circuit.

Detector microwave receivers are being applied mainly in the decimetric band and are based on cryogenically cooled bolometers and semiconductor detectors. In the centimeter and millimeter ranges (up to a frequency $f = 230$ GHz) in most cases uncooled detectors are used. Shorter microwave receivers, which are often cooled, are used only in research.

However, studies have shown that microwave techniques can be successfully used not only to control the moisture in solids but also for studies of atmospheric water vapor. Already, weather satellites using IR sensors have widened our knowledge of the mechanisms of cloud formations, and through the use of IR spectrometers the vertical profiles of temperature and water vapor above the clouds have been probed. Thus, a great volume of important information was gathered. However, this information does not provide the necessary data concerning the lower troposphere, which are necessary for numerical weather prediction. The cloud-top limit effectively shields the majority of the lower atmosphere from IR sensors. Even thin clouds are opaque to IR radiation and restrict the region of the troposphere, which is to be

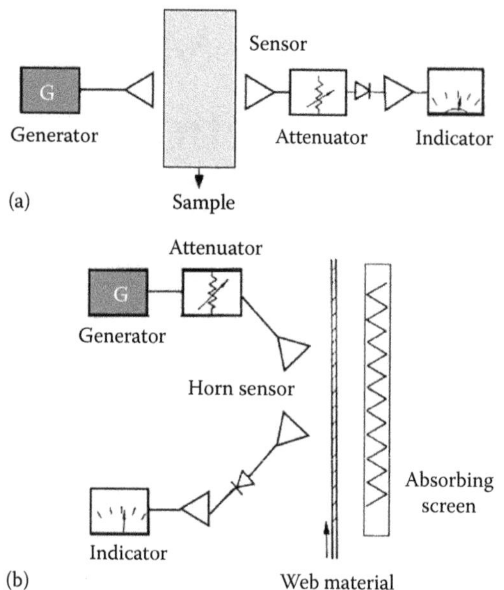

FIGURE 5.1 Two simple microwave moisture meters: (a) transmission measurement and (b) reflection measurement. (Idea from Kraszewski, A., *J. Microw Power*, 8, 323–335, 1973.)

explored by IR sensors. Analysis of hemispheric pictures of the Earth taken from space for percentage of cloud cover has revealed how much of a handicap this is.

Experiment and theoretical simulations have shown that the use of microwave region, which extends arbitrarily from millimeter to meter wavelengths, gives possibility to circumvent this limitation. Just in this range (22.235 and 183.31 GHz), it is being observed atmospheric absorption arising from pure rotational spectral lines of the water vapor (Van Vleck 1947; Payne et al. 2011). Research has shown that in the microwave spectral region, the principal sources of thermal radiation are atmospheric oxygen, the water vapor, and the liquid water within clouds. In the range between 20 and 200 GHz, emission is dominated by the oxygen complex from 50 to 70 GHz, the isolated oxygen line at 118.75 GHz, the water vapor lines indicated before at 22.235 and 183.31 GHz, and the so-called water vapor continuum arising from higher frequency lines contribution. Hydrometeors, forming clouds, contribute with the emission, absorption, and scattering, although for lower frequencies and for nonprecipitating clouds, the scattering effects can be considered negligible (Janssen 1993a).

The theoretical substantiation of possibilities of microwave radiometry for sounding of the atmosphere is represented in Gunn and East (1954), Staelin et al. (1973), and Karmakar (2012, 2014). It was established that the microwave region of the spectrum has a zone of frequencies within which clouds at the lower frequencies are essentially transparent, and at the higher frequencies opaque. Below approximately 30 GHz (1 cm) the clouds rapidly lose the ability to absorb and scatter radiation; at higher frequencies they rapidly become opaque. Quite fortuitously, the water-vapor resonance of lowest frequency is centered near 1.35 cm, and therefore is little affected by ordinary cloud cover. A total atmospheric attenuation of this band does not exceed 1.5 dB at resonance. Thus, there are important advantages in passively probing the atmosphere in this part of the electromagnetic spectrum rather than in the IR: (1) clouds are not opaque at these frequencies, thereby making it possible to study the cloud itself and the region below the cloud; and (2) instrumentation exists whose bandwidth is much smaller than the widths of spectral lines arising from several of the most important atmospheric gases. This last fact allows detailed analysis of line shape, which in turn facilitates the inference of the atmospheric conditions in which the lines arise. At the same time, the 183.3 GHz resonance line, which is a very strong spectral line, is subjected to strong attenuation in the atmosphere. In the moist tropical regions, the peak one-way attenuation through the atmosphere reaches more than 200 dB. In dry Arctic regions the peak attenuation falls below 20 dB, still, however, it is optically thick. This means that the measurements on two frequencies 22 and 183 GHz will contain information about the clouds. It is important to note that the resonant absorption bands related to the water vapor do not coincide with the absorption bands of oxygen (Figure 5.2), which simplifies the monitoring of the atmosphere. As it is seen in Figure 5.2, oxygen has an absorption band around 60 GHz. Although other atmospheric molecules have spectral lines in this frequency region, their expected strength is too small to affect propagation significantly (Meeks 1976; Raghavan 2003).

Also, due to the properties of the water vapor absorption bands in the microwave portion of the spectrum, it is possible to retrieve the profiles of humidity, rather than the column integrated humidity quantities that are derived from IR radiances (e.g., Blankenship et al. 2000). It is known that the intensity of radiation emitted at any altitude is proportional to the concentration of gases and hydrometeors, and to the local temperature. Thus, the principle of radiometric retrieval of the temperature and humidity profiles is based on the measurement of radiation generated at different atmospheric levels. This can be accomplished in part by measuring the emitted spectrum at frequencies conveniently distributed along the wing of an absorption line/complex, which correspond to different absorption and penetration depth (Rosenkranz 1998). For instance, temperature profiles can be estimated from

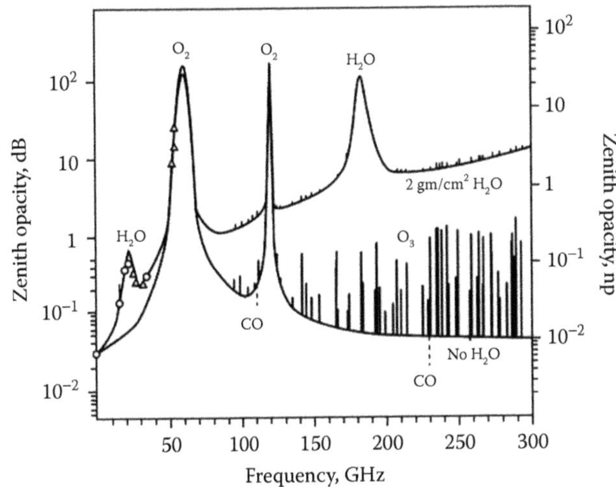

FIGURE 5.2 Atmospheric zenith opacity for H_2O, O_2, O_3, and CO. (Data extracted from Meeks, M.L., *Methods of Experimental Physics: Astrophysics*, Academic Press, New York, 1976, from Raghavan, S., *Radar Meteorology*, Kluwer Academic Publishers, Dordrecht, the Netherlands, 2003, and from Karmakar, P.K. *Microwave Propagation and Remote Sensing: Atmosphere Influence with Models and Applications*, CRC Press, Boca Raton, FL, 2012.)

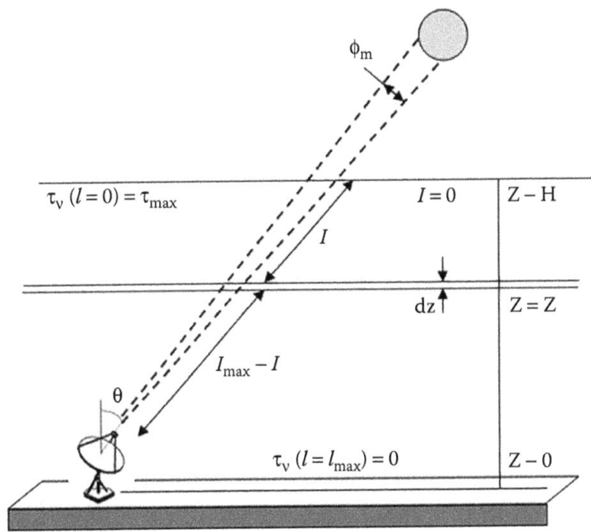

FIGURE 5.3 Schematic illustration of the passive microwave technique application for the study of atmospheric water vapor using solar radiation.

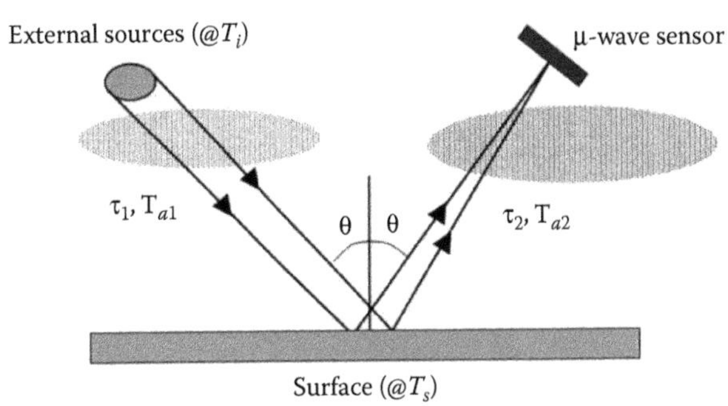

FIGURE 5.4 Combined influence of active (external) sources and atmosphere: T_i = brightness temperature from external sources. τ_1 = transmittance of the whole atmosphere in direction θ. τ_2 = transmittance of atmosphere between surface and sensor. T_{a1} = mean temperature of the whole atmosphere. T_{a2} = mean temperature of atmosphere between surface and sensor.

spectral measurements in the 50–60 GHz band, whereas the measurements around 22 GHz yield information on the water vapor profile (Cimini et al. 2006).

This is an important advantage of microwave radiometry, because the use of this approach for measuring atmospheric vertical profiles of temperature and humidity eliminates the need for probes. The probes are expensive devices and moreover they give a finite space/time resolution (routinely once or twice per day). So, they are inadequate for a detailed study of the diurnal evolution of the near surface atmosphere. At the same time, the study of the atmosphere using microwave radiometry can be carried out round-the-clock. Moreover, these studies can be organized using both satellites and aircrafts with generators or receivers of mircowave radiation, as well as passive microwave techniques, when the light source is the Sun (Figures 5.3 and 5.4), which significantly simplifies and reduces the cost of the research process. Advantages and disadvantages of such approach to the study of the atmosphere are shown in Table 5.1. Moreover, microwave radiometers can be operated in long-term unattended mode in nearly all weather conditions, with temporal resolution of the order of seconds. These features make a microwave radiometry very appealing for planetary boundary layer research (Ruffieux et al. 2006), where atmospheric processes can develop in a time scale of the order of minutes. Thus, a microwave technology can significantly enhance the ability to study atmospheric phenomena. Research has shown that microwave technique can also be used for analyzing the reflected radiation from the Earth's surface using satellites and aircrafts (Guan et al. 2011). It was found that the precipitation and the rain rates can also be estimated using microwave radiation (Chwala et al. 2012, 2014). However, accurate retrieval of humidity from the satellite microwave sensor requires accurate knowledge of the surface emissivity at the wavelength(s) to which the sensor is sensitive. Unfortunately, this information is only available for the oceans and therefore definitions

TABLE 5.1
Features of the Passive Microwave Remote Sensing from Space

Advantages	Disadvantages
• Penetration through nonprecipitating clouds • Radiance is linearly related to temperature (i.e., the retrieval is nearly linear) • Highly stable instrument calibration • Global coverage and wide swath	• Larger field of views (10–50 km) compared to VIS/IR sensors • Variable emissivity over land • Polar orbiting satellites provide discontinuous temporal coverage at low latitudes (need to create weekly composites)

for surface temperature, atmospheric water vapor, cloud liquid water, and rainfall rate when using Passive Microwave Remote Sensing from Space are possible only over the oceans. Other examples of using a microwave radiation for the study of the atmosphere and mathematical tools used for processing the information received one can find in Janssen (1993b), Westwater et al. (2003), Matzler (2006), Noam et al. (2011), David et al. (2012), Karmakar (2012, 2014), Kampfer (2013), Massaro et al. (2015), Roy et al. (2016), and Grankov and Milshin (2016). Unfortunately, a detailed discussion of these issues is beyond the scope of this book.

It should be noted that the use of microwave radiometry for studying the atmosphere increased sharply only after the appearance of commercial microwave radiometers, among which the most widely used are microwave radiometers, made by Radiometer Physics GmbH (Table 5.2). The appearance of these hygrometers is shown in Figure 5.5a. The microwave radiometers developed by Argonne National Laboratory (Figure 5.5b) (Candlish et al. 2012), are also widely used. Other radiometers were designed for large international projects related to the study of the atmosphere.

TABLE 5.2
Microwave Radiometers Which Have Been Used for a Humidity Measurement in the Atmosphere

Microwave Radiometers	Producer
RPG–HATPRO, RPG–HUMPRO, RPG–LHUMPRO (Humidity and temperature profilers)	Radiometer Physics GmbH
MWRP, MWR3C (ARM microwave radiometers)	Argonne National Laboratory
RA-2 (Microwave Radiometer, MWR)	Envisat (European Space Agency)
AMSR-E (Advanced Microwave Scanning Radiometer)	Japan Aerospace Exploration Agency
Ground-based water vapor millimeter-wave spectrometer (WVMS)	The Naval Research Laboratory (NRL)
The MPAE ground-based microwave spectrometer (WASPAM)	Max Planck Institut für Aeronomie (MPAE)

(a)

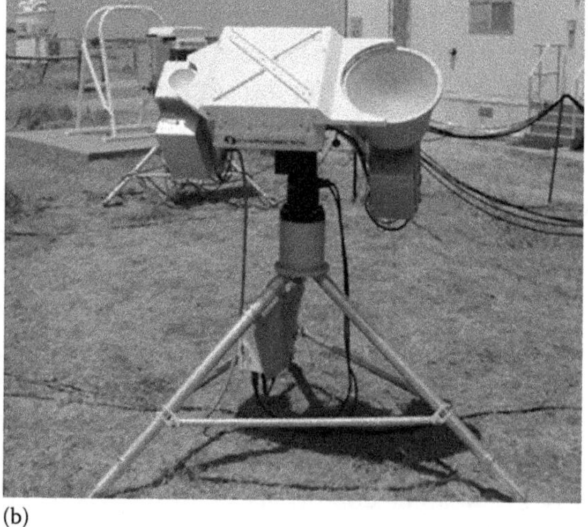
(b)

FIGURE 5.5 (a) The humidity and temperature profilers RPG–HATPRO (Radiometer Physics GmbH) (From http://www.radiometer-physics.de/); (b) 3 Channel Microwave Radiometer MWR3C. (From the U.S. Department of Energy, http://www.arm.gov/.)

5.2 EXAMPLES OF MICROWAVE HYGROMETERS REALIZATION

Usually ground-based microwave spectrometers use passive microwave techniques and carry out measurements at 22.2 GHz, which corresponds to the emission frequency of the 6_{16}–5_{23} rotational line of H_2O. With regard to these differences in hygrometers, they are mostly associated with the use of either a variety of detectors, or various software (Kley et al. 2000). The ground-based Water Vapor Millimeter-wave Spectrometer (WVMS), designed by the Naval Research Laboratory (NRL), was based on a High Electron Mobility Transistors amplifier, which was refrigerated to 20 K and provided 30 dB gained more than a 500 MHz bandwidth. Radiation entered the radiometer through a scalar feedhorn that had a full width at half-maximum (FWHM) beam size of ≈8°, after being reflected from an elliptical aluminium plate. A stepper motor rotated the aluminium plate that was inclined at a 45° angle to the axis of the motor and the horn, thus allowing variation of the measurement elevation angle. The WVMS2 and WVMS3 instrument spectrometers consisted of a filter bank with thirty 14 MHz wide filters, thirty 2 MHz filters, twenty 200 kHz filters, and ten 50 kHz filters.

An estimate of the optical depth of the troposphere (τ_{trop}) was obtained by tipping the instrument through 11 angles from 45° to 75° from zenith, using a high density of angles near 75° where the rate of the change of the air mass (μ) as a function of angle is the highest. To solve for the tropospheric optical depth NRL used the measured system temperature from the widest filters (T_{sys}), estimated the receiver temperature and the atmospheric temperature (T_{rx} and T_{atm}), and then solved for the tropospheric optical depth using the best fit to the equation,

$$T_{sys} = T_{rx} + T_{atm}\left(1 - \exp\left(-\mu \cdot \tau_{trop}\right)\right) \tag{5.1}$$

Details of the measurement and retrieval technique are discussed in Rodgers (1976) and Nedoluha et al. (1995, 1996).

The WVMS instruments have been in operation, providing nearly continuous measurements of water vapor from ≈40 to 80 km (Kley et al. 2000). An integration time of ≈one week was required in order to achieve adequate signal-to-noise for retrievals at 80 km, while in the stratosphere daily retrievals were possible. For weekly retrievals the random error was estimated to be 4%–7%, with the largest error at the highest altitudes. The altitude resolution was ≈10 km, and systematic uncertainties were ≈5%–10%. The systematic uncertainties were arised primarily from errors in the calibration and pointing.

The ground-based microwave spectrometer of the Max Planck Institut für Aeronomie (MPAE), the WASPAM (Wasserdampf und Spurengasmessungen mit Mikrowellen) (Hartogh and Jarchow 1995; Kley et al. 2000) in addition to 22.235 GHz frequency observed also the water vapor continuum at 31.5 GHz. A rotating mirror reflected the microwave radiation of the atmosphere, a cold load and a hot load alternately to a polarization grid. While one polarization passed the grid into the horn antenna of the 31.5 GHz receiver, the other polarization was reflected via an elliptic mirror to the horn antenna of the 22 GHz receiver. The signal level was amplified by about 30 dB using three High Electron Mobility Transistors. The amplifier covered the frequency range of 20–24 GHz and was cooled to 20 K by a closed Helium loop. A steep low-pass filter at the output of the amplifier is cut off all frequencies above 22.5 GHz. The single sideband (SSB) receiver temperature was about 100 K. A Chirp Transform Spectrometer (CTS) with 2048 channels of 21 kHz resolution (i.e., 43 MHz bandwidth) performs the spectral analysis.

The atmospheric radiation was received with a fixed elevation angle of 18° and a beam width of 7° FWHM. The tropospheric transmission was derived from the measured brightness temperature in the line wings of the water vapor line, assuming that the troposphere is a single layer with a constant temperature. It can be shown that the error of this approach is negligible for low brightness temperatures of the atmosphere and small line amplitudes (Jarchow 1998). The retrieval techniques, which have been applied to determine altitude profiles of water vapor from the measured spectra, were described in Backus and Gilbert (1970) and Rodgers (1976). While the approach described by Rodgers (1976) gives the best results when retrieving single spectra, for instance for a weekly or monthly mean, the approach described in Backus and Gilbert (1970) has advantages for retrieving continuous data sets, especially when there are large variabilities in the tropospheric transmission (Jarchow and Hartogh 1995).

The altitude coverage of the data observed by WASPAM was 35–85 km (Kley et al. 2000). For weekly retrievals in the altitude range of 40–80 km the random error is estimated to be 0.15 ppmv. The altitude resolution was about 10 km. The systematic errors ranged between 0.1 ppmv at 80 km and 0.5 ppmv at 40 km. The main source of systematic errors, the so-called baseline, occurred due to reflections in the receiver. The contribution of pointing errors to the total systematic error was negligible for the system. A detailed error analysis was described in Jarchow (1998).

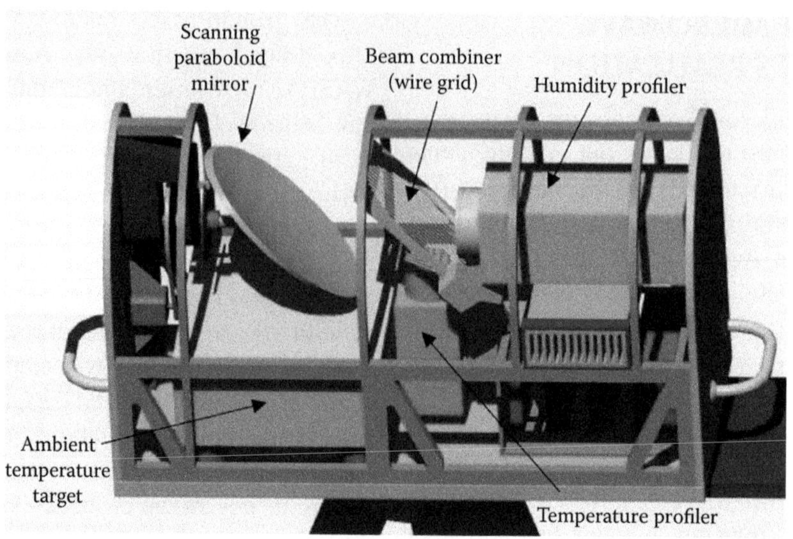

FIGURE 5.6 Internal structure of the RPG–HATPRO/RPG–LHATPRO radiometers designed by Radiometer Physics GmbH. (From ftp://ftp.radiometer-physics.de.)

The same frequency range of 22.24–31.4 GHz was used by Radiometer Physics GmbH when elaborating the RPG–HATPRO и RPG–HUMPRO Humidity and Temperature Profilers. A schematic drawing of the inner components of these radiometers is shown in Figure 5.6. The following functional blocks can be identified in this figure:

- Receiver optics comprising a corrugated feed-horn (encapsulated in thermal insulation) for each frequency band and off-axis paraboloid (scanning mirror).
- Two receiver units (22.24–31.4 GHz and 51.3–59 GHz) 51.3–59 GHz range were used for temperature profiling. Each channel has its own detector diode.
- The ambient load as part of the calibration system.
- The internal scanning mechanism.
- The instrument electronics sections.
- Data acquisition system.

One should note that discussed radiometers are based on the direct detection technique without using mixers and local oscillators for signal down conversion. Instead the input signal is directly amplified, filtered, and detected. The receivers are integrated with their feedhorns and are thermally insulated to achieve a high thermal stability. The zero bias highly doped GaAs Schottky detector diodes used in hygrometers can handle frequencies up to 110 GHz with a virtually flat detection sensitivity from 10 to 35 GHz. In addition, the detector diode offers superior thermal stability when compared to silicon zero bias Schottky diodes.

According to Operating Manual, RPG–HATPRO provides the construction of the tropospheric humidity profiles with the following vertical resolution:

- 200 m (range 0–2000 m)
- 400 m (range 2000–5000 m)
- 800 m (range 5000–10,000 m)

Accuracy of measurements is ~0.4 g/m^3 RMS (absolute humidity) or ~5% RMS (relative humidity).

As for the microwave hygrometers, operating at frequency of 183.31 GHz, they were developed for aircraft and satellite applications (Kley et al. 2000). The transition at 183.310 GHz is much better suited for these applications because its line strength is stronger by two orders of magnitude than the 22.235 transition. The brightness temperature for the 183 GHz line observed from aircraft is approximately 100 K in contrast to that measured from the ground at 23 GHz that is only a few tenths of a degree. Therefore, the integration time of a microwave receiver will be much shorter for the stronger line. This offers the opportunity to deduce latitudinal variations of water vapor if observed from a moving platform.

The microwave radiometer designed at the University of Bern, Switzerland is one of the examples of such devices (Peter et al. 1988; Peter 1998). The University of Bern system consists of a heterodyne receiver with an uncooled Schottky diode mixer that converts the high frequency signals to an intermediate frequency of 3.7 GHz. The radiometer has a single sideband system temperature of 4000 K. A Martin–Puplett interferometer acts as a filter to suppress the unwanted sideband. This sideband filter can be tuned to measure alternately in either of the two sidebands around 183.3 and 175.9 GHz that enables measurement of radiation in the very far wing of the water vapor line. Side band switching is controlled by the onboard computer. A rotating mirror switches the instrument's field of view approximately every 1.5 s between the atmosphere at an elevation angle of 15° and two microwave absorbers at temperatures of 77 K and 312 K that serve as calibration loads. The intermediate frequency signal is analyzed simultaneously with two acousto-optical spectrometers (AOS). One AOS has a large bandwidth of 1 GHz and a resolution of 1.2 MHz. It is well suited for the observation of spectral lines from the stratosphere. The second AOS has a total bandwidth of 50 MHz at a high resolution of 50 kHz. It is used for the analysis of narrow spectral lines emitted from the mesosphere. Furthermore, the instrument has its own independent global positioning system (GPS) receiver to record position and altitude for each measurement.

Water vapor profiles are retrieved between 15 and 75 km with an altitude resolution of approximately 8 km in the lower stratosphere and approximately 15 km in the upper mesosphere. The root sum square of all error contributions is below 0.6 ppmv.

Another example of a microwave hydrometers operating at frequency of 183.31 GHz is a LHUMPRO radiometer designed by Radiometer Physics GmbH. This device was designed for ultra-low humidity sites (IWV < 1.0 kg/m^2) such as high altitudes or arttic/Antarctic areas. LHUMPRO has a layout similar to the layout of the RPG-HATPRO/RPG-LHATPRO radiometers showed in Figure 5.5. The 183 GHz water vapor receiver is a DSB (double side-band) heterodyne radiometer with LO (local oscillator) tuned to the line center at 183.31 GHz. The mixer's lower/upper sideband response has been characterized by using a Rhode and Schwarz network analyzer plus frequency extension for the 170–220 GHz range.

REFERENCES

Backus G.E., Gilbert J.F. (1970) Uniqueness in the inversion of inaccurate gross Earth data. *Philos. Trans. R. Soc. London, Ser. A* 266, 123–192.

Blankenship C.B., Al-Khalaf A., Wilheit T.T. (2000) Retrieval of water vapor profiles using SSM/T-2 and SSM/I data. *J. Atmos. Sci.* 57, 939–955.

Candlish L.M., Raddatz R.L., Aaplin M.G., Barber D.G. (2012) Atmospheric temperature and absolute humidity profiles over the Beaufort Sea and Amundsen Gulf from a microwave radiometer. *J. Atmos. Oceanic Technol.* 29, 1182–1201.

Chwala C., Gmeiner A., Qiu W., Hipp S., Nienaber D., Siart U., Eibert T., Pohl M., Seltmann J., Fritz J., Kunstmann H. (2012) Precipitation observation using microwave backhaul links in the alpine and pre-alpine region of Southern Germany. *Hydrol. Earth Syst. Sci.* 16, 2647–2661.

Chwala C., Kunstmann H., Hipp S., Siart U. (2014) A mono-static microwave transmission experiment for line integrated precipitation and humidity remote sensing. *Atmosph. Res.* 144, 57–72.

Cimini D., Hewison T.J., Martin L., Guldner J., Gaffard C., Marzano F.S. (2006) Temperature and humidity profile retrievals from ground-based microwave radiometers during TUC. *Meteorolog. Z.* 15 (5), 45–56.

David N., Alpert P., Messer H. (2012) Novel method for fog monitoring using cellular networks infrastructures. *Atmos. Meas. Tech. Discuss.* 5, 5725–5752.

Grankov A.G., Milshin A.A. (2016) *Microwave Radiation of the Ocean-Atmosphere*, 2nd ed. Springer, New York.

Guan L., Zou X., Weng F., Li G. (2011) Assessments of FY-3A microwave humidity sounder measurements using NOAA-18 microwave humidity sounder. *J. Geophys. Res.* 116, D10106.

Gunn K.L.S., East T.W.R. (1954) The microwave properties of precipitation particles. *Quart. J. Roy. Meteor. Soc.* 80, 522–545.

Hartogh P., Jarchow C. (1995) Ground based detection of middle atmospheric water vapor. In: *Global Process Monitoring and Remote Sensing of the Ocean and Sea Ice*, Deering D.W., Gudmandsen P. (Eds.), *Proc. SPIE* 2586, 188–195.

Janssen M.A. (1993a) An introduction to the passive microwave remote sensing of atmospheres. In: Janssen M. (Ed.) *Atmospheric Remote Sensing by Microwave Radiometry*, John Wiley & Sons, New York.

Janssen M. (Ed.) (1993b) *Atmospheric Remote Sensing by Microwave Radiometry*, John Wiley & Sons, New York.

Jarchow C. (1998) Bestimmung atmosphärischer Wasserdampf- und Ozonprofile mittels bodengebundener Millimeterwellen-Fernerkundung. PhD thesis, MPAE-W-016-99-06.

Jarchow C., Hartogh P. (1995) Retrieval of data from ground-based microwave sensing of the middle atmosphere: Comparison of two inversion techniques. *Proc. SPIE* 2586, 196–205.

Kampfer N.K. (2013) *Monitoring Atmosphere Water Vapors: Ground-based Remote Sensing and In-situ Methods.* Vol. 10, Springer, New York.

Karmakar P.K. (2012) *Microwave Propagation and Remote Sensing. Atmosphere Influence with Models and Applications.* CRC Press, Boca Raton, FL.

Karmakar P.K. (2014) *Ground-Based Microwave Radiometry and Remote Sensing: Methods and Applications.* CRC Press, Boca Raton, FL.

Kley D., Russell III J.M., Phillips C. (Eds.) (2000) SPARC assessment of upper tropospheric and stratospheric water vapour. World Climate Research Programme, SPARC Report No. 2.

Kraszewski A. (1973) Microwave instrumentation for moisture content measurement. *J. Microw Power* 8 (3/4), 323–335.

Massaro G., Stiperski I., Pospichal B., Rotach M.W. (2015) Accuracy of retrieving temperature and humidity profiles by ground-based microwave radiometry in truly complex terrain. *Atmos. Meas. Tech.* 8, 3355–3367.

Matzler C. (Ed.) (2006) *Thermal Microwave Radiation: Applications for Remote Sensing*. The Institution of engineering and Technology, London, UK.

Meeks M.L. (1976). *Methods of Experimental Physics: Astrophysics*. Academic Press, New York.

Nedoluha G.E., Bevilacqua R.M., Gomez R.M., Thacker D.L., Waltman W.B., Pauls T.A. (1995) Ground-based measurements of water vapor in the middle atmosphere. *J. Geophys. Res.* 100, 2927–2939.

Nedoluha G.E., Bevilacqua R.M., Gomez R.M., Waltman W.B., Hicks B.C., Thacker D.L., Matthews W.A. (1996) Measurements of water vapor in the middle atmosphere and implications for mesospheric transport. *J. Geophys. Res.* 101, 21183–21193.

Noam D., Pinhas A., Hagit M. (2011) Humidity measurements using commercial microwave links. In: Mutamed K. (Ed.), *Advanced Trends in Wireless Communications*, Intech, Rijeka, Croatia, pp. 65–78.

Payne V.H., Mlawer E.J., Cady-Pereira K.E., Moncet J.-L. (2011) Water vapor continuum absorption in the microwave. *IEEE Trans. Geosci. Remote Sens.* 49 (6) 2194–2208.

Peter R., Künzi K., Hartmann G.K. (1988) Latitudinal survey of water vapor in the middle atmosphere using an airborne millimeter-wave sensor. *Geophys. Res. Lett.* 15, 1173–1176.

Peter R. (1998) Stratospheric and mesospheric latitudinal water vapor distributions obtained by an airborne millimeter-wave spectrometer. *J. Geophys. Res.* 103, 16275–16290.

Raghavan S. (2003). *Radar Meteorology*. Kluwer Academic Publishers, Dordrecht, the Netherlands.

Rodgers C.D. (1976) Retrieval of atmospheric temperature and composition from remote measurements of thermal radiation. *Rev. Geophys. Space Phys.* 14, 609–624.

Rosenkranz P.W. (1998) Water vapor microwave continuum absorption: A comparison of measurements and models. *Radio Sci.* 33, 919–928.

Roy V., Gishkori S., Leus G. (2016) Dynamic rainfall monitoring using microwave links. *EURASIP J. Adv. Signal Proc.* 2016, 1–17.

Ruffieux D., Nash J., Jeannet P., Agnew J.L. (2006) The COST 720 temperature, humidity, and cloud profiling campaign: TUC. *Meteorol. Z.* 15, 5–10.

Staelin D.H., Barrett A.H., Waters J.W., Barath F.T., Johnston E.J., Rosenkranz P.W., Gaut N.E., Lenoir W.B. (1973) Microwave spectrometer on the nimbus 5 satellite: Meteorological and geophysical data. *Science* 182 (4119), 1339–1341.

Tsui J.B.-Y. (2008) Microwave Receivers and Related Components–Electronic Engineering Series, Published by Wexford College Press, PA.

Van Vleck J.H. (1947) The absorption of microwaves by uncondensed water vapor. *Phys. Rev.* 71 (7), 425–433.

Westwater E.R., Stankov B.B., Cimini D., Han Y., Shaw J.A., Lesht B.M., Long C.N. (2003) Radiosonde humidity soundings and microwave radiometers during Nauru99. *J. Atmosph. Ocean. Technol.* 20, 953–971.

6 Global Positioning System Monitoring of Atmospheric Water Vapor

6.1 GLOBAL POSITIONING SYSTEM

Measurement of the concentration of water vapor in the atmosphere using the global positioning system (GPS), also called the global navigation satellite system (GNSS), is a variety of other microwave hygrometry (Bevis et al. 1992, 1994; Elgered et al. 2005; De Haan 2006). This ability to control the humidity of the air came after deployment the satellite GPS constellation (Figure 6.1) that filled the atmosphere with microwave signals based on atomic clocks.

The GPS is a satellite navigation system, which can provide both position and time information (Figure 6.2). GPS is working under in principle all weather conditions and anywhere as long as there is an unobstructed line of sight to four or more GPS satellites. It is developed and maintained by the U.S. government and is freely accessible for civilian applications and scientific research. Since the first experimental Block-I GPS satellite was launched in 1978, GPS has been in operation for more than 30 years. It consists of more than 30 satellites orbiting the Earth at an altitude close to 20,200 km (above the surface of the Earth), and in six nearly circular orbital planes with an inclination angle of 55°. With approximately 11 h, 58 min of the orbital period, each satellite can make two revolutions in one sidereal day. Currently, GPS uses two frequencies 1.575 GHz (L1) ($\lambda \sim 19$ cm) and 1.228 GHz (L2) ($\lambda \sim 24$ cm) to carry the signals. A third frequency 1.176 GHz (L5), which is proposed for use as a civilian safety-of-life (SoL) signal, is being implemented.

Currently, such satellite navigation systems are developed by Russia (GLONASS), Europe (Galileo), and China (BeiDou). This means that it is really possible to create a global system for in situ monitoring of the atmosphere when capabilities of these systems are combined.

There are numerous books describing the concepts of GPS. Notable to this list are works by Hofmann-Wellenhof et al. (1992) and Kaplan (1996). It is important that there are three components to the GPS system: a space segment (the satellite constellation), a control segment (the network of monitoring and tracking stations, which are operated by the military), and a user segment. For the user segment, GPS is designed to be a passive system. There is no interaction from the user segment to the satellite or control segment and there can be an infinite number of users of the system. Users of GPS system are jointed by the International GPS Service (IGS). IGS is a volunteer collection of research groups, mainly consisting of government agencies and participating universities. Their goal is to provide high quality data and products for use in high precision science applications. The key components of the IGS include a network of continuously operating GPS stations, data centers who collect and distribute data through the Internet, analysis centers who produce products based on data collected from the network, and a governing body (including working groups and a central bureau) to guide the development and progress of the IGS. The IGS collects data from more than 360 stations worldwide. A map of the *global* stations is plotted in Figure 6.3. This network is the backbone of the IGS and is the basis for which it generates its products. Individual institutions voluntarily operate and contribute the data collected by their station to the IGS for analysis.

6.2 HISTORY

As it was indicated before, GPS was originally designed as a navigational and time transfer system, but since its implementation it has revolutionized the field of geodetic positioning (very precise surveying) and in this context it is now widely employed by geodesists, geophysicists, and surveyors. In GPS geodesy, the distances between GPS satellites and GPS receivers are determined either by measuring the time of flight of the time-tagged radio signals that propagate from satellites to receivers (*pseudoranging*) or, over a more extended period of time, by finding the associated path lengths by an interferometric technique (*phase measurement*). Both of these approaches are complicated by the presence of the Earth's atmosphere, which increases the optical path length between satellite and receiver and the corresponding time of flight of the GPS radio signal. One of the key tasks of geodetic GPS processing software is to *correct* the ranges between receivers and satellites so as

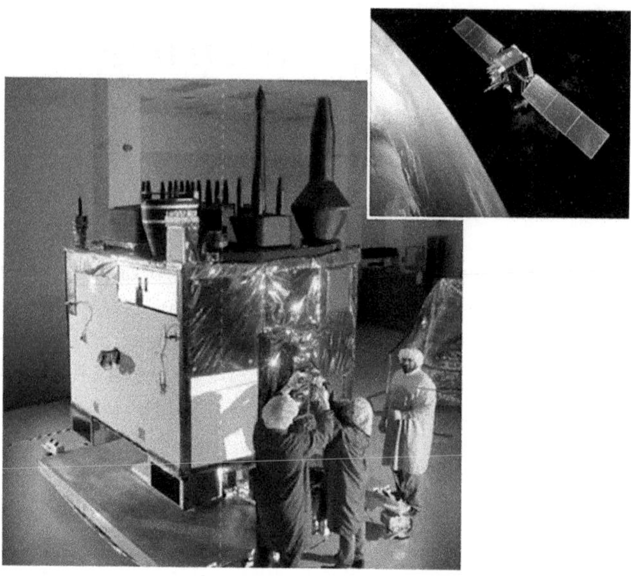

FIGURE 6.1 A GPS 2F satellite. (From http://www.nasa.gov.)

FIGURE 6.2 GPS positioning system. (From http://www.nasa.gov.)

to remove the effects of the Earth's atmosphere, thereby reducing all optical path lengths to straight-line path lengths. These corrections can be formulated in terms of excess path lengths, signal delays, or phase advances.

Geodesists and geophysicists have spent a great deal of effort learning how to model these delays to be able to remove them from GPS observations (e.g., Herring et al. 1990; Tralli and Lichten 1990).

As a result of research and modeling techniques, the methods, which allowed with sufficient accuracy estimation of the effect of the atmosphere on the optical path length and taking into account the hydrostatic delay and wet delay in the results of the GPS observations, have been found. However, the GPS geodesists estimated ionospheric and tropospheric delays only to eliminate them. Furthermore, the techniques used to estimate hydrostatic and wet delays were focused on correcting the signals recorded at an isolated receiver and not on studying the atmosphere for its own sake. In the same time, analysis of the results of these studies shows that the GPS observations can be successfully used for atmospheric research (Bevis et al. 1992).

6.3 PRINCIPLES AND PROCESSING OF GPS MONITORING OF ATMOSPHERIC WATER VAPOR

As it was shown in previous Section 6.2, the microwave signals are influenced by the ionosphere and the atmosphere along the signal path. Like visible light, the radio waves are refracted when passing through the atmosphere. If we define the index of refraction, n, as the ratio between the speed of light in a vacuum and the speed of light in the atmosphere, it is found that at microwave wavelengths the dependency of the phase refractivity on atmospheric variables can be approximated by the following empirical expression (Kursinski et al. 2000):

$$N = 77.6\frac{P}{T} + 3.73\cdot 10^5\frac{P_w}{T^2} - 4.03\frac{n_e}{f^2} \tag{6.1}$$

where:

refractivity $N = (n - 1) \times 10^6$
T is temperature in degrees kelvin
P and P_w are total pressure and partial pressure of water vapor in hectopascals, respectively
n_e is the free electron density in electrons per cubic meter
f is the signal frequency in hertz

Above 90 km, the pressure and water vapor terms are negligible, so N is therefore directly proportional to the electron density. In the stratosphere and upper troposphere, water vapor is negligible and N can be used to deduce accurate temperatures with the application of the hydrostatic equation. At the same time, the influence of water cannot be ignored in the lower troposphere. In this area, water vapor can contribute as much as 30% of N, which is the basis for the determination of water vapor content in the atmosphere (Bengtsson et al. 2003).

It was found that the atmosphere affects microwave transmissions from space in two ways. First, the waves travel slower than they would in a vacuum. Second, they travel in a curved path instead of in a straight line. Both of these effects are due to a variable index of refraction

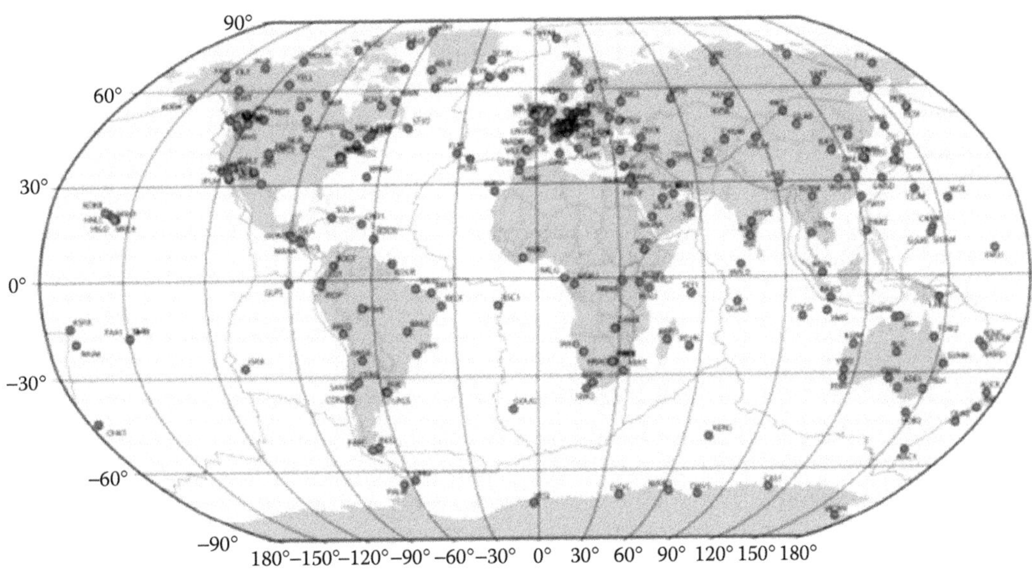

FIGURE 6.3 The international GNSS (Global Navigation Satellite System) stations, which consist of GPS stations. (From http://www.nasa.gov.)

along the ray path. At that microwave signals are delayed by water vapor, dry air, hydrometeors, and other particulates (Niell 1996; Solheim et al. 1999). This means that the measurement of the delay due to water vapor can be used for sensing water vapor with GPS (Bevis et al. 1992, 1994; Rocken et al. 1993, 1995; Businger et al. 1996; Duan et al. 1996). Thus, the primary observable during GPS monitoring atmospheric water vapor is the time of arrival of the transmitted signal. However, signal delays at a given receiver should be measured from multiple radio sources that differ in their angles of elevation. This atmospheric delay is usually mapped to zenith in the processing and is called zenith total delay (ZTD) expressed generally in meters.

It should be noted that the definition of the integrated water vapor (IWV) content and the integrated liquid water (ILW) when using data of GPS observations, is quite a complex procedure, which requires taking account of a large number of parameters and assumptions (Bevis et al. 1992, 1994). At first, it is necessary to determine the position of a receiver very accurately. One of the GPS sites is shown in Figure 6.4. But for precise positioning of the receiver (i.e., with millimeter accuracy), the atmospheric term is an error source, which needs to be estimated to obtain the acquired accuracy. At second, it is necessary to take into account the influence on a signal delay such factors as the Earth's ionosphere and neutral atmosphere (hydrostatic delay) (Rocken et al. 1995). It is important because GPS measures the total neutral delay. Ionospheric delays are highly variable, ranging from 1 to 15 m in the zenith direction. However, these delays can be corrected with millimeter accuracy, because GPS signals are transmitted at two frequencies, and because it is known that the ionospheric delay is approximately proportional to the inverse square of the signal frequency (Spilker 1978). Thus, combined wet and hydrostatic zenith delay can be estimated for each station equipped with GPS receiver. The hydrostatic delay is proportional to the total mass of atmosphere along the radio path, and hence to the surface pressure. Estimations have shown that the hydrostatic delay of a zenith GPS signal traveling

FIGURE 6.4 The GPS site (ONSA) located at the Onsala Space Observatory is a site in the International Global Navigation Satellite System (GNSS) Service (IGS), formerly the international GPS service, network (http://www.igscb.jpl.nasa.gov/network/site/onsa.html). The antenna is protected by a hemispheric radome and microwave absorbing material is attached to the antenna. (From http://www.nasa.gov.)

to an atmosphere depth of 1000 mb is approximately 230 cm. This delay can also be predicted to better than several millimeters with the surface pressure measurement accuracies of 0.5 mb. This means that if the station is also equipped with good barometer for independent estimation of the hydrostatic delay, it is possible to estimate zenith wet delay (ZWD). Unfortunately, the surface pressure measurements are available only at a limited amount of GPS stations. Thus, the surface pressure has to be obtained from a different location. The wet delay is nearly proportional to the total amount of water vapor along the radio path. Experiment has shown that the GPS signal delay ranges from 0 to 40 cm in the zenith direction. ZWD is highly variable and cannot be accurately predicted from surface observations.

There are of course other source terms, which play also a crucial role with respect the accuracy of exact positioning (Rocken et al. 1995; Hagemann et al. 2002). Other sources of error are: satellite positions, satellite clock error, receiver clock error, ionosphere, and noise. Data from each GPS satellite usually are weighted according to the goodness of fit to a previous day global tracking data, provided by the International group controlling GPS satellites. However, satellite maneuvers are unpredictable and can cause large orbit errors. Therefore, usually real-time analysis that should detect and eliminate such errors prior to the column water vapor estimation is used. As a rule, such analysis is based on the goodness of fit to the hourly data from the Forecast Systems Laboratory (FSL) GPS network. Thus, the accuracy of the atmospheric observable depends on (1) the size and the network configuration (the network should be large enough to guarantee an absolute value of the GPS ZTD estimate and stable); (2) quality and location of the receivers; (3) accuracy of the orbits; and (4) method of processing and processing parameters, such as a processing time window, elevation cut-off angel, and update frequency of a priori coordinates.

One should note that water vapors determined during the process of GPS observations can be represented as the IWV or the total precipitable water (TPW). IWV is a total amount of water vapor present in a vertical atmospheric column. TPW is the amount of water in a column of the atmosphere, if all the water and water vapor in that column were condensed to liquid or precipitated as rain. IWV and TPW are equivalent parameters, but the IWV expressed in kg/m^2, and TPW or the ILW expressed in mm of liquid water. But strictly speaking, not all the water vapor is actually precipitable.

Virtually all high-accuracy GPS data processing software packages in use today include among their functions the estimation of the atmospheric correction or tropospheric delay parameters (Bevis et al. 1992, 1994; Rocken et al. 1995). In most cases, these parameters are estimated at the same time as the station and satellite coordinates are being estimated. While most software packages eliminate clock errors through double differencing (Rocken et al. 1997; Hernandez-Pajares et al. 2001; Iwabuchi et al. 2006), a technique in which one combines observables in such a way that clock errors are canceled out, other software packages process the undifferenced data by introducing an explicit model for clock drift (Zumberge et al. 1997; Gendt et al. 2004; Karabatic et al. 2011). Several packages are capable of modeling tropospheric delays in more than one way. For example, one may choose to predict the zenith hydrostatic delay (ZHD) from surface meteorological measurements and model only the ZWD. In this case, the ZWD model tends to absorb errors in the modeling of the ZHD. Another approach is to measure the wet delay, using an upward-looking water vapor radiometers (WVR), and to model only the residual wet delay due to the instrumental and calibration errors of the WVR. In some situations, it is reasonable to model the total zenith delay as a single entity.

Thus, from above-mentioned discussions, it is clear that the atmospheric parameter, which is measured during GPS observation is thus ZTD (or IWV), and the observation technique involves the following data sources/techniques (Figures 6.5 and 6.6):

- Gathering RINEX (Receiver Independent Exchange format) data from a network
- Obtain orbit and clock information (in case of a regional network)
- Process the network
- Disseminate ZTD/IWV

Signal delay due to the dry atmosphere can be removed using surface barometric pressure measurements and a mapping function (Niell 1996). From ZTD data, an estimate of the IWV can be computed, when the surface pressure and temperature observations are available (Bevis et al. 1992). The ZWD can be converted to zenith column water vapor using the linear relationship,

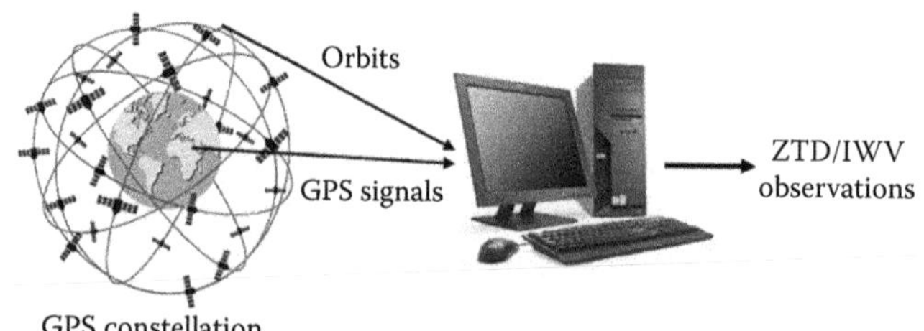

FIGURE 6.5 Processing GPS.

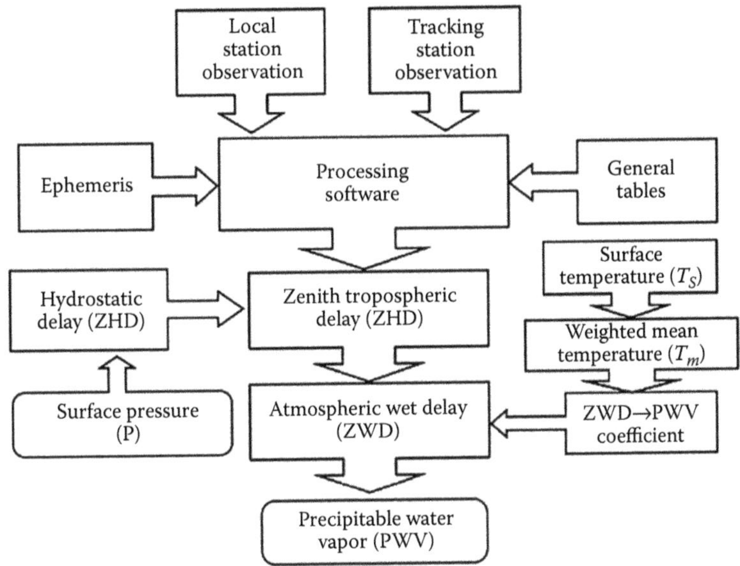

FIGURE 6.6 GNSS observation processing flowchart. (From Liang, H. et al., *Geodesy Geodynam.*, 6, 135–142, 2015. Published by Institute of Seismology, China Earthquake Administration as open access.)

published by Bevis et al. (1994). If the vertical IWD overlying a receiver is stated in term of (TPW), then this quantity can be related to the ZWD at the receiver; thus

$$\text{TPW} = \Pi \times \text{ZWD} \tag{6.2}$$

Here the ZWD is given in units of length, and the dimensionless constant of proportionality Π, which is given by

$$\Pi = \frac{10^6}{\rho R_v \left[\left({k_3}/{T_m} \right) + k_2' \right]} \tag{6.3}$$

where

$$k_2' = k_2 - mk_1 \tag{6.4}$$

and m is M_w/M_d, the ratio of molar masses of water vapor and dry air, and T_m is the vertically integrated mean temperature within an atmospheric column represented by N levels. The physical constants k_2', k_2, and k_3 are from the widely used formula for atmospheric refractivity N (Smith and Weintraub 1953; Boundouris 1963).

For these calculations, Bernese 4.0 software (Beutler et al. 1996) can be used. The current standard processing software (e.g., Bernese, GAMIT/CLOBK, or GPSY–OASIS) for processing GPS observations are capable of estimating ZTD with good accuracy in near real time. Coordinates of GPS station are computed using GPS satellite orbits generated by the Center for Orbit Determination (CODE) in Berne, Switzerland. Bernese 4.0 uses a batched least squares algorithm, which stores the normal equations for each processed time interval. As a rule, the accuracy of the GPS ZTD estimates in the order of 10–15 mm. The accuracy is dependent on the used processing scheme and software. The accuracy of IWV derived from ZTD is around 2 kg/m^2 deduced from collocated GPS and radiosonde observations (De Haan 2006). Comparison of GPS data with data received by WVR and other instruments such as radiosonds showed that these results are in good agreement (Revercombe et al. 2003; Liang et al. 2015; Guerova et al. 2016). Comparisons between GPS and microwave radiometers (MWRs) data indicated their agreement to approximately 5% (Revercombe et al. 2003).

It was established that 1 cm of TPW causes approximately 6.5 cm of GPS wet signal delay. This 6.6-ford *amplification* effect is important for accurate TPW measurement with GPS (Rocken et al. 1995). Delay estimation at every GPS station is called absolute tropospheric estimations. Estimation of differences in delay between two stations is called relative of differential estimation.

One should note that equipment of GPS observation is constantly being improved by incorporating the best possible statistical description of water vapor (Kuo et al. 1996; Fang et al. 1998; Guo et al. 2000; Bengtsson et al. 2003; Ge et al. 2008; Geng et al. 2012; Li et al. 2013, 2014). For example, techniques have been developed to obtain IWV along slant paths between ground-based GPS receivers and the GPS satellites (Figure 6.7).

As a result, the opportunity to improve the accuracy of measurement (Braun et al. 2001, 2003) and simulate the three-dimensional distribution of water vapor

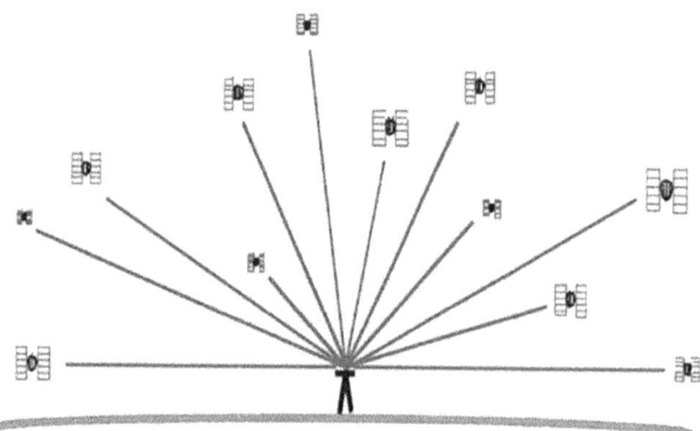

FIGURE 6.7 Signals from up to 12 GPS satellites can be tracked by a ground-based GPS receiver. SW can be simultaneously sensed along all of the GPS ray paths. (Reprinted from *J. Atmos. Solar-Terr. Phys.*, 63, Ware, R.H. et al., Real-time national GPS networks for atmospheric sensing, 1315–1330, Copyright 2001, With permission from Elsevier.)

in the atmosphere appeared (Macdonald et al. 2002; Champollion et al. 2005). Higher spatial resolution can be obtained by solving for the IWV or *slant water* (SW) along each GPS ray path. SW is obtained by solving for the total slant delay along each ray path, and then subtracting the dry component of the slant delay. The dry slant delay can be estimated from surface pressure measurements or from three-dimensional numerical weather models (Chen and Herring 1996).

6.4 PROSPECTS OF GPS OBSERVATION—GLOBAL MONITORING

Analysis of opportunities of GPS observation for monitoring water vapor in the atmosphere shows that despite the fact that under good operating conditions, ground-based MWRs may constitute a better means of sensing IWV than does recovering the wet delay from GPS observations, formed GPS system has great potential for development. GPS receivers established for geodetic applications are now available in different regional and local networks around the world. The new technology of GPS atmospheric remote sensing in comparison with conventional water vapor observing system is low cost with reliable and stable results and high measurement accuracy. Moreover, the price of GPS receivers drops 25%–50% every year. This GPS technology is characterized by all-weather operability. For comparison, MWRs are not *all-weather* instruments. The resulting IWV data are of excellent quality and are practically equivalent to radiosonde measurements. In addition, GPS observation operates on the radio frequencies that can penetrate clouds and dusts (Gutman et al. 2004b; Awange and Grafarend 2005). However, perhaps the most important potential advantage of GPS over ground-based MWRs is that GPS receivers are going to exist in huge numbers over wide areas in just a few years, and the use of this resource for meteorological purposes could occur at little incremental cost.

All this indicates that the GPS can really serve as a basis for global atmospheric monitoring (Lee et al. 2001, 2013). GPS-based measurements offer new and promising possibilities. One of these is the capability to provide data at similar quality under all weather conditions. GPS could be used to supplement the Geostationary Operational Environmental Satellite (GOES) and other satellite observations that are degraded in cloudy regions. GPS column water vapor can also be used for calibration and validation of satellite radiometers. Numerous studies have shown that this technology already now plays a major role in complementing the existing techniques, for example, radiosonde and ground-based WVR.

GPS column water vapor that is available close to real-time can be included in operational numerical weather models that provide the primary input for climate studies (Yuan et al. 1993; Lee et al. 2001, 2013). Thus, real-time GPS estimation is valuable for climate monitoring and short-term weather forecasting (Kuo et al. 1996; McPherson et al. 1997). Experiment has shown that the GPS data have very high temporal resolution (typically a few minutes), which so far has not been explored systematically, but it is expected that in the future this ability will provide detailed information on fronts, squall lines, and other small-scale meteorological systems. It could be used even more extensively than at present in numerical weather prediction (NWP). With comprehensive data assimilation, there are interesting possibilities of better determining the high-frequency exchange of water between the atmosphere and the land surface (Bengtsson et al. 2003).

Measurements of SW vapor with GPS networks (Ware et al. 1997) could potentially be applied to water vapor tomography, prediction of convection, and related aviation wind shear hazard. These applications are also most valuable if conducted close to real-time. GPS-sensed PW data can be used to improve a storm system analysis (Rocken et al. 1995; Businger et al. 1996). In addition, improved vertical structure of water-vapor and short-term precipitation forecasts can be obtained by assimilating the surface humidity and PW data into mesoscale models (Kuo et al. 1996). Park and Droegemeier (1996) showed that simulations of thunderstorms can be quite sensitive to the distribution of water vapor in their near environment.

It should be noted that at present, the base has already been created for the global atmosphere monitoring. As we mentioned earlier, the users of GPS system are jointed in a global network by the IGS, which formally started on January 1, 1994. IGS collects ZPD data from more than 260 GPS stations. IGS is a member of the Astronomical and Geophysical Data Analysis Services (FAGS) and it operates in close cooperation with the International Earth Rotation and Reference System Services (IERS). One of the objectives of the IGS is to strive to provide the highest quality, reliable GNSS data, and products, openly and readily available to all. The IGS provides products such as (De Haan 2006):

- RINEX (Receiver Independent Exchange Format) data from IGS stations
- Satellite orbits and clocks with various latencies and qualities
- Earth rotation parameters
- Atmospheric parameters (with a latency of more than 2 h)

At present the RINEX data available through IGS has currently three types of latency, where the latency determines the length of the window:

- Daily data
- Hourly data
- 15-min data

The daily datasets are valuable for climatology purposes, while the hourly data, when the latency is small enough, can be used as input for GPS processing which is dedicated for NWP. The 15-min data set can be valuable for processing more than once per hour which in turn can be valuable for nowcasting purposes by for example, presenting two dimensional IWV maps to forecasters. The coverage of the 15-min data set is not good enough on its own to create a valuable nowcasting product.

In addition to the global system currently there are many regional networks. As a notable example, in the framework of the European EUMETNET EIG GNSS water vapor program (http://www.egvap.dmi.dk/) project more than 2400 GNSS sites are continuously operated, providing hourly updated tropospheric ZTD for assimilation into NWP models. A positive impact of GPS-derived tropospheric products on NWP has been demonstrated in several studies (e.g., Haan et al. 2004; Gutman et al. 2004a, b; Karabatic et al. 2011; Shoji et al. 2011). Several European projects such as WAVEFRONT, MAGIC, and European Cooperation in Science and Technology (COST-716) (Elgered et al. 2005) have all demonstrated the ability of GNSS to serve as an accurate atmospheric water vapor sensor for meteorological applications. GPS networks for atmospheric monitoring are also formed in Canada, Denmark, France, Germany, Finland, Japan, and many other countries (Ware et al. 2001; Hagemann et al. 2002; Bai and Feng 2003; Benevides et al. 2015; Liang et al. 2015; Guerova et al. 2016). These national networks of GPS receivers are maintained by commercial companies, government agencies (such as survey departments), or universities. Some of these networks are set up in cooperation with national meteorological services (e.g., Switzerland, United Kingdom, the Netherlands, and United States). These GPS networks are usually involved in many international programs such as EUREF, NOAA and many others (De Haan 2006).

REFERENCES

Awange J.L., Grafarend E.W. (2005) *Solving Algebraic Computational Problems in Geodesy and Geoinformatics*. Springer, New York.

Bai Z., Feng Y. (2003) GPS Water vapor estimation using interpolated surface meteorological data from Australian automatic weather stations. *J. Global Position. Syst.* 2 (2), 83–89.

Benevides P., Catalao J., Miranda P.M.A. (2015) On the inclusion of GPS precipitable water vapour in the nowcasting of rainfall. *Nat. Hazards Earth Syst. Sci.* 15, 2605–2616.

Bengtsson L., Robinson G., Anthes R. et al. (2003) The use of GPS measurements for water vapor determination. *Bull. Am. Meteorol. Soc.* 1249–1258.

Beutler G., Brockman E., Frankhauser S. et al. (1996) Bernese GPS Software Version 4.0, University of Berne, September.

Bevis M., Businger S., Herring T.A., Rocken C., Anthes A., Ware R. (1992) GPS Meteorology: Remote sensing of atmospheric water vapor using the global positioning system. *J. Geophys. Res.* 97, 15787–15801.

Bevis M., Businger S., Chiswell S., Herring T., Anthes R., Rocken C., Ware R. (1994) GPS meteorology: Mapping zenith wet delays onto precipitable water. *J. Appl. Meteorol.* 33, 379–386.

Boundouris G. (1963) On the index of refraction of air, the absorption and dispersion of centimeter waves by gases. *J. Res. Natl. Bur. Stand.* 67D, 631–684.

Braun J.J., Rocken C., Ware R.H. (2001) Validation of single slant water vapor measurements with GPS. *Radio Sci.* 36, 459–472.

Braun J.J., Rocken C., Liljegren J. (2003) Comparisons of line-of-sight water vapor observations using the global positioning system and a pointing microwave radiometer. *J. Atmos. Oceanic Technol.* 20, 606–612.

Businger S., Chiswell S., Bevis M., Duan J., Anthes R., Rocken C., Ware R., Van Hove T., Solheim F. (1996) The promise of GPS in atmospheric monitoring. *Bull. Am. Meteorol. Soc.* 77, 5–18.

Champollion C., Masson F., Bouin M. N., Walpersdorf A., Doerflinger E., Bock O., Van Baelen J. (2005) GPS water vapour tomography: Preliminary results from the ESCOMPTE field experiment. *Atmos. Res.* 74, 253–274.

Chen G., Herring T. (1996) Effects of atmospheric azimuthal asymmetry on the analysis of space geodetic data. *J. Geophys. Res.* 102, 20489–20502.

De Haan S. (Ed.) (2006) National/regional operational procedures of GPS water vapour networks and agreed international procedures: Instruments and observing methods. Report No. 92, WMO.

Duan J., Bevis M., Fang P. et al. (1996) GPS meteorology: Direct estimation of the absolute value of precipitable water. *J. Appl. Meteorol.* 35, 830–838.

Elgered G., Plag H.P., Van der Marel H., Barlag S., Nash J. (2005) COST Action 716—Exploitation of ground-based GPS for operational numerical weather prediction and climate applications. EC/COST, Official Publications of the European Communities, Luxembourg.

Fang P., Bevis M., Bock Y., Gutman S., Wolfe D. (1998) GPS meteorology: Reducing systematic errors in geodetic estimates for zenith delay. *Geophys. Res. Lett.* 25, 3583–3586.

Ge M., Gendt G., Rothacher M., Shi C., Liu J. (2008) Resolution of GPS carrier-phase ambiguities in precise point positioning (PPP) with daily observations. *J. Geod.* 82 (7), 389–399.

Gendt G., Dick G., Reigber C., Tomassini M., Liu Y., Ramatschi M. (2004) Near real time GPS water vapor monitoring for numerical weather prediction in Germany. *J. Meteorol. Soc. Jpn.* 82, 361–370.

Geng J., Shi C., Ge M., Dodson A.H., Lou Y., Zhao Q., Liu J. (2012) Improving the estimation of fractional-cycle biases for ambiguity resolution in precise point positioning. *J. Geod.* 86, 579–589.

Guerova G., Jones J., Douša J. et al. (2016) Review of the state-of-the-art and future prospects of the ground-based GNSS meteorology in Europe. *Atmos. Meas. Tech. Discuss.* 9, 5385–5406.

Guo Y.-R., Kuo Y.-H., Dudhia J., Parsons D. (2000) Fourdimensional variational data assimilation of heterogeneous mesoscale observations for a strong convective case. *Monthly Weather Rev.* 128, 619–643.

Gutman S.I., Sahm S.R., Benjamin S.G., Smith T.L. (2004a). GPS water vapor observation error. *Bull. Am. Meteorol. Soc.* 2004, 6205–6211.

Gutman S.I., Sahm S.R., Benjamin S.G., Schwartz B.E., Holub L., Stewart J.Q., Smith T.L. (2004b) Rapid retrieval and assimilation of ground based GPS precipitable water observations at the NOAA forecast systems laboratory: Impact on weather forecasts. *J. Meteorol. Soc. Jpn.* 82, 351–360.

Haan S., Barlag S., Baltink H., Debie F. (2004) Synergetic use of GPS water vapor and meteosat images for synoptic weather forecasting. *J. Appl. Meteorol.* 43, 514–518.

Hagemann S., Bengtsson L., Gendt G. (2002) On the determination of atmospheric water vapour from GPS measurements. MPI Report 340, Max-Planck-Institut für Meteorologie, Hamburg, Germany.

Hernandez-Pajares M., Juan J.M., Sanz J., Colombo O.L., Van Der Marel H. (2001) A new strategy for real-time integrated water vapor determination in WADGPS networks. *Geophys. Res. Lett.* 28 (17), 3267–3270.

Herring T., Davis J.L., Shapiro I.I. (1990) Geodesy by radio interferometry: The application of Kalman filtering to the analysis of very long baseline interferometry data. *J. Geophys. Res.* 95, 12561–12581.

Hofmann-Wellenhof B., Lictenegger H., Collins J. (1992) *GPS Theory and Practice*. Springer-Verlag, New York.

Iwabuchi T., Rocken C., Lukes Z., Mervart L., Johnson J., Kanzaki M. (2006) PPP and network true real-time 30 sec estimation of ZTD in dense and giant regional GPS network and the application of ZTD for nowcasting of heavy rainfall. In: *Proceedings of the Institute of Navigation, 19th international technical meeting of the Satellite Division*, ION GNSS 2006, Vol. 4, pp. 1902–1909.

Kaplan E.D. (1996) *Understanding GPS: Principles and Applications*. Artech House, Norwood, MA.

Karabatic A., Weber R., Haiden T. (2011) Near real-time estimation of tropospheric water vapour content from ground based GNSS data and its potential contribution to weather now-casting in Austria. *Adv. Space Res.* 47, 1691–1703.

Kuo Y.-H., Zou X., Guo Y.-R. (1996) Variational assimilation of precipitable water using a nonhydrostatic mesoscale adjoint model. *Monthly Weather Rev.* 124, 122–147.

Kursinski E.R., Hajj G.A., Leroy S.S., Herman B. (2000) The GPS radio occultation technique. *Terr. Atmos. Oceanic Sci.* 11, 53–114.

Lee L.C., Kursinski R., Rocken C. (Eds.) (2001) *Applications of Constellation Observing System for Meteorology, Ionosphere & Climate*. Springer-Verlag, New York.

Lee S.-W., Kouba J., Schutz B., Kim D.H., Lee Y.J. (2013) Monitoring precipitable water vapor in real-time using global navigation satellite systems. *J. Geod.* 87, 923–934.

Li X., Ge M., Zhang H., Wickert J. (2013), A method for improving uncalibrated phase delay estimation and ambiguity-fixing in real-time precise point positioning. *J. Geod.* 87(5), 405–416.

Li X., Dick G., Ge M., Heise S., Wickert J., Bender M. (2014) Real-time GPS sensing of atmospheric water vapor: Precise point positioning with orbit, clock, and phase delay corrections. *Geophys. Res. Lett.* 41(10), 3615–3621.

Liang H., Cao Y., Wan X., Xu Z., Wang H., Hu H. (2015) Meteorological applications of precipitable water vapor measurements retrieved by the national GNSS network of China. *Geodesy Geodynam.* 6(2), 135–142.

Macdonald A.E., Xie Y., Ware R.H. (2002) Diagnosis of three-dimensional water vapor using a GPS network. *Monthly Weather Ren.* 130, 386–397.

McPherson R., Kalnay E., Lord S. (1997) The potential role of GPS observations in operational numerical weather prediction. In: *Proceedings of the National Research Council Workshop on GPS in Geoscience*, National Academy Press, Washington, DC, pp. 111–113.

Niell A. (1996) Global mapping functions for the atmosphere delay at radio wavelengths. *J. Geophys. Res.* 101, 3227–3246.

Park S., Droegemeier K. (1996) Sensitivity of 3-D convective storm evolution to water vapor and implications for variational data assimilation. In: *Proceedings of 11th Conference on NWP*, Norfolk, VA, pp. 137–139.

Revercombe H.E., Turner D.D., Tobin D.C. et al. (2003) The ARM program's water vapor intensive observation periods: Overview, initial accomplishments, and future challenges. *Bull. Am. Meteorol. Soc.* 84 (2), 217–236.

Rocken C., Ware R., Van Hove T., Solheim F., Alber C., Johnson J. (1993) Sensing atmospheric water vapor with GPS. *Geophys. Res. Lett.* 20, 2631–2634.

Rocken C., Van Hove T., Johnson J., Solheim F., Ware R., Bevis M., Businger S., Chiswell S. (1995) GPS storm—GPS sensing of atmospheric water vapor for meteorology. *J. Ocean. Atmos. Technol.* 12, 468–478.

Rocken C., Van Hove T., Ware R. (1997) Near real-time sensing of atmospheric water vapor. *Geophys. Res. Lett.* 24, 3221–3224.

Shoji Y., Kunii M., Saito K. (2011) Mesoscale data assimilation of Myanmar cyclone nargis part II: Assimilation of GPS-derived precipitable water vapor. *J. Meteorol. Soc. Jpn.* 89, 67–88.

Smith E.K., Weintraub S. (1953) The constants in the equation for atmospheric refractive index at radio frequencies. *J. Res. Natl. Bur. Stand.* 50, 39–41.

Solheim F., Vivekanandan J., Ware R., Rocken C. (1999) Propagation delays induced in GPS signals by dry air, water vapor, hydrometeors and other particulates. *J. Geophys. Res.* 104(D8), 9663–9670.

Spilker J.J. (1978) GPS signal structure and performance characteristics. *Navigation* 25 (2), 121–146.

Tralli D.M., Lichten S.M. (1990) Stochastic estimation of tropospheric path delays in global positioning system geodetic measurements. *Bull. Geod.* 64, 127–159.

Ware R., Alber C., Rocken C., Solheim F. (1997) Sensing integrated water vapor along GPS ray paths. *Geophys. Res. Lett.* 24, 417–420.

Ware R.H., Fulker D.W., Stein S.A. et al. (2001) Real-time national GPS networks for atmospheric sensing. *J. Atmos. Solar-Terr. Phys.* 63(12), 1315–1330.

Yuan L., Anthes R. A., Ware R.H., Rocken C., Bonner W., Bevis M., Businger S. (1993) Sensing climate change using the global positioning system. *J. Geophys. Res.* 98 (D8), 14925–14937, 1993.

Zumberge J.F., Heflin M.B., Jefferson D.C., Watkins M.M., Webb F.H. (1997) Precise point positioning for the efficient and robust analysis of GPS data from large networks. *J. Geophys. Res.* 102 (B3), 5005–5017.

7 Atmosphere Monitoring Using Methods of Absorption of Electromagnetic Radiation—Terahertz Absorption

7.1 TERAHERTZ RADIATION

Terahertz (THz) radiation is the electromagnetic spectrum with the frequency defined from 0.1 to 10 THz (1 THz = 10^{12} Hz). This spectral range corresponds to a wavelength between 30 μm and 3 mm. THz frequencies lay between the operation ranges of classical microwave and infrared spectroscopy (Figure 7.1) and thus cannot be effectively covered by any of these techniques. However, THz region plays an important role for the Earth's radiation budget; for example, (1) up to 50% outgoing long-wave radiation (OLR) is below 650 cm^{-1} and (2) up to 50% of basic greenhouse effect is in THz/far-IR range. But, until recently, because of the difficulty of generating and detecting techniques in this region, THz frequency band remains unexplored compared to other range and tremendous effort has been made to fill in *THz gap* (Zhang and Xu 2009). Significant progress in the development of THz region has been achieved only in the last 30 years, when suitable for use sources and detectors of THz radiation have been developed (Gallerano and Biedron 2004; Krishna et al. 2012; Yin et al. 2012; Lewis 2014).

7.2 ABSORPTION OF THz RADIATION

Currently there are large amounts of research related to the study of the propagation of THz radiation in the atmosphere. But the main efforts are focused on the study of the atmospheric transparency in the THz range and searching the air transmission windows for communicating and sensing system (Yuan et al. 2003; Foltynowicz et al. 2005; Yang et al. 2011, 2012; Yao et al. 2012). As is known, THz communication will benefit from the high-bit-rate wireless technology which takes advantage of higher frequency and broader information bandwidth allowed in this range than microwave. It is possible for such a system to achieve data rate in tens of gigabits per second (Lee 2008). However, the atmospheric opacity severely limits the communication applications at this range (Siegel 2002). Water vapor in the air has a high absorption coefficient in the THz frequency range. The humidity affects the shape and amplitude of pulsed T-rays, especially in long distance propagation. Water vapor absorption lines are extremely ubiquitous and result in relatively short atmospheric propagation paths through most of the THz range. In addition, as the relative humidity increased, the *windows*, which present on the THz range, got narrower and shallower. Therefore, the definition of conditions for THz radiation propagation over long distances will undoubtedly determine whether THz communication will be carried out into practical application. That is why the characteristics of THz atmospheric propagation now rank among the most critical issues in the principal application of space communication and atmospheric remote sensing (Tonouchi 2007).

7.3 ATMOSPHERE MONITORING USING THz SPECTROSCOPY

Research has shown that the absorption in THz region is due mainly to intense rotational transitions of the water molecules (Hall and Dowling 1967; Burch 1968; van Exter et al. 1989). In gases like water vapor, the relative isolation of the molecules leads to sharp resonant peaks of absorption centered at specific frequencies (Figure 7.2). The intensity of these peaks is strongly dependent on the concentration of water vapor in the atmosphere. Furthermore, the absorption strength of water vapor lines is proportional to the concentration of water molecules, and therefore to the relative humidity. This means that THz spectroscopy can be applied for measurement of atmospheric humidity (Xin et al. 2006). However, despite the encouraging prospects, it should be recognized that there is no significant progress in the use of THz radiation for monitoring the atmosphere.

Unfortunately, despite the significant progress in theoretical predictions of atmospheric absorption

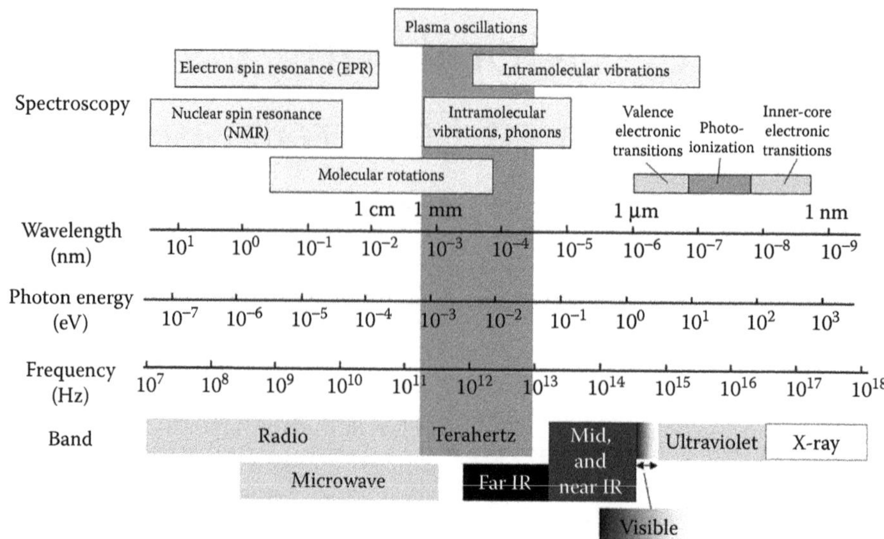

FIGURE 7.1 Electromagnetic spectrum from radio to X-ray wavelengths, including spectral ranges for probing different excitations in materials.

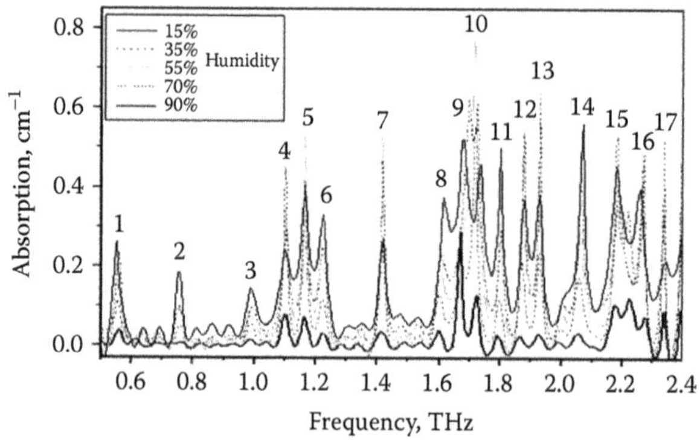

FIGURE 7.2 Water vapor absorption profiles at different humidities at room temperature. Air with a relative humidity of 10%, 30%, 50%, and 90% used in these experiments. (Reprinted with permission from Xin, X. et al., *J. Appl. Phys.*, 100, 094905, 2006. Copyright 2006 American Institute of Physics.)

(Liebe 1989; Rosenkranz 1998; Pardo et al. 2001b), we did not yet reach a good understanding of the behavior of THz-wave in the atmosphere, and the knowledge and technology for the atmospheric remote sensing are still poor. There still exist large difficulties in developing appropriate observation technology for THz-wave that lies in the transition region between optical and microwave regions.

As a result, now we do not have yet an adequate model that could describe the atmospheric propagation THz radiation in different kinds of climatic conditions. Therefore, in order to develop hygrometers, working in THz region, and on the basis of their readings to make predictions on the climate change, we need a detailed studies aimed at (1) the construction of radiative transfer algorithm; (2) the collection of accurate spectral parameters, such as linear and continuum absorption and complex refractive index in THz region, that is, to collect the spectroscopic fingerprinting of atmospheric molecules for THz atmospheric monitoring; and (3) the standardization of measurement procedures with purpose to improve both the signal-to-noise ratio (SNR), and the restoration of the original signal from the observed signal by the process of deconvolution (Foltynowicz et al. 2005; Ryu and Kong 2010; Zhan et al. 2015). It is essential to understand the actual effects on the amplitude and phase of THz radiation propagating through the atmosphere, which depends on the frequency of incident wave, gas components, and ambient temperature or barometric pressure in different atmospheric conditions (Yao et al. 2012).

Xin et al. (2006) have found that the line amplitude variations of three orders of magnitude can be accurately measured with THz spectroscopy, providing a large dynamic range for humidity measurements. However, the lines used for humidity measurement must be carefully selected and the variation of their amplitude with humidity must be calibrated. As it is seen in Figure 7.3, the influence of humidity strongly depends on the spectral range used for measurements.

It should also be borne in mind that to build the correct model the propagation of THz radiation in the atmosphere, more completed information is needed on the absorption spectra of water vapor. Three spectroscopic parameters: (1) center of the line, (2) oscillator intensity,

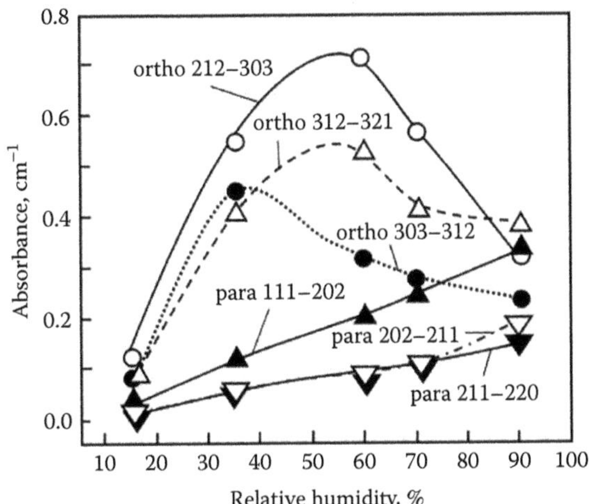

FIGURE 7.3 Color online the peak intensity of three para triangles and three ortho squares rotational transitions for different humidities. While para transitions increase, ortho transitions increase and then decrease with increasing humidity. (Reprinted with permission from Xin, X. et al., *J. Appl. Phys.*, 100, 094905, 2006. Copyright 2006, American Institute of Physics.)

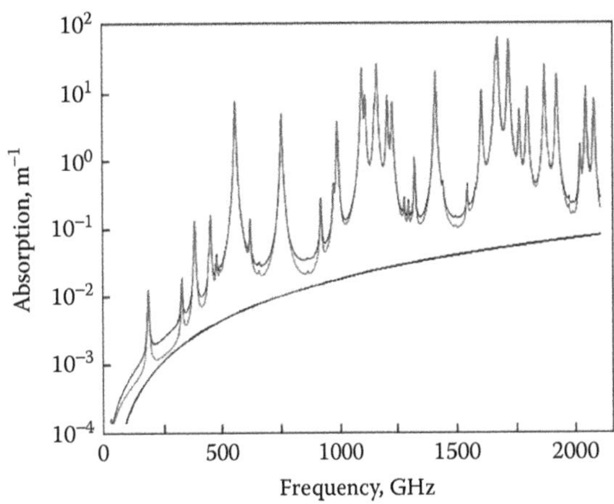

FIGURE 7.4 The calculated air-broadened spectrum (light gray), continuum absorption (black), and air-broadened spectrum with continuum absorption (gray). All plots are 15.23 Torr of water vapor and 746.77 Torr of air. (Reprinted from *J. Quant. Spectrosc. Rad. Transfer*, 127, Slocum, D.M. et al. (2013) Atmospheric absorption of terahertz radiation and water vapor continuum effects, 49–63, Copyright 2013, With permission from Elsevier.)

and (3) pressure broadening coefficient fully describe the absorption lines (Yasuko and Takamasa 2008; Rothman et al. 2009). While the first parameter is well determined for all of the lines because it depends only on the molecule inner structure, the other two are much more difficult to predict because they depend on the molecule interactions. It has been known for a long time that the line contribution is not enough to reproduce the experimentally measured atmospheric absorption, with calculated absorption being as much as an order of magnitude too small. In order to explain the discrepancy, the so-called continuum absorption was introduced (Rice and Ade 1979; Ma and Tipping 1990; Pardo et al. 2001a). The atmospheric absorption spectrum doesn't correspond to the accumulation of water vapor absorption lines. The continuum absorption is what remains after subtraction of linear contributions from the total absorption that can be measured directly (Rosenkranz, 1998) (Figure 7.4). It may be observed in wide electromagnetic spectrum (from microwave to infrared) and cannot be described by water vapor absorption lines. The physical mechanisms of the continuum absorption are still not well explained while several theories have been proposed (Burch and Gryvnak 1979; Clough et al. 1989; Liebe 1989; Ma and Tipping 1992; Yasuko and Takamasa 2008; Yang et al. 2014).

Certain difficulties connected with the interpretation of the results are also related to the fact that the absorption cross section for H_2O vapor depends nonlinearly on the pressure and temperature leading to an absorption profile that can skew from a Beer–Lambert–Bouguer model with respect to environmental conditions (Xin et al. 2006). It was also established that the absorption spectra for water vapor are a mixture of the contribution of two components, *para*-H_2O and *ortho*-H_2O with an *ortho* to *para* ratio (OPR) of 3:1, which even more complicates their structure. *Para*-H_2O and *ortho*-H_2O appear when considering the nuclear spin effect of hydrogen atoms in the water molecule (Townes and Schawlow 1975). At that Xin et al. (2006) have shown (Figure 7.3) that water molecules that undergo *ortho*-level rotational transitions increase, then decrease in absorption strength with increasing humidity, whereas most *para* transitions simply increase as expected for higher concentrations in the THz beam path. In terms of nuclear spin statistics, the *ortho* levels possess populations three times larger than the *para* levels at room temperature due to the presence of two symmetric hydrogen nuclei.

It is not clear yet how can we separate desired signals from the thermal blackbody-background created by any warm object present in the THz (Foltynowicz et al. 2005). Research has shown that background is a severe issue in this frequency range, much more so than for other frequency ranges due to blackbody radiation. The spectral density of background radiation from any room temperature object increases at the long wavelengths by about four orders of magnitude per decade. Since this is so prevalent, if we wish to get clean signals, one has to think about coherent detection, very narrow line-width sources, and very narrow line-width detectors, to minimize saturation of the detector.

In addition, one should take into account the technical immaturity of the technology in this frequency range, including the lack of compact, tunable sources, detectors, and other standard optical components (Foltynowicz et al. 2005), which restrains the development of devices suitable for wider application.

Furthermore, it should be noted that the measurements themselves in the field of THz are not trivial. For example, THz spectroscopy operates with pulses of electromagnetic radiation which are too short to be resolved by conventional electronic display instruments such as oscilloscopes. Usually, for research in this area it is used so-called the transmission time-domain THz spectroscopy (TTDTS). The basic idea of the TTDTS can be described in the following way: a sub-picosecond pulse of electromagnetic radiation passes through a sample and gets its time profile changed compared to the one of the reference pulse. The last can be either a freely propagating pulse or a pulse transmitted through a medium with known properties. Through an analysis of changes in the *complex* Fourier spectrum, which are introduced by the sample, the spectrum of the refractive index of the sample's material is obtained.

TTDTS is being realized by so-called gated-detection technique (Khazan 2002). Typical gated-detection scheme usually used for studies is shown in Figure 7.5. The THz spectrometer is powered by a laser which emits a train of pulses each of several tens of femtoseconds (1 fs = 10^{-15} s) in duration. The initial laser beam is split by a beam splitter in two parts called pump and probe beams. A pump pulse hits an emitter which in response releases a short (few picoseconds) pulse of electromagnetic radiation. The spectrum of the radiation is centered at several hundreds of GHz (1 GHz = 10^9 Hz) so that one- or even half-cycle pulses are released. This THz radiation then comes to a detector which is gated by a probe pulse. The output signal of the detector is proportional to the magnitude and the sign of the field of the THz pulse in every certain moment of time. Thus, by the variation of the delay between pump and probe optical pulses one can trace the whole time profile of the THz pulse.

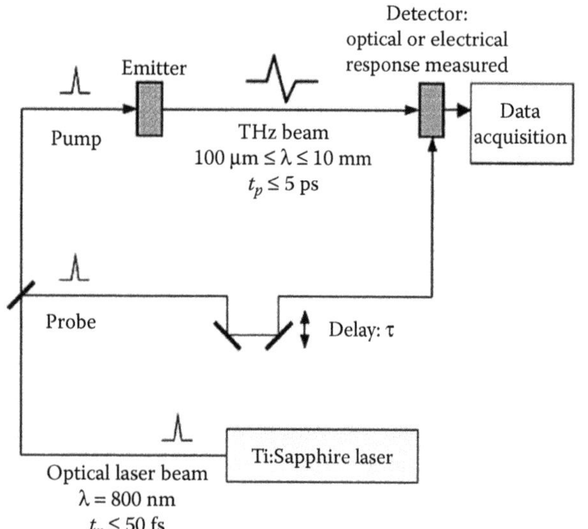

FIGURE 7.5 Typical gated-detection scheme: λ is wavelength and t_p is pulse duration (of either laser or THz pulse). (Idea from Khazan, M.A., Time-domain Terahertz spectroscopy and its application to the study of high-T_c superconductive this films, PhD Thesis, Universität Hamburg, Germany, 2002.)

Main sources and receivers of THz radiation, which can be used in the development of devices suitable for monitoring the atmosphere, are listed in Tables 7.1 and 7.2. Their advantages, shortcomings, and opportunities for the applications are considered in detail in Gallerano and Biedron (2004), Krishna et al. (2012), Yin et al. (2012), and Lewis (2014). For example, large scale free electron lasers (FEL) facilities are capable of providing a high power *continuous-wave* (CW) THz radiation while low-cost solid state sources are expected to drive the realization of more portable THz systems for a variety of applications (Gallerano and Biedron 2004). According to Yin et al. (2012) most advanced solid state T-ray sources which are primarily used for current popular THz research include (1) pulsed T-rays based on ultrafast laser sources; (2) high-frequency electronic sources, for their integration into existing electronic technology, to achieve low-power *continuous-wave* (CW) operation; and (3) *Quantum Cascade Laser*s (QCLs), for their small size and tunability, to realize CW operation. The pulsed nature of ultrafast T-ray systems provides high SNRs, broad bandwidth, and low average power, making them ideal tools to study biological and medical materials. However, QCLs as THz narrowband laser sources show deeper penetration, due to higher average power, which complements Tuned Port Injection (TPI) systems.

As for the receivers of THz radiation, there is also no definite opinion about what receiver is preferred for use. Everything is determined by the radiation power and frequency range. So, traditionally used in the microwave and IR radiation detectors that operate at temperatures close to room temperature (piezoelectric detectors, bolometers, thermocouples, etc.), have good sensitivity in the THz range, but their use in the THz systems is limited by low operation speed which at best is tens of milliseconds.

Detectors based on the Schottky barrier diodes perform significantly better frequency properties. For example, Microtech Inc. (USA) manufactures detectors based on GaAs, which can operate in the range of 0.2–1.0 THz. The

TABLE 7.1
Some Sources of Terahertz Radiation

Source Type	Example
Thermal	Cosmic background radiation
	Globar
	Mercury lamp
Vacuum electronic	Backward-wave oscillator
	Extended-interaction klystron
	Traveling-wave tube
	Gyrotrons
	Free electron lasers (FELs)
	Synchrotrons
Solid-state electronic	Gunn diodes
	Transistors
	Frequency multiplication
	Superconductor
	Resonant tunnel diodes
Lasers	Molecular gas lasers
	Ti:Sapphire-based Lasers
	Semiconductor lasers (*p*-type Ge laser)
	Quantum cascade lasers
Sources pumped by lasers (continuous)	Photomixer
	Mechanical resonance
Sources pumped by lasers (pulsed)	Photoconductive switches
	Air
	Magnetic dipoles
	Terahertz parametric oscillator
Optical rectification	Bulk electro-optic rectification
	Nonresonant optical rectification
Transient currents	Diffusion
	Drift
Mechanical excitation	Peeling tape
	Surface formation

Source: Data extracted from Lewis, R.A., *J. Phys. D: Appl. Phys.*, 47, 374001, 2014.

TABLE 7.2
Detection of Terahertz Radiation

Detector Type	Example
Electronic	• Schottky diode • Backward diode • Rectifying transistor (Tera-FET) • Superconductive mixers
Thermal	• Bolometer • Golay cell • Pyroelectric • Thermopile
Optical	• Photoconductive switches or antenna/photomixers • Electro-optic sampling • Biased-air coherent detection • Synchronous and asynchronous optical sampling

Source: Data extracted from Yin, X. et al., *Terahertz Imaging for Biomedical Applications: Pattern Recognition and Tomographic Reconstruction*, Springer Science+Business Media, New York, pp. 9–26, 2012, and from Zouaghi, W. et al., *Eur. J. Phys.* 34, S179–S199, 2013.

main advantages of these detectors are speed of response (up to 20 GHz), and operation at room temperature.

However, their sensitivity (10^{-9}–10^{-10}) is an order of magnitude worse than the sensitivity of superconducting receivers such as hot electron bolometer (HEB). Their work is based on the effect of heating of electrons in thin films of superconductors, when they absorb electromagnetic radiation. This means that the superconducting receivers are preferable when detecting very weak signals.

At the same time, experience shows that when working with a pulse radiation, it is better to use detectors based on electro-optic sampling and photoconductivity switch antenna/photomixer. There are reports that as the receivers, a quantum dots-based single electron transistors can also be used. However, their use is limited because the operating temperature is close to zero, 0.3–1.5 K (Lewis 2014).

One should realize that optical components such as mirrors, lenses, and polarizers are also components of THz spectrometers (Lewis 2014). In contrast to visible optical systems, where lenses and similar transmitting elements predominate, THz systems tend to employ reflecting elements, which have minimal loss and no dispersion. THz mirrors have conventionally been made of metal. Other materials have been recently trialed, for example, doped and undoped GaAs and a hybrid of polypropylene and high-resistivity silicon. Tunable mirrors, based on one-dimensional photonic crystals, have also been developed. Lenses are typically made of plastics. It is advantageous if the lens transmits visible radiation, as this facilitates optical alignment. Traditionally, plastic THz lenses were made by machining on a lathe. Recently, lenses have been manufactured by compressing various micropowders in metal moulds using a tabletop hydraulic press. Diffractive elements, cavities, and waveguides for THz range have also been constructed using micromachining technology. On the other hand, temporary, or reconfigurable, components may be formed by optical modulation, using visible light projected onto a silicon chip; aperture arrays and polarizers have been made in this manner.

As the TDTS is a new measurement technique, today's market of scientific instruments cannot offer a reasonably priced time-domain THz spectrometer. As a rule, such systems throughout the world are normally hand-made and are a subject of constant development and improvement. Therefore, modern plants designed to operate in THz region are quite expensive and their operation requires highly skilled professionals.

Figure 7.6 shows a diagram of the ZnTe crystal-based THz–TDS spectrometer used in Sandia Lab for investigations in THz range. This is photomixer-based spectrometer. The maintaining the stability of the entire system, including the ultrafast laser pulse shape, repetition rate, and power in the THz pumping and THz sampling beams, the generated THz pulse shapes, bandwidth and power, coupling into and out of the sample tube, and the temperature of sample is also important problem during experimental studies (Yang et al. 2011).

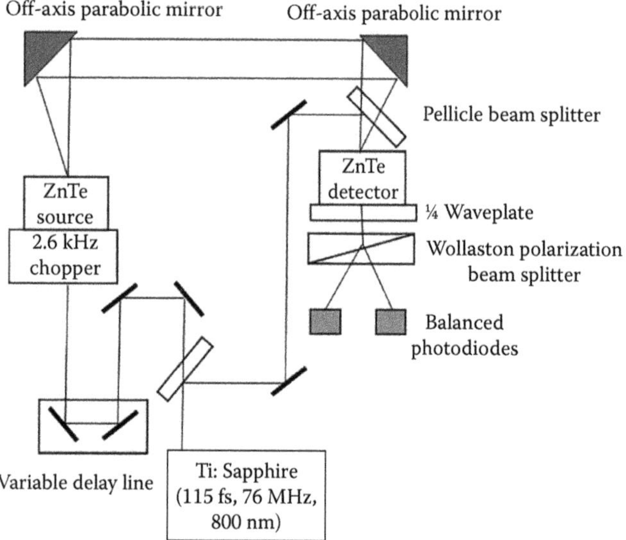

FIGURE 7.6 Terahertz time-domain spectroscopy overview. A mode-locked, Ti:Sapphire oscillator (Coherent Mira 900) with a 115 fs pulsewidth, 76 MHz repetition rate and centered at 800 nm is split into a high energy pump and low energy probe beam. The THz beam that is generated on the output face of the ZnTe crystal is collected and collimated by a 10 cm focal length, off-axis parabolic mirror with a 90° reflection angle. The THz beam traverses a path length of 94 cm and is focused onto a 10 × 10 × 0.9 mm thick <110> ZnTe crystal sensor. The polarization change is analyzed by allowing the probe beam to traverse through a ¼ wave-plate oriented at 45°. The polarization components are split by a Wollaston prism with a 10° divergence angle and the resulting split beam is weakly focused onto a large area balanced photoreceiver (New Focus Model 2307) with a gain setting of 10^5 V/A. The average probe optical power of each polarization component present at the balanced photodiode is 3.8 mW. The THz generation is modulated at 2.6 KHz and the differential signal is measured by a SRS Model 830 DSP lock in amplifier with an integration time of 300 ms and a low pass filter roll off of 12 dB/Oct. The data acquisition is controlled by LabVIEW 6.0 on a desktop PC. (From Foltynowicz, R.J. et al., Atmospheric propagation of THz radiation, Sandia National Laboratories, Report SAND2005-6389, New Mexico, 2005.)

The effect of atmospheric turbulence also needs to be taken into account. In parallel, scattering effect also results in the energy attenuation along the optical path. It comprises the molecular Rayleigh scattering and the Mie scattering by aerosols and the water vapor coagulum. As the wavelength of THz radiation lies in the order of aerosols, only Mie scattering should be taken into consideration. Aerosol particles mainly refer to the solid and liquid particles suspending in the atmosphere, for example, dusts, salts, ice particles and water droplets, and the Mie scattering effect mainly depends on their size-distribution, complex refractive index, and the wavelength of incident radiation. It is difficult to simulate the scattering by aerosols due to their large scale change in time and space domain.

This means that to build a theoretical model able to give information about the state of the atmosphere, basing on data of THz spectroscopy, an additional study of THz atmospheric transmittance as a function of water vapor content is required. However, data of these types are very limited in the open literature. Only a few broadband studies have been performed in the THz region, most of which were performed in a laboratory environment using either pure nitrogen or oxygen gas as the foreign broadening gas (Podobedov et al. 2005, 2008; Slocum et al. 2015). This method allows for control over the experimental conditions at the expense of using a foreign gas that is not of the same composition as the Earth's atmosphere. Other studies have been performed using atmospheric air (Pardo et al. 2001a, b), however these studies lack of control of the experimental parameters and are often performed at remote high altitude locations. No doubts that new technologies and approaches that will develop instruments suitable for wide application, are required (Krishna et al. 2012).

After solving all problems, the THz remote sensing technology can certainly be a powerful tool for the accurate prediction of the natural disaster, such as localized torrential rain and extraordinarily heavy snowfall, due to the climate change, because THz-radiation is suitable to observe humidity, ice cloud, and the temperature simultaneously with high spatial resolution (Liebe et al. 1993; Yasuko 2008). On account of the fact that water is highly absorptive in the THz range, while most materials which absorb moisture are either very transparent (e.g., paper, plastics) or reasonably transparent to THz, the THz spectroscopy can also become very prospective method for

THz imaging: a high contrast in the image between *moist* and *dry* regions could be a good tool for detecting materials with different properties (Federici 2012). Experimental studies confirmed this statement. THz imaging is shown to be a viable nondestructive evaluation tool not only for point measurements of water content, but also as a tool to visualize the spatial dependence of moisture.

Bidgoli et al. (2014) have shown that THz spectroscopy also can be successfully used for technological process control. For water vapor monitoring they used frequency scan in the range of 300–500 GHz (0.3–0.5 THz). The choice of this range (f < 1 THz) Bidgoli et al. (2014) explained by the fact that in this spectral range an absorption and scattering by the dust and tar components in the industrial gases are much less as compared to the THz waves at above 1 THz. In addition, within this spectral range, two strong rotational lines, at 448 and 383 GHz, have the potential to determine the concentration of water in gaseous mixtures. Other stronger lines (i.e., at 556 GHz) have a tendency to saturate at the high water vapor concentrations. The CO lines at 346 and 461 GHz can be used to quantify the concentration of CO in a complex gas mixture. A compact THz transmitter with an output power of 50–100 μW launched the EM waves into the gas-cell. Bidgoli et al. (2014) used a frequency multiplier-based Tx unit (Virginia Diodes, Inc., USA), which up-converted the microwave signal (8–14 GHz) to the desired range for this application (300–500 GHz). The frequency of the microwave source was controlled by a DC source that was integrated into a lock-in amplifier (Stanford Research SR830). The THz wave was amplitude-modulated (AM) at 18 Hz. To minimize the effects of instabilities in the frequency control voltage, the THz wave was also frequency-modulated (FM) at 333 Hz (which was much higher than the AM frequency). Through long operating time of the system with no service, Bidgoli et al. (2014) clearly proved that the THz radiation meets the optimal level of sensitivity requirement to not only overcome the hurdles facing IR measurement systems in an excellent way, but also to yield high-quality data with an acceptable temporal resolution. The replacement of slow conventional water measurement techniques with this new method allows rapid and stringent monitoring of gaseous steams from the biomass gasifiers, thereby providing better control over the whole process. The statistical analysis shows that with the electronic devices applied in the present work and the current design of the gas cell, the H_2O fractions in hot gaseous streams can be determined with an absolute precision close to 0.2% within a timescale of 1 min. This might be further improved on through optimization of the structural design of the gas cell and optimizing the THz transmitter/receiver system.

A more detailed analysis of the features of THz spectroscopy and possible fields of application one can find in Mittleman (2003), Schmuttenmaer (2004), Dexheimer (2007), Liu et al. (2007). Baxter and Guglietta (2011), Jepsen et al. (2011), Haddad et al. (2013), and Zouaghi et al. (2013).

REFERENCES

Baxter J.B., Guglietta G.W. (2011) Terahertz spectroscopy. *Anal. Chem.* 83, 4342–4368.

Bidgoli H., Cherednichenko S., Nordmark J., Thunman H., Seemann M. (2014) Terahertz spectroscopy for real-time monitoring of water vapor and CO levels in the producer gas from an industrial biomass gasifier. *IEEE Trans. Terahertz Sci. Technol.* 4 (6), 722–733.

Burch D.E. (1968) Absorption of infrared radiant energy by CO_2 and H_2O. Absorption by H_2O between 0.5 and 36 cm^{-1}. *J. Opt. Soc. Am.* 58 (10), 1383–1394.

Burch D.E., Gryvnak D.A. (1979) Continuum absorption by water vapor in the infrared and millimeter regions. In: *Proceedings of the International Workshop on Atmospheric Water Vapor*, Vail, CO, pp. 47–76, September 11–13, 1979.

Clough S., Kneizys F., Davies R. (1989) Line shape and the water vapor continuum. *Atmos. Res.* 23 (3), 229–241.

Dexheimer S.L. (Ed.) (2007) *Terahertz Spectroscopy: Principles and Applications.* CRC Press, Boca Raton, FL.

Federici J.F. (2012) Review of moisture and liquid detection and mapping using Terahertz imaging. *J. Infrared Milli. Terahz. Waves* 33, 97–126.

Foltynowicz R.J., Wanke M.C., Mangan M.A. (2005) Atmospheric propagation of THz radiation. Sandia National Laboratories, Report SAND2005-6389, New Mexico.

Gallerano G.P., Biedron S. (2004) Overview of Terahertz radiation sources. In: *Proceedings of the 26th international Free Electron Laser Conference*, Trieste, Italy, pp. 216–221, August 29–September 3, 2004.

Haddad J.El, Bousquet B., Canioni L., Mounaix P. (2013) Review in terahertz spectral analysis. *Trends Anal. Chem.* 44, 98–105.

Hall R.T., Dowling J.M. (1967) Pure rotational spectrum of water vapor. *J. Chem. Phys.* 47 (7), 2454.

Jepsen P.U., Cooke D.G., Koch M. (2011) Terahertz spectroscopy and imaging modern techniques and applications. *Laser Photon. Rev.* 5, 124–166.

Khazan M.A. (2002) Time-domain Terahertz spectroscopy and its application to the study of high-T_c superconductive this films. PhD Thesis, Universität Hamburg, Germany.

Krishna M.G., Kshirsagar S.D., Tewari S.P. (2012) Terahertz emitters, detectors and sensors: Current status and future prospects. In: Gateva S. (Ed.), *Photodetectors.* InTech, Rijeka, Croatia, pp. 115–144.

Lee Y. (2008) *Principles of Terahertz Science and Technology.* Springer Science+Business Media, New York.

Lewis R.A. (2014) A review of terahertz sources. *J. Phys. D: Appl. Phys.* 47, 374001.

Liebe H.J. (1989) MPM–An atmospheric millimeter-wave propagation model. *Int. J. Infrared Millimeter Waves* 10(6), 631–650.

Liebe H.J., Hufford G.A., Cotton M.G. (1993) Propagation modeling of moist air and suspended water/ice particles at frequencies below 1000 GHz. *AGARD Conference Proceedings*, Vol. 542, pp. 3.1-3.11 (SEE N94-30495 08-32)

Liu H.-B., Zhong H., Karpowicz N., Chen Y., Zhang X.-C. (2007) Terahertz spectroscopy and imaging for defense and security applications. *Proc. IEEE* 95 (8), 1514–1527.

Ma Q., Tipping R.H. (1990) Water vapor continuum in the millimeter spectral region. *J. Chem. Phys.* 93, 6127–6139.

Ma Q., R.H. Tipping R.H. (1992) A far wing line shape theory and its application to the water vibrational bands (II). *J. Chem. Phys.* 96, 8655–8663

Mittleman D.M. (Ed.) (2003) *Sensing with Terahertz Radiation.* Springer Series in Optical Sciences, Vol. 85. Springer, New York.

Pardo J.R., Serabyn E., Cernicharo J. (2001a) Submillimeter atmospheric transmission measurements on Mauna Kea during extremely dry El Niño conditions: Implications for broadband opacity contributions. *J. Quant. Spectrosc. Radiat. Transfer* 68, 419–433.

Pardo J.R., Cernicharo J., Serabyn E. (2001b) Atmospheric transmission at microwaves (ATM): An improved model for millimeter/submillimeter applications. *IEEE Trans. Anten. Propag.* 49 (12), 1683–1694.

Podobedov V.B., Plusquellic D.F., Fraser G.T. (2005) Investigation of the water-vapour continuum in the THz region using a multipass cell. *J. Quant. Spectrosc. Radiat. Transfer* 91, 287–295.

Podobedov V.B., Plusquellic D.F., Siegrist K.E., Fraser G.T., Ma Q., Tipping R.H. (2008) New measurements of the water vapour continuum in the region from 0.3 to 2.7 THz. *J. Quant. Spectrosc. Radiat. Transfer* 109, 458–467.

Rice D.P., Ade P.A.P. (1979) Absolute measurements of the atmospheric transparency at short millimeter wavelengths. *Infrared Phys.* 19, 575–584.

Rosenkranz P.W. (1998) Water vapor microwave continuum absorption: A comparison of measurements and models. *Radio Sci.* 33, 919–928.

Rothman L., Gordon I., Barbe A. (2009) The HITRAN 2008 molecular spectroscopic database. *J. Quant. Spectrosc. Rad. Transfer* 110(9), 533–572.

Ryu C., Kong S. (2010) Atmospheric degradation correction of terahertz beams using multiscale signal restoration. *Appl. Optics* 49 (5), 927–935.

Schmuttenmaer C.A. (2004) Exploring dynamics in the far-infrared with terahertz spectroscopy. *Chem. Rev.* 104, 1759.

Siegel P. (2002) Terahertz technology. *IEEE Trans. Microwave Theory Techn.* 50 (3), 910–928.

Slocum D.M., Slingerland E.J., Giles R.H., Goyette T.M. (2013) Atmospheric absorption of terahertz radiation and water vapor continuum effects. *J. Quant. Spectrosc. Rad. Transfer* 127, 49–63.

Slocum D.M., Giles R.H., Goyette T.M. (2015) High-resolution water vapor spectrum and line shape analysis in the Terahertz region. *J. Quant. Spectrosc. Rad. Transfer* 159, 69–79.

Tonouchi M. (2007) Cutting-edge terahertz technology. *Nat. Photonics* 1 (2), 97–105.

Townes C.H., Schawlow A.L. (1975) *Microwave Spectroscopy.* Dover, New York, p. 104.

Van Exter M., Fattinger Ch., Grischkowsky D. (1989) Terahertz time-domain spectroscopy of water vapor. *Opt. Lett.* 14 (20), 1128–1130.

Yang Y., Shutler A., Grischkowsky D. (2011) Measurement of the transmission of the atmosphere from 0.2 to 2 THz. *Opt. Express* 19 (9), 8830–8838.

Yang Y., Shutler A., Grischkowsky D. (2012) Understanding THz pulse propagation in the atmosphere. *IEEE Trans. Terahertz Sci. Technol.* 2 (4), 406–415.

Yang Y., Mandehgar M., Grischkowsky D. (2014) Determination of the water vapor continuum absorption by THz-TDS and molecular response theory. *Opt. Express* 22(4), 4388–4403.

Yao J., Wang R., Cui H., Wang J. (2012) Atmospheric propagation of Terahertz radiation. In: Escalante B. (Ed.) *Remote Sensing–Advanced Techniques and Platforms.* InTech, Rijeka, Croatia, pp. 371–386.

Yasuko K. (2008) Terahertz-wave remote sensing. *J. Nat. Inst. Inform. Commun. Technol.* 55(1), 79–81.

Yasuko K., Takamasa S. (2008). Atmospheric propagation model of Terahertz-wave. *J. Nat. Inst. Inform. Commun. Technol.* 55 (1), 73–77.

Yin X., Ng B.W.-H., Abbott D. (2012) *Terahertz Imaging for Biomedical Applications: Pattern Recognition and Tomographic Reconstruction*, Chapter Terahertz Sources and Detectors. Springer Science+Business Media, New York, pp. 9–26.

Yuan T., Liu H.B., Xu J.Z., Al-Douseri F., Hu Y., Zhan X.-C. (2003) THz time-domain spectroscopy of atmosphere with different humidity. In: Hwu R.J., Woolard D.L. (Eds.) Terahertz for military and security applications. *SPIE Proceedings*, 5070, 28–37.

Xin X., Altan H., Saint A., Matten D., Alfano R.R. (2006) Terahertz absorption spectrum of *para* and *ortho* water vapors at different humidities at room temperature. *J. Appl. Phys.* 100, 094905.

Zhan H.L., Sun S.N., Zhao K., Leng W.X., Bao R.M., Xiao L.Z., Zhang Z.W. (2015) Less than 6 GHz resolution THz spectroscopy of water vapor. *Sci. China. Technol. Sci.* 58 (12), 2104–2109.

Zhang X., Xu J. (2009) *Introduction to THz Wave Photonics.* Springer Science+Business Media, New York.

Zouaghi W., Thomson M.D., Rabia K., Hahn R., Blank V., Roskos H.G. (2013) Broadband terahertz spectroscopy: Principles, fundamental research and potential for industrial applications. *Eur. J. Phys.* 34, S179–S199.

8 LIDAR Systems for Atmosphere Monitoring

8.1 INTRODUCTION

Light detection and ranging (LIDAR) hygrometry is a modern kind of optical methods for control of water vapor in the atmosphere. LIDAR is an optical remote sensing technology that measures properties of scattered light to obtain information about atmospheric composition, clouds, and aerosols (Grant 1991; Turner and Whiteman 2002). It is known that remote control based on absorption measurements can be divided into two main types: passive, for which no special light sources are needed; and active, for measurements that utilize additional sources of electromagnetic radiation. In the first type, the Sun functions as the light source. Light from the Sun is reflected and reemitted from the ground as it passes through a gas plume. As all molecules absorb light in very discrete and narrow wavelength bands of ultraviolet (UV), visible, and infrared (IR) light, by focusing on these bands of absorption, one can observe and predict how much less energy reaches the sensor over a gas plume as opposed to an unaffected area. Using atmosphere modeling software, it is possible to obtain a reasonable value for the concentration of gas. As shown previously, passive methods for studying the atmosphere can be based on using optical, microwave, and terahertz spectroscopy. Using these methods of testing and the methods of processing the information obtained, one can receive information about the temperature and humidity distribution in the height of the atmosphere (e.g., Table 8.1), and on the Earth's surface, which is essential for the development of models suitable for adequate weather forecasting. As is known, humidity, in particular, is a critical variable in the initialization of these models. Profiles of the water vapor and temperature height distribution can be obtained by using radiosondes. However, it is well known that radiosonde measurements are often not reliable at upper-tropospheric temperatures. Furthermore, the temporal resolution of routine observations performed by weather services is rather low, with typically two radiosonde launches per day. Therefore, important weather phenomena such as the development of the convective boundary layer (CBL) and the passage of cold and warm fronts cannot be resolved. Radiosondes take only one data point per measurement height. Thus, the measured relative humidity (RH) profiles are often not representative. For these reasons, alternatives to routine radiosonde observations are required.

Studies have shown that many of these problems can be solved with the help of LIDAR technique, which refers to the methods of active remote sensing (Grant 1991; Turner and Whiteman 2002). The advantage of LIDAR methods is that range-resolved water vapor measurements can be made at a high-resolution during day or night. LIDAR systems can be operated in spectral regions that avoid interference from absorption by other species and LIDAR retrievals require fewer assumptions than those for passive techniques and do not require complex inversion algorithms (Kley et al. 2000). There is only one limitation. LIDAR measurements are possible whenever the laser beam is not blocked by clouds.

8.2 PRINCIPLES OF OPERATION

LIDAR originated in the early 1960s, shortly after the invention of the laser, and combined laser-focused imaging with radar's ability to calculate distances by measuring the time for a signal to return. Its first applications came in meteorology, where the National Center for Atmospheric Research used it to measure clouds (Goyer and Watson 1963). The physical principles of LIDAR operation are based on the measurement of the properties of the radiation that is returned either from molecules and particles in the atmosphere or from the Earth's surface when illuminated by a laser source with purpose to find information about remote object (Cracknell and Hayes 2007). The principle of LIDAR operation is identical to the principle of the radar operation. However, whereas radar uses radio waves, LIDAR uses much shorter wavelengths of the electromagnetic spectrum, typically in the UV, visible, or near IR region.

In its basic form a LIDAR transmitter produces a pulse of optical radiation, which is directed into the medium of interest by the system optics. As the transmitted optical energy propagates, it is affected by the characteristics of the region of atmosphere through which it passes. Gas molecules and particles or droplets

TABLE 8.1
The Change of the Atmosphere Parameters with the Altitude

Altitude, km	Air Density, kg/m³	Air Pressure, Pa	Temperature		Water Vapor Content, Sub-Artic, ppmv	
			K	°C	Summer	Winter
0	1.22	101.3	288	15	11,900	1410
1	1.11	89.9	282	8, 5	8700	1620
2	1.00	79.5	275	2	6750	1430
3	0.91	70.1	269	−5	4820	1170
4	0.82	61.6	263	−11	3380	790
5	0.74	54.0	256	−18	2220	431
6	0.66	47.2	250	−24	1330	237
7	0.59	41.1	244	−30	797	147
8	0.53	35.6	236	−37	400	33.8
9	0.47	30.7	229	−44	130	29.8
10	0.41	26.4	223	−50	42	20

Source: ICAO, Manual of ICAO Standard Atmosphere, International Civil Aviation Organization Doc. 7488/2, ICAO, Montreal, Canada, 1964; Abreu L.W., Anderson G.P. (Eds.). MODTRAN Report, Ontar Corporation, North Andover, MA, 1996.

cause some of the energy to be scattered. A small fraction of this scattered energy is backscattered, that is, directed back toward the LIDAR system. This energy can be detected at the LIDAR receiver. Since energy which is not directed back along the path of propagation is lost, scattering also produces attenuation in the optical field. Additional attenuation occurs from absorption by the gases and particles, which occurs along the path. Absorption of energy increases sharply when the operational wavelength coincides with a resonance absorption line of an atmospheric gas.

In LIDAR applications, the signal of interest is usually the energy backscattered by the atoms and particles within some segment of the propagation medium. In some cases, when the atmospherically backscattered signal is too weak to be detected, the energy reflected from a solid target at one extreme of the propagation medium is used. The returned energy is collected by the receiver optics and directed onto a photodetector, which produces an electrical signal proportional to the intensity of the optical radiation incident at the detector. In a pulsed LIDAR, the time elapsed between the transmit pulse and the return signal is proportional to the range from which the return originates. This means that by measuring the time delay between transmission of a pulse and detection of the reflected signal, the distance to an object can be determined.

Since LIDAR system performance characteristics are affected by particular properties of the propagation medium and target, analysis of the received signal yields information on these properties. The magnitude of the signal from a given range is a function of the backscatter characteristic at that range as well as the attenuative properties of the intervening medium. In order to estimate one parameter, for example, backscatter, information on the other parameters (propagation-path *characteristics*) must be known or estimated. For the most part, both backscatter and transmission properties are a strong function of operating wavelength.

In addition, it is necessary to have in mind that in general the atmosphere is a spatially and temporally random medium with respect to its scattering and absorptive properties. Although it can be characterized in a statistical sense, its exact characteristics during the instant in which it interacts with the optical energy are not precisely known. This random characterization also applies to refractive index. Inhomogeneity in refractive index perturb the phase of a propagating wave, producing bending, decoherence, and scintillations in the wave intensity. Backscatter and turbulence effects also vary with wavelength; at shorter wavelengths a molecular scattering and turbulence effects are both enhanced. By appropriately selecting the operating wavelength one can often deliberately emphasize the effects of one parameter relative to another in order to perform a particular function.

Although the term *LIDAR* originally applied to any system which employed an optical source, almost all present-day systems use lasers as the system transmitter. The monochromatic radiation produced by laser transmitters enables the operating wavelength to be exactly specified, and provides transmit beams which remain collimated over long distances. In remote sensing applications, the minimal dispersion of the beam permits measurements to be made at low elevation angles where typical radar systems would be hampered by ground clutter effects.

Experiments and simulations have shown that different types of scattering can be used for different LIDAR applications, most common are Rayleigh scattering, Mie scattering (Hua and Kobayashi 2005), and Raman scattering (Refaat et al. 2008) as well as fluorescence. Therefore, at present there are different types of LIDAR instrument:

- The backscatter LIDAR, in which the laser beam is backscattered, reflected or reradiated by the target, gives information on the scattering and extinction coefficients of the various atmospheric layers being probed.
- The differential absorption LIDAR analyses (DIAL) the returns from a tunable laser at different wavelengths to determine densities of specific atmospheric constituents, as well as water vapor and temperature profiles.
- Doppler LIDAR measures the Doppler shift of the light backscattered from particles or molecules moving with the wind, thereby allowing the determination of wind velocity.
- The ranging and altimeter LIDAR provides accurate measurements of the distance from a reference height to precise locations on the Earth's surface.

In atmospheric physics, it is considered that for humidity monitoring and for measuring aerosol and cloud properties the DIAL technique (Shipley et al. 1983; Measures 1984; Cha et al. 1991; Grund and Eloranta 1991; Grund et al. 1995; Wulfmeyer and Walther 2001) is the most appropriate. This technique was proposed by Schotland (1966). DIAL transmits two wavelengths: an *online* wavelength that is absorbed by the gas of interest and an *off-line* wavelength that is not absorbed. The differential absorption between the two wavelengths is a measure of the concentration of the gas as a function of range. So, DIAL measures the atmospheric backscattered radiation from two laser pulses transmitted at slightly different wavelengths. This process is applied to determine the relation of a water vapor mixture in the atmosphere. Using the DIAL method, the average molecular number density between ranges R_1 and R_2 is calculated using the relation 8.1. A limitation on DIAL measurement accuracy is a lack of wavelength stability, which is required to keep the laser held within a picometer of the absorption-line center.

$$n = \frac{1}{2\Delta\sigma(R_2 - R_1)} ln \frac{S_{on}(R_1) \cdot S_{off}(R_2)}{S_{on}(R_2) \cdot S_{off}(R_1)} \qquad (8.1)$$

where:

n is the average water vapor number density between the ranges R_1 and R_2

S is the LIDAR signal

$\Delta\sigma$ is the differential absorption cross section between the on- and off-lines

The $\Delta\sigma$ values are computed from the line parameters for water vapor lines in the near visible region measured using high-resolution spectroscopic techniques (Grossmann and Browell 1989a, b; Ponsardin and Browell 1997).

The advantages of DIAL measurements are their relatively high spatial resolution, high measurement specificity, lack of dependence on external light sources, and relatively simple inversion methods compared to retrieval methods used in passive remote sensing (Measures 1984; Grant 1991). With the DIAL method, LIDAR measurements can be made during day or night and in between and up to cloudy regions in the atmosphere. The DIAL method provides a self-calibrated absolute water vapor measurement capability. The DIAL technique can be used to directly obtain the concentration profile of water vapor using the LIDAR signals and $\Delta\sigma$. The advantage of the DIAL method is also that in addition to measuring gas concentration profiles, high spatial resolution aerosol backscattering distributions are simultaneously can be obtained as part of the DIAL measurement using the off-line LIDAR signals. The vertical and/or horizontal resolution can be adjusted in the DIAL analysis by changing how the high-resolution LIDAR data are averaged.

The main requirements for DIAL measurements are knowledge of the absorption spectra of the measured species, a tunable pulsed laser source, and a suitable detection element. The detector converts the optical power of the DIAL return signal into an electrical signal, thereby directly affecting the instrument measurement sensitivity. Therefore, both the laser source and the detector must be spectrally compatible with the absorption spectra, with minimal influence from other species.

This method for water vapor detection uses single absorption lines that are typically 10 pm wide. DIAL measurement simulations involving these narrow absorption lines showed that the quality of laser spectral radiation has to be controlled very well to avoid undesirable systematic influences (Ismail and Browell 1989). As it is the usual dominant noise source in the DIAL instrument, the detector affects the system minimum detectable signal and correspondingly the minimum detectable concentration at a given range. These issues drive the need for a detector of narrow spectral bandwidth, high quantum efficiency, and low noise. Semiconductor quantum detectors are an attractive solution for DIAL receivers due to their ruggedness, compact size, and small sensitive areas, which lead to lower noise (Refaat et al. 2003). It was also shown that the signals and $\Delta\sigma$ are influenced by a number of atmospheric and instrumental effects that need to be taken into consideration during the operation of DIAL and during data analysis (Ismail and Browell 1989). For example, gradients in the atmospheric aerosol scattering, Doppler broadening, and the near range LIDAR overlap region present challenges in DIAL data reduction (Bosenberg 1998).

The first detection of water vapor using the DIAL technique was made in 1966 with a laser system that had a fortuitous coincidence with a water vapor absorption line. The DIAL technique for water vapor measurements using tunable lasers was demonstrated from the ground by Browell et al. (1979), and subsequently by Cahen et al. (1982) and Bosenberg (1987). The first airborne measurements of water vapor came in 1981 (Browell et al. 1984), and this was followed by subsequent airborne measurements (Ehret et al. 1993; Higdon et al. 1994). The results of last experiments carried out using DIAL technique were published by Vogelmann et al. (2015).

LIDAR systems based on Raman scattering are most commonly used for monitoring the atmosphere and building profiles of temperature and humidity (Sherlock et al. 1999a, b; Turner et al. 2002; Venable et al. 2005; Refaat et al. 2008; Navas-Guzmán et al. 2014). Examples of such Raman LIDAR systems which have been operated from the ground more than 30 years are listed in Table 8.2.

Widespreading of Raman LIDAR systems are connected with its relative simplicity compared to competing techniques such as differential absorption LIDAR (Grant 1991). In contrast to the DIAL technique, the laser source in Raman LIDAR does not have to be tuned to a water vapor absorption line and generally does not require a very narrow linewidth; that the Stokes Raman scattering is essentially temperature independent for tropospheric water vapor; and that high spatial resolution measurements are possible (Cooney et al. 1985). This has permitted to develop an automated Raman LIDAR system (Turner et al. 2002). The technique is relatively free of systematic errors (e.g., aerosol effects) and can be operated from the ground because the radiation is not absorbed by water vapor in the lower troposphere. In addition, Raman LIDAR technique is versatile, allowing studies of cloud liquid water and droplet radius, and direct measurement of aerosol extinction (Whiteman and Melfi 1999). Raman LIDARs are able to profile the water vapor *in the same parcels* as the aerosol measurements. Di Girolamo et al. (2009) have shown that the characterization of atmospheric processes involving cirrus clouds can also be organized using the capabilities of Raman LIDAR systems. Cirrus clouds represent a fundamental component of the global climate system,

TABLE 8.2
Raman LIDAR Systems Used for Atmosphere Monitoring

Developer	Working Wavelength	References
NASA Goddard Space Flight Center, USA	371, 382, and 403 nm	Melfi and Whiteman (1985), Whiteman et al. (1992), Whiteman and Melfi (1999)
U.S. Department of Energy (DOE), USA	355, 387, and 408 nm	Goldsmith et al. (1998), (http://www.arm.gov/instruments/rl)
Jet Propulsion Laboratory, California Institute of Technology	355, 387, and 407 nm	Leblanc et al. (2012)
The Observatoire de Haute Provence (OHP), France	532.1 nm	Sherlock et al. (1999a, b)
Swiss Federal Institute of Technology (EPFL) (MeteoSwiss Payerne), Switzerland	355, 387, and 408 nm	Dinoev (2009), Dinoev et al. (2013)
CEILAP (CITEDEF–CONICET) in the frame of the Argentinean Cherenkov Telescope Array (CTA), Argentina	355, 532, and 1064 nm	Pallotta et al. (2013)
The Institute for Tropospheric Research (German aerosol LIDAR network), Germany	355, 532, and 1064 nm	Mattis et al. (1998, 2002), Wandinger (2005b)
Ecole Polytechnique (Meteorologie Nationale, EERM, Observatoire de Magny-les-Hameaux), France	278, 284, and 295 nm	Renaut et al. (1980)
University College of Wales, UK	355 and 532 nm	Vaughan et al. (1988)
The Howard University, Washington, USA	355, 387, and 408 nm	Adam et al. (2010)

primarily through their radiative effects. Cirrus radiative properties are important modulators of the incoming solar and the outgoing planetary radiation (Liou 1986; Stephens et al. 1990). The role played by cirrus clouds in humidifying the underlying atmosphere through evaporation of falling ice crystals is also important. Its disadvantage is the very weak Raman scattering cross section, which has traditionally meant a long averaging time (several hours) for a precise measurement near the tropopause (e.g., Vaughan et al. 1988). In addition, atmospheric aerosols and molecular species, such as ozone and oxygen, interfere with the measurement. However, recent systems based on excimer lasers have extended the Raman technique to the tropopause.

The basics of Raman scattering theory are described by Long (2002). The calculation of frequency shifts and line intensities for Raman LIDAR applications is summarized in Wandinger (2005a). The standard single scattering Raman LIDAR equation, which can be used for atmosphere monitoring and building profiles of RH, can be expressed as in Equation 8.2:

$$P(\lambda_{X,r}) = \frac{O_X(r)\cdot P_o(\lambda_L)\cdot N_X(r)\left[\dfrac{d\sigma_X(\lambda_L,\pi)}{d\Omega}\right]A\cdot\xi(\lambda_X)}{r^2} \times \exp\left\{-\int_0^r[\alpha(\lambda_L,r')+\alpha(\lambda_X,r')]dr'\right\} \quad (8.2)$$

Here $P(\lambda_{X,r})$ is the background subtracted power received at the Raman-shifted wavelength appropriate for molecular species X as a function of range, r. $Q_X(r)$ is the channel overlap function, and $P_0(\lambda_L)$ is the output power of the laser at the laser wavelength, λ_L. $N_X(r)$ is the number density of molecules and $d\sigma_X(\lambda_L,\pi)/d\Omega$ is the Raman differential backscatter cross section at the laser wavelength. $\xi(\lambda_X)$ is the total LIDAR receiver optical efficiency at the wavelength of the Raman species of interest and includes factors such as the reflectivity of the telescope, the transmission of any conditioning optics, the transmission of ant filters, and the quantum efficiency of the detector. A is the receiver telescope area. The exponential factor gives the two-way atmospheric transmission, where $\alpha(\lambda_{X,r})$ is the total extinction coefficient due to scattering and absorption by molecules and aerosols at the Raman-shifted wavelength as a function of range along the path of the laser beam. In this context, the term *aerosols* may be used to describe any nonmolecular atmospheric constituent such dust, water droplets, ice crystals, and so on. This equation uses the simplifying assumption that the Raman scattered signals are monochromatic.

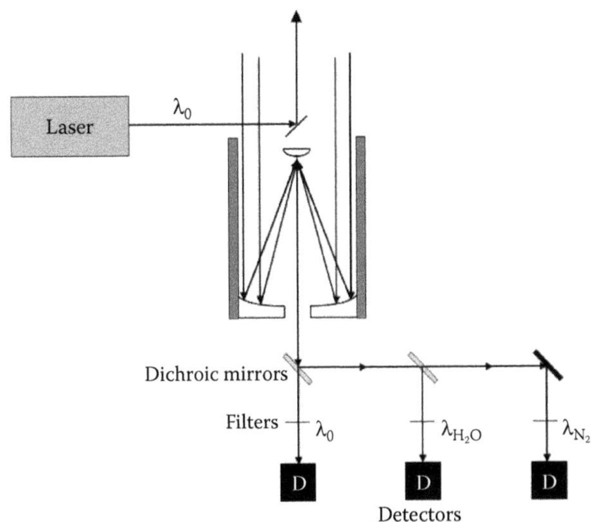

FIGURE 8.1 Basic Raman LIDAR concept. (Idea from Goldsmith, J. et al., *Appl. Opt.*, 37, 4979–4990, 1998. With permission.)

As mentioned earlier, one of the most important tasks of Raman LIDAR measurements of water vapor is building of the atmospheric extinction profile. Simulation has shown that these profiles can be determined using analysis of molecular and aerosol scattering. However, these scattering can be modeled if atmospheric pressure and temperature are known. Therefore, for resolving these problems LIDAR systems usually have additional LIDAR channels, such as one at the laser wavelength and one at either the N_2 or O_2 Raman shifted wavelength (Figure 8.1). The rotational Raman spectra of nitrogen and water permit to determine atmospheric temperature and pressure with good accuracy (Cooney and Pina 1976; Endemann and Byer 1981).

As it is known, Raman LIDAR exploits inelastic scattering to single out the gas of interest from all other atmospheric constituents. A small portion of the energy of the transmitted light is deposited in the gas during the scattering process, which shifts the scattered light to a longer wavelength by an amount that is unique to the species of interest, thus giving high selectivity of the method. For example, water vapor, nitrogen, and oxygen, molecules of interest in the normal atmosphere, exhibit convenient vibrational Raman shifts, that are approximately 3657, 2331, and 1555 cm^{-1}, respectively. This means that if Raman LIDAR uses an XeF excimer laser (λ = 351 nm), during the measurement process one should control a backscattered light at the laser wavelength as well as a Raman scattered light from water vapor (403 nm), nitrogen (382 nm), and oxygen (371 nm) molecules. The same case for Nd:YAG laser is shown in Figure 8.2. At that the

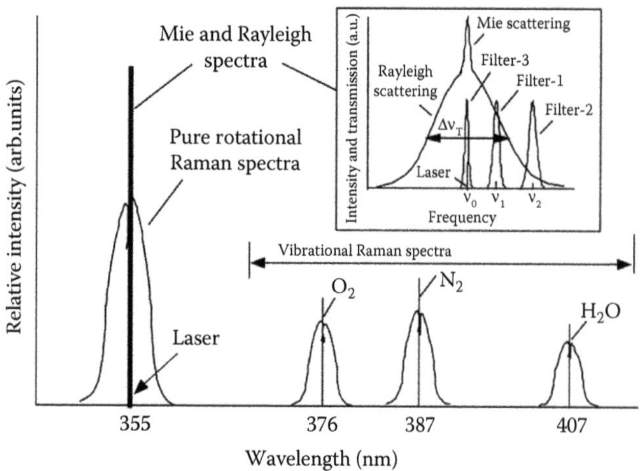

FIGURE 8.2 Overview of LIDAR backscattering spectra for a laser wavelength of 355 nm. Mie and Rayleigh scattering spectra and transmission of filters are shown in the upper inset right panel. (From Hua, D. and Kobayashi, T., *Jpn. J. Appl. Phys.* 44, 1287, 2005. Copyright 2005: The Japan Society of Applied Physics. With permission.)

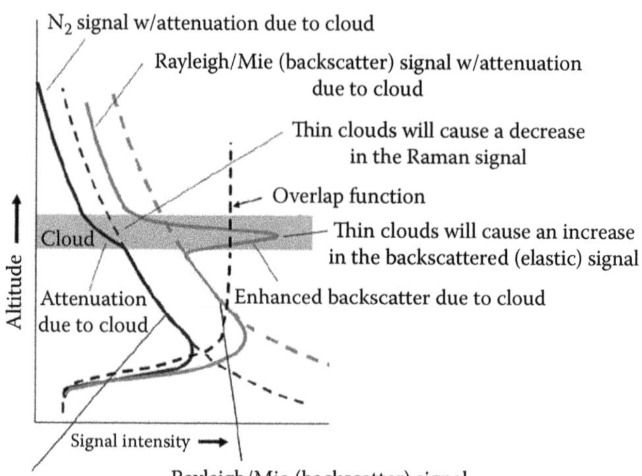

FIGURE 8.3 Schematic representation of simplified LIDAR returns. (Data extracted from Wandinger, U., Raman lidar, In Weitkamp C. (Ed.), *Lidar*, Springer, New York, 2005a.)

higher the concentration of the gas is, the greater will be the magnitude of the backscattered signal.

In this case, the profiles of water vapor mixing ratio, RH, aerosol backscattering, and aerosol extinction are derived usually routinely using a set of automated algorithms (Turner et al. 2002). As a rule, the water vapor mixing ratio $q(R)$ are computed from the ratio of the return signals due to Raman scattering by water vapor $P_{H_2O}(R)$ to Raman scattering by nitrogen $P_{N_2}(R)$, using the equation:

$$q(R) = k_{H_2O} \cdot \frac{P_{H_2O}(R)}{P_{N_2}(R)} \cdot \tau(R) \tag{8.3}$$

where $\tau(R)$ is a term correcting the differences in the atmospheric transmission at the water vapor and nitrogen Raman wavelengths. The coefficient of proportionality k_{H_2O}, commonly denoted as the LIDAR calibration constant, can be presented as

$$k_{H_2O} = \frac{M_{H_2O} \cdot n_{N_2}}{M_{air} \cdot n_{air}} \times \frac{\int [T_{N_2}(R,\lambda) \cdot \eta_{N_2}(R,\lambda) \cdot \sigma_{N_2}(\lambda)] d\lambda}{\int [T_{H_2O}(R,\lambda) \cdot \eta_{H_2O}(R,\lambda) \cdot \sigma_{H_2O}(\lambda)] d\lambda} \tag{8.4}$$

where:

M_X and n_X is the molecular mass and the number density of species X (air denotes dry air)

$\sigma_X(\lambda)$ is the Raman cross section of species X

$T_X(\lambda,R)$ and $\eta_X(\lambda,R)$ are the instrument transmission function

PMT is the efficiency of the respective Raman channel X

The calibration constant is commonly obtained by comparison of a LIDAR profile to data from a reference instrument: collocated radiosonde, microwave radiometer, and global positioning system (GPS) (Whiteman et al. 1992; Kley et al. 2000), but the calibration constant (determined by the average LIDAR/sonde ratio) varied from sonde to sonde. There is a clear need for further work in this area to characterize the performance of both LIDARs and sondes.

It is clear that the possibilities of building water vapor profiles are determined by the value of returned signal and the ability to process it. A simplified model of the behavior of the LIDAR equation reveals that without the overlap term, the returned signal would primarily decrease as $1/r^2$ as the range (altitude) increases (Wandinger 2005b). This is indicated by the dashed lines (gray—Rayleigh–Mie scattering [H_2O] and black—Raman scattering [N_2]) in Figure 8.3. Signals close to the aperture of the telescope (altitude approaches zero) are attenuated due to incomplete overlap of the telescope's field of view and the signal source (Measures 1984). This region of incomplete overlap is primarily impacted by shadowing of the telescope's secondary mirror and nonoptimum focusing of the near field source. Here it is assumed that the telescope is designed for focusing of a source located at infinity. An idealized overlap function ranging from a value of 0 at ground level to a value of 1 at an altitude that corresponds to complete overlap of the source and the telescope's field of view is represented by the dashed black line. On account of the incomplete overlap the actual detected signal starts near 0 in the near field and increases to a local maximum value near the region of complete overlap. The described behavior is the same for both the Rayleigh–Mie and Raman channels. The overlap factor in Equation 8.3 is correct for this behavior. The solid gray and black lines at low altitudes indicate the behavior of the idealized signal at altitudes near 0. In a region of optically thin clouds or aerosol layers the Rayleigh–Mie and Raman signals behave differently. The former shows a spike in the return signal due to the impact of aerosol. The Raman signal shows enhanced attenuation of the return signal. Both show possible attenuation of the laser by the aerosol layer. The resulting signals are depicted by the solid green (Rayleigh–Mie) and solid black (Raman) lines.

Experiment has shown that an agreement between LIDAR and radiosondes was generally good up to 8 km. For a 1 min profile, the random error in the profiles of water vapor mixing ratio is less than 10% for altitudes below 7.5–8.5 km. A vertical resolution can be varied from 100 m (boundary layer) to 500–1000 m (free troposphere) (Mattis et al. 2002; Whiteman et al. 2007). By averaging for longer periods of time and/or by reducing the vertical resolution, the profiles up to 10 km and more can be obtained (Ferrare et al. 2003). Examples of profiles of water vapor mixing ratio and RH built basing on data received by LIDAR systems are shown in Figure 8.4. For comparison, at the same figure it is given a profile of water vapor mixing ratio measured with radiosonde. It is seen that there is a very good agreement between the data presented in Figure 8.4. The relative uncertainties in mixing ratio and RH are of the order of 5%–10%, and 5%–25%, respectively, for the Raman LIDAR that was used.

One should note that vertical profiling of RH with high spatial and temporal resolution may allow to study in detail the processes of cloud formation at the top of the boundary layer. Furthermore, Raman LIDARs are powerful tools for aerosol monitoring (Ferrare et al. 1998). Aerosol and RH observations with one LIDAR and thus in the same air volume are an attractive approach to studying aerosol–climate interactions, because the optical properties of particles depend strongly on RH. Carbon dioxide measurements using Raman LIDAR can also be performed (Whiteman et al. 2007). Wulfmeyer et al. (2010) have shown that Raman LIDAR system is able to observe turbulent processes present in the CBL as well. As for other possibilities of Raman LIDAR systems, they are listed in Table 8.3. It is seen that the

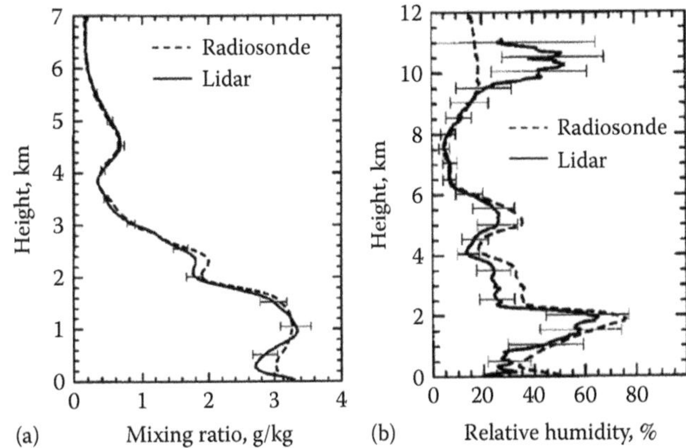

FIGURE 8.4 Mixing ratio (a) and relative humidity (b) measured with radiosonde and determined with a calibrated Raman LIDAR (k_{H_2O} = 1 in Equation 8.3). The LIDAR signals were smoothed before calculation of the mixing ratio to reduce a signal noise (Signal smoothing: 0–3 km–120 m; 3–5 km–480 m; 5–12 km–1200 m). The same smoothing lengths were applied to the radiosonde data. Error bars indicate the standard deviation as a function of Δk_{H_2O} (=0.0006) and a signal noise. (Form Mattis, I. et al., *Appl. Opt.*, 41, 6451, 2002. With permission of Optical Society of America.)

TABLE 8.3
Overview of Raman LIDAR Techniques

Measured Quantity	Interacting Molecule, Raman Band Used	Typical Achievable Measurement Range
Water-vapor mixing ratio	H_2O (vapor), VRR (+ reference gas, VRR)	0–12 km (night) 0–5 km (day)
Temperature	N_2 and/or O_2, RR	0–40 km (night) 0–12 km (day)
Extinction coefficient Backscatter coefficient LIDAR ratio	N_2 or O_2, VRR or RR (+ elastic signal for backscatter and LIDAR ratio)	0–30 km (night) 0–10 km (day)
Ozone concentration (Raman DIAL)	N_2 and/or O_2, VRR–VRR or RR–VRR	3–20 km (night) 0–3 km (day SB)
Other trace-gas concentrations	Specific gas, VRR (+ reference gas, VRR)	0–1 km (day and night)
Liquid water	H_2O (liquid), VRR (+ reference gas, VRR)	0–4 km (night)

Source: Data extracted from Wandinger, U., *Proc. SPIE*, 5830, 307–316, 2005b.

VRR—vibration–rotation Raman band, RR—rotational Raman band, SB—solar-blind spectral region.

Raman LIDAR systems allow the measurement of atmospheric trace gases, temperature, and aerosol properties (Wandinger 2005a, b).

While the literature on errors and corrections to Raman LIDAR measurements of water vapor is not as extensive as it is for DIAL measurements, there are indeed a number of random and systematic error sources that must be contended with, either in the system design or in its operation. A main limitation of Raman LIDAR measurements of water vapor is the limited signal strength, necessitating more multipulse averaging than for the DIAL technique (Grant 1991). This effect ultimately limits a system performance. The use of photon-counting techniques improves the measurement results compared with analog signal detection. If photons are counted, the measurement random error can be determined from the counting statistics. The wavelength dependence of atmospheric extinction is another limitation. This effect makes the measurements and analysis more complex even when the signal strength is adequate. However, as Raman LIDAR systems become more widespread, these effects will be better understood, and the measurements will be improved.

8.3 APPROACHES TO LIDAR SYSTEMS FABRICATION

The most important decision for the development of a new water vapor DIAL or Raman LIDAR system is the proper choice of the operating wavelength of the laser transmitter. This choice is determined by compromises among the following requirements (Wulfmeyer and Walther 2001): (1) the presence of suitable water-vapor absorption lines within the tuning range of the laser transmitter; (2) availability of detectors with high efficiency and low noise; (3) wavelength dependence of the atmospheric backscatter and extinction coefficients; (4) wavelength dependence of the atmospheric background signal, and, as far as possible; and (5) eye safety of the laser transmitter. Thus, the laser transmitter has to be operated in a wavelength region where suitable water-vapor absorption lines exist. The absorption lines must be strong enough to permit measurements to be made in the lower stratosphere and the upper troposphere from airborne platforms.

Experiment has shown that suitable water-vapor absorption lines present in all spectral ranges and therefore, in principle, LIDAR systems can be designed for the UV (Hua and Kobayashi 2005), visible (Werner and Herrmann 1981; Wilkerson and Schwemmer 1982), or IR spectral range (Rosenberg and Hogan 1981; Cha et al. 1991). So, now there are devices using the laser radiation at 266 nm (Froidevaux et al. 2013), 351 and 355 nm (Ferrare et al. 1998, 2003; Hua and Kobayashi 2005; Whiteman et al. 2007), 693 nm (Werner and Herrmann 1981), 724 nm (Ehret et al. 1993), 815 nm (Browell et al. 1997), 940 nm (Wilkerson and Schwemmer 1982), 1.7 μm (Rosenberg and Hogan 1981), 2.1 μm (Cha et al. 1991), and 9.2–10.7 μm (Murray et al. 1976; Baker 1983; Grant et al. 1987).

It is believed that IR LIDAR, unlike the UV and visible versions, provide the potential for increasing the transmitted laser energy while maintaining eye safety requirements, which allows for longer operating range with higher sensitivity (Refaat et al. 2011). At the same time, Froidevaux et al. (2013) believe that the UV spectral range is better suited for working with solar radiation. The UV spectral band known as the *solar-blind* region (wavelengths shorter than 300 nm) where nearly all solar radiation is absorbed by stratospheric ozone. In the visible spectrum, solar background is much greater than any Raman scattering.

The first DIAL measurement of water vapor was made using a temperature-tuned ruby laser operating from 693.7 to 694.5 nm (Schotland 1966; Zuev et al. 1983). These systems were used for ground-based measurements of water vapor, and made measurements at ranges up to above 10 km, presumably at night. However, problems with the low laser pulse repetition frequency, frequency chirp, and difficulty in determining the water vapor absorption coefficients in that spectral region, coupled with improved tunable laser sources in other spectral regions, have tended to shift interest away from the ruby laser for water vapor measurements (Grant 1991).

The next DIAL systems for water vapor were based on using single CO_2 lasers with direct detection receiver systems (Murray et al. 1976; Baker 1983; Grant et al. 1987). The CO_2 lasers, which are discretely tunable in the 9.2–10.7 μm spectral region, have accidental overlaps with a number of vibrational and rotational water vapor lines of differing line strengths. However, it was established that the sensitivity of conventional LIDAR systems, which employ a direct detection of backscattered or reflected radiation, is less at IR wavelengths than in the visible or UV spectral regions (Grant 1991). This degradation at longer wavelengths is a result of the limitations imposed by current detector technology, as well as the increased level of optical background which exists in the middle IR. Detectors in the IR range have high level of thermal noise and these detectors does not have high intrinsic gain in order to minimize the effect of electrical preamplifier noise (as photomultipliers do that for visible and UV spectral ranges). Reduced detection sensitivity of LIDAR systems in the IR is especially significant because atmospheric backscatter coefficients in that region are substantially lower than at visible or UV wavelengths. Also, spectral interference from other molecular or from aerosol spectral properties is more likely to occur here than in the near-IR spectral region. As a result of this combination of detrimental effects, the LIDAR systems operating at wavelengths in the region around $\lambda \sim 10$ μm require much more powerful radiation source to obtain backscattered signals comparable to those obtained with the system, which uses shorter wavelengths. Finally, CO_2 laser DIAL systems also suffer from having a long tail after the gain-switched spike, which can be corrected by modeling or by using some method to eliminate the tail before transmitting, such as reducing the amount of N_2 in the gas mixture or chopping the beam electro-optically (Grant 1991).

It was established that an alternate technique to improve a DIAL measurement sensitivity of LIDAR in IR region without constructing larger lasers is to use optical heterodyne detection on the backscatter radiation (Menzies 1985). Differences of heterodyne or coherent detection system from direct or incoherent detection system are shown in Figure 8.5 and listed in Table 8.4. Here the electric field of the backscatter signal is mixed with the field of a local oscillator (LO). In the case of a coherent LIDAR system this LO is another frequency-stable continuous-wave (CW) laser (analogous to the stable LO in a radar system).

Experiments and simulations have shown that in principle, a heterodyne (also known as coherent) DIAL LIDAR system can provide great increase in sensitivity over a direct detection system because the detector can be operated in a quantum-noise limited mode (Menzies 1985; Zhao et al. 1990; Holmes and Rask 1995; Belmonte 2003). In addition, the frequency and phase of the optical field can be measured using coherent detection. The capability to measure frequency is extremely valuable for remote monitoring. As a result, coherent DIAL IR LIDAR systems extended the range of monitoring from 2–3 km to 5–7 km and shown capability of operation to the 10–15 km horizontal range. Today however, coherent LIDAR systems are not widely used. It was established that coherent LIDAR systems have never reached the reliability and resolution of direct-detection instruments. Probably this primarily connected with complexity of constructing a pulsed coherent system. Other reasons for this condition can be found in Wulfmeyer and Walther (2001). As the state-of-the-art technology required to make such a system performing adequately entails high costs, research on coherent IR LIDAR systems has advanced slowly in the absence of a well-defined benefit or application. Therefore, for the most part, LIDAR systems used for monitoring the atmosphere are Raman LIDARs, working in the UV or visible spectral ranges (Table 8.2) and using *incoherent* or direct energy detection.

However, the DIAL systems can operate day or night and should be scalable for operation of the LIDAR

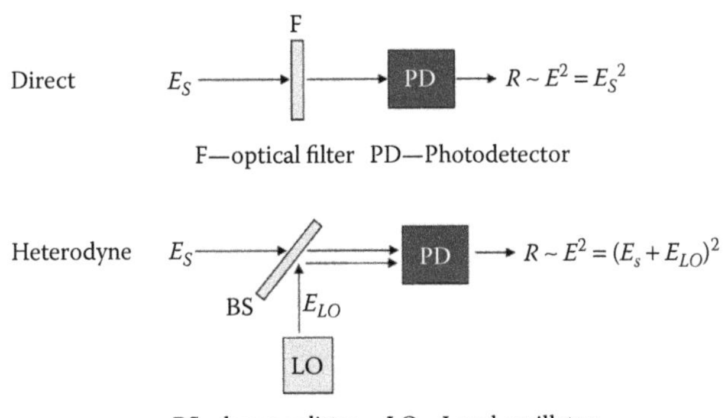

FIGURE 8.5 Differences between incoherent and coherent detection systems. (Idea from Menzies, R.T., Coherent and incoherent Lidar–An overview. In Byer R.L., Gustafson E.K., Trebino R. (Eds.), *Tunable Solid State Lasers for Remote Sensing*, Vol. 51 of the Springer series in Optical Sciences, Springer, Berlin, Germany, pp. 17–21, 1985.)

TABLE 8.4
Differences between Coherent and Incoherent Detection Systems

Incoherent Detection System	Coherent Detection System
Spectral resolution determined by optical filter, typically 0.1–10 cm^{-1}	Spectral resolution determined by IF bandpass filter, typically 10^{-5} cm^{-1}–10^{-1} cm^{-1}
Detection *Gain* may exist, for example, photoemissive dynodes, avalanche effect	Detection *Gain* provided by photomixing with local oscillator
Noise sources: background radiation fluctuations, thermal fluctuations in photodiode, and amplifier	Thermal noise source effects can be minimized with large enough local oscillator power. Local oscillator fluctuations become limiting noise

Source: Data extracted from Menzies, R.T., Coherent and incoherent Lidar–An overview. In Byer R.L., Gustafson E.K., Trebino R. (Eds.), *Tunable Solid State Lasers for Remote Sensing*, Vol. 51 of the Springer series in Optical Sciences, Springer, Berlin, Germany, pp. 17–21, 1985.

TABLE 8.5
Lasers in Raman LIDAR Systems Used or Proposed for Water Vapor Measurements

Laser	Wavelength, nm
KrF	245.5
4xNb:YAG	266
*KrF+D_2	268.5
XeCl	308
N_2	337.1
2xRuby	347.2
3xNb:YAG	354.7
XeF	351
2xNb:YAG	532
Ruby	694.4
Nb:YAGs	1064

Source: Data extracted from Grant, W.B., *Opt. Eng.*, 30, 40–49, 1991.

system at great distances from the region of interest, such as from a space-based or airborne platforms (Ismail and Browell 1989; Wulfmeyer and Walther 2001). After solving the technological problems, the DIAL systems undoubtedly will find wider application.

The optimal laser source for Raman scattering measurements of water vapor has been pursued during the past decades. The basic considerations are that the wavelength should be short to take advantage of the inverse fourth power of wavelength dependence of the Raman scattering cross section, and that the laser should have a high laser pulse energy and a high pulse repetition frequency (PRF) (Grant 1991). The first Raman LIDAR measurements of water vapor used the second harmonic of a ruby laser or the N_2 laser. More recent measurements used the first, second, third, or fourth harmonics of a Nd:YAG laser (Table 8.5), a ruby laser, or a KrF, XeC1, or XeF excimer lasers (Grant 1991). One should note that the most successful Raman LIDAR instruments use Nd:YAG lasers, which can operate in either a CW or pulsed mode (Ferrare et al. 2003; Hua and Kobayashi 2005; Whiteman et al. 2007; Froidevaux et al. 2013).

Another source of optical radiation in the IR for remote sensing applications can be the tunable diode laser. However, because of the low power output of these devices, their use is generally restricted to applications as LOs in heterodyne systems. A potentially promising new source of optical radiation for remote sensing is also a tunable laser made from transition-metal-doped crystals. Although still in a research stage these devices are continuously tunable and can be Q-switched to generate a high-peak-power output. The examples of such devices are the Co:MgF_2, Co:ZnF_2, and Cr:ZnS lasers (Budgor et al. 1986; Sorokina and Vodopyanov 2003), which are tunable across the 1.0–2.3 μm region. This spectral region contains strong absorption bands due to H_2O and CO_2.

Undoubtedly, Laser transmitters are an important element of LIDAR system, since these are the principal elements which distinguish one LIDAR system from another. However, other characteristics, such as optical configuration and principle of operation are also important for LIDAR systems. For example, most single-ended LIDARs operate in either a monostatic or bistatic configuration. In a monostatic system the transmitted and backscattered signals follow a common optical path, hence alignment is simplified. However, in the case where the transmitter and receiver also share the same optics, some kind of transmit/receive switch is usually necessary to prevent saturation or damage to the signal detector from leakage of the transmit energy back along the receiver path. This problem can be reduced by the use of a two-mirror coaxial system. Mounting the transmit mirror coaxially in the middle of the receiver provides isolation as well as monostatic operation, although alignment difficulties are increased. Isolation is also improved for single transmit/receive mirror monostatic systems when the telescope employs an off-axis configuration, such that no transmit energy reflects back into the detector from the system primary mirror. Bistatic configurations alleviate isolation problems at a cost of increased alignment difficulty.

Receiver is another important part of LIDAR system (Wulfmeyer and Walther 2001). The returned energy in LIDAR systems is collected by the receiver optics and directed onto a photodetector, which produces an electrical signal proportional to the intensity of the optical radiation incident at the detector. In the receiver, the choice of a system detector and electronic bandwidth is a strong function of both the transmitter characteristics and the system performance requirements. To get a high signal-to-noise ratio (SNR) of the return signal, one needs a product of quantum efficiency and gain of the detector of at least 100 to be able to discriminate efficiently between the atmospheric signal and the dark current background. Furthermore, the detector noise current should be as low as possible. The response of the detector must be linear over a dynamic range of 10^8. Research shows that only photomultipliers and avalanche photodiodes (APDs) meet these requirements. Photomultipliers are applicable for observations in the UV, visible and near IR range up to 1000 nm. Photomultiplier tubes provide high sensitivity from the UV region of the spectrum through the visible to the near IR range up to 1000 nm. However, the best performance in visible and near IR range was achieved by silicon avalanche photodiodes, as they have both high quantum efficiencies of more than 50% up to a wavelength of 1000 nm and a gain of the order of 100. APDs are less prone than photomultipliers to ringing after saturation of the detector. This is an important issue in a LIDAR system, as backscatter signals from the clouds often cause saturation of the detector. If a ringing of the detector occurs, signals can be contaminated for a certain time after the cloud disappears. InGaAs p-i-n detectors are commonly used at wavelengths longer than 1000 nm. However, although they have high quantum efficiency, the gain is only of the order of 1. If applications necessitate operation at longer wavelengths in the IR, photoconductive and photovoltaic semiconductors are employed. The materials from which these detectors are fabricated include silicon, germanium, InSb, PbSnTe, PbSnSe, InAsSb, and HgCdTe. In general, HgCdTe has become the preferred material because detector response can be peaked at any wavelength from 0.9 to 40 μm by using the appropriate alloy of HgTe and CdTe. As photons at longer wavelengths possess less energy, detectors in the IR portion of the spectrum must be cooled to reduce the effect of thermal excitation. Since this property is often a disadvantage in certain applications, room temperature detectors have been developed. However, their performance is at least 10 dB poorer than appropriately cooled IR detectors (Spears 1983).

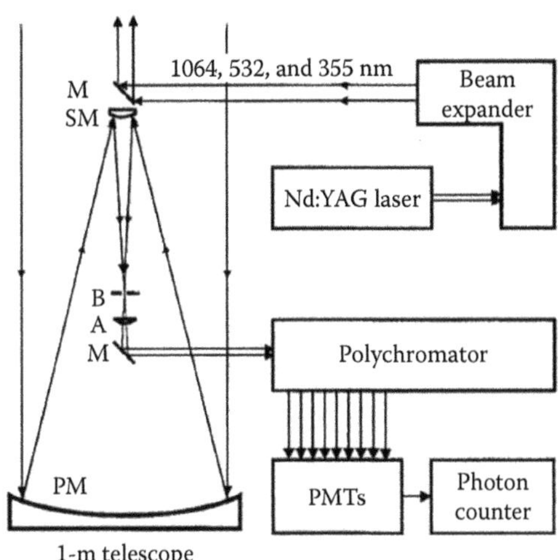

FIGURE 8.6 Schematic view of the Raman LIDAR. Three laser beams, at 355, 532, and 1064 nm, are transmitted vertically into the atmosphere by way of a mirror (M). Laser shots are fired at a repetition rate of 30 Hz. The outgoing beam is collimated with a beam expander that has a magnification factor of 10 before it is directed to the atmosphere by a beam-folding mirror. The Cassegrain telescope consists of a primary mirror (PM), a secondary mirror (SM), a field stop (B), an achromatic lens (A), and a mirror to direct the backscattered photons to the beam-separation unit. PMT, photomultiplier tube. (From Mattis, I. et al., *Appl. Opt.*, 41, 6451, 2002. With permission of Optical Society of America.)

It was indicated earlier that in general there are two kinds of LIDAR detection schemes: (1) *incoherent* or direct energy detection, and (2) *coherent* or heterodine detection, which can operate in continuous or impulse mode. At that both coherent and incoherent LIDARs, operated in a pulse mode, can be designed in two variants: (1) micropulse and (2) high-energy systems (Wilkerson and Schwemmer 1982). Micropulse systems have been developed as a result of more powerful computers with greater computational capabilities. These lasers are lower powered and are classed as *eye-safe* allowing them to be used with little safety precautions. Typically, such systems are used for measurements near the ground, where the concentration of water vapor is high. High-energy systems are more commonly used for atmospheric research aimed at the water vapor, aerosol, and the temperature profiling (Wilkerson and Schwemmer 1982; Ferrare et al. 1998). Better resolution can be achieved with shorter pulses provided the receiver detector and electronics have sufficient bandwidth to cope with the increased data flow.

An important element of LIDAR systems are telescopes, the size, and design of which can be sufficiently different, depending on the tasks to be resole (Mattis et al. 2002; Hua and Kobayashi 2005; Dinoev et al. 2013; Froidevaux et al. 2013). For example, the device shown in Figure 8.6 uses a 1-m-diameter Cassegrain telescope, which collected radiation that has been backscattered by atmospheric molecules and particles. In addition, the telescope provides shielding from unwanted light. Mattis et al. (2002), who designed this Raman-LIDAR, employed the elastic backscattering at the three emitted wavelengths, including information on the depolarization at 532 nm, to study the height stratification and some optical and microstructure properties of the atmospheric aerosol. In addition, they recorded Raman-LIDAR returns from the main molecular constituents of the atmosphere: nitrogen and water vapor, to acquire the profiles of temperature and moisture content as well as particle extinction coefficients of the atmosphere. To obtain the water vapor profile, Mattis et al. (2002) used radiation at 355 nm to excite LIDAR returns from the ν_1 vibrational–rotational Raman band of water vapor, centered at 407 nm and from nitrogen at 387 nm. To obtain the temperature profile they isolated four portions from the Stokes and anti-Stokes branches of the pure rotational Raman spectrum (PRRS) of nitrogen excited by radiation of the second harmonic of the Nd:YAG laser.

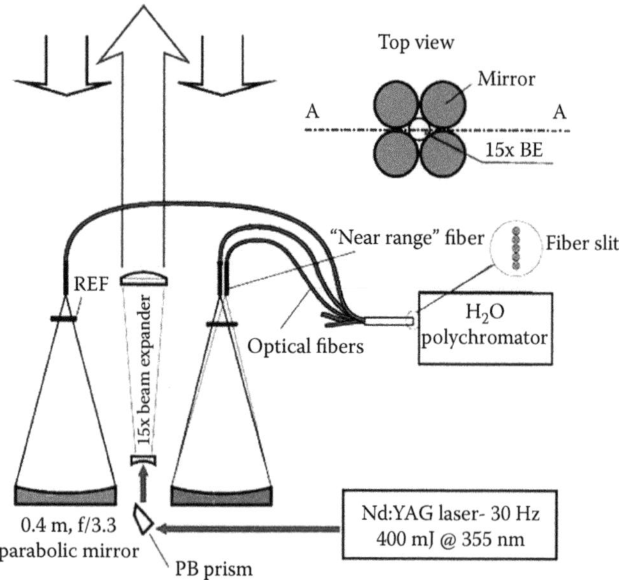

FIGURE 8.7 LIDAR optical scheme: A 15× beam expander is centered between four 30 cm in diameter, f/3.33 parabolic mirrors as illustrated by the top view shown in the upper right corner. The optical scheme is a cross section along the line A–A of the top view, therefore only two of the telescope mirrors can be seen; REF—Razor edge filter (Semrock®); PB—Pellin–Broca prism; PMT—photomultiplier. (From Dinoev, T. et al., *Atmos. Meas. Tech.* 6, 1329, 2013. Published by the European Geosciences Union as open access.)

TABLE 8.6
System Parameters of Raman LIDAR Designed by Federale de Lausanne–EPFL with Swiss Meteorological Service (MeteoSwiss) and Supported by the Swiss National Foundation

Transmitter	Receiver
Nd:YAG laser—third harmonic:	*Four fiber coupled parabolic mirrors:*
Wavelength: 354.7 nm	Focal length: 1 m
Spectral line width: 0.7 cm^{-1}	Diameter: 0.3 m
Rep. rate: 30 Hz	Axial displacement: 235 mm (to expander axis)
Pulse energy: 300 mJ	FOV: 3 × 0.20 and 1 × 0.22 mrad
Pulse duration: 8 ns	*Polychromator—Diffraction grating-based:*
Beam expander—Galilean type:	Bandwidth (FWHM): 0.33 nm (H_2O and N_2)
Expansion ratio 15×	Central wavelength:
Transmitted beam:	H_2O—407.45 nm
Divergence:	N_2—386.7 nm
calculated 0.06 mrad	Efficiency: 33% (peak at H_2O)
measured 0.09 ± 0.02 mrad	*Photodetectors* (Hamamatsu)
Diameter: 140 mm	H_2O—R7600U-200
	N_2 and O_2—H6780

Source: Data extracted from Dinoev, T. et al., *Atmos. Meas. Tech.* 6, 1329, 2013.

Raman LIDAR designed by Dinoev et al. (2013) has another configuration (Figure 8.7). The main LIDAR parameters are summarized in Table 8.6. The LIDAR uses a narrowband, narrow field-of-view (NFOV) configuration, and an excitation laser operating at 355 nm to ensure daytime operation. The receiver telescope consists of four, 30 cm in diameter mirrors, fiber coupled to a grating polychromator. Extension of the operational range in the lower part of the profile down to 100 m is achieved by using incomplete overlap signals and two fibers in one of the telescopes. The second fiber is collecting signals from distances close to the LIDAR. The need for range-dependent corrections is eliminated by the careful design of the receiver, which removes the range dependence of the instrumental part of the LIDAR calibration constant (Equation 8.2). Elimination of the range-dependent corrections not only simplifies the data treatment and calibration but also allows accurate humidity measurements for cases of transmitter–receiver misalignment, but at the price of reduced distance range (or precision). To achieve long-term data consistency, a grating-based (instead of interference-filter-based) polychromator is used. The compact and rigid mechanical design of the LIDAR, together with the fiber-optic connection between the LIDAR telescope and the polychromator, ensures long-term alignment stability and eliminates the need for frequent transmitter–receiver alignments.

Hua and Kobayashi (2005) used another approach to the design of the LIDAR system (Figure 8.8). They have built a Rayleigh–Mie Raman LIDAR with an eye safe 355 nm laser wavelength and a 25-cm diameter telescope. In response to 355 nm radiation, the centers of the nitrogen and water vapor Raman spectra were 387 and 407 nm, respectively. The nitrogen Raman scattering signal, which was simultaneously measured with the water vapor Raman scattering in conventional Raman LIDAR, was applied to compute the water vapor mixing ratio. The rotational Raman scattering and Rayleigh scattering were used for temperature profiling by making use of the temperature dependence of spectral intensity and linewidth. The temperature was determined from the molecular Rayleigh scattering linewidth and the water vapor density was determined from the intensity of the water vapor vibration Raman line centered at 407.5 nm for an excitation laser wavelength of 354.7 nm. Aerosol

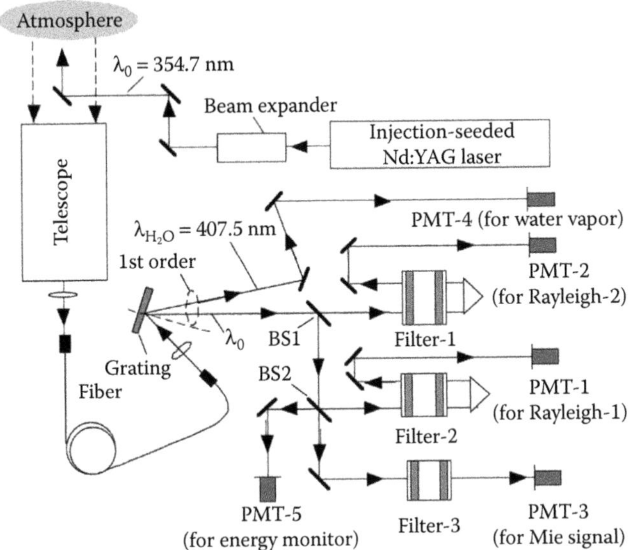

FIGURE 8.8 Schematic of the Rayleigh–Mie Raman LIDAR for simultaneous profiling of atmospheric temperature, humidity and aerosol optical properties: PMT—photomultiplier tube; BS—beam splitter. (From Hua, D. and Kobayashi, T., *Jpn. J. Appl. Phys.* 44, 1287, 2005. Copyright 2005, The Japan Society of Applied Physics. With permission.)

Mie scattering and molecular rotational Raman scattering as well as nitrogen Raman scattering signals were considered to be interference signals that should be rejected to achieve an accurate temperature and water vapor profiling. Since the Raman scattering is about three orders of magnitude weaker than the Rayleigh scattering and distributed much more widely in frequency, the effect of rotational Raman scattering can be removed easily by using a high-resolution grating. Moreover, the grating also functions as a solar band-pass filter that blocks a major portion of solar background during a daytime temperature measurement. Two Rayleigh filters were designed with a dual-pass optical layout so that high rejection for aerosol Mie scattering and large transmittance for Rayleigh scattering can be realized. The results obtained using this device showed that the uncertainty of water vapor in RH was less than 10% up to an altitude of 2.5 km with a 60 m vertical resolution and a 3.5 min temporal resolution with a temperature error of less than 1 K.

8.4 SOLAR-BLIND RAMAN LIDAR OPERATION

It is important to note that most of the measurements were performed for research purposes at night time (Whiteman et al. 1992; Balin et al. 2004). The primary problem is, of course, that at laser wavelengths longer than about 270 nm (putting the water vapor Raman shifted return at 300 nm), solar background radiation swamps the Raman LIDAR signal, while for laser wavelengths shorter than 270 nm, ozone interferes, and for laser wavelengths shorter than 265 nm, oxygen also interferes. There are a few reports of solar background flux as a function of wavelength below 300 nm (Renault and Capitini 1988). However, solar radiance is highly variable, depending on solar angle and atmospheric conditions, so it is difficult to generalize as to the exact wavelength for which solar blind operation is possible.

Some of the competing effects are discussed in Renault and Capitini (1988), where it is suggested that on the basis of ozone attenuation of laser radiation and solar radiation and atmospheric scattering that the optimum laser wavelength is about 265 nm. Thus, the fourth harmonic of the Nd:YAG laser was adjudicated optimal for solar-blind Raman LIDAR measurements of water vapor. It is proposed here that when the laser pulse energy and PRF are considered, a Raman-shifted KrF laser may be a better choice. However, according to Grant (1991), the KrF laser has the highest laser pulse energy and PRF of the three, but is strongly affected by ozone and oxygen absorption. The Nd:YAG laser has a favorable wavelength, but has low laser pulse energy and PRF; a D_2-shifted KrF laser has a high PRF and a moderate pulse energy approximately 30% of the KrF laser (Grant 1991) but has a wavelength (268.5 nm) that puts the water vapor Raman signal at 297.7 nm, which is just outside the true solar-blind region. Grant (1991) believes that further research and development may result in a better laser source near 266 nm with higher average power, perhaps by using another Raman-shifting medium.

Optimization design of LIDAR, the use of a NFOV, narrow band (NB) receiver also allows daytime operation at visible and near UV wavelengths with an operational range up to the mid-troposphere (Werner and Herrmann 1981; Cooney et al. 1985; Goldsmith et al. 1998; Dinoev et al. 2013). However, at daytime operation the measurement accuracy is lower. For example, the precision for ground-based measurements of the Goddard Space Flight Center (GSFC) Raman LIDAR (Whiteman and Melfi 1999) was typically quoted as <10% up to 8 km with 1 min averaging and 75 m vertical resolution (nighttime). The day-time precision was 10% random error at 4 km using 10 min average. Also, Raman LIDARs operated during the daytime did not have possibility to conduct measurements at high altitudes.

8.5 LIDAR SYSTEMS FOR AIRCRAFT AND SATELLITE APPLICATIONS

Besides Earth-based systems, the DIAL-based hygrometers for aircraft and satellite applications have been developed (Kley et al. 2000; Wulfmeyer and Walther 2001). Raman LIDAR instruments are not suitable for these applications. The first autonomously operating DIAL-based devices intended for these applications were LASE (LIDAR Atmospheric Sensing Experiment) system designed by the NASA Langley Research Center (Browell et al. 1997), and DLR (Deutsches Zentrum für Luft- und Raumfahrt) airborne water vapor DIAL systems designed in Oberpfaffenhofen, Germany (Ehret et al. 1993). Pulses of laser light were fired vertically below the aircraft. A small fraction of the transmitted laser light was reflected from the atmosphere back to the aircraft and collected with a telescope receiver.

LASE operated in the 815 nm absorption band of water vapor and it used a Ti:sapphire laser that was locked onto a strong and temperature insensitive water vapor line. This line was chosen to have a residual temperature sensitivity of less than a few percent (Ismail and Browell 1989). Water vapor over the entire troposphere was measured by electronically tuning the online to operate at any spectral location of the water vapor absorption line profile. This was done by first locking to the line center and then adjusting the diode current a known amount from the line center before the Ti:sapphire laser was fired. Since the actual water vapor line was used to lock the diode and the known spectral properties of the absorption line, the LASE water vapor measurement was self-calibrated. LASE has demonstrated the capability to measure water vapor over the entire troposphere (Browell et al. 1997) with an accuracy of 6% or 0.01 g/kg. The horizontal resolution of LASE water vapor measurements is about 24 km (or 2 min of data averaging) and vertical resolutions of 300 m and 500 m in the lower and upper troposphere, respectively.

The first DLR H_2O DIAL system was operated in the near IR spectral region at 724 nm. This system took part in several field experiments to investigate processes in the boundary layer. In order to increase both accuracy and spatial resolution, and to be sensitive for water vapor measurements in the stratosphere, a new water vapor DIAL system has been developed at DLR. Based on a high peak and average power optical parametric oscillator, its transmitter fulfilled the spectral requirements for water vapor measurements in the troposphere as well as in the stratosphere (Ehret et al. 1998). The residual systematic error caused by the spectral properties of the new radiation source amounted to only 2%. The transmitter was designed to be operated at either the weak 4v vibrational absorption band of water vapor near 830 nm, suitable for tropospheric measurements, or at the 3v vibrational absorption band lying in the 940 nm spectral region, which is one order-of-magnitude stronger. The latter has the advantage of much higher measurement sensitivity in regions of low water vapor content at upper tropospheric and lower stratospheric heights. The new DLR H_2O DIAL system has shown possibility to measure the water vapor in the range from 5 to 100 ppmv (Ehret et al. 1999). In the downward looking mode it could scan vertical cross sections of up to 6 km in range, with a resolution of about 20 km in the horizontal and 250–500 m in the vertical.

Wulfmeyer and Walther (2001) believe the realization of all advantages of the DIAL technique requires that the system be a direct-detection system that operates in the band of water vapor near 940-nm wavelength. The setup for a direct-detection LIDAR system is straightforward (Hinkley 1976; Measures 1984). The optical path of such a system is easy to align and highly stable even in a hostile environment. Another important advantage is that the bandwidth of the receiver channel is large enough for molecular backscattering to be detected. Consequently, measurements are possible in clear air without any aerosol particles, which is particularly important in the free troposphere and the lower stratosphere.

According to Wulfmeyer and Walther (2001), great potential of DIAL for aircraft application has three causes. First, the DIAL technique is flexible and has the unique property that it can be applied for measurements in the boundary layer as well as in the lower stratosphere and from tropical to arctic regions. The same low relative error in a water-vapor measurement can be maintained by choice of different strengths of water-vapor absorption lines under different atmospheric conditions. Second, the DIAL technique is currently the most accurate remote-sensing technique. That this is so is due to its intrinsic self-calibrating measurement method. Unknown system constants cancel out in the measurement of the water vapor profile, as the derivative of the ratio of two signals, the online and the off-line signals, is calculated. Typically, a systematic error of 5% in water-vapor profiles from the ground to the lower stratosphere can be expected. Third, a DIAL system can perform measurements during either daytime or nighttime with outstanding resolution under clear-air conditions because this system measures elastic backscatter signals, which provide high SNRs. At the same time Wulfmeyer and Walther (2001) recognized that the full potential of the DIAL technique has not been explored yet and its realization requires time.

When a LIDAR sensor is mounted on a mobile platform such as satellites, airplanes or automobiles, the absolute position and the orientation of the sensor usually are controlled by GPS, which provide accurate geographical information regarding the position of the sensor, and by an inertia measurement unit (IMU), which records the precise orientation of the sensor at that location. These two devices provide the method for translating sensor data into static points for use in a variety of systems.

REFERENCES

Abreu L.W., Anderson G.P. (Eds.) (1996) MODTRAN Report, Ontar Corporation, North Andover, MA.

Adam M., Demoz B.B., Venable D.D., Joseph E., Connell R., Whiteman A., Fitzgibbon J. (2010) Water vapor measurements by Howard University Raman Lidar during the WAVES 2006 campaign, *J. Atmos. Oceanic Technol.* 27, 42–60.

Baker P.W. (1983) Atmospheric water vapor differential absorption measurements on vertical paths with a CO_2 lidar. *Appl. Opt.* 22(15), 2257–2264.

Balin I., Serikov I., Bobrovnikov S., Simeonov V., Calpini B., Arshynov Y., van den Bergh H. (2004) Simultaneous measurement of atmospheric temperature, humidity, and aerosol extinction and backscatter coefficients by a combined vibrational–pure-rotational Raman lidar. *Appl. Phys. B* 79, 775–782.

Belmonte A. (2003) Analyzing the efficiency of a practical heterodyne lidar in the turbulent atmosphere: Telescope parameters. *Opt. Express* 11(17), 2041–2046.

Bosenberg J. (1987) A DIAL system for remote sensing: Instrumentation and techniques. *OSA Tech. Digest Series*, 18, 22–25.

Bosenberg J. (1998) Ground-based differential absorption LIDAR for water vapor and temperature profiling. *Appl. Opt.* 37, 3845–3860.

Browell E.V., Wilkerson T.D., Mcilrath T.J. (1979) Water vapor differential absorption Lidar development and evaluation. *Appl. Opt.* 18, 3474–3483.

Browell E.V., Goroch A.K., Wilkerson T.D., Ismail S., Markson R. (1984) Airborne DIAL water vapor measurements over the Gulfstream. In: *Proceedings of 12th International Laser Radar Conference*, Aix-en-Provence, France, pp. 151–155, August 13–17.

Browell E.V., Ismail S., Hall W.M. et al. (1997) LASE validation experiment. In: Ansmann A., Neuber R., Rairoux P., Wandinger U. (Eds.), *Advances in Atmospheric Remote Sensing with Lidar.* Springer-Verlag, Berlin, Germany, pp. 289–295

Budgor A.B., Esterowitz L., DeShazer L.G. (Eds.) (1986) *Tunable Solid-State Lasers II.* Springer-Verlag, Berlin, Heidelberg.

Cahen C., Megie G., Flamant P. (1982) Lidar monitoring of water vapor cycle in the troposphere. *J. Appl. Meteorol.* 21, 1506–1515.

Cha S., Chan K., Killinger D. (1991) Tunable 2.1-μm Ho lidar for simultaneous range-resolved measurements of atmospheric water vapor and aerosol backscatter profiles. *Appl. Opt.* 30(27), 3938–3943.

Cooney J.A., Pina M. (1976) Laser radar measurements of atmospheric temperature profiles by use of Raman rotational backscatter. *Appl. Opt.* 15, 602–603.

Cooney J., Petri K., Salik A. (1985) Measurements of high resolution atmospheric water vapor profiles by use of a solar blind Raman lidar. *Appl. Opt.* 24(1), 104–108.

Cracknell A.P., Hayes L. (2007) *Introduction to Remote Sensing*, 2nd ed. Taylor & Francis Group, London, UK.

Di Girolamo P., Summa D., Lin R.-F., Maestri T., Rizzi R., Masiello G. (2009) UV Raman lidar measurements of relative humidity for the characterization of cirrus cloud microphysical properties. *Atmos. Chem. Phys.* 9, 8799–8811.

Dinoev T. (2009) Automated Raman lidar for day and night operational observation of tropospheric water vapor for meteorological applications. PhD Thesis, Federal Institute of Technology EPFL, Lausanne, Switzerland.

Dinoev T., Simeonov V., Arshinov Y., Bobrovnikov S., Ristori P., Calpini B., Parlange M., van den Bergh H. (2013) Raman Lidar for meteorological observations, RALMO–Part 1: Instrument description. *Atmos. Meas. Tech.* 6, 1329–1346.

Ehret G., Fix A., Weiß V., Poberaj G., Baumert T. (1998) Diode-laser-seeded optical parametric oscillator for airborne water vapor DIAL application in the upper troposphere and lower stratosphere. *Appl. Phys. B* 67, 427–431.

Ehret G., Kiemle C., Renger W., Simet G. (1993) Airborne remote sensing of tropospheric water vapor using a near infrared DIAL system. *Appl. Opt.* 32, 4534–4551.

Ehret G., Hoinka K.P., Stein J., Fix A., Kiemle C., Poberaj G. (1999) Low stratospheric water vapor measured by an airborne DIAL. *J. Geophys. Res.* 104(D24), 31,351–31,359.

Endemann M., Byer R.L. (1981) Simultaneous remote measurements of atmospheric temperature and humidity using a continuously tunable IR lidar. *Appl. Opt.* 20(18), 3211–3217.

Ferrare R.A., Melfi H., Whiteman D.N., Evans K.D., Leifer R. (1998) Raman lidar measurements of aerosol extinction and backscattering. 1. Methods and comparisons. *J. Geophys. Res.* 103(D16), 19663–19672.

Ferrare R.A., Turner D.D., Tooman T.P. et al. (2003) Vertical variability of aerosols and water vapor over the southern great plains. In: *Proceedings of Thirteenth ARM Science Team Meeting Proceedings*, Broomfield, CO, pp. 1–13, March 31–April 4, 2003.

Froidevaux M., Higgins C.W., Simeonov V., Ristori P., Pardyjak E., Serikov I., Calhoun R., van den Bergh H., Parlange M.B. (2013) A Raman lidar to measure water vapor in the atmospheric boundary layer. *Adv. Water Resour.* 51, 345–356.

Goldsmith J., Blair F.H., Bisson S.E., Turner D.D. (1998) Turn-key Raman lidar for profiling atmospheric water vapor, clouds, and aerosols. *Appl. Opt.* 37, 4979–4990.

Goyer G.G., Watson R. (1963) The laser and its application to meteorology. *Bull. Am. Meteorol. Soc.* 44(9), 564–575.

Grant W.B. (1991) Differential absorption and Raman lidar for water vapor profile measurements: A review. *Opt. Eng.* 30(1), 40–49.

Grant W.B., Margolis J.S., Brothers A.M., Tratt D.M. (1987) CO_2 DIAL measurements of water vapor. *Appl. Opt.* 26(15), 3033–3042.

Grossmann B.E., Browell E.V. (1989a) Spectroscopy of water vapor in the 720-nm wavelength region: Line strengths, self-induced pressure broadenings and shifts, and temperature dependence of line widths and shifts. *J. Mol. Spectrosc.* 136, 264–294.

Grossmann B.E., Browell E.V. (1989b) Water vapor line broadening and shifting by air, nitrogen, oxygen, and argon in the 720-nm wavelength region. *J. Mol. Spectrosc.* 138, 562–595.

Grund C.J., Eloranta E.W. (1991) University of Wisconsin high spectral resolution lidar. *Opt. Eng.* 30, 6–12.

Grund C.J., Hardesty R.M., Rye B.J. (1995) Feasibility of tropospheric water vapor profiling using infrared heterodyne differential absorption Lidar. In: *Proceedings of 5th Atmospheric Radiation Measurement (ARM) Science Team Meeting*, San Diego, CA, March 19–23. http://www.arm.gov/publications/proceedings/conf05/

Higdon N.S., Browell E.V., Ponsardin P., Grossmann B.E., Butler C.F., Chyba T.H. et al. (1994) Airborne differential absorption lidar system for measurements of atmospheric water vapor and aerosols. *Appl. Opt.* 33, 6422–6438.

Hinkley E.D. (Ed.) (1976) *Laser Monitoring of the Atmosphere.* Springer-Verlag, Berlin, Germany.

Holmes J.F., Rask J. (1995) Optimum optical local-oscillator power levels for coherent detection with photodiode. *Appl. Optics* 34, 927–933.

Hua D., Kobayashi T. (2005) UV Rayleigh–Mie Raman Lidar for simultaneous measurement of atmospheric temperature and relative humidity profiles in the troposphere. *Jpn. J. Appl. Phys.* 44(3), 1287–1291.

ICAO (1964) Manual of ICAO Standard Atmosphere. International Civil Aviation Organization Doc. 7488/2, ICAO, Montreal, Canada.

Ismail S., Browell E.V. (1989) Airborne and spaceborne Lidar measurements of water vapor profiles: A sensitivity analysis. *Appl. Opt.* 28, 3603–3615.

Kley D., Russell III J.M., Phillips C. (Eds.) (2000) SPARC assessment of upper tropospheric and stratospheric water vapour. World Climate Research Programme, SPARC Report No.2.

Leblanc T., McDermid I.S., Walsh T.D. (2012) Ground-based water vapor Raman lidar measurements up to the upper troposphere and lower stratosphere for long-term monitoring. *Atmos. Meas. Tech.* 5, 17–36.

Liou K.N. (1986) Influence of cirrus clouds on weather and climate processes: A global perspective. *Mon. Weather Rev.* 114, 1167–1199.

Long D. (2002) *The Raman Effect.* John Wiley & Sons, Cambridge.

Mattis I., Wandinger U., Muller D., Ansmann A., Althausen D. (1998) Routine dual-wavelength Raman lidar observations at Leipzig as part of an aerosol lidar network in Germany. In: *Proceedings of 19th International Laser Radar Conference*, NASA/CP-1998-207671/PT1, pp. 29–32.

Mattis I., Ansmann A., Althausen D. et al. (2002) Relative-humidity profiling in the troposphere with a Raman lidar. *Appl. Opt.* 41(30), 6451–6462.

Measures R. (1984) *Laser Remote Sensing: Fundamental and Applications.* John Wiley & Sons, New York.

Melfi S., Whiteman D. (1985) Observations of lower-atmospheric moisture structure and its evolution using a Raman lidar. *Bull. Am. Meteorol. Soc.* 66, 1288–1292.

Menzies R.T. (1985) Coherent and incoherent Lidar–An overview. In: Byer R.L., Gustafson E.K., Trebino R. (Eds.), *Tunable Solid State Lasers for Remote Sensing*, Vol. 51 of the Springer series in Optical Sciences. Springer, Berlin, Germany, pp. 17–21.

Murray E.R., Hake R.D., Van der Laan J.E., Hawley J.G. (1976) Atmospheric water vapor measurements with a 10 micrometer DIAL system. *Appl. Phys. Lett.* 28(9), 542–543.

Navas-Guzmán F., Fernández-Gálvez J., Granados-Muñoz M.J., Guerrero-Rascado J.L., Bravo-Aranda J.A., Alados-Arboledas L. (2014) Tropospheric water vapour and relative humidity profiles from lidar and microwave radiometry. *Atmos. Meas. Tech.* 7, 1201–1211.

Pallotta J., Ristori P., Otero L., Chouza F., Raul D., Gonzalez F., Etchegoyen A., Quel E. (2013) Argentinian multi-wavelength scanning Raman lidar to observe night sky atmospheric transmission. In: *Proceedings of 33rd International Cosmic Ray Conference*, Rio de Janeiro. The Astroparticle Physics Conference.

Ponsardin P.L., Browell E.V. (1997) Measurements of $H_2{}^{16}O$ linestrengths and air–induced broadenings and shifts in the 815-nm region. *J. Mol. Spectrosc.* 185, 58–70.

Refaat T.F., Abedin M.N., Koch G.J., Ismail S., Singh U.N. (2003) Infrared detectors characterization for CO_2 DIAL measurement. *Proc. SPIE* 5154, 65–73.

Refaat T.F., Ismail S., Abedin N.N., Spuler S.M., Mayor S.D., Singh U.N. (2008) Lidar backscatter signal recovery from phototransistor systematic effect by deconvolution. *Appl. Opt.* 47(29), 5281–5295.

Refaat T.F., Ismail S., Koch G.J. et al. (2011) Backscatter 2-μm Lidar validation for atmospheric CO_2 differential absorption Lidar applications. *IEEE Trans. Geosci. Remote Sensing* 49(1), 572–580.

Renault D., Capitini R. (1988) Boundary-layer water vapor probing with a solar-blind Raman lidar: Validations, meteorological observations, and prospects. *J. Atmos. Oceanic Tech.* 5(5), 585–601.

Renaut D., Pourney J.C., Captini R. (1980) Daytime Raman-lidar measurements of water vapor. *Opt. Lett.* 5, 233–235.

Rosenberg A., Hogan D.B. (1981) Lidar technique of simultaneous temperature and humidity measurements: Analysis of Mason's method. *Appl. Opt.* 20(19), 3286–3288.

Schotland R.M. (1966) Some observations of the vertical profile of water vapor by means of a ground based optical radar. In: *Proceedings of Fourth Symposium on Remote Sensing of the Environment*, Ann Arbor, MI, pp. 271–273.

Sherlock V.J., Lenoble J., Hauchecorne A. (1999a) Methodology for the independent calibration of Raman backscatter water vapour lidar systems. *Appl. Opt.* 38, 5838–5850.

Sherlock V.J., Garnier, A., Hauchecorne, A., Keckhut, P. (1999b) Implementation and validation of a Raman lidar measurement of middle and upper tropospheric water vapour. *Appl. Opt.* 38(27), 5838–5850.

Shipley S.T., Tracy D.H., Eloranta E.W., Trauger J.T., Sroga J.T., Roesler F.L., Weinman J.A. (1983) High spectral resolution lidar to measure optical scattering properties of atmospheric aerosols: 1. Theory and implementation. *Appl. Opt.* 22, 3716–3724.

Sorokina I.T., Vodopyanov K.L. (2003) *Solid-State Mid-Infrared Laser Sources*. Springer, Berlin, Germany.

Spears D.C. (1983) IR detectors: heterodyne and direct. In: Killinger D.K., Morradian A. (Eds.), *Optical and Laser Remote Sensing*. Springer-Verlag, Berlin, Germany, pp. 287–286.

Stephens G.L., Tsay S.-C., Stackhouse P.W., Flatau P.J. (1990) The relevance of the microphysical and radiative properties of cirrus clouds to climate and climatic feedback. *J. Atmos. Sci.* 47, 1742–1754.

Turner D.D., Whiteman D.N. (2002) Remote Raman spectroscopy. Profiling water vapor and aerosols in the troposphere using Raman lidars. In: Chalmers J.M., Griffiths P.R. (Eds.), *Handbook of Vibrational Spectroscopy,*. John Wiley & Sons, Chichester, UK. doi:10.1002/0470027320.s6803.

Turner D.D., Ferrare R.A., Heilman L.A., Feltz W.F., Tooman T. (2002) Automated retrievals of water vapor and aerosol profiles over Oklahoma from an operational Raman lidar. *J. Atmos. Oceanic Tech.* 19, 37–50.

Vaughan G., Wareing D.P., Thomas L., Mitev V. (1988) Humidity measurements in the free troposphere using Raman backscatter. *Q. J. Royal Meteorol. Soc.* 114, 1471–1484.

Venable D., Joseph, E., Whiteman, D., Demoz, B., Connell, R., Walford, S. (2005) The development of the Howard University Raman lidar. In: *Proceedings of 85th AMS Annual Meeting*, American Meteorological Society, San Diego, CA, P1.1, pp. 5227–5232, January 9–13, 2005.

Vogelmann H., Sussmann R., Trickl T., Reichert A. (2015) Spatiotemporal variability of water vapor investigated using lidar and FTIR vertical soundings above the Zugspitze. *Atmos. Chem. Phys.* 15, 3135–3148.

Wandinger U. (2005a) Raman lidar. In: Weitkamp C. (Ed.) *Lidar*. Springer, New York.

Wandinger U. (2005b) Raman lidar techniques for the observation of atmospheric aerosols, temperature, and humidity. *Proc. SPIE* 5830, 307–316.

Werner Ch., Herrmann H. (1981) Lidar measurements of the vertical absolutr humidity distribution in the boundary layer. *J. Appl. Meteorol.* 20, 476–481.

Whiteman D.N., Melfi S.H. (1999) Cloud liquid water, mean droplet radius and number density measurements using a Raman lidar. *J. Geophys. Res.* 104, 31411–31419.

Whiteman D.N., Melfi S.H., Ferrare R.A. (1992) Raman lidar system for the measurement of water vapor and aerosols in the Earth's atmosphere. *Appl. Opt.* 31(16), 3068–3082.

Whiteman D.N., Veselovskii I., Cadirola M., Rush K., Comer J., Potter J.R., Tola R. (2007) Demonstration measurements of water vapor, cirrus clouds, and carbon dioxide using a high-performance Raman Lidar. *J. Atmos. Oceanic Technol.* 24, 1377–1388.

Wilkerson T.D., Schwemmer G.K. (1982) Lidar techniques for humidity and temperature measurement. *Opt. Eng.* 21(6), 1022–1024.

Wulfmeyer V., Walther C. (2001) Future performance of ground-based and airborne water-vapor differential absorption lidar. I. Overview and theory. *Appl. Opt.* 40, 5304–5320.

Wulfmeyer V., Pal S., Turner D.D., Wagner E. (2010) Can water vapour Raman Lidar resolve profiles of turbulent variables in the convective boundary layer? *Boundary-Layer Meteorol.* 136, 253–284.

Zhao Y., Post M.J., Hardesty R.M. (1990) Receiving efficiency of monostatic pulsed coherent lidars. 1: Theory. *App. Opt.* 29(28), 4111–4119.

Zuev V.V., Zuev V.E., Makushkin Yu. S., Marichev V.N., Mitsel A.A. (1983) Laser sounding of atmospheric humidity: Experiment. *Appl. Opt.* 22(23), 3742–3746.

9 Upper Tropospheric and Stratospheric Water Vapor Control

9.1 THE ROLE OF WATER VAPORS IN THE UPPER TROPOSPHERE AND LOWER STRATOSPHERE

As it was indicated in Section 2.8, the water vapor distribution in the upper troposphere (UT) and lower stratosphere (LS) is of central importance in several ways (Kley et al. 2000): it plays a major role in the balance of planetary radiation; it influences and responds to atmospheric motions; and it plays a key role in many aspects of UT/LS chemistry. Effects on the radiation balance are especially important because the water vapor molecule is strongly polar in shape giving it a strongly absorbing infrared (IR) spectrum. Water vapor is the dominant atmospheric greenhouse gas, and because the amount of water vapor which can be held by the atmosphere increases strongly with temperature, it is very sensitive to changes in climate. For example, the inclusion of water vapor and other climate feedbacks within general circulation models (GCMs) produces ~3°C of warming for a doubling of CO_2 from preindustrial times, compared to 1.2°C of warming calculated without feedbacks (IPCC 2007). Thus, it is clear that changes in the water vapor concentration have important significance for the formation of the Earth's climate.

Although water vapor in the stratosphere is present at levels much lower than those in the troposphere, these very changes in stratospheric water vapor have a profound impact on radiative forcing and surface temperatures. It was established that longwave radiation at the wavelengths of water absorption is fully attenuated in the lower troposphere, such that increasing water vapor concentrations at those levels will have no impact on radiative forcing. However, surface temperatures are very sensitive to changes in water vapor in the UT and LS, where water vapor absorption lines are unsaturated. De F. Forster and Shine (1999), extending earlier work of Rind and Lonergan (1995), calculated that if the increase in the lower stratospheric H_2O mixing ratio reported over Boulder, Colorado, from 1979 to present, by Oltmans and Hofmann (1995) is occurring globally, the contribution to surface warming would be 40% of that from the CO_2 increase over the same time period. Solomon et al. (2010) determined that a 1 ppmv increase (~1% per year) in stratospheric water vapor between 1980 and 2000, though uncertain, would have enhanced the rate of surface warming over that period by ~30% compared to estimate without this change. The 0.4 ppmv decrease in water vapor after 2001 acted to reduce the rate of global surface warming from 2000–2009 by ~25%. The combination of these trends likely contributed to the observed flattening of the global surface warming trend observed in the past decade. De F. Forster and Shine (1999) also emphasized that an increase of water vapor causes a cooling of the LS that is comparable to the contribution due to ozone changes.

It was found that the increased stratospheric water vapor levels could also enhance ozone destruction. Water vapor is a source of HO_x radicals in the stratosphere, which catalytically destroy ozone. At the poles, elevated water vapor also increases the formation and persistence of polar stratospheric clouds (PSCs); a 1 ppmv increase in water vapor increases the threshold temperature for PSC activation by ~1°C, which would increase the spatial and temporal extent of polar ozone loss (Kirk-Davidoff et al. 1999). Elevated water vapor also increases activation of sulfate-water aerosols for ozone depletion, both at the poles and at middle attitudes. Deep convective outflow has been observed to produce plumes with up to 12 ppmv of water vapor in the middle attitude stratosphere, which could potentially activate halogen-catalyzed ozone loss on sulfate-water aerosols, leading to destruction of up to 25% of local ozone more than seven days (Anderson et al. 2012). Deep convection can also transport very short-lived bromine and iodine species into the stratosphere on short timescales, which would amplify this ozone destruction. If increasing global surface temperatures lead to a greater frequency of such deep convective events, substantial local ozone destruction could be observed in the summer over populated areas. This means that the increase in the stratospheric concentration of water vapor presents a continuing threat to the global ozone layer.

Increased climate forcing by greenhouse gases is also predicted to increase continental convective activity at middle attitudes. In particular, climate models show that increased atmospheric moisture content in a warmer climate leads to an increase in the frequency and intensity of severe storms over the United States (Trapp et al. 2009; Van Klooster and Roebber 2009). Such changes could increase the amount of ice transported into the stratosphere by convection, increasing water vapor in both the lowermost stratosphere and overworld. As the lowermost stratosphere is a region of descent, increases in water vapor there would have an important local radiative impact, but would not impact the global stratosphere. However, elevated water vapor above the 380 K isentrope could be isentropically transported into the tropics where it would enter the mean circulation and moisten the overworld stratosphere while bypassing the cold point (Dessler and Sherwood 2004).

It is therefore critical to better understand how stratospheric water vapor will be affected by a changing climate to predict future changes and their feedback on the climate system. The climate feedbacks of tropospheric water vapor are well represented in climate models because water vapor at these levels closely follows temperature changes. On the other hand, the processes that control water vapor in the stratosphere are not well understood, which makes predictions of stratospheric water vapor levels and their radiative feedbacks much more uncertain. From the results presented by Harries (1997), one can realize that it is very important to have reliable information about the state of the atmosphere. Harries (1997) pointed out that, because the effects of water vapor on the Earth radiative balance are so large, small errors in spectroscopic parameters and in radiative-dynamical models used to model the energy balance, can produce potentially large uncertainties in the prediction of climatic change. He presented evidence from sensitivity studies showing that the humidity concentration, particularly in the UT may need to be known with accuracy in the range of 3%–10% to avoid uncertainties in calculated radiative forcing that are of the order of the effect due to doubling CO_2 concentrations.

These facts emphasize the urgent need to understand the behavior and to assess the concentration of the water vapor in the UT and LS, to detect short- and long-term trends in stratospheric water vapor, and to establish the mechanisms, which control upper tropospheric and stratospheric humidity.

9.2 WATER VAPOR MONITORING IN TROPOSPHERE AND STRATOSPHERE

The previous analysis has shown that atmospheric water vapor is abundant in the atmosphere and possesses absorption features spread over a broad range of the electromagnetic spectrum. Therefore, it should be easy to measure the concentration of water vapor in UT and stratosphere. However, in practice, water vapor measurements in these areas have proved to be difficult. Water sticks to surfaces, thereby providing challenges for *in situ* techniques. Sharp vertical and horizontal gradients present difficulties for remote sensing techniques. A lack of understanding of the fundamental physics behind the observed spectrum complicates the analysis of remote sensing measurements. The radiative effects of clouds are yet another complicating factor in making water vapor measurements. In addition, comparisons between satellite and direct water vapor measurements from *in situ* observations present a difficulty. This was due mainly to problems with the *in situ* methods, especially the radiosondes, and also because of difficulties in making comparisons in an inhomogeneous atmosphere when instruments have very different spatial coverage and altitude resolution. As a result, our understanding of the distribution of upper tropospheric and lower stratospheric water vapor is not as thorough as it should be.

As follows from Chapters 4 through 8, the study of UT and stratosphere can be carried out using either a remote sensing instruments, installed at ground stations, or *in situ* monitoring using instruments mounted on balloons (Figure 9.1a), satellites (Figure 9.1b), and aircrafts (Kley et al. 2000; Pankine et al. 2009). The advantage of most *in situ* techniques is their higher precision and spatial resolution compared to remote sensing instruments; thus *in situ* instruments are well suited particularly for case studies on smaller scales. Their absolute calibration either in the laboratory or in flight, tracing to calibration standards can be repeated at regular intervals and possible instrument drifts are easier to detect. There is no need for additional algorithms for the measurement geometry. At the same time, the retrieval of atmospheric water vapor profiles from ground-based measurements is a difficult task due to its large vertical gradient and large temporal variability; and standard retrieval methods are often not

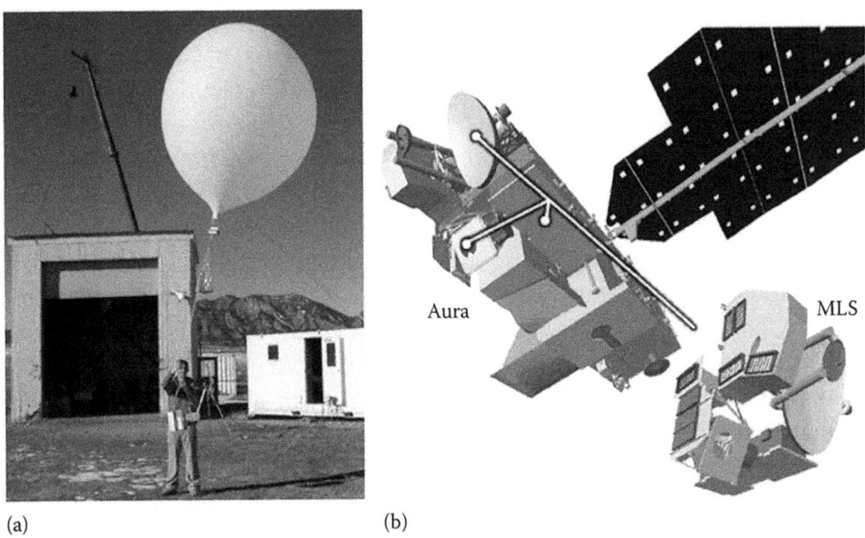

FIGURE 9.1 (a) Launching of a balloon for meteorological observations (From http://www.esrl.noaa.gov/gmd/ozwv/wvap/) and (b) Satellite with AURA-Microwave Limb Sounder (MLS) (From http://www.mls.jpl.nasa.gov/). The Earth Observing System (EOS) MLS is one of four instruments on the NASA's EOS Aura satellite, launched on July 15, 2004. MLS makes measurements of atmospheric composition, temperature, humidity, and cloud ice. The MLS measurements are made globally day and night. A feature of the MLS technique is that its measurements can be obtained in the presence of ice clouds and aerosol that prevent measurements by shorter-wavelength infrared, visible, and ultraviolet techniques. The EOS MLS measures thermal emission from broad spectral bands centered near 118, 190, 240, 640, and 2500 GHz are measured continuously (24 h per day) by 7 microwave receivers (2 each at 118 and 2500 GHz) using a limb viewing geometry, which maximizes signal intensities and vertical resolution. MLS is radiometrically calibrated after each 25 s limb scan. The EOS MLS instrument contains three modules: (1) The GHz radiometer module, which includes the 118–640 GHz receivers and a scanning offset antenna with 0.8 × 1.6 m primary mirror; (2) The THz radiometer module which contains the 2500 GHz receivers and the THz telescope and scan mirror, whose scan is synchronized with that of the GHz antenna; and (3) The Spectrometer module which receives signals from the GHz and THz radiometer modules, detects and digitizes them, and passes the digitized signals to the spacecraft for telemetry to the ground.

suited (Schneider et al. 2013). The development of water vapor retrievals has recently made substantial progress, but still no common water vapor retrieval method is applied at all NDACC sites. *In situ* instruments are therefore often used for the validation of space-borne remote sensing experiments. Since the operation of *in situ* instruments in the UT/LS requires the use of platforms such as balloons or aircraft, which are either expensive to operate, have limited availability, or both, the data sets cover only limited regions and time periods. This disadvantage is partly compensated by the large number of measurements made to date.

With regard to the comparison of opportunities for research using balloons and aircrafts, it should be noted that these studies tend to solve various tasks. Of course, more complicated equipment, which can allow conducting more complex experiments can be installed at aircrafts. The ability of the aircraft to go when and where observations are needed makes it also very valuable for atmosphere monitoring. Moreover, it is possible with long range jet aircraft to reach any spot on the globe. This means that it is able to cover a large geographical range with a number of flights. In addition, jet aircraft can reach the UT or LS depending on the latitude, and thus, are above most of the water vapor that prevents observation from the ground at many wavelengths. However, the number of aircrafts that can be used for research is very limited. In addition, the use of aircraft has weather restrictions. Another disadvantage of using aircrafts as observing platforms is their inability to reach the altitudes of greatest interest for stratospheric chemistry. The vibration level which can cause noise problems is also a shortcoming of aircraft platform.

Regarding satellite possibilities for controlling the water vapor distribution in the atmosphere, then of course they are used (Noël et al. 1999; Sherwood et al. 2010). But of course, these measurements may not be *in situ* measurements. Weather satellites are operated by agencies in China, France, India, Japan, Korea, the Russian Federation, the United States and Eumetsat for Europe, with international coordination by the World Meteorological Organization (WMO). In addition to geostationary weather satellites around 160 environmental satellite missions in low-earth orbit are currently measuring selected climate parameters. As a rule, for remote monitoring of water vapor in the atmosphere satellites equipped with microwave radiometers (Engelen and Stephens 1999; Susskind et al. 2003; Froidevaux et al. 2006; Waters et al. 2006) or with multichannel spectrometers (Schmetz et al. 2002; Buehler et al. 2008; Fetzer et al. 2008; Chou et al. 2009) (Table 9.1). For example, the Atmospheric Infrared Sounder (AIRS) contains hundreds of channels in the water vapor absorption bands and can resolve vertical layers of a few kilometers. However, even this sounder is still affected by cloud cover and tends to underestimate wet and dry relative humidity extremes (Fetzer et al. 2008; Chou et al. 2009). As it was shown in Chapter 5, the microwave radiation is less affected by clouds and thus offers a useful alternative method of moisture sounding from space. In particular, advanced microwave sounding unit (AMSU) is able to detect humidity averaged over several broad layers, especially in the UT, with significant interference only from thick clouds (Engelen and Stephens 1999; Susskind et al. 2003). Similar instruments that observe limb emission, the two Microwave Limb Sounders (MLSs, first flown in 1991), observe moisture above ~350 hPa (Froidevaux et al. 2006; Waters et al. 2006). We should also not forget possibilities of the global positioning system (GPS) technology, considered in Chapter 6. This technology can be used to estimate a total column water vapor over suitably equipped surface stations (Wang and Zhang 2008). The GPS technique has the important advantages of being an absolute measurement that does not need an independent

TABLE 9.1
Estimates of Random Errors, Systematic Errors and Vertical Resolution of Stratospheric H_2O Profiles Derived from Satellite Instrumentation

Instrument and Data Set	Random Error	Systematic Error	Vertical Resolution, km
LIMS (version 5)	20–15% (1–5 hPa)	31–24% (1–5 hPa)	~5
(Limb IR emission)	15–10% (5–10 hPa)	24–20% (5–10 hPa)	
	10% (10–50 hPa)	20–37% (10–50 hPa)	
SAGE II (version 5.9)	10–5% (3–10 hPa)	6–13% (3–7 hPa)	~3
(IR solar occultation)	5–14% (10–25 hPa)	13% (7–25 hPa)	
	14% (25–300 hPa)	13–27% (25–100 hPa)	
		27% (100–300 hPa)	
ATMOS (version 3)	9–11% (1–300 hPa)	6% (1–300 hPa)	3–6
(IR solar occultation)			
HALOE (version 19)	9–7% (1–10 hPa)	10–14% (1–10 hPa)	2.3
(IR solar occultation)	7–13% (10–40 hPa)	14–19% (10–40 hPa)	
	13% (40–100 hPa)	19–24% (40–100 hPa)	
MLS (version 0104)	4% (1–10 hPa)	6–9% (1–10 hPa)	~3
(Limb microwave emission)	3% (10–50 hPa)	9–16% (10–50 hPa)	
	3–8% (50–100 hPa)	16–50% (50–100 hPa)	
MAS (Limb microwave emission)	5–10% (1–50 hPa)	10–15% (1–50 hPa)	~5
ILAS (version 4.20)	More than 10% above 2 hPa	30% (1–2 hPa)	1–2
(IR Solar occultation)	10–5% (2–300 hPa)	30–10% (2–7 hPa)	
		10% (7–300 hPa)	
POAM III (version 2)	5% (3–100 hPa)	15% (3–100 hPa)	1–3
(IR solar occultation)			

Source: Kley D. et al., SPARC Report No. 2. World Climate Research Programme, Geneva, Switzerland, 2000.

ATMOS—Atmospheric Trace Molecule Spectroscopy, ILAS—Improved Limb Absorption Spectrometer, LIMS—Limb Infrared Monitor of the Stratosphere, MAS—Microwave Atmospheric Sounder, MLS—Microwave Limb Sounder (on UARS), POAM—Polar Ozone and Aerosol Measurements, SAGE—Stratospheric Aerosol and Gas Experiment.

calibration and of not being affected at all by clouds or other absorbers (although temperature must be known fairly well to achieve good accuracy).

9.3 PECULIARITIES OF AIRCRAFT OBSERVATIONS—FACTORS INFLUENCING THE IN-FLIGHT PERFORMANCE

Experiments have shown that numerous factors can influence the in-flight performance of airborne water vapor measurements. In particular, of crucial importance are the sticking of water vapor at surfaces and the appropriate use of sampling systems (Kley et al. 2000; Bange et al. 2013).

Water molecules are highly polar, such that water molecules attach themselves tenaciously to surfaces. Particularly, at low temperatures, this can lead to large memory effects of the water vapor measurements. Additional heating may eliminate any such memory effects. Therefore, the selection of hygrophobic materials is an important part of the sampling and measuring systems design of the hygrometer. In general, to avoid any water vapor contamination or memory effect from the aircraft skin, it is most favorable that the air inlets are sampling air outside the aerodynamic boundary layer at the aircraft skin. In addition, moisture must not be allowed to leak into the measurement system or interface with the measurements. This is the most critical in dry regions or at low humidities in the UT and stratosphere. Therefore, most airborne hygrometers use extractive sampling systems that force the samples ambient air through an appropriate inlet system into a closed-path detector system installed inside the aircraft. Usually the sideward or backward facing type of air inlets are deployed by using a pump, and the forward directed type of air inlets use the ram pressure caused the moving aircraft, such as fast Lyman–alpha or IR absorption hygrometers, or open-path TLAS systems. These open-path systems have the advantages of virtually eliminating contamination. The heating of the instrument and its inlet before measurements can also be used for decrease of contamination influence.

It should be noted that similar problems exist in the measurements carried out with the use of balloons. Therefore, proper selection of wall materials, optimization of probe location (increasing the distance between a measurement unit and balloon), heating of the instrument and its inlet, large flow rates, or open-cell designs have to be used to avoid or minimize measurement artifacts (Kley et al. 2000).

9.4 COMPARATIVE CHARACTERISTICS OF THE DEVICES USED FOR THE UPPER TROPOSPHERIC AND STRATOSPHERIC MEASUREMENTS

Tropospheric and stratospheric water vapor has been measured more than the past 50 years by a large number of individuals and institutions using a variety of *in situ* and remote-sensing measurement techniques (Mastenbrook 1968; Bertaux and Delannoy 1978; Kley and Stone 1978; Kley et al. 1979; Buck 1985; Goutail and Pommereau 1987; Vömel et al. 2007; Leblanc et al. 2012; Hurst et al. 2011, 2014). Operating principles and measurement specifications of most *in situ* research-type instruments currently in use are presented in Table 9.2 along with their estimated measurement accuracy. These instruments provide point measurements in time and space with high vertical resolution, typically in the range of a few hundred meters or better. Accuracy estimates range from 5% to 10% based on known or estimated random and systematic uncertainties inherent in the instrument system, calibration procedures and retrieval

TABLE 9.2
***In Situ* and Remote-Sensing Techniques for Measurements of H_2O from Ground-Based, Balloon-Borne and Airborne Platforms, along with Their Typical Measurement Range and Overall Accuracy, That Is, the Sum of Systematic and Random Errors**

Technique	Range	Altitude Range	Accuracy
Lyman–alpha fluorescence	500–0.2 ppmv	5–35 km	6–7%
Tunable diode laser spectrometry	>0.1 ppmv	0–30 km	5–10%
Microwave spectrometry	20–0.2 ppmv	20–80 km	0.6–0.2 ppmv
LIDAR	>4 ppmv	0–20 km	5–10%
IR and FTIR spectrometry	>1 ppmv	5–40 km	5–13%
Frost point hygrometry	10,000–0.5 ppmv	5–30 km	5–10%
MOZAIC sensor	>20 ppmv	Troposphere	5–7% RH
Radiosonde	100 ≈ 5% RH	Middle and lower troposphere	not assessed

Source: Kley D. et al., SPARC Report No. 2. World Climate Research Programme, Geneva, Switzerland, 2000.

FIGURE 9.2 LIDAR system designed by Jet Propulsion Laboratory (JPL) Table Mountain Facility (TMF) atmospheric LIDAR group. (From http://www.tmf.jpl.nasa.gov/tmf-lidar/.)

algorithms. Remote sensing instruments deployed on ground-based (Figure 9.2), balloon-borne and airborne platforms provide vertical profile measurements with stated accuracy similar to *in situ* instruments, although with coarser vertical resolution. Such vertical resolutions range from several hundred meters in the case of light detection and ranging (LIDAR), to a few kilometers for the IR and far-infrared (FIR) spectrometers, and approximately 10 km for microwave instruments.

A survey of different techniques designed for satellite application and their performance is presented in Table 9.1. The vertical resolution of satellite instruments depends on the individual measurement concept (e.g., occultation or emission) and the specific instrument implementation. Horizontal resolution is typically on the order of 50–300 km depending on whether the experiment is nadir or limb viewing.

Despite the significant research experience gained over the past decades, measurements of atmospheric humidity in upper tropospheric and stratospheric regions have been and continue to be a challenging task. Measurements of water vapor in the UT and in the stratosphere require tremendous technical effort due to the large gradients around the tropopause and the low stratospheric mixing ratios of a few ppmv in contrast to the moist tropospheric air masses. Further, in the stratosphere, the spatial and temporal variability of the H_2O abundance is relatively small, that is, changes of a few tenths of 1 ppmv need to be detected with a similar accuracy of the measurement. Therefore, till now there is no sensor available that can cover the full dynamic range of water vapor levels from a few percentages near the surface down to a few of ppmv in 15–20 km altitude. This means that there is no single existing instrument which is capable of H_2O measurements at all altitudes, with adequate global and temporal coverage. Figure 9.3 shows a diagram of the altitude range and platforms for the most important techniques. It is seen, while a broad

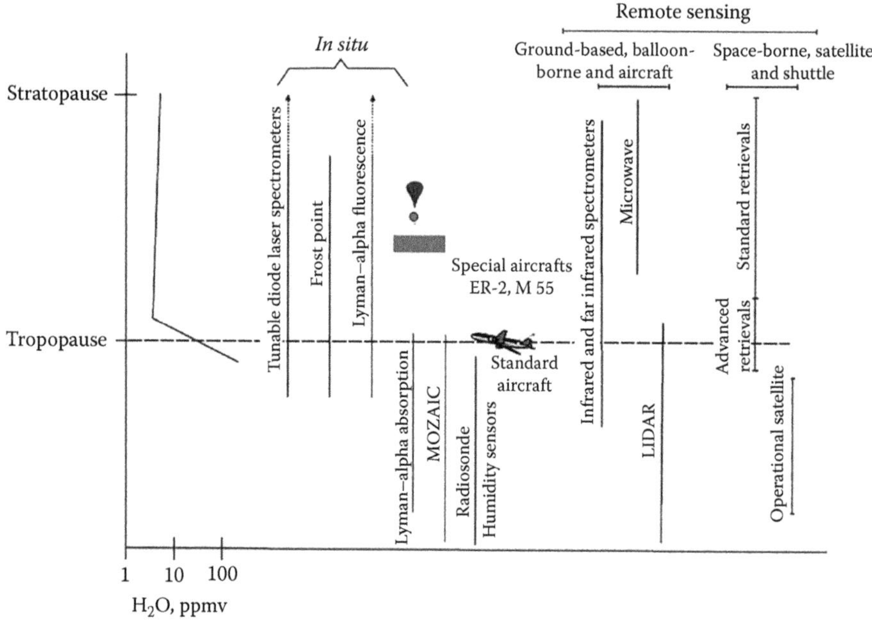

FIGURE 9.3 Typical vertical profile of water vapor in the upper troposphere and the stratosphere, the altitude range where the techniques described in this Assessment can be applied, and the carriers available for integration of the different instruments. (From Kley D. et al., SPARC Report No. 2. World Climate Research Programme, Geneva, Switzerland, 2000.)

spectrum of instruments exists today, all of them have their limitations (Beaton and Spowart 2012).

For example, the chilled-mirror hygrometer, commonly used for aircraft-based water vapor measurements, directly measures the dew point or frost point of the air. While these instruments are inexpensive, easy to deploy, and accurate, the need to adjust the temperature of the mirror makes them slow to respond, requiring seconds or even tens of seconds, and they often overshoot or oscillate when subjected to large, abrupt humidity changes.

Earlier in Chapters 3 and 4, it was shown that optical water vapor absorption-measuring methods are a powerful tool for monitoring the atmosphere. They are based on tunable diode lasers (May 1998; Zondlo et al. 2010), with data rates of 10–25 samples per second, equivalent to a bandwidth of 5–12.5 Hz (Shannon 1949). These instruments provide an independent absolute measurement of the water vapor number density in the troposphere and LS, but their using require very careful calibration, with consideration to the pressure, temperature, and homogeneous and heterogeneous broadening coefficients of water vapor. They also require skilled field support because troubleshooting and repair can be very difficult. Another limitation for the use of TDLAS hygrometers for these purposes is a significant shift in the central wavelength of absorption when the altitude is changed (Table 9.3). This means that the coordination of the laser radiation and the absorption spectrum by water vapor, required for this method, can be violated.

Nondispersive infrared (NDIR) instruments such as the LI-COR LI-7500 have also been deployed. This instrument has a bandwidth of up to 20 Hz; however, the large sample volume, roughly 30 cm^3 or larger, is designed for ground-based flux measurements and is not suitable for mounting outside an aircraft without significantly strengthening the optical assembly, even at airspeeds below 50 m/s (Hall et al. 2006). If mounted inside the aircraft, it requires a very large airflow to exchange the air inside the large sample volume at 20 times per second. For comparison, the Zeeman-split argon emission hygrometer described by Kebabian et al. (2002) has a sample volume of 300 cm^3, and they employed a 600 L/min flow to sufficiently clear the cell for measurements in a 10-Hz bandwidth.

TABLE 9.3
Effect on the Altitude of the Measurement on the Position of Central Wavelength of Water Vapor Absorption

Central Wavelength of Absorption, μm	The Area of Measurement
7.43	Lower-tropospheric humidity
7.02	Mid-tropospheric humidity
6.51	Upper-tropospheric humidity

Source: Menzel, W.P. and Purdom, J.F.W., *Bull. Am. Meteo. Soc.*, 75, 757–781, 1994.

Until recently, Lyman–alpha absorption hygrometers were the preferred high-rate hygrometers for water vapor measurements from research aircraft. These instruments were inexpensive, provided data with a bandwidth of 25 Hz, and required no optical alignment or skilled field support, only occasional cleaning of the lamp and detector windows. However, because of pressure- and temperature-dependent interference from atmospheric oxygen, and to a lesser extent carbon dioxide, which also absorbs at this wavelength, and drift in the electronics, lamp output, and detector responsivity, the Lyman–alpha hygrometer often was used as a high rate signal only, with the long-term output forced to follow that of a slower, more accurate sensor, such as the chilled-mirror dew point sensors by a loose-coupling algorithm such as described by Schanot (1987). There have been attempts to remove the oxygen interference through variable pathlength (Buck 1976), multiwavelength (Weinheimer and Schwiesow1992), and reference path techniques (Zuber and Witt 1987), but for a variety of reasons these have not been widely deployed on aircraft. Recent developments made in Harvard University (Cambridge, Massachusetts) (Hintsa et al. 1999) and in the National Center for Atmospheric Research (Boulder, Colorado) (Beaton and Spowart 2012) contributed to the improvement of parameters of Lyman–alpha absorption hygrometers. A schematic of the Harvard Water Vapor (HWV) instrument based on the Harvard Lyman–alpha hygrometer is shown in Figure 9.4. This instrument was designed for aircraft applications. However, the main disadvantages of Lyman–alpha absorption hygrometers, indicated earlier, still exist.

Similar hygrometers (Campbell and Tanner 1985) have been designed around krypton lamps, which have emission lines near the Lyman–alpha line—a strong emission line at 123.6 nm and a weaker line at 116.5 nm. While the krypton lamps are stable, long-lived, inexpensive sources, they are notably inferior from a spectroscopic standpoint. At 123.6 nm, the water vapor absorption strength is down by a factor of almost 2, whereas the oxygen absorption is stronger by a factor of about 40 (Tillman 1965; Yoshino et al. 1996), making the oxygen interference much more severe and variable

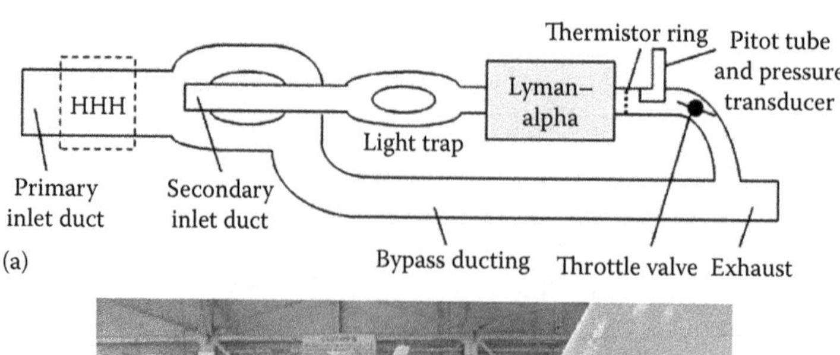

FIGURE 9.4 (a) A schematic of the HWV instrument. The Lyman–alpha detection axis and the subsystems that measure temperature and pressure and control the velocity through the instrument duct are labeled; the location of the HHH detection axis, which was included in the payload for the first time in the 2011 MACPEX campaign, is also indicated. Harvard Herriott Hygrometer (HHH) is an independent water vapor measurement. HHH uses a tunable diode laser (TDL) to measure water vapor via direct absorption in the infrared range. (b) The HWV instrument as it was flown in 2011 using a 3.5 × 4″ rectangular primary inlet which incorporates the HHH detection axis. (From http://www.nasa.gov.)

with air density, a particular problem for measurements from aircraft.

Thus, one can see that an ideal instrument capable of measuring the concentration of water vapor in all possible conditions does not exist. Moreover, one should recognize that there is no single technique or instrument platform that is recognized as a standard to which other techniques should be compared. This means that the development and operation of precise and accurate hygrometers for the atmosphere monitoring as well as their accurate calibration is still at the research level and is far from being routine. For example, in recent years there have been studies aimed at developing methods of water vapor measurements based on the use of airborne mass spectrometers (AIMS–H_2O). In particular, Kaufmann et al. (2016) presented a new setup based on linear quadrupole mass spectrometers (LQMS) to measure water vapor by direct ionization of ambient air (Figure 9.5). Air was sampled via a backward facing inlet that includes a bypass flow to assure short residence times (<0.2 s) in the inlet line, which allows the instrument to achieve a time resolution of ~4 Hz, limited by the sampling frequency of the mass spectrometer. From the main inlet flow, a smaller flow is extracted into the novel pressure controlled gas discharge ion source of the mass spectrometer. Developed instrument had possibility to quantify low water-vapor mixing ratios typical for the UT and LS with a high accuracy between 7% and 15% in the measurement range between 1 and 500 ppmv, depending on specific humidity and time resolution of the measurement. Of course, this powerful tool is not simple, cheap, and portable, which can be used for example, in the measurements carried out with the use of balloons. But its application can enhance the possibilities of scientific research in this field.

The same situation takes place when comparing methods, used for remote monitoring. For example, the comparison of the total precipitable water vapor (PWV) from air measured using FTIR spectroscopy with data, obtained by other methods, has shown the following (Schneider et al. 2009, 2010a, b): when it was compared with instruments such as a Multifilter Rotating Shadow-band Radiometer (MFRSR), a Cimel sun photometer, a GPS receiver, and daily radiosondes (Vaisala RS92), it was estimated that the FTIR spectrometer provides very precise trophospheric water vapor data, but when the area-wide coverage and real-time data availability is very important, the GPS and the RS92 data

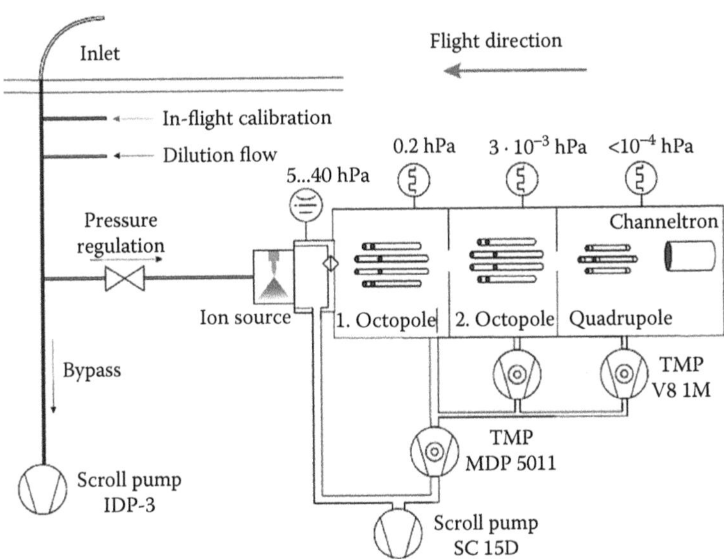

FIGURE 9.5 Schematic of the flight configuration of AIMS. Ambient air enters via a backward faced inlet and passes through a pressure regulation valve before entering the ion source. The ion beam is then focused by two adjacent octopoles and finally separated by mass-to-charge ratio in the quadrupole. In addition, connections for an optional dilution of ambient air and background measurements and for addition of trace gases for in-flight calibration are mounted right beneath the inlet. (From Kaufmann S. et al., *Atmos. Meas. Tech.* 9, 939, 2016. Published by the European Geoscience Union as open access.)

are more appropriate. At the same time, the FTIR spectroscopy can be used as a reference when assessing the accuracy of the other techniques, but those who use this technique have to be aware of the FTIR's significant clear sky bias.

Palm et al. (2010) also did comparisons between total water vapor column amounts measured by the ground-based FTIR, the ground-based microwave radiometer for atmospheric measurments (RAM), and by the satellite sensors Scanning Imaging Absorption Spectrometer for Atmospheric Chartography (SCIAMACHY) and Advanced Microwave Sounding Unit B (AMSU-B). It was shown that the microwave sensors, RAM and AMSU-B, operate best in winter, when the IWV is low. They are partly independent of weather conditions; that is, light clouds do not distort the measurements beyond recovery. The IR sensor, FTIR, depends on clear sight to the Sun, that is, no clouds between the instrument and the Sun. The optical sensors, SCIAMACHY and The Global Ozone Monitoring Experiment (GOME), also depend on solar light. Due to the short wavelength of the radiation recorded by the SCIAMACHY and the GOME instrument, clouds disturb the measurements considerably. This fact makes it necessary to cloud filter those measurements. The GPS IWV is derived from the zenith path delay (ZPD) standard product of the International GNSS Service (IGS) global network processing. It measures all year round and is independent of weather.

It was also established that measurements from ground-based remote sensors, Solar-FTIR and GPS (IWV higher than 10×10^{21} mol/cm^2), were of superior quality (Palm et al. 2010). The satellite instruments, AMSU-B, SCIAMACHY, and GOME, also performed well if errors are taken into account. For the AMSU-B measurements this included IWV higher than 30×10^{21} mol/cm^2. The higher variance of the satellite-based instruments may be explained by the spatial coverage and by the high variance of IWV. The Lunar-FTIR exhibited a low SNR and very sparse measurements necessitating long coincidence times to enable radiosonde comparisons. The Lunar absorption measurements using the FTIR are very sparse and also rather noisy. However, the Lunar-FTIR gives possibility to make measurements during the night. The measurements of the RAM were of fair quality. It has to be taken into account; however, that the RAM is not designed for measuring the IWV but this is a by-product of the O_3-measurements. The IWV values derived from GPS ZPD measurements exhibited high relative scatter for low IWV (smaller than 10×10^{21} mol/cm^2). For lowest IWV (smaller than 5×10^{21} mol/cm^2), they were even inferior to the RAM data. This means that the gap during night time, cloudy weather conditions, and in the polar winter can however be filled using the GPS, except for very low IWV. Similar conclusions can be drawn for the satellite-based instruments SCIAMACHY, GOME, and AMSU-B. The strength of the GOME and the SCIAMACHY is the ability to measure high IWV but rely on solar irradiation of the atmosphere and on little interference by clouds. AMSU-B, however, does not rely on external irradiation of the atmosphere and delivers data throughout polar winter.

Thus, our analysis shows that, despite efforts to develop various methods and tools for monitoring the concentration of water vapor in the atmosphere, our possibilities for measuring the complete seasonal cycle are still very limited (Palm et al. 2010). The typical surface station instruments commonly provide only very local, point, observations, and therefore suffer from low spatial resolution. Compounding this problem is the limited accessibility to position humidity gauges in heterogeneous terrain, or areas with complex topography. In addition, because of surface perturbation a point measurement close to the surface (e.g., 2 m from the ground as in a standard meteorological surface station) is not satisfactory for model initialization. What is ideally required for meteorological modeling purposes is an area average measurement of near-surface moisture over a box with the scale of the model's grid and at an altitude of a few tens of meters. Sensors installed on both commercial and research aircraft to measure water vapor also can be used for *in situ* atmosphere monitoring. Of course, this approach to humidity measurement is a promising. However, this approach would only provide measurements when and where scheduled flights occur and thus the researcher might not have data where they are needed. In addition, the cost is high for dedicated aircraft flights. Another approach is to fly small *in situ* packages on balloons filled with helium or hydrogen. However, radiosondes, which are typically launched only 2–4 times a day, also provide very limited information. In addition, these monitoring methods are costly for implementation, deployment, and maintenance. In addition, it was established that data from radiosondes have biases that lead to underestimation of relative humidity, particularly in the mid and UT (e.g., Soden et al. 1994; Ferrare et al. 1995; Lesht and Liljegren 1997; Kley et al. 2000; Miloshevich et al. 2001; Wang et al. 2002; Turner et al. 2003; Soden et al. 2004). Wang et al. (2002) identified the source of some of the biases, including contamination of the sensor due to inappropriate packaging of the radiosondes prior to use, errors in the model relating the change of capacitance of the humidity sensor with changes in relative humidity, and errors in the model relating the change of capacitance of the humidity sensor with changes in temperature. Soden et al. (2004) applied the bias corrections of Wang et al. (2002) and found that upper-tropospheric humidity is still underestimated (up to 40%) by radiosondes compared with LIDAR and IR satellite measurements, which means the cause (or causes) of the radiosonde dry bias is (or are) still largely unknown. In recent years, much attention has been paid to the use of satellites for these purposes. Satellites allow for a large area to be covered, but results obtained by satellites are frequently not accurate enough in measuring surface-level moisture, whereas this near-surface moisture is, in most cases, the important variable for convection. In addition, the application of satellites for these purposes is possible primarily for only cloud-free zones. Microwave sounders have the ability to sound through cloud and hence offer nearly all-weather capability. However, their spatial resolution (both vertical and horizontal) is generally lower than that of the IR instruments. So, further research aimed at finding and developing new methods of monitoring the atmosphere is still required.

REFERENCES

Anderson J.G., Wilmouth D.M., Smith J.B., Sayres D.S. (2012) UV dosage levels in summer: Increased risk of ozone loss from convectively injected water vapor. *Science* 337, 835–839.

Bange J., Espositi M., Lenschow D.H. et al. (2013) Measurement of aircraft state and thermodynamic and dynamic variables. In: Wendisch M., Brenguier J.-L. (Eds.), *Airborne Measurements for Environmental Research. Methods and Instruments.* Wiley-VCH, Weinheim, Germany, Chapter 2.

Beaton S.P., Spowart M. (2012) UV absorption hygrometer for fast–response airborne water vapor measurements. *J. Atmos. Oceanic. Technol.* 29, 1295–1303.

Bertaux J.-L., Delannoy A. (1978) Vertical distribution of H_2O in the stratosphere as determined by UV fluorescence in-situ measurements. *Geophys. Res. Lett.* 5, 1017–1020.

Buck A.L. (1976) The variable–path Lyman–alpha hygrometer and its operating characteristics. *Bull. Amer. Meteor. Soc.* 57, 1113–1118.

Buck A.L. (1985) The Lyman-alpha absorption hygrometer. In: *Proceedings, Moisture and Humidity Symposium*, Washington, DC, Instrument Society of America, Research Triangle Park, NC, p. 411.

Buehler S.A., Kuvatov M., John V.O., Milz M., Soden B.J., Jackson D.L., Notholt J. (2008), An upper tropospheric humidity data set from operational satellite microwave data. *J. Geophys. Res.* 113, D14110.

Campbell G.S., Tanner B.D. (1985) A krypton hygrometer for measurement of atmospheric water vapor concentration. In: Moisture and Humidity: Measurement and Control in Science and Industry. Instrument Society of America, Research Triangle Park, *Proceedings of the 1985 International Symposium on Moisture and Humidity*, Instrument Society of America, pp. 609–614.

Chou C., Neelin J.D., Chen C.A., Tu J.Y. (2009) Evaluating the "rich-get-richer" mechanism in tropical precipitation change under global warming. *J. Clim.* 22, 1982–2005.

De F. Forster P.M., Shine K.P. (1999) Stratospheric water vapour changes as a possible contributor to observed stratospheric cooling. *Geophys. Res. Lett.* 26, 3309–3312.

Dessler A.E., Sherwood S.C. (2004) Effect of convection on the summertime extratropical lower stratosphere. *J. Geophys. Res.* 109, D23301.

Engelen R.J., Stephens G.L. (1999) Characterization of water-vapour retrievals from TOVS/HIRS and SSM/T-2 measurements. *Q. J. Royal Meteorol. Soc.* 125, 331–351.

Ferrare R.A., Melfi S.H., Whiteman D.N., Evans K.D., Schmidlin F.J., O'C Starr D. (1995) A comparison of water vapor measurements made by Raman Lidar and radiosondes. *J. Atmos. Ocean. Technol.* 12, 1177–1195.

Fetzer E.J., Read W.R., Waliser D. et al. (2008) Comparison of upper tropospheric water vapor observations from the microwave limb sounder and atmospheric infrared sounder. *J. Geophys. Res.* 113, D22110.

Froidevaux L., Livesey N.J., Read W.G. et al. (2006), Early validation analyses of atmospheric profiles from EOS MLS on the Aura satellite. *IEEE Trans. Geosci. Remote Sens.* 44, 1106–1121.

Goutail F., Pommereau J.-P. (1987) Stratospheric water vapor in situ measurements from infra-red montgolfier. *Adv. Space Res.* 7(7), 111–114.

Hall P., Dumas E., Senn D. (2006) NOAA ARL mobile flux platform instrumentation integration on University of Alabama Sky Arrow Environmental Aircraft. NOAA Technical Memorandum, ARL-257, 51 p.

Harries J.E. (1997) Atmospheric radiation and atmospheric humidity. *Q. J. Royal Meteorol. Soc.* 123, 2173–2186.

Hintsa E.J., Weinstock E.M., Anderson J.G., May R.D., Hurst D. (1999) On the accuracy of *in situ* water vapor measurements in the troposphere and lower stratosphere with the Harvard Lyman-α hygrometer. *J. Geophys. Res.* 104, 8183–8189.

Hurst D.F., Oltmans S.J., Vömel H., Rosenlof K.H., Davis S.M., Ray E.A., Hall E.G., Jordan A.F. (2011) Stratospheric water vapor trends over Boulder, Colorado: Analysis of the 30 year Boulder record. *J. Geophys. Res.* 116, D02306.

Hurst D.F., Lambert A., Read W.G., Davis S.M., Rosenlof K.H., Hall E.G., Jordan A.F., Oltmans S.J. (2014) Validation of Aura microwave limb sounder stratospheric water vapor measurement by the NOAA frost point hygrometer. *J. Geophys. Res. Atmos.* 119, 1612–1625.

IPCC (2007) Climate Change 2007. Fourth assessment report: Synthesis report. http://www.ipcc.ch/publications_and_data/ar4/syr/en/contents.html.

Kaufmann S., Voigt C., Jurkat T., Thornberry T., Fahey D.W., Gao R.-S., Schlage R., Schäuble D., Zöger M. (2016) The airborne mass spectrometer AIMS—Part 1: AIMS-H_2O for UTLS water vapor measurements. *Atmos. Meas. Tech.* 9, 939–953.

Kebabian P., Kolb C.E., Freedman A. (2002) Spectroscopic water vapor sensor for rapid response measurements of humidity in the troposphere. *J. Geophys. Res. Atmos.* 107(23), 4670.

Kirk-Davidoff D.B., Hintsa E.J., Anderson J.G., Keith D.W. (1999) The effect of climate change on ozone depletion through changes in stratospheric water vapour. *Nature* 402(6760), 399–401.

Kley D., Stone E. (1978) Measurements of water vapor in the stratosphere by photo dissociation with Ly-α (1216 Å) light. *Rev. Sci. Instrum.* 49(6), 691–697.

Kley D., Stone E.J., Henderson W.R., Drummond J.W., Harrop W.J., Schmeltekopf A.L., Thompson T.L., Winkler R.H. (1979) *In situ* measurements of the mixing ratio of water vapor in the stratosphere, *J. Atmos. Sci.* 36, 2513–2524.

Kley D., Russell J.M. III, Phillips C. (Eds.) (2000) SPARC assessment of upper tropospheric and stratospheric water vapor. World Meteorological Organization Technical Document 143. SPARC Report No.2. World Climate Research Programme, Geneva, Switzerland.

Leblanc T., McDermid I.S., Walsh T.D. (2012) Ground-based water vapor Raman lidar measurements up to the upper troposphere and lower stratosphere for long-term monitoring. *Atmos. Meas. Tech.* 5, 17–36.

Lesht B.M., Liljegren J.C. (1997) Comparison of precipitable water vapor measurements obtained by microwave radiometers and radiosondes at the southern great plains cloud and radiation Testbed Site. In: *Proceedings of the Sixth Atmospheric Radiation Measurement Science Team Meeting*, San Antonio, TX, pp. 165–168.

Mastenbrook H.J. (1968) Water vapor distribution in the stratosphere and high troposphere. *J. Atmos. Sci.* 25, 299–311.

May R.D. (1998) Open–path, near–infrared tunable diode laser spectrometer for atmospheric measurements of H_2O. *J. Geophys. Res.* 103(D15), 19161–19172.

Menzel W.P., Purdom J.F.W. (1994) Introducing *GOES-I:* The first of a new generation of geostationary operational environmental satellites. *Bull. Am. Meteo. Soc.* 75, 757–781.

Miloshevich L.M., Vomel H., Paukkunen A., Heymsfield A.J., Oltmans S.J. (2001) Characterization and correction of relative humidity measurements from Vaisala RS80A radiosondes at cold temperatures. *J. Atmos. Oceanic Technol.* 18, 135–155.

Noël S., Buchwitz M., Bovensmann H., Hoogen R., Burrows J.P. (1999) Atmospheric water vapor amounts retrieved from GOME satellite data. *Geophys. Res. Lett.* 26(13), 1841–1844.

Oltmans S.J., Hofmann D.J. (1995) Increase in lower-stratospheric water vapor at a mid-latitude Northern Hemisphere site from 1981 to 1994. *Nature* 374, 146–149.

Palm M., Melsheimer C., Noel S., Heise S., Notholt J., Burrows J., Schrems O. (2010) Integrated water vapor above Ny Alesund, Spitsbergen: A multi-sensor intercomparison. *Atmos. Chem. Phys.* 10, 1215–1226.

Pankine A., Nock K., Li Z., Parsons D., Purucker M., Wiscombe W., Weinstock E. (2009) Stratospheric satellites for earth observations. *BAMS* 97(8), 1109–1119.

Rind D., Lonergan P. (1995) Modeled impacts of stratospheric ozone and water vapour perturbations with implications for high speed civil transport aircraft. *J. Geophys. Res.* 1007381–1007396.

Schanot A.J. (1987) An evaluation of the uses and limitations of Lyman-alpha hygrometer as an operational airborne humidity sensor. In: *Proceedings of 6th Symposium Meteorological Observations and Instrumentation*, New Orleans, LA., American Meteorological Society, pp. 257–260.

Schmetz J., Pili P., Tjemkes S., Just D., Kerkmann J., Rota S., Ratier A. (2002) An introduction to Meteosat second generation (MSG). *Bull. Am. Meteorol. Soc.* 83, 977–992.

Schneider M., Romero P.M., Hase F., Blumenstock T., Cuevas E., Ramos R. (2009) Quality assessment of Izana's upper-air water vapour measurement techniques: FTIR, Cimel, MFRSR, GPS, and Vaisala RS92. *Atmos. Meas. Tech. Discuss.* 2, 1625–1662.

Schneider M., Toon G.C., Blavier J.-F., Hase F., Leblanc T. (2010a) H_2O and D profiles remotely-sensed from ground in different spectral infrared regions. *Atmos. Meas. Tech.* 3:1599–1613.

Schneider M., Romero P.M., Hase F., Blumenstock T., Cuevas E., Ramos R. (2010b) Continuous quality assessment of atmospheric water vapour measurement techniques: FTIR, Cimel, MFRSR, GPS, and Vaisala RS92. *Atmos. Meas. Tech.* 3, 323–338.

Schneider M., Demoulin P., Sussmann R., Notholt J. (2013) Fourier transform infrared spectrometry. In: Kampfer, N. (Ed.), *Monitoring Atmospheric Water Vapor.* Springer, New York, pp. 95–112.

Shannon C.E. (1949) Communication in the presence of noise. *Proc. Inst. Radio Eng.* 37, 10–21.

Sherwood S.C., Roca R., Weckwerth T.M., Andronova N.G. (2010) Tropospheric water vapor, convection, and climate. *Rev. Geophys.* 48, RG2001.

Soden B.J., Ackerman S.A., O'C Starr D., Melfi S.H., Ferrare R.A. (1994) Comparison of upper tropospheric water vapor from GOES, Raman lidar, and CLASS sonde measurements. *J. Geoph. Res.* 99, 21005–21016.

Soden B.J., Turner D.D., Lesht B.M., Miloshevich L.M. (2004) An analysis of satellite, radiosonde, and lidar observations of upper tropospheric water vapor from the ARM program. *J. Geoph. Res.* 109, D04105.

Solomon S., Rosenlof K.H., Portmann R.W., Daniel J.S., Davis S.M., Sanford T.J., Plattner G.K. (2010) Contributions of stratospheric water vapor to decadal changes in the rate of global warming. *Science* 327(5970), 1219–1223.

Susskind J., Barnet C.D., Blaisdell J.M. (2003) Retrieval of atmospheric and surface parameters from AIRS/AMSU/HSB data in the presence of clouds. *IEEE Trans. Geosci. Remote Sens.* 41, 390–409.

Tillman J.E. (1965) Water vapor density measurements utilizing the absorption of vacuum ultraviolet and infrared radiation. In: Ruskin R.E. (Ed.), *Humidity and Moisture*, Vol. 1, Principles and Methods of Measuring Humidity in Gases. Reinhold, Wisconsin, WI, pp. 428–443.

Trapp R.J., Diffenbaugh N.S., Gluhovsky A. (2009) Transient response of severe thunderstorm forcing to elevated greenhouse gas concentrations. *Geophys. Res. Lett.* 36, L01703.

Turner D.D., Lesht B.M., Clough S.A., Liljegren J.C., Revercomb H.E., Tobin D.C. (2003) Dry bias and variability in Vaisala radiosondes: The ARM experience. *J. Atmos. Oceanic Technol.* 20, 117–132.

Van Klooster S.L., Roebber P.J. (2009), Surface-based convective potential in the contiguous United States in a business-as-usual future climate. *J. Climate* 22, 3317–3330.

Vömel H., Yushkov V., Khaykin S., Korshunov L., Kyro E., Kivi R. (2007) Intercomparison of stratospheric water vapor sensors: FLASH-B and NOAA/CMDL frost point hygrometer. *J. Atmos. Oceanic Tech.* 27, 941–952.

Wang J.H., Zhang L.Y. (2008) Systematic errors in global radiosonde precipitable water data from comparisons with ground-based GPS measurements. *J. Clim.* 21, 2218–2238.

Wang J., Cole H.L., Carlson D.J., Miller E.R., Beirle K., Paukkunen A., Laine T.K. (2002) Corrections of humidity measurement errors from the Vaisala RS80 radiosonde—application to TOGA-COARE data. *J. Atmos. Oceanic Technol.* 19, 981–1002.

Waters J.W., Froidevaux L., Harwood R.S., Jarnot R.F., Pickett H.M., Read W.G. et al. (2006) The Earth observing system microwave limb sounder (EOS-MLS) on the Aura satellite. *IEEE Trans. Geosci. Remote Sens.* 44, 1075–1092.

Weinheimer A.J., Schwiesow R.L. (1992) A two–path, two–wavelength ultraviolet hygrometer. *J. Atmos. Oceanic Technol.* 9, 407–419.

Yoshino K., Esmond J.R., Parkinson W.H., Ito K., Matsui T. (1996) Absorption cross section measurements of water vapor in the wavelength region 120 to 188 nm. *Chem. Phys.* 211, 387–391.

Zondlo M.A., Paige M.E., Massick S.M., Silver J.A. (2010) Vertical cavity laser hygrometer for the National Science Foundation Gulfstream-V aircraft. *J. Geophys. Res.* 115, D20309.

Zuber A., Witt G. (1987) Optical hygrometer using differential absorption of hydrogen Lyman–α radiation. *Appl. Opt.* 26(15) 3083–3089.

Section III

Optical and Fiber-Optic Humidity Sensors

10 Introduction in Humidity Measurement by Optical and Fiber-Optic Sensors

10.1 INTRODUCTION

In Chapters 3, 4, and 8, we also have considered optical-based instruments designed for humidity measurements. The operation of these optical hygrometers is based on the absorption and fluorescence properties of water molecules. This is a so-called direct sensing scheme. However, studies have shown that for development of the optical humidity sensors one can use another approach. Instead of studying the properties of air, containing water vapor, we can investigate the effect of water vapor on the physical properties of humidity-sensitive materials.

Experiment has shown that water can have a comprehensive effect on the properties of different materials (Section 2.4). Under the influence of the water volume, properties of materials, such as the refractive index (RI), density, and the phase composition can change. Surface properties such as the composition of the adsorbed molecules, activation energies of adsorption and desorption are also sensitive to the presence of water vapor in the atmosphere. Condensation of water modifies the dielectric properties of the environment and thus affects the interaction between the nanoparticles. This means that for building the solid-state optical humidity sensors one can use almost all the variety of optical effects observed in the solid bodies—transmittance, reflectance, luminescence (fluorescent), interference, scattering, and the plasmon resonance (Wang et al. 1991; Xu et al. 2004; Wolfbeis 2006, 2008). This approach uses the so-called indirect or indicator-based sensing scheme. In many cases, this approach proved to be more effective, because it allowed reducing significantly the size of devices and increase their sensitivity. In addition, fiber-optic and planar optical humidity sensors, despite of the optical nature of measurements, are quite different structurally from the optical hygrometers discussed in Chapters 3 through 5. For this reason, a consideration of these sensors we have allocated in the individual chapters.

It should be noted that the appearance of solid-state planar optical and fiber-optic sensors is a result of three critical factors that have been instrumental toward development of this technology (Davis et al. 1986; Udd 1991; Culshaw 2008; Sharma and Wei 2013). First, it was the invention of laser that provided a high-intensity light source such as light-emitting diodes (LED) and laser diodes (LD) with strong spatial and temporal coherence properties. Second, it was the advancements in optoelectronics industry that facilitated the development of optical detectors to realize sophisticated low noise detection techniques to record optical interference effects with high sensitivity. And finally, it was the development of low-loss single-mode fibers that made it possible to guide light in a flexible fiber-optic waveguide and realize compact, low cost, robust, and versatile fiber-optic interferometers that would otherwise be impractical to achieve with bulk-optic components. The availability of rapid data acquisition and data-processing techniques also contributed to the fast development of this sensing technology.

As it was indicated earlier, solid-state optical and fiber-optic sensors can be described as devices, which can be employed for the detection and determination of physical and chemical parameters through measurements of optical property of humidity-sensitive materials. In general, solid-state optical sensors incorporate optical fibers as light guides. That is, in optical sensors optical fibers perform a passive role. At the same time in the fiber-optic sensors, optical fibers serve as the active element. Although we must admit that in many cases this separation is very conditional, because mechanisms, providing the sensitivity of the majority of solid-state optical and fiber-optic sensors to humidity are the same.

The current approach based on the development of solid-state optical and fiber-optic sensors is dominant in various fields of analytical sciences, for example, in the development of a variety of chemical sensors and biosensors (Narayanaswamy and Wolfbeis 2004). It was shown that various chemical and biochemical analytes can be qualitatively detected by spot tests using spectroscopic techniques, including colorimetry and photometry. Optical sensors in analytical sciences offer several advantages (Grattan and Sun 2000; Alwis et al. 2013). These sensors are capable of observing a sample in its dynamic environment, no matter how distant, difficult to reach, or hostile this environment is. As it is known, optical fibers allow transmission of light over the great distances; and for chemical sensing typical monitoring distances required the range from 1 to 100 m. These devices are intrinsically safe, involving a low optical power and are nonelectrical at the sensing point. These sensors are electrically passive and immune to electromagnetic disturbances, geometrically flexible and corrosion resistant, capable of being miniaturized, and compatible with telemetry. Furthermore, the possibility of multiplexing several sensors to a single instrumentation unit can afford substantial economic advantages to this type of sensor. In particular, this possibility allows developing devices capable of carrying out a simultaneous control of humidity and the presence of toxic gases.

More detailed information on the features of the fiber-optic sensor functioning can be found in reviews (Davis et al. 1986; Dakin and Culshaw 1988; Grattan and Sun 2000; Yu and Yin 2002; Narayanaswamy and Wolfbeis 2004; Kara 2011; Lou et al. 2014; Rajan 2015). With regard to additional information relating to the use of fiber-optic sensors for measuring humidity, one can refer to the reviews prepared by Moreno-Bondi et al. (2004), Yeo et al. (2008), Alwis et al. (2013), Kolpakov et al. (2014), Noor et al. (2015), Sikarwar and Yadav (2015), Jung et al. (2016).

10.2 TYPES OF OPTICAL HUMIDITY SENSORS—SENSOR CLASSIFICATION

On the basis of the properties of the material, changing under the influence of humidity, currently there can be marked out three basic approaches to the building of optical humidity sensors. The first approach is based on the color change of certain metal–ion complexes, when water is participating in the coordinating sphere (Dacres and Narayanaswamy 2006). So, the absorption band of the humidity-sensitive material undergoes a dramatic hypsochromic shift (Sharkany et al. 2005). This is so-called absorption, photometric, or colorimetric-based humidity sensors (Chapter 12). Other optical humidity sensors, that is, refractometric-based sensors, are based on the physi- or chemisorption of water directly onto the surface of the optical device or in a very thin and porous film, leading to a detectable change in the RI (Chapter 14). The third approach is derived from the phenomenon that water molecules are able to influence the fluorescence emission of certain compounds. Therefore, this type of sensors is called fluorescence-based sensors (Chapter 15). With increasing relative humidity (RH), either an enhancement or a quenching of the fluorescence intensity of an indicator dye is observed (Wang et al. 1991).

In optical sensors, the light may be modulated either inside or outside the optical waveguide, that is, sensing location may be inside or outside the waveguide. Hence, based on the sensing location, optical sensors are classified broadly as intrinsic and extrinsic sensors (read Chapter 11). In intrinsic fiber-optic sensor, the interaction occurs within an element of optical fiber itself and light never leaves the waveguide. External environment acts on the fiber and the fiber in turn changes some characteristics of the light inside the fiber that is measured using the detector. In the extrinsic fiber-optic sensor, the optical fiber is used to couple the light, usually to and from the region where the light beam is influenced by the measurand (or external environment). In this case, the fiber just acts as a mean of getting the light to the sensing location and to detector.

Besides the previously mentioned types of humidity sensors using different properties of the humidity-sensitive materials, it is also necessary to mark the humidity sensors, which are different from others by the specific method of measurement or configuration of the sensor element. These sensors must first of all include interferometric humidity sensors. Most of the components of interferometric fiber sensors use either all-fiber or integrated optic material to provide better stability and compactness. An interferometric sensor works on the modulation in the phase of light emerging from a single mode fiber. The variation in phase is converted into an intensity shift using interferometric schemes (Sagnac forms, ring resonators, Mach–Zehnder, Michelson, Fabry–Perot or dual mode, polarimetric, grating, and Etalon-based interferometers). This type of sensors will be considered in Chapter 16. Developers of the humidity sensors also mark out microfiber-based humidity sensors (Chapter 17), humidity sensors based on special fibers (Chapter 18), surface Plasmon resonance-based humidity sensors (Chapter 19), lossy-mode resonance humidity sensors (Chapter 20), and ellipsometry-based humidity sensors (Chapter 21).

One can also select the intensity modulated, spectrometric (wavelength modulated), polarization modulated, and phase modulated sensors (Krohn et al. 2015). Intensity-modulated sensors are based on the modulation of light intensity (amplitude) in the fiber. This kind of sensor can measure any parameter that can cause intensity losses in the guided light beam. Sensors that vary the intensity of light are the simplest and cheapest method of detecting different parameters using optical fiber as only a source and detector are required. Intensity-modulated sensors are analog in nature and have significant usage in digital applications for switches and counters. As a rule, such principle of measurements is used in photometric and refractometry- or evanescent wave-based sensors (Chapters 12 and 14). Other types of sensors also can use this method of measurements. The advantages of these sensors are the following: simplicity of implementation, low cost, possibility of being multiplexed, and ability to perform as real distributed sensors. However, as a rule, such sensors have decreased sensitivity. In addition, relative measurements and variations in the intensity of the light source may lead to false readings, unless a reference system is used. Potential error sources include variable losses due to connectors and splices, microbending loss, macrobending loss, and mechanical creep and misalignment of light sources and detectors. To circumvent these problems, many of the successful higher-performance, intensity-based fiber sensors employ dual wavelengths. One of the wavelengths is used to calibrate out all of the errors due to undesired intensity variations by bypassing the sensing region. The intensity-based sensor requires more light and therefore usually uses multimode large core fibers or bundles of fibers.

The spectrometric or wavelength modulated sensors monitor changes in the wavelength of the light. Fluorescence sensors and fiber grating sensors are examples of wavelength-modulated sensors. These sensors are not as sensitive as the interferometric sensors but their configuration, installation, and data processing are extremely easy. An advantage of these sensors is that the sensed information (shift in wavelength) is an absolute parameter, and thus obtained absolute measurements are more reliable as compared to relative ones. The wavelength-encoded nature of the output also permits an ease in multiplexing. However, here intermittent segments of fiber act as sensors and such segmental formation is achieved by fiber gratings. Intensive study on fiber gratings began after a controllable and effective method for their fabrication was devised by Meltz et al. (1989). More detailed information about the fiber Bragg gating (FBG) and long-period grating (LPG)-based sensors one can find in Chapter 14.

Phase-modulated sensors use the modulation in the phase of the light being transmitted in an optical fiber for detection. The optical phase of the light passing through the fiber is changed by the physical quantity to be measured. This phase modulation is then detected interferometrically by comparing the phase of the light in the signal fiber to that in a reference fiber. Once the beams are recombined, they interfere with each other. This measurement principle is implemented in interferometric-based sensors (Chapter 16). In the interferometer, the light is split into two beams, where one beam is exposed to the sensing environment and undergoes a phase shift and the other is isolated from the sensing environment and is used for as a reference. Phase-modulated optical sensors are highly sensitive to measurand. Hence, they are employed when extreme sensitivity is required.

Standard single-mode fibers transmit light without regards to polarization. But some special fibers known as polarization mainlined fibers, maintain the input polarization or transmit only one polarization state. It was found that the output polarization state is changed according to the perturbation caused by various external influences. Hence, by analyzing the output polarization state at the exit end of the fiber, the external perturbation can be detected (Yu and Yin 2002; Lecler and Meyrueis 2012). Ellipsometry uses this approach for studying the processes of absorption and characterization of surface coverings (Chapter 21).

Fiber-optic sensors can be also be classified in response to their measurements points (Chapter 24). Three important types are the following: (1) point-to-point sensors, (2) multiplex sensors, and (3) distributed sensors. In point-to-point type, there is a single measurement point at the end of the fiber-optic connection cable, similarly to most electrical sensors. Multiplexed sensors allow the measurement at multiple points along a single fiber line and distributed sensors are able to sense at any point along a single fiber line, typically every meter over many kilometers of length (Grattan and Sun 2000; Yu and Yin 2002).

10.3 HUMIDITY-SENSITIVE MATERIALS

10.3.1 Polymers

As it will be shown in the following Chapters 11–24, a variety of materials, which reversibly change their optical properties with the humidity, can be used as humidity-sensitive materials. The most popular are organic and inorganic polymers, including silicones, poly(acrylic acid) (PAA), poly(vinyl chloride) (PVC), polytetrafluoroethylene (PTFE), poly(methyl methacrylate) (PMMA), polyvinyl alcohol (PVA), polyaniline (PANI), Nafion, nylon, agarose, and so on. (Harsanyi 1995; Brook and Narayanaswamy 1998; Raimundo and Narayanaswamy 1999; Bariain et al. 2000; McMurtry et al. 2000; Fuke et al. 2009). These materials usually give a good robustness to the sensor design, especially if optical properties are changed in the visible or near-infrared region. This permits the use of the low-cost instrumentation and standard fibers.

Except for a few, polymers are organic macromolecules made of carbon and hydrogen atoms as the major percentage, with some heteroatoms such as nitrogen, oxygen, sulfur, phosphorus, and halogens as minor constituents. A polymer molecule is formed by the repetitive union of a large number of reactive small molecules in a regular sequence. Polymers in optical sensors can have both conducting and insulating properties. Reversible change of optical properties such as refraction, transmittance, and fluorescence during interaction with water vapor is more important for polymers aimed at applications in optical humidity sensors. Optical properties of the polymers such as a RI, change due to water absorption and swelling effect (read following Section 10.3.2). Figure 10.1 shows typical humidity-dependent swelling and deswelling of a PAH and PAA films (Secrist and Nolte 2011). PAA was observed to approximately double in film thickness near 100% RH, whereas the PAH film swelled on average to four times of its original film thickness. When the water content increases, the polymer film swelling generates a reduction in the RI (Figure 10.2). Depending on the effect used to influence their optical properties, the polymers may have hydrophilic or hydrophobic properties.

Taking into account the mechanism of the effect of water on the parameters of the polymers, it becomes clear that the superabsorbent polymers are the most suitable material for the development of humidity sensors. Superabsorbent polymers (SAPs) are the materials that have the ability to absorb and retain large volumes of water and aqueous solutions (Buchholz and Graham 1997; Zohuriaan-Mehr and Kabiri 2008). Exactly these materials have excellent water swelling capacity. In many cases, such materials are

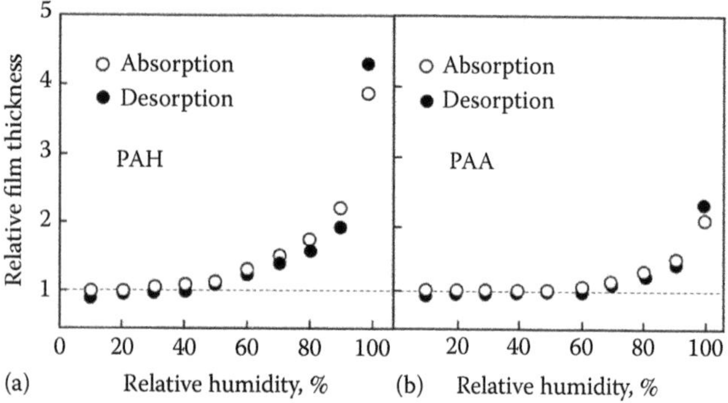

FIGURE 10.1 Humidity-dependent swelling and deswelling of (a) poly(allylamine hydrochloride) (PAH) and (b) poly(acrylic acid) (PAA) films. (Reprinted with permission from Secrist, K.E. and Nolte, A.J., *Macromolecules*, 44, 2859–2865, 2011. Copyright 2011 American Chemical Society.)

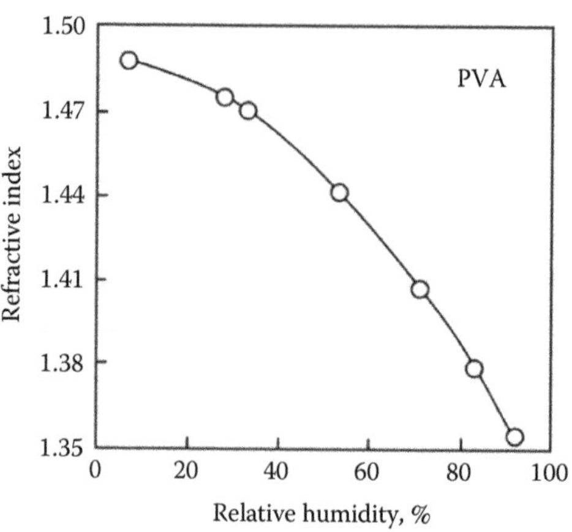

FIGURE 10.2 Refractive indexes of PVA film versus RH. $d \sim 82\ \mu m$. (Adapted from *Opt. Commun.*, 311, Su D. et al., A fiber Fabry–Perot interferometer based on a PVAcoating for humidity measurement, 107–110, Copyright 2013, with permission from Elsevier.)

called hydrophilic gels or hydrogels (Bajpai 2001; Laftah et al. 2011). The most famous superabsorbents were made from chemically modified starch and cellulose and other polymers such as poly(vinyl alcohol) PVA, poly(ethylene oxide) PEO all of which are hydrophilic and have a high affinity for water. When lightly cross-linked, chemically or physically, these polymers became water-swellable but not water-soluble.

Such polymers are widely used in medicine, purification systems, and so on. Usually SAPs are scientifically irreversible in the process of interaction. However, unlike the superabsorbent polymers intended for such applications, superabsorbent polymers developed for humidity sensors must have one feature—the process of interaction with water vapor should be reversible and fast as possible. In addition, they have to be optically transparent in the desired optical wavelengths, capable of forming films with good adhesion and be stable. The absence of hysteresis in the swelling/deswelling process is also an important requirement for polymers used in humidity sensors. According to Secrist and Nolte (2011), the hysteresis mechanism for humidity-swollen polymer films is tied to both the ability of a film to restructure on exposure to increasing water activity as well as the frustration of chain movements due to the presence of electrostatic cross-linking. All this requires a specific approach to the choice of polymers for optical humidity sensors and their synthesis technology. For example, the use of polyelectrolyte multilayers (PEMs) films formed by layer-by-layer (LbL) adsorption, significantly increases a swelling/deswelling hysteresis in comparison with films formed by one material (Figures 10.1 and 10.3).

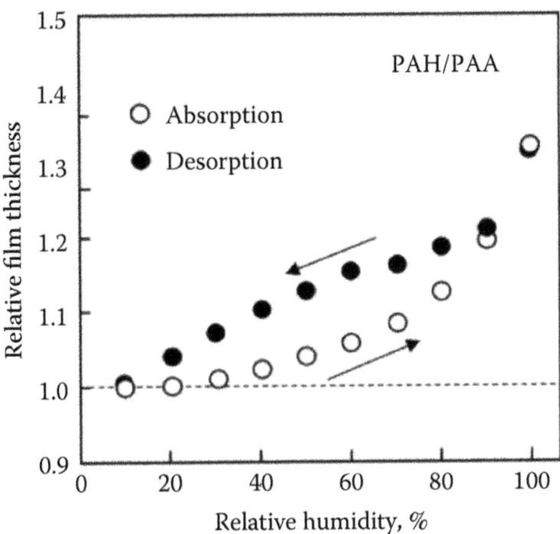

FIGURE 10.3 Humidity-dependent swelling and deswelling of a PAH/PAA 50-bilayer film deposited by LbL method. The squares and circles indicate measurements during absorption and desorption, respectively. (Reprinted with permission from Secrist, K.E. and Nolte, A.J., *Macromolecules*, 44, 2859–2865, 2011. Copyright 2011 American Chemical Society.)

Moreover, Secrist and Nolte (2011) have shown that this hysteresis was dependent on the parameters of the fabrication process.

10.3.2 Mechanisms of Swelling in Superabsorbent Polymers

There are several mechanisms to the process of swelling, which include hydration and the formation of hydrogen bonds (Buchholz and Graham 1997; Kenkare et al. 2000; Bajpai 2001). Figure 10.4 is a diagrammatic representation of the part of the polymer network. The polymer backbone in SAP is hydrophilic, that is, *water loving* because it contains water loving carboxylic acid groups (–COOH). When water is added to SAP there is a polymer/solvent interaction; hydration and the formation of hydrogen bonds are two of these interactions. Hydration is the interaction of ions of a solute with molecules of a solvent, that is, COO^- and Na^+ ions attract the polar water molecules, while hydrogen bonds are electrostatic interactions between molecules, occurring in molecules that have hydrogen atoms attached to small electronegative atoms such as N, F, and O. The hydrogen atoms are attracted to the nonbonding electron pairs (lone pairs) on other neighboring electronegative atoms. These effects decrease the energy and increase the entropy of the system. Due to the hydrophilic nature of SAP, the polymer chains have a tendency to disperse in a given amount of water (i.e., they are trying to dissolve in the water), which leads to a higher number of configurations for the system and also increases entropy. At the same time cross-links between polymer chains form a three-dimensional network and prevent the polymer swelling to infinity, that is, dissolving. This is due to the elastic retraction forces of the network, and is accompanied by a decrease in entropy of the chains, as they become stiffer from their originally coiled state (Figure 10.5).

It is important that there is a balance between the forces of retraction and the tendency for the chains to swell to infinite dilution. The degree of cross-linking has a direct effect on the level of swelling of the polymer and

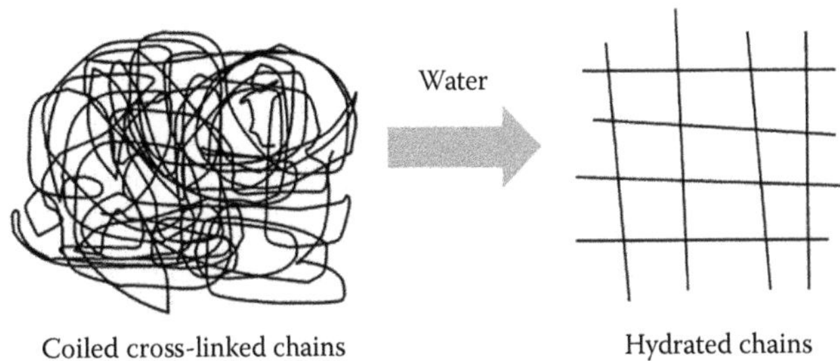

FIGURE 10.5 Changes in a three-dimensional network of polymer chains during swelling. (From Buchholz, F.L. and Graham, A.T., (Eds.): *Modern Superabsorbent Polymer Technology*. Copyright Wiley-VCH Verlag GmbH & Co. KGaA. Reproduced with permission.)

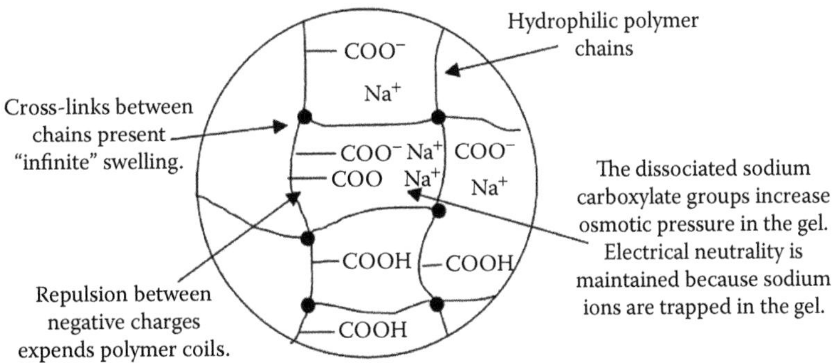

FIGURE 10.4 Diagrammatic representation of the part of the polymer network. (From Buchholz, F.L. and Graham, A.T., (Eds.): *Modern Superabsorbent Polymer Technology*. Copyright Wiley-VCH Verlag GmbH & Co. KGaA. Reproduced with permission.)

the strength of the network, that is, increased cross-link density = decreased swelling capacity = increased gel strength. This means that the swelling effect has strong dependence on the technology of polymer synthesis.

For ionic polymers there is another solvent/polymer interaction beyond simple mixing (Buchholz and Graham 1997). The neutralized chains contain charges that repel each other (Figure 10.6). Overall electrical neutrality is maintained as the negative carboxylate groups are balanced by the positive sodium ions. Upon contact with water the sodium ions are hydrated (Figure 10.6) which reduces their attraction to the carboxylate ions (due to the high dielectric constant of water). This allows the sodium ions to move freely within the network, which contributes to the osmotic pressure within the gel. The mobile positive sodium ions, however, cannot leave the gel because they are still weakly attracted to the negative carboxylate ions along the polymer backbone and so behave like they are trapped by a semipermeable membrane. So, the driving force for swelling is the difference between the osmotic pressure inside and outside the gel (Bajpai 2001). Increasing the level of sodium outside of the gel will lower the osmotic pressure and reduce the swelling capacity of the gel. The maximum swelling of the gel will occur in deionized water.

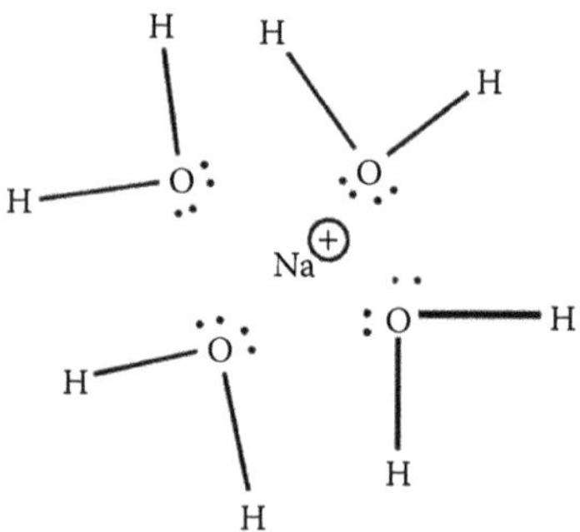

FIGURE 10.6 Solvent/polymer interaction in ionic polymer. (From Buchholz, F.L. and Graham, A.T., (Eds.): *Modern Superabsorbent Polymer Technology*. Copyright Wiley-VCH Verlag GmbH & Co. KGaA. Reproduced with permission.)

The synthesis of polymeric hydrogels may be organized in a number of *classical* chemical ways (Enas 2015). These include one-step procedures like polymerization and parallel cross-linking of multifunctional monomers, as well as multiple step procedures involving synthesis of polymer molecules having reactive groups and their subsequent cross-linking, possibly also by reacting polymers with suitable cross-linking agents, that is, modification or functionalization of existing polymers. The polymer engineer can design and synthesize polymer networks with molecular-scale control over the structure such as cross-linking density and with tailored properties, such as biodegradation, mechanical strength, and chemical and biological response to stimuli. Various reviews have discussed in detail the synthetic methods of hydrogels (Peppas 1987; Mathur et al. 1996; Laftah et al. 2011; Enas 2015). Table 10.1 describes the commonly used monomers for synthesis of hydrogels.

Regarding a swelling rate, experimental studies have shown that for a rapid effect, the polymer should be a macroporous (the large pores with size in the range of 0.1–1 μm), or even super-porous, that is, along with high porosity, polymer should have

TABLE 10.1
Commonly Used Synthetic Monomers for Hydrogel Synthesis

Monomer	Abbreviation	Formula	Chemical Structure
Hydroxyethyl methacrylate	HEMA	$C_6H_{10}O_3$	CH_2=C(CH_3)$COOCH_2CH_2OH$
Methoxyethyl methacrylate	MEMA	$C_7H_{12}O_3$	CH_2=C(CH_3)$COOCH_2CH_2OCH_3$
Ethylene glycol dimethacrylate	EGDMA	$C_{10}H_{14}O_4$	CH_2=C(CH_3)$COOCH_2CH_2OCOC$(CH_3)=CH_2
Acrylic acid	AA	$C_3H_4O_2$	CH_2=CHCOOH
Methacrylic acid	MA	$C_4H_6O_2$	CH_2=C(CH_3)COOH
Methyl methacrylate	MMA	$C_5H_8O_2$	CH_2=C(CH_3)$COOH_3$
N-vinyl-2-pyrrolidone	NVP	C_6H_9NO	CH_2=$CHNCOCH_2CH_2CH_2$
Vinyl acetate	VAc	$C_4H_6O_2$	CH_2=$CHCOOCH_3$
Acrylamide	AAam	C_3H_5NO	CH_2=CH-$CONH_2$
N-isopropylacrylamide	NIPA Aam	$C_6H_{11}NO$	CH_2=CHCONHCH(CH_3)$_2$

Source: Data extracted from Bajpai, S.K., *J. Sci. Ind. Res.*, 60, 451–462, 2001.

interconnected open-cell structure (Ganji et al. 2010). In this case, the swelling is very fast and this effect is sample size-independent.

10.3.3 Other Materials

Some salts and complexes of transition metals also have humidity-sensitive properties. For example, $CoCl_2$ is one of the most commonly used materials for humidity sensor design because it can absorb water and form coordination complexes, which change its characteristic color (Boltinghouse and Abel 1989). More detailed information about $CoCl_2$-based humidity sensors one can find in Chapter 12.

Fluorescent and phosphorescent materials such as rhodamine (Otsuki et al. 1998; Choi and Ling 1999), Perylene dyes (Porsh and Wolfbeis 1988; Zhang et al. 2016) or Ruthenium-based complex (Bedoya et al. 2006) are another big group of humidity-sensitive materials. The main advantage of these sensors is that the lifetime-based measurement can be used, which does not depend on the light-source intensity changes caused by the sensing molecule, the transducer, or the optical path. Such sensors will be considered in Chapter 15.

Porous metal oxides such as Co_3O_4 (Ando et al. 1996), Indium Tin oxide (ITO) (Zamarreno et al. 2010) and Titanium dioxide (TiO_2) (Hawkeye and Brett 2011; Socorro et al. 2016), ZnO, $BaTiO_3$, Al_2O_3, SnO_2 (Ansari et al. 1997), silica (Viegas et al. 2011), and carbon-based materials such as carbon nanotubes and graphene (Chi et al. 2015) are also acceptable for humidity sensors design.

The metal films such as Au or Ag are also used in the manufacture of humidity sensors. Interaction between the light and thin metal film is used in the surface Plasmon resonance-based humidity sensors, considered in Chapter 19. Metal nanoparticles can also be incorporated in the polymer matrix for increasing sensitivity. For example, Fuke et al. (2009) found that Ag–polyaniline nanocomposite was more sensitive to humidity in comparison to polyaniline alone.

As in the polymers, the change of optical properties of materials mentioned above is due to the adsorption and condensation of water vapor on the solid surface. As the air and water have different RIs, the appearance of water on the surface and then filling the pores with water will be accompanied by a change in the effective RI of this structure. This means that the water adsorption and the water capillary condensation in the film deposited on the surface of the fiber can influence the interaction between the light guided in the fiber and environment. The amount of water, adsorbed by humidity-sensitive materials, depends on their porosity. Therefore, humidity-sensitive materials are typically highly porous.

10.3.4 Adsorption and Diffusion in Porous Media

Adsorption processes of water vapor taking place in porous media usually are described by an adsorption isotherm (Brunauer et al. 1938, 1940; Brunauer 1943; Pavlík et al. 2012). Although, in the literature many types of isotherms for different absorbents and adsorbed media can be found, most of them can be classified into one of the five or six isotherms classes originally formulated by Brunauer (1943). The original types of isotherms are presented in Figure 10.7. Type I is characteristic for purely microporous adsorbents such as zeolites or activated carbon, type II for nonporous or purely macroporous materials, type III for the rare cases of nonporous adsorbents with very small interactions between the adsorbent and the adsorbed medium. Type-IV isotherm

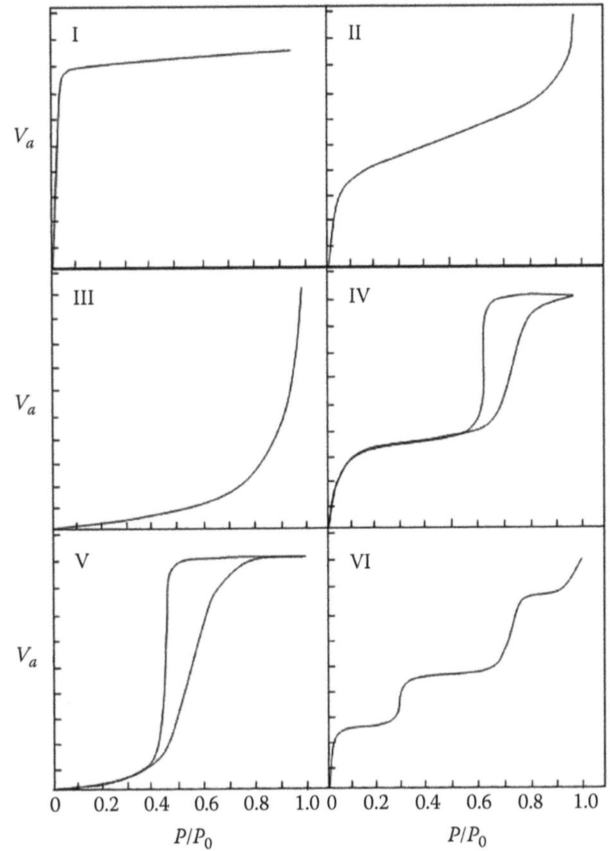

FIGURE 10.7 Basic shapes of sorption isotherms according to Brunauer S. (Data from Brunauer, S. et al., *J. Am. Chem. Soc.*, 60, 309–319, 1938; Brunauer, S. et al., *J. Am. Chem. Soc.*, 62, 1723–1732, 1940; Brunauer, S., *The Adsorption of Gases and Vapors*, Princeton University Press, Princeton, NJ, 1943. with permission.)

characterizes mesoporous adsorbents and is very common, contrary to the type V, which presents a relatively rare case of mesoporous adsorbents with small interactions with the adsorbed medium. Type VI is very rare; the adsorption process is realized here in several separated steps.

In porous materials used in humidity sensors, the adsorption isotherms are almost exclusively of type II, that is, with an S-shape (Hansen 1986). At low RH, the molecules of water are bound in one layer to the surface of pores by van der Waals forces. This phase of adsorption expresses the relatively fast moisture increase in the initial regions of the curve. The molecules of the first adsorbed layer have a character similar to the liquid phase. Before the end of the adsorption in the first layer, further layers appear. This process is characteristic for a plane surface of the adsorbent and bigger pores where the curvature of the adsorbed film does not affect the gas pressure of the adsorbent above it. This phase is characterized by the linear part of the isotherm. The final phase is the capillary condensation. This phenomenon can be explained by the laws of reversible thermodynamics, on the basis of the assumption that on the interface between water and water vapor a thermodynamic equilibrium is established (Gregg and Sing 1982).

According to the basic theory of adsorption on a porous matrix (Adamson and Gast 1997), when the vapor molecules are first physicosorbed onto the porous material, capillary condensation will occur if the micropores are narrow enough. The critical size of pores for a capillary condensation effect is characterized by the Kelvin radius. In the case of water, the condensation of vapor into the pores can be expressed by a simplified Kelvin equation (Ponec et al. 1974):

$$r_K = \frac{2\gamma V_M \cos\theta}{\rho RT ln(\%RH/100)} \tag{10.1}$$

where:

γ is the surface tension of vapor in the liquid phase
V_M is molecular volume
θ is contact angle
ρ is the density of vapor in the liquid phase

In this equation, the thickness of the adsorbed layer has been ignored. While V_M and surface tension γ are constants at room temperature, the possibility of controlling the condensation by simply changing the contact angle θ becomes attractive.

The values of Kelvin radius, calculated for water vapors in accordance with Equation (10.2), are presented in Table 10.2. For these calculations, the following data and assumptions were used: the contact angle between the liquor and the wall of concave (θ) is constant; the V_M and γ are 1.8×10^{-2} L/mol and 72.8 mN/m (20°C), respectively; the temperature (T) is about 293 K; and $x = \cos\theta$. It is seen that the Kevin radius increases with the RH, and the rate of change (the slope) also increases with RH. This means that the pores with smaller diameter are filled first, whereas bigger pores are filled later.

TABLE 10.2
The Values of Kelvin Radius Calculated for Various Air Humidity

RH (%)	*r* (nm)
11	0.4906 *x*
23	0.7369 *x*
33	1.095 *x*
43	1.283 *x*
52	1.656 *x*
67	2.704 *x*
75	3.764 *x*
86	7.180 *x*
97	35.55 *x*

Source: Chen, W.-P. et al., *Sensors*, 9, 7431, 2009.

Of course, this is somewhat simplified case, as in many cases during the interaction of water vapor with a solid, in addition to physisorption may also participate chemisorption. As a rule, the physisorption of the water follows after chemisorption. If the physical adsorption took place under the influence of forces of the van der Waals type, then the water molecule chemisorption takes place via a dissociative mechanism. On the surface of metal oxides, this process usually takes place as follows. The coordinative unsaturated metal cations, M^+, and oxygen anions, O^{2-}, at the surface facilitate the dissociation of water vapor in the vicinity of the surface into a hydroxyl group, which binds to a metal cation site, and proton, which binds to an oxygen anion site. Once surface hydroxylation is present, water vapor is physisorbed through double hydrogen bonds shared with two surface hydroxyl groups. On completion of this first physisorbed layer, subsequent layers are adsorbed via single hydrogen bonds. Thus the initial chemisorbed layer is tightly attached to the surface grains. Therefore, desorption of the chemisorbed layer only takes place at higher temperatures. Once forms, chemisorbed layer is no further affected by exposure to humidity. As RH increases, an additional layer of water molecules starts to be formed on the chemisorption one. It is important that many more physisorption layers can be joined as humidity gets higher. Moreover, with more than one layer of physisorption water molecules, water can start to be condensed into the capillary pores. The filling of the pores depends on its radius and the thickness of the physisorption layer, which depends on the magnitude of the RH. Increasing RH, the thickness of the physisorption layer of water gets bigger, which leads to the filling of successively higher diameter pores, according to the prediction of Kelvin equation.

It is important to take into account that at desorption, we sometimes get a different dependence of the water content on the RH than at adsorption (Pavlík et al. 2012). The desorption isotherm always lies above the adsorption isotherm. This effect is called hysteresis. It is typical for types-IV and V isotherms (Figure 10.7), but often it is observed also for type II isotherms. Hysteresis is caused by two main mechanisms. The first is the hysteresis of the contact angle, which at adsorption is significantly smaller than at desorption. The probable causes of this fact are water contamination on the solid surface, surface roughness, or immobility of the surface layer of water (Adamson 1990). The capillary hysteresis, which is the second main mechanism in this respect, is a consequence of the appearance of pores with alternating wide and narrow parts. In the narrow pore necks, evaporation is significantly slower than in wider parts because the partial pressure of saturated water vapor is lower there. The phenomenon, which accelerates capillary condensation, thus slows down the drying process at desorption. Theoretically, the hysteresis can also be induced by a third factor: by the irreversible processes taking place at water adsorption, but it is assumed that in most cases this factor is much less significant than the previous two (Everett 1967). The presence of this effect shows that the parameters of humidity sensors may also have the hysteresis, that is, the sensor signal, when measurements, occurring against a background of increasing and decreasing humidity content may vary.

The position and the shape of an adsorption isotherm also strongly depend on the temperature (Pavlík et al. 2012). At higher temperatures, the transport of water molecules is faster, the bonds can be released more easily, and therefore both adsorption and desorption isotherms corresponding to higher temperatures are lower (or shifted to the right) as compared to those corresponding to lower temperatures (some water molecules already bound on the solid surface can be released, polymolecular layers on the pore walls are thinner). Therefore, the capillary condensation occurs (all other conditions being the same) at higher RH.

Since the most drastic changes in the optical parameters of the humidity-sensitive materials occur at the water vapor capillary condensation (Israelachvili 1992), it becomes obvious that the pore size becomes a factor controlling the sensitivity of the sensor and the measuring range. In particular, a decrease in the pore size in the humidity-sensitive material should promote to increase sensitivity toward water vapor; the smaller the pore radius is, the lower is the partial pressure at which condensation can occur at a given temperature (Adamson 1990). However, it should be borne in mind that the reduction of pore diameter will influence simultaneously the kinetics and hysteresis of sensor response, that is, the response time and the magnitude of hysteresis will increase. Thus, the performance of a humidity sensor based on a porous sensing layer is determined by its nano- and microscopic dimensions, including pore size, thickness of the porous layer, size distribution of the surface structural unit, and regularity of the surface morphology (Kim et al. 2000; Di Francia et al. 2002). This fact indicates that the regularity and controllability of porous structures are of great importance in sensor applications. For example, due to water condensation in nanosize pores and the decrease in the gas penetration, it is difficult to prepare rapid humidity sensors on the basis of nanoporous material (Bjorkqvist et al. 2004). Usually transport in nanoporous material is described by Knudsen diffusion (Reinecke and Sleep 2002). The Knudsen diffusivity (D_{iK}) of gas species i can be estimated in Equation 10.2 (Cunningham and Williams 1980):

$$D_{iK} = \frac{d_p}{3}\sqrt{\frac{8RT}{\pi M_i}} \tag{10.2}$$

where:

M_i represents the molecular weights of gas species i

d_p is the mean pore size of the porous media.

D_{iK} can be simplified further as Equation 10.3. In Equation 10.3, d_p has the unit of cm, M_i has the unit of g/mol, and temperature T has the unit of K. It is seen that with a decrease in the pore diameter the Knudsen diffusivity decreases.

$$D_{iK} = 4850 d_p \sqrt{\frac{T}{M_i}} \tag{10.3}$$

In addition, experiment has also shown that the characteristics of humidity sensors are strongly affected by the hydrophilic/hydrophobic properties of the walls (Yarkin 2003). On the basis of the Kelvin equation, the Kelvin radius becomes smaller when the surface becomes more hydrophobic.

All this means that the structural engineering and the surface functionalizing of humidity-sensitive materials should be the target of the humidity sensor developers. In this case, it is important to control not only the pore size, but also their dispersion. Unfortunately, this approach is not practiced in most studies.

REFERENCES

Adamson A.W. (1990) *Physical Chemistry of Surfaces*, 5th ed. Wiley & Sons, New York.

Adamson A.W., Gast A.P. (1997) *Physical Chemistry of Surface*. Wiley & Sons, New York.

Alwis L., Sun T., Grattan K.T.V. (2013) Optical fibre-based sensor technology for humidity and moisture measurement: Review of recent progress. *Measurement* 46, 4052–4074.

Ando M., Kobyashi T., Haruta M. (1996) Humidity-sensitive optical absorption of Co_3O_4 film. *Sens. Actuators B* 32, 157–160.

Ansari Z.A., Karekar R.N., Aiyer R.C. (1997) Humidity sensor using planar optical waveguides with claddings of various oxide materials. *Thin Solid Films* 305, 330–335.

Bajpai S.K. (2001) Swelling studies on hydrogel networks—A review. *J. Sci. Ind. Res.* 60, 451–462.

Bariain C., Matias I.R., Arregui F.J., Lopez-Amo M. (2000) Optical fibre humidity sensor based on a tapered fibre coated with agarose gel. *Sens. Actuators B* 69, 127–131.

Bedoya M., Diez M.T., Moreno-Bondi M.C., Orellana G. (2006) Humidity sensing with a luminescent Ru(II) complex and phase-sensitive detection. *Sens. Actuators B* 113, 573–581.

Bjorkqvist M., Salonen J., Paski J., Laine E. (2004) Characterization of thermally carbonized porous silicon humidity sensor. *Sens. Actuators A* 112, 244–247.

Boltinghouse F., Abel K. (1989) Development of an optical relative humidity sensor. Cobalt chloride optical absorbency sensor study. *Anal. Chem.* 61(17), 1863–1866.

Brook T.E., Narayanaswamy R. (1998) Polymeric films in optical gas sensors. *Sens. Actuators B* 51(1–3), 77–83.

Brunauer S. (1943) *The Adsorption of Gases and Vapors*. Princeton University Press, Princeton, NJ.

Brunauer S., Emmet P.H., Teller W.E. (1938) Adsorption of gases in multimolecular layers. *J. Am. Chem. Soc.* 60, 309–319.

Brunauer S., Deming L.S., Deming E.W., Teller E. (1940) On a theory of the van der Waals adsorption of gases. *J. Am. Chem. Soc.* 62, 1723–1732.

Buchholz F.L., Graham A.T. (Eds.) (1997) *Modern Superabsorbent Polymer Technology*, John Wiley & Sons, New York.

Chen W.-P., Zhao Z.-G., Liu X.-W., Zhang Z.-X., Suo C.-G. (2009) A capacitive humidity sensor based on multi-wall carbon nanotubes (MWCNTs). *Sensors* 9, 7431–7444.

Chi H., Liu Y.J., Wang F.K., He C. (2015) Highly sensitive and fast response colorimetric humidity sensors based on graphene oxides film. *ACS Appl. Mater. Interfaces* 7, 19882–19886.

Choi M.M.F., Ling T.O. (1999) Humidity-sensitive optode membrane based on a fluorescent dye immobilized in gelatin film. *Anal. Chim. Acta* 378(1–3), 127–134.

Culshaw B. (2008) Fiber-optic sensing: a historical perspective. *J. Lightwave Technol.* 26(9), 1064–1078.

Cunningham R.E., Williams R.J.J. (1980) *Diffusion in Gases and Porous Media*. Plenum Press, New York.

Dacres H., Narayanaswamy R. (2006) Highly sensitive optical humidity probe. *Talanta* 69, 631–636.

Dakin J., Culshaw B. (Eds.) (1988) *Optical Fiber Sensors-Principles and Components*, Vol. I, Artech House, Boston, MA.

Davis C.M., Carome E.F., Weik M.H., Ezekiel S., Einzig R.E. (1986) *Fiber-optic Sensor Technology Handbook*. Optical Technologies, Herndon, VA.

Di Francia G., Noce M.D., Ferrara V.L., Lancellotti L., Morvillo P., Quercia L. (2002) Nanostructured porous silicon for gas sensor application. *Mater. Sci. Technol.* 18, 767–771.

Enas M.A. (2015) Hydrogel: Preparation, characterization, and applications: A review. *J. Adv. Res.* 6, 105–121.

Everett D.H. (1967) Adsorption hysteresis. In: Flood E.A. (Ed.), *The Solid Gas Interface*, Vol. 2, Chapter 36. Dekker, New York.

Fuke M.V., Vijayan A., Kanitkar P., Kulkarni M., Kale B.B., Aiyer R.C. (2009) Ag-polyaniline nanocomposite cladded planar optical waveguide based humidity sensor. *J. Mater. Sci.: Mater. Electron.* 20, 695–703.

Ganji F., Vasheghani-Farahani S., Vasheghani-Farahani E. (2010) Theoretical description of hydrogel swelling: A review. *Iran. Polym. J.* 19(5), 375–398.

Grattan K.T.V., Sun T. (2000) Fiber optic sensor technology: An overview. *Sens. Actuators* 82, 40–61.

Gregg S.J., Sing K.S.W. (1982) *Adsorption, Surface Area, Porosity*, 2nd ed. Academic Press, London, UK.

Hansen K.K. (1986) Sorption Isotherms—A Catalogue. Technical report 162/86. TU Denmark, Lyngby.

Harsanyi G. (1995) Polymeric sensing films: New horizons in sensorics? *Sens. Actuators A* 46–47, 85–88.

Hawkeye M.M., Brett M.J. (2011) Optimized colorimetric photonic-crystal humidity sensor fabricated using glancing angle deposition. *Adv. Funct. Mater.* 21, 3652–3658.

Israelachvili J.N. (1992) *Intermolecular and Surface Forces*, 2nd ed. Academic Press, London, UK.

Jung H.S., Verwils P., Kim W.Y., Kim J.S. (2016) Fluorescent and colorimetric sensors for the detection of humidity or water content. *Chem. Soc. Rev.* 45, 1242–1256.

Kara P. (2011) Polymer optical fiber sensors—A review. *Smart Mater. Struct.* 20, 013002.

Kenkare N.R., Hall C.K., Khan S.A. (2000) Theory and simulation of the swelling of polymer gels. *J. Chem. Phys.* 113, 404–418.

Kim S.J., Park J.Y., Lee S.H., Yi S.H. (2000) Humidity sensors using porous silicon layer with mesa structure. *J. Phys. Appl. Phys.* 33, 1781–1784.

Kolpakov S.A., Gordon N.T., Mou C., Zhou K. (2014) Toward a new generation of photonic humidity sensors. *Sensors* 14, 3986–4013.

Krohn D.A., MacDougall T.W., Mendez A. (2015) *Fiber Optic Sensors: Fundamentals and Applications*. SPIE, Bellingham, Washington, DC.

Laftah W.A., Hashim S., Ibrahim A.N. (2011) Polymer hydrogels: A review. *Polymer-Plastics Technol. Eng.* 50(14), 1475–1486.

Lecler S., Meyrueis P. (2012) Intrinsic optical fiber sensor. In: Yasin M., Harun S.W., Arof H. (Eds.), *Fiber Optic Sensors*. InTech, Rijeka, Croatia, pp. 53–76.

Lou J., Wang Y., Tong L. (2014) Microfiber optical sensors: A review. *Sensors* 14, 5823–5844.

Mathur A.M., Moorjani S.K., Scranton A.B. (1996) Methods for synthesis of hydrogel networks: A review. *J. Macromol. Sci. C* 36(2), 405–430.

McMurtry S., Wright J.D., Jackson D.A. (2000) A multiplexed low coherence interferometric system for humidity sensing. *Sens. Actuators B* 67, 52–56.

Meltz G., Morey W.W., Glenn W.H. (1989) Formation of Bragg gratings in optical fibers by a transverse holographic method. *Opt. Lett.* 14(15), 823–825.

Moreno-Bondi M.C., Orellana G., Bedoya M. (2004) Fiber-optic sensors for humidity monitoring. In: Narayanaswamy R., Wolfbeis O.S. (Eds.), *Optical Sensors: Industrial Environmental and Diagnostic Applications*. Springer, Berlin, Germany, pp. 251–280.

Narayanaswamy R., Wolfbeis O.S. (Eds.) (2004) *Optical Sensors: Industrial Environmental and Diagnostic Applications*. Springer, Berlin, Germany.

Noor M.Y.M., Peng G.-D., Rajan G. (2015) Optical fiber humidity sensors. In: Rajan G. (Ed.), *Optical Fiber Sensors: Advanced Techniques and Applications*. CRC, Boca Raton, FL, pp. 413–454.

Otsuki S., Adachi K., Taguchi T. (1998) A novel fibre-optic gas-sensing configuration using extremely curved optical fibres and an attempt for optical humidity detection. *Sens. Actuators B* 53(1–2), 91–96.

Pavlík Z., Žumár J., Medved I., Cerný R. (2012) Water vapor adsorption in porous building materials: Experimental measurement and theoretical analysis. *Transport Porous Med.* 91, 939–954.

Peppas N.A. (Ed.) (1987) *Hydrogels in Medicine and Pharmacy*, Vols. 1 and 3. CRC Press, Boca Raton, FL.

Ponec V., Knor Z., Cerný S. (1974) *Adsorption on Solids*. Butterworth, London, UK, p. 405.

Porsh H.E., Wolfbeis O.S. (1988) Optical sensors, 13: Fibre-optical humidity sensor based on fluorescence quenching. *Sens. Actuators* 15, 7–83.

Raimundo I.M. Jr., Narayanaswamy R. (1999) Evaluation of Nafion-crystal violet films for the construction of an optical relative humidity sensor. *Analyst* 124(11), 1623–1627.

Rajan G. (Ed.) (2015) *Optical Fiber Sensors: Advanced Techniques and Applications*. CRC, Boca Raton, FL.

Reinecke S.A., Sleep B.E. (2002) Knudsen diffusion, gas permeability, and water content in an unconsolidated porous medium. *Water resources Res.* 38(12), 1280.

Secrist K.E., Nolte A.J. (2011) Humidity swelling/deswelling hysteresis in a polyelectrolyte multilayer film. *Macromolecules* 44, 2859–2865.

Sharkany J., Korposh S.O., Batori-Tarci Z.I., Trikur I.I., and Ramsden J.J. (2005) Bacteriorhodopsin-based biochromic films for chemical sensors. *Sens. Actuators B* 107, 77–81.

Sharma U. and Wei X. (2013) Fiber optic interferometric devices. In: Kang J.U. (Ed.), *Fiber Optic Sensing and Imaging*. Springer Science+Business Media, New York, pp. 29–53.

Sikarwar S., Yadav B.C. (2015) Opto-electronic humidity sensor: A review. *Sens. Actuators A* 233, 54–70.

Socorro A.B., Hernaez M., Del Villar I., Corres J.M., Arregui F.J., Matias I.R. (2016) Single-mode—multimode—single-mode and lossy mode resonance-based devices: A comparative study for sensing applications. *Microsyst. Technol.* 22, 1633–1638.

Su D., Qiao X., Rong Q., Sun H., Zhang J., Bai Z., et al. (2013) A fiber Fabry–Perot interferometer based on a PVAcoating for humidity measurement. *Opt. Commun.* 311, 107–110.

Udd E. (1991) *Fiber Optic Sensors*. Wiley & Sons, New York.

Viegas D., Hernaez M., Goicoechea J., Santos J.L., Araújo F.M., Arregui F.J., Matias I.R. (2011) Simultaneous measurement of humidity and temperature based on an SiO_2 nanospheres film deposited on a Long-Period Grating in-line with a fiber Bragg grating. *IEEE Sensors J.* 11(1), 162–166.

Wang K., Seiler K., Haug J.-P., Lehmann B., West S., Hartman K., and Simon W. (1991) Hydration of trifluoroacetophenones as the basis for an optical humidity sensor. *Anal. Chem.* 63, 970–974.

Wolfbeis O.S. (2006) Fiber-optic chemical sensors and biosensors. *Anal. Chem.* 78, 3859–3874.

Wolfbeis O.S. (2008) Fiber-optic chemical sensors and biosensors. *Anal. Chem.* 80, 4269–4283.

Yarkin D.G. (2003) Impedance of humidity sensitive metal/porous silicon/n-Si structures. *Sens. Actuators A* 107, 1–6.

Yeo T.L., Sun T., Grattan K.T.V. (2008) Fibre-optic sensor technologies for humidity and moisture measurement. *Sens. Actuators A* 144, 280–295.

Yu F., Yin S. (Eds.) (2002) *Fiber Optic Sensors*. Marcel-Dekker, New York.

Xu L., Fanguy J.C., Soni K., Tao S. (2004) Optical fiber humidity sensor based on evanescement-wave scattering. *Opt. Lett.* 29(11), 1191–1193.

Zamarreno C.R., Hernaez M., Del Villar I., Matias I.R., Arregui F.J. (2010) Tunable humidity sensor based on ITO-coated optical fiber. *Sens. Actuators B* 146, 414–417.

Zhang S., Zhou F., Peng H., Liu T., Ding L., Fang Y. (2016) Fabrication and humidity sensing performance studies of a fluorescent film based on a cholesteryl derivative of perylene bisimide. *Spectrochim. Acta A* 165, 145–149.

Zohuriaan-Mehr M.J., Kabiri K. (2008) Superabsorbent polymer materials: A review. *Iranian Polymer J.* 17(6), 451–477.

11 Optical and Fiber-Optic Humidity Sensors—General Consideration

As it was shown in Chapter 10, there are many different types of optical and fiber-optic sensors (FOS) which can be used for humidity measurement. Naturally, each direction in the development of humidity sensors requires humidity indicators with specific properties. But at the same time it should be noted that in many cases optical humidity sensors independently on the type have the same configuration.

11.1 INSTRUMENTATION

The instrumentation associated with optical humidity sensors is similar to those associated with conventional spectrophotometric techniques. It includes components such as a light source, a wavelength selector, a photodetector, and a display of the output (Davis et al. 1986; Wolfbeis 1991, 1992; Narayanaswamy and Wolfbeis 2004). In addition, other optical components such as a lens, optical couplers, and connectors are used for coupling light into optical fibers. The use of solid-state optoelectronic components has enabled the development of commercial portable systems and a review discusses the instrumentation systems specifically used in conjunction with optical sensors (Taib and Narayanaswamy 1995). A schematic of a typical instrumentation system employed in optical sensing is shown in Figure 11.1. The optical radiation supplied by the light source is modulated and launched into optical fibers, and the detected light from the sensor is passed through a wavelength selector, for example, monochromator, and directed to the photodetector. The detected optical radiation results in an electrical signal, which is processed and displayed by a suitable read-out system. A wavelength selector can also be used in a position between the light source and the sensor to provide increased specificity in detection.

The fiber-optic sensor simplified form, can be represented as shown in Figure 11.2. It consists of an optical source, optical fiber, sensing, or modulator element (which transduces the measurand to an optical signal), an optical detector, and processing electronics.

The light source used in optical sensing systems, generates an intense and stable optical radiation that will be required to probe the optical property of the molecular-recognition element in the sensor (Chapter 22). The optical signal generated in the sensor, will encode information on the analyte interaction with the molecular recognition element and its concentration in the analytical sample. Light sources such as incandescent (filament) lamps, gas discharge lamps, lasers, light-emitting diodes (LEDs), and laser diodes (LDs) are widely used. Incandescent lamps provide a broad and moderate spectral radiance and are used in short-range optical sensors. The miniature tungsten–halogen lamp is a commonly used source in such sensors. Gas discharge lamps, for example, xenon arc lamp, provide high optical radiance and are ideal sources for use in absorbance and fluorescence measurements. In the case of using lamps, the required wavelength of radiation is achieved using optical filters or monochromators (Figure 11.3).

Among discontinuously emitting sources, LEDs and LDs are particularly attractive for sensor applications. The reasons are their small dimensions, their low-energy demand and their manufacturing technology, which is compatible with the sensor production. In addition, lasers emit highly intense and coherent optical radiation making them ideal for long-range sensing systems. Description of these devices can be found in the book by Bass and Van Stryland (2002). At present, LEDs and LDs, sources with relatively intense radiation of narrow bandwidth, are available over a range of wavelengths. LDs are made from the same materials as LEDs, but their structure is somewhat more complex, and they are processed differently.

The wavelength selector sorts different optical signals according to wavelengths which are directed to the photodetector. It contributes to decoding of the chemical information associated with the optical signal and to sensitive measurements of absorbance or a selective excitation of molecules to generate fluorescence. Optical filters and grating-based monochromators are among

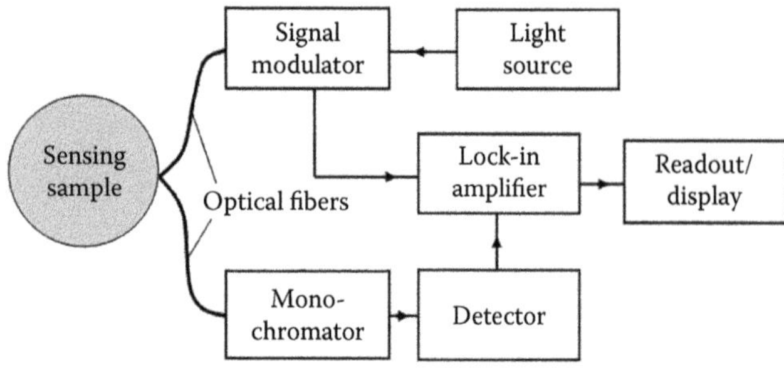

FIGURE 11.1 Schematic of a typical instrumentation associated with optical sensors. (From Korotcenkov, G. et al., Optical and fiber-optic chemical sensors. In Korotcenkov G. (Ed.), *Chemical Sensors: Comprehensive Sensor Technologies.* Vol. 5: *Electrochemical and Optical Sensors*, Momentum Press, New York, pp. 311–476, 2011. With permission.)

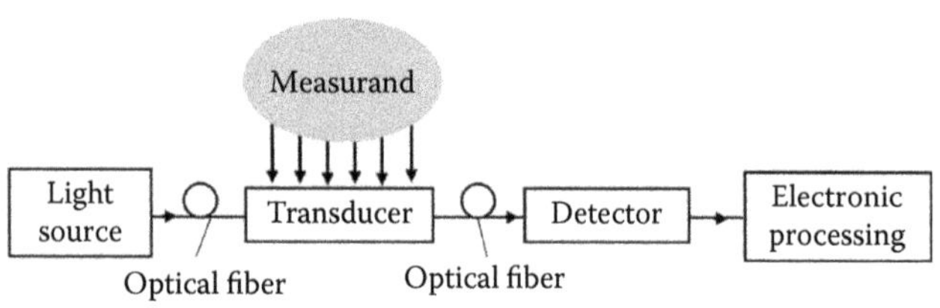

FIGURE 11.2 Basic components of an optical fiber sensor system.

typical wavelength selectors employed in optical chemical sensors. When using optical filters, the unwanted wavelengths are removed through either absorption of radiation (in absorption filters) or the destructive interference of out-of-phase radiation (in interference filters). Monochromators enable the dispersion of the constituents of polychromatic radiation and focus the desired wavelength band onto a slit in the system. Furthermore, the integration of monochromators is readily achieved through the use of optoelectronic components along with optical fibers.

The photodetection system converts the optical signal into an electrical signal, which preserves the original chemical information. The optical characteristics of the photodetector should match those of the light source to prevent any loss of information at the measurement wavelengths. Typically, the photodetection system must exhibit a high signal-to-noise ratio in order that the measured signals are highly sensitive. Photomultiplier tubes, photodiodes, photodiode arrays, phototransistors and charge-coupled devices (CCD) are among common photodetection devices. The photomultiplier tubes are the most sensitive photodetection systems and are preferred for the low-level light detection. The phototransistors and photodiodes (which are also known as quantum detectors) offer compactness and miniaturization of analytical systems, and are useful for measurements in the ultraviolet, visible, and near-infrared spectral regions. Photodiode arrays and CCDs are multichannel detectors that can do simultaneous detection of the dispersed optical radiation.

The display provides a visual representation of the output data, usually in numerical form, and common display systems include analog voltmeter, digital panel meters, chart recorders, and computer monitors. Optical couplers or beam splitters are sometimes utilized in optical fiber-based instrumentation systems to distinguish and separate the incident and detected radiation traveling through a single optical fiber. Refracting and reflecting optical components such as lenses and mirrors are also employed in the instrumentation systems in order to manipulate the optical signals effectively in the fibers.

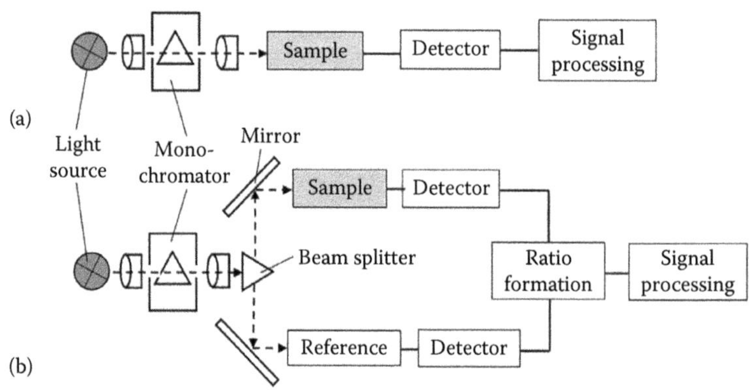

FIGURE 11.3 (a) Single- and (b) double-beam arrangements of spectrometers used in optical sensors.

11.2 CONFIGURATION OF OPTICAL HUMIDITY SENSORS

A multitude of configurations have been adopted in optical and fiber-optic humidity sensors. At the same time, these configurations do not differ from the configuration of optical sensors used for the detection of toxic or flammable gases. The main configurations of a typical optical sensor are schematically shown in Figure 11.4. Light from the excitation light source (L) passes through the excitation filter (F) and is absorbed by the sensor spot (S), containing an indicator. The light from the indicator (transmitted, reflected, or emitted) is registered by the detector (D). In case of a luminescent sensor, an emission filter (F′) is placed between the sensor spot and the detector to avoid interception of excitation light by the detector.

Instead of filters the instrument can contain a monochromator. The angle between the excitation light source and the detector may vary. For absorption-based sensors a linear scheme of the optical components is usual (Figure 11.4a), while an angular setup (Figure 11.4b) is preferred for luminescent sensors. For luminescent optical sensors, however, either 90° or 180° is preferred.

As we indicated before, the most important component of an optical sensor system is the sensor spot (S), which contains an analyte-specific indicator (humidity-sensitive material). The configuration of the light source, excitation and emission filters, and the modulation frequencies are optimally fitted for

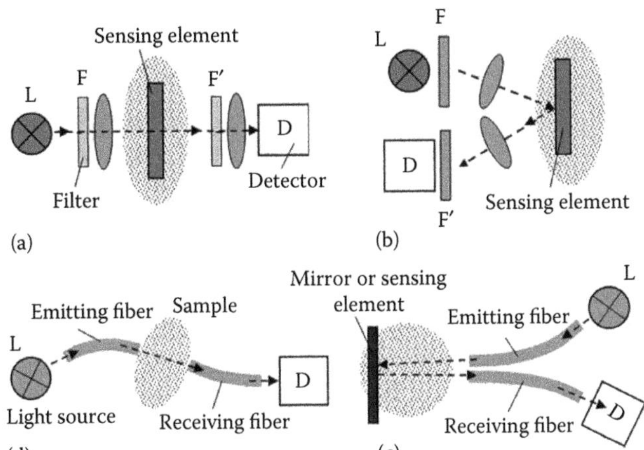

FIGURE 11.4 Scheme of two principal setups for optical sensing: (a,c) a linear configuration convenient for colorimetric, spectroscopic and photometric-based sensors, and (b,d) an angular optical setup optimal for luminescence or fluorescence-based sensors.

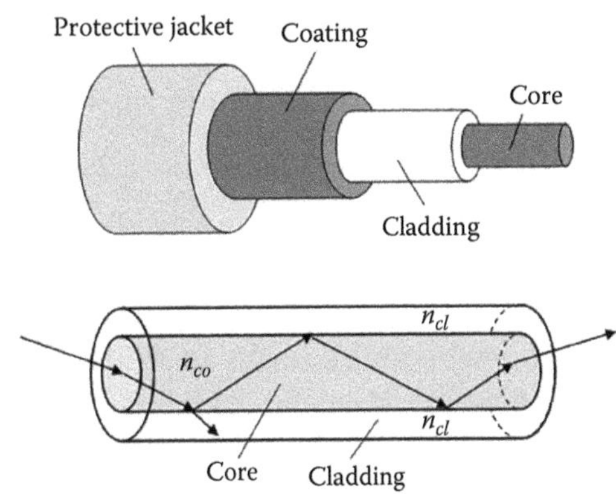

FIGURE 11.5 Schematic diagram of fiber.

the best sensing performance of an indicator (Gansert et al. 2006). In optical sensors, using given configuration, the use of fibers as simple light pipes to transmit spectroscopic information to an instrument detector is the easiest way of remote sensing (Culshaw and Kersey 2008). Generally, an optical fiber consists of a fiber as the core surrounded by a cladding layer and a light-impermeable jacket (Figure 11.5). The refractive index (RI) of the core (n_1) is always higher than that of the cladding (n_2) layer. As a result, the light beams are reflected toward the inside of the guide by internal total reflection.

This means that the light generated by a light source and sent through an optical fiber can be transmitted over a long distance. The light then in the same way returns through the optical fiber and is captured by a photodetector. One of possible variants of optical sensors based on transmission or reflectance measurements is shown in Figure 11.4. As noted earlier, these sensors are also called the extrinsic optical fiber sensors. In the extrinsic type, the optical fiber is only used as a means of light transport to an external sensing system, that is, the fiber structure is not modified in any way for the sensing function (Elosua et al. 2006). It should be noted that many hygrometers discussed in Chapter 12 refer to extrinsic fiber-optic humidity sensors.

Another option is the construction of the humidity sensor is using a planar waveguide. The typical experimental configuration used to study the spectroscopic properties of the planar waveguide sensors is shown in Figure 11.6. This set up is practically the same one used in *m-line* spectroscopy, where the light is injected in the film by prism coupling (Tien 1997). Light is totally reflected numerous times at the interfaces due to the low thickness of the active layer. This means that an

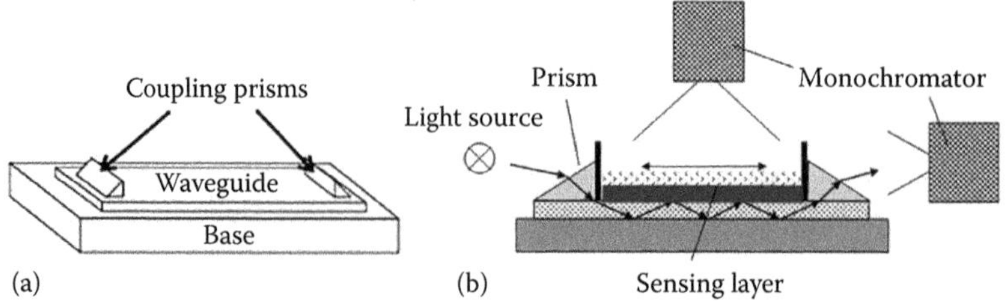

FIGURE 11.6 (a) Planar waveguide configuration; (b) Experimental setup used for measurements of Raman and Brillouin scattering, absorbance and luminescence in planar waveguide configuration. The laser light is injected into the guide by prism coupling. (From Korotcenkov, G. et al., Optical and fiber optic chemical sensors. In Korotcenkov G. (Ed.), *Chemical Sensors: Comprehensive Sensor Technologies.* Vol. 5: *Electrochemical and Optical Sensors*, Momentum Press, New York, pp. 311–476, 2011. With permission.)

extraordinary intensive interaction with the interface, that is, with the sample medium, takes place. This configuration allows an appreciable increase of the contrast as well as selectivity of both mode and polarization. Detailed discussions about the wave-guiding geometry are reported in several books and review articles (Martellucci et al. 1994; Ferrari et al. 1996).

It is important to note that when using this technique, a humidity-sensitivity is achieved either by the humidity-sensitive properties of the waveguide, or due to humidity-sensitive properties of the material deposited on the surface of the planar waveguide. It turns out that with proper selection of the material capable to interact with water vapors, it is possible to create the conditions, providing a much stronger interaction of the radiation distribution in the waveguide with the environment.

Experiment has shown that the resulting optical planar sensor can be compact, miniature, and robust, making it ideal for rugged field measurements. We need to note that the research of sensors based on planar waveguide platforms exhibits a rapid growth in terms of innovative design and integration of multiple functionalities onto a single sensor chip. This is largely due to the compatibility of the planar geometry with a range of advanced microfabrication technologies and the ease with which such sensor geometry can be integrated with microfluidic and lab-on-a-chip systems (McDonagh et al. 2008). The generic design of an integrated planar optical waveguide sensor and a waveguide surface Plasmon resonance sensor is shown Figure 11.7. Currently, such an approach to the construction of optical sensors is of interest because in this case planar waveguides can be integrated with solid-state radiation sources and phototransducers.

In the development of fiber-optic sensors, the so-called intrinsic fiber-optic sensors, is used the same approach as in the development of planar waveguide-based sensors. However, as the waveguide an optical fiber is being used.

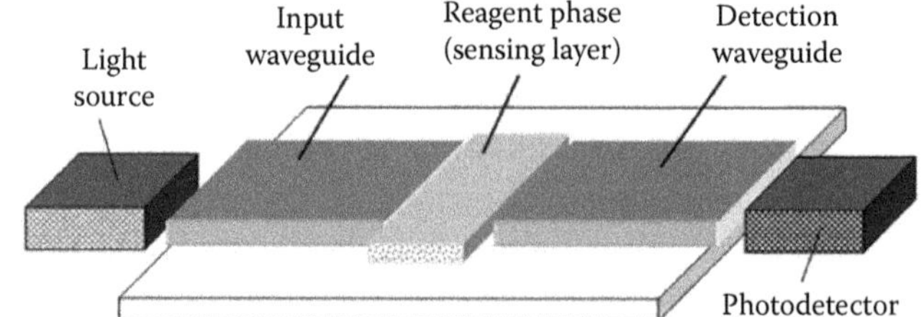

FIGURE 11.7 Diagram of an integrated optical waveguide sensor.

The using of optical fibers to sense chemical concentration has been reported in literature since 1946 (Culshaw and Kersey 2008). The optical fiber sensors or fiber-optic sensors (FOS) since then found applications in chemical (Chan et al. 1984; Wolfbeis and Posch 1986; Wolfbeis 1991, 1992; Stewart et al. 1997), biochemical (Ferguson et al. 1996; Healy et al. 1997; Rowe-Taitt and Ligler 2001), biomedical, and environmental sensing (Mizaikoff et al. 1995; Dietrich 1996; Schwotzer 1997; Holst and Mizaikoff 2001), including humidity sensing (Yeo et al. 2008; Sikarwar and Yadav 2015). There are four basic types of light property modulation used in optical fiber sensors: an intensity, phase, wavelength, and polarization. These light properties can be modulated by changing the boundary conditions through different types of physical perturbations. FOS are mostly amplitude-modulations sensors, that is, sensors in which the intensity of the light transported by the fiber is directly modulated by the parameter being investigated which itself has optical properties (spectrophotometric sensors); or by a special reagent connected to the fiber, whose optical properties vary with the variation in the concentration of the parameter being studied (transducer sensors). Only in a few special cases a phase modulation is used for the optical modulation, because the chemical species being investigated modify the optical path of the light transported by the fiber (Brenci and Baldini 1991).

Now we will consider the types of sensors indicated previously in more detail. Since in recent years the designers in the development of optical-based humidity sensors gave preference to FOS, our discussion will start with this type of optical sensors.

11.3 OPTICAL FIBERS

An optical fiber is a cylindrical, flexible, and transparent long strand made either from a glass, plastic, silica, or other ceramic materials. An optical fiber consists of at least two components: (1) a core and (2) a cladding (Figure 11.6). The core is the thin center of fiber where the light travels, whereas the cladding is the outer optical material, which surrounds the core and function to reflect the light back into the core by total internal reflection principle (Lacey 1982; Cheo 1985). This occurs when the RI of the core is higher than the surrounding cladding. As it is known, when $n_2 < n_1$, where n_2 and n_1 are refractive indices of two contacting medium, any ray impinging at the interface with an incident angle greater than θ_c Equation 11.1 is totally reflected inside the first medium.

$$\theta_c = \sin^{-1}\frac{n_2}{n_1} \tag{11.1}$$

Figure 11.8 illustrates this situation. In this figure, it was shown that the light beam enters from air to the optical fiber, a less dense to the denser medium, with an external angle θ_{max}. In these cases, the light refracted toward the normal at an angle θ_1. To propagate the light beam down the optical fiber, the light beam at the core and cladding interface must take an angle greater than the critical angle θ_c.

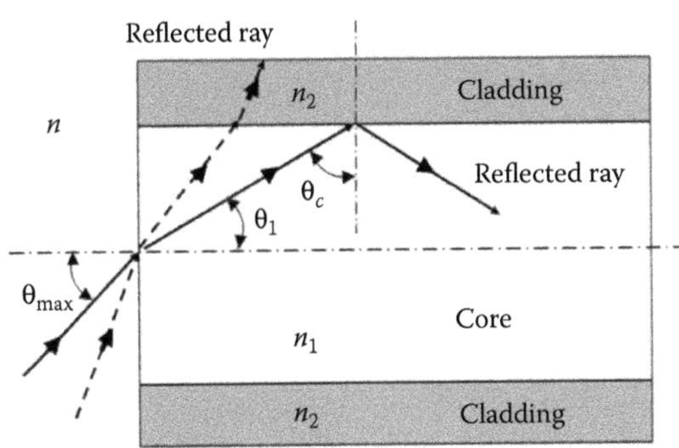

FIGURE 11.8 The light wave propagation in the fiber. (From Korotcenkov, G. et al., Optical and fiber optic chemical sensors. In Korotcenkov G. (Ed.), *Chemical Sensors: Comprehensive Sensor Technologies*. Vol. 5: *Electrochemical and Optical Sensors*, Momentum Press, New York, pp. 311–476, 2011. With permission.)

Thus, the cladding performs the following necessary functions: (1) reduces loss of light from the core into the surrounding air, (2) reduces scattering loss at the surface of the core, (3) protects the fiber from absorbing surface contaminants, and (4) adds mechanical strength.

Optical fibers are most often (though not always) fabricated from a very pure silica, or SiO_2 (Syms and Cozens 1993). This has a RI of $n \sim 1.458$ at $\lambda_0 = 850$ nm. Dopants useful for fabrication of silica-based fiber include Germania (GeO_2) and phosphorus pentoxide (P_2O_5), both of which increase the RI of silica, boric oxide (B_2O_3), and fluorine (F_2), which reduce it. Thus, a typical fiber might consist of a GeO_2:SiO_2 core, with a SiO_2 cladding. It was established that germanosilicate optical fibers are photosensitive (Hill et al. 1978) and this property can be used for the reflection filter fabrication (Section 22.6). Alternatively, a pure SiO_2 core could be used, with a B_2O_3:SiO_2 cladding. The boundary between the core and cladding may either be abrupt (step-index fiber) or gradual (graded-index fiber). Other low-loss fiber materials that have been investigated include low melting-point silicate glasses (soda-lime silicates, germanosilicates, and borosilicates) and halide crystals (TlBrI, TlBr, KCl, CsI, and AgBr). Most recently, attention has been transferred to the fluoride glasses (mixtures of GdF_3, BaF_2, ZrF_4, and AlF3). Plastic-based fibers were designed as well (Yang et al. 2014). Plastics for optic fiber have large losses in comparison with silica fibers, and therefore they are not suitable for long distance communications. But for short links of a few tens of meters and for sensor design plastic fibers are satisfactory and simple to use (Yang et al. 2014). They can be handled without special tools or techniques. Selected parameters of several optical fibers used in chemical sensors are presented in Table 11.1.

Optical fiber is mainly used to transmit light along its length. Therefore, optical fibers are also called the light guides. They are widely used in the fiber-optic communication, which permits transmission over longer distances and at higher data transfer rates than other forms of wired and wireless communications. Remarkable technological advances in the fabrication of optical fibers over the past two decades allowed light to be guided through 1 km of the glass fiber with a loss as low as ≈0.15 dB (≈3.4%) at the wavelength of maximum transparency. Optionally, an optical fiber also consists of the third component, a jacket (coating), which gives further protection to the surface of the fiber.

The spectral transmission of an optical fiber is determined by the material used in its core and cladding. The *numerical aperture* (NA) is an important parameter for the familiar light beam approach.

TABLE 11.1
Selected Properties for Optical Fibers Used as Sensors

Fiber/Parameter	Wavelength	Attenuation	Refractive Index, Core	Maximum Use Temperature
Silica	0.2–4 μm	0.5 dBm^{-1} (1.5 μm)	1.458	800°C
Chalcogenide	3–10 μm	0.5 dBm^{-1} (6 μm)	2.9	300°C
Fluoride	0.2–4.3 μm	0.02 dBm^{-1} (2.6 μm)	1.51	250°C
Sapphire	0.2–4 μm	20 dBm^{-1} (3 μm)	1.7	>1500°C
Single-mode photonic crystal	0.4–3 μm	<1 dBm^{-1} (16 μm)	1.0 (core) 1.46 (cladding)	800°C
AgBr/Cl	3.3–15 μm	0.7 dBm^{-1} (10.6 μm)	2.0	400°C
PMMA	0.4–0.8 μm	0.1 dBm^{-1} (600 nm) 30 dBm^{-1} (800 nm)	1.492	80°C

Source: Fernando, G.F. and Degamber, B., *Int. Mater. Rev.* 51, 65–106, 2006.
PMMA—Poly(methyl methacrylate).

It provides a connection between the maximum angle of incidence θ_{max}, at which light will still be propagated within the core, and the refractive indices of corresponding materials:

$$NA = n_0 \cdot \sin(\theta_{max}) = \frac{\sqrt{n_1^2 - n_2^2}}{n_0} \quad (11.2)$$

The RI n_0 belongs to the medium surrounding of the fiber. For air, $n_0 = 1$. Table 11.2 shows a list of typical acceptance angles, NA, and transmission ranges for commercially available fibers.

TABLE 11.2
Acceptance Angles (θ_c) and Numerical Apertures (NA) of Different Types of Optical Fibers

Fiber Material and Transmission Range	θ_{max} (°)	NA (λ < 587 nm, $n_0 = 1$)
Glass (400–1600 nm)	26.5	0.45
	32.5	0.54
	35.0	0.57
	40.0	0.64
	60.0	0.86
Quartz (250–1000 nm)	12.5	0.22
	15.0	0.26
PMMA (400–800 nm)	27.5	0.46
	30.0	0.50

Source: Haus, J., *Optical Sensors: Basics and Applications*, Wiley-VCH Verlag, Weinheim, Germany, 2010.
PMMA—Poly(methyl methacrylate) (acrylic glass).

The acceptance angle θ_{max} of the fiber determines the cone of external rays that are guided by the fiber. Rays incident at angles greater than θ_{max} are refracted into the fiber but are guided only for a short distance because they do not undergo total internal reflection. The NA therefore describes the light-gathering capacity of the fiber, as illustrated in Figure 11.9. When the guided rays arrive at the terminus of the fiber, they are refracted back into a cone of angle θ_{max}. The acceptance angle is thus a crucial design parameter for coupling light into and out of a fiber.

Optical fibers can be divided into two basic types based on the transmission characteristics: (1) as multimode fiber and (2) single-mode fiber. Multimode fibers simply refer to the fact that numerous modes or light rays are carried simultaneously through the light guide, whereas those which can only support a single

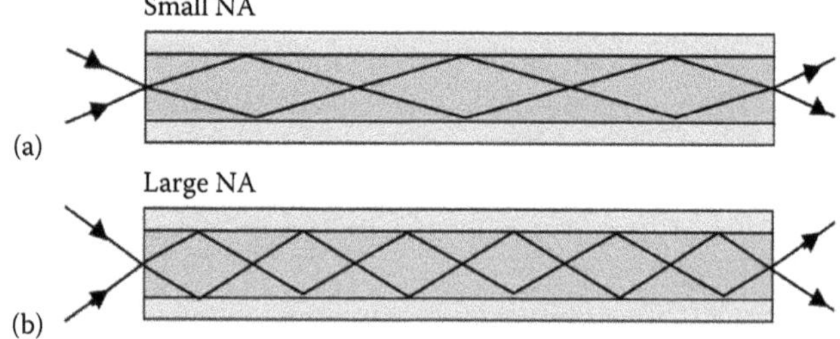

FIGURE 11.9 Light propagation in the fiber with different NA. The light-gathering capacity of a large NA fiber (b) is greater than that of a small NA fiber (a). (From Korotcenkov, G. et al., Optical and fiber optic chemical sensors. In Korotcenkov G. (Ed.), *Chemical Sensors: Comprehensive Sensor Technologies.* Vol. 5: *Electrochemical and Optical Sensors*, Momentum Press, New York, pp. 311–476, 2011. With permission.)

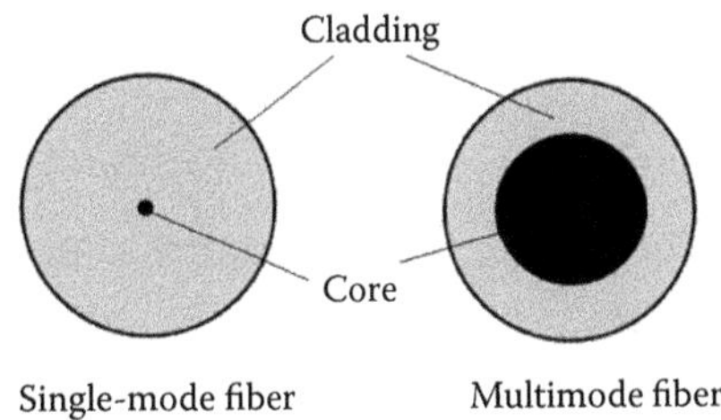

FIGURE 11.10 Fibers of single-mode and multimode.

mode are called single-mode fibers. Optical fibers of the *single-mode type* have a core diameter of some micrometers at maximum, whereas multimode fibers can have a diameter of up to 1 mm (Lacey 1982; Cheo 1985) (Figure 11.10).

The number of modes of multimode fiber cable depends on the wavelength of light, the core diameter, and the material composition. This can be determined by the Normalized frequency parameter (V). The V is expressed as (Buck 1995):

$$V = \frac{\pi d}{\lambda}\sqrt{n_1^2 - n_2^2} = \frac{\pi d}{\lambda}(\text{NA}) \tag{11.3}$$

where:
d is a fiber core diameter
λ is a wavelength of light
NA is a numerical aperture

For a single-mode fiber, $V \leq 2.405$ and for multimode fiber, $V \geq 20$. Mathematically, the number of modes for a step index, fiber is given by Keiser (1991)

$$N_{SI} = \frac{V^2}{2} \tag{11.4}$$

For a graded index fiber, the number of mode is given by

$$N_{GI} = \frac{V^2}{4} \tag{11.5}$$

Optical fibers can be used as single fibers or in the form of fiber bundles. Fiber bundles often are used as carriers for chemical receptor layers in optodes. Single fibers become increasingly meaningful for miniature chemical sensors.

11.4 CONFIGURATION OF FIBER-OPTIC HUMIDITY SENSORS

Currently, there are many approaches to the classification of FOS's configuration. But all these classifications are not universal. Therefore, we consider the sensor types typically useful for the most sensing applications.

Broadly speaking, FOS can be divided into two categories (Sharma and Wei 2013):

- *Intrinsic fiber sensors:* The light is modified inside the fiber and the fiber itself acts as a transducer, in part or as a whole. Often the fiber is attached to a material that acts as a transducer in tandem with fiber. The intrinsic fiber sensors have the advantage of compact and efficient design with high sensitivity.
- *Extrinsic or hybrid fiber sensors:* The light is carried by a fiber to a location where the light is modified by environmental perturbations or measurand, and the modified light is then collected back by the same or another fiber and directed to a detector where it is processed and analyzed. One of the major advantages of the extrinsic fiber sensors is that the fiber acts as a flexible and rugged dielectric conduit of light and enables delivery and collection of light for measurement purposes, which otherwise would had been prohibitive due to harsh environmental conditions.

In the intrinsic fiber sensors, these environmental perturbations impart a change in the physical property of the fiber itself; such as temperature- or strain-induced changes in length or RI, which in turn imparts a change in the optical phase of the light traveling through the fiber. On the other hand, extrinsic fiber sensors could have much wider range of applications because it is then feasible to expand the transducing mechanisms as the light is not necessarily confined within the fiber when interacting with the perturbation field.

With regard to the position of the humidity-sensitive materials in fiber-optic humidity sensors, we can mark out three versions: (1) end-of-fiber, (2) side-of-fiber, and (3) porous or interrupted fiber configurations (Rogers and Poziomek 1996).

11.4.1 End-of-Fiber Configuration

For end-of-fiber sensors, the optical fiber acts as a conduit to carry light to and from the sample (Figure 11.11a). This method involves sending a light source directly through the optical fiber and analyzing the light that is reflected or emitted by the contaminant. The RI of the material at the tip of the optical fiber can be used to determine what phases (vapor, water, or any solution) are present. The modulation of intensity for a given range of wavelengths is dependent on the absorbance or fluorescence of the analyte, indicator, or analyte–indicator

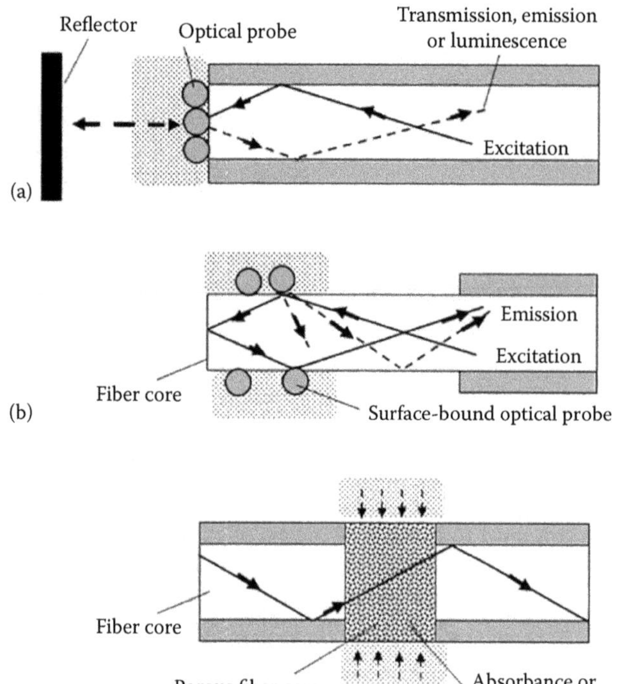

FIGURE 11.11 Fiber-optic sensor configurations. (a) End of fiber, (b) Side of fiber, and (c) Porous or interrupted fiber. (Idea from Rogers, K.R. and Poziomek, E., *Chemosphere*, 33, 1151–1174, 1996.)

complex. The indicator compound can be trapped behind a membrane, in a polymer, or covalently immobilized to the end of the fiber. In some end-of-fiber sensors, there is no molecular recognition element, because the analyte possesses optical properties through which it can be measured directly. The intrinsically small dimensions of fiber-optic and waveguide experiments mean that the observable quantities of sample contained within the indicator phase are rather small, which make sensitive measurement approaches, especially appealing.

The above-mentioned configuration is suitable for remote analysis, analysis of materials within small spaces, and, importantly, for novel ways to combine optical response-producing reagents with the sample medium. In this case, the quantity of light absorbing or emitting material observed by the fiber depends on a number of factors including the depth of light penetration into and/or the dimension of the indicator phase and the optical input and/or collection characteristics of the fiber/indicator phase interface (Murray 1989). Chemiluminescence and fluorescence measurements are typical approaches used for design of end-of-fiber sensors. For fluorescence measurements, the fiber optics design must provide for admitting the exciting wavelengths and collecting the emitted wavelengths.

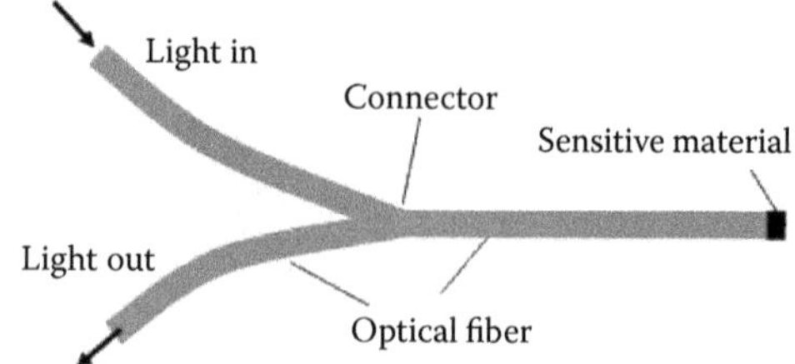

FIGURE 11.12 Fiber-optic optode for chemical sensing.

The optode is the most widespread version of end-of-fiber sensors. Since the probe is practically the fiber itself, a compact, highly miniaturized sensing structure can be attained. The basic optode design consists of a source fiber and a receiver fiber connected to a third optical fiber by a special connector as shown in Figure 11.12. The tip of the third fiber is coated with a sensitive material, mostly by the dip-coating procedure.

However, the optode can be fabricated using more simple approaches (Figure 11.13). They can be fabricated either by using some combination of two different fibers (bifurcated) (Figure 11.13b), or by using a single fiber (Figure 11.13a) and by exploiting differences between the collimation of the exciting source (typically a laser) and the emanating emitted light at the instrument end of the fiber.

In the single-fiber system, the incident light from the light source is directed by an optical fiber or a fiber bundle to the reagent phase, which consequently reflects the radiation back to the optical fiber. The reflected radiation

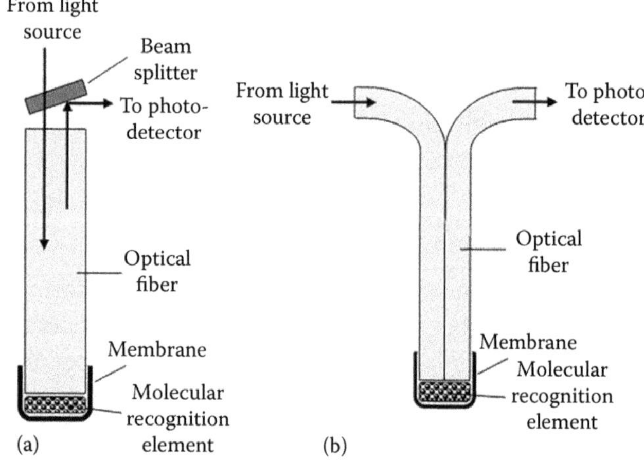

FIGURE 11.13 Typical configuration of optodes: (a) employing a single fiber and (b) employing a bifurcated optical fiber bundle. (From Korotcenkov, G. et al., Optical and fiber optic chemical sensors. In Korotcenkov G. (Ed.), *Chemical Sensors: Comprehensive Sensor Technologies*. Vol. 5: *Electrochemical and Optical Sensors*, Momentum Press, New York, pp. 311–476, 2011. With permission.)

is then led by the same optical fiber to the phototransducer system. The incident radiation and the reflected radiation are discriminated through by the use of a beam splitter in the instrumentation, which isolates the returning light and focuses it on the detector (Figure 11.13a). In the bifurcated system, two fibers or two fiber bundles are involved: one to feed the incident light to the reagent phase and the other to direct the reflected radiation to the photodetector (Figure 11.13b). The tip of the third fiber is coated with a sensitive material, mostly by the dip-coating procedure. The chemical to be sensed may interact with the sensitive tip by changing the absorption, reflection, scattering properties, change in luminescence intensity, change in RI, or a change in polarization behavior; hence, changing the reflected light properties. The fiber in this case acts as a light pipe transporting light to and from the sensing region.

The single-fiber configuration has high optical collection efficiency, because there is maximum overlap of the incident and collected radiation. On the other hand, the bifurcated optical fiber bundle arrangement requires a simpler instrumentation, not needing any beam splitter that could make the optics complicated. It also has a high optical throughput, because several fibers are used to transport the radiation.

11.4.2 Side-of-Fiber Configuration

If the optics-chemistry coupling is done longitudinally (side-of-fiber configuration), this involves either removing the cladding from a commercial optic fiber (Figure 11.14), or using a hollow fiber, or planar geometries, and coating the indicator phase, that is, humidity-sensitive material, along the length of the optically transmitting phase. Such sensors may also be referred to sensors with active coating or cladding-based humidity FOS. In this design, a small section of the optical fiber passive cladding is replaced by an active coating (Yuan et al. 2001). Humidity indicator is usually a microporous material, sensitive to air humidity (Ogawa et al. 1988). The water vapors react with the coating to change the optical properties of the coating, that is, RI, absorbance, fluorescence, and so on, which is then coupled to the core to change the transmission through the optical fiber. In particular, these changes can take place due to interaction between the indicator phase and the internally reflected light, which depends on the penetration of the radiation field into the chemical phase around the core (Murray 1989). This mechanism based on the evanescent field will be discussed in the Section 14.1 in relation to the evanescent spectroscopicy in planar waveguide-based sensors.

11.4.3 Hybrid Intrinsic Sensors

Attempts to improve parameters of FOS usually are accompanied by designing hybrid configurations (Elosua et al. 2006). This configuration of FOS results from the combination of the different approaches as described before. For example, one possibility consists of an evanescent wave sensor incorporated into a reflection setup: this can be accomplished by modifying the cladding of an optical mirror-ended pigtail (Figure 11.15). This configuration guarantees that the optical power is modulated twice, first, when the light passes through the sensing area to the mirrored end, and second, when the signal travels back from the mirror at the end of the fiber to the detector, passing again through the modified cladding region (Elosua et al. 2006).

It is also possible to use a bent evanescent wave mirror-ended sensor. Another option for hybrid sensors is the combination of an active core fiber with a reflection scheme; as in both cases the modulated light is

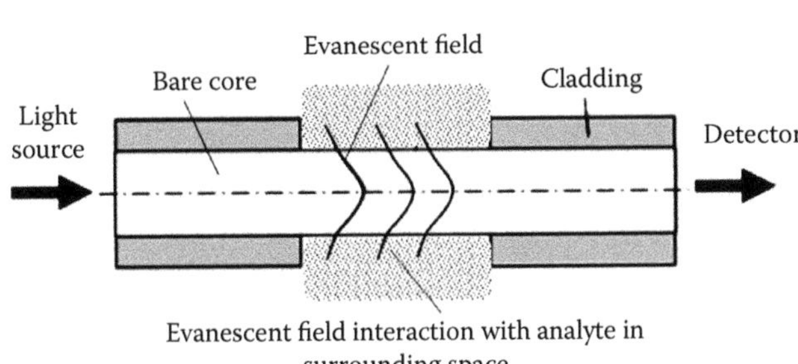

FIGURE 11.14 Schematic showing of an evanescent spectroscopic fiber-optic sensor design. Removal of cladding of part of optical fiber allows evanescent wave to interact with the material under test.

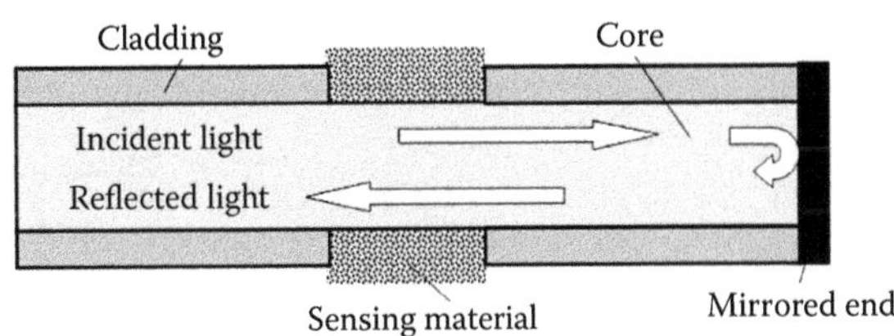

FIGURE 11.15 This configuration, compared to the classic transmission one, enhances the sensing performance of the sensors because the incident light passes twice through the sensing zone due to the mirrored end. (From Korotcenkov, G. et al., Optical and fiber optic chemical sensors. In Korotcenkov G. (Ed.), *Chemical Sensors: Comprehensive Sensor Technologies. Vol. 5: Electrochemical and Optical Sensors*, Momentum Press, New York, pp. 311–476, 2011. With permission.)

the one that travels through the fiber, the effect of the active core region is duplicated and that of the deposition onto the cut-end fiber is added to it. A hybrid sensor can also combine a sensing signal and an architecture not typically used with it. For example, there are some experimental transmissive configurations using the fluorescence produced by a chemical dye replacing the cladding (Potyrailo and Hieftje 1998).

Hybrid configurations can also be used to solve the technical problems associated with the use of FOS. For example, Gu et al. (2011) have designed relative humidity (RH) sensor based on a thin-core fiber modal interferometer (Chapter 16) with a fiber Bragg grating (FBG) (Section 22.8) between, where FBG was used for measurement of temperature and interferometer for RH sensing. The knowledge of the temperature is necessary for compensating the temperature effects of the sensor performance.

Hybrid humidity sensors can also be designed on the base of devices using different principles of humidity detection. For example, Shinbo et al. (2009) proposed a hybrid humidity sensor, in which the optical waveguide was formed on the surface of the quartz crystal microbalance (QCM). Thus, two independent humidity measurement techniques have been implemented in one unit. Fluorinated polyimide was used as waveguide and polyvinyl alcohol (PVA) film with $CoCl_2$ was used as humidity-sensitive material. The authors have developed device structures of two types shown in Figure 11.16a and b. Figure 11.16a shows the hybrid sensor of a slab optical waveguide type (SOWG), and Figure 11.16b shows a ridge optical waveguide type (ROWG). However, only sensors of the first type were tested. Testing has shown efficiency of this approach. Shinbo et al. (2009) believed that by observing both the mass and the optical transmittance or the optical spectrum, the hybrid sensors can realize not only high sensitivity but also accurate discrimination of adsorbed chemical species. Moreover, they can also be used for the evaluation of sorption phenomena in liquids—even in colored solutions.

11.4.4 Porous or Interrupted Fiber Configuration

For porous or interrupted fiber configuration (Figure 11.11), the humidity indicator is typically incorporated directly into the structure. Sensors with such configuration are often called sensors with active core. For several reasons, this configuration has the potential to be extremely versatile. For example, because of the large surface area provided by porous fiber core, this method is particularly well suited for absorbance measurements. Further, because the porous regions are intrinsically coupled with the fiber (i.e., they are part of the fiber), measurements can be made at multiple locations along a single fiber. Porous fibers may exhibit very high gas permeability. Thus, vapors permeating into the porous

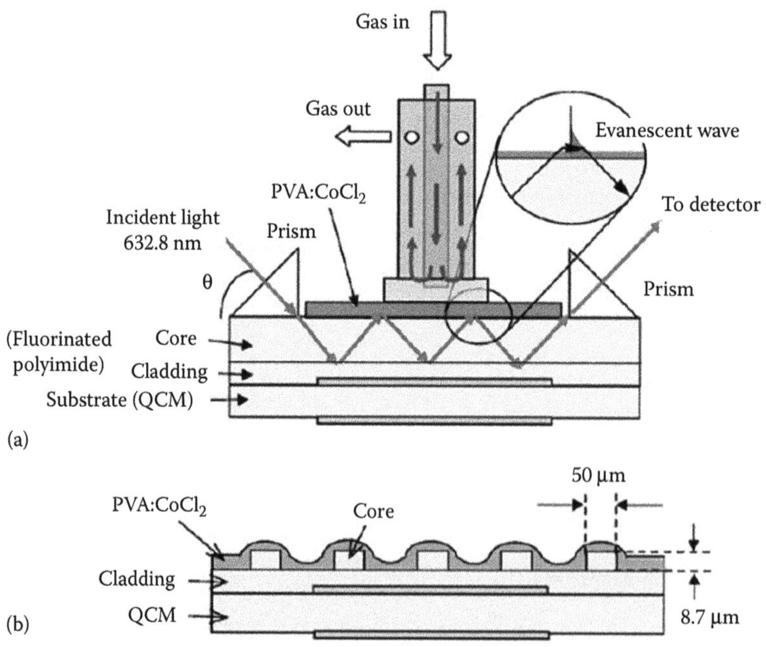

FIGURE 11.16 The structures of the fabricated sensors: (a) the QCM with the slab waveguide. and (b) the QCM with the ridge waveguide. (Reprinted from *Thin Solid Films*, 518, Shinbo, K. et al., A hybrid humidity sensor using optical waveguides on a quartz crystal microbalance, 629–633, Copyright 2009, with permission from Elsevier.)

zone can produce a spectral change in transmission. For better sensitivity and selectivity, colorimetric reagents can be trapped in the pores. However, the response time of these sensors is very long. So, this technique finds applications where very fast response is less important than sensitivity.

Regarding an interrupted fiber configuration, this version, in respect to planar waveguide-based sensors is shown in Figure 11.7. It can also be realized by using a fiber-optic technology. In this case, we get the device for absorption or spectroscopic measurements but in micro-miniaturized performance.

11.5 PLANAR WAVEGUIDE-BASED PLATFORMS FOR HUMIDITY SENSORS

The description of a planar waveguide-based chemical sensor (PWCS) in detail, one can find in the review papers of McDonagh et al. (2008), Lambeck (2006), Gauglitz (2005), Sparrow et al. (2009), and Campbell (2010). It should be noted that the humidity sensor is a form of chemical sensors because they tend to use the same platform. The only difference is in the use of different materials to ensure sensitivity to the specified analyte.

In its core, a PWCS comprises a planar substrate (e.g., glass, plastic, or silicon) that forms the basis of the sensor chip. In some cases, this substrate acts as the waveguide, whereas in others an additional waveguide layer is deposited onto the substrate. The basic requirements of a thin-film optical guide material are that it be transparent to the wavelength of interest and that it has a RI higher than that of the medium in which it is embedded (Selvarajan and Asoundi 1995). Usually, a layer of the film on the top of a substrate (of lower index) serves as the guide. In channel optical waveguides, light is confined in the film not only across its thickness but also across its lateral direction. To realize this, the waveguide in its cross section should be surrounded in all the directions by media having lower indices of refraction. Thus, in terms of total internal reflection concept, guidance can be considered to be the result of the rays, which propagate within the high RI region, suffering total internal reflection at the boundary interfaces with the media of lower refractive indices. Thus, materials that enable a light mode propagation are characterized by a high RI and a low attenuation <3 dB/cm. Especially, metal oxides or nitrides meet these optical requirements, and hence SiO_2, SiO_xN_y, Si_3N_4, TiO_2, Ta_2O_5, and Nb_2O_5 are the typical waveguide materials. With regard to the waveguide layer, several configurations have been employed that impart various optical functionalities to the sensor platform, and some of these are reported in the following Chapters 12, 15, 16, 19 and 23. In many cases, the light that propagates within the waveguide facilitates the operation of the platform as a sensor through the interaction of its evanescent field with the sensing environment above the waveguide.

There are two approaches acceptable for a planar waveguides fabrication (Schmitt and Hoffmann 2010). In the first approach, the substrate material is modified, for example, by ion exchange in SiO_2 or $LiNbO_3$ (Syms and Cozens 1993; Korishko et al. 2000; Hu et al. 2001). For example, titanium metal can be diffused into lithium niobate ($LiNbO_3$) or lithium tantalate substrates, by first depositing the metal in patterned strips of ~100 nm thickness and then carrying out an in-diffusion at a high temperature (~1000°C) for several hours (3–9). This is known as the *Ti: $LiNbO_3$ process*. The additional impurities cause a change in RI that is approximately proportional to their concentration, with a typical maximum value of $\Delta n \sim 0.01$. Other treatments can also be used for ion exchange. For example, protons (H^+ ions) can be exchanged with Li^+ ions in $LiNbO_3$. Silver ions can be exchanged with Na^+ ions in soda-lime glass by immersing the substrate in molten $AgNO_3$, at temperatures of 200°C–350°C. Ion implantation can also be used for these purposes. As this process is ballistic, it causes damage to the lattice as well as a change in RI. Ion implantation is therefore normally followed by an annealing step, which shakes out the lattice damage. Waveguides, fabricated using the first approach, have the disadvantage that their RI is increased only to a minor extent, yet they show less porosity than coated waveguide layers.

In the second approach, a waveguide layer is deposited on the top of a substrate material (coating) as it is done with metal oxides or nitrides (Worhoff et al. 1999). Here, high-RI waveguides can be achieved more easily, but often they are porous and have insufficient chemical resistance. These waveguides are mainly fabricated using chemical vapor deposition techniques (CVD). The need for low porosity in combination with high chemical resistance requires optimized coating techniques to fabricate extremely dense waveguide films and also to reduce a surface roughness to minimize light losses due to scattering (Schmitt and Hoffmann 2010).

Like fiber-based sensors, planar optical waveguides can be subdivided into two classes: (1) single-mode (small waveguide thickness) and (2) multimode (comparably large thickness). A light wave reproducing itself after two reflections in the waveguide is called an eigenmode, or simply, the mode of a waveguide (Saleh and Teich 2007). In single-mode waveguides, only one light mode can be guided, whereas thicker waveguides allow several modes. For multimode waveguides, a description based on the ray-optics approach is adequate. Yet, it is not enough to describe thin-film (thin core) waveguides, where the electromagnetic approach is more suitable. These two approaches basically follow the description of the optical waveguide theory described in Snyder and Love (1983) and Schmitt and Hoffmann (2010).

We need to note that up to now most of the work, which has been done for the development of planar evanescent wave chemical sensors, was based on single-mode waveguides. Due to the small waveguide core diameter in the range of a few microns they allow using only LDs as a light source (Burck et al. 1996). However, research has shown that planar multimode waveguides can also be fabricated (Klein and Voges 1994; Burck et al. 1996). As is known, a multimode waveguide enormously facilitates the fiber coupling to the planar structure and allows the use of white-light sources.

According to Burck et al. (1996), the application of planar evanescent wave absorbance sensors with smaller waveguide dimensions compared with the existing cylindrical FOS offers the following advantages: (1) due to the smaller waveguide dimensions a higher number of reflections and thus an increased sensitivity per unit length and a higher fraction of light intensity in the evanescent field are obtained; (2) in contrast to a cylindrical fiber geometry, a planar substrate allows easy deposition of tailor-made polymer superstrates, which act as sensing layer; and (3) the planar structure provides a much higher mechanical stability compared to a sensing fiber coiled up close to its minimum bend radius on a support.

At the same time, Burck et al. (1996) believe that besides above-mentioned advantages planar configuration is faced with the problem of low light transmission intensities in such small surface waveguide structures. On account of the small dimensions of thin-film waveguides, the coupling of light into the planar structure is not very efficient. Therefore, the coupling of light requires special attention. There are different methods for inputting light into the active layer. Several methods used for these purposes are shown in Figures 11.6, 11.7, and 11.17. Commonly, light is coupled to the device by means of prisms (prism coupler sensor). This approach was discussed earlier in Section 11.2 (Figure 11.6). Alternatively, the light can be coupled by miniature diffraction gratings at the sensor ends (grating coupler sensor, GCS) (Figure 11.17a). At present various gratings with different design have been developed for the energy transfer of a light beam into or out of an optical waveguide. They are characterized by their geometrical dimensions and the grating period. The different shapes encompass triangular, rectangular, and trapezoidal profiles, but asymmetric profiles are also possible (Tamir and Peng 1977). With grating couplers, preferably laser beams are used, which are coupled to the planar medium by diffraction at the grating with a precisely defined angle of incidence. Front-face coupling (Figure 11.17b) can be accomplished either by a lens focusing a collimated light beam onto the waveguide or by a waveguide illumination with an optical fiber (Figure 11.17c).

After comparison of above-mentioned approaches, one can conclude that a prism coupling requires no waveguide structuring and is a relatively simple method (Campbell 2010; Schmitt and Hoffmann 2010). Compared to front-face coupling, the alignment of the light beam is not as critical as long as the correct coupling angle is used. The coupling angle depends on the effective RI of the waveguide. The use of prisms can find application when the sensing is conducted in air. However, when the measurement is taken with a liquid flow cell, the flow cell and the gasket between the flow cell and the waveguide become a major problem.

An effective front-face coupling via lens and fibers requires well-prepared (e.g., polished) square edges of the waveguide and an exact alignment of the optical elements focusing the light beam. This ensures a maximal overlap integral, which is determined by the intensity distribution of the light and the distribution of the waveguide modes. Although, being technically not demanding, the front-face coupling has the disadvantage of low robustness against vibrations in the sensor system, and it requires extensive alignment procedures to minimize variations in the coupling efficiency.

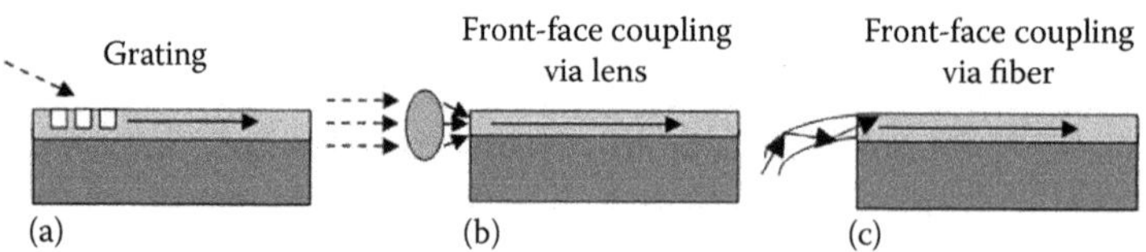

FIGURE 11.17 Light coupling methods into waveguides: (a) Grating coupling, (b) front-face coupling via optical lens, and (c) front-face coupling via optical fiber. (From Korotcenkov, G. et al., Optical and fiber-optic chemical sensors. In Korotcenkov G. (Ed.), *Chemical Sensors: Comprehensive Sensor Technologies.* Vol. 5: *Electrochemical and Optical Sensors*, Momentum Press, New York, pp. 311–476, 2011. With permission.)

The choice of whether to input light into a waveguide using either a fiber to couple into the end of a waveguide or to couple a focused beam through a grating seems to be driven by the individual researcher's equipment and experience without any fundamental reason behind the choice.

For grating coupling, a defined grating structure in the waveguide is necessary, which is technically more elaborate than the other methods. As with prism coupling, the coupling angle is determined by the effective RI of the waveguide (sensing principle of the GCS). Gratings provide the option and advantage of having the light come in from the bottom of the waveguide, thus allowing for direct sensing of the environment, or the placement of the test cell and related fluidics on the sensing side. However, the grating structure allows only a very narrow range of coupling angles, depending on the spectral bandwidth of the light source. The position of the light beam on the grating influences the coupling efficiency to a large extent, along with grating parameters, that is, a structure depth. An advantage of this coupling method over the previously mentioned methods is a good reproducibility of coupling conditions because no further optical elements are directly involved. The grating design allows waveguide chips to be routinely replaced, producing a *plug and play* sensor. Gratings work well for slab waveguides but pose a more difficult task for channel waveguides. The measurement principle of the grating coupler for chemical sensing was first discovered and published by Lukosz and Tiefenthaler (1984).

11.6 SUMMARY

A crucial problem in developing any sensor, including humidity sensors, is to create a new intermediate-class device, which must combine such qualities as portability, off-line operation, small price, long operating life, and possibility to measure concentrations in a wide range with high accuracy in various environmental conditions. Summing up the consideration of the optical sensors, it is necessary to state that the present approach to the development of humidity sensors is really progressive and promising for widespread use. On the basis of this review, one can conclude that primarily it refers to FOS.

REFERENCES

Bass M. and Van Stryland E.W. (Eds.) (2002) *Fiber Optic Handbook: Fiber, Devices, and Systems for Optical Communications*. McGraw-Hill, New York.

Brenci M. and Baldini F. (1991) Fiber optic optrodes for chemical sensing. In: *Proceedings of the 8th IEEE Optical Fiber Sensors Conference*, New York, pp. 313–319.

Buck J.A. (1995) *Fundamentals of Optical Fibers*. John Wiley & Sons, Chichester, UK.

Burck J., Zimmermann B., Mayer J., and Ache H.-J. (1996) Integrated optical NIR-evanescent wave absorbance sensor for chemical analysis. *Fresenius J. Anal. Chem.* 354, 284–290.

Campbell D.P. (2010) Planar-waveguide interferometers for chemical sensing. In: Zourob M. and Lakhtakia A. (Eds.), *Chemical Sensors and Biosensors: Methods and Applications*. Springer-Verlag, Berlin, Germany, pp. 55–113.

Chan K., Ito H., and Inable H. (1984) An optical fiber based gas sensor for remote adsorption measurement of low level methane gas in near infrared region. *J. Lightwave Technol.* 2, 234–237.

Cheo P.K. (1985) *Fibre Optics: Devices and Systems*. Prentice-Hall, Englewood Cliffs, NJ.

Culshaw B. and Kersey A. (2008) Fiber-optic sensing: A historical perspective. *J. Lightwave Technol.* 26(9), 964–1078.

Davis C.M., Carome E.F., Weik M.H., Ezekiel S., and Einzig R.E. (1986) *Fiber-optic Sensor Technology Handbook*. Optical Technologies, Herndon, VA.

Dietrich A.M. (1996) Measurement of pollutants: Chemical species. *Water Environ. Res.* 68, 391–406.

Elosua C., Matias I.R., Bariain C., and Arregui F.J. (2006) Volatile organic compound optical fiber sensors: A review. *Sensors*. 6, 1440–1465.

Ferguson J.A., Boyles T.C., Adams C.P., and Walt D.R. (1996) Fiber optic DNA biosensor microarray for the analysis of gene expression. *Nature Biotechnol.* 14, 1681–1684.

Fernando G.F. and Degamber B. (2006) Process monitoring of fibre reinforced composites using optical fibre sensors. *Int. Mater. Rev.* 51(2), 65–106.

Ferrari M., Gonella F., Montagna M., and Tosello C. (1996) Detection and size determination of Ag nanoclusters in ion-exchanged soda-lime glasses by waveguided Raman spectroscopy. *J. Appl. Phys.* 79, 2055–2059.

Gansert D., Arnold M., Borisov S., Krause C., and Muller A. (2006) Hybrid optodes (HYBOP). In: Popp J. and Strehle M. (Eds.), *Biophotonics: Visions for Better Health Care*. Wiley-VCH Verlag, Weinheim, Germany, pp. 477–518.

Gauglitz G. (2005) Direct optical sensors: Principles and selected applications. *Anal. Bioanal. Chem.* 381, 141–155.

Grundler P. (2006) *Chemical Sensors: An Introduction for Scientists and Engineers*. Springer-Verlag, Berlin, Germany.

Gu B., Yin M., Zhang A.P., Qian J., and He S. (2011) Optical fiber relative humidity sensor based on FBG incorporated thin-core fiber modal interferometer. *Opt Express*. 19, 4140–4146.

Haus J. (2010) *Optical Sensors: Basics and Applications*. Wiley-VCH Verlag, Weinheim, Germany.

Healy B.G., Li L., and Walt D.R. (1997) Multianalyte biosensors on optical imaging bundles. *Biosens. Bioelectron.* 12, 521–529.

Hill K.O., Fujii Y., Johnson D.C., and Kawasak B.S. (1978) Photosensitivity in optical fiber waveguides: Application to reflection filter fabrication. *Appl. Phys. Lett.* 32, 647–649.

Holst G. and Mizaikoff B. (2001) Fiber optic sensors for environmental sensing. In: Lopez-Higuera J.M. (Ed.), *Handbook of Optical Fiber Sensing Technology.* John Wiley & Sons, New York, pp. 729–749.

Hu H., Lu F., Chen F., Shi B.-R., Wang K.-M., and Shen D.-Y. (2001) Monomode optical waveguide in lithium niobate formed by MeV Si^+ ion implantation. *J. Appl. Phys.* 89, 5224–5226.

Keiser G. (1991) *Optical Fiber Communications.* McGraw-Hill, New York.

Klein R. and Voges E. (1994) Integrated-optic ammonia sensor. *Fresenius J. Anal. Chem.* 349, 394–398.

Korishko Y.N., Fedorov V.A., and Feoktistova O.Y. (2000) $LiNbO_3$ optical waveguide fabrication by high-temperature proton exchange. *J. Lightwave Technol.* 18, 562–568.

Korotcenkov G., Cho B.K., Sevilla III F., and Narayanaswamy R. (2011) Optical and fiber optic chemical sensors. In: Korotcenkov G. (Ed.), *Chemical Sensors: Comprehensive Sensor Technologies.* Vol. 5: *Electrochemical and Optical Sensors*, Momentum Press, New York, pp. 311–476.

Lacey E.A. (1982) *Fibre Optics.* Prentice-Hall, Englewood Cliffs, NJ.

Lambeck P.V. (2006) Integrated optical sensors for the chemical domain. *Meas. Sci. Technol.* 17, R93–R116.

Leung A., Shankar P.M., and Mutharasan R. (2007) A review of fiber-optic biosensors. *Sens. Actuators B.* 125, 688–703.

Lukosz W. and Tiefenthaler K. (1983) Directional switching in planar waveguides effected by adsorption-desorption processes. In: *Proceedings of 2nd European conference on Integrated Optics*, Florence, Italy. IEE Conference Publication No. 227, London, UK, pp. 152–155.

Martellucci S., Chester A.N., and Bertolotti M. (Eds.) (1994) *Advances in Integrated Optics.* Plenum Press, New York.

McDonagh C., Burke C.S., and MacCraith B.D. (2008) Optical chemical sensors. *Chem. Rev.* 108, 400–422.

Mignani A.G., Falciai R., and Ciaccheri L. (1998) Evanescent wave absorption spectroscopy by means of bi-tapered multimode optical fibers. *Appl. Spectrosc.* 52(4), 546–551.

Mizaikoff B., Taga K., and Kellner R. (1995) Infrared fiber optic gas sensor for chlorofluorohydrocarbons. *Vib. Spectrosc.* 8, 103–108.

Murray R.W. (1989) Chemical sensors and microinstrumentation: An overview. In: Murray R., Dessy R.E., Heineman W.R., Janata J., and Seitz W.R. (Eds.), *Chemical Sensors and Microinstrumentation.* ACS Symposium Series. American Chemical Society, Washington, DC, pp. 1–19.

Narayanaswamy R. and Wolfbeis O.S. (Eds.) (2004) *Optical Sensors—Industrial, Environmental and Diagnostic Applications*, Springer Series on Chemical Sensors and Biosensors, Vol. 1. Springer-Verlag, Berlin, Germany.

Ogawa K., Tsuchiya S., Kawakami H., and Tsutsui T. (1988) Humidity-sensing effects of optical fibers with microporous SiO_2 cladding. *Electron. Lett.* 24(1), 42–43.

Potyrailo R.A. and Hieftje G.M. (1998) Oxygen detection by fluorescence quenching of tetraphenylporphyrin immobilized in the original cladding of an optical fiber. *Anal. Chim. Acta.* 370, 1–8.

Rogers K.R. and Poziomek E. (1996) Fiber optic sensors for environmental monitoring. *Chemosphere.* 33, 1151–1174.

Rowe-Taitt C.A. and Ligler F.S. (2001) Fiber optic biosensors. In: Lopez-Higuera J.N. (Ed.), *Handbook of Optical Fiber Sensing Technology.* John Wiley & Sons, New York, pp. 687–700.

Saleh B.E.A. and Teich M.C. (2007) *Fundamentals of Photonics.* John Wiley & Sons, New York.

Schmitt K. and Hoffmann C. (2010) High-refractive-index waveguide platforms for chemical and biosensing. In: Zourob M. and Lakhtakia A. (Eds.), *Chemical Sensors and Biosensors: Methods and Applications.* Springer-Verlag, Berlin, Germany, 21–54.

Schwotzer G. (1997) Optical sensing of hydrocarbons in air or in water using UV absorption in the evanescent field of fibers. *Sens. Actuators B.* 38, 150–153.

Selvarajan A. and Asoundi A. (1995) Photonics, fiber optic sensors and their applications in smart structures. *J. Non-Destruct. Eval.* 15(2), 41–56.

Sharma U. and Wei X. (2013) Fiber optic interferometric devices. In: Kang J.U. (Ed.), *Fiber Optic Sensing and Imaging.* Springer Science+Business Media, New York, pp. 29–53.

Shinbo K., Otuki S., Kanbayashi Y., Ohdaira Y., Baba A., Kato K., Kaneko F., and Miyadera N. (2009) A hybrid humidity sensor using optical waveguides on a quartz crystal microbalance. *Thin Solid Films.* 518, 629–633.

Sikarwar S. and Yadav B.C. (2015) Opto-electronic humidity sensor: A review. *Sens. Actuators A.* 233, 54–70.

Snyder A.W. and Love J.D. (1983) *Optical Waveguide Theory.* Chapman & Hall, London, UK.

Sparrow I.J.G., Smith P.G.R., Emmerson G.D., Watts S.P., and Riziotis C. (2009) Planar Bragg grating sensors—Fabrication and applications: A review. *J. Sensors.* 2009, 607647.

Stewart G., Jin W., and Culshaw B. (1997) Prospects for fiber optic evanescent field gas sensors using absorption in the near infrared. *Sens. Actuators B.* 38, 42–47.

Syms R.R.A. and Cozens J.R. (1993) *Optical Guided Waves and Devices.* Academic Press, London, UK.

Taib M.N. and Narayanaswamy R. (1995) Solid-state instruments for optical fibre chemical sensors. A review. *Analyst.* 120, 1617–1625.

Tamir T. and Peng S.T. (1977) Analysis and design of grating couplers. *Appl. Phys.* 14, 235–254.

Tien P.K. (1997) Integrated optics and new wave phenomena in optical waveguides. *Rev. Modern Phys.* 49, 361–420.

Wolfbeis O.S. (1991) *Fiber Optic Chemical Sensors and Biosensors*, Vol. 1. CRC Press, Boca Raton, FL.

Wolfbeis O.S. (1992) *Fiber Optic Chemical Sensors and Biosensors*, Vol. 2. CRC Press, Boca Raton, FL.

Wolfbeis O.S. and Posch H.E. (1986) Fiber optic fluorescing sensor for ammonia. *Anal. Chim. Acta.* 185, 321–327.

Worhoff K., Lambeck P.V., and Driessen A. (1999) Design, tolerance analysis, and fabrication of silicon oxynitride based planar optical waveguides for communication devices. *J. Lightwave Technol.* 17, 1401–1407.

Yang H.Z., Qiao X.G., Luo D., Lim K.S., Chong W.Y., and Harun S.W. (2014) A review of recent developed and applications of plastic fiber optic displacement sensors. *Measurement.* 48, 333–345.

Yeo T.L., Sun T., and Grattan K.T.V. (2008). Fibre-optic sensor technologies for humidity and moisture measurement. *Sens. Actuators A.* 144(2008) 280–295.

Yuan J., El-Sherif A., MacDiarmid M.A., and Jones W. (2001) Fiber optic chemical sensors using modified conducting polymer cladding. *Proc. SPIE.* 4204, 170–179.

12 Absorption or Colorimetric-Based Optical Humidity Sensors

12.1 PRINCIPLES OF OPERATION

Absorption or colorimetric-based *optical humidity sensors*, as the name would suggest, are based upon detection of a humidity-induced absorption or color change in the sensor material. Examples of such sensors are presented in Table 12.1.

In absorption or colorimetric-based optical humidity sensors, the change in the intensity of radiation, measured at a particular wavelength, is the main index of concentration of water vapor in the air. To implement these sensors on the base of optical fibers, configurations shown in Figure 12.1, are usually used. Humidity indicators are located on either end of the fiber or between the two fibers. For example, Otsuki et al. (1998) have discussed an air-gap design, shown in Figure 12.2, utilizing the in-line absorption configuration. The sensor demonstrated was formed with an air-gap between two sections of a large core fiber positioned on the same axis. One end of the fiber was dip-coated with a dye solution containing Rhodamine B (RB) and hydroxypropyl cellulose (HPC). To measure the optical signal as a function of humidity, light was coupled into the sensor from one end of the fiber, passing through the dye-doped film and collected by the other fiber. The sensor discussed was able to operate between 0% and 95% RH and had a response time of approximately 2 min.

In principle, the same approach was used to develop the planar moisture absorption-based sensors. One of the most prevalent examples of the absorption-based the partial pressure of the water vapour (PW) humidity sensor (PWHS) platform is the one, based on evanescent-wave absorption (Section 14.1.1). Sensing layers that are doped with colorimetric, analyte-sensitive indicators can be deposited onto the upper surface of the waveguide, and any analyte-induced color changes can be probed by the evanescent field of a suitable light source. In the cases where the analyte is detected using direct spectroscopic, such a sensing layer may not be necessary, although transparent enrichment layers with high permeability coefficients for the analyte(s) of interest are sometimes employed. Such sensors for water vapor detection were developed by Skrdla et al. (2002). Platform, shown in Figure 12.3, was used for fabrication of these sensors.

These platforms typically incorporate design features intended to enhance the interaction of the interrogating light with the sensing environment, thereby improving platform sensitivity. The integrated waveguide absorbance optode (IWAO) developed by Puyol et al. (1999) composes of a single-mode integrated optic waveguide (as the input waveguide), a bulk optode (as the recognition region), and a wider multimode waveguide (as the collection waveguide) in a continuous arrangement. Optode incorporates a poly(vinyl chloride) (PVC)-based sensor membrane that is located between two antiresonant reflecting optical waveguides (ARROWs). The sensor membrane also serves as a waveguide and transports the interrogating radiation from the input ARROW to the output waveguide. The increased optical path length within the sensing layer provides enhanced sensitivity. The input and output waveguides are rib ARROW structures made by dielectric materials compatible with the standard integrated circuit and micromachining technology. ARROW structures are technologically easier to fabricate than conventional waveguides, and therefore, less costly and more robust fabrication processes are implied.

It is interesting that the sensors which utilize this technique have been modified to facilitate the interaction of the light guided within the fiber with the color-changing indicator or the water vapor itself. The main advantages of these sensors include the ability to transduce environmental changes into visual color changes. In this case, it is possible to eliminate the use of detectors by making environmental changes visible to the unaided eye. Basing on this principle of visualization, indicators have been developed and they will be discussed in Chapter 13.

TABLE 12.1
Colorimetric Humidity Sensors

Sensing Material	RH, %	Response Time	References
$CoCl_2$/borosilicate optical fiber	20–50	<5 min	Zhou et al. (1988)
$CoCl_2$/cellulose	4–60	5–20 min	Boltinghouse and Abel (1989)
$CoCl_2$–gelatin	20–80		Kharaz and Jones (1995)
$CoCl_2$–PVA	60–100	3–5 s	Hypszer and Wierzba (1995)
$CoCl_2$/porous sol–gel fiber	2–10		Tao et al. (2004)
$CoCl_2$/PEO	20–80		Tsigara et al. (2007)
Co_3O_4/glass plate	10–90	<5 min	Ando et al. (1996)
NiO/glass plate	0–90	<3 min	Ando et al. (1999)
Silica gel	0–95	1.5–2 min	Fong and Hieftje (1995)
Silica gel–crystal violet dye	20–90	~10 min	Carmona et al. (2007)
Cu(I)–MOF	30–90	>3 h	Yu et al. (2014)
Saponite clay incorporating Ni(II)–chelate complexes	0–100		Hosokawa and Mochida (2015)
Methylene blue–TiO_2 (Al_2O_3)	0–90	30 ms	Ishizaki and Katoh (2016)
HPC film	0–91	5–17 min	Otsuki and Adachi (1993)
Rhodamine B-doped HPC film	0–95	~2 min	Otsuki et al. (1998)
Trifluoroacetophenones-PVC	1–100	<0.5 min	Wang et al. (1991)
N,N-dioctylaminophenyl-4′-trifluoroacetyl-azobenzene (ETH^T 4001)-PVC	1–100	>1 h	Mohr and Spichiger-Keller (1998)
Nafion film with crystal violet dye	40–82		Brook et al. (1997)
PMMA or PEO films containing Reichardt's dye	0–71	2 min	Sadaoka et al. (1992a)
Nafion with various triphenylmethane or cyanide dyes	0–90	<2 min	Sadaoka et al. (1991)
TiO_2 films deposited by GLAD	0–90	>50 min	Hawkeye and Brett (2011)
Graphene oxide multilayers	0–98	250 ms	Chi et al. (2015)
PCs based on polystyrene-*b*-poly(2-vinyl pyridine) in the form of gel film	20–100		Kang et al. (2007), Kang (2012)
PCs based on dipentaerythritol hydroxypentaacrylate	40–95	~2 s	Shi et al. (2008)
PCs based on poly(styrenesulfonate-methylbutylene)	15–95	5 s	Kim et al. (2012)
PVA thin-film doped with Rh6G coated onto a silver surface	5–75	<5 s	Kumari et al. (2016)

HPC—hydroxypropyl cellulose; GLAD—glancing angle deposition; MAcoBMA—poly(methacrylic acid-co-*tert* butyl methacrylate); MOF—metal-organic frameworks; PC—photonic crystal; PEO—poly(ethylene oxide); PMMA—poly(methyl methacrylate); PVA—poly(vinyl alcohol); PVC—poly(vinyl chloride); Rh6G—Rho-damine 6G.

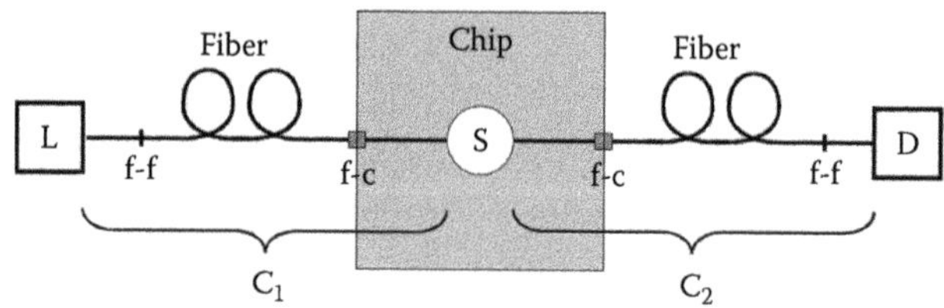

FIGURE 12.1 Schematic representation of a fiber-connected intensity modulated IO sensing scheme: L—light source; D—detector; S—sample; f-f—fiber-fiber connection; f-c—fiber-chip connection; C_1—emitting part; C_2—receiving part.

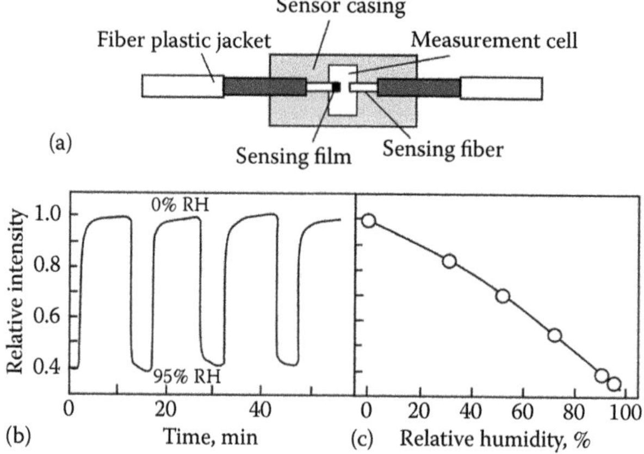

FIGURE 12.2 (a–c) Sensing characteristics of a fiber-optic humidity sensor using the air-gap configuration. (Adapted from Otsuki, S. et al., *Anal. Sci.* 14, 633, 1998. Copyright 1998: Japan Society for Analytical Chemistry. With permission.)

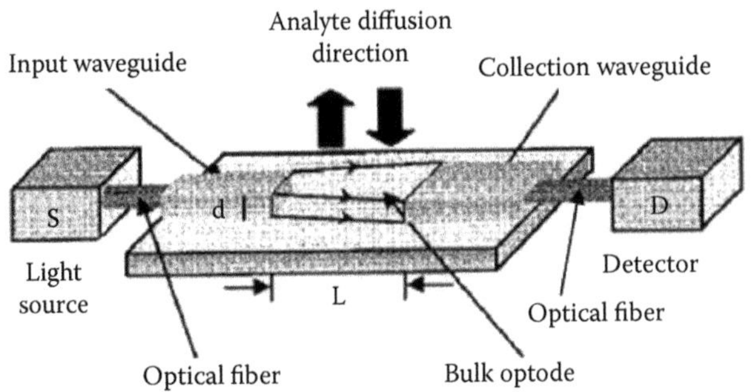

FIGURE 12.3 Integrated waveguide absorbance optode (IWAO) configuration developed by Puyol et al. (1999). (Reprinted with permission from Puyol, M. et al., *Anal. Chem.*, 71, 5037–5044, 1999. Copyright 1999 American Chemical Society.)

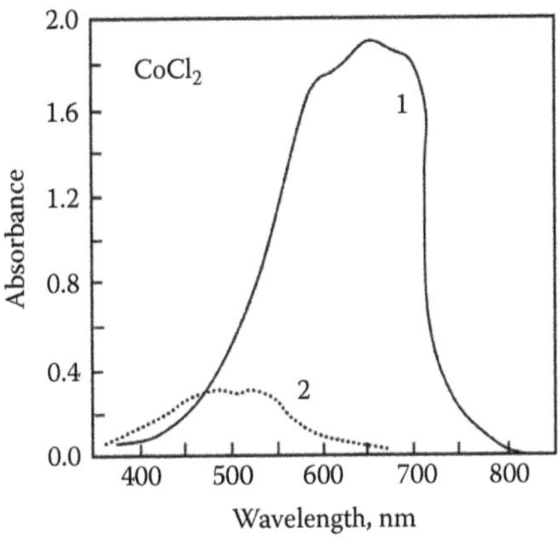

FIGURE 12.4 Absorbency spectra of $CoCl_2$ on cellulose filter paper: solid line (1), anhydrous: broken line (2), hexahydrate. (Reprinted with permission from Boltinghouse, F. and Abel, K., *Anal. Chem.*, 61, 1863–1866, 1989. Copyright 1989 American Chemical Society.)

12.2 MATERIALS USED IN ABSORPTION-BASED OPTICAL HUMIDITY SENSORS

As it was shown in Table 12.1, many different materials can change their optical properties under the influence of moisture. A summary of the color and λ_{max} changes, for the wet and dry forms for some of the major dye-based humidity sensors is presented in Table 12.2.

12.2.1 Salts and Oxides

Tables 12.1 and 12.2 show that the most common material used in the development of colorimetric humidity sensors is cobalt chloride, which shows a strong change in the absorption spectrum at hydration (Figure 12.4). The anhydrous salt $CoCl_2$ is blue, and exhibits a strong absorption peak at ~660 nm. When hydrated, the absorption peak shifts to 510 nm, characteristic of the $CoCl_2{\cdot}H_2O$ complex. So, the absorption band of the salt $CoCl_2$ undergoes a dramatic hypsochromic shift (Sharkany et al. 2005). Moreover, this behavior is completely reversible and is well suited for probing with light-emitting diodes (LEDs) that produce 635–675 nm wavelength of light.

To construct colorimetric humidity sensors a configuration shown in Figures 12.2 and 12.3 is commonly used. In the first case, the cobalt chloride is being applied to any carrier, which is optically transparent in the spectral region under study. For example, Boltinghouse and Abel (1989) immobilized cobalt chloride directly on cellulose and on acetylated cellulose paper. According to their results, acetylation of cellulose affords a support with reduced interaction with water that allows an important decrease of the light scattering and improves a reproducibility of the measurements, shortens the response time, and lowers the sensor hysteresis. In the second case,

TABLE 12.2
Humidity Sensitive Dyes

System	Color (λ_{max}, nm)		$\Delta\lambda_{max}$ (nm)	References
	Dry	**Wet**		
$CoCl_2$/cellulose	Sky blue (650)	Purple (525)	−125	Boltinghouse and Abel (1989)
7-diethylamino-4′-dimethylaminoflavylium perchlorate/k-carrageenan	Purple (550)	Red (520)	−30	Matsushima et al. (2002)
Thionine/k-carrageenan	Purple (600)	Red (520)	−80	Matsushima et al. (2000)
Methylene blue/urea/hydroxyethyl cellulose	Purple (550)	Blue (660)	+110	Mills et al. (2010)
Methylene blue/HPC	Purple (496)	Purple (505)	+9	Mills et al. (2017)

cobalt chloride is being applied directly to the fiber (Zhou et al. 1988). Zhou et al. (1988) and Hypszer and Wierzba (1995) have shown that the measuring range and sensitivity could be controlled by changing the concentration of the $CoCl_2{\cdot}H_2O$ solution used to treat the fiber. Cobalt chloride can also be applied to the surface of the fiber as the composite $CoCl_2$/gelatin (Kharaz and Jones 1995), $CoCl_2$/poly(vinylpyrrolidone) (Ballantine and Wohtjen 1986), $CoCl_2$/PVA (Hypszer and Wierzba 1995), $CoCl_2$/PEO, and $CoCl_2$/MAcoBMA (Tsigara et al. 2007). Experiment has shown that all indicated polymers can be used in combination with $CoCl_2$. However, according to Tsigara et al. (2007), PEO/$CoCl_2$ systems are the most promising hybrids for photonic humidity sensors. PEO oxide is absolutely stable over continuous and long-term measurements. Only MAcoBMA is not acceptable, because in the case of MAcoBMA, a performance degradation effect was observed after some cycles of exposure at high RH% (Tsigara et al. 2007). This may possibly be attributed to adsorption and permanent trapping of an amount of water molecules within the copolymers. After several cycling tests, the $CoCl_2$/MAcoBMA composites turned to become milky in appearance and showed a slight decrease in their dynamic range and sensitivity.

Colorimetric humidity sensors can also operate in the reflecting mode if they are manufactured in configuration of optodes. The selectivity of these devices seems to be quite good, whereas generally the response time as well as in some cases the hysteresis is not yet satisfactory. Thus, $CoCl_2$ is a humidity-sensitive material, suitable for the development of humidity sensors. The main shortcoming of $CoCl_2$-based sensors is the Co(II) toxicity.

It was established that spectral characteristics of crystal violet (CV) have also strong hypsochromic shift (Figures 12.4 and 12.5). Therefore, numerous materials, including Nafion films doped with CV have been investigated for humidity sensing using reflectance measurements. Dacres and Narayanaswamy (2006) have established that CV-based sensors responded to relative humidity (RH) with high sensitivity in the range 0% and 0.25% with detection limit as low as 0.018% RH (~4 ppm). The response was fully reversible in dry nitrogen with little hysteresis, and was capable of detection of moisture in process gases such as nitrogen and HCl.

Ando et al. (1996) have established that optical absorbance of nanosized metal oxide films was humidity sensitive as well. Cobalt oxide (Co_3O_4) films, prepared by pyrolysis of an organometallic precursor, has shown a humidity-sensitive absorbance change in the visible wavelength region around 350–400 nm at room temperature.

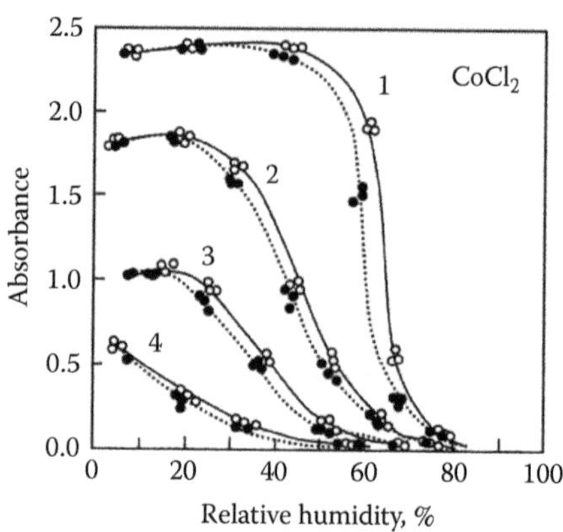

FIGURE 12.5 Effects of cobalt chloride concentration (mg $CoCl_2$/ml MeOH) and hysteresis effects on acetylated cellulose: 1-0.173; 2-0.155; 3-0.061; 4-0.042. (Reprinted with permission from Boltinghouse, F. and Abel, K., *Anal. Chem.*, 61, 1863–1866, 1989. Copyright 1989 American Chemical Society.)

The reversible absorbance changes and the relatively fast response time (within a few minutes) make the Co_3O_4 film a potential candidate for optical humidity detection. Nanosized NiO films prepared by plasma oxidation of nickel–carbon composite films (Ando et al. 1999) and TiO_2 films fabricated by sol–gel and thermal evaporation methods (Yadav et al. 2007) also showed humidity-sensitive absorbance changes at room temperature. At that NiO films display the largest absorbance changes in the Vis/NIR region with better sensitivity that the Co_3O_4 films at the measuring wavelength. With regard to TiO_2, a comparative study carried out by Yadav et al. (2007) established that humidity sensors with a TiO_2 film prepared by the thermal evaporation method were more sensitive in the range of 5%–80% than the sensor with the sol–gel method, whereas sensors on the base of sol–gel films were sensitive in the higher range of humidity, that is, from 80% to 95%. However, the sensitivity of the above-mentioned films to the humidity change was much worse than that of $CoCl_2$ films. In addition, the use of these films did not improve the sensor operation speed and did not expand the dynamic range of humidity measurement.

12.2.2 Polymers

As the humidity-sensitive materials for colorimetric humidity sensors organic molecules were also tested, Wang et al. (1991) described the use of trifluoroacetophenones immobilized in plasticized PVC for RH monitoring. On hydration, the absorption spectra of these molecules in the ultraviolet (UV) region changed drastically due to formation of a ketal. It was shown that when using two isologous ligands it was possible to measure RH in the range 1–100% with high stability, good

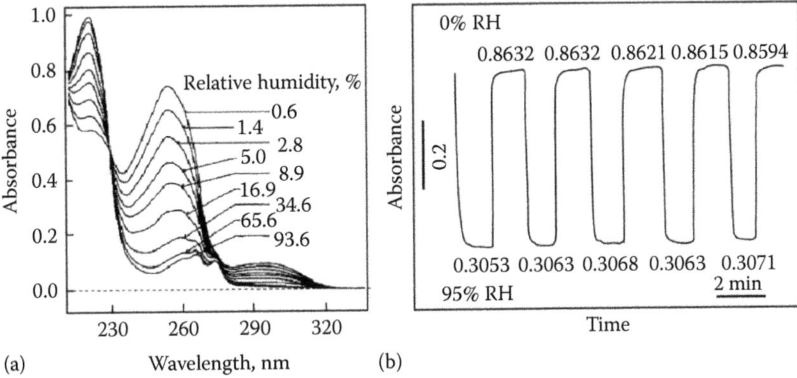

FIGURE 12.6 (a) Absorption spectra of two 4-μm-thick optode membranes after equilibration with air streams of different relative humidity. Plasticized poly(vinyl chloride) (PVC) was used as the membrane matrix. The dehydrated form of 4-(*n*-dodecylsulfonyl)-1-(trifluoroacetyl)benzene (ETH 6019) shows an absorbance maximum at 253 nm, the hydrated form at 220 nm; (b) Short-term reproducibility of the absorbance response of two 4-μm-thick optode membranes at 261 nm for relative humidity changes between 95% RH and 0% RH. (Reprinted with permission from Wang, K. et al., *Anal. Chem.*, 63, 970–974, 1991. Copyright 1991 American Chemical Society.)

reproducibility, and short response time (Figure 12.6). One important limitation to the broad application of these indicator dyes for humidity sensing is associated with their low absorption wavelength maxima (253–251 nm). In addition, ethanol behaves as an important interferent of the optode. To overcome this drawback, Mohr and Spichiger-Keller (1998) have prepared a new indicator, *N,N*-dioctylaminophenyl-4′-trifluoroacetyl-azobenzene (ETH^T 4001), with a similar sensitivity and selectivity to the previous indicator but with the absorption maximum shifted by more than 200 nm into visible region. For optode development, the reagent has been immobilized in polyurethane to afford a broad response range with highest sensitivity in the 5%–40% RH region. However, the response time of this sensor was long (in order of hours), preventing their use for continuous monitoring applications. The same approach was used by Sharkany et al. (2005). Sharkany et al. (2005) have designed miniaturized optical vapor sensors that utilized bacteriorhodopsin (BR)-based biochromic films. Films were fabricated by dispersing nanosize fragments of BR in gelatin. When the ambient RH was increased from 12% to 85%, the optical absorption of the films doubled.

Sadaoka et al. (1992a) have observed that a poly(methyl methacrylate) (PMMA) or poly(ethylene oxide) (PEO) films containing Reichardt's dye also changed their absorption spectrum under the influence of air humidity. It was established that absorption spectrum in the 400–10,000 nm range was changed considerably and the intensity of light reflected at 750 nm could be used to monitor RH. Sadaoka et al. (1991, 1992b) have also prepared composite films of hydrolyzed Nafion with various triphenylmethane or cyanide dyes containing terminal *N*-phenyl groups, working in the reflecting mode (Figure 12.7). They studied various dyes such as CV (hexamethyl-*p*-rosaniline chloride, CV), ethyl violet,

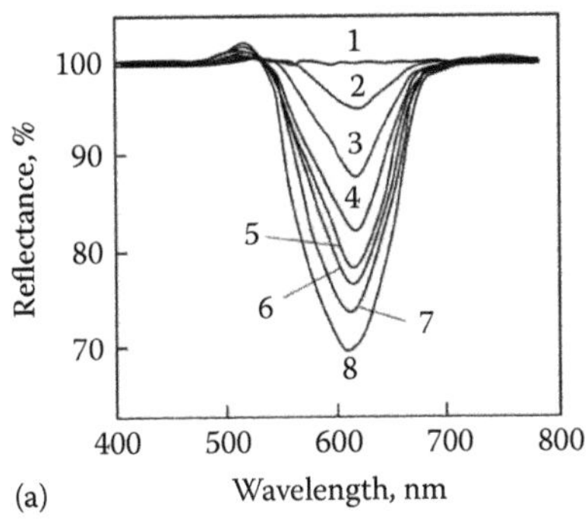

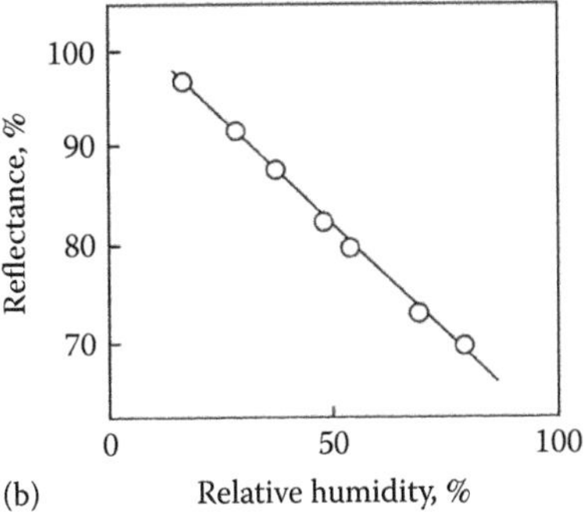

FIGURE 12.7 Humidity dependences of (a) spectrum and (b) reflectance at 610 nm for the Nafion film with malachite green: 1-0; 2-17; 3-29; 4-38; 5-44; 6-50; 7-63, and 8-71% RH. (Reprinted from *Sens. Actuators A*, 25–27, Sadaoka, Y. et al. Optical-fibre and quartz oscillator type gas sensors: Humidity detection by Nafion^R film with crystal violet and related compounds, 489, Copyright 1991, with permission from Elsevier.)

malachite green, aizen cathilon pinl FGH, and so on. Acceptable results were obtained for CV and malachite green dyes. For these dyes entrapped on Nafion, the color changed from yellow to blue or green with an increase in the sorbed water molecules and good reversibility with changes of humidity was confirmed. In particular, it was established that the optical intensity, with a minimum at 615 nm in the reflection mode, decreased with RH for a Nafion film with malachite green. The rise and recovery times were less than 1 min and the hysteresis was only +l% RH at 50% RH. For the film with CV, a similar characteristic was confirmed at 630 nm. Carmona et al. (2007) also compared two organic dyes: CV and chlorophenol red (3′,3″-dichlorophenolsulfonephthalein, Cr) encapsulated in silica gel and found that in contrast to Cr the CV dye really could be used for a humidity sensor design. According to Carmona et al. (2007), the mechanism that drives the CV change of color by the action of RH is based on a mixed phenomenon of neutralization–dilution. Water from RH is able to produce the dilution of the acid-basic chemical species (H_3O^+ and OH^-), which can react with CV molecules. The results obtained by (Carmona et al. 2007) showed the possibility to estimate by silica gel–CV sensors the environmental RH in the range between 20% and 90% (λ = 590 nm). However, the response was slow since a color change took place after approximately 10 min exposure.

Park et al. (2016) have found that polydiacetylenes (PDAs) also possess hydrochromic properties. The incorporation of headgroups composed of hygroscopic ions such as cesium or rubidium and carboxylate counterions enables PDAs to undergo a blue-to-red colorimetric transition as well as a fluorescence turn-on response to water. Kumari et al. (2016) established that PVA thin-film doped with Rhodamine 6G (Rh6G) also shows good sensitivity to changes in humidity (Figure 12.8). Kumari et al. (2016) found that via optimization of the thickness of the sensing layer (t ~ λ/4n), the spectral shifts of the Fabry–Perot (FP) resonance induced by the changes in RH, can be increased in ~10 times. As it is seen in Figure 12.8, the FP resonances in the dyed thin-film cavity can have a spectral shift up to $\Delta\lambda$ = 61 nm when the RH changes from 10% to 73%. It is important that the enhanced sensitivity of the dyed thin-film cavity can also be monitored by measuring the changes in the magnitude of the absorption measured at different RH environments.

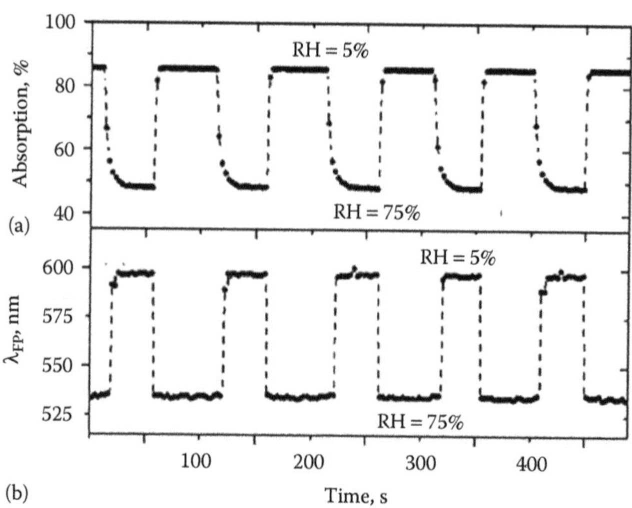

FIGURE 12.8 (a) Variation of absorption magnitude at λ = 527 nm and (b) spectral shift of FP resonances (λ_{FP}) in the absorption spectra as a function of time for PVA_{Rh6G}–Ag, C = 0.002 wt% under s-polarized illumination at θ = 81 when RH cycles between 5% and 75%. (Reprinted from *Sens. Actuators B*, 231, Kumari, M. et al., Enhanced resonant absorption in dye-doped polymer thin-film cavities for water vapour sensing, 88–94, Copyright 2016, with permission from Elsevier.)

12.2.3 Photonic Crystals

In recent years, increased interest has been shown to humidity-sensitive materials based on photonic crystals (PCs) (Kang et al. 2007; Tian et al. 2008; Wang et al. 2010; Hawkeye and Brett 2011; Kang 2012; Wang and Zhang 2013; Lu et al. 2016). A PC is a periodic optical nanostructure that affects the motion of photons in much the same way that ionic lattices affect electrons in solids. A PC can be defined as a structure whose dielectric constant periodically varies in one or more spatial directions. That allows classifying the PCs in one, two, and three dimensions (1D, 2D, and 3D) depending on the spatial directions in which the dielectric constant varies periodically (Joannopoulos et al. 2008). Figure 12.9 shows three different kinds of PCs (1D, 2D, and 3D from left to right) where the dielectric constant varies periodically in one, two, or

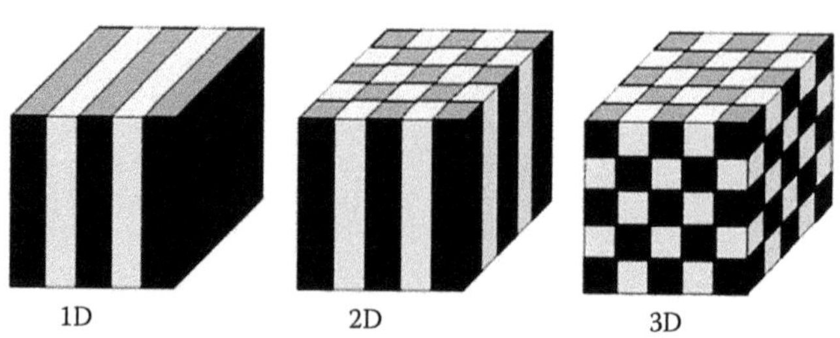

FIGURE 12.9 Simple examples of 1D, 2D, and 3D photonic crystals. Light and dark areas represent materials of different dielectric constant. (From Joannopoulos, J.D. et al., *Photonic Crystals. Molding the Flow of Light*, Princeton University Press, Princeton, NJ, 2008.)

three spatial directions. The periodical distribution of its dielectric constant can prohibit light propagation of a certain range of electromagnetic wave frequencies inside the crystal.

PCs have increasingly attracted the interest of researchers due to their unique structural color properties (Zhao et al. 2012). PCs have a band gap that forbids propagation of a certain frequency range of light. This property enables one to control light with amazing facility and produce effects that are impossible with conventional optics. The colorful appearances of the PCs materials can be ascribed to interference and reflection, which can be described by Bragg's and Snell's laws (Ozin and Arsenault 2008; Zhao et al. 2012). The law is given by

$$\lambda = 2D\left(n_{\mathrm{eff}}^2 - \cos^2\theta\right)^{1/2} \tag{12.1}$$

where:

- λ is the wavelength of the reflected light
- n_{eff} is the average refractive index of the constituent photonic materials
- D is the distance of diffracting plane spacing
- θ is the Bragg angle of incidence of the light falling on the nanostructures

Based on the equation, there are several methods for tuning structural color, such as changing the diffracting plane spacing D, the average refractive index n_{eff}, Bragg glancing angle θ, and changing the n_{eff} and D simultaneously. Taking into account that under certain conditions, an environment, in particular the air humidity, can affect indicated parameters; it becomes clear that the PCs can become humidity indicators in colorimetric humidity sensors. For example, the changes in the refractive index or the lattice spacing can be modified by the water absorption and condensation in PCs. However, many believe that this approach is well-suited to the realization of low-cost and low-power sensors operated in the visible region of the spectrum (Hawkeye and Brett 2011).

Of course, to satisfy the increasing number of requirements for actual application of colorimetric PC-based humidity sensors, it is critical to develop smart artificial photonic materials with excellent sensitivity, response rate, durability, and selectivity to water vapors. At that it is considered that to use PCs as sensors, diffractions that fall into the visible range are usually preferred, as the optical output can be directly observed by the naked eye without the need of complicated and expensive apparatuses to read the signals. Photonic structures capable of producing structural colors include 1D multilayer interference, 2D diffraction grating, and 2D/3D PCs. Compared to 1D photonic structures, 2D and 3D photonic structures in nature provide richer color, but the fabrication of such structures, of course, is more complicated.

At present, when developing PCs for humidity sensors the solid-state materials and polymers are being used (Kang et al. 2007; Shi et al. 2008; Tian et al. 2008; Wang et al. 2010; Hawkeye and Brett 2011; Wang and Zhang 2013; Lu et al. 2016). For inorganic humidity sensors, structural color changes are often caused by changes in the effective RI. Hawkeye and Brett (2011) have developed a mesoporous TiO_2 PCs with high- and low-density structural layers constituting of high- and low-RI layers. It was shown that the structural color changes of TiO_2 PCs can be sensitively observed despite the fact that the RH changes are smaller than 1%. The colorful response of the sensor was stable more than hundreds of hours. However, due to a large thickness and high film density the response was very slow. Hydrogel-based sensors generally induce a diffraction wavelength shift in response to humidity changes, owing to the volume change of polymer networks. Tian et al. (2008) have developed a humidity sensor by infiltrating acrylamide (AAm) solution into a P(St–MMA–AA) PC template and subsequently photopolymerizing. The colors of such sensors could reversibly vary from transparent to violet, blue, cyan, green, and red under various humidity conditions, covering the whole visible range. Furthermore, the color response showed exceptional stability under cyclic humidity experiments. Kang et al. (2007) reported a hydrophobic block–hydrophilic polyelectrolyte block polymer (Polystyrene-*b*-poly(2-vinyl pyridine) (PS-*b*-P2VP)) that formed a simple one-dimensional periodic lamellar structure. They showed very large reversible optical changes caused by swelling effect. This effect was observed for air humidity above 70% RH. Latter Kang (2012) have shown that the sensitivity of the photonic gels to humidity can be easily modulated simply by exchanging the counter anions pairing with the pyridinium groups in the QP2VP gel layers (Figure 12.10).

Wang et al. (2010) reported an organic/inorganic hybrid 1D PCs consisting of alternating thin films of titania and poly(2-hydroxyethyl methacrylate-co-glycidyl methacrylate) (PHEMA-co-PGMA) by the simple, reproducible, and low-cost approach of spincoating. Kim et al. (2012) developed fast responsive polymeric humidity sensors from a series of self-assembled poly(styrenesulfonate–methylbutylene) (PStS–b–PMB) block copolymers with tailored hygroscopic properties (Figure 12.11). Under different humidity, the PStS–b–PMB thin films displayed discernible reflective color changes covering almost entire visible light regions from violet (RH = 20%) to red (RH = 95%).

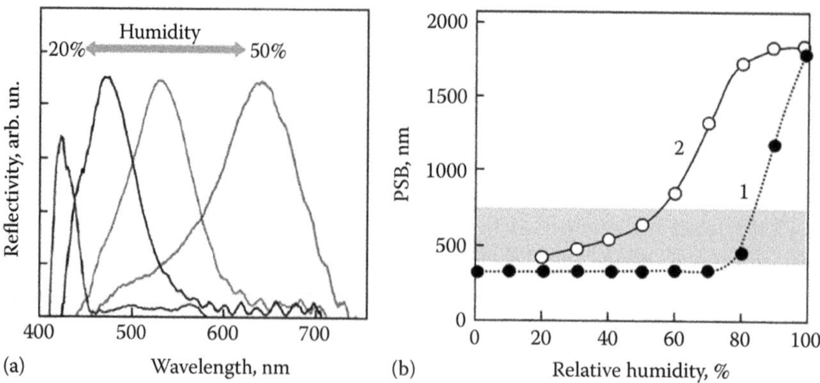

FIGURE 12.10 (a) UV–vis spectra of photonic gel film modified with acetate as a function of humidity and (b) variation of PSBs in the photonic gel film modified with (1) chloride and (2) acetate primary measured on UV–vis. The shaded area represents the visible regime. (With kind permission from Springer Science+Business Media: *Macromol. Res.*, Colorimetric humidity sensors based on block copolymer photonic gels, 20, 2012, 1223–1225, Kang, Y.)

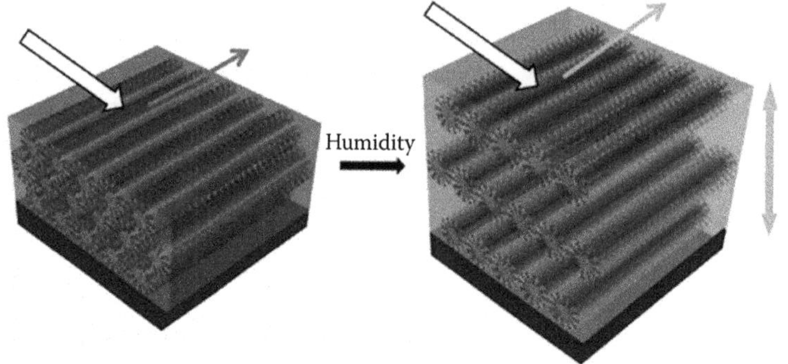

FIGURE 12.11 A schematic illustration of the structure of the PSS–b–PMB humidity sensor and the mechanism of the color changes between low and high RH conditions. The hygroscopic PSS chains spontaneously absorb water from the moist air and the swelling changes the film thickness to reflect visible light with different wavelengths. (Reprinted with permission from Kim, E. et al., *ACS Appl. Mater. Interfaces*, 4, 5179–5187, 2012. Copyright 2012 American Chemical Society.)

Interesting, but unsuitable for wide use method of manufacturing colorimetric humidity sensors was offered by Lu et al. (2016). They have developed a new kind of humidity responsive hierarchical structured photonic material by chemically coating polyacrylamide (PAAm) onto the surface of butterfly wing scales. The combination of the strong water absorption properties of the PAAm and the PC structures of the butterfly wing scales demonstrated excellent humidity responsive properties and a tremendous color change.

It should be noted that the graphene oxide (GO) multilayers studied by Chi et al. (2015), can be also referred to PCs, coloration of which is caused by the optical interference. Chi et al. (2015) have shown that due to the water adsorption on GO, two main reflection peaks at 386 and 526 nm, which were observed in an interference spectrum in the relatively dry atmosphere (RH = 12%), shifted at RH = 98% to 474 and 645 nm, respectively (Figure 12.12a). The ultrafast response ($\tau \sim 250$ ms) was the main advantage of these sensors (Figure 12.12b). It was established that the humidity induced color change was attributed to the swelling of GOs multilayers film at higher RH, and the ultrafast sensing was attributed to the super permeability of GOs. Chi et al. (2015) believe that the easy preparation, low cost, high sensitivity, and fast response of the GO-based humidity sensors provide a promising opportunity for the design of new humidity sensing systems for industry and domestic applications.

12.2.4 Other Materials

Many other materials such as HPC (Otsuki and Adachi 1993), cellulose acetate butyrate (CAB) (Li et al. 2015), silica gel (Fong and Hieftje 1995), porous metal–organic frameworks (MOFs) (Yu et al. 2014; Ohira et al. 2015), saponite clay incorporating Ni(II)–chelate complexes (Hosokawa and Mochida 2015) have been also tested.

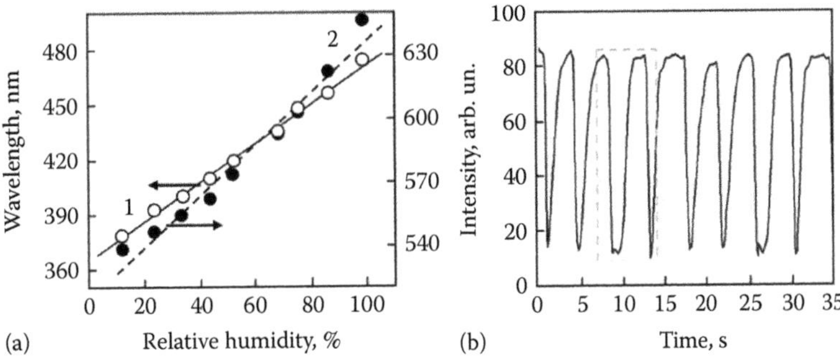

FIGURE 12.12 (a) The linear wavelength change of the reflection spectra peaks (1, 2) of a GOs film under different RHs. (b) Response and recovery properties of the GOs film based sensors under alternate atmosphere and high humidity at 25°C. (Reprinted with permission from Chi, H. et al., *ACS Appl. Mater. Interfaces*, 7, 19882–19886, 2015. Copyright 2015 American Chemical Society.)

In particular, Fong and Hieftje (1995) have proposed to control the concentration of water vapor through a direct measurement of absorption of radiation by water molecules, adsorbed on silica gel. They have shown that virtually any near-IR or even visible wavelength could be used to detect water adsorbed on silica gel plates; however, the 1940 nm combination band was the most sensitive to changes in humidity.

Version of optical sensor with MOF filter is shown in Figure 12.13a. MOF filter was prepared using copper benzene-1,3,5-tricarboxylate (Cu–BTC) (Ohira et al. 2015). The main feature of Cu–BTC is relatively rapid water uptake even with low concentrations of water. One study found a water uptake of 285 cm^3/g at a water pressure fraction (P/P_0) of 0.1, and no significant difference in the repeatability for the water uptake when P/P_0 was less than 0.05 (Furukawa et al. 2014). As a result, Cu–BTC-based sensors had possibility to detect very low concentration of water vapor in air (Figure 12.13b). The dynamic range of the sensor response developed by Ohira et al. (2015) was five orders of magnitude, from several tens of ppbv to thousands of ppmv (approximately 2000 ppmv). At that the sensor showed a reversible response to trace water, did not require heating to remove the adsorbed water molecules. In addition, the sample gas flow rate did not affect the sensitivity. The response time for sample gas containing 2.5 ppmv H_2O was 23 s. Studies have shown that the main drawback of these sensors is interference with ammonia and possibly with volatile organic compounds (VOCs).

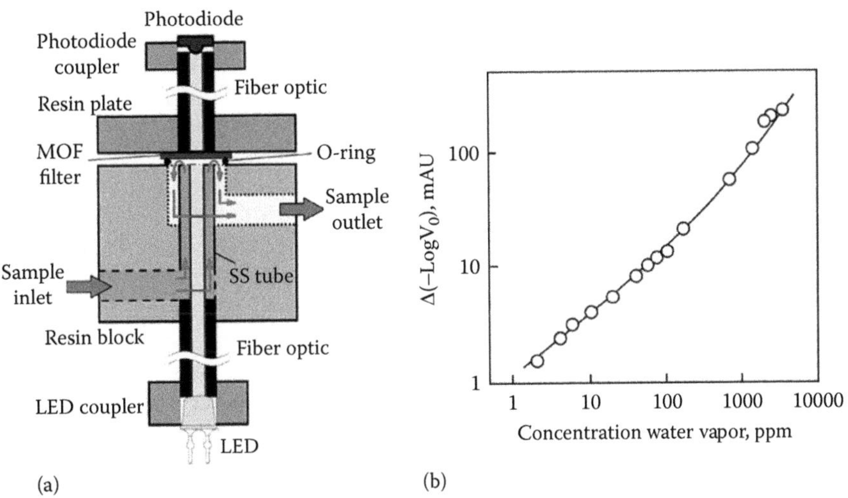

FIGURE 12.13 (a) Schematic of the fiber-optic gas sensing device. Absorbance detection was achieved by passing light through the filter. The light source was a LED (λ_{max} = 468 nm, driven at 20 mA, L-7113QBC-D, http://www.kingbright.com) and the photodetector was a current to voltage convertor integrated miniature PD (TSL 257, http://www.taosinc.com). (b) Responses of MOF-based humidity sensor. (Reprinted from *Anal. Chim. Acta*, 886, Ohira, S.I. et al., A fiber optic sensor with a metal organic framework as a sensing material for trace levels of water in industrial gases, 188–193, Copyright 2015, with permission from Elsevier.)

Bridgeman et al. (2014) believe that liquid composite materials can be also used for humidity monitoring. The sensing material was a liquid composite that comprised a hygroscopic medium for environmental humidity capture and a color indicator that translated the humidity level into a distinct color change. Sodium borohydride was used to form a liquid composite medium, and DenimBlu30 dye was used as a redox indicator. Thus, the liquid composite medium provided a hygroscopic response to the RH, and DenimBlu30 translated the chemical changes into a visual change from yellow to blue. Test results indicated that this new sensing material can detect RH in the range of 5%–100% in an irreversible manner with good reproducibility and high accuracy. At the same time, it was established that required stability was achieved only after aging during nine days. Apparently, slow response also is a characteristic for this material.

12.3 SUMMARY

When analyzing the results concerning colorimetric humidity sensors, it is possible to conclude that the main disadvantages of the most developed sensors of this type are (1) nonlinearity (linearity is possible only in a narrow range of the humidity change); (2) a notable temperature dependence; and (3) slow response and recovery. However, studies using PCs showed that the speed of response of colorimetric humidity sensors can be significantly improved. Apparently, such improvement is associated with the high permeability and small film thickness used in sensors. The same explanation of the fast response of humidity sensors based on methylene blue aggregates formed on nanoporous TiO_2 and Al_2O_3 films give Ishizaki and Katoh (2016). They found that the response began to occur within 10 ms. Ishizaki and Katoh (2016) believed that the response was fast because all the methylene blue molecules attached to the nanoporous semiconductor surface were directly exposed to the environment. They also deduced that the color changes were caused by structural changes of the methylene blue aggregates on the metal oxide surface.

The lack of selectivity is also a significant drawback of many humidity sensors. For example, a blue-to-red color transition in PDAs can be triggered by various stimuli, including VOCs (Park et al. 2014), surfactants (Shimogaki and Matsumoto 2011), ligand–receptor binding (Charych et al. 1993), heat (Ampornpun et al. 2012), and mechanical stress. A similar situation can also be observed for other hydrochromic materials.

REFERENCES

Ampornpun S., Montha S., Tumcharern G., Vchirawongkwin V., Sukwattanasinitt M., Wacharasindhu S. (2012) Odd-even and hydrophobicity effects of diacetylene alkyl chains on thermochromic reversibility of symmetrical and unsymmetrical diyndiamide polydiacetylenes. *Macromolecules* 45, 9038–9045.

Ando M., Kobyashi T., Haruta M. (1996) Humidity-sensitive optical absorption of Co_3O_4 film. *Sens. Actuators B* 32, 157–160.

Ando M., Sato Y., Tamura S., Kobayashi T. (1999) Optical humidity sensitivity of plasma-oxidized nickel oxide films. *Solid State Ionics* 121, 307–311.

Ballantine D.S., Wohltjen H. (1986) Optical waveguide humidity detector. *Anal. Chem.* 58(13), 2883–2885.

Boltinghouse F., Abel K. (1989) Development of an optical relative humidity sensor. Cobalt chloride optical absorbency sensor study. *Anal. Chem.* 61, 1863–1866.

Bridgeman D., Corral J., Quach A., Xian X., Forzani E. (2014) Colorimetric humidity sensor based on liquid composite materials for the monitoring of food and pharmaceuticals. *Langmuir* 30, 10785–10791.

Brook T.E., Taib M.N., Narayanaswamy R. (1997) Extending the range of a fiberoptic relative humidity sensor. *Sens. Actuators B* 38–39, 272–276.

Carmona N., Herrero E., Llopis J., Villegas M.A. (2007) Chemical sol–gel-based sensors for evaluation of environmental humidity. *Sens. Actuators B* 126, 455–460.

Chi H., Liu Y.J., Wang F.K., He C. (2015) Highly sensitive and fast response colorimetric humidity sensors based on graphene oxides film. *ACS Appl. Mater. Interfaces* 7, 19882–19886.

Charych D.H., Nagy J.O., Spevak W., Bednarski M.D. (1993) Direct colorimetric detection of a receptor-ligand interaction by a polymerized bilayer assembly. *Science* 261, 585–588.

Dacres H. and Narayanaswamy R. (2006) Highly sensitive optical humidity probe. *Talanta* 69, 631–636.

Fong A., Hieftje G.M. (1995) Near infrared measurement of relative and absolute humidity through detection of water adsorbed on a silica gel layer. *Anal. Chem.* 67, 1139–1146.

Furukawa H., Gandara F., Zhang Y.-B., Jianf J., Queen W.L., Hudson M.R., Yaghi O.M. (2014) Water adsorption in porous metal–organic frameworks and related materials. *J. Am. Chem. Soc.* 136 (11), 4369–4381.

Hawkeye M.M., Brett M.J. (2011) Optimized colorimetric photonic-crystal humidity sensor fabricated using glancing angle deposition. *Adv. Funct. Mater.* 21, 3652–3658.

Hosokawa H., Mochida T. (2015) Colorimetric humidity and solvent recognition based on a cation-exchange clay mineral incorporating Nickel(II)–Chelate complexes. *Langmuir* 31, 13048–13053.

Hypszer R., Wierzba I.J. (1995) Fiber optic technique for relative humidity sensors. *Proc. SPIE* 3054, 145–150.

Ishizaki R., Katoh R. (2016) Fast-response humidity-sensing films based on methylene blue aggregates formed on nanoporous semiconductor films. *Chem. Phys. Lett.* 652, 36–39.

Joannopoulos J.D., Jonson S.G., Winn J.N., Meade R.D. (2008) *Photonic Crystals. Molding the Flow of Light.* Princeton University Press, Princeton, NJ.

Kang Y., Walish J.J., Gorishnyy T., Thomas E.L. (2007) Broad-wavelength-range chemically tunable block-copolymer photonic gels. *Nat. Mater.* 6, 957–960.

Kang Y. (2012) Colorimetric humidity sensors based on block copolymer photonic gels. *Macromol. Res.* 20(12), 1223–1225.

Kharaz A., Jones B.E. (1995) A distributed optical-fiber sensing system for multi-point humidity measurement. *Sens. Actuators B* 46–47, 491–493.

Kim E., Kim S., Jo G., Kim S., Park M. (2012) Colorimetric and resistive polymer electrolyte thin films for real-time humidity sensors. *ACS Appl. Mater. Interfaces* 4, 5179–5187.

Kumari M., Ding B., Blaikie R. (2016) Enhanced resonant absorption in dye-doped polymer thin-film cavities for water vapour sensing. *Sens. Actuators B* 231, 88–94.

Li G., Xu W., Huang X. (2015) A simple fiber optic humidity sensor based on water-absorption characteristic of CAB. *Proc. SPIE* 9446, 944617.

Lu T., Zhu S., Chen Z., Wang W., Zhang W., Zhang D. (2016) Hierarchical photonic structured stimuli-responsive materials as high-performance colorimetric sensors. *Nanoscale* 8, 10316.

Matsushima R., Atsushi O., Fujimoto S. (2000) Thermochromic properties of flavylium salts in agar-gel matrix. *Chem. Lett.* 6, 590–591.

Matsushima R., Ogiue A., Kohno Y. (2002) Humidity-sensitive color changes of ionic dyes in solid thin film of sugar gel. *Chem. Lett.* 4, 436–437.

Mills A., Grosshans P., Hazafy D. (2010) A novel reversible relative-humidity indicator ink based on methylene blue and urea. *Analyst* 135, 33–35.

Mills A., Hawthorne D., Burns L., Hazafy D. (2017) Novel temperature-activated humidity-sensitive optical sensor. *Sens. Actuators B* 240, 1009–1015.

Mohr G.J., Spichiger-Keller U.E. (1998) Development of an optical membrane for humidity. *Microchim. Acta* 130, 29–34.

Ohira S.I., Miki Y., Matsuzaki T., Nakamura N., Sato Y.K., Hirose Y., Toda K. (2015) A fiber optic sensor with a metal organic framework as a sensing material for trace levels of water in industrial gases. *Anal. Chim. Acta* 886, 188–193.

Otsuki S., Adachi K. (1993) Reversible opacity change of hydroxypropyl cellulose films as a basis for optical humidity measurements. *Anal. Sci.* 9, 299–301.

Otsuki S., Adachi K., Taguchi T. (1998) A novel fiber-optic gas sensing arrangement based on an air gap design and an application to optical detection of humidity. *Anal. Sci.* 14, 633–635.

Ozin G., Arsenault A.C. (2008) P-ink and elast-ink from lab to market. *Mater. Today* 11, 44–51.

Park D.-H., Hong J., Park I.S., Lee C.W., Kim J.-M. (2014) A colorimetric hydrocarbon sensor employing a swelling-induced mechanochromic polydiacetylene. *Adv. Funct. Mater.* 24, 5186–5193.

Park D.-H., Park B.J., Kim J.-M. (2016) Hydrochromic approaches to mapping human sweat pores. *Acc. Chem. Res.* 49, 1211–1222.

Puyol M., del Valle M., Garces I., Villuendas F., Dominguez C., Alonso J. (1999) Integrated waveguide absorbance optode for chemical sensing. *Anal. Chem.* 71(22), 5037–5044.

Sadaoka Y., Matsuguch M., Sakai Y. (1991) Optical-fibre and quartz oscillator type gas sensors: Humidity detection by Nafion^R film with crystal violet and related compounds. *Sens. Actuators A* 25–27, 489–492.

Sadaoka Y., Sakai Y., Murata Y.-U. (1992a) Optical humidity and ammonia gas sensors using Reichardt's dye-polymer composites. *Talanta* 39(12), 1675–1679.

Sadaoka Y., Matsuguch M., Sakai Y., Murata Y.-U. (1992b) Optical humidity sensing characteristics of Nafion–dyes composite thin films. *Sens. Actuators B* 7, 443–446.

Sharkany J., Korposh S.O., Batori-Tarci Z.I., Trikur I.I., and Ramsden J.J. (2005) Bacteriorhodopsin-based biochromic films for chemical sensors. *Sens. Actuators B* 107, 77–81.

Shi J., Hsiao V.K.S., Walker T.R., Huang T.J. (2008) Humidity sensing based on nanoporous polymeric photonic crystals. *Sens. Actuators B* 129, 391–396.

Shimogaki T., Matsumoto A. (2011) Structural and chromatic changes of host polydiacetylene crystals during interaction with guest alkylamines. *Macromolecules* 44, 3323–3327.

Skrdla P.J., Armstrong N.R., Saavedra S.S. (2002) Starch-iodine films respond to water vapor. *Anal. Chim. Acta* 455(1), 49–52.

Tao S., Winstead C.B., Jindal R., Singh J.R. (2004) Optical-fiber sensor using tailored porous sol-gel fiber core. *IEEE Sens. J.* 4, 322–328.

Tian E., Wang J., Zheng Y., Song Y., Jiang L., Zhu D. (2008) Colorful humidity sensitive photonic crystal hydrogel. *J. Mater. Chem.* 18, 1116–1122.

Tsigara A., Mountrichas G., Gatsouli K., Nichelatti A., Pispas S., Madamopoulos N., Vainos N.A., Du H.L., Roubani-Kalantzopoulou F. (2007) Hybrid polymer/cobalt chloride humidity sensors based on optical diffraction. *Sens. Actuators B* 120, 481–486.

Wang K., Seiler K., Haug J.-P., Lehmann B., West S., Hartman K., Simon W. (1991) Hydration of trifluoroacetophenones as the basis for an optical humidity sensor. *Anal. Chem.* 63, 970–974.

Wang Z., Zhang J., Xie J., Li C., Li Y., Liang S., Tian Z., Wang T., Zhang H., Li H. (2010) Bioinspired water-vapor-responsive organic/inorganic hybrid one-dimensional photonic crystals with tunable full-color stop band. *Adv. Funct. Mater.* 20, 3784–3790.

Wang H., Zhang K.-Q. (2013) Photonic crystal structures with tunable structure color as colorimetric sensors. *Sensors* 13, 4192–4213.

Yadav B.C., Pandey N.K., Srivastava A.K., Sharma P. (2007) Optical humidity sensors based on titania films fabricated by sol–gel and thermal evaporation methods. *Meas. Sci. Technol.* 18, 260–264.

Yu Y., Zhang X.-M., Ma J.-P., Liu Q.-K., Wang P., Dong Y.-B. (2014) Cu(I)-MOF: Naked-eye colorimetric sensor for humidity and formaldehyde in single-crystal-to-single-crystal fashion. *Chem. Commun.* 50, 1444–1446.

Zhao Y., Xie Z., Gu H., Zhu C., Gu Z. (2012) Bio-inspired variable structural color materials. *Chem. Soc. Rev.* 41, 3297–3317.

Zhou Q., Shahriari M.R., Kritz D., Sigel G.H. Jr. (1988) Porous fiber-optic sensor for high-sensitivity humidity measurements. *Anal. Chem.* 60, 2317–2320.

13 Moisture Indicators

In the truest sense of the word, moisture indicators are not the sensors. However, their concept of operation is based on the same physical and chemical principles that have been discussed in Chapter 12. Therefore, we decided to include a description of moisture indicators in the current chapter.

13.1 PRINCIPLES AND MATERIALS

Sometimes it is desirable to detect only a change in the amount of moisture in the sample, whereas the exact concentration is of only secondary importance. Such may be the case on the outlet of drying beds where the moisture breakthrough will indicate the need for regeneration. In other cases, for example, we must be sure that the relative humidity (RH) does not exceed 30%–35%. RH in the range of 30%–35% was the concern because this is when corrosion can begin.

Research has shown that indicated problem can be solved by using certain salt crystals, which change color with the appearance of moisture. It has been found that the use of this effect is the most economical method for detecting changes of moisture level (Blinn 1965). Experiments have shown that cobalt chloride, discussed in the previous Chapter 12, is the material most suitable for such applications (Katzin and Ferraro 1952). One should note that *chemical compounds of the element cobalt have been used for thousands of years as coloring agents in paint, ink, ceramics, and glass.* The color change is a result of chemical reactions taking place with the participation of cobalt chloride and water. As the humidity increases, and water is absorbed by $CoCl_2$, the crystal structure rearranges itself to make room for water molecules. The *hydration* reaction may be represented by the chemical reaction in Equation 13.1. Two water molecules surround each cobalt atom, forming the *dehydrate*. Cobalt chloride dehydrate is purple.

$$CoCl_2\ (\text{blue}) + 2H_2O \rightarrow CoCl_2 \cdot 2H_2O\ (\text{purple}) \quad (13.1)$$

Chemists use the raised dot "·" symbol before the H_2O to indicate the number of water molecules that have become incorporated into a compound at the atomic level. As the humidity increases further, the crystal structure again changes, this time rearranging itself to let four more water molecules in to surround each cobalt atom, forming the *hexahydrate*:

$$CoCl_2 \cdot 2H_2O\ (\text{purple}) + 4H_2O \rightarrow CoCl_2 \cdot 6H_2O\ (\text{pink}) \quad (13.2)$$

In reality, cobalt chloride has six states of hydration and exhibits progressive color changes with corresponding changes in hydration state. Several reports indicate that with excess exposure to moisture cobalt chloride can exist in even higher states of hydration (Russell and Fletcher 1985), although no further color change can be noted. As the initially anhydrous cobalt salt bonds with each water molecule, it exhibits a color change from blue to a fully hydrated pink (Katzin and Ferraro 1952). Heating the hydrated forms of cobalt chloride reverses the above-mentioned reactions, returning cobalt chloride to the blue, water-free, or *anhydrous*, state. Water is *liberated* in these reactions, known as *dehydration* reactions. Thus, indicated property allows creation of a color-change element, which turns from blue to lavender to pink as the humidity increases, and turns back to blue when the humidity decreases. It was established that this characteristic enables these elements to be reused as long as they have not been exposed to humidity in excess of 90% for periods of 36 h or longer, which tend to wash out the element.

It was found that on the basis of cobalt chloride nonreversible humidity indicators can be made. For these purposes, ammonia salts are added in the element. In the presence of vapors of ammonia, the elements lose the reversibility of the color. Such nonreversible elements or maximum humidity indicator cards are needed to prevent the fact when certain humidity levels have been exceeded, or when the desiccant may need to be recharged or replaced. For example, these cards can indicate the highest level of humidity experienced by cargo during its voyage, regardless of current (potentially lower) humidity levels. Maximum humidity indicator cards provide a clear, unmistakable means of determining if goods have been exposed to damaging humidity levels during their journey. If the card indicates high levels of humidity, users know how to check their products to avoid a possible damage or modify their packaging regimen accordingly.

Humidity indicators designed on the basis of above-mentioned principle are a special kind of hygrometer (humidity indicator strips) that use the blotting paper impregnated with cobalt salts, blue cobalt(II)chloride or cobalt thiocyanate (Balköse et al. 1998, 1999). These indicators are inexpensive, easily stored, not easily damaged, and may be used in many variants. Humidity indications can be determined by comparison with a color guide printed next to indicating color spots for easy reading. As a rule, a series of patches are labeled with RH, usually in 10% increments, from 10% to 80% RH, as well as various combinations of these humidities. Emerson reported that Emerson's moisture indicator can detect moisture at 3% RH. Usually the humidity-sensitive elements have long-term calibration and accuracy to within ±5% when read at 24°C. Indicators have a temperature-caused error of 2.5% per 5°C deviation from a base of 24°C (Spomer and Tibbitts 1997). At other temperatures a small correction factor for each 5.5°C higher or lower than 24°C must be taken into consideration. For example, at high temperatures the element will indicate a lower humidity than is actually the case, and conversely, a higher humidity will be indicated at low temperatures. As indicated above, these elements are water soluble and should not be placed in direct contact with water or steam or exposed to extremely high humidity for protracted periods of time. If these indicators are used in a moist environment, they can become inaccurate. They are damaged by excessive humidity or condensation, which dilutes the spots and results in nonuniform chemical distribution and inconsistent color response (Spomer and Tibbitts 1997). Strong solvents such as ammonia fumes, which cause the bleaching of the color change dye and lose color reversibility, should not be present in the atmosphere.

Other substrates such as cellulose, cellulose acetate, polyvinyl pyrollidon, wool, calcium sulphate, silica gel, zeolites, and alumina can also be used. Usually the substrate type does not affect the color change: the color is blue at low RH levels and pink at high RH levels. But wool impregnated with cobaltous chloride is blue in dry form and converts to yellow on moisture adsorption (Balköse and Ülkü 2007). It was established that decreasing the amount of cobalt chloride in the matrix decreased both hysteresis effects and the effective RH range over which the salt/substrate combination could be used for RH measurement. It was also found that acetylation of the cellulose extended the effective RH range, increased reproducibility, and decreased hysteresis effects (Boltinghouse and Abel 1989). Moreover, optimizing the type of the substrate and the type and concentration of the cobalt salt, it is possible to produce indicators, which show a sharp color change in a small humidity range (Balköse et al. 1998).

13.2 MARKET OF MOISTURE INDICATORS

Currently various companies (Table 13.1) offer a big variety of humidity indicators. Humidity indicators can be manufactured as strips, cards, and in the form of elements embedded in the gas main. In particular, the crystals and humidity-sensitive elements, which change the color under the influence of water, can be packed into a small chamber with a glass window for observation of the color. A sample side stream is piped through the chamber, and increase of moisture in the sample is indicated to the operator by a color change.

At that both types of elements, reversible, and nonreversible, may be installed in the same plug-type housing, which is leak-proof and will withstand shock and vibration. Examples of humidity indicator cards and components from different manufacturers are shown in Figures 13.1 and 13.2. Humidity indicators are often incorporated into plug-type desiccators for small enclosures. The plugs can be quickly installed with common hand tools, and are suitable for almost all applications. Of course, these cards are not precision indicators. In addition, these cards indicate only the RH. However, they are very good at telling you the RH in a container. For example, if all of the dots are deep blue on a card which is a 10%–60% card then you can be assured that the humidity in that container is below 10%. Humidity indicator strips or cards are also designed for mounting in moisture-sensitive electronic or electro-optic equipment, or any other humidity controlled enclosure.

TABLE 13.1
Examples of Manufacturers of Humidity Indicators

Company	Internet Site
Desiccare, Inc.	http://www.desiccare.com
Emerson Climate Technologies	http://www.emersonclimate.eu
Castel S.r.l.	http://www.castel.it
Parker Hannifin Corporation. Sporlan Division	http://www.sporlan.com
Henry Technologies Ltd.	http://www.henrytech.com
Texas Instruments Incorporated	http://www.ti.com
AGM Container Controls, Inc.	http://www.agmcontainer.com
IGLOO	http://www.igloo-refrigerazione.com
IMPAK Corporation. Sorbent Systems	http://www.sorbentsystems.com
Brownell Ltd.	http://www.brownell.co.uk
Clariant International Ltd.	http://www.clariant.com

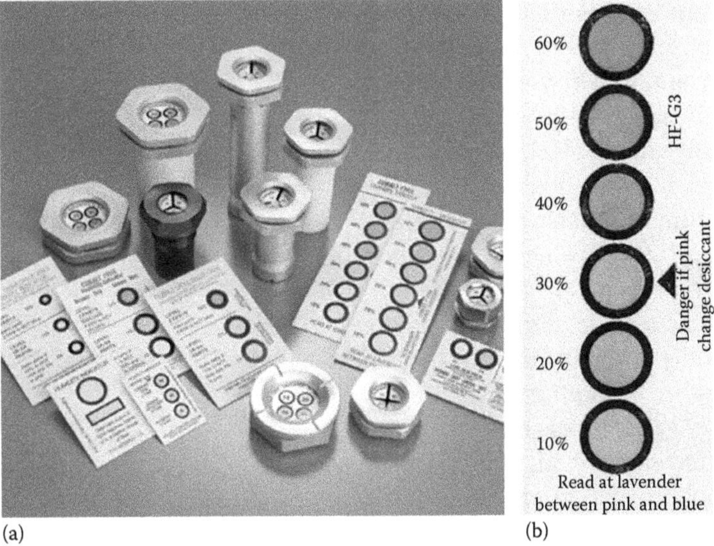

FIGURE 13.1 (a) Humidity indicators designed by Clariant International Ltd. (From http://www.clariant.com/.) (b) Reversible humidity indicator cards fabricated by Brownell Ltd. In this figure the bottom image shows a new card that is directly out of a sealed bag from the supplier. A lavender color usually indicates the relative humidity of the environment the card is placed within. (From http://www.brownell.co.uk/.)

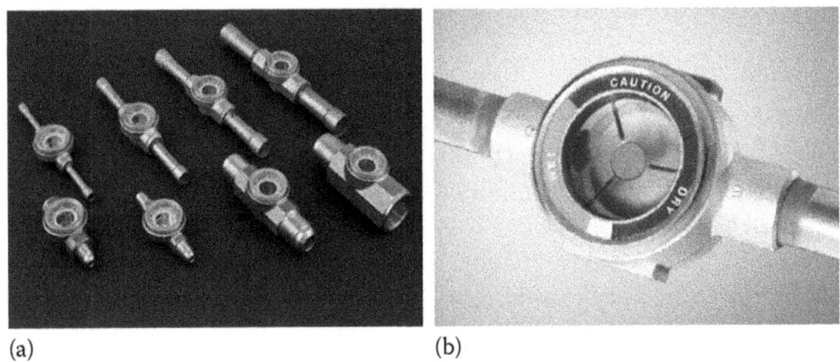

FIGURE 13.2 Indicators with moisture sensitive element designed by different companies to monitor the condition of the refrigerant: (a) from http://www.igloo-refrigerazione.com and (b) from http://www.emersonclimate.eu.

For example, humidity indicator cards are present on many small electronic devices, ranging from cellular phones to laptop computers, for the purpose of alerting the manufacturer that the device has been exposed to high levels of moisture. In many cases, this voids or changes the terms of warranty coverage for the device.

One should note that in addition to cobalt salt indicators other salts can be also applied in humidity indicators. The need for such developments was due to the fact that in 1998, the European Community (EC) issued a directive which classified the items containing cobalt (II) chloride from 0.01% to 1% w/w as T (Toxic). Studies have shown that there is a sufficiently large set of materials that can also change color when exposed to moisture (Moreton 2002; Matsushima et al. 2000, 2002, 2003; Carmona et al. 2006; Fueda et al. 2007; Mackenzie and O'Leary 2008; Mills et al. 2010, 2017; Zhang et al. 2013). For example, Matsushima et al. (2000, 2002, 2003) presented a series of reports on reversible humidity-sensitive indicator films, formed from a combination of a dye with a sugar-based hydrogel (such as agarose or k-carrageenan), which undergo a reversible color change due to the formation of an aggregated form of the dye at high humidities, and its subsequent breakdown at low humidities. In these RH indicators, the color change is attributed to the humidity-promoted aggregation of the dye. In 2010 and 2017, Mills et al. (2010, 2017) published details of a humidity indicator based on the thiazine dye, (esp. methylene blue, MB), encapsulated in a film of hydroxyethyl cellulose (HEC) containing urea. In this system, the urea was present in a quantity that was roughly 20 times that of the dye. Under both dry (i.e., RH = 0%) and ambient conditions (RH~60%, at 23°C), the film was bright purple, but turned a deep blue after exposure to an ambient atmosphere with RH ≥ 85%. This discovery is unusual in such that it yields a sharp evident reversible color change at high RH, whereas this has been a problem in other investigations. Mills et al. (2017) believe that such a film is suitable for use as an indicator in the packaging of goods which cannot tolerate highly humid environments. Copper(II) chloride and potassium lead iodide can also be used in humidity indicators. Yellow lead iodide is obtained from potassium lead iodide upon moisture adsorption. Experiment has shown that *polymer–dye* systems also changed their color with air humidity (Kunzelman et al. 2007). Some studies have shown that polymer-based photonic crystals were also promising for these applications (Shi et al. 2008). The latter are interesting in view of the ways that ordered materials can interact with light to produce a variety of optical effects. A high sensitivity optical humidity probe was developed by using Nafion-crystal violet films. Reversible change in absorbance of the films at 650 nm with RH was linear in the 0% to 1% range (Dacres and Narayanaswamy 2006).

REFERENCES

Balköse D., Ulutan S., Özkan F.C., Çelebi S., Ülkü S. (1998) Dynamics of water vapor adsorption on humidity-indicating silica gel. *Appl. Surf. Sci.* 134, 39–46.

Balköse D., Köktürk U., Yilmaz H. (1999) A study of cobaltous chloride dispersion on the surface of the silica gel. *Appl. Surf. Sci.* 147, 77–84.

Balköse D., Ülkü S. (2007) Water vapour adsorption and desorption on cobaltous chloride impregnated wool in packed column. In: Pethrick R.A., Zaikov G.E., and Horak D. (Eds.), *Polymers and Composites: Synthesis, Properties, and Applications.* Nova Science Publishers, New York, pp. 35–40.

Blinn B.G. (1965) Properties and uses of color change humidity indicators. In: Wexler A., Ruskin R.E. (Eds.), *Humidity and Moisture*, vol. 1. Principles and Methods of Measuring Humidity in Gases, Reinhold Publishing Cooperation, New York, pp. 602–605.

Boltinghouse F., Abel K. (1989) Development of an optical relative humidity sensor. Cobalt chloride optical absorbency sensor study. *Anal. Chem.* 61, 1863–1866.

Carmona N., Herrero E., Llopis J., Villegas M.A. (2006) Chemical sol-gel-based sensors for evaluation of environmental humidity. *Sens. Actuators B* 126(2), 455–460.

Dacres H., Narayanaswamy H. (2006) Highly sensitive optical humidity probe. *Talanta* 69, 631–636.

Fueda Y., Matsumoto J., Shiragami T., Nobuhara K., Yasuda M. (2007) Porphyrin/$MgCl_2$/silica gel composite as a cobalt-free humidity indicator. *Chem. Lett.* 36(10), 1246–1247.

Katzin L.I., Ferraro J.R. (1952) The system cobaltous chloride—water—acetone at 25°. *J. Am. Chem. Soc.* 74, 2752–2754.

Kunzelman J., Crenshaw B.R., Weder C. (2007) Self-assembly of chromogenic dyes—a new mechanism for humidity sensors. *J. Mater. Chem.* 17, 2989–2991.

Mackenzie K., O'Leary B. (2008) Inorganic polymers (geopolymers) containing acid-base indicators as possible colour-change humidity indicators. *Mater. Lett.* 63(2), 230–232.

Matsushima R., Nishimura N., Goto K., Kohno Y. (2003) Vapochromism of ionic dyes in thin films of sugar gels. *Bull. Chem. Soc. Jpn.* 76, 1279–1283.

Matsushima R., Atsushi O., Fujimoto S. (2000) Thermochromic properties of flavylium salts in agar-gel matrix. *Chem. Lett.* 6, 590–591.

Matsushima R., Ogiue A., Kohno Y. (2002) Humidity-sensitive color changes of ionicdyes in solid thin film of sugar gel. *Chem. Lett.* 4, 436–437.

Mills A., Grosshans P., Hazafy D. (2010) A novel reversible relative-humidity indicator ink based on methylene blue and urea. *Analyst* 135, 33–35.

Mills A., Hawthorne D., Burns L., Hazafy D. (2017) Novel temperature-activated humidity-sensitive optical sensor. *Sens. Actuators B* 240, 1009–1015.

Moreton S. (2002) Silica gel impregnated with iron(III) salts: A safe humidity indicator. *Mater. Res. Innov.* 5(5), 226–229.

Russell A.P., Fletcher K.S. (1985) Optical sensor for the determination of moisture. *Anal. Chim. Acta* 170, 209–216.

Shi J., Hsiao V.K.S., Walker T.R., Huang T.J. (2008) Humidity sensing based on nanoporous polymeric photonic crystals. *Sens. Actuators B* 129, 391–396.

Spomer L.A., Tibbitts T.W. (1997) Humidity. In: Langhans R.W., Tibbitts T.W. (Eds.), *Plant Growth Chamber Handbook*. Iowa State University, Ames, IA, pp. 43–64.

Zhang Y.P., Chodavarapu V.P., Kirk A.G., Andrews M.P. (2013) Structured color humidity indicator from reversible pitch tuning in self-assembled nanocrystalline cellulose films. *Sens. Actuators B* 176, 692–697.

14 Refractometry-Based Optical Humidity Sensors

14.1 PRINCIPLES OF OPERATION

As noted previously, the sensing mechanism of refractometric optical humidity sensors is based on the change of effective refractive index of the porous coating formed on the surface of the fiber or waveguide. In *refractometric optical humidity sensors,* due to interaction with water vapor adsorbed on the core or on the cladding of the fiber or planar waveguide, a perturbation of the corresponding refraction index (RI) is obtained. As a result, a modification in the intensity of the light transported by the fiber or planar waveguide takes place and the adsorbed species can thus be detected. Thus, the principle of these sensors rely on the use of the humidity-sensitive materials to generate secondary effects such as RI change or strain on the sensing fiber that result in the shift of output spectra or change in the intensity. The change of the RI and polarization in an optical humidity sensor are generally explained by the adsorption water and also by the capillary water condensation within the pores of a humidity-sensitive material (Section 10.3).

14.1.1 Evanescent Spectroscopy

There are different approaches to design of refractometric optical humidity sensors. As a rule these sensors are based on the principles of evanescent spectroscopy. The principle of measurement is noted in Figure 14.1. The evanescent field is the basis of such sensors (Gründler 2006). The evanescent spectroscopic sensor design uses the evanescent field associated with the propagation of light in optical waveguide. This effect occurs when light is propagated down an optical waveguide. The geometric ray approach shows no distinction between a dielectric and metallic surfaces. In a metallic conductor, the tangential electric field component is zero, whereas there is no such restriction in the tangential electric field at the dielectric surface. This makes the boundary conditions at the two surfaces different. The boundary condition results in the extension of the electric field across the boundary into the medium of lower index of refraction and shift in the phase of the reflected wave. So, an evanescent wave is a wave field outside the dielectric boundary. The amplitude of the evanescent wave decays (falls) exponentially with distance from the boundary and can be presented as

$$E(x) = E_0 \exp\left(\frac{x}{d_p}\right) \tag{14.1}$$

where:

E_0 is the initial value evanescent field
$E(x)$ is the intensity of the exponentially decaying evanescent field with the distance x perpendicular to the interface
d_p is the penetration depth of the evanescent field

This means, that in the case of planar optical waveguides, the electromagnetic fields do not abruptly reduce to zero at the interface between core and cladding. The light also extends to the cladding of a lower RI into which it cannot penetrate according to the geometric ray approach. So, the sensitivity of evanescent wave-based sensors depends on the optical power transferred into the evanescent field and also by the penetration depth of these waves into the sensing cladding. The penetration depth of the evanescent field in the cladding is given by the Equation 14.2 (Wise and Wingard 1991).

$$d_p = \frac{\lambda}{2\pi\sqrt{\left(n_1^2 \sin^2\theta - n_2^2\right)}} \tag{14.2}$$

where:

λ is the wavelength of light
θ is the angle of incidence
n_1 is the RI of the waveguide
n_2 is the RI of the cladding

This evanescent zone, which is usually limited to less than 100 nm, can be used to detect the presence of optical indicators or changes in RI at the surface of the unclad waveguide (Murray 1989). As it will be shown in the following Sections of this Chapter, the evanescent field can be considered as an advantage for many sensing applications. The increase in evanescent field forms the basis for increased sensitivity. Now, if a thin-film of

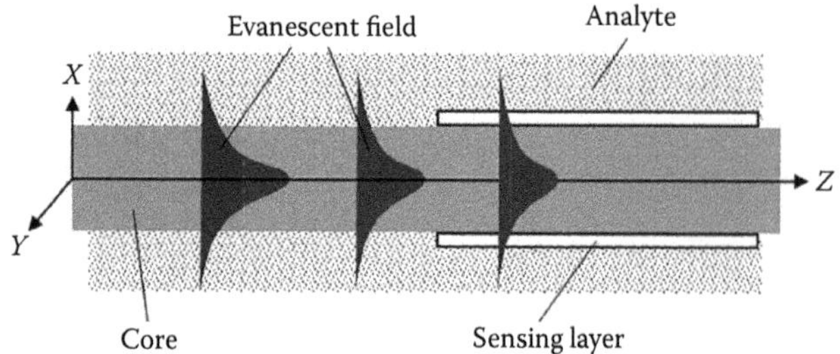

FIGURE 14.1 Schematic of the evanescent field energy distribution along the sensing fiber. In many cases, evanescent wave interactions with the analyte form the basis for the sensor. In some cases, a sensing layer may be employed to facilitate transduction by imparting colorimetric or fluorometric properties to the sensor.

a certain material is deposited onto the silica substrate, its behavior will be influenced by the evanescent field of the waveguide. Indeed, part of the energy propagated in these modes will be transferred to the thin film. Under these conditions, some wavelengths can suffer strong attenuation, whereas the opposite is true for the rest. This leads to a resonant phenomenon called electromagnetic resonance (EMR). Most of sensing technologies use just these phenomena to increase the sensitivity of sensors developed. In particular, the humidity sensors, developed on the basis of the surface Plasmon resonance (SPR) and lossy-mode resonance (LMR) will be considered in Chapters 19 and 20.

The properties of light in the waveguide are determined by the number of modes N. As it is known, light propagates as a combination of propagation modes. In a multimode waveguide, there is a one-to-one relationship between modes and angles of incidence. The number of modes is directly related to a dimensionless parameter known as the V number. In a uniform-diameter single-mode waveguide, the light only propagates in one mode. However, if V changes along the fiber due to geometry or local RI changes, the single mode waveguide becomes multimode, and the coupling of modes takes place so that transmission by many modes occur, which increases the evanescent portion. This means that each of these propagation modes generates its own evanescent field. Consequently, the evanescent field of the waveguide is actually a superposition of these propagation modes, each one contributing with its own evanescent field in a very short distance out of the waveguide.

The evanescent spectroscopic sensor undoubtedly can be realized on the base of the fiber. The side-of-fiber configuration discussed before should be used for this purpose. According to this mechanism, if the modified cladding has a lower RI than the core, then the total reflection condition is met. In this case, the sensor response is governed by the intensity modulation caused by light absorption of the evanescent wave, which is guided through the cladding; this interaction results in the attenuation of the guided light in the fiber core (Yuan and El-Sherif 2003). On the other hand, if the modified cladding has a higher RI than the core, part of the optical power is refracted into the cladding, and another part is reflected back into the core (Khalil et al. 2004). Both the optical power reflected in the core and the light passing through the cladding depend on the optical properties of the cladding, which change in presence of the analyte to be detected. The total optical power loss is characterized mainly by the light absorption of the chemical dye that replaces the cladding, the number of interactions between the core and the cladding, the diameter of the core, and the numeric aperture of the fiber, among other parameters. However, there is a trade-off between these parameters to achieve the desired result (Messica et al. 1996).

As for the distinctions, characteristic for fiber-optic technology, then, in contrast to planar waveguide technology, it gives possibility to enhance the interaction of the evanescent field with the environment. For example, several investigators have attempted to increase the penetration depth of the evanescent field and facilitate mode coupling by bending (Khijwania and Gupta 1999), tapering (Mackenzie and Payne 1990; Mignani et al. 1998), altering the light launching angle (Ahmad and Hench 2005), and increasing the wavelength (Moar et al. 1999). It was found that tapering not only exposed the evanescent field to the surroundings but also increased the evanescent field magnitude and penetration depth. Thus, tapering provides strong interaction with the surrounding area allowing efficient excitation of the species in the sensing region by the propagating light and possible coupling back via evanescent field. Tapering can be performed by removing the cladding and then tapering the core, or keeping both the core and cladding in place and tapering the entire fiber. Figure 14.2 illustrates the profile of a tapered fiber where both the core and the cladding were tapered. Other approaches have been discussed by Leung et al. (2007). Experiment has shown that tapered fiber-based sensors had suitable sensitivity, but had limited application as distributed sensors; it is only possible to build a quasi-distributed system on the tapered fiber basis. In addition, the tapered regions are delicate and required special packaging.

It is also necessary to take into account that the evanescent fields due to interaction with the analyte can create fluorescence in the region outside the core (evanescent excitation) or couple fluorescence from the surrounding medium into the fiber core (evanescent collection). For

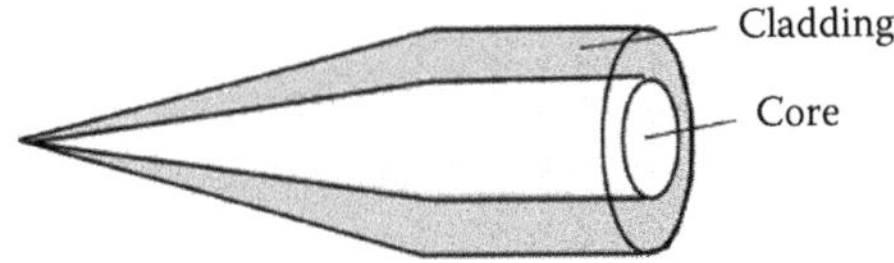

FIGURE 14.2 The profile of a tapered fiber.

example, Sadaoka et al. (1993) have shown that such situation could be realized in sensors when using calcein, (3,3′-bis[*N*,*N*-di(carboxymethyl)aminomethyl]fluorescein), as a humidity-sensitive indicator. Calcein has a high fluorescence quantum yield and a humidity-sensitive absorption band in the wavelength range 400–600 nm. These effects, of course, influence the light in the optical fiber and can be used for humidity sensors design. Sensors of such type will be discussed latter in Chapter 15.

14.1.2 A Bend Loss-Based Humidity Sensor

With regard to the practical implementation of refractometric optical humidity sensors, then the use of a simple macrobent optical fiber coated with a hygroscopic material is the simplest and affordable method of such sensors manufacture. This possibility was first demonstrated by Bownass et al. (1997). A bend loss-based humidity sensor is attractive because the fabrication of the sensor head does not require special treatment of the fiber itself. The example of such fiber refractometer is shown in Figure 14.3a. In this design, the core of the optical fiber is exposed and the fiber is bent in a U shape. When an uncoated fiber is bent, the effective RI profile is being changed (Taylor 1984; Love and Durniak 2007; Yao et al. 2009). The light propagating on the outside of the bend has further to travel

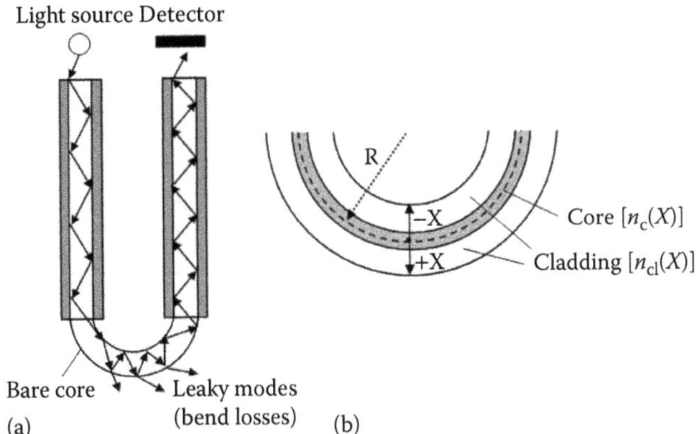

FIGURE 14.3 (a) Schematic showing of an optical fiber refractometer sensor design. (b) Schematic diagram of a fiber macrobend. (From Korotcenkov, G. et al., Optical and fiber optic chemical sensors, in *Chemical Sensors*, Vol. 5: Electrochemical and Optical Sensors, Korotcenkov, G. (Ed.), Momentum Press, New York, 2011. With permission.)

and this effect can be modeled as a tilt applied to the RI profile of the fiber (Taylor 1984), given as

$$n(x) = n_0(x)\left[1+(1+\gamma)x/R\right] \tag{14.3}$$

where:

$n_0(x)$ is the RI profile when the fiber is straight
R is the bend radius
x is a transverse coordinate along a line joining the center of curvature and the center of the fiber, with the origin in the center of the fiber and with x increasing in a positive fashion toward the fiber outer surface (Figure 14.3b)

The factor $\gamma = -0.22$ for silica fiber (Taylor 1984) accounts for the strain-optic effect. With increasing bend curvature the RI profile of the bend fiber becomes more tilted and the effective index for the cladding modes increases, reaching a maximum at the outer edge. At some critical bend radius, the effective index of the cladding modes equals the effective index for the core-guided mode. If there is a sufficient overlap between the core mode and a cladding mode at this point, the core mode and a cladding mode will be coupled together. At tighter bends the core mode will couple to other cladding modes. In the straight portion of the fiber following the bend region, the conventional fiber buffer coating suppresses the cladding modes so that the power coupled to the cladding mode is absorbed and is not recoupled to the core mode. Therefore, at certain bend radii and wavelengths, there are dips in the transmission spectrum of the macrobend fiber that are a function of the degree of coupling of the fundamental mode to the cladding modes (Mathew et al. 2009, 2011, 2012). The phase-matching condition for mode coupling also depends on the ambient RI at the bend. If the surrounding RI of the bent fiber is changed, this will lead to a change in the coupling conditions and will result in a shift in the resonant wavelength. To utilize such a bent fiber as a humidity sensor, the buffer-stripped fiber bend can be coated with a hygroscopic material, whose RI changes with respect to the ambient humidity. The presence of the water vapor in the vicinity of the bend causes a change in the RI of the surrounding, which changes the bend losses through the optical fiber causing a change in the transmission efficiency through the optical fiber.

In practice, the simplest mechanical configuration for such a sensor is a simple fiber bend where the number of turns is less than one. Such configuration was used by Bownass et al. (1998). They have shown that because of the low-bend sensitivity of standard fiber such as SMF28 (http://www.corning.com), a sensor fabricated from such

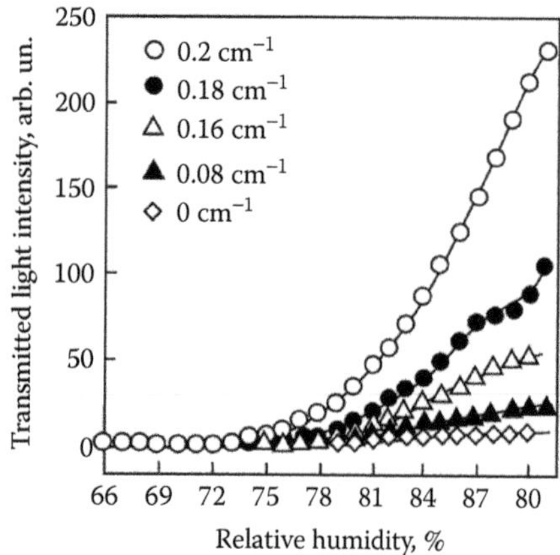

FIGURE 14.4 The changes in the transmitted light intensity with respect to the humidity for a variety of curvatures. POF was coated by humidity-sensitive PVP–PNDF layer. (Data extracted from Morisawa, M. et al., Improvement of POF humidity sensor with swelling polymer cladding via bending. In: *Proceedings of IEEE Sensors*, Orlando, FL, 16597073, October 30–November 3, 2016.)

a fiber should have very low values of bend radius ~6 mm to achieve a reasonable sensitivity, increasing the risk of stress-induced fiber breakage. To improve sensitivity and increase the bend radius so as to reduce the risk of fiber breakage, a high bend loss fiber such as 1060XP (http://www.nufern.com) is much more suitable.

The results presented in Figure 14.4 show how strong is the influence of the bending of fibers on the sensitivity to humidity. Morisawa et al. (2016) have found that the sensitivity of the bent plastic optical fiber (POF)-based sensor with a curvature of 0.20 cm^{-1} was 30 times greater than that of the straight POF sensor.

Mathew et al. (2012) have noted that in developing the fiber bend-based humidity sensor it should be also borne in mind that to achieve a linear wide range of operation, apart from selecting a material with a linear RI change with respect to humidity, it is also important to select a bend radius such that there is no resonant dip in the transmission spectrum. Even though the RI of the polymer can be greater than the RI of the fiber cladding, the effective RI of the coating will be less because its thickness is very small (~μm), so at a certain bend radius and operating wavelengths the power coupling from the fundamental core mode to the cladding mode will be high (resonant mode coupling at phase matched condition) (Mathew et al. 2011). Therefore, to achieve a linear spectral response in required spectral range it is necessary to suitably select some bend radii for which there are no resonant dips in the spectral range studied. For example, Mathew et al. (2012) have found that for the agarose-coated fiber bend (ACFB) sensors operated in the range of 1500–1600 nm, there were no resonant dips in the transmission spectrum of the ACFB with bend radii circa 13.2 and 16.7 mm.

With regard to the limitations of the fiber bend-based humidity sensor, they can be formulated as follows (Mathew 2013). The optical transmission of a fiber bend depends on the bend radius and the wavelength of the propagating light. In practice, there exists a difficulty in reproducing the exact bend radius of the fiber bend. Hence, the fabrication repeatability of the humidity sensor based on the fiber bend is also difficult. Furthermore, the mechanical stability of the fiber bend-based RH sensor is poor and the airflow or a strong acoustic noise or some displacement to the fiber bend could result in erroneous measurement of RH. The optical transmission level of the fiber bend itself is temperature dependent, so the temperature compensation is required for the fiber bend-based humidity sensor. It is also recognized that a limitation of a fiber bend-based humidity sensor is that the sensor size is limited to the diameter of the bend. A small fiber bend will cause the stress-induced breakage of the sensor head. Thus, the main limitations of this type of sensors are associated with the long-term mechanical reliability of the fiber and the ability to fabricate sensors with identical characteristics. The effect of the interfering species also requires evaluation.

14.1.3 Microbent Sensors

Microbent chemical sensors proposed and demonstrated in 1980 by Fields (1980), are a variety of refractometric optical sensors. One should note that microbend loss has always been an undesirable effect that caused problems in fiber-optic communication links. The bending of optical fiber results in loss in light. It was established that the fiber bending created higher order modes, which resulted in the evanescent field having greater penetration depth (Leung et al. 2007). Later a description of some chemical sensors based on microbending has appeared in the literature. These sensors essentially used an indirect method to determine the chemical concentration: a transducer converted the value of the chemical concentration to the pressure applied to microbend the fiber (MacLean et al. 2000). However, Lee et al. (2001) have shown that a microbend optical fiber could be directly used to detect any chemical species that had absorption at the transmitting wavelength. The reader is invited to see references of Bertold (1995) and Thomas et al. (2002) for additional discussions on the theory of microbend fiber-optic sensor (FOS). One version of microbent-based humidity sensor is shown in Figure 14.5. As it is seen, the swelling of polymer influences the fiber bending and thus affects the intensity of the radiation detected by the photoreceiver.

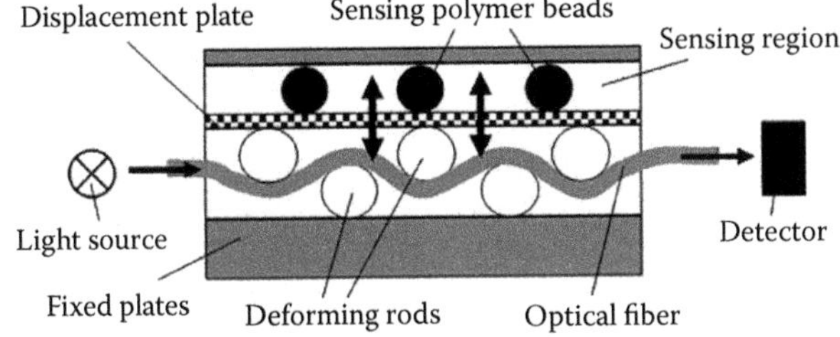

FIGURE 14.5 Microbend humidity sensor based on the polymer swelling in humid atmosphere. (Reprinted from Grossman, B.G. et al., In situ device for salinity measurements [chloride detection] of ocean surface, *Optics Laser Technol.*, 37, 217–223, Copyright 2005, with permission from Elsevier.)

14.2 EVANESCENT WAVE-BASED HUMIDITY SENSORS

Examples of evanescent wave-based humidity sensors using evanescent spectroscopy are listed in Table 14.1.

In principle, several approaches can be used to construct refractometry-based humidity sensors. The simplest and cheapest method of sensor construction is when the plastic cladding is removed over a suitable fiber length, exposing the core, and allowing access to the evanescent field. Another arrangement is a fiber surrounded by a cladding that has been impregnated with humidity-sensitive material. Both approaches are widely presented in the literature related to sensing technology; advantages and problems are discussed and solutions suggested (Ruddy et al. 1990; MacCraith et al. 1991; Börner et al. 2009). Moreover, this type of optical fiber construction provides a platform to develop truly distributed sensing systems. The problem is that the low percentage of total power carried by the evanescent field results in poor sensitivity. One way to overcome this problem is, for example, the tapered fiber (Figure 14.2), which was developed to maximize interaction with the sensing region (Mackenzie and Payne 1990). The use of a tapered fiber for humidity sensing based on the RI change has been demonstrated by Bariain et al. (2000). The sensor consisted of a tapered single mode telecommunication grade fiber, coated with agarose gel (Figure 14.6). The sensing mechanism was based on

TABLE 14.1
Evanescent Wave-Based Humidity Sensors

Sensing Method	Sensing Material	Range (% RH)	Response Time	References
Absorption measurement using U-bend fiber	$CoCl_2$-doped gelatin film	50–80	<1 min	Russell and Fletcher (1985)
	Gelatin/PEO film	–	–	Bownass et al. (1998)
	Rhodamine B-doped HPC film	0–95	~2 min	Otsuki et al. (1998)
	$CoCl_2$-doped HEC (H) and gelatin (G) films	H: 30–96 G: 40–80	–	Kharaz et al. (2000)
	Phenol red-doped PMMA film	20–80	–	Gupta and Ratnanjali (2001)
	$CoCl_2$-doped PVA film	–	–	Jindal et al. (2002)
	Co NPs dispersed in PANI	20–95	8 s	Vijayan et al. (2008)
	Methylene blue doped Silica sol–gel film	1–70	0.3–3 min	Zhao and Duan (2011)
	PEO	80–95	~1 s	Mathew et al. (2011)
	Agarose	25–90	~0.1 s	Mathew et al. (2012)
Absorption measurement using coated fiber	PAH/SiO_2	40–90	~10 s	Gomez et al. (2016)
	PEGDMA hydrogel	65–90	–	Zhang and Zhang (2015)
	Graphene oxide or TiO_2 NPs	30–95	~0.9 s	Ghadiry et al. (2016)
OTDR reflectometry	Porous SiO_2 film	25–95	–	Ogawa et al. (1988)
	Water swellable polymer			Michie et al. (1995)
	$CoCl_2$-doped gelatin film	20–80	1 s	Kharaz and Jones (1995)
Attenuation measuring using tapered fiber	Agarose gel	30–80	<1 min	Bariain et al. (2000)
	PDDAMAC/Poly R-478	75–100	–	Corres et al. (2006)
	Gelatin film	9–94	79 ms	Zhang et al. (2008)
	PVA film	30–95	~2 s	Li et al. (2012)
	HEC/PVDF composite	50–80	–	Lokman et al. (2014)
	Agarose, seeded ZnO or HEC/PVDF composite	50–80		Batumalay et al. (2015)

(Continued)

TABLE 14.1 (*Continued*)
Evanescent Wave-Based Humidity Sensors

Sensing Method	Sensing Material	Range (% RH)	Response Time	References
Attenuation measuring using side-polished fiber	PEO overlay	–	<2 h	Bownass et al. (1997)
	HEC/PVDF film	20–80	<5 s	Muto et al. (2003)
	Agarose gel	10–100	~90 s	Arregui et al. (2003)
	PVA film	50–90	1 min	Gaston et al. (2004)
	Polyethylene glycol	10–95	–	Acikgoz et al. (2008)
	Polyallylamin hydrochloride/PAA film	20–80	–	Sanchez et al. (2011)
	Porous sol–gel cladding	3–90	<1 min	Xu et al. (2004)
	TiO_2 film	0–80	–	Alvarez-Herrero et al. (2004)
	TiO_2 doped sol–gel film	24–95	10 ms	Aneesh and Khijwania (2012)
	WS_2	35–85	few seconds	Luo et al. (2016)

HEC—hydroxyethyl cellulose; HPC—hydroxypropyl cellulose; NPs—nanoparticles; OTDR—optical time-domain reflectometer; PAA—Poly(acrylic acid); PAH—Poly(allylamine hydrochloride); PANI—Polyaniline; PEGDMA—Poly(ethylene glycol) dimethacrylate; PDDA—poly(diallyldimethylammonium); PDDAMAC—poly(diallyldimethylammonium chloride); PMMA—poly(methyl methacrylate); PVA—polyvinyl alcohol; PEO—polyethylene oxide; Poly R-478—the polymeric dye polyvinyl amine sulfonate anthrapyridone; PVDF—polyvinylidene fluoride.

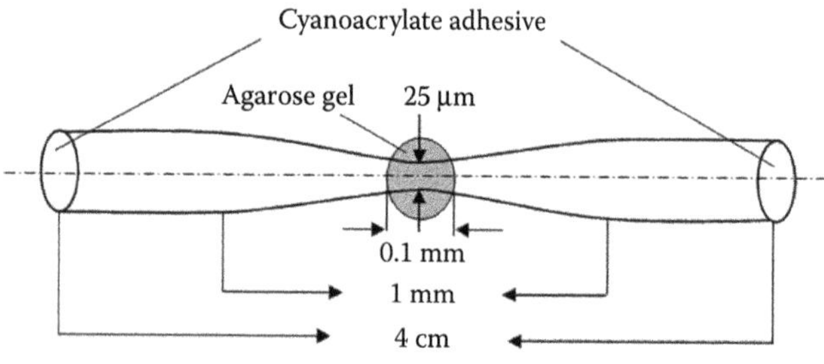

FIGURE 14.6 Geometrical parameters of a bioconical tapered single-mode optical fiber used for relative humidity monitoring. (Reprinted from Bariain, C. et al., Optical fiber humidity sensor based on a tapered fiber coated with agarose gel, *Sens. Actuators B*, 69, 127–131, Copyright 2000, with permission from Elsevier.)

the variation of the optical power transmitted through the taper as a function of the RI changes of the aragose gel at different RH values. Variations of 6.5 dB of the transmitted optical power were obtained for changes between 30% and 80% RH with a response time of less than 1 min and low hysteresis. The material was checked for more than half a year and showed no degradation. At the same time, Arregui et al. (2003) designed an agarose-based humidity sensor using a declad fiber. Comparison of these sensors has shown that the tapered sensor was reported to have a similar operating range and time response to the sensor using a declad fiber. Tapered fibers were also used by Corres et al. (2007). They have shown that the sensitivity of the sensor to the humidity can be optimized by controlling the thickness of the coating film and a faster time response was also possible due to the thin sensing film used.

Employing a side-polished optical fiber with a humidity-sensitive overlay represents another scheme for humidity sensing using refractometric approach (Gaston et al. 2003, 2004; Alvarez-Herrero et al. 2004). To expose the evanescent field and create such a sensor, the flat surface parallel to the fiber axis was polished back to remove the cladding. In particular, Gaston et al. (2003, 2004) have proposed a humidity sensor based on a single mode, side-polished fiber with a polyvinyl alcohol (PVA) overlay. The fiber block, with an exposed interaction length of about 2–3 mm, was covered by a PVA layer with thickness in the order of 100 μm. The humidity response was examined using two different laser sources emitting at 1550 and 1310 nm, respectively. Better results were obtained for $\lambda = 1550$ nm; an operating range of ~50%–90% RH, dynamic range: 2.2 dB. A sensor designed for low humidity detection and based on side-polished fiber, using a titanium oxide (TiO_2) overlay, was also demonstrated by Alvarez-Herrero et al. (2004). The nanostructure overlay was deposited over the polished fiber block by using the electron beam evaporation method. Side polishing can be realized by first immobilizing the optical fiber in a rigid material, forming a rectangular block with fiber extending out from the two end faces of the block orthogonal to the fiber axis. The advantage of using this scheme is that the sensing element can be fabricated using inexpensive components and a variety of coating materials can be deposited onto the flat surface of the fiber block. However, the fabrication procedure is very time-consuming and depending on the design of the fiber block and the exposed interaction length can be limited.

14.3 FIBER GRATING

Humidity sensors using fiber gratings are more advanced options of refractometric-based sensors (Kersey 1996; Grattan and Sun 2000). In contrast to conventional sensors shown in Figures 14.1 and 14.3, they involve addition of RI-sensitive optical structures to the optical fiber. Fiber grating is one of such structures (Bass and Van Stryland 2002). Examples of such structures are shown in Figure 14.7. Fiber gratings are broadly classified into fiber Bragg gratings (FBGs) and long-period gratings (LPGs). The period of an FBG is approximately half a micrometer, whereas the period of an LPG is typically several hundred micrometers. The theory of the optical fiber gratings one can find in Kogelnik and Shank (1972), Yariv (1973), and Erdogan (1997a, b).

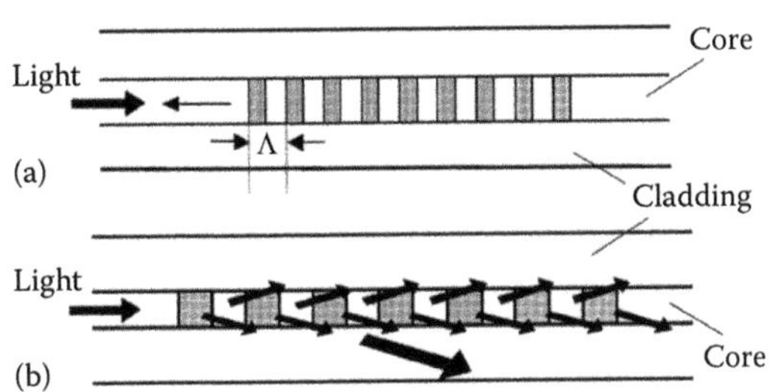

FIGURE 14.7 Types of fiber gratings. (a) Fiber Bragg grating and (b) long-period fiber grating. (Reprinted Lee, B., Review of the present status of optical fiber sensors, from *Optic. Fiber Technol.*, 9, 57–79, Copyright 2003, with permission from Elsevier.)

14.3.1 A Fiber Bragg Grating

A *Bragg grating* is a permanent periodic modulation of the RI in the core of a single-mode optical fiber. The change of the core RI is between 10^{-5} and 10^{-3}, and the length of a Bragg grating is usually around 10 mm. The principle of a FBG is described as follows: when light within a fiber impinges on Bragg grating, constructive interference between the forward wave and the contrapropagating light wave leads to narrowband back-reflection of light when the Bragg condition is satisfied (Figure 14.8). On account of this, a FBG can serve as an intrinsic sensor. In other words, Bragg gratings are optical filters, which allow the transmission of some wavelengths and reflect others. In FBGs the Bragg wavelength λ_B, or the wavelength of the light that is reflected, is given by

$$\lambda_B = 2n_{\text{eff}}\Lambda \tag{14.4}$$

where:

n_{eff} is the effective RI of the fiber core
Λ is the grating period (Figures 14.7a and 14.8)

In Equation 14.4, it can be seen that the Bragg wavelength is changed with a change in the grating period or the effective RI (Lee 2003). This means that the perturbation of the grating caused by environment change will result in a shift in the Bragg wavelength of the device, which can be detected in either the reflected or transmitted spectrum (Kwesey 1996). Thus, changes in the intensity of reflected light may accurately represent physical as well as chemical events that occur.

According to Sparrow et al. (2009), the ability to detect changes in the RI depends on two key aspects when using a Bragg grating device. First, and most obviously, the rate at which the Bragg wavelength varies with RI, the intrinsic sensor sensitivity, must be considered.

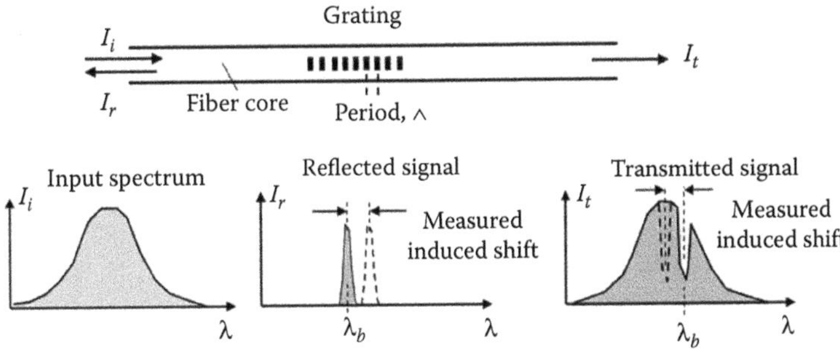

FIGURE 14.8 Basic transmission and reflection properties of Bragg gratings. (From Korotcenkov, G. et al., Optical and fiber optic chemical sensors, in *Chemical Sensors*, Vol. 5: Electrochemical and Optical Sensors, Korotcenkov, G. (Ed.), Momentum Press, New York, 2011. Copyright 2011, with permission from Momentum Press.)

Second, the resolution and accuracy to which the Bragg wavelength can be determined must be taken into account. Other factor in determining a RI resolution is the measurement technique used.

FBG sensors have a unique property: they encode the wavelength, which is an absolute parameter and does not suffer from disturbances of the light paths. Wavelength changes can be detected by an interrogator, which employs edge filters, tunable narrowband filters, or charge-coupled device (CCD) spectrometers. The bandwidth of the reflected signal depends on several parameters, particularly the grating length, but typically is ~0.05 to 0.3 nm in most sensor applications.

In FBG-based sensors it is possible to have an evanescent wave that only penetrates very short distances (<100 nm) into the medium, allowing study of localized surface phenomena. In other applications, it may be desirable to have the light penetrate as deeply as possible into the medium.

Multiplexing is another advantage of FBG-based sensors. Many tens of gratings can be written on the same optical fiber, and several hundred can be simultaneously interrogated by one multichannel instrument. This permits to use FBGs-based arrays to arrange a simple sensing system with a high number of sensing points distributed along a wide area. Moreover, these gratings can be configured to measure various parameters. This means that the humidity measurement can be performed simultaneously with the measurement of pressure, temperature, and concentration of toxic gases.

14.3.2 Long-Period Gratings

There is other type of Bragg gratings in which the pitch of the refractive variation is much longer, around hundreds of micrometers. These devices are known as LPG (Othenos and Kalli 1999; Grattan and Sun 2000; Tan et al. 2005). Indicated types of fiber gratings are shown in Figure 14.7b. A LPG is also a wavelength filter, but in this case the filtered wavelengths are not propagated back but rather coupled in evanescent fields through the cladding (Maier et al. 2006). It may couple the fundamental core mode to a forward propagating cladding mode when the phase-matching condition is satisfied by Vengsarkar et al. (1996)

$$\lambda_{res} = \left(n_{core} - n_{cl}^{i}\right) \cdot \Lambda \tag{14.5}$$

where:

λ_{res} is the LPG resonance wavelength

n_{core} and n_{cl}^{i} are the effective indexes of the core mode and the cladding mode

Similarly to FBG-based sensors, parameters that can affect the phase-matching condition will result in a shift in the LPG resonance wavelength.

The sensitivity of an LPG is also typically defined as a shift of the resonance wavelength induced by a measurand. The sensitivity characteristic of a bare LPG to surrounding RI changes has an increasing (in modulus) nonlinear monotone trend. The result is that the maximum sensitivity is achieved when the external index is close to the cladding index, whereas for lower RIs (around 1.33), the LPG is scarcely sensitive (Gouveia et al. 2013).

Figure 14.9 shows operating characteristics of one of LPG-based humidity sensors. The loss-peak wavelength shift is proportional to the level of RH. From this figure, one can clearly observe that the resonant wavelength is shifted toward the shorter wavelength side as the RH increases. From the experimental results shown

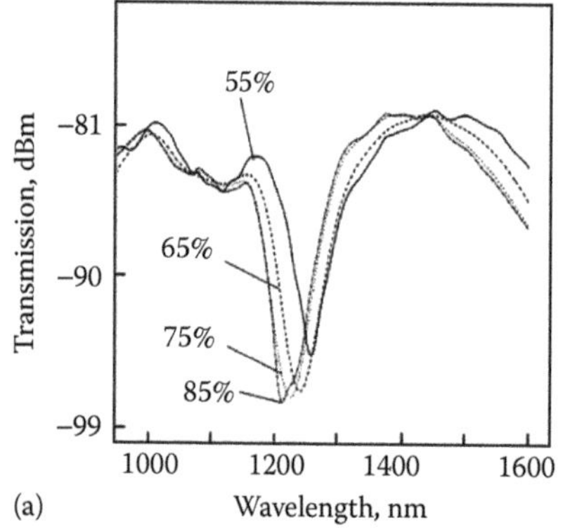

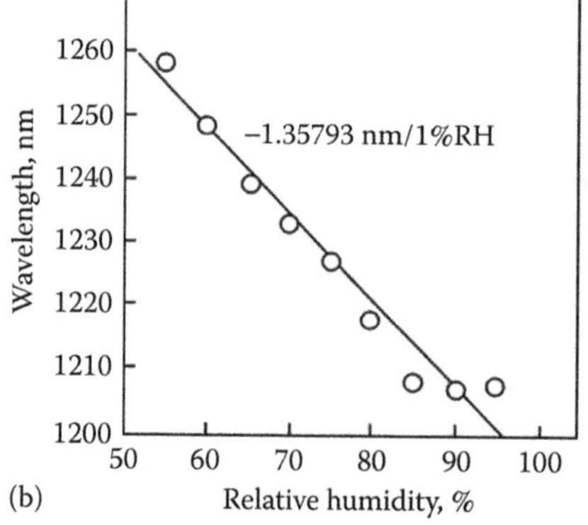

FIGURE 14.9 (a) Transmission spectra of fiber-optic humidity sensor, based on a calcium chloride thin film coated on an air-gap long-period grating fabricated by combining the fiber side-polishing, measured with different RH values at temperature of 30°C. (b) Curve of grating wavelength shift versus relative humidity at 30°C. (Adapted from Fu, M.Y. et al., Fiber-optic humidity sensor based on an air-gap long period fiber grating. *Opt. Rev.*, 18, 93, 2011. Copyright 2011, with permission from the Optical Society of Japan.)

in Figure 14.9 it is seen that the humidity sensitivity of 1.36 nm/1% RH is experimentally demonstrated.

Of course, fiber grating-based sensors can be realized in planar variant. The possibility of planar configuration realization has been demonstrated in a variety of different material platforms such as in polymers, sol–gel systems, silicon-on-insulator (SOI), lithium niobate, and silica-on-silicon.

14.3.3 Tilted Fiber Gratings

In the case of tilted gratings, as shown in Figure 14.10, the mode coupling becomes more complex.

The resonant wavelengths are Erdogan and Sipe (1996) and Lee and Erdogan (2000):

$$\lambda_{\text{co-cl}} = (n_{\text{co}}^{\text{eff}} \pm n_{\text{cl},m}^{\text{eff}}) \cdot \frac{\Lambda_g}{\cos\theta} \tag{14.6}$$

where $n_{\text{co}}^{\text{eff}}$ and $n_{\text{cl},m}^{\text{eff}}$, respectively, are the effective indices of core and the mth cladding mode. The grating period along the fiber axis is simply

$$\Lambda = \frac{\Lambda_g}{\cos\theta} \tag{14.7}$$

One of the unique characteristics of tilted fiber gratings (TFGs) is the strong polarization dependent loss (PDL) effect when the tilt angle becomes large. A near-ideal in-fiber polarizer based on 45° TFG has been reported by Zhou et al. (2005a, b), exhibiting a polarization-extinction ratio higher than 33 dB more than 100 nm range and an achievement of 99.5% degree of polarization for the unpolarized light.

Tilted fiber Bragg gratings (TFBGs) are a suitable option for refractometric sensing in terms of performance and robustness of the fiber structure. However, a TFBG couples the core mode to a number of cladding modes in a large wavelength bandwidth, which renders difficulty in signal readout and multiplexing. In addition, the fact that the measurement must be made in transmission, requiring access to the sensor from both sides, can represent a difficulty in some applications. Recently, a few authors have been exploring the possibility to excite the cladding modes of standard FBG by transferring power from the fundamental core mode to the cladding modes in the upstream of the FBG. Thereby, the FBG will couple back the light to the fundamental core mode. This arrangement enables the possibility to read the cladding mode of the Bragg grating in the reflected spectrum (Gouveia et al. 2013).

Of course, fiber grating-based sensors can be realized in planar variant. The possibility of the planar configuration realization has been demonstrated in a variety of different material platforms such as in polymers, sol–gel systems, SOI, lithium niobate, and silica-on-silicon.

14.3.4 Fiber Grating-Based Humidity Sensors

When developing moisture sensors Bragg grating is usually coated with an appropriate hygroscopic material (Table 14.2). When sensor exposed to the surrounding with humidity change, the volume expansion or compression of the humidity-sensitive polymer, the so-called swelling effect taking place due to absorption or desorption of water molecules, induces strain effect on the grating, and this results in a shift of the Bragg wavelength, which can be measured. Examples of refractometric-based sensor are humidity sensor for breath monitoring (Kang et al. 2006).

14.3.5 Limitations of Fiber-Grating Sensors

As for disadvantages of given sensors, according to Claus et al. (2002), Bragg grating and LPG sensors have the following limitations:

- The primary limitation of Bragg grating sensors is the complex and expensive fabrication technique.

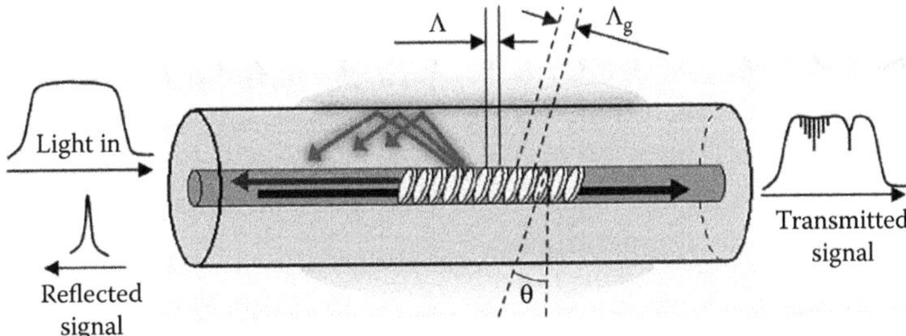

FIGURE 14.10 Schematic diagram of tilted grating in fibre core and refractometer based on a tilted fiber Bragg grating. (From Gouveia, C.A.J. et al., Refractometric optical fiber platforms for label free sensing, in *Current Developments in Optical Fiber Technology*, InTech, Rijeka, Croatia, 2013. Published as open access.)

TABLE 14.2
In-Fiber Grating-Based Humidity Sensors

Sensing Method	Sensing Material	Range (% RH)	Response Time	References
Strain-induced Bragg wavelength measurement	Polyimide	10–90	–	Giaccari et al. (2001), Kronenberg et al. (2002), Ding et al. (2011)
	Polyimide	22–97	18–25 min	Yeo et al. (2005), Venugopalan et al. (2009)
	Polyimide	11–98	5 s	Huang et al. (2007)
	Bragg grating recorded in silica and polymer fibers	50–95	~30 min	Zhang et al. (2010)
	Di-ureasil (formed of polyether chains covalently linked to a siliceous inorganic skeleton by urea bridges)	5–95	~8 min	Correia et al. (2012)
	Pyralin	15–95	~3 s	David et al. (2012)
	PMMA	30–90	13–25 min	Zhang and Webb (2014)
	MWCNTs	20–90	~10 min	Shivananju et al. (2014)
FBG resonant band intensity measurement	PVA	30–90	~3 s	Liang et al. (2015)
	Graphene oxide	10–80	–	Wang et al. (2016)
LPG resonant band wavelength measurement	Carboxy methyl cellulose	0–95		Luo et al. (2002)
	$CoCl_2$-doped PEO film	40–80	<1 s	Konstantaki et al. (2006)
	$CaCl_2$ film	55–95	–	Fu et al. (2011)
	TiO_2 sol–gel	0–75	–	Berruti et al. (2014)
LPG resonant band intensity measurement	Gelatin	90–99		Tan et al. (2005)
	PVA	33–97	50–90 s	Venugopalan et al. (2008)
	Silica nanospheres film	20–80	30–200 ms	Viegas et al. (2009, 2011)

LPG—long-period grating; MWCNTs—multiwalled carbon nanotubes; PEO—poly(ethylene oxide); PMMA—poly(methyl methacrylate); PVA—Polyvinyl alcohol.

- The second primary limitation of gratings is their limited bandwidth. The typical value of the full width at half-maximum (FWHM) for Bragg grating is between 0.1 and 1 nm. The limited bandwidth requires high-resolution spectrum analysis to monitor the grating spectrum.
- In addition, FBG- and LPG-based sensors are not very sensitive (Elosua et al. 2006). This means that sensors may not be adequate for certain applications.
- Finally, the cross-sensitivity to temperature leads to erroneous displacement measurements in applications where the ambient temperature has a temporal variation. A Bragg wavelength shift occurs in presence of temperature variations, due to the intrinsically dependence of λ_B to temperature (Figure 14.11). Moreover, it was demonstrated in the literature (Kronenberg et al. 2002) that the response behavior of a polymer-coated FBG was a linear superposition of relative humidity and temperature effects.

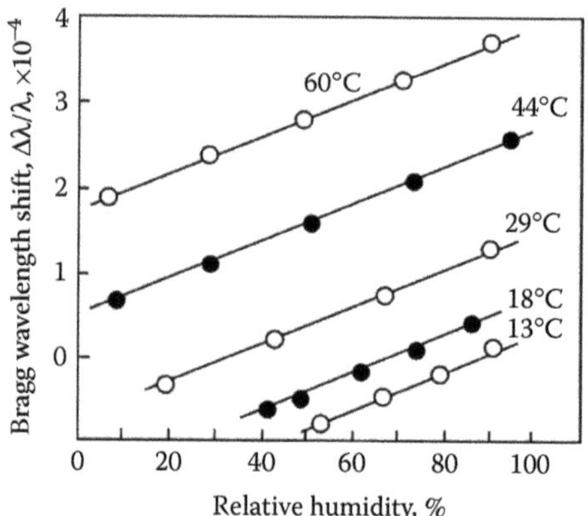

FIGURE 14.11 Bragg wavelength shift for several values of temperatures reported by Kronenberg P. and coworkers. (Data extracted from Kronenberg, P. et al., Relative humidity sensor with optical fiber Bragg grating. *Opt. Lett.*, 27, 1385–1387, 2002.)

This means that in real-sensing applications, in presence of both temperature and humidity variations, one should contemplate the application of temperature compensation techniques. A precise deconvolution of the temperature and humidity effects from the sensor signal requires therefore an independent temperature reading as close as possible to the humidity sensor position. In addition, in order to protect the FBG-based sensor and to avoid unwanted additional stress (not due to humidity variations), appropriate holders needs to be used to package the optical gratings and keep them in a tension-free state. Examples of humidity sensors, where temperature compensation is realized, can be found in Korenko et al. (2015). In particular, Korenko et al. (2015) show that a thermal compensation scheme is possible if birefringent fibers are used. Such fibers result in double reflection peaks where the distance of the peaks and the individual shift of the peaks can be used as separate information parameters. Another approach was proposed by Urrutia et al. (2016). For simultaneous measurement of humidity and temperature, they used semicoated LPG (Figure 14.12a) and analyzed the position of the bands formed by coated and uncoated parts of LPG (Figure 14.12b). The dual-wavelength-based measurement provided a simultaneous monitoring of RH and temperature, with sensitivity ratios of 63.23 pm/% RH and 410.66 pm/°C for the attenuation band corresponding to the coated contribution, and 55.22 pm/% RH and 405.09 pm/°C for the attenuation band corresponding to the uncoated grating.

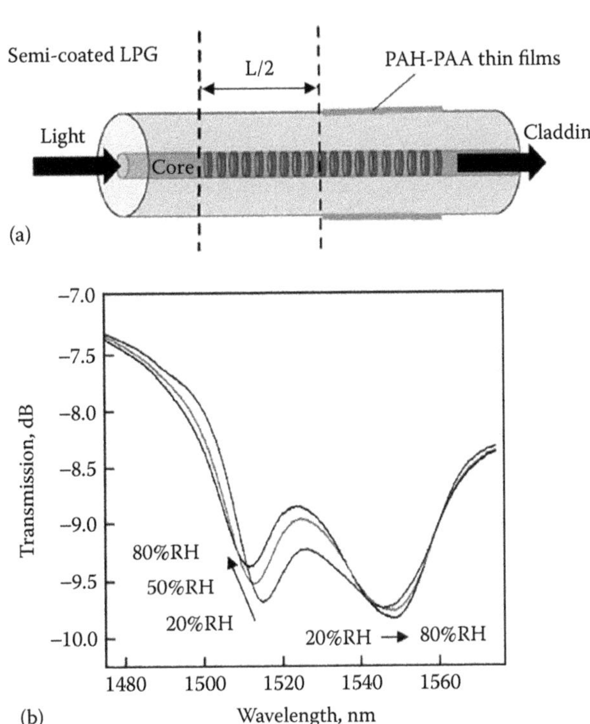

FIGURE 14.12 (a) Schematic diagram of LPG sensor designed for detection of RH and temperature and (b) transmission spectra evolution of the partially-coated LPG as the RH is changed. (Reprinted from Urrutia, A. et al., Simultaneous measurement of humidity and temperature based on a partially coated optical fiber long period grating, *Sens. Actuators B*, 227, 135–141, Copyright 2016, with permission from Elsevier.)

14.4 WHISPERING GALLERY MODE-BASED MICRORESONATORS

Sensors based on optical whispering gallery mode (WGM)-based microresonators are another promising direction in humidity sensor development. WGM-based microresonators have attracted significant attention from researchers for applications in various photonic devices and sensors due to their ultrahigh quality factors (Q), low absorption loss, and easy and inexpensive fabrication methods. A significant amount of research work has been carried out on various microresonator shapes, such as microspheres (Matsko and Ilchenko 2006; Mallik et al. 2016), microdisks (Eryürek et al. 2017), microtubes (Ling and Guo 2007), and microrings (Ramachandran et al. 2008). Inside such a circular-shaped resonator the light propagates in the form of WGMs as a result of total internal reflection. Optical microresonators are extensively described in literature (Vahala 2003; Heebner et al. 2008; Foreman et al. 2015; Van 2016).

Photograph of microsphere resonator-based sensor is presented in Figure 14.13. Typical configuration of planar microring resonator (MRR) is shown in Figure 14.14. In general, a ring resonator consists of a looped optical waveguide and a coupling mechanism to access the loop. In its simplest form, a ring resonator can be constructed by feeding one output of a directional coupler back into its input, the so-called all-pass filter (APF) or notch filter configuration (Figure 14.14a). The term ring resonator is typically used to indicate any looped resonator, but in the narrow sense it is a circular ring. When the shape

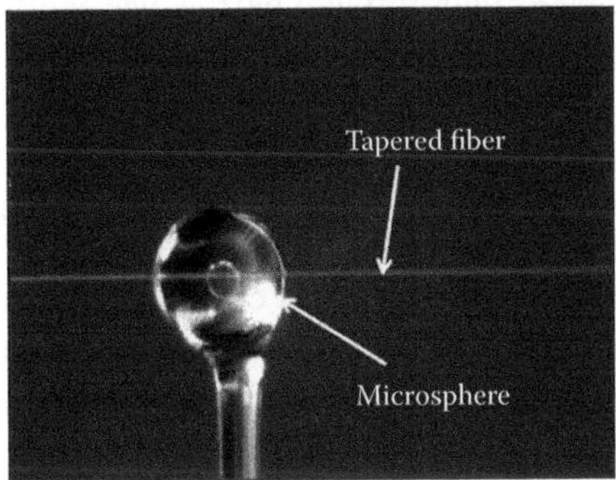

FIGURE 14.13 Photograph of coupled sphere-tapered fiber. (From Ioppolo, T. et al., Development of whispering gallery mode polymeric micro-optical electric field sensors, *J. Vis. Exp.*, 71, e50199, 2013. Published by Jove as open access.)

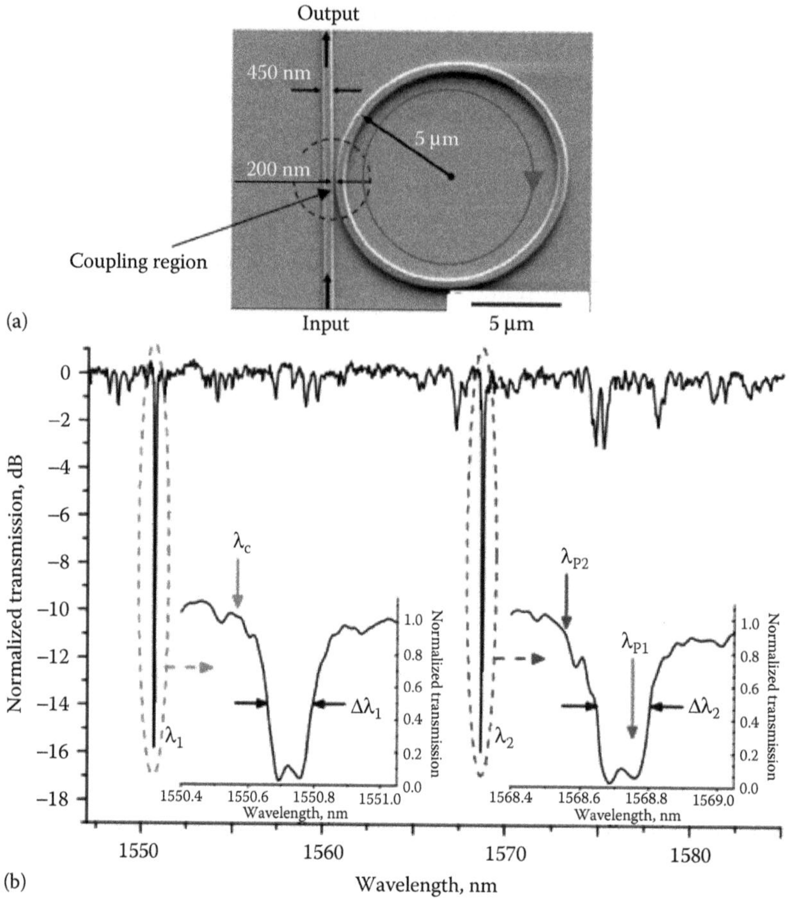

FIGURE 14.14 (a) Top-view SEM picture of the silicon microring resonator before the SiO_2-cladding deposition. (b) The transmission spectrum of the resonator. The insets show the zoom-ins around each resonance. The wavelengths of the pump (λ_C) and the probes for the AND (λ_{P1}) and NAND (λ_{P2}) gates are marked in the insets. (From Xu, Q., and Lipson, M., All-optical logic based on silicon micro-ring resonators. *Opt. Exp.*, 15, 925–929, 2007. Published by the Optical Society as open access.)

is elongated with a straight section along one direction (typically along the coupling section) the term racetrack resonator is also used.

The functioning of MRRs is based on waveguiding phenomena specified by boundary conditions. Propagation of light through narrow waveguides generates an evanescent field, a local region of light *leakage* that extends a distance of 100–300 nm (at near-infrared wavelength) at the surface of the waveguide. In MRRs, the linear waveguides and microrings are precisely aligned and positioned, either vertically or laterally, so as to achieve evanescent field coupling. This facilitates light passing through the input waveguide to enter into the microring structure, as well as light to escape from the microring structure into the output waveguide. The continuous recirculation of light enhances the effective path length several fold over the actual optical path length of the microring. Thus, the intensity of light within the microring structure builds up significantly as the continuous light from the input waveguide gets coupled in and adds up in phase, thereby generating an amplified evanescent field that can be exploited for sensing applications. When the waves in the loop build up a round trip phase shift that equals an integer times 2π, the waves interfere constructively and the cavity is in resonance (Figure 14.14b). Optical resonance in the microring is attained when the condition in Equation 14.8 is satisfied

$$\lambda_{res} = \frac{n_{eff} 2\pi R}{m}, \qquad m = 1, 2, 3..... \qquad (14.8)$$

where:

R is the radius of the microring
n_{eff} the effective RI of the microring
λ is the wavelength
m is any integer

As only a select few wavelengths will be at resonance within the loop, the optical ring resonator functions as a filter. For ideal cavities with zero attenuation, the transmission at resonance drops to zero.

Due to λ_{res} extreme sensitivity to the size of the resonator and also to the RIs of the resonator and the surrounding medium (Ilchenko et al. 1998; Krioukov et al. 2002), WGMs can be used for humidity sensing using the variation of the RI and stress, caused by interaction with water vapors. A water vapor that settles near the microring surface changes the effective RI and alters the resonance properties of the microring. This effect causes a shift in resonance wavelength, which can be determined by monitoring the optical spectrum of the output waveguide. A perturbation to a cavity can also produce changes in the linewidth of the resonance.

Examples of humidity sensors based on WGM microresonators are listed in Table 14.3. As a rule these sensors comprise ring-shaped waveguide structures (microrings or microdisks) optically coupled to one or more linear waveguides, patterned on a planar surface. Microring and microdisk resonators are made using dielectric materials such as Si_3N_4 or SiO_2. This type of sensors offers several advantages for use as humidity-sensitive platforms. They provide label-free, sensitive, and real-time detection capabilities. As can be seen from the data in Table 14.3, the best WGM-based humidity sensors have the sensitivity greater than 1000 pm/% RH and response time is less than 1 s. As a rule, the sensitivity increases with an increase in the thickness of the humidity-sensitive material. For example, Mallik et al. (2016) reported that the

TABLE 14.3
Humidity Sensors Based on Optical Microresonators

Configuration	Sensing Material	Range (% RH)	Maximum Sensitivity	Response Time	References
Microsphere resonator	Agarose-coated silica microsphere	30–70	518 pm/% RH	–	Mallik et al. (2016)
	Microsphere-coated with SiO_2 nanoparticles	0–10	1–5 pm/% RH	3–5 min	Ma et al. (2010)
	A liquid optical glycerol droplet resonator doped with rhodamine 6G	40–65	~2000 pm/% RH	<1 s	Labrador-Paez et al. (2017)
Microdisk resonator	SU-8 polymer microdisk	0–10	108 pm/% RH	50–80 s	Eryürek et al. (2017)
Microring resonator	SU-8 ring coated by sol–gel silica	30–85	16 pm/% RH	<0.2 s	Bhola et al. (2009)
	Uncoated	4–100	–	~1 s	Alali and Wang (2016)
	Uncoated	3–100	10–50 pm/% RH	–	Tao et al. (2016)
	SiO_2 microring coated with pNIPAAm	0–3	~13 pm/% RH	<10 s	Mehrabani et al. (2013)
	Resonator with three coupling points coated with PI	0–100	1700 pm/% RH	–	Guo et al. (2014)
Knot resonator	PMMA microfibers	10–95	~1 pm% RH	<1 s	Wu et al. (2011)
Tubular microcavity	PAA/PEI modified microcavity	0–95	130 pm/% RH	–	Zhang et al. (2014)

PAA—poly(acrylic acid); PEI—poly(ethylenimine); PI—polyimide; PMMA—poly(methyl methacrylate); pNIPAAm—poly(*N*-isopropylacrylamide).

increase of the agarose layer thickness from one layer to six layers deposited by layer-by-layer (LbL) method was accompanied by increase in sensitivity more than 100 times (Figure 14.15). However, simultaneously with the increase in the coating thickness, the response time also increases (Figure 14.16). As shown by Bhola et al. (2009) on the example of SU-8 sensors coated with silica sol–gel thin film, increasing the coating thickness from 0.4 to 4.0 μm resulted in an increase in response time from 0.2 to 35 s. This time is required for diffusion of water vapor into sol–gel film.

In addition, owing to their small footprint, arrays of MRRs can be fabricated on a small chip, enabling multiplexing capabilities. Moreover, MRRs can be ultimately mass produced at low cost. They can be fabricated by standard silicon wafer processes using photolithographic techniques similar to those used for manufacture of integrated circuit (IC) chips and are amenable to very large-scale integration (VLSI) (Tan et al. 2002). As for the shortcomings of WGM-based sensors, they are common to all optical sensors.

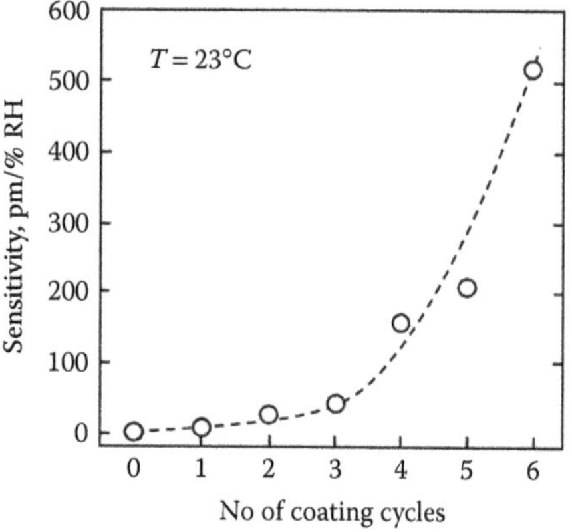

FIGURE 14.15 RH sensitivity of microsphere resonators (D ~100 μm) with a different number of agarose-coating layers. Agarose was deposited using LbL method. (Data extracted from Mallik, A.K. et al., Agarose coated spherical micro resonator for humidity measurements. *Opt. Exp.*, 24, 21216–21227, 2016.)

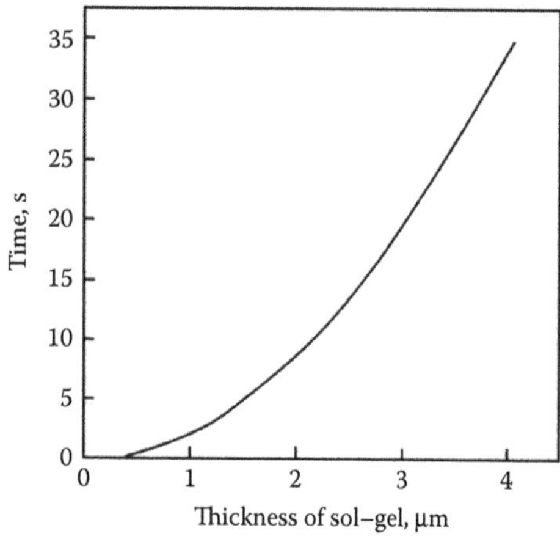

FIGURE 14.16 A plot of time required to for a 5% change in relative humidity to propagate homogeneously inside the sol–gel thin film. (Data extracted from Bhola, B. et al., Sol–gel-based integrated optical microring resonator humidity sensor. *IEEE Sens.*, 9, 740–747, 2009.)

Optical fibers are also used for fabrication of WGM microresonator-based sensors Yan et al. (2011). Moreover, Yan et al. (2011) have shown that the usage of tapered fibers in such sensors increases the efficiency of coupling. In this coupling method, tapered fiber was positioned approaching the microcavity using a high-resolution (generally 10–20 nm) stage adjusting coupling efficiency. The gap between the two parts at a range from several nanometers to several hundred nanometers is preferable and it is vital for coupling efficiency. However, at the same time Yan et al. (2011) noted that to excite a certain WGM, the taper need to reside in the equatorial plane of the cavity with minimal tilt angle. These rigorous demanding of the separate coupling structures can be achieved in laboratory, but is unimaginable in device development. Therefore, the separate structures still challenge stability of physical structure, although it's the highest coupling efficiency at present.

The RH sensor, made only of optical fibers, was developed by Alali and Wang (2016). Configuration of such sensor is shown in Figure 14.17. The sensing signal (ringdown time) is affected by the length of the fiber loop and the total optical loss of the light pulse inside the fiber loop (Wang 2014). Alali and Wang (2016) have shown that this sensor did not need coatings and allowed to measure humidity in wide range from 4% to 100% RH. Moreover, the sensor was easy to fabricate and of low cost; no complex instruments were needed for sensor operation. The time response to switch from

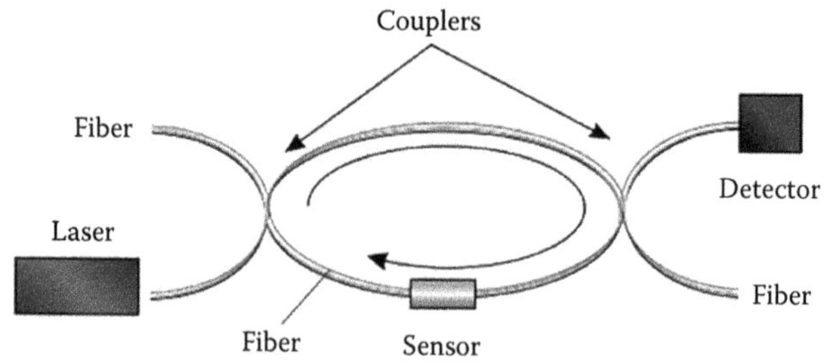

FIGURE 14.17 Setup of a fiber loop cavity. The fiber is bent into a loop to form a ring cavity. Fiber couplers with high splitting ratios may be used to inject light into the loop and to direct a small fraction of the light to the detector. (From Waechter, H. et al., Chemical sensing using fiber cavity ring-down spectroscopy. *Sensors*, 10, 1716–1742, 2010. Published by MDPI as open access.)

lower humidity to higher humidity was between 15 s and ~3 min, depending on the RH variation, whereas the sensor's response to RH was ~1 s.

14.5 MATERIALS ACCEPTABLE FOR REFRACTOMETRIC OPTICAL HUMIDITY SENSORS

14.5.1 Polymers

Experiment has shown that in general a suitable coating for refractometric optical humidity sensors must satisfy a number of criteria (Mathew et al. 2011; Mathew 2013): (1) it should be hygroscopic and should have a suitable RI change with respect to ambient humidity, (2) the preparation of the coating solution and the coating process should be simple, (3) the material should also possess a good adhesion to silica, and finally (4) the coating material should have a low evaporation tendency at higher humidity and should exhibit good long-term stability. Approximately the same requirements imposed for the materials used in the grating-based sensors, one of the types of refractometric sensors. However, the value of the swelling effect is more important parameter for such sensors, because it determines the magnitude of the Bragg wavelength shift. In addition, the requirement of good adhesion is stronger, because the adhesion of the coating onto the grating may affect the sensor performances.

Materials tested in refractometric optical humidity sensors are listed in Tables 14.1 through 14.3. As can be seen from the table, polymers are the most commonly used humidity-sensitive materials in refractometric humidity sensors. This is due to the fact that polymers, when interacting with water, at the expense of swelling effect can provide the maximum RI change and create the strain effect induced onto the grating. In addition, polymers provide useful mechanical properties; they are compatible with oxides and ceramics, low-cost, flexible, light weight, easy to process, and can be used at room temperatures. At that, very often the polymers are used in the form of hydrogels (Arregui et al. 2003). Hydrogel is a kind of polymeric material known for its high swelling effect due to excellent water absorption properties. However, when choosing polymers for FBG- and LPG-based sensors it is necessary to borne in mind that not all materials that have a high moisture swelling property may induce a detectable strain on the fiber. Therefore, there should be a trade-off between these factors when selecting a suitable material for an optimized FBG- and LPG-based relative humidity sensor.

As it is seen in Tables 14.1 through 14.3, hygroscopic polymers such as PANI (polyaniline), PVA, Chitosan, PEG (polyethylene glycol), PDMAA (poly(dimethylacrylamide)), PMMA poly(methyl methacrylate), Polyimide (PI), and many other polymers can be used in optical humidity sensors. For example, Gaston et al. (2004) used PVA as a coating in humidity sensors. PVA dissolves in water and offers a simple coating procedure but it is not suitable for sensing in a wide humidity range. The reason is that the PVA film dries completely when it reaches a RH value of ~50% so that a subsequent decrease in the environmental humidity has no effect on the coating. Poly(ethylene oxide) (PEO) (Konstantaki et al. 2006; Mathew et al. 2011) and PEG (Acikgoz et al. 2008) are also a possible candidates and are highly sensitive. But it was shown that PEO does not provide a linear response, and because PEO is soluble in the water, a long-term exposure to high humidity levels is likely to damage the coating. In addition, the use of PEO does not provide a wide range of humidity measurements (Mathew et al. 2011). Hydroxyethyl cellulose (HEC)/polyvinylidene fluoride (PVDF) composite also did not provide the required sensitivity and the range of humidity measurement (Lokman et al. 2014). The sensitivity of this FOS based on a tapered fiber was estimated to be around 0.0228 dB/% RH. Sensors, which used hydrogel Poly(ethylene glycol) dimethacrylate (PEGDMA), had better sensitivity (Zhang and Zhang 2015). However, the sensing range of theses sensors was still very narrow, from 70% to 90%, due to the moisture absorption behavior of the used hydrogel PEGDMA.

The materials agarose and gelatin are other potential candidates (Bariain et al. 2000; Tan et al. 2005; Mathew et al. 2007, 2012; Lee et al. 2007; Zhang et al. 2008); both of them offer a wide operating humidity range with a simple coating procedure. However, compared to gelatin, agarose is a more consistent product with fewer impurities and less ethical concerns. It has a higher melting point than the gelatin and also is a more stable material with respect to temperature. Agarose shows a linear change in its RI with respect to humidity (Lee et al. 2007). Agarose is an unbranched polysaccharide obtained from the cell walls of some species of red algae or seaweed. Chemically, agarose is a polymer made up of subunits of the sugar galactose. Agarose has an additional advantage of the low material degradation compared with the materials used in Bownass et al. (1998). Since agarose is soluble in hot water (Arregui et al. 2003; Mathew et al. 2007), the preparation and coating procedures are simple. Agarose also has a good adhesion to silica and easily forms a thin coating film on silica fiber. Therefore, many believe that agarose is a suitable choice as a coating for the fiber-optic refractometric-based humidity sensors (Bariain et al. 2000; Lee et al. 2007; Mathew et al. 2007, 2012). For example, Arregui et al. (2003) compared various polymers such as poly-hydroxyethyl metharyclate (poly-HEMA), polyacrylamide, poly-*N*-vinyl pyrrolidinone (poly-*N*-VP) and agarose gel and found that among the materials tested, the sensor with agarose gel gave the best overall performance, achieving an operating range of 10%–90% RH. It was also reported that these sensors exhibited a better response time and stability than the rest of the materials evaluated. Figure 14.18 shows typical response of the ACFB humidity sensors developed by Mathew et al. (2012). At the same time, Batumalay et al. (2015) compared the parameters of the tapered fiber humidity sensor coated with agarose gel, HEC/PVDF composite and nanostructured ZnO and found that all coating materials were sensitive and efficient for humidity sensing.

Alwis et al. (2013a, b) have also made a comparison of PI and PVA coverings. As it is known, both the PVA and PI swell with the increase of RH in its surroundings. For this purpose, they used LPGs. The RI of PVA and PI is 1.53 and 1.7, respectively. Since the cladding RI is around 1.44, PVA lies closer to the cladding RI than that of PI and therefore it would experience a greater RI change than PI. This latter material has been coated to create a moisture related strain induced RH sensor and PVA is used to induce a RH related external RI variation on the LPG. A comparison between the performance characteristics of these two different polymer-coated self-interfering LPGs (SILPGs) are shown in Figure 14.19. It can be seen that PVA offers higher sensitivity of the two polymers used, although its sensing region is limited and the performance is nonlinear, but these disadvantages may be overcome for certain applications. PI on the other hand, offers a linear performance that is easy to process, but with

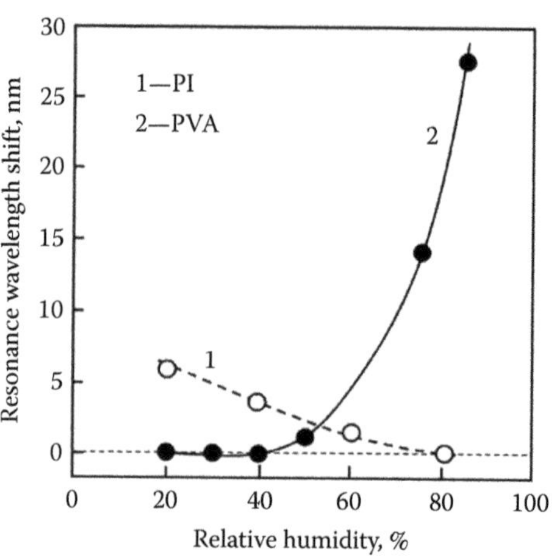

FIGURE 14.19 Comparison between the performance of PI- and PVA-coated LPG-based RH sensor probes. (Data extracted from Alwis, L. et al., Design and performance evaluation of PVA/PI coated optical fibre grating-based humidity sensors. *Rev. Sci. Instrum.*, 84, 025002–025008, 2013.)

overall less sensitivity in most of the target range compared to that achieved with PVA.

At the same time Huang et al. (2007) and Guo et al. (2014) believe that PI is more effective material for humidity sensor application. For example, Guo et al. (2014) reported that MRR-based RH sensors with PI coating had extremely high sensitivity, which reached 1700 pm/% RH. PI-recoated FBGs-based sensors also have acceptable parameters. Huang et al. (2007) have shown that these sensors had linear response over a wide humidity range. The sensor was reported to respond well to a humidity range of 10%–90% RH and display good repeatability. Typical sensitivity of FBGs-based sensors varied in the range from 1.7 pm/% RH (Ding et al. 2011) to 6.8 pm/% RH (Venugopalan et al. 2009). At that, the sensitivity increased with the increasing of the film thickness (Figure 14.20). However, this increase was accompanied by the increase in hysteresis and response time (Figure 14.21). It was also found that the highly alkali environment degrades the PI coating and hence reduces the performance of the PI-coated FBG-based fiber-optic humidity sensor with time. However, experiments performed by Bremer et al. (2014) have shown that sensor packaging using a polytetrafluoroethylene (PTFE) membrane protects the PI coating of the fiber-optic humidity sensor against alkaline solutions. Some approaches to the optimization of PI-based FBG humidity sensors have been also proposed by Swanson et al. (2016). They suggest that the improvement of the sensor parameters can be achieved through chemical structural modifications: to add more water associating sites and limit carbon chain lengths.

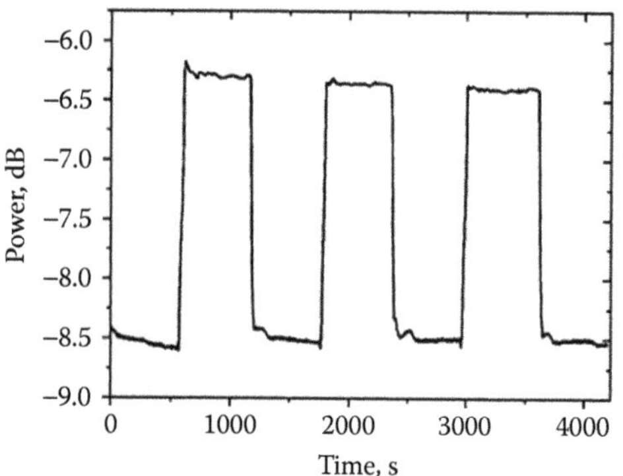

FIGURE 14.18 Humidity cycling (60%–80% RH) response to show the repeatability and stability of the ACFB. (Reprinted from Mathew, J. et al., A fiber bend-based humidity sensor with a wide linear range and fast measurement speed, *Sens. Actuators A*, 174, 47–51, Copyright 2012, with permission from Elsevier.)

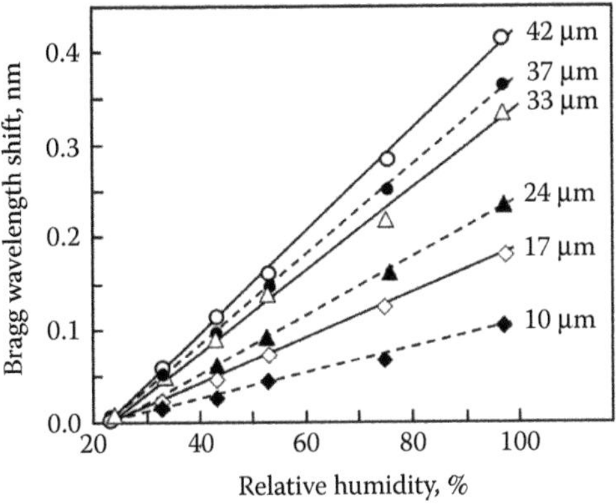

FIGURE 14.20 RH response of FBG-based sensors with different coating thicknesses at 23°C reported by Yeo, T.L. and coworkers. (Data extracted from Yeo, T.L. et al., Characterisation of a polymer-coated fiber Bragg grating sensor for relative humidity sensing. *Sens. Actuators B*, 110, 148–156, 2005.)

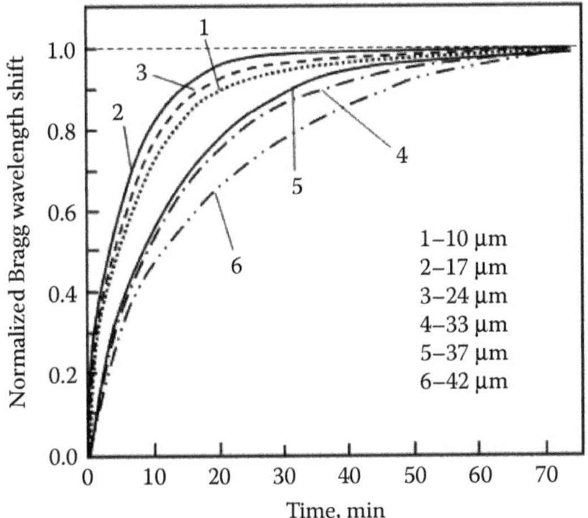

FIGURE 14.21 Response time of FBG-coated relative humidity sensors reported by Yeo, T.L. and coworkers. (Data extracted from Yeo, T.L. et al., Characterisation of a polymer-coated fiber Bragg grating sensor for relative humidity sensing. *Sens. Actuators B*, 110, 148–156, 2005.)

It is observed that the presence of metal oxide ions also increases the humidity sensitivity of hydrophilic polymers. Khijwania et al. (2005) have reported on a RH sensor based on U-shaped probe coated with anhydrous $CoCl_2$ and PVA. Sensors had a response time of 1 s with a sensing range of 2%–92% RH. Recently, an optical fiber humidity sensor based on TiO_2-nanoparticle (NP) doped nanostructured thin film as the fiber-sensing cladding was reported by Aneesh and Khijwania (2012). SnO_2 NPs suspended polyvinyl alcohol matrix was used for humidity sensing by Hatamie et al. (2009). Enhancement of the humidity-sensing properties of PMMA with the addition of alkali salts (KOH and K_2CO_3) was also reported by Su et al. (2006).

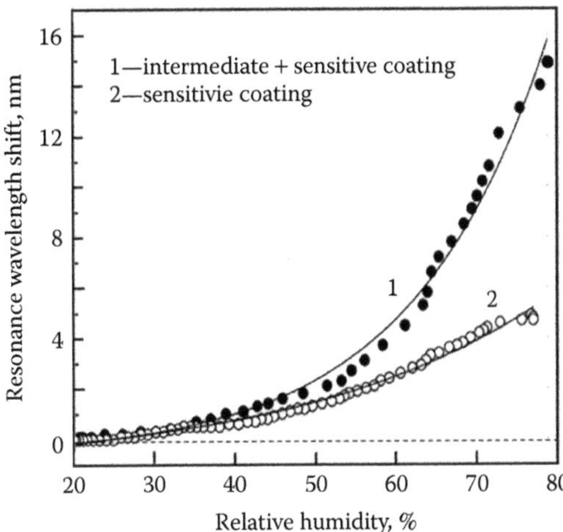

FIGURE 14.22 Resonant wavelength shift dependence with relative humidity for both coatings. (Adapted from Viegas, D. et al., Sensitivity improvement of a humidity sensor based on silica nanospheres on a long-period fiber grating. *Sensors*, 9, 519, 2009. Published by MDPI as open access.)

It follows from the foregoing analysis, the RI and the coating thickness of the sensing material are parameters that require careful selection when elaborating refractometric humidity sensors, as their response, that is, the wavelength or intensity variation will depend strongly on these parameters. Therefore, it is quite natural to look for the optimal coatings that will ensure an increase in the sensitivity of the RH sensors. In particular, Corres et al. (2008b) and Viegas et al. (2009) basing on their theoretical and experimental investigations of LPG-based sensors have established that one can achieve an improvement of their parameters at the expense of using a *double-coating*: (1) the LPG is initially coated with a thin overlay of a material with a particular RI leading to an increase in the sensitivity of the LPG to external RI variations, and (2) then a second overlay is deposited with the species-specific material targeting the particular analyte. As an example of such *double-coating*, Viegas et al. (2009) have suggested to use PDDA/PolyR-478 as an initial coating (intermediate coating) and PAH/SM30 as a humidity-sensitive material. The results for varying the RH over the range of 20%–80% at room temperature for RH sensors fabricated using single and double coatings are presented in Figure 14.22. It can be clearly seen that the initial coating of the higher RI polymer layer has greatly increased the overall sensitivity of the sensor to varying RH.

14.5.2 Other Materials

Porous sol–gel silica (Arregui et al. 2003; Zhao and Duan 2011) and TiO_2 (Berruti et al. 2014; Ghadiry et al. 2016) can be also coated on to the surface of a silica

TABLE 14.4
The Effect of the Coating Thickness on the Sensitivity and Response Time of the Graphene Oxide (GO) and TiO_2-Based Humidity Sensors

	Thickness (µm)		Sensitivity (dB/RH%)		Response Time (s)		Recovery Time (s)	
Sample	TiO_2	GO	TiO_2	GO	TiO_2	GO	TiO_2	GO
1	0.2	0.5	0.09	0.11	0.23	0.17	0.33	0.23
2	0.7	1.1	0.23	0.23	0.74	0.63	0.91	0.76
3	0.9	2.3	0.39	0.33	1.83	1.71	2.07	1.93
4	1.1	3.9	0.505	0.43	2.44	2.33	2.81	2.64

Source: Ghadiry, M. et al., *PLoS One*, 2016, 0153949, 2016.

optical fiber core in humidity sensors. The evanescent waves that penetrate the coating layer are increasingly scattered as the humidity of the surrounding air increases (Xu et al. 2004; Corres et al. 2008a). Berruti et al. (2014) have established that LPG-based humidity sensors with TiO_2 coating retain their characteristics even after high radiation exposures. Ghadiry et al. (2016) have reported that evanescent-based humidity sensors, which used TiO_2 coating, also had high sensitivity and fast response. It was observed that the device response was almost linear over a wide dynamic range (35%–98% RH) with high sensitivity of around 0.21 dB/%RH. Sensor was characterized by short response time of ~0.7 s for humidification and 0.95 s for desiccation. It is important to note that the sensors sensitivity, as well as the response time and recovery time increased with increasing layer thickness (Table 14.4), which is quite a natural phenomenon.

Corres et al. (2008a) have shown that the low thickness of the SiO_2 film (760 nm) and the highly porous structure obtained using the electrostatic self-assembly (ESA) deposition technique allowed obtaining fiber-optic-based humidity sensors with low hysteresis and a very fast response. Sol–gel silica also offers a wide humidity operating range. But the preparation and the coating process for this material is difficult and there is a tendency for cracks to develop in the coating, reducing the uniformity of the coating and thus compromising the operation of the sensor. In addition, other gases, especially volatile organic compounds (VOCs) can also influence the RI and therefore can heavily contribute to the response signal of humidity sensors (Zhao and Duan 2011).

It was established that one-dimensional materials such as carbon nanotubes (CNTs) (Shivananju et al. 2014), tungsten disulfide (WS_2) (Luo et al. 2016), and graphene (Ghadiry et al. 2016; Wang et al. 2016) can be also used for humidity sensor design. Luo et al. (2016) have demonstrated a fiber-optic humidity sensor, comprising a WS_2 film overlay on a side polished fiber, while Wang et al. (2016) and Shivananju et al. (2014) proposed to use CNTs and graphene oxide in FBG-based sensors. According to Luo et al. (2016), WS_2-based humidity sensors had a linear correlation coefficient of 99.39%, sensitivity of 0.1213 dB/% RH, and a humidity resolution of 0.475% RH in a RH range of 35%–85%. Furthermore, this sensor showed good repeatability and reversibility, and fast response. Shivananju et al. (2014) reported that CNTs-based sensors also were highly sensitive over a wide range of RH (20%–90%). A sensitivity of ~31 pm/% RH, which is achieved, is one of the highest values compared to the existing FBG-based humidity sensors. The limit of detection observed in carbon nanotubes coated etched fiber Bragg grating (CNT-EFBG) sensor has been found to be about 0.03 RH. No doubts these are good parameters for refractometric humidity sensors. However, it should be kept in mind that technologies used in the manufacture of these sensors do not provide the necessary adhesion and mechanical strength of the coatings made from these materials. This means that the stability, the reproducibility of the parameters, and the lifetime of these sensors may be unsatisfactory for actual application.

14.6 SUMMARY

Conducted analysis demonstrates that refractometric approach has great potential for the development of humidity sensors. The sensors can provide a measurement of humidity in a wide range with acceptable accuracy. However, this method has the problem of lacking selectivity since, whatever the substance adsorbed may be, there is a variation in the RI and, therefore, in the intensity of the light. This problem can be overcome by combining the measurement via evanescent spectroscopy with the analysis by absorption or by fluorescence; for example, by setting on the fiber or planar waveguide a chromophore selective for the species being investigated.

Reproducibility of the parameters is another problem of refractometric humidity sensors. As it was mentioned earlier, parameters of refractometric humidity sensors depend on the processes controlling diffusion, adsorption, and condensation of water vapor in the pores of humidity-sensitive materials. Since these processes, as well as the swelling effect, depend on the structure of the films and the conditions of their formation, it becomes clear that establishing the relationship between the structure of the films, the conditions of their deposition and the parameters of the sensors, and the use of these regularities for the subsequent formation of humidity-sensitive films with the required structure should become the main approach for development of the technology of sensor fabrication. Only in this case one can provide an acceptable reproducibility of the parameters of devices developed. Unfortunately, research in this direction is practically not carried out. Most studies end at the stage of the demonstration effect of the possibility of measuring humidity. Stability and durability are also the important issues for polymer-based humidity sensors. It was established that some polymers degrade under the influence of ultraviolet (UV) radiation and oxidizers such as ozone and NO_2 (Razumovskii and Zaikov 1982). It is clear that environment stability of sensors depends on the type of polymer material and therefore, this feature of the polymers should be taking into account when selecting the polymers for humidity sensor design.

However, we need to note that all the above-mentioned problems, including longtime instability, which is accompanied by the temporal drift and degradation of sensor performance, are a big drawback of all types of sensors based on polymers (Kumar and Sharma 1998; Kondratowicz et al. 2001; Bai and Shi 2007).

REFERENCES

Acikgoz S., Bilen B., Demir M.M., Menceloglu Y.Z., Skarlatos Y., Aktas G., Inci M.N. (2008) Use of polyethylene glycol coatings for optical fiber humidity sensing. *Opt. Rev.* 15(2), 84–90.

Ahmad M., Hench L.L. (2005) Effect of taper geometries and launch angle on evanescent wave penetration depth in optical fibers. *Biosens. Bioelectron.* 20(7), 1312–1319.

Alali H., Wang C. (2016) Fiber loop ringdown humidity sensor. *Appl. Opt.* 55(31), 8938–8945.

Alvarez-Herrero A., Guerrero H., Levy D. (2004) High-sensitivity sensor of a low relative humidity based on overlay on side-polished fibers. *IEEE Sens. J.* 4(1), 52–56.

Alwis L., Sun T., Grattan K.T.V. (2013a) Analysis of polyimide-coated optical fiber long-period grating-based relative humidity sensor. *IEEE Sens.* 13, 767–771.

Alwis L., Sun T., Grattan K.T.V. (2013b) Fibre optic long period grating-based humidity sensor probe using a Michelson interferometric arrangement. *Sens. Actuators B* 178, 694–699.

Alwis L., Sun T., Grattan K.T.V. (2013c) Design and performance evaluation of PVA/PI coated optical fibre grating-based humidity sensors. *Rev. Sci. Instrum.* 84, 025002–025008.

Aneesh R., Khijwania S.K. (2012) Titanium dioxide nanoparticle based optical fiber humidity sensor with linear response and enhanced sensitivity. *Appl. Opt.* 51, 2164–2171.

Arregui F.J., Ciaurriz Z., Oneca M., Matias I.R. (2003) An experimental study about hydrogels for the fabrication of optical fiber humidity sensors. *Sens. Actuators B* 96, 165–172.

Bai H., Shi G. (2007) Gas sensors based on conducting polymers. *Sensors* 7, 267–307.

Bariain C., Matias I.R., Arregui F.J., Lopez-Amo M. (2000) Optical fiber humidity sensor based on a tapered fiber coated with agarose gel. *Sens. Actuators B* 69, 127–131.

Bass M. and Van Stryland E.W. (eds.) (2002) Fiber Optic Handbook: *Fiber, Devices, and Systems for Optical Communications*. McGraw-Hill, New York.

Batumalay M., Harun S.W., Irawati N., Ahmad H., Arof H. (2015) A study of relative humidity fiber-optic sensors. *IEEE Sens. J.* 15(3), 1945–1950.

Berruti G., Consales M., Cusano A. et al. (2014) Fiber optic sensors for relative humidity monitoring in high energy physics applications. In: *Proceedings of Italian Conference on Photonics Technologies, 2014 Fotonica AEIT*, Naples, Italy, 14415606, May 12–14.

Bertold III J.W. (1995) Historical review of microbend fiber optic sensors. *J. Lightwave Technol.* 13(7), 1193–1199.

Bhola B., Nosovitskiy P., Mahalingam H., Steier W.H. (2009) Sol–gel-based integrated optical microring resonator humidity sensor. *IEEE Sens.* 9, 740–747.

Börner S., Orghici R., Waldvogel S. R., Willer U., Schade, W. (2009) Evanescent field sensors and the implementation of waveguiding nanostructures. *Appl. Opt.* 48(4), B183–B189.

Bownass D.C., Barton J.S., Jones D.C. (1997) Serially multiplexed point sensor for the detection of high humidity in passive optical networks. *Opt. Lett.* 22(5), 346–348.

Bownass D.C., Barton J.S., Jones D.C. (1998) Detection of high humidity by optical fiber sensing at telecommunications wavelengths. *Opt. Commun.* 146, 90–94.

Bremer K., Wollweber M., Guenther S., Werner G., Sun T., Grattan K.T.V., Roth B. (2014) Fibre optic humidity sensor designed for highly alkaline environments. *Proc. SPIE* 9157, 9157A4-1.

Claus R.O., Matias I., Arregui F. (2002) Optical fiber sensors. In: Bass M. and Van Stryland E.W. (Eds.), *Fiber Optic Handbook: Fiber, Devices, and Systems for Optical Communications*. McGRAW-HILL, New York, Chapter 14.

Correia S.F., Antunes P., Pecoraro E. et al. (2012) Optical fiber relative humidity sensor based on a FBG with a Di-Ureasil coating. *Sensors* 12(7), 8847–8860.

Corres J.M., Bravo J., Matias I.R. (2006) Nonadiabatic tapered single-mode fiber coated with humidity sensitive nanofilms. *IEEE Photon. Technol. Lett.* 18(8), 935–937.

Corres J.M., Bravo J., Matias I.R. (2007) Sensitivity optimisation of tapered optical fiber humidity sensors by means of tuning the thickness of nanostructured sensitive coatings. *Sens. Actuators B* 122, 442–449.

Corres J.M., Matias I.R., Hernaez M., Bravo J., Arregui F.J. (2008a) Optical fiber humidity sensors using nanostructured coatings of SiO_2 nanoparticles. *IEEE Sens. J.* 8(3), 281–285.

Corres J.M., Villar I.D., Matias I.R., Arregui F.J. (2008b) Two-layer nanocoatings in long-period fiber gratings for improved sensitivity of humidity sensors. *IEEE Trans. Nanotechnol.* 7, 394–400.

David N.A., Wild P.M., Djilali N. (2012) Parametric study of a polymer-coated fiber-optic humidity sensor. *Measur. Sci. Technol.* 23(3), 035103.

Ding F., Wang L., Fang N., Huang Z. (2011) Experimental study on humidity sensing using a FBG sensor with polyimide coating. *Proc. SPIE* 7990, Optical Sensors and Biophotonics II, 79900C.

Elosua C., Matias I.R., Bariain C., Arregui F.J. (2006) Volatile organic compound optical fiber sensors: A review. *Sensors* 6, 1440–1465.

Erdogan T. (1997a) Fiber grating spectra. *J. Lightwave Technol.* 15, 1277–1294.

Erdogan T. (1997b) Cladding-mode resonances in short- and long-period fiber grating filters. *J. Opt. Soc. Am. A* 14, 1760–1773.

Erdogan T., Sipe J.E. (1996) Tilted fiber phase gratings. *J. Opt. Soc. Am. A* 13, 296–313.

Eryürek M., Tasdemir Z., Karadag Y., Anand S., Kilinc N., Alaca B.E., Kiraz A. (2017) Integrated humidity sensor based on SU-8 polymer microdisk microresonator. *Sens. Actuators B* 242, 1115–1120.

Fields J.N. (1980) Attenuation of a parabolic index fiber with periodic bends. *Appl. Phys. Lett.* 36(10), 799–801.

Foreman M.R., Swaim J.D., Vollmer F. (2015) Whispering gallery mode sensors. *Adv. Opt. Photon.* 7, 168–240.

Fu M.Y., Lin G.R., Liu W.F., Wu C. (2011) Fiber-optic humidity sensor based on an air-gap long period fiber grating. *Opt. Rev.* 18(1), 93–95.

Gaston A., Lozano I., Perez F., Auza F., Sevilla J. (2003) Evanescent wave optical fiber sensing (temperature, relative humidity and pH sensors). *IEEE Sens. J.* 3(6), 806–811.

Gaston A., Perez F., Sevilla J. (2004) Optical fiber relative humidity sensor with polyvinyl alcohol film. *Appl. Opt.* 43(21), 4127–4132.

Ghadiry M., Gholami M., Lai C.K., Ahmad H., Chong W.Y. (2016) Ultra-sensitive humidity sensor based on optical properties of graphene oxide and nano-anatase TiO_2. *PLoS One* 2016, 0153949. doi:10.1371/journal.pone.0153949.

Giaccari P., Limberger H.G., Kronenberg P. (2001) Influence of humidity and temperature on polyimide-coated fiber Bragg gratings. In: *Proceedings of the Trends in Optics and Photonics Series: Bragg Gratings, Photosensitivity, and Poling in Glass Waveguides*, Washington, DC, Vol. 61, pp. BFB2.

Gomez D., Morgan S.P., Gill B.R.H., Korposh S. (2016) Polymeric fibre optic sensor based on a SiO_2 nanoparticle film for humidity sensing on wounds. *Proc. SPIE* 9916, 991623-1

Gouveia C.A.J., Baptista J.M., Jorge P.A.S. (2013) Refractometric optical fiber platforms for label free sensing. In: Harun S.W., Arof H. (Eds.), *Current Developments in Optical Fiber Technology*. InTech, Rijeka, Croatia, pp. 345–373.

Grattan K.T.V., Sun T. (2000) Fiber optic sensor technology: An overview. *Sens. Actuators* 82, 40–61.

Grossman B.G., Yongphiphatwong T., Sokol M. (2005) In situ device for salinity measurements (chloride detection) of ocean surface. *Optics Laser Technol.* 37, 217–223.

Gründler P. (2006) *Chemical Sensors: An Introduction for Scientists and Engineers.* Springer-Verlag, Berlin, Germany.

Guo S.-L., Wang W.-J., Hu C.-H. (2014) A high sensitivity humidity sensor based on micro-ring resonator with three coupling points. *Proc. SPIE* 9297, 92972P-1.

Gupta B.D., Ratnanjali (2001) A novel probe for a fiber optic humidity sensor. *Sens. Actuators B* 80, 132–135.

Hatamie S., Dhas V., Kale B.B., Mulla I.S., Kale S.N. (2009) Polymerembedded stannic oxide nanoparticles as humidity sensors. *Mater. Sci. Eng. C* 29, 847–850.

Heebner J., Grover R., Ibrahim T. (2008) *Optical Microresonators: Theory, Fabrication and Applications*, 1st ed. Springer Series in Optical Sciences. Springer, London, UK.

Huang X.F., Sheng D.R., Cen K.F., Zhou H. (2007) Low-cost relative humidity sensor based on thermoplastic polyimide-coated fiber Bragg grating. *Sens. Actuators B* 127, 518–524.

Ilchenko V.S., Volikov P.S., Velichansky V.L., Treussart F., Lefèvre-Seguin V., Raimond J.-M., Haroche S. (1998) Strain tunable High-Q optical microsphere resonator. *Opt. Commun.* 145(1–6), 86–90.

Ioppolo T., Ötügen V., Ayaz U. (2013) Development of whispering gallery mode polymeric micro-optical electric field sensors. *J. Vis. Exp.* 71, e50199, 3–6.

Jindal R., Tao S., Singh J.R., Gaikwad P.S. (2002) High dynamic range fiber optic relative humidity. *Opt. Eng.* 41(5), 1093–11093.

Kang Y., Ruan H., Wang Y., Arregui F.J., Matias I.R., Claus R.O. (2006) Nanostructured optical fibre sensors for breathing airflow monitoring. *Meas. Sci. Technol.* 17, 1207–1210.

Kersey A.D. (1996) A review of recent developments in fiber optic sensor technology. *Opt. Fiber Technol.* 2, 291–317.

Khalil S., Bansal L., El-Sherif M. (2004) Intrinsic fiber optic chemical sensor for the detection of dimethyl methylphosphonate. *Opt. Eng.* 43, 2683–2688.

Kharaz A., Jones B. (1995) A distributed fiber optic sensing system for humidity measurement. *Meas. Contr.* 28, 101–103.

Kharaz A., Jones B.E., Hale K.F., Roche L., Bromley K. (2000) Optical fiber relative humidity sensor using a spectrally absorptive material. In: *SPIE Proceedings of International Conference on Optical Fiber Sensors*, Venice, Italy, Vol. 4185, pp. 370–373.

Khijwania S.K., Gupta B.D. (1999) Fiber optic evanescent field absorption sensor: Effect of fiber parameters and geometry of the probe. *Opt. Quant. Electron.* 31(8), 625–636.

Khijwania S.K., Srinivasan K.L., Singh J.P. (2005) Performance optimized optical fiber sensor for humidity measurement. *Opt. Eng.* 44, 34401(1–7).

Kogelnik H., Shank C.W. (1972) Coupled wave theory of distributed feedback lasers. *J. Appl. Phys.* 43, 2327–2335.

Kondratowicz B., Narayanaswamy R., Persaud K.C. (2001) An investigation into the use of electrochromic polymers in optical fibre gas sensors. *Sens. Actuators B* 74, 138–144.

Konstantaki M., Pissadakis S., Pispas S., Madamopoulos N., Vainos N.A. (2006) Optical fiber long-period grating humidity sensor with poly(ethylene oxide)/cobalt chloride coating. *Appl. Opt.* 45(19), 4567–4571.

Korenko B., Rothhardt M., Hartung A., Bartelt H. (2015) Novel fiber-optic relative humidity sensor with thermal compensation. *IEEE Sens. J.* 15(10), 5450–5454.

Korotcenkov G., Cho B.K., Sevilla III F., Narayanaswamy R. (2011) Optical and fiber optic chemical sensors. In: Korotcenkov G. (Ed.), *Chemical Sensors: Comprehensive Sensor Technologies.* Vol. 5: Electrochemical and Optical Sensors. Momentum Press, New York, pp. 311–476.

Krioukov E., Klunder D.J.W., Driessen A., Greve J., Otto C. (2002) Sensor based on an integrated optical microcavity. *Opt. Lett.* 27(7), 512–514.

Kronenberg P., Rastogi P.K., Giaccari P., Limberger H.G. (2002) Relative humidity sensor with optical fiber Bragg grating. *Opt. Lett.* 27, 1385–1387.

Kumar D., Sharma R.C. (1998) Advances in conductive polymers. *Eur. Polym. J.* 34, 1053–1060.

Kwesey A.S. (1996) A review of recent developments in fiber optic sensor technology. *Opt. Fiber Technol.* 2, 291–317.

Labrador-Paez L., Soler-Carracedo K., Hernandez-Rodriguez M., Martin I.R., Carmon T., Martin L.L. (2017) Liquid whispering-gallery-mode resonator as a humidity sensor. *Opt. Exp.* 25(2), 1165–1172.

Lee B. (2003) Review of the present status of optical fiber sensors. *Optic. Fiber Technol.* 9, 57–79.

Lee K., Erdogan T. (2000) Fiber mode coupling in transmissive and reflective tilted fibre gratings. *Appl. Opt.* 39, 1394–1404.

Lee T.S., George N.A., Sureshkumar P., Radhakrishnan P., Vallabhan C.P.G., Nampoori V.P.N. (2001) Chemical sensing with microbent optical fiber. *Opt. Lett.* 26, 1541–1543.

Lee K.J., Wawro D., Priambodo P.S., Magnusson R. (2007) Agarose-gel based guided-mode resonance humidity sensor. *IEEE Sens. J.* 7(3), 409–414.

Leung A., Shankar P.M., Mutharasan R. (2007) A review of fiber-optic biosensors. *Sens. Actuators B* 125, 688–703.

Li T., Dong X., Chan C.C., Zhao C.L., Zu P. (2012) Humidity sensor based on a multimode-fiber taper coated with polyvinyl alcohol interacting with a fiber Bragg grating. *IEEE Sens. J.* 12(6), 2205–2208.

Liang Y., Yan G., He S. (2015) Enlarged-taper tailored fiber Bragg grating with polyvinyl alcohol coating for humidity sensing. *Proc. SPIE* 9620, 962007-1.

Ling T., Guo L.J. (2007) A unique resonance mode observed in a prism-coupled micro-tube resonator sensor with superior index sensitivity. *Opt. Exp.* 15(25), 17424–17432.

Lokman A., Nodehi S., Batumalay M., Arof H., Ahmad H., Harun S.W. (2014) Optical fiber humidity sensor based on a tapered fiber with hydroxyethylcellulose/polyvinyldenefluoride composite. *Microwave Opt. Technol. Lett.* 56(2), 380–382.

Love J.D., Durniak C. (2007) Bend loss, tapering, and cladding-mode coupling in single-mode fibers. *IEEE Photonics Technol. Lett.* 19(16), 1257–1259.

Luo S., Liu Y., Sucheta A., Evans M., van Tassell R. (2002) Applications of LPG fiber optical sensors for relative humidity and chemical warfare agents monitoring. *Proc. SPIE*, Advanced Sensor Systems and Applications, 4920, 193–204.

Luo Y., Chen C., Xia K. et al. (2016) Tungsten disulfide (WS_2) based. *Opt. Exp.* 24(8), 8956–8966.

Ma Q., Huang L., Guo Z., Rossmann T. (2010) Spectral shift response of optical whispering-gallery modes due to water vapor adsorption and desorption. *Meas. Sci. Technol.* 21, 115206.

MacCraith B.D., Ruddy V., Potter C., O'Kelly B., McGilp J.F. (1991) Optical waveguide sensor using evanescent wave excitation of fluorescent dye in sol-gel glass. *Electron. Lett.* 27(14), 1247–1248.

Mackenzie H.F., Payne F.P. (1990) Evanescent field amplification in a tapered single-mode optical fibre. *Electron. Lett.* 26(2), 130–132.

MacLean A., Moran C., Johnstone W., Culshaw B., Marsh D., Watson V., Andrews G. (2000) A distributed fiber optic sensor for hydro-carbon detection. In: *Proceedings of 14th International Conference on Optical Fiber Sensors, OFS 2000*, Venice, Italy, pp. 382–385, October 11–13.

Maier R.R.J., Barton J.S., Jones J.D.C., McCulloch S., Jones B.J.S., Burnell G. (2006) Palladium-based hydrogen sensing for monitoring of ageing materials. *Meas. Sci. Technol.* 17, 1118–1123.

Mallik A.K., Liu D., Kavungal V., Wu Q., Farrell G., Semenova Y. (2016) Agarose coated spherical micro resonator for humidity measurements. *Opt. Exp.* 24(19), 21216–21227.

Mathew J. (2013) Development of novel fiber optic humidity sensors and their derived applications. PhD Thesis, Dublin Institute of Technology.

Mathew J., Rajan G., Semenova Y., Farrell G. (2009) All fiber tunable loss filter. *Proc. SPIE* 7503, 75037G.

Mathew J., Semenova Y., Farrell G. (2012) A fiber bend based humidity sensor with a wide linear range and fast measurement speed. *Sens. Actuators A* 174, 47–51.

Mathew J., Semenova Y., Rajan G., Wang P., Farrell G. (2011) Improving the sensitivity of a humidity sensor based on fiber bend coated with a hygroscopic coating. *Opt. Laser Technol.* 43(7), 1301–1305.

Mathew J., Thomas K.J., Nampoori P.N., Radhakrishnan P. (2007) A comparative study of fiber optic humidity sensors based on chitosan and agarose. *Sens. Trans. J.* 84(10), 1633–1640.

Matsko A.B., Ilchenko V.S. (2006) Optical resonators with Whispering-gallery modes—Part I: Basics. *IEEE J. Sel. Top. Quant. Electron.* 12, 3–14.

Mehrabani S., Kwong P., Gupta M., Armani A.M. (2013) Hybrid microcavity humidity sensor. *Appl. Phys. Lett.* 102, 241101.

Messica A., Greenstein A., Katzir A. (1996) Theory of fiber-optic, evanescent-wave spectroscopy and sensors. *Appl. Opt.* 35, 2274–2284.

Michie W.C., Culshaw B., Konstantaki M., McKenzie I., Kelly S., Graham N.B., Moran C. (1995) Distributed pH and water detection using fiber-optic sensors and hydrogels. *Lightwave Technol. J.* 13(7), 1415–1420.

Mignani A.G., Falciai R., Ciaccheri L. (1998) Evanescent wave absorption spectroscopy by means of bi-tapered multimode optical fibers. *Appl. Spectrosc.* 52(4), 546–551.

Moar P.N., Huntington S.T., Katsifolis J., Cahill L.W., Roberts A., Nugent K.A. (1999) Fabrication, modeling, and direct evanescent field measurement of tapered optical fiber sensors. *J. Appl. Phys.* 85(7), 3395–3398.

Morisawa M., Yamaoka H., Suzuki Y. (2016) Improvement of POF humidity sensor with swelling polymer cladding via bending. In: *Proceedings of IEEE Sensors*, Orlando, FL, 16597073, October 30–November 3. doi:10.1109/ICSENS.2016.7808454.

Murray R.W. (1989) Chemical sensors and microinstrumentation: An overview. In: Murray R., Dessy R.E., Heineman W.R., Janata J., Seitz W.R. (Eds.), *Chemical Sensors and Microinstrumentation, ACS Symposium Series*. American Chemical Society, Washington, DC, pp. 1–19.

Muto S., Suzuki O., Amano T., Morisawa M. (2003) A plastic optical fiber sensor for real-time humidity monitoring. *Meas. Sci. Technol.* 14, 746–750.

Ogawa K., Tsuchiya S., Kawakami H. (1988) Humidity sensing effects of optical fibers with microporous SiO_2 cladding. *Electron. Lett.* 24(1), 42–43.

Othenos A., Kalli K. (1999) *Fiber Bragg Gratings*. Artech House, Norwood, MA.

Otsuki S., Adachi K., Taguchi T. (1998) A novel fiber optic gas sensing configuration using extremely crived optical fibers and an attempt for optical humidity detection. *Sens. Actuators B* 53, 91–96.

Ramachandran A., Wang S., Clarke J. et al. (2008) A universal biosensing platform based on optical micro-ring resonators. *Biosens. Bioelectron.* 23, 939–944.

Razumovskii S.D., Zaikov G.Y. (1982) Effect of ozone on saturated polymers. *Polym. Sci. USSR* 24(10), 2305–2325.

Ruddy V., MacCraith B.D., Murphy J.A. (1990) Evanescent wave absorption spectroscopy using multimode fibers. *J. Appl. Phys.* 67(10), 6070–6074.

Russell A.P., Fletcher K.S. (1985) Optical sensor for the determination of moisture. *Anal. Chim. Acta* 170, 209–216.

Sadaoka Y., Sakai Y., Murata Y.U. (1993) Optical humidity and ammonia gas sensor using calcein-based films. *Sens. Actuators B* 13, 420–423.

Sanchez P., Zamarreño C.R., Hernaez M., Del Villar I., Fernandez-Valdivielso C., Matias I.R., Arregui F.J. (2011) Lossy mode resonances toward the fabrication of optical fiber humidity sensors. *Meas. Sci. Technol.* 23(1), 014002.

Shivananju B.N., Yamdagni S., Fazuldeen R., Kumar A.K.S., Nithin S.P., Varma M.M., Asokan S. (2014) Highly sensitive carbon nanotubes coated etched fiber Bragg grating sensor for humidity sensing. *IEEE Sens. J.* 14(8), 2615–2619.

Sparrow I.J.G., Smith P.G.R., Emmerson G.D., Watts S.P., Riziotis C. (2009) Planar Bragg grating sensors—Fabrication and applications: A review. *J. Sens.* 2009, 607647.

Su P.-G., Sun Y.-L., Lin C.-C. (2006) Humidity sensor based on PMMA simultaneously doped with two different salts. *Sens. Actuators B* 113, 883–886.

Swanson A.J., Raymond S.G., Janssens S., Breukers R.D., Bhuiyan M.D.H., Lovell-Smith J.W., Waterland M.R. (2016) Development of novel polymer coating for FBG based relative humidity sensing. *Sens. Actuators A* 249, 217–224.

Tan F.S., Klunder D.J.W., Kelderman H., Hoekstra H.J.W.M., Driessen A. (2002) High finesse vertically coupled waveguide-microring resonators based on Si_3N_4-SiO_2 technology. In: *Proceedings of 2002 IEEE/LEOS Workshop on Fibre and Optical Passive Components*, Glasgow, Scotland, pp. 228–232, June 5–6.

Tan K.M., Tay C.M., Tjin S.C., Chan C.C., Rahardjo H. (2005) High relative humidity measurements using gelatine coated long-period grating sensors. *Sens. Actuators B* 110, 335–341.

Tao J., Luo Y., Wang L., Cai H., Sun T., Song J., Liu H., Gu Y. (2016) An ultrahigh-accuracy miniature dew point sensor based on an integrated photonics platform. *Sci. Rep.* 6, 29672.

Taylor H.F. (1984) Bending effects in optical fibers. *J. Lightwave Technol.* LT-2(5), 617–628.

Thomas L.S., Geetha K., Nampoori V.P.N., Vallabhan C.P.G., Radhakrishnan P. (2002) Microbent optical fibers as evanescent wave sensors. *Opt. Eng.* 41(12), 3260–3264.

Urrutia A., Goicoechea J., Ricchiuti A.L., Barrera D., Sales S., Arregui F.J. (2016) Simultaneous measurement of humidity and temperature based on a partially coated optical fiber long period grating. *Sens. Actuators B* 227, 135–141.

Vahala K.J. (2003) Optical microcavities. *Nature* 14, 839–846.

Van V. (2016) *Optical Microring Resonators: Theory, Techniques, and Applications*. CRC Press, Boca Raton, FL.

Vengsarkar A., Pedrazzani J., Judkins J., Lemaire P., Bergano N., Davidson C. (1996) Long-period fiber-grating-based gain equalizers. *Opt. Lett.* 21(5), 336–338.

Venugopalan T., Yeo T.L., Basedau F., Henke A.S., Sun T., Grattan K.T.V., Habel W. (2009) Evaluation and calibration of FBG-based relative humidity sensor designed for structural health monitoring. *Proc. SPIE* 7503, 750310-1.

Venugopalan T., Yeo T.L., Sun T., Grattan K.T.V. (2008) LPG-based PVA coated sensor for relative humidity measurement. *IEEE Sens. J.* 8(7), 1093–1098.

Viegas D., Goicoechea J., Santos J.L., Araújo F.M., Ferreira L.A., Arregui F.J., Matias I.R. (2009) Sensitivity improvement of a humidity sensor based on silica nanospheres on a long-period fiber grating. *Sensors* 9(1), 519–527.

Viegas D., Hernaez M., Goicoechea J., Santos J.L., Araújo F.M., Arregui F.J., Matias I.R. (2011) Simultaneous measurement of humidity and temperature based on an SiO_2 nanospheres film deposited on a long-period grating in-line with a fiber Bragg grating. *IEEE Sens. J.* 11(1), 162–166.

Vijayan A., Fuke M., Hawaldar R., Kulkarni M., Amalnerkar D., Aiyer R.C. (2008) Optical fiber based humidity sensor using Co-polyaniline clad. *Sens. Actuators B* 129, 106–112.

Waechter H., Litman J., Cheung A.H., Barnes J.A., Loock H.P. (2010) Chemical sensing using fiber cavity ring-down spectroscopy. *Sensors* 10, 1716–1742.

Wang C. (2014) Fiber loop ringdown sensors and sensing. In: Gagliardi G., Loock H.-P. (Eds.), *Cavity-Enhanced Spectroscopy and Sensing*. Springer, Berlin, Germany, pp. 411–461.

Wang Y., Shen C., Lou W., Shentu F., Zhong C., Dong X., Tong L. (2016) Fiber optic relative humidity sensor based on the tilted fiber Bragg grating coated with graphene oxide. *Appl. Phys. Lett.* 109, 031107.

Wise D.L., Wingard L.B. Jr. (1991) *Biosensors and Fiberoptics*. Humana Press, Clifton, NJ.

Wu Y., Zhang T., Rao Y., Gong Y. (2011) Miniature interferometric humidity sensors based on silica/polymer microfiber knot resonators. *Sens. Actuators B* 155, 258–263.

Xu L., Fanguy J.C., Soni K., Tao S. (2004) Optical fiber humidity sensor based on evaenscent wave scattering. *Opt. Lett.* 29(11), 1191–1193.

Xu Q., Lipson M. (2007) All-optical logic based on silicon micro-ring resonators. *Opt. Exp.* 15(3), 925–929.

Yan Y., Yan S., Zhang Y., Wang L., Liu J., Xue C., Xiong J. (2011) Packaging and isolating microsphere coupling system. In: *Proceedings of the 6th IEEE International Conference on Nano/Micro Engineered and Molecular Systems*, Kaohsiung, Taiwan, pp. 928–931, February 20–23.

Yao L., Birks A., Knight J.C. (2009) Low bend loss in tightly-bend fibers through adiabatic bend transitions. *Opt. Exp.* 17(4), 2962–2967.

Yariv A. (1973) Coupled-mode theory for guided-wave optics. *IEEE J. Quantum Electron.* 9, 919–933.

Yeo T.L., Sun T., Grattan K.T.V., Parry D., Lade R., Powell B.D. (2005) Characterisation of a polymer-coated fiber Bragg grating sensor for relative humidity sensing. *Sens. Actuators B* 110, 148–156.

Yuan J., El-Sherif A. (2003) Fiber-optic chemical sensor using polyaniline as modified cladding material. *IEEE Sens. J.* 3, 5–12.

Zhao Z., Duan Y. (2011) A low cost fiber-optic humidity sensor based on silica sol–gel film. *Sens. Actuators B* 160, 1340–1345.

Zhang C., Zhang W., Webb D.J., Peng G.D. (2010) Optical fiber temperature and humidity sensor. *Electron. Lett.* 46(9), 643–644.

Zhang L., Gu F., Lou J., Yin X., Tong L. (2008) Fast detection of humidity with a subwavelength diameter fiber taper coated with gelatin film. *Opt. Exp.* 16(17), 13349–13353.

Zhang W., Webb D.J. (2014) Humidity responsivity of poly(methyl methacrylate)-based optical fiber Bragg grating sensors. *Opt. Lett.* 39(10), 3026–3029.

Zhang Z.F., Zhang Y. (2015) Humidity sensor based on optical fiber attached with hydrogel spheres. *Opt. Laser Technol.* 74, 16–19.

Zhang J., Zhong J., Fang Y.F., Wang J., Huang G.S., Cui X.G., Mei Y.F. (2014) Roll up polymer/oxide/polymer nanomembranes as a hybrid optical microcavity for humidity sensing. *Nanoscale* 6, 13646–13650.

Zhou K., Chen X., Simpson A.G., Zhang L., Bennion I. (2005a) High extinction ratio infiber polarizer based on a 45°-tilted fiber Bragg grating. In: *Proceedings of Optical Fibre Communication Conference*, Anaheim, CA, paper OME22.

Zhou K., Simpson G., Chen X., Zhang L., Bennion I. (2005b) High extinction ratio infiber polarizers based on 45°-tilted fiber Bragg gratings. *Opt. Lett.* 30, 1285–1287.

15 Luminescence (Fluorescence)-Based Humidity Sensors

15.1 BASIC PRINCIPLES

As noted earlier, the third main type of optical humidity sensors was derived from the phenomenon that water molecules are able to influence the fluorescence emission of certain compounds. With increasing relative humidity (RH), either an enhancement or a quenching of the fluorescence intensity of a humidity-indicator dye was observed (Wang et al. 1991; Jung et al. 2016).

In fluorescence-based optical chemical sensors, two facts can be distinguished: (1) the substance being investigated is itself fluorescent; (2) the substance is not fluorescent: in this case, it can be labeled with a fluorophore or else it can react chemically with a reagent, giving rise to a fluorescent product, or even interact with a fluorophore, causing a variation of its emission of fluorescence. One should note that humidity sensors can be designed using only second approach. Thus, luminescence (fluorescence) optical humidity sensors are based on the humidity influence on the spontaneous light emission of a fluorophore when it is excited with light at a wavelength (excitation wavelength) located in the absorption spectral region of such fluorophore. Usually excitation wavelength differs from the emission wavelength (Gansert et al. 2006; Orellana 2006). In that regard, of particular interest is the phenomenon known as fluorescence *quenching*, in which the fluorescence intensity decreases, following interaction with the water vapors under examination, which can thus be detected. Quenching of the indicator dye fluorescence by the water vapors is one of the most widespread methods adopted to develop humidity optosensors. It is also possible to fabricate fluorescent sensors based on excited state *energy transfer* quenching. *Fluorescence resonance energy transfer* (FRET) has also been used very often to design optical sensors. In this case, the sensitive layer contains the fluorophore and an analyte-sensitive dye, the absorption band of which overlaps significantly with the emission of the former. Reversible interaction of the absorber with the analyte species leads to a variation of the absorption band so that the efficiency of energy transfers from the fluorophore changes. In this way, both emission intensity- and lifetime-based sensors may be fabricated. Photo-induced *electron transfer* (PET) quenching processes can also be employed to develop fluorescence optosensors (Figure 15.1). Due to the electron promotion on absorption of light, every electronic excited state is both a better oxidant and a better reducing agent than its corresponding ground state. Therefore, the analyte will undergo a photoredox reaction with the appropriate (luminescent) indicator dye (intermolecular quenching) or will interfere with an existing intramolecular PET quenching. For a more in-depth discussion of this phenomenon one can refer to the overview (de Silva et al. 2009).

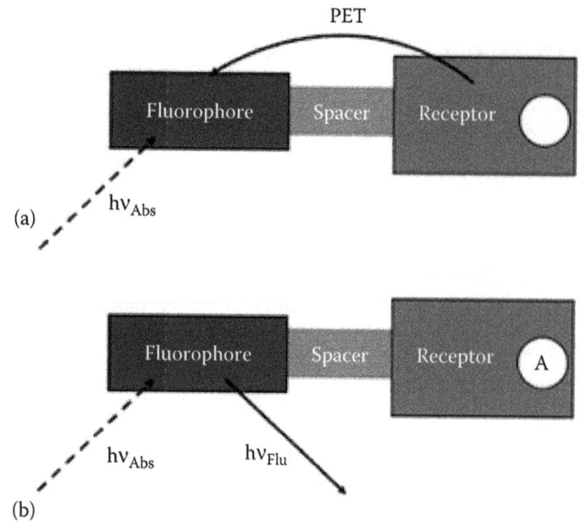

FIGURE 15.1 (a) An electron transfer from the analyte-free receptor to the photo-excited fluorophore creates the *off* state of the sensor. (b) The electron transfer from the analyte-bound receptor is blocked resulting in the *on* state of the sensor. (de Silva, A.P. et al., *Analyst*, 134, 2385–2393, 2009. Reproduced by permission of The Royal Society of Chemistry.)

15.2 PRINCIPLES OF OPERATION

As a rule, the probe molecule, fluorophore, in humidity sensors is immobilized onto a (thin) polymer support (sometimes even the waveguide surface itself) and placed at the distal end or in the evanescent field of an optical fiber (Figure 15.2) or integrated optic sensor (Figure 15.3).

Planar waveguide sensors usually employ the technique of evanescent wave excitation of fluorescence (Figure 15.3). The sensor configuration is relatively straightforward, consisting of a planar waveguide to which light is coupled using one of the techniques discussed before. The evanescent field of the guided light extends into the superstrate (i.e., the sensing environment) and induces fluorescence in susceptible molecules located within a distance of typically 100–200 nm from the waveguide surface. In some cases, the excited fluorescence is detected either above or below the sensor platform, but it is also possible to use the waveguide itself to capture fluorescence for detection at the output end the face of the waveguide (Burke et al. 2005) (Figure 15.3). With such a configuration, it is possible to utilize direct excitation of the fluorescence as opposed to evanescent wave excitation, which results in higher signal levels and improved sensor performance, while improving the inherent ability of the platform to discriminate between the excitation light and the fluorescence (Burke et al. 2005).

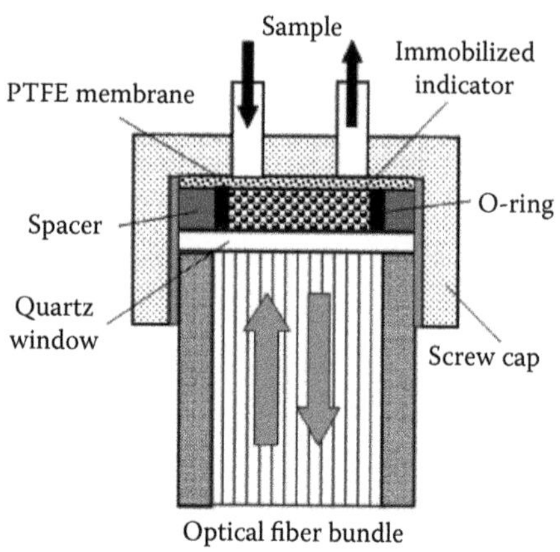

FIGURE 15.2 Custom-made flow-through cell set up used for fiber-optic measurements using fluorescence method. (Reprinted with permission from Orellana, G. et al., *Anal. Chem.*, 67, 2231–2238, 1995. Copyright 1995 American Chemical Society.)

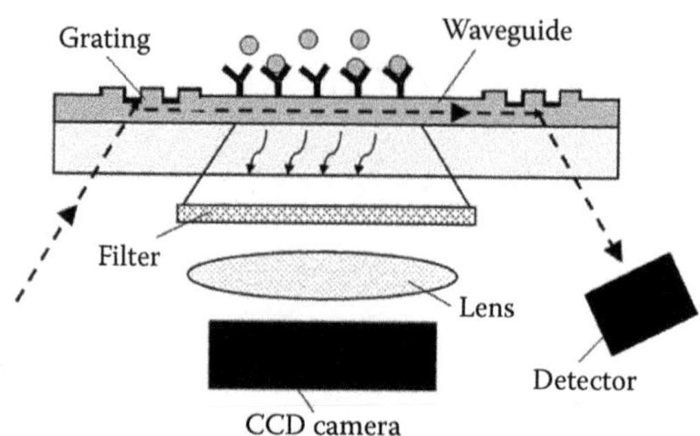

FIGURE 15.3 Detection principle of evanescent-field fluorescence on planar waveguides: excitation light is coupled into a thin-film waveguide; surface confined fluorescence of bound labeled molecules is detected by a CCD camera. The fluorescence could be also separated via the outcoupling grating and detected by a photodiode. (From Korotcenkov, G. et al., Optical and fiber optic chemical sensors, in *Chemical Sensors*, Vol. 5: Electrochemical and Optical Sensors, Korotcenkov, G. (Ed.), Momentum Press, New York, 2011. With permission.)

As it follows from the principle of the sensor operation, fluorescence spectra are composed of both excitation and emission components, which should, in principle, give this spectroscopic approach a great potential for selective analysis. The difference in wavelength between both emission (usually longer wavelength) and excitation wavelengths (shorter wavelength, thus higher photon energy) is called a Stokes shift (Figure 15.4). In addition to specific excitation and emission measurements, selectivity can be enhanced for fluorescence by utilizing matrix effects (micelles, surfaces, and heavy atoms), temperature effects (Shpol'skii effect), fluorescence and phosphorescence lifetime filtering, and synchronous

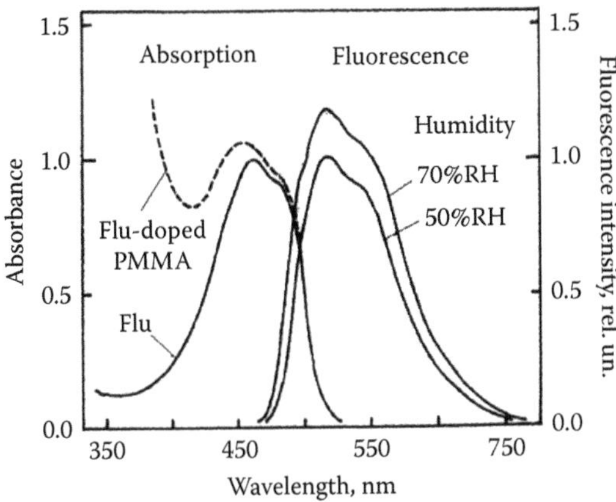

FIGURE 15.4 Absorption and fluorescence spectra of fluorescein-dye-doped PMMA film. (From Muto, S. et al., *Jpn. J. Appl. Phys.*, 33, 6060–6064, 1994. Copyright 1994, The Japan Society of Applied Physics. With permission.)

scanning. For a detailed review of fluorescence spectroscopy, the reader should be referred to Lakowicz (1982, 1999), Fleming (1986), and Valeur (2002). One should note that luminescence (fluorescent)-based sensors occupy nowadays a prominent place among the optical devices due to their superb *sensitivity* (just a *single* photon sometimes sufficient for quantifying luminescence compared to detecting the intensity difference between two beams of light as in absorption techniques) (Orellana 2006).

In *fluorescence-based optical chemical sensors*, different measurement schemes have been proposed. Among them the most popular are fluorescence intensity sensors, fluorescence lifetime sensors and fluorescence phase-modulation sensors (Mitsubayashi et al. 2006). If the sensing dye interacts with the water vapors, a change in the emission of the dye will be used as the sensing response. Looking ahead to maximize the emitted light coupled to the fiber, some special terminations are built onto the cut-end pigtails, usually tapers (Jorge et al. 2004; Moreno et al. 2005) (Figure 15.5). The measurement of luminescence intensity would be simple in terms of instrumentation but its accuracy is often compromised by adverse effects such as drifts of the optoelectronic system, due to variations in the intensity of the light sources and the sensitivity of the detector, and variations in the optical properties of the sample, including fluorophore concentration, turbidity, coloration, and refractive index, caused by the change in the surrounding temperature.

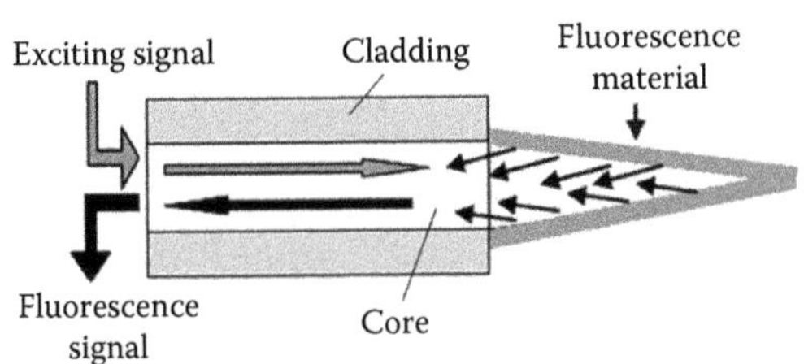

FIGURE 15.5 Reflection fluorescence sensor based on a tapered-end fiber. (From Elosua, C. et al., *Sensors*, 6, 1440–1465, 2006. Published by MDPI as open access.)

To solve these problems, there are different approaches. For example, Muto et al. (1994), who designed RH fiber-optic sensors based on a transparent plastic core and fluorescein-dye-doped polymer-cladding layer, suggested using the measuring circuit shown in Figure 15.6. To eliminate the influence of the fluorescence decay caused by thermal quenching, another fluorescence plastic optical fiber with humidity-proof layer was also set inside the jacket as a compensator.

Another possibility for overcoming above-mentioned problems is to use a luminescence decay time of chromophore indicator for sensor purposes (Figure 15.7), which is hardly affected by fluctuations of the overall fluorescence intensity, and the time-resolved fluorescence spectroscopy has been successfully applied in chemical sensors (Draxler and Lippitsch 1996). According to Gansert et al. (2006), the measurement of the luminescence decay time, an intrinsically referenced parameter, has the following advantages compared to conventional intensity measurements:

- The decay time does not depend on fluctuations of the intensity of the light source and the sensitivity of the detector.
- The decay time is not influenced by signal loss caused by fiber bending or by intensity changes caused by changes in the geometry of the sensor.
- The decay time is, to a great extent, independent of the concentration of the indicator dye in the sensitive layer. Therefore, photobleaching and leaching of the indicator dye have no influence on the measured signal.
- The decay time is not influenced by variations in the optical properties of the sample, including turbidity, refractive index, and coloration.

Discussions of the methods, which can be used for design an optical sensor using decay time measurements have been presented in the literature (Gansert et al. 2006). For example, a fluorescent lifetime filtering can be used to selectively measure luminescence from a specific set of chromophores. This method can be utilized for resolution of mixtures of various compounds which are of environmental interest. In the case when several species interact with the reagent, causing emission of fluorescence characterized by different

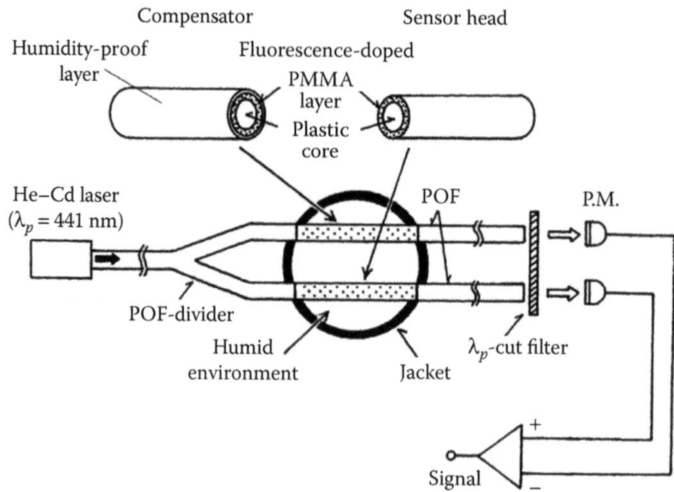

FIGURE 15.6 Configuration of RH sensor developed by Muto S. and coworkers: P.M. is photomultiplier. (From Muto, S. et al., *Jpn. J. Appl. Phys.*, 33, 6060–6064, 1994. Copyright 1994, The Japan Society of Applied Physics. With permission.)

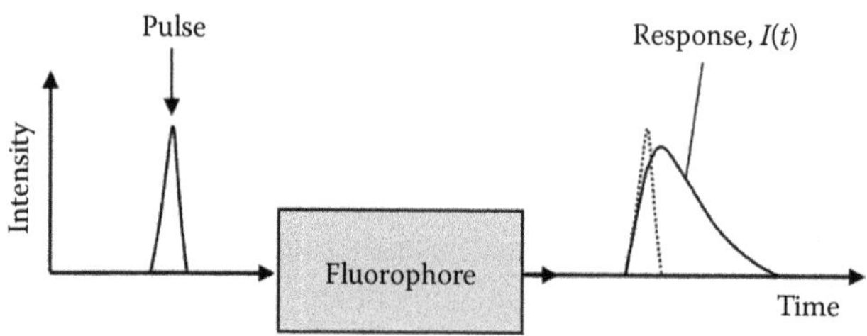

FIGURE 15.7 Principle of time-resolved fluorometry.

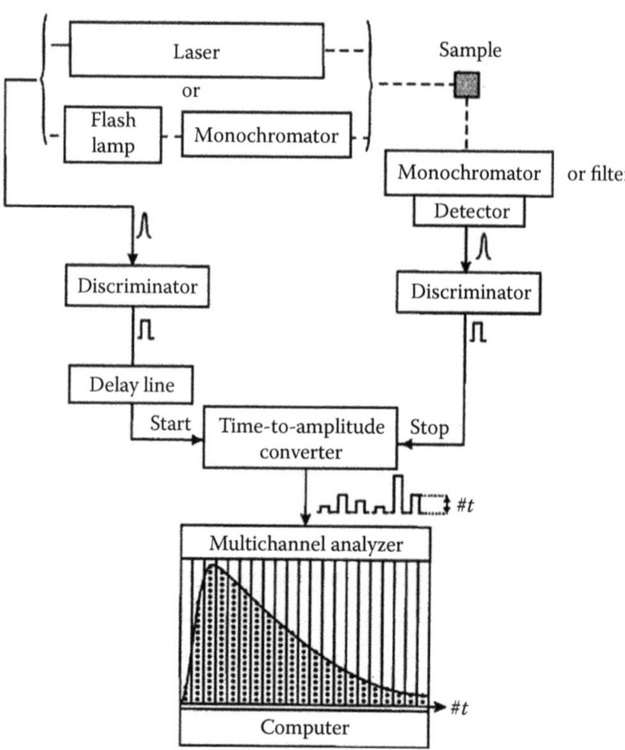

FIGURE 15.8 Scheme diagram of a single-photon timing fluorimeter. (From Valeur, B., *Molecular Fluorescence. Principles and Applications*, Wiley-VCH Verlag, Weinheim, Germany, 2002.)

decay times, these can be detected simultaneously, by using time-resolution devices (Brenci and Baldini 1991). One of variants of experimental setup used for fluorescence decay-time measurements is shown in Figure 15.8. A detailed discussion about luminescence measurements is given in the literature (Righini and Ferrari 2005; Ryskevic et al. 2010). Importantly, this method is not simple and the realization of this method requires the ultrafast pulse excitation and fast photomultiplier detection.

Phase modulation is also frequently employed in fluorescence sensors to avoid the jittering in the fluorescence and also changes in fluorescence produced by movements of the own fiber (Jorge et al. 2004). Fluorescence is excited with sinusoidally modulated light at high variable frequency. Polarization is another physically observable property of the luminescence, which can be used in humidity sensors (Passaro et al. 2007). It is caused by unique symmetries and orientations of electric moment vectors and wave functions involved in electronic transitions. The electric dipole moment determines the direction along which the charge is displaced in a molecule, undergoing an electronic transition. It is possible to study the polarization of the transition using polarized light to excite and detect luminescence. Exciting the sample with polarized light and measuring the luminescence intensity along two perpendicular directions, it is possible to define the degree of polarization P as

$$P = \frac{I_{EE} - I_{EB}}{I_{EE} + I_{EB}} \tag{15.1}$$

where I_{EE} and I_{EB} are luminescence intensities measured along the direction parallel and perpendicular to the excitation electric vector, respectively.

In practice, fluorescence experiments are performed under continuous or pulsed excitation. In the first case, the system is considered in equilibrium, that is, the population density of the excited state is constant (Righini and Ferrari 2005). Under pulsed excitation, interesting information about the relaxation mechanisms can be obtained. In particular, the fluorescence decay time or lifetime of the emitting state can be determined (Righini and Ferrari 2005). To transform a standard optical fiber into a fluorescence-based optical chemical sensor, it is necessary to impart analyte-sensitive fluorescence to the fiber in some manner. This may be achieved by replacing the cladding of the fiber over a portion of its length with a solid matrix that contains the fluorescent compound. This process involves removing a portion of the original cladding of the fiber and coating the decladded region with a sensing material, which is subsequently cured to form a solid, fluorescent cladding. A variation of the fluorescent cladding configuration was employed by Ahmad et al. (2005).

One should note that the configuration of fluorescence-based humidity sensors generally produces a rapid response because it requires only a very thin reagent coating (about a few μm thick) and promotes an efficient mass transfer during the interaction with the *analyte*. In the case of fluorescent indicators, the method is very sensitive. This design has found applications in *in situ* humidity diagnostics (Muto et al. 1994; Willer et al. 2002).

15.3 FLUORESCENCE DYES FOR HUMIDITY SENSORS

Humidity sensitive materials acceptable for design of fluorescence-based humidity sensors are listed in Table 15.1. It can be seen that there is a sufficiently large number of fluorescence dyes, which can be used to develop humidity sensors. We do not indicate the concentration of fluorescence dyes, used in the manufacture of sensors, because in most papers it is not optimized for use in humidity sensors. Generally, this concentration may vary in the range of 1 wt%–10 wt%. An additional information about the materials that can be used as fluorescence dyes for humidity sensors one can find in Jung et al. (2016).

TABLE 15.1
Fluorescence-Based Humidity Sensors

Sensing Method	Sensing Material	Range, % RH	Response Time	References
Fluorescence quenching	Perylene dyes	0–100	3–10 min	Posch and Wolfbeis (1988), Zhang et al. (2016)
	Aluminum/morin metal ion-organic complex doped polyvinyl pyrrolidone membrane	0–80		Raichur and Pederse (1995)
	2′,7′-Dibromo-50-[hydroxymercuri]fluorescein (mercurochrome)	0–100	3–5 min	Costa-Fernández and Sanz-Medel (2000)
	4-[2-(pyrazin-2-yl)-1,3-oxazol-5-yl] benzenamine immobilized in hydroxypropyl cellulose (HPC)	0–100	1–2 min	Bedoya et al. (2001)
	4′,7-dihydroxyflavylium perchlorate	0–100	~5 min	Galindo et al. (2005)
	$[Ru(phen)_2dppz]Cl_2$ in polymer	0–100	<5–10 min	Xu et al. (2007)
	Eu(III)-doped mesoporous silica	0–100	~20 min	Park et al. (2010)
Fluorescence increase	Rhodamine 6G-impregnated in a perfluorinated polymer matrix	5–40	~1 s	Zhu et al. (1989)
	Fluorescein-dye doped PMMA	0–80	~1 s	Muto et al. (1994)
	Platinum(II) double salt materials	0–100		Drew et al. (2004)
	Anionic fluorescein dyes and 1-butanesulfonate (C4S) ions hybridized with layered double hydroxides	20–90		Sasai and Morita (2017)

(Continued)

TABLE 15.1 (*Continued*)
Fluorescence-Based Humidity Sensors

Sensing Method	Sensing Material	Range, % RH	Response Time	References
Stokes shift	5-dimethylaminonaphthalene-l-sulphonic acid (DNSA)	0–100	13–20 min	Otsuki and Adachi (1993)
	4-[4-(dimethylamino)styryl]pyridine	0–100	2–20 s	Gao et al. (2016)
	Dapoxyl sulfonic acid incorporated into hydrogel films, agarose, and a copolymer	0–100	~20 min	Tellis et al. (2011)
Fluorescence lifetime	Rhodamine 6G-impregnated Nafion film			Litwiler et al. (1991)
	Aminonaphthalenesulphonate (ANS) derivatives	0–90		Otsuki and Adachi (1994)
	Pd- and Pt-porphyrin dyes	40–100		Papkovsky et al. (1994, 1995)
	Lithium-treated Nafion membrane			Glenn et al. (2001)
	Ruthenium-based complex doped poly(tetrafluoroethylene) membrane	4–100	~2 min	Bedoya et al. (2006)
	Ruthenium(II) diphenylphenanthroline–dipyridophenazinehexafluorophosphate	0–100		McGaughey et al. (2006)

Perylene dyes such as perylenedibutyrate and perylenetetracarboxylic acid bis-imides (PTCABs) were among the first fluorescence dyes used for the development of humidity sensors (Posch and Wolfbeis 1988) (Figure 15.9). These materials were chosen as indicators because of their high stability and their long-wave excitation maxima of 470 and ~490 nm, respectively. Perylenedibutyrate can be excited between 450 and 490 nm (max. at 468 run) and fluorescence is strong between 480 and 560 nm. The other dyes have excitation and emission maxima of, respectively, 490 ± 5 nm and 545 ± 5 nm. They are therefore susceptible to fluorescence excitation by tungsten halogen lamps or blue light-emitting diode (LEDs), which offers cost advantages over xenon arc lamps. The indicators have large Stokes' losses (50–70 nm), which facilitates the separation of fluorescence (which is humidity dependent) from reflected excitation light (which is humidity independent). Silica gel sheets were chosen as supports because the dyes can be easily immobilized thereon by adsorption and then display strong fluorescence, which is efficiently quenched by humidity. Other support materials were also studied. It is found, however, that the quenching efficiency of water vapor is distinctly smaller in this case. With regard to the shortcomings relating to the data of fluorescence dyes, they are as follows. Recovery of a sensor becomes annoyingly slow once it has been exposed to 100% RH. Also, oxygen can act as a quencher with a response time much shorter than that of water vapor. Thus, when the oxygen content is lowered from 21% (as normal) to 18%, a signal increase of typically 10% is observed, a fact that limits the range of application. There are also observed hysteresis and nonlinearity of sensor characteristics in the measurement of humidity in a wide range.

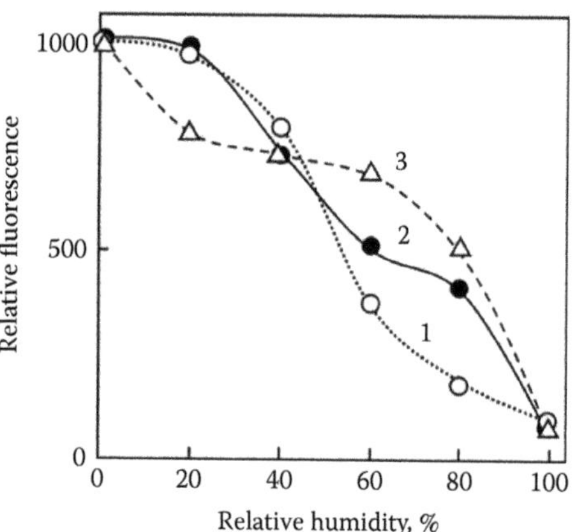

FIGURE 15.9 Humidity dependence of the fluorescence of PTCABs 1–3 when adsorbed on silica gel. Excitation and emission wavelength: 1—487/550 nm; 2—495/545 nm; 3—495/545 nm. (Reprinted from *Sens. Actuators*, 15, Posch, H.E., and Wolfbeis, O.S., Optical sensors, 13: Fibre-optical humidity sensor based on fluorescence quenching, 77–83, Copyright 1988, with permission from Elsevier.)

Perylene dyes in the form of perylene bisimides (PBI) were studied further by Zhang et al. (2016). For studies nanofiber-based films formed by spinning were used. It was established that the fluorescence emission of the film decreased significantly with increasing RH. At that there was a linear relationship between the $(I_o–I)/I_o$ within a range of 6.4%–97.3% RH. Zhang et al. (2016) have found that at room temperature, the vapors such as alcohol, dichloromethane, and acetone, showed some quenching effect to the fluorescence emission of the film, but they were much weaker than that quenched by a RH of 55%. The response was also fast as the equilibrium took less than a few minutes.

Other fluorescence dyes such as 2′,7′-Dibromo-50-[hydroxymercuri]fluorescein (mercurochrome) (Costa-Fernández and Sanz-Medel 2000), 4′,7-dihydroxyflavylium and 7-hydroxy-4-methylflavylium in poly (hydroxyethyl methacrylate) (PHEMA) (Galindo et al. 2005), dapoxyl sulfonic acid (Tellis et al. 2011), which are characterized by the fluorescence-quenching effect in the presence of water vapor, have approximately the same characteristics. The differences are largely connected with the influence of the matrix used (Tellis et al. 2011). At that most people agree with Posch and Wolfbeis (1988), that sol–gel glasses offer several advantages over the organic polymer resin supports, including chemical inertness, low cost, high thermal stability, excellent optical transparency, and low intrinsic fluorescence.

One should note that in addition to the quenching effect, the opposite effect is possible, when the humidity stimulates fluorescence (Figures 15.10 and 15.11). This effect was observed for rhodamine 6G dye incorporated into a perfluorinated polymer matrix (Zhu et al. 1988), platinum(II) double salt materials such as polycrystalline [(phen)Pt(CNcyclohexyl)$_2$] [Pt(CN)$_4$] and [Pt(CN-*n*-tetradecyl)$_4$][Pt(CN)$_4$], [(phen)-Pt(CN-cyclododecyl)Cl)]$_2$[(phen)Pt(CNcyclododecyl)$_2$]$_2$[Pt(CN)$_4$]$_3$ deposited from a diethyl ether suspension or a perfluorohexane/Teflon AF suspension (Drew et al. 2004), and fluorescein-dye-doped poly(methyl methacrylate) (PMMA) (Muto et al. 1994). In particular, Zhu et al. (1988) have shown that for rhodamine 6G dye the fluorescence intensity increases strongly and linearly with increasing a water vapor partial pressure even though the lifetime of fluorescence is simultaneously lowered. The response time of the optrode was approximately 1 s. It is a very good rate for response in comparison with other dyes. In addition, during measurements the effect of photodegradation of 6G was not detectable. Degradation was observed only when the optrode was exposed to relatively strong laser radiation (>200 mW) for several hours. Zhu et al. (1988) believe that the observed changes in fluorescence intensity can be attributed to the formation of a complex between immobilized rhodamine 6G and H_2O. The complex appears to have a higher absorbance and consequently greater excitation efficiency than the immobilized dye by itself.

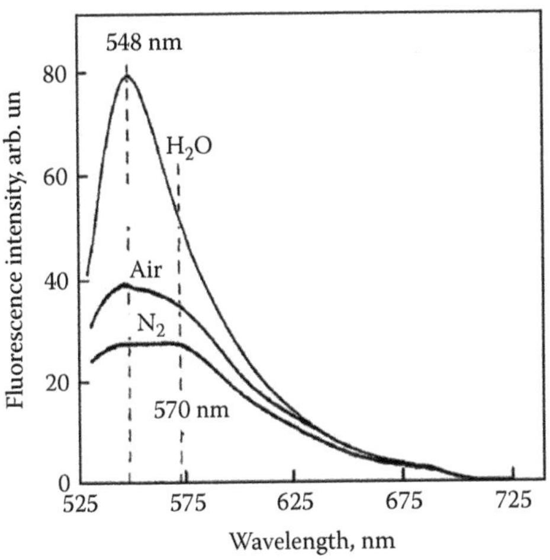

FIGURE 15.10 Emission spectra of the optrode in the indicated environment (excitation wavelength: 514.5 nm). N_2 refers to a dry-nitrogen atmosphere, air to laboratory environment, and H_2O to a location just above the water level in an open vessel. (Adapted from Zhu, C. et al., A new fluorescence sensor for quantification of atmospheric humidity, Office of Naval Research Contract N14-86-K-0366 R&T Code 4134006, Technical Report No. 35, July 11, 1988.)

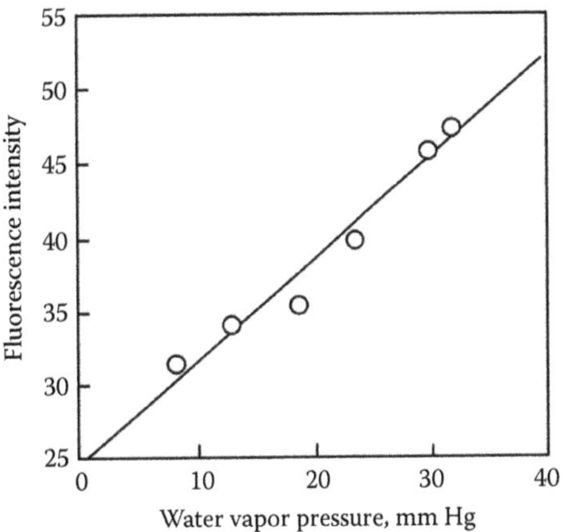

FIGURE 15.11 Calibration curve for the water partial pressure measurement. (Adapted from Zhu, C. et al., A new fluorescence sensor for quantification of atmospheric humidity, Office of Naval Research Contract N14-86-K-0366 R&T Code 4134006, Technical Report No. 35, July 11, 1988.)

It is interesting that rhodamine 6G dye immobilized in gelatin film also has a linear dependence of the intensity from the humidity (Choi and Tse 1999). However, in contrast to the results reported by Zhu et al. (1988), Choi and Tse (1999) observed a decrease in fluorescence intensity with increasing humidity. The response and recovery times were also longer (>2 min). But with decreasing of the thickness, the time constants decreased, indicating that the response time is controlled by equilibrium time between the humidity-sensitive matrix and water vapor. It is important to note that oxygen, nitrogen, and carbon dioxide exerted weak influence on the fluorescence intensity. Only slight relative fluorescence signal changes were also observed for acetone, toluene vapors, and NO_x. However, serious intereferences were found for ethanol, chloroform, acetic acid vapors, and SO_2. Choi and Tse (1999) believed that gases possessing acidic protons have some chemical interactions, possibly hydrogen bond formation, between gelatin and these interferents. Many fluorescence dyes were also sensitive to ammonia vapors. For example, in the presence of ammonia vapor, the film of 4′,7-dihydroxyflavylium changes its color from yellow to orange–brown, and finally to pink, on a few minutes (Galindo et al. 2005). In contrast, the recovery from the pink films (on being treated with ammonia vapor) to yellow by contact with HCl vapor was a much slower process. The process of 4′,7-dihydroxyflavylium drying is also slow and requires ~30 min.

Otsuki and Adachi (1993) have used another approach. They analyzed the shift of the fluorescence spectra under influence of the air humidity. For this study, Otsuki and Adachi (1993) used 5-dimethylaminonaphthalene-l-sulphonic acid (DNSA), embedded in a hydrophilic polymer film, a hydroxypropyl cellulose (HPC). Nafion, which is usually used as support, is not an appropriate matrix for physicochemical studies of fluorophores because of its very strong acidity. The fluorescence spectra of DNSA in an HPC film at various RHs are shown in Figure 15.12. I_{max} is plotted in reverse for ease of comparison of the curves for λ_{max} and I_{max}. The authors notice good reproducibility of fluorescence spectrum and the intensity of fluorescence. Only small decrease in the relative intensity of the fluorescence spectrum was observed at each RH after repeated humidification and desiccation. In contrast, the λ_{max} values and the spectral shape at each RH were not varied after repeated RH changes. The fluorophore was also not degraded by exposure to the analyzing light, because the light intensity was so small.

Gao et al. (2016) have shown that the spectra of 4-[4-(dimethylamino)styryl]pyridine (DSP) chromophore incorporated into a polyvinylpyrrolidone (PVP) host were also characterized by a pronounced Stokes shift. DSP was chosen because it contains a pyridine group that can potentially form hydrogen-bonding interactions with water molecules, which can influence the electronic structure of DSP. Different extents of humidity can also change the aggregation states of the hydrophobic DSP, thus allowing tailoring of its fluorescence properties. The maximum intensity of luminescence of DSP chromophore, depending on the humidity, can be shifted in the range of 462–489 nm. The main characteristic feature of obtained results is fast and reversible humidity response. Apparently, high performance is the result of using the structure of humidity-sensitive material. Humidity indicator was prepared in the form of nanofibers using an electrospinning process, which provided a high porosity of the material and the rapid establishing of equilibrium with the water vapor.

Aminonaphthalenesulphonate (ANS) derivatives (Figure 15.13), which include 5-dimethylamino-l-naphthalenestdphonamide (DNSM), DNSA, and 8-anilino-1-naphthalenesulphonic acid (ANSA) are representatives of fluorescent probes, which allow us to get information about the air humidity, basing on the

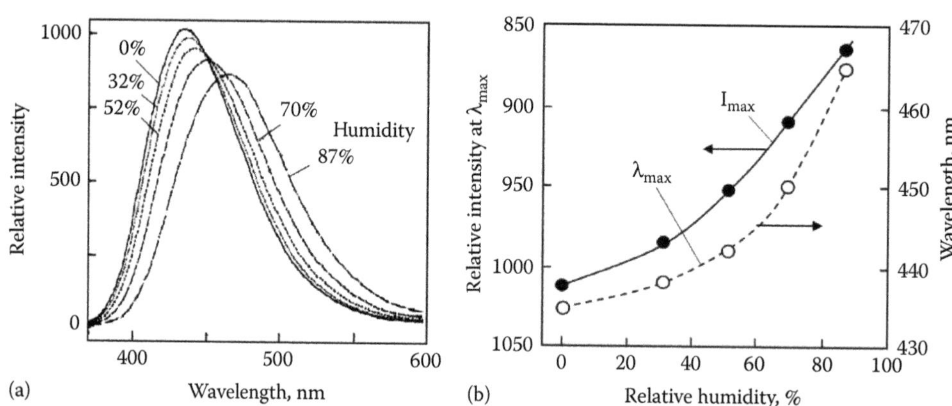

FIGURE 15.12 (a) Effect of RH on the corrected fluorescence spectrum of a DNSA–HPC film and (b) Relation of λ_{max} and I_{max} to RH for a DNSA–HPC film. (Reprinted from *Photochem. Photobiol. A: Chem.*, 71, Otsuki, S., and Adachi, K., Effect of humidity on the fluorescence properties of a medium sensitive fluorophore in a hydrophilic polymer film, 169–173, Copyright 1993, with permission from Elsevier.)

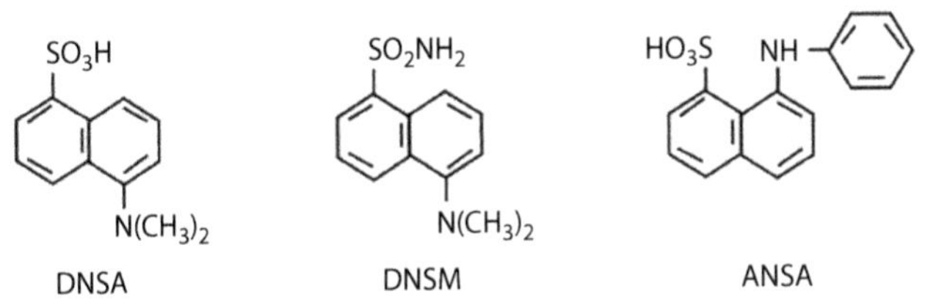

SO_3H $N(CH_3)_2$ DNSA

SO_2NH_2 $N(CH_3)_2$ DNSM

HO_3S NH ANSA

FIGURE 15.13 The formulae of ANS derivatives.

measurement of the fluorescence decays (Otsuki and Adachi 1994). It is believed that lifetime-based detection methods are virtually independent of external perturbations, as the lifetime is an intrinsic property of the fluorophore. A HPC films containing a fluorophore, which were studied in Otsuki and Adachi (1994), had a thickness of about 17 μm. The fluorescence decay curves of the fluorophore in the HPC film were measured between 405 and 640 nm at various RHs and are shown in Figure 15.14. It is expected that the

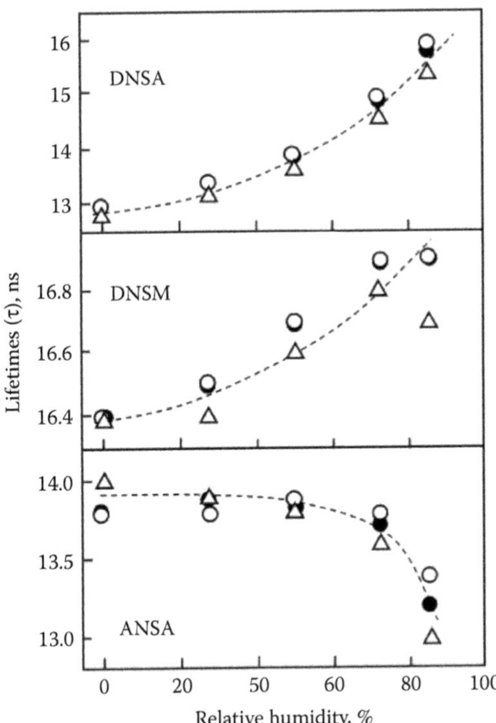

FIGURE 15.14 Effect of RH on the lifetimes (τ) of the fluorophores in the HPC film. The fluorescence decays were fitted to single exponential kinetics. Concentration: DNSA: 0.0040 mol/kg (○); 0.011 mol/kg (●); 0.022 mol/kg (Δ); DNSM: 0.0067 mol/kg (○); 0.013 mol/kg (●); 0.027 mol/kg (Δ); ANSA: 0.0058 mol/kg (○); 0.012 mol/kg (●); 0.023 mol/kg (Δ). (Reprinted from *J. Photochem. Photobiol. A: Chem.*, 84, Otsuki, S. and Adachi, K., Effect of humidity on the dynamic fluorescence properties of aminonaphthalenesulphonate derivatives in a hydrophilic polymer film, 91–96, Copyright 1994, with permission from Elsevier.)

TABLE 15.2
Fluorescence Lifetimes of ANS Derivatives in Alcohol and Water

Fluorophore	τ (ns)	
	Alcohol	Water
DNSA	11.9[a]	16.5
DNSM	18.2[a]	3.3
ANSA	11.1[b]	0.25

Source: Otsuki, S. and Adachi, K., *J. Photochem. Photobiol. A: Chem.*, 84, 91–96, 1994.

[a] 2-propanol.

[b] ethanol.

microenvironment of HPC becomes more polar when the film adsorbs water vapor. Consequently, τ depends on the polarity of the medium and increases with polarity for DNSA and DNSM but decreases for ANSA. Otsuki and Adachi (1994) believe that the spectral relaxation is probably dependent on the local viscosity around the fluorophore molecules: the local viscosity around the fluorophore immediately after excitation increases with increasing polarity of the excited state. This means that these measurements really can give information about air humidity. However, these measurements are complicated and require specific equipment. The principles of the measurements of fluorescence lifetimes and time-resolved fluorescence spectra one can find in Fleming (1986), Lakowicz (1999), and Valeur (2002). In addition, τ depends strongly on the presence of solvents vapors in the atmosphere (Table 15.2), which significantly limits the field of application of such sensors. The range of the fluorescence lifetime change is also quite narrow, making it difficult to conduct accurate measurements.

Phosphorescent water-soluble Pt-coproporphyrin III and Pd-coproporphyrin III dyes, especially Pd-porphyrin one, have longer decay times (Table 15.3), and, therefore, seem to be more suitable for practical use in lifetime-based RH sensors (Papkovsky et al. 1994). In addition, they are more compatible with LED excitation (e.g., with pure green 557 nm LEDs) and frequency-modulated optoelectronics, and they showed decreased sensitivity to quenching with molecular oxygen. However, this does not help to avoid the difficulties existing in such measurements.

Interferences of other widespread species, such as gas quenchers (SO_2 NO_x) and polar gases (HCl, NH_3, and

TABLE 15.3
Luminescent Properties of the Metalloporphyrin LB Coatings

LB Structure	Excitation Maximum (nm)	Emission Maximum (nm)	Lifetime (μs)		Relative Intensity, Water
			Dry Air	Water	
Pt–CP3	380; 535	647	96	45	7.0
Pt–CP3–PEI	380; 535	647	93	41	10.0
Pd–CP#	393; 547	667	980	53	5.0
Pd–CP#–PEI	393; 547	667	952	50	6.5

Source: Papkovsky, D.B. et al. *Anal. Lett.*, 28, 2027–2040, 1995.
CP3—coproporphyrin III; PEI—polyethyleneimine.

organic vapors) (Papkovsky et al. 1995), they also hinder the use of such materials.

15.4 LUMINESCENCE HUMIDITY SENSORS

Luminescence humidity sensor is a type of sensor which operation is based on principles other than those discussed earlier in Chapters 12, 13 and 14. For the development of such sensors, one can use the materials, which have fluorescent properties, such as semiconductors and luminophores. It is known that photoluminescence (PL) intensity depends on the surface properties of semiconductor, particularly on the diffusion potential on the surface, controlling surface recombination of generated charge carriers. Due to the fact that the water adsorption on the surface of semiconductors is one of the most powerful instruments of influence on the surface properties of semiconductors, it becomes clear that the photoluminescence properties are humidity-sensitive parameter. However, the same experiment showed that too many factors affect the intensity of photoluminescence (the surface condition, conditions of the synthesis, the intensity and spectrum of the excitation, the concentration of oxygen, the presence of active gases in the atmosphere, a prehistory of the samples, etc.). Therefore, this approach is practically not used for the development of humidity sensors for real-world applications.

Despite the fact that the studies related to the effect of humidity on the luminescent properties of semiconductors and luminophores are being conducted intensively (Nagai et al. 1979; Appleby et al. 2011; Zhang et al. 2011; Smyntyna et al. 2014; Jeseentharani et al. 2017), in the literature one can find only several works where the authors position their research as the development of humidity sensors (Hu et al. 2013; Kayahan 2015; Nayef and Khudhair 2017). The results of some of these studies are presented in Figures 15.15 and 15.16.

There are also studies that have shown the ability to manufacture luminescence-based humidity sensors based on Ln^{3+}-metal–organic frameworks (MOFs)

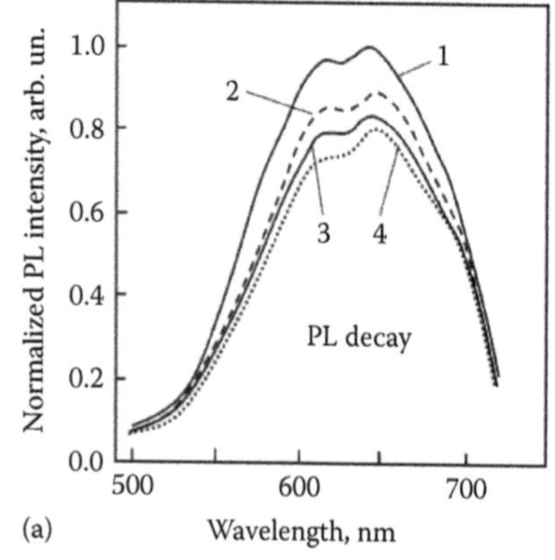

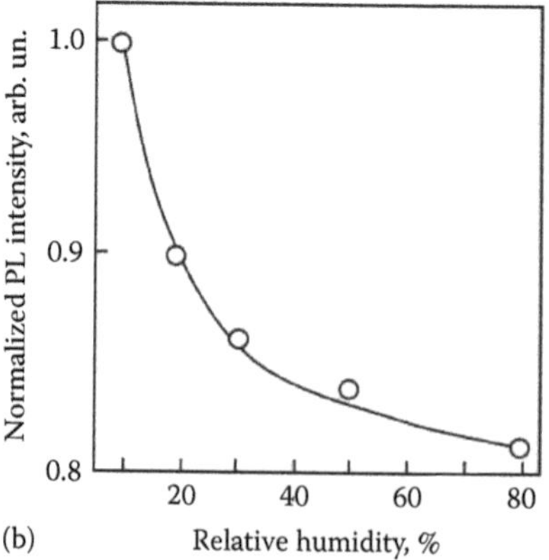

FIGURE 15.15 (a) The PL spectra change of porous silicon with increasing RH levels. (PL_{exc} = 254 nm): 1—10% RH, 2—19% RH, 3—50% RH, 4—80% RH; (b) The PL intensity changes with RH humidity levels. (Adapted from Kayahan, E., *Acta Phys. Polonica A*, 127, 1397–1399, 2015. Published by Institute of Physics of the Polish Academy of Sciences as open access.)

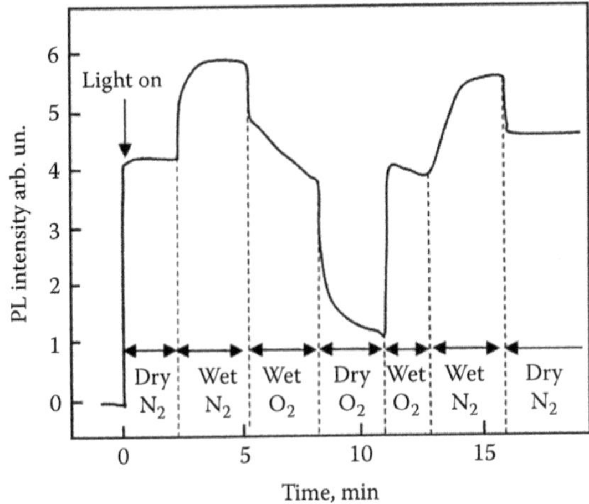

FIGURE 15.16 Reversible PL intensity response from n-InP:Sn (n ~2×10^{18} cm^{-3}) crystal surface with a change in humidity of gaseous species (N_2 and O_2). The PL was excited by the 632.8 nm red emission from a He–Ne laser with an excitation density of ~2 W/cm^2. (Adapted from Nagai, H. et al., *Jpn. J. Appl. Phys.*, 18, 377–381, 1979. Copyright 1979, The Japan Society of Applied Physics. With permission.)

(Yu et al. 2012). According to Yu et al. (2012), Ln^{3+}-MOFs are of interest because of their narrow emission and high color purity resulting from the Ln^{3+} emitters. Furthermore, Ln-based emissions are sensitive to the chemical environment; for example, the high energy O–H, N–H, and C–H vibrations around them. So, porous Ln^{3+}–MOFs could be useful fluorescent scaffolds to probe or sense the encapsulated guest species depending on their guest-driven emissions. Experiments conducted by Yu et al. (2012) have shown that the humidity sensing by this material could be realized by controlling O–H oscillators on the Ln^{3+} emitters through host–guest H-bonding interactions. However, it is too early to make any conclusions about the real prospects of given material for the development of humidity sensors. Although Lin et al. (2016) believe that luminescent MOFs and related devices are very promising for sensor applications, the versatile structures and chemistry, easy utilization of encapsulated guest luminophores, and many viable types of sensing mechanisms are unique advantages of MOFs-based sensors.

Coutino-Gonzalez et al. (2016) have shown that zeolite-based materials can also be used for humidity sensor development. It was established that the emission of small silver clusters confined inside zeolite matrices has high external quantum efficiencies (EQE) and spans the whole visible spectrum. Moreover, it has been reported that the ultraviolet (UV) excited luminescence of partially Li-exchanged sodium Linde type A zeolites (LTA[Na]) containing luminescent silver clusters can be controlled by adjusting the water content of the zeolite. These samples showed a dynamic change in their emission color from blue to green and yellow upon an increase of the hydration level of the zeolite, showing the great potential that these materials can have as luminescence-based humidity sensors at the macro and micro scale. The sensor can be fabricated by suspending the luminescent Ag-zeolites in an aqueous solution of polyethylenimine (PEI) to subsequently deposit a film of the material onto a selected substrate.

One should note that luminescence-based sensors are significantly inferior to previously considered devices in their parameters, and they have insignificant prospects for wide application.

15.5 SUMMARY

Despite the significant number of fluorescence dyes, tested in the development of fluorescence-based humidity sensors, their sufficiently high sensitivity and acceptable speed of response, we must admit that none of these materials is suitable for manufacturing the sensors for the sensor market. Short lifetime, interference effects, especially to oxygen (Figure 15.17) (McGaughey et al. 2006), ammonia and acetic acid, the hysteresis of characteristics, the influence of the temperature (Figure 15.18) (Bedoya et al. 2001), the need for using sophisticated equipment, etc., significantly limit their applications.

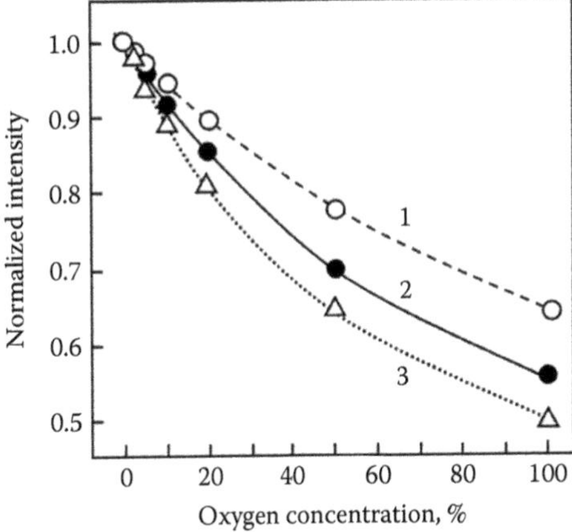

FIGURE 15.17 Intensity response of a ruthenium compound fluorescence to oxygen. 1—aging time 30 min, 2—aging time 1 h, 3—agining time 6 h. (Reprinted from *Anal. Chim. Acta*, 570, McGaughey, O. et al., Development of a fluorescence lifetime-based sol-gel humidity sensor, 15–20, Copyright 2006, with permission from Elsevier.)

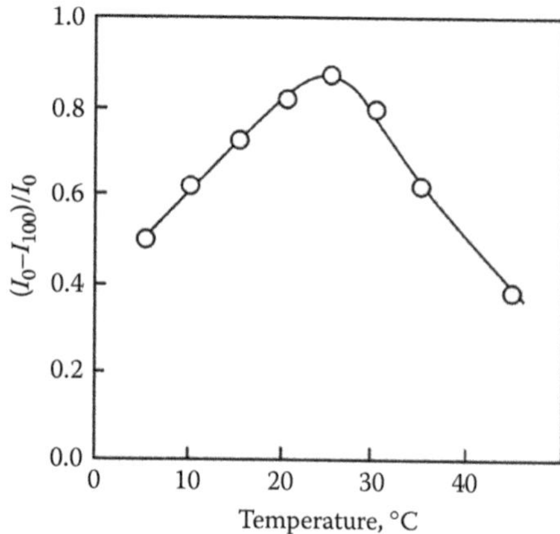

FIGURE 15.18 Influence of temperature on the normalized fluorescence intensity of 4-[2-(pyrazin-2-yl)-1,3-oxazol-5-yl]benzenamine immobilized in hydroxypropylcellulose obtained with the optode for the determination of a 100% relative humidity gas sample. λ_{exc} = 390 nm, λ_{em} = 496 nm. (From Bedoya, M. et al.: Fluorescent optosensor for humidity measurements in air. *Helvetica Chimica Acta*. 2001. 84. 2628–2639. Copyright Wiley-VCH Verlag GmbH & Co. KGaA. Reproduced with permission.)

REFERENCES

Ahmad M., Chang K.P., King T.A., Hench L.L. (2005) A compact fibre-based fluorescence sensor. *Sens. Actuators A* 119, 84–89.

Appleby G.A., Kroeber P., Zimmermann J., von Seggern H. (2011) Influence of oxygen doping and hydration on photostimulated luminescence of CsBr and CsBr:Eu^{2+}. *J. Appl. Phys.* 109, 073510.

Bedoya M., Diez M.T., Moreno-Bondi M.C., Orellana G. (2006) Humidity sensing with a luminescent Ru(II) complex and phase-sensitive detection. *Sens. Actuators B* 113, 573–581.

Bedoya M., Orellana G., Moreno-Bondi M.C. (2001) Fluorescent optosensor for humidity measurements in air. *Helvetica Chim. Acta* 84, 2628–2639.

Brenci M., Baldini F. (1991) Fiber optic optrodes for chemical sensing. In: *Proceedings of the 8th IEEE Optical Fiber Sensors Conference*, New York, pp. 313–319.

Burke C.S., McGaughey O., Sabattie J.M., Barry H., McEvoy A.K., McDonagh C., MacCraith B.D. (2005) Development of an integrated optic oxygen sensor using a novel, generic platform. *Analyst* 130, 41–45.

Choi M.M.F., Tse O.L. (1999) Humidity-sensitive optode membrane based on a fluorescent dye immobilized in gelatin film. *Anal. Chim. Acta* 378, 127–134.

Costa-Fernández J.M., Sanz-Medel A. (2000) Air moisture sensing materials based on the room temperature phosphorescence quenching of immobilized mercurochrome. *Anal. Chim. Acta* 407, 61–69.

Coutino-Gonzalez E., Baekelant W., Dieu B., Roeffaers M.B.J., Hofkens J. (2016) Nanostructured Ag-zeolite composites as luminescence-based humidity sensors. *J. Vis. Exp.* 117, e54674 (1–7).

de Silva A.P., Moody T.S., Wrighta G.D. (2009) Fluorescent PET (Photoinduced Electron Transfer) sensors as potent analytical tools. *Analyst* 134, 2385–2393.

Draxler S., Lippitsch M.E. (1996) Time-resolved fluorescence spectroscopy for chemical sensors. *Appl. Opt.* 35, 4117–4123.

Drew S.M., Mann J.E., Marquardt B.J., Mann K.R. (2004) A humidity sensor based on vapoluminescent platinum(II) double salt materials. *Sens. Actuators B* 97, 307–312.

Elosua C., Matias I.R., Bariain C., Arregui F.J. (2006) Volatile organic compound optical fiber sensors: A review. *Sensors* 6, 1440–1465.

Fleming G.R. (1986) *Chemical Applications of Ultrafast Spectroscopy*. Oxford University Press, New York.

Galindo F., Lima J.C., Luis S.V., Melo M.J., Parola A.J., Pina F. (2005) Water/humidity and ammonia sensor, based on a polymer hydrogel matrix containing a fluorescent flavylium compound. *J. Mater. Chem.* 15, 2840–2847.

Gansert D., Arnold M., Borisov S., Krause C., Muller A. (2006) Hybrid optodes (HYBOP). In: Popp J., Strehle M. (Eds.), *Biophotonics: Visions for Better Health Care*. Wiley-VCH Verlag, Weinheim, Germany, pp. 477–518.

Gao R., Cao D., Guan Y., Yan D. (2016) Fast and reversible humidity-responsive luminescent thin films. *Ind. Eng. Chem. Res.* 55, 125–132.

Glenn S.J., Cullum B.M., Nair R.B., Nivens D.A., Murphy C.J., Angel S.M. (2001) Lifetime-based fiber-optic water sensor using a luminescent complex in a lithium-treated Nafion membrane. *Anal. Chim. Acta* 448, 1–8.

Hu J., Wu P., Deng D., Jiang X., Hou X., Lv Y. (2013) An optical humidity sensor based on CdTe nanocrystals modified porous silicon. *Microchem. J.* 108, 100–105.

Jeseentharani V., Dayalan A., Nagaraja K.S. (2017) Co-precipitation synthesis, humidity sensing and photoluminescence properties of nanocrystalline Co^{2+} substituted zinc(II)molybdate ($Zn_{1-x}Co_xMoO_4$; x = 0, 0.3, 0.5, 0.7, 1). *Solid State Sci.* 67, 46–58.

Jorge P.A.S., Caldas P., Rosa C.C., Oliva A.G., Santos J.L. (2004) Optical fiber probes for fluorescente based oxygen sensing. *Sens. Actuators B* 130, 290–299.

Jung H.S., Verwilst P., Kim W.Y., Kim J.S. (2016) Fluorescent and colorimetric sensors for the detection of humidity or water content. *Chem. Soc. Rev.* 45, 1242–1256.

Kayahan E. (2015) Porous silicon based humidity sensor. *Acta Phys. Polonica A* 127(4), 1397–1399.

Korotcenkov G., Cho B.K., Sevilla III F., Narayanaswamy R. (2011) Optical and fiber optic chemical sensors. In: Korotcenkov G. (Ed.), *Chemical Sensors: Comprehensive Sensor Technologies*. Vol. 5: Electrochemical and Optical Sensors. Momentum Press, New York, pp. 311–476.

Lakowicz J. (1982) *Principles of Fluorescence Spectroscopy*. Plenum Press, New York.

Lakowicz J.R. (1999) *Principles of Fluorescence Spectroscopy*, 2nd ed. Kluwer/Plenum, New York.

Lin R.-B., Liu S.-Y., Ye J.-W., Li X.-Y., Zhang J.-P. (2016) Photoluminescent metal–organic frameworks for gas sensing. *Adv. Sci.* 3, 1500434.

Litwiler K.S., Kluczynski P.M., Bright F.V. (1991) Determination of the transduction mechanism for optical sensors based on rhodamine 6G impregnated perfluorosulfonate films using steady-state and frequency-domain fluorescence. *Anal. Chem.* 63, 797–802.

McGaughey O., Ros-Lis J.V., Guckian A., McEvoy A.K., McDonagha C., MacCraith B.D. (2006) Development of a fluorescence lifetime-based sol–gel humidity sensor. *Anal. Chim. Acta* 570, 15–20.

Mitsubayashi K., Minamide T., Otsuka K., Kudo H., Saito H. (2006) Optical bio-sniffer for methyl mercaptan in halitosis. *Anal. Chim. Acta* 573–573, 75–80.

Moreno J., Arregui F.J., Matias I.R. (2005) Fiber optic ammonia sensing employing novel thermoplastic polyurethane membranes. *Sens. Actuators B* 105, 419–424.

Muto S., Sato H., Hosaka T. (1994) Optical humidity sensor using fluorescent plastic fiber and its application to breathing-condition monitor. *Jpn. J. Appl. Phys.* 33, 6060–6064.

Nagai H., Noguchi Y., Mizushina Y. (1979) Influence of ambient gas on the PL intensity from InP and GaAs. *Jpn. J. Appl. Phys.* 18(Suppl. 18-1), 377–381.

Nayef U.M., Khudhair I.M. (2017) Study of porous silicon humidity sensor vapors by photoluminescence quenching for organic solvents. *Optik* 135, 169–173.

Orellana G., Gomes-Carneros A.M., De Dios C., Garcia-Martinez A.A., Moreno-Bondi M.C. (1995) Reversible fiber-optic fluorosensing of lower alcohols. *Anal. Chem.* 67, 2231–2238.

Orellana G. (2006) Fluorescence-based sensors. In: Baldini F., Chester A.N., Homola J., Martellucci S. (Eds.), *Optical Chemical Sensors*. Springer-Verlag, Dordrecht, the Netherlands, pp. 99–116.

Otsuki S., Adachi K. (1993) Effect of humidity on the fluorescence properties of a medium sensitive fluorophore in a hydrophilic polymer film. *J. Photochem. Photobiol. A: Chem.* 71, 169–173.

Otsuki S., Adachi K. (1994) Effect of humidity on the dynamic fluorescence properties of aminonaphthalenesulphonate derivatives in a hydrophilic polymer film. *J. Photochem. Photobiol. A: Chem.* 84, 91–96.

Papkovsky D.B., Desyaterik I.V., Ponomarev G.V., Kurochkin I.N., Korpela T. (1995) Phosphorescence lifetime-based sensing of SO_2 in the gas phase. *Anal. Lett.* 28(11), 2027–2040.

Papkovsky D.B., Ponomarev G.V., Chernov S.F., Ovchinnikov A.N., Kurochkin I.N. (1994) Luminescence lifetime-based sensor for relative air humidity. *Sens. Actuators B* 22, 57–61.

Park J.-Y., Suh M., Kwon Y.-U. (2010) Humidity sensing by luminescence of Eu(III)-doped mesoporous silica thin film. *Micropor. Mesopor. Mater.* 127, 147–151.

Passaro V.M.N., Dell'Olio F., Casamassima B., De Leonardis F. (2007) Guided-wave optical biosensors. *Sensors* 7, 508–536.

Posch H.E., Wolfbeis O.S. (1988) Optical sensors, 13: Fibre-optical humidity sensor based on fluorescence quenching. *Sens. Actuators* 15, 77–83.

Raichur A., Pedersen H. (1995) Fiber optic moisture sensor for baking and drying process control. In: *Proceedings of the Food Processing Automation Conference, FPAC, IV*, Chicago, IL, pp. 180–189, November 3–5.

Righini G.C., Ferrari M. (2005) Photoluminescence of rare-earth-doped glasses. *Rivista del Nuovo Cimento* 28, 1–53.

Ryskevic N., Jursenas S., Vitta P., Bakiene E., Gaska R., Zukauskas A. (2010) Concept design of a UV light-emitting diode based fluorescence sensor for real-time bioparticle detection. *Sens. Actuators B* 148, 371–378.

Sasai R., Morita M. (2017) Luminous relative humidity sensing by anionic fluorescein dyes incorporated into layered double hydroxide/1-butanesulfonate hybrid materials. *Sens. Actuators B* 238, 702–705.

Smyntyna V., Semenenko B., Skobeeva V., Malushin N. (2014) Photoactivation of luminescence in CdS nanocrystals. *Beilstein J. Nanotechnol.* 5, 355–359.

Tellis J.C., Strulson C.A., Myers M.M., Kneas K.A. (2011) Relative humidity sensors based on an environment-sensitive fluorophore in hydrogel films. *Anal. Chem.* 83, 928–932.

Valeur B. (2002) *Molecular Fluorescence. Principles and Applications*. Wiley-VCH Verlag, Weinheim, Germany.

Wang K., Seiler K., Haug J.-P., Lehmann B., West S. Hartman K., Simon W. (1991) Hydration of trifluoroacetophenones as the basis for an optical humidity sensor. *Anal. Chem.* 63, 970–974.

Willer U., Scheel D., Kostjucenko I., Bohling C., Schade W., Faber E. (2002) Fiber-optic evanescent-field laser sensor for in-situ gas diagnostics. *Spectrochim. Acta A* 58, 2427–2432.

Xu W., Wittich F., Banks N., Zink J., Demas J.N., Degraff B.A. (2007) Quenching of luminescent Ruthenium(II) complexes by water and polymer-based relative humidity sensors. *Appl. Spectrosc.*, 61(11), 1238–1245.

Yu Y., Ma J.-P., Dong Y.-B. (2012) Luminescent humidity sensors based on porous Ln^{3+}-MOFs. *Cryst. Eng. Commun.* 14, 7157–7160.

Zhang Q., Jiao F., Chen Z., Xu L., Wang S., Liu S. (2011) Effect of temperature and moisture on the luminescence properties of silicone filled with YAG phosphor. *J. Semicond.* 32(1), 012002.

Zhang S., Zhou F., Peng H., Liu T., Ding L., Fang Y. (2016) Fabrication and humidity sensing performance studies of a fluorescent film based on a cholesteryl derivative of perylene bisimide. *Spectrochim. Acta A* 165, 145–149.

Zhu C., Bright F.V., Wyatt W.A., Hieftje G.M. (1988) A new fluorescence sensor for quantification of atmospheric humidity, Office of Naval Research Contract N14-86-K-0366 R&T Code 4134006, Technical Report No. 35, July 11.

16 Interferometric Humidity Sensors

16.1 PRINCIPLES OF OPERATION

Physically interferometric methods are based on the superposition of two coherent light beams, where one beam has interacted with the sample and the other served as the reference. The superposition of both beams leads to constructive or destructive interferences. When light beams recombine in the detection stage, the interferometric pattern will vary, providing the desired changes in the wavelength, phase, intensity, frequency, and other parameters. Depending on the modulation of the light source, these interferences can be detected simply by a photodiode or laterally resolved by a charge-coupled device (CCD) detector.

Thus, optical interferometry is a technique used to measure various changes that may occur along an optical path. These changes may be the result of a change in the wavelength of the light, a change in the length of the path the light is traveling through, or a change in the refractive index (RI) of the medium through which the light is passing (Campbell 2010). In the case of humidity sensing we have last variant. A variation of the RI results in a change of the effective RI of the waveguide, and hence in a phase shift $\Delta\varphi$, which can be described as follows (Schmitt and Hoffmann 2010):

$$\Delta\varphi = \frac{2\pi}{\lambda_0} L \Delta n_{\text{eff}} \quad (16.1)$$

where:

λ_0 denotes the vacuum wavelength
L the interaction length, that is, the distance the evanescent field is in contact with the sample

RI changes can occur from the reaction or interaction of a chemical species with the sensing medium, a humidity-sensitive material in our case. Depending on the experimental approach, the reference branch can compensate intrinsic instabilities due to the light source or ambient temperature fluctuations. Measuring the apparent shift in this pattern provides a measure of the change in RI occurring in either arm of the interferometer. With calibration, this shift yields the concentration of the analyte present.

In a typical optic interferometric sensor, the light is divided in at least two parts and at least one part of the light interacts with the measurand of humidity-sensitive materials. The interaction of the measurand with the light field would result in a phase shift or phase modulation, which can be detected when the modified light field interferes with the reference light.

At present there are four interferometer architectures such as (1) Michelson, (2) Mach–Zehnder, (3) Sagnac, and (4) Fabry–Perot (FP) configuration, which can be used for sensor design (Sharma and Wei 2013). However, different configurations may present different design, cost, and performance trade-offs and one or another architecture may be preferred depending on the unique requirements of a specific application.

16.1.1 Mach–Zehnder Interferometers

Mach–Zehnder interferometers (MZIs) have been commonly used in diverse sensing applications because of their flexible configurations. In an MZI, the incident light field is divided into two parts (sample and reference) by a beam splitter or fiber-optic coupler (Lee et al. 2012). Unlike Michelson, the light in the reference and sample paths is not reflected; rather it is directed to a second splitter/combiner where the two lights combine and the interference patterns are measured using detectors (Figure 16.1). For sensing applications, the reference arm is kept isolated from external variation and only the sensing arm is exposed to the variation.

The scheme of using two separated arms in the MZIs has been rapidly replaced with the scheme of in-line waveguide interferometer since the advent of long-period gratings (LPGs). As shown in Figure 16.2a, a part of the beam guided as the core mode of a single-mode fiber (SMF) is coupled to cladding modes of the

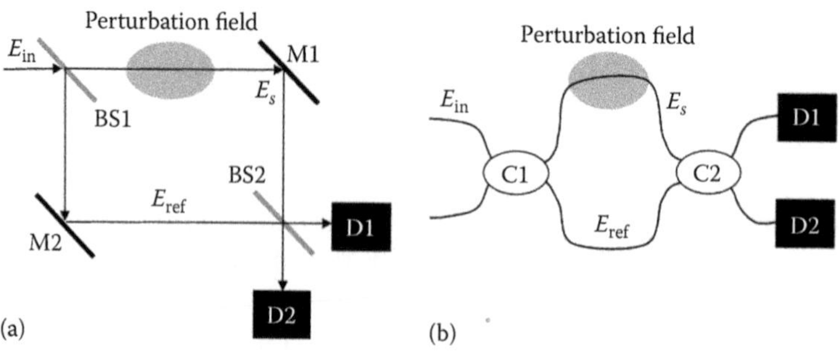

FIGURE 16.1 (a,b) Schematic design for a bulk-optic and fiber-optic Mach–Zehnder interferometer architecture. The electric fields, optical, and fiber-optic components shown in the figure are as follows: E_{in}—incident electric field; E_s—electric field in the sample arm; E_{ref}—electric field in the reference arm; BS—beam splitter; M—mirrors; C—fiber optic coupler; and D—detector. (Idea from Sharma, U. and Wei, X., Fiber optic interferometric devices, in *Fiber Optic Sensing and Imaging*, Kang, J.U. (Ed.), Springer Science+Business Media, New York, 29–53, 2013.)

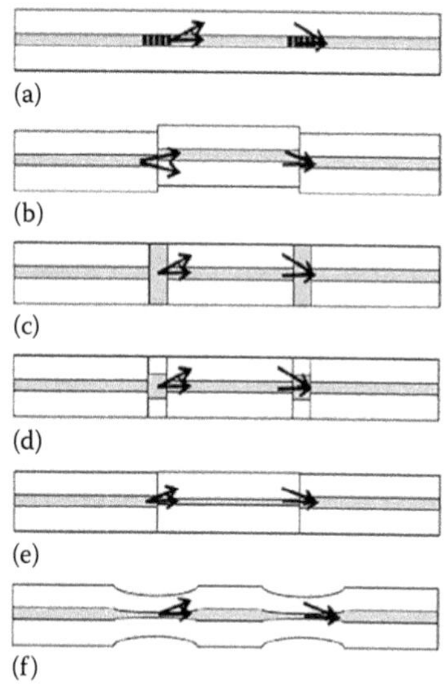

FIGURE 16.2 Configuration of various types of MZIs; the methods of using (a) a pair of long-period fiber gratings (LPGs), (b) core mismatch, (c) air-hole collapsing of phonic crystal fiber (PCF), (d) multimode fiber (MMF) segment, (e) small core single-mode fiber (SMF), and (f) fiber tapering. (From Lee, B.H. et al., Interferometric fiber optic sensors. *Sensors*, 12, 2467, 2012. Published by MDPI as open access.)

same fiber by an LPG, and then recoupled to the core mode by another LPG. The combined beam and the uncoupled beam in the core make interference, which gives a compact but very effective MZI. This in-line type of MZIs has the same physical lengths in both the reference arm and the sensing arm, but has the different optical path lengths due to the modal dispersion; the cladding mode beam has a lower effective index than the core mode beam.

16.1.2 Michelson Interferometer

Michelson interferometer (MI) is probably the most commonly known optical configuration that is used for interferometry. Figure 16.3 shows the typical schematics of bulk-optic and fiber-optic realizations of MIs. In an MI, an incident light field is divided into two parts (sample and reference) by a beam splitter or fiber-optic coupler. The light in the reference and sample paths are reflected back and combined using the same splitter and the interference patterns of the two lights are measured using a detector. The light in the sample arm accrues perturbation field-dependent phase shifts that can be measured using interference with the reference light. In fact, an MI is like a half of an MZI in configuration. Thus, the fabrication method and the operation

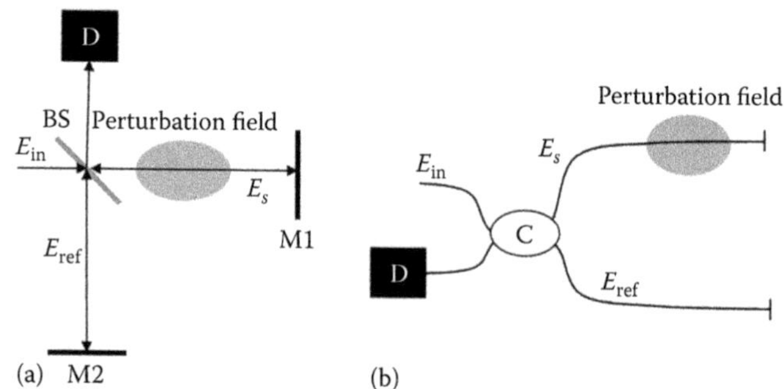

FIGURE 16.3 (a,b) Schematic design for a bulk-optic and fiber-optic Michelson interferometer architecture. The designations are identical to those shown in Figure 16.1. (Idea from Sharma, U. and Wei, X., Fiber optic interferometric devices, in *Fiber Optic Sensing and Imaging*, Kang, J.U. (Ed.), Springer Science+Business Media, New York, 29–53, 2013.)

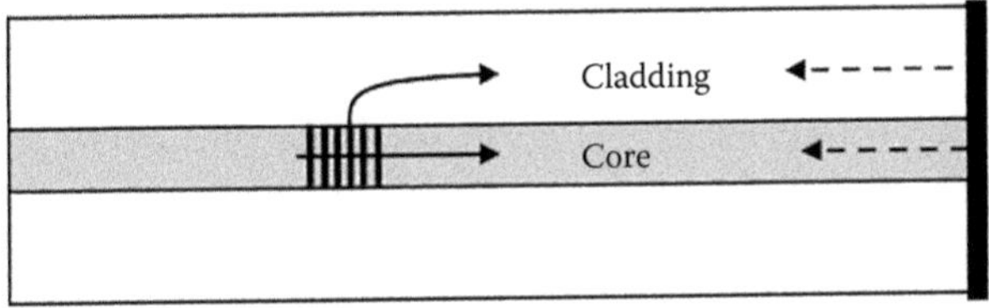

FIGURE 16.4 Schematic of a compact in-line Michelson interferometer. (From Lee, B.H. et al., *Sensors*, 12, 2467–2486, 2012. Published by MDPI as open access.)

principle of MIs are almost the same as MZIs. The main difference is the existence of a reflector(s). Since MIs use reflection modes, they are compact and handy in practical uses and installation. Multiplexing capability with parallel connection of several sensors is another beneficial point of MIs. However, it is essential to adjust the fiber length difference between the reference arm and the sensing arm of an MI within the coherence length of the light source. An in-line configuration of MI is also possible as shown in Figure 16.4. A part of the core mode beam is coupled to the cladding mode(s), which is reflected along with the uncoupled core mode beam by the common reflector at the end of the fiber (Lee et al. 2012).

16.1.3 A Fabry–Perot Interferometer

A Fabry–Perot (FP) interferometer (FPI) consists of two optically parallel reflectors separated by a cavity of length L (Figure 16.5). Sometimes FPI is called an etalon. Reflectors can be mirrors, interface of two dielectrics, or fiber Bragg gratings (FBGs). The cavity may be an optical fiber or any other optical medium. There are two different types of optical fiber FPIs. One is based on the light transmission through an FP, while the other is based on the reflection. The incident light is reflected back and forth and transmitted multiple times at the two partially reflective surfaces. Due to multiple reflections and interference effects, the reflected and transmitted spectrums are functions of cavity length, medium index of refraction, and mirrors reflectivity, in the modulus of 2π. The reflection or transmission spectrum of an FPI can be described as the wavelength-dependent intensity modulation of the input light spectrum, which is mainly caused by the optical-phase difference between two reflected or transmitted beams. The maximum and the minimum peaks of the modulated spectrum mean that both beams, at that particular wavelength, are in phase and out-of-phase, respectively. On account of energy conservation law, the transmitted spectrum is opposite to the reflected spectrum. The phase difference of the FPI is simply given as follows:

$$\delta_{FPI} = \frac{2\pi}{\lambda} n2L \qquad (16.2)$$

where:

- λ is the wavelength of incident light
- n is the RI of cavity material or cavity mode
- L is the physical length of the cavity

When perturbation is introduced to the sensor, the phase difference is influenced with the variation in the optical path difference (OPD) of the interferometer.

Optical fiber FPs are classified as intrinsic and extrinsic types (Rao 2006). In the intrinsic fiber FP interferometer (IFFPI), the two mirrors are separated by an SMF, whereas in the extrinsic fiber FP interferometer (EFFPI), the two mirrors are separated by an air gap or by some solid material other than fiber. In both IFFPI and EFFPI, the light from the emitter to the FP and from FP to the detector are transmitted by an SMF. In the intrinsic fiber-optic FP sensor the reflective surfaces inside the fiber can be created by micromachining, FBG, or thin film deposition. Other embodiments of FPI are shown in Figure 16.6. The optical cavity, shown in Figure 16.6b and c, is formed by the air gap between two uncoated fiber faces and the fibers used may be held using glue or epoxy resin.

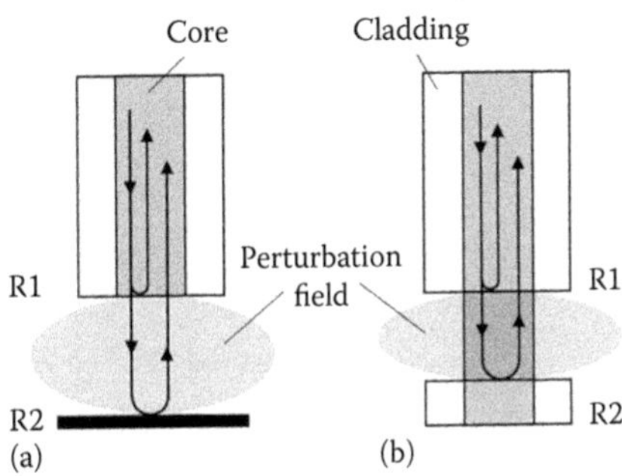

FIGURE 16.5 Schematic design for a bulk-optic (a) extrinsic fiber-optic (middle), and (b) intrinsic fiber-optic Fabry–Perot interferometer architecture. R1 and R2 are the two reflective surfaces forming the Fabry–Perot cavity. (Idea from Sharma, U. and Wei, X., Fiber optic interferometric devices, in *Fiber Optic Sensing and Imaging*, Kang, J.U. (Ed.), Springer Science+Business Media, New York, 29–53, 2013.)

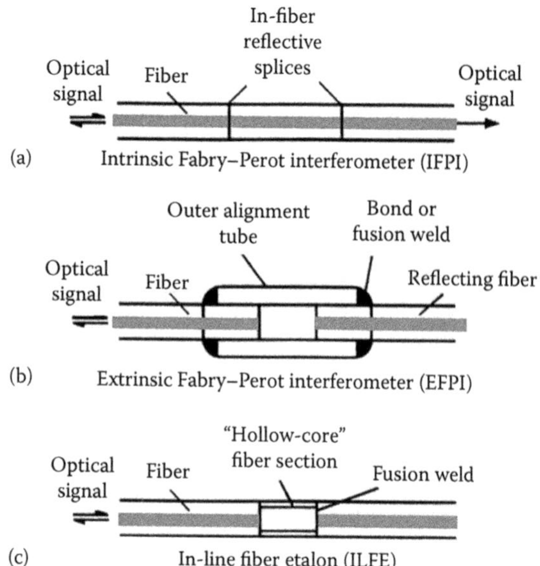

FIGURE 16.6 Interferometric fiber F–P implementations: (a) intrinsic fiber F–P, (b) extrinsic fiber F–P, and (c) in-line fiber etalon. (Reprinted from *Sens. Actuators*, 82, Grattan, K.T.V., and Sun, D.T., Fiber optic sensor technology: An overview, 40–61, Copyright 2000, with permission from Elsevier.)

16.1.4 Sagnac Interferometers

Sagnac interferometers (SIs) are recently in great interest in various sensing applications owing to their advantages of simple structure, easy fabrication, and environmental robustness. In Sagnac or ring interferometer, the incident light field is split into two parts (Vali and Shorthill 1976). While the path traveled by the two beams is the same, the two beams travel in opposite angular directions (i.e., clockwise and anticlockwise). After completing the loop trajectory, the two beams combine at the point of entry and undergo interference (Figure 16.7). As the two beams move in opposite angular directions, the interference signal at the coupler is highly sensitive to the angular motion of Sagnac loop itself.

16.1.5 Planar Interferometers

Typical configurations of planar interferometric sensors are shown in Figure 16.8. The most investigated approaches are the so-called MZI (Figure 16.8a) and the Young interferometer (Figure 16.8b) configurations (Schmitt and Hoffmann 2010). MZI consists of one light beam that is split into a sensing and a reference branch (Heideman and Lambeck 1999). One passes through the area where the chemical is present, whereas the other is isolated from the chemical that serves as a reference. While the sensing branch interacts with the sample, the reference branch is in contact with the reference medium, for example, air, buffer, solvent, and so on, and eventually both branches are combined to interfere.

Applying this MZI configuration to a sensor chip means that this structure has to be deposited with a high-RI material on a substrate. When monochromatic light is guided through this waveguide structure, the ratio of the beam power is exactly one and the phase difference is exactly zero if the sensor chip is perfectly manufactured. Applying a sample to the sensing branch, the local RI will change at the waveguide surface inducing a phase shift, and the output power will change obeying the following equation:

$$P_{out} = \frac{1}{2} P_{in}(1 + \cos\Delta\varphi) \tag{16.3}$$

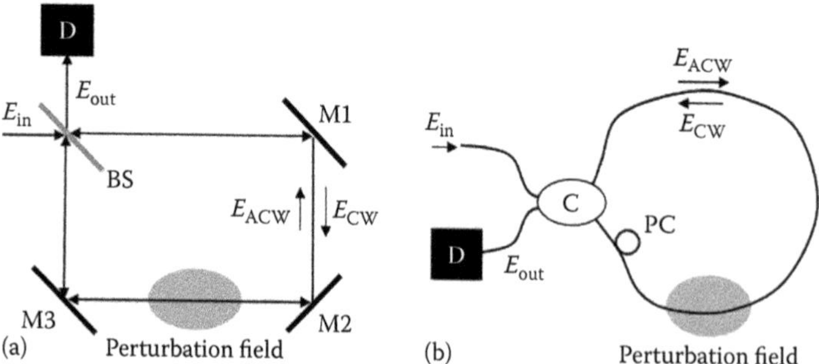

FIGURE 16.7 Schematic design for a bulk-optic (a), and intrinsic fiber-optic (b) Sagnac interferometer architecture. ECW and EACW are the electric field traveling in the clockwise and anticlockwise direction in the Sagnac ring or loop. (From Sharma, U. and Wei, X., Fiber optic interferometric devices, in *Fiber Optic Sensing and Imaging*, Kang, J.U. (Ed.), Springer Science+Business Media, New York, 29–53, 2013.)

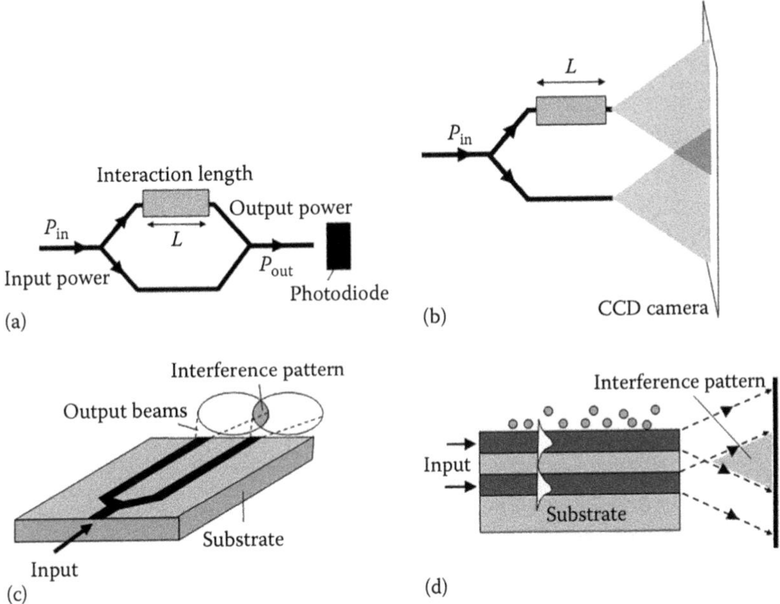

FIGURE 16.8 Schematic drawings of interferometric arrangements: (a) Mach–Zehnder configuration: the sensing and the reference branch are combined after the interaction of the sample and a photodiode that records the light intensity time-resolved; (b) Young interferometer. The sensing and the reference branch interfere in the far field. The interference pattern is recorded by a CCD. Either the end face of the waveguide structure approximates two point light sources or a double slit can be set in the beam path to generate the interference pattern; (c) interferometer using dispersion for beam overlap in a side-by-side configuration; and (d) a sandwich configuration. (From Schmitt, K. and Hoffmann, C., High-refractive-index waveguide platforms for chemical and biosensing, in *Chemical Sensors and Biosensors: Methods and Applications*, Zourob, M., and Lakhtakia, A. (Eds.), Springer, Berlin, Germany, 21–54, 2010; Campbell, D.P., Planar-waveguide interferometers for chemical sensing, in *Chemical Sensors and Biosensors: Methods and Applications*, Zourob, M. and Lakhtakia, A. (Eds.), Springer, Berlin, Germany, 55–113, 2010.)

where:

- P_{out} denotes the output power
- P_{in} is the input power
- $\Delta\varphi$ is the sample-induced phase shift

The MZI is generally designed as a system using a monomode wave propagating through the waveguide.

The Young interferometers, Figure 16.8b, can be very similar to the MZIs. However, the Young interferometer overcomes the problem of sensitivity fading in the extremum areas because it resolves the signal laterally using a CCD sensor, whereas the photodetector of a MZI counts the fringes in a time-resolved manner (Schmitt and Hoffmann 2010). But the Young interferometer approach has the same problem of ambiguity as the MZI, unless a semicoherent (or incoherent) light source is used. In this case, the interference pattern is superimposed on the envelope function known from a single slit. With an algorithm resolving the peak of the interference pattern only, the absolute phase value can be determined.

It should be noted that the various methods of interferometric spectroscopy began to develop rapidly simultaneously with the development of semiconductor electronics and the appearance of lasers, high-speed photodetectors, CCD cameras, and computerized data acquisition systems (Campbell 2010). Fabrication techniques and instruments, including photolithography, ion etching, and material deposition to very controlled geometries and thicknesses, which are necessary for the manufacture of optical bulk interferometers, were developed in the semiconductor industry. Devices such as diode lasers and megapixel array cameras provide inexpensive and high-quality light sources and detectors. Computers on a chip and small, inexpensive liquid-crystal displays, which appeared in recent years, take the images and signals produced by the interferometers and then convert them fast into displayed concentrations of the analytes. All this considerably improved the versatility, range, and resolution of interferometers, and as a result promoted expansion of the fields of their applications.

16.2 HUMIDITY SENSORS REALIZATION

With regard to the humidity sensors, the interferometers of FP configuration received the most prevalence (Grattan and Sun 2000), although other configurations such as Michelson (Kersey et al. 1991; Rao and Jackson 2000), Mach–Zehnder (Lee et al. 2012; Shao et al. 2013, 2014), Sagnac (Vali and Shorthill 1976; Sun et al. 2016), and difference interferometer (Lukosz et al. 1997) are being used as well (Table 16.1).

A chemical film placed on the waveguide surface can select and concentrate various analytes through a number of mechanisms that will produce measureable signals on the interferometer. These interactions and reactions can be categorized as either passive or active in their mode of sensing (Campbell 2010). Passive mechanisms can occur alone or in combination with an active moiety added to the passive sensing film. Passive sensing involves physical changes in the film caused by adsorption and concentration of a chemical analyte from the surroundings causing the film to swell, filling in void space within the film, or dissolving into the sensing film. Active sensing mechanisms can change

TABLE 16.1
Interferometric-Based Humidity Sensors

Sensing Method	Sensing Material	Range (%RH)	Response Time	References
Fabry–Perot interferometer	SiO_2–TiO_2–SiO_2 cavity	0–80	~1 min	Mitschke (1989)
	SiO_2 nanoparticles	40–98	150 ms	Corres et al. (2008)
	Microstructured SnO_2 overlay	2–40	–	Consales et al. (2011)
	Porous SiO_2 layer	0–100	5–36 s	Pevec and Donlagic (2015)
	Porous Al_2O_3 film	20–90	18 min	Huang et al. (2015b)
	Porous TiO_2–SiO_2–TiO_2 cavity	0–70	–	Huang et al. (2015a)
	SnO_2 deposited on the MOF	20–90	–	Lopez-Aldaba et al. (2016)
	SiO_2–[Au:PDDA+/poly(sodium 4-styrene-sulfonate)(PSS)]–air cavity	11–100	1.5 s	Arregui et al. (1999)
	SiO_2-[PDDA+/PolyS119-]-air cavity	0–97	3 s	Yu et al. (2001)
	Poly R-478-PDDA	11–97	<1.5 s	Arregui et al. (2002)
	Chitosan film	20–95	380 ms	Chen et al. (2012a)
	Nanocomposite polyacrylamine	38–98	250 ms	Yao et al. (2012)
	Polyvinyl alcohol (PVA)	7–91	~6 min	Su et al. (2013)
	Poly(sodium-*p*-styrenesulfonate) NPs	5–90	2–6 s	Sui et al. (2014)
	Nafion film	22–80	<1 s	Santos et al. (2014)
	Agarose	20–80	5–10 s	Wang et al. (2016a)
	Hygroscopic polymer NOA 61	20–90	~5 s	Lee et al. (2016)
Michelson interferometer	Polyurethane urea-PEO/poly(propylene oxide) hydrogel	–	–	Kronenberg et al. (1999)
	Polyvinyl alcohol	38–95	–	Kronenberg et al. (2000)
		30–90	300 ms	Wong et al. (2012)
	Chitosan	60–90	3–5 s	Hu et al. (2014)
Mach–Zehnder interferometer	Without special treatment	50–100	–	Shao et al. (2013, 2014)
	Without special treatment	30–90	–	Liu et al. (2016)
	Graphene oxide/PVA composite	25–80	–	Wang et al. (2016b)
	ZnO nanowires	35–60	–	Lokman et al. (2016)
	Chitosan	10–90		Ni et al. (2017)
	Carboxy-methyl cellulose	65–85		Ma et al. (2017)
Sagnac interferometer	Chitosan	20–95		Chen et al. (2012b)
	Without special treatment	20–90	60 ms	Sun et al. (2016)
Difference interferometer	Porous SiO_2–TiO_2 waveguide	25–90	–	Lukosz et al. (1997)
TCFMI	P4VP·HCl/ PVS multilayer	20–90	~2 s	Gu et al. (2011)

MOF—microstructured optical fiber; P4VP·HCl—poly(*N*-ethyl-4-vinylpyridinium chloride); PVS—poly(vinylsulfonic acid, sodium salt); SMF—single-mode fiber; TCFMI—thin-core fiber modal interferometer.

the RI to a large degree through the creation of a new compound within the sensing film and can provide enhanced responses to small amounts of material present. A combination of the two mechanisms: passive and active, allows adsorption-driven response by the partitioning to be followed by a chemical reaction within the film and further increases the interferometric signal produced, assuming that the RI changes occurring possess the same sign. In the case of humidity sensors, as a rule, passive sensing mechanisms are being used: swelling when using polymers (Arregui et al. 2002; Wong et al. 2012; Yao et al. 2012) and the change in the RI when using metal oxide films (Mitschke 1989; Corres et al. 2008; Consales et al. 2011).

16.2.1 Fabry–Perot Interferometers

FPI is an important technique used for humidity sensing (Table 16.1). Different from other fiber interferometric sensors (Match–Zehnder, Michelson, and Sagnac), the FP structure is stable and robust since there is no fiber couples and movable components. In addition, fiber F–PI sensors are extremely sensitive to perturbations.

One of the earliest fiber-optic interferometric humidity sensors was demonstrated by Mitschke (1989). The proposed sensor design consists of a thin film FPI formed at the tip of the optical fiber as shown in Figure 16.9. The interference between the optical signals reflected by the mirror at both ends of the cavity gives rise to a spectral response, which gives a maximum intensity output (resonances) at specific wavelengths. As shown in Figure 16.9, the FP cavity in the proposed design was created by a layer of TiO_2 sandwiched in between two partially reflecting mirrors, with the thickness of the cavity optimized to match the operation at the wavelength of the input diode laser source. The use of multilayer coatings has enabled high reflectivities to be achieved. As the RI of the cavity material has a dependence on humidity, the resonance was therefore shifted in response to a humidity change and can be conveniently detected by performing intensity measurement at a fixed wavelength. The sensor demonstrated suffers from cross-sensitivity to temperature, which can be corrected using a suitable compensation scheme. Nevertheless, it showed a good response between 0% and 80% RH and a response time of less than a minute.

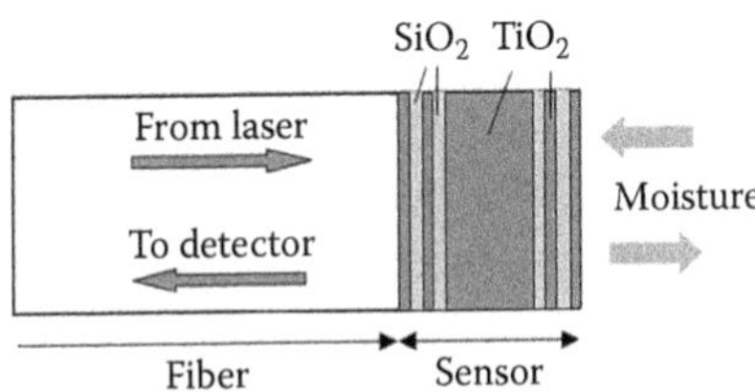

FIGURE 16.9 Cross section of a thin-film Fabry–Perot resonator (not to scale). The overall thickness of the stack of films is approximately 2 μm. (From Mitschke, F., *Opt. Lett.*, 14, 967, 1989. With permission of Optical Society of America.)

Pevec and Donlagic (2015) also used FPI for humidity measurement. However, as a humidity-sensitive material the porous SiO_2 layer was used. It has been shown that the thickness of the film is the main parameter controlling the sensor performance. The thicker layer provided higher sensor sensitivity. But at the same time the increase in the film thickness was accompanied by a significant increase in the response time. Response time was increased from 4.5 to 36 s by increasing the thickness of the SiO_2 layer from 240 to 600 nm. This behavior is quite natural, because the kinetics of the sensor response is largely controlled by the diffusion of water vapor in the humidity-sensitive film.

Huang et al. (2015b) have suggested using porous anodic Al_2O_3 (PAA) as a humidity-sensitive material. The sensing probes were made by attaching a thin porous alumina film, prepared by anodical etching, on the fiber tip. Experimental result showed that the sensitivity of 0.31 nm/% RH can be realized with very good repeatability in the range of 20%–90% RH (Figure 16.10). It can be concluded that fiber-optic RH sensors with PAA film of thinner thickness, smaller aperture size, and aspect ratio would present promising sensing performance. However, the response of these sensors was rather slow. In addition, parameters of these sensors were strongly dependent on anodizing conditions.

Similar FP interferometric humidity sensors with a submicron cavity length were reported by Arregui et al. (1999, 2000) and Yu et al. (2001). A typical multilayer thin film interferometric cavity was formed by stacking bilayers of alternating cationic and anionic polymers at the fiber tip. This was achieved by using the ionic self-assembled monolayer (ISAM) technique, which provided good control over the cavity length as well as the material composition of each coating layer. Sensors with a cavity length (or the number of bilayers) optimized at a specific operating wavelength were shown to be able to operate over a wide humidity range. A very fast response time, in the order of less than a few seconds, was reported and the sensor has been recommended as a possible diagnostic tool to monitor human breathing (Chen et al. 2002).

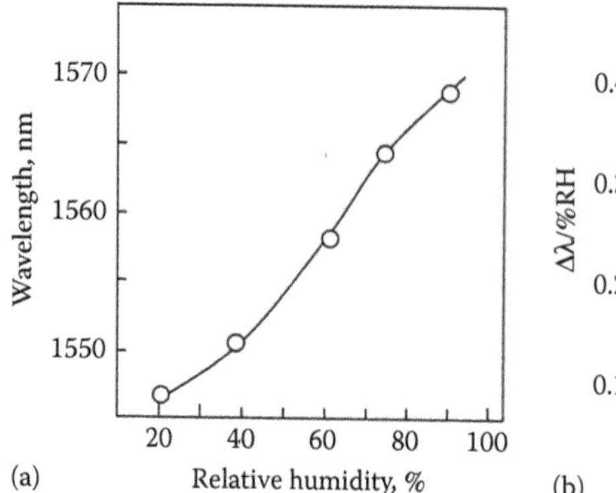

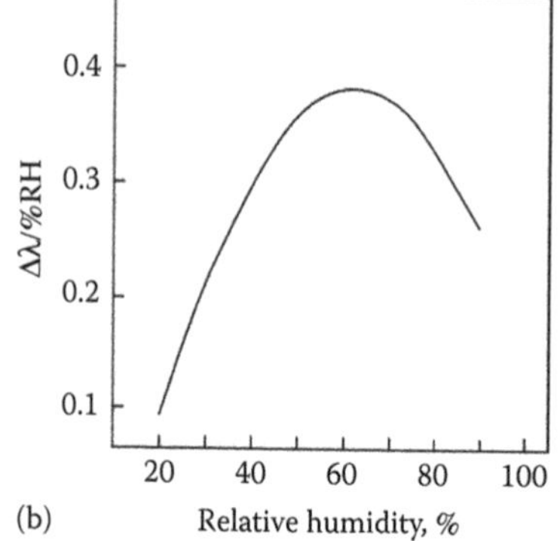

FIGURE 16.10 Characteristics of Fabry–Perot interferometer-based humidity sensor with porous Al_2O_3 layer: (a) the shift of characteristic wavelength and (b) the sensitivity curve to different RH levels. (Data extracted from Huang, C. et al., Optical fiber Fabry–Perot humidity sensor based on porous Al_2O_3 film. *IEEE Photon. Technol. Lett.*, 27(20), 2127–2130, 2015b.)

Santos et al. (2014) have shown that the fast response can be achieved when using thick enough humidity-sensitive films. Using the Nafion films with thickness 35–76 μm, Santos et al. (2014), they have developed sensors with high sensitivity, large dynamic range with good linearity of characteristics and fast response ($\tau < 1$ s). The configuration of the sensors and their sensitivity to humidity are shown in Figures 16.11 and 16.12. Apparently, the authors were able to form a porous Nafion films with a large pore size. Santos et al. (2014) believe that Nafion, presenting high hydrophilicity, chemical and thermal stability, high conductivity, and mechanical toughness, is a suitable material for humidity sensor design. Su et al. (2013) have tested the FPI using polyvinyl alcohol (PVA) films with a thickness of ~82 μm. However, unlike sensor developed by Santos et al. (2014), these sensors had a much lower sensitivity. Besides, the sensor response was too long.

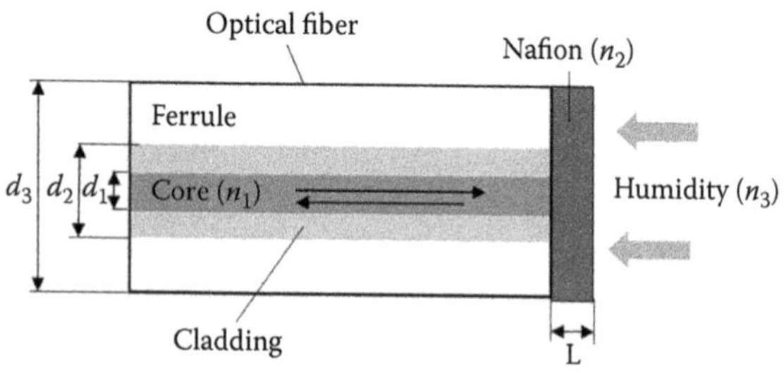

FIGURE 16.11 Schematic showing the structure of the fiber sensor probe ($d_1 = 9$ μm, $d_2 = 125$ μm, and $d_3 = 2$ mm). The sensor probe was assembled by depositing a drop of Nafion solution on the tip of a standard SMF28 single mode fiber. (Reprinted from *Sens. Actuators B*, 196, Santos, J.S. et al., Characterisation of a Nafion film by optical fibre Fabry–Perot interferometry for humidity sensing, 99–105, Copyright 2014, with permission from Elsevier.)

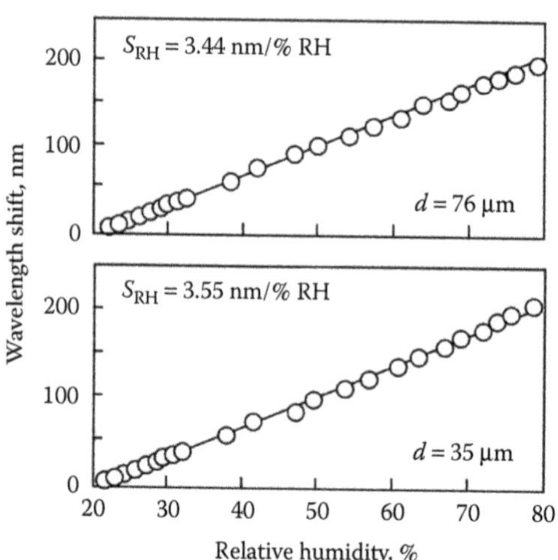

FIGURE 16.12 Analytical curves for the proposed FPI-based humidity sensor obtained with Nafion thicknesses of 35 and 76 μm. (Reprinted from *Sens. Actuators B*, 196, Santos, J.S. et al., Characterisation of a Nafion film by optical fibre Fabry–Perot interferometry for humidity sensing, 99–105, Copyright 2014, with permission from Elsevier.)

16.2.2 Michelson Interferometric Sensors

A description of Michelson interferometric humidity sensors one can find in Kronenberg et al. (1999), Wong et al. (2012), and Hu et al. (2014). In particular, Kronenberg et al. (1999) have explored the use of a low coherence interferomertric system arranged in tandem configuration for humidity detection. The system consisted of two MIs—(1) an interrogation system to perform the measurements and (2) a bi-fiber-optic sensor based on Michelson configuration formed by a reference uncoated fiber and a hydrogel-coated fiber. In a similar way to the FBG-sensing mechanism described

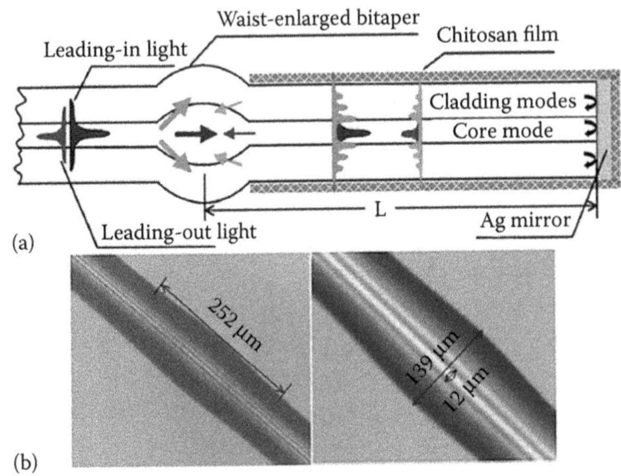

FIGURE 16.13 Schematic diagram of the proposed WEBMI-based humidity sensor. The inset (a) and (b) are microscope photos of the waist-enlarged bitaper with different magniscales. (Reprinted from *Sens. Actuators B*, 194, Hu, P. et al., 180–184, Copyright 2014, with permission from Elsevier.)

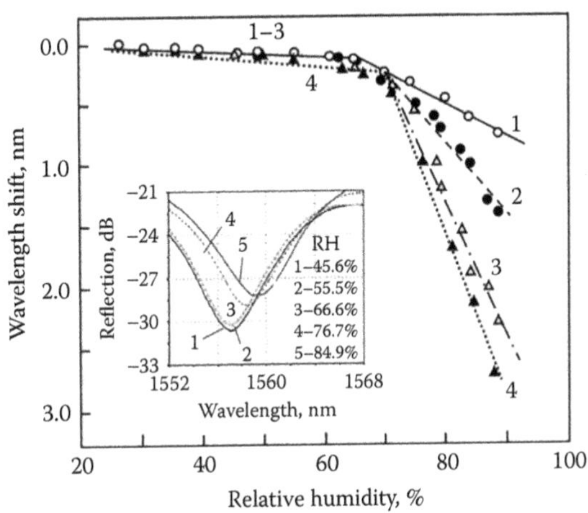

FIGURE 16.14 Wavelength shift versus surrounding humidity for fiber sensor with different time of etching and chitosan oxidization: 1, nonetching; 2, 40-min etching; 3, 60-min etching; 4, 60-min etching and oxidation. The inset shows optical spectra of the sensor with 60-min etching and chitosan oxidization at various humidity levels. (Reprinted from *Sens. Actuators B*, 194, Hu, P. et al., 180–184, Copyright 2014, with permission from Elsevier.)

in the preceding Chapter 14, the fiber sensor relies on a humidity-induced swelling to stretch the fiber, which in turn creates an interferometric sensor with an unbalanced path length.

Another approach was used by Hu et al. (2014). Schematic diagram of the proposed chitosan-coated, waist-enlarged bitaper-based Michelson interferometer (WEBMI)-based humidity sensor is shown in Figure 16.13. The WEBMI is formed between the bitaper and a silver mirror at the fiber end. The chitosan film was deposited on the fiber from solution by using a dip-coating method. Hu et al. (2014) have shown that the sensitivity can be strongly increased by reducing the fiber cladding thickness and oxidizing the chitosan film (Figure 16.14). They believe that this improvement in sensitivity may be related to the enhanced hydrophilicity of the oxidized chitosan sensing film: (1) the periodate ions attack vicinal diols to cleave the carbon–carbon bond by an oxidation reaction, leading to the formation of new functional groups (such as dialdehyde), which are capable of absorbing additional water molecules, and (2) partial depolymerization of the chitosan permits more chain flexibility, resulting in higher swelling index to enhance the RH sensitivity of the proposed sensor. Experimentations showed that the measurement of RH in the range of 25%–90% RH by WEBMI-based sensors was fully reversible and repeatable with fast response and recovery time of ~5 s and 3 s, respectively. However, wavelength shift acceptable for real application was observed only at RH > 60%.

16.2.3 Mach–Zehnder Interferometers

One of the variants of MZI humidity sensors is shown in Figure 16.15. Such configuration was developed by Shao et al. (2013, 2014). The interferometer consists of two arc-induced ellipsoid fiber tapers and a section of SMF between them. The cladding modes are excited from the core mode by the first taper, and then enter in a section of SMF as the interferometer arm. Before the core mode and cladding modes reach the second taper, different modes possess different phase thus making the

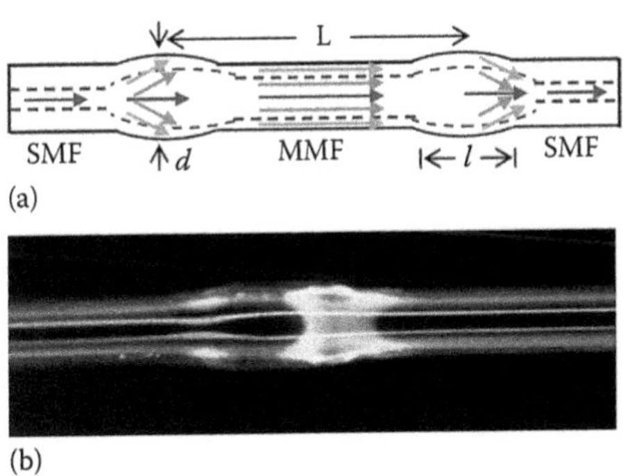

FIGURE 16.15 (a) Schematic diagram of the MZI constructed by two ellipsoid fiber tapers. (b) Photograph of the waist-enlarged fiber taper. (Reprinted from *Opt. Laser Eng.*, 52, Shao, M. et al., A Mach–Zehnder interferometric humidity sensor based on waist-enlarged tapers, 86–90, Copyright 2014, with permission from Elsevier.)

modal phase difference. The core mode and cladding modes is recouped by the second taper into the SMF, which makes the modal interferences. Experimental results showed that this kind of MZI is easily fabricated by commercial splicer in an SMF and it does not need special fibers or coated film. When the interferometer length was 47 mm, the humidity sensitivity of MZI, determined as the wavelength shift, was −0.047 nm/% RH. Such sensitivity is about 10 times higher than that of the common FBG humidity sensors. However, in comparison with other interferometric humidity sensors, this sensitivity to humidity is not high. In addition, this response was dependent on the temperature. But it should be taken into account that this Mach–Zehnder interferometric humidity sensor was designed on the basis of multimode fiber without a coating of sensitive film. In the same time, Shao et al. (2014) have established that the measurement of the power variation of the dip in the transmission spectra was of interest to monitor humidity, because the power of the dip remained unchanged with temperature. When measuring this parameter, the linear humidity response with enhanced sensitivity of 0.119 dB/% RH was achieved over the humidity range of 35%–90% RH.

Another version of the MZI-based humidity sensor, which also did not use additional humidity-sensitive films deposited on the fiber, was proposed by Liu et al. (2016). They proposed a simple in-fiber MZI comprised of a hollow-core fiber (HCF) sandwiched between two SMFs. The interferometer is sensitive to changes in RI near 1.33 and thus can be further employed to measure the relative humidity (RH). A schematic diagram of the sensor is shown in Figure 16.16. An 8 mm-long HCF is sandwiched between two SMFs by aligning splices. The core/cladding diameters of the SMFs are 9/125 μm, and the HCF's diameters are 2/125 μm. The HCF performs as the interference arms. As the device is formed, the part of the input core mode is coupled to the cladding modes within the HCF cladding via the core-mismatch between the SMF and HCF. The excited cladding modes are transmitted along the HCF cladding and then are partially recoupled to the output SMF via the second splice interface. On account of the effective differences in the RIs of the core and cladding modes, phase delays occur between them that result in a well-defined interference spectrum.

A well-defined interference pattern was obtained as a result of a fiber-core mismatch and core-cladding mode interference. The selected interference dip was sensitive to the ambient RI and provided a stable RI response. However, the response was less than that of sensors developed by Shao et al. (2013, 2014).

Wang et al. (2016b) showed that the use of graphene oxide/PVA composite incorporated into the MZI can significantly increase the sensitivity of the humidity sensor. The MZI was consisted of two waist-enlarged tapers, and the length between two waist-enlarged tapers was 20 mm (Figure 16.17b). The waist-enlarged taper was made easily by using a commercial fusion splicer (Fujikura FSM-60s). To make the end of the SMF to be spherical, the fiber was arc is discharged for 4–5 times under the manual splicing mode. And then it was arc discharged with a flat end face, where it was used a much larger overlap distance of 130 μm to replace the common setting of 10 μm. Figure 16.17a shows the structure of the waist-enlarged taper, in which the black and the white areas refer to the fiber cladding and fiber core, respectively. Obviously, an elliptical core area is formed in the center of the waist-enlarged taper, which means that light can be coupled into the cladding easily and the fringes visibility of interference pattern can be enhanced.

Figure 16.17b shows that when the light is launched into SMF 1, massive light energy is coupled from the core of SMF 1 into the cladding of SMF 2 and a large number of cladding modes were excited by the waist-enlarged taper. The remaining light still propagates in

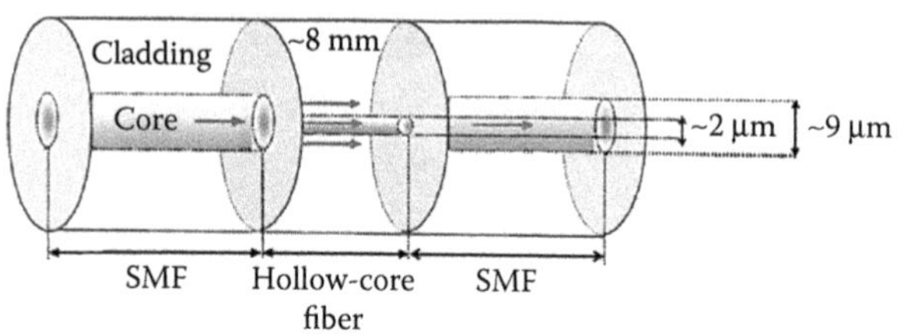

FIGURE 16.16 Schematic diagram of the MZI-based humidity sensor. (Reprinted from *Opt. Commun.*, 367, Liu, N. et al., A fiber-optic refractometer for humidity measurements using an in-fiber Mach–Zehnder interferometer, 1–5, Copyright 2016, with permission from Elsevier.)

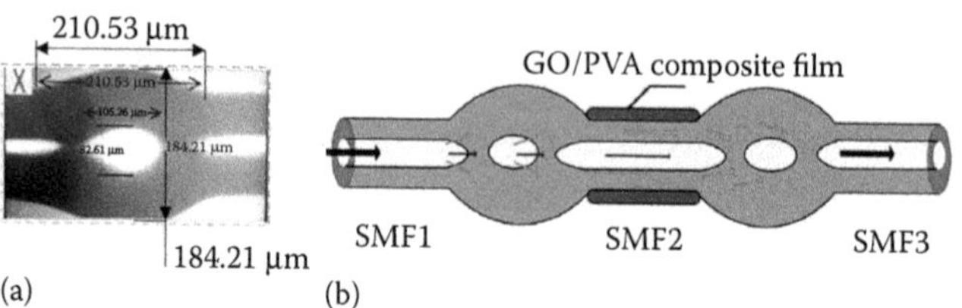

FIGURE 16.17 (a) The partially enlarged drawing of the sensing head and (b) schematic diagram of the structure of the waist-enlarged taper. (Reprinted from *Opt. Commun.*, 372, Wang, Y. et al., Fiber optic humidity sensor based on the grapheme oxide/PVA composite film, 229–234, Copyright 2016, with permission from Elsevier.)

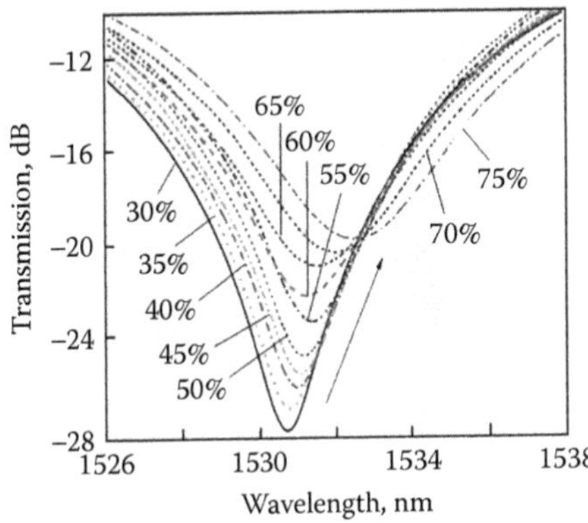

FIGURE 16.18 Changes of interference spectra at the resonant wavelength of $\lambda = 1531$ nm coated with the composite film formed by 0.3 g PVA + 10 mL GO dispersion. (Reprinted from *Opt. Commun.*, 372, Wang, Y. et al., Fiber optic humidity sensor based on the graphene oxide/PVA composite film, 229–234, Copyright 2016, with permission from Elsevier.)

the core of SMF 2 as the fundamental core mode. Due to the formation of the elliptical core area in the center of the waist-enlarged taper, there will be much light coupled into the cladding of SMF 2, and recoupled into the fiber core of SMF 3 by the second taper to interfere with the fundamental core mode. The lengths of this kind of MZI's two arms are exactly the same, but the optical lengths are different because of the different effective indices of each optical length. As measurand Wang et al. (2016b) like Shao et al. (2014) used the power variation of the dip in the transmission spectra (Figure 16.18). It was found that MZI-based sensor had sensitivity of 0.193 dB/% RH with a good stability and linearity at the RH range 25%–80%. A linear correlation coefficient was of 99.1%.

Lokman et al. (2016) also tried to improve the characteristics of MZI-based humidity sensors by using ZnO nanowire coating. However, significant progress has not been achieved. A wavelength shift, not exceeding 0.490 nm was achieved for a humidity variation from 35% to 60% RH. At the same time, the use of Chitosan as a humidity-sensitive material deposited on the cladding has yielded positive results (Ni et al. 2017). After preparation of chitosan solutions, the deposition was achieved by dipping the surface-treated SMF–MZI into the solution. This proposed sensor had a sensitivity of ~0.18 nm/% RH. Even more sensitive sensors were developed by Ma et al. (2017). MZI-based sensors were coated by a carboxymethyl cellulose. However, a noticeable sensitivity of these sensors was observed only at RH > 60%. Apparently, when RH > 60%, a capillary condensation of water vapor becomes possible.

16.2.4 Sagnac Interferometers

SIs are extremely sensitive to rotations and therefore these devices find application mainly in the development of inertial guidance systems such as gyroscopes (Anderson et al. 1994). Sun et al. (2016) have shown that humidity sensors based on SIs also can be designed. Figure 16.19 shows the schematic diagram of the humidity sensor based on the high-birefringence (Hi–Bi) elliptical microfiber SIs. A section of Hi–Bi microfiber that contains two transition regions and a central uniform waist region, together with a Hi–Bi Panda fiber, is fusion spliced into a fiber loop mirror. The microfiber waist has a micron-meter size so that light could be extended in the whole silica region. High birefringence of the microfiber is mainly induced by the elliptical structure cross section. The polarization states of light can be adjusted by the polarization controllers of PC_1 and PC_2. The Panda fiber is applied to enhance the RH sensitivity of the device. The spectral characteristic is measured by using a broad band light source (BBS) and an optical spectrum analyzer (OSA). Light from the BBS splits into clockwise and counterclockwise beams via 3 dB coupler as it enters the loop. The polarimetric interference fringes can be achieved by the recombination of the counter propagating beams at the coupler. The optical path difference between the polarized light beams is determined by the relative azimuths of the Hi–Bi fibers.

As shown in Figure 16.19, the sensing is realized due to the interaction between the evanescent mode field of the Hi–Bi elliptic microfiber and external physical quantities. Sun et al. (2016) have fabricated the elliptic profile of the fiber cross section by cutting away the part of the silica cladding on the opposite

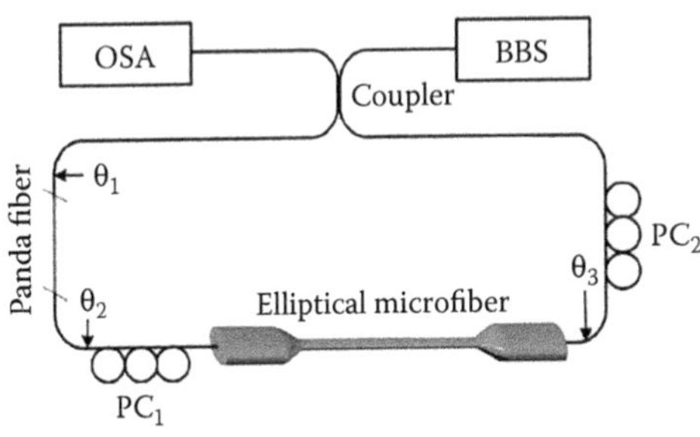

FIGURE 16.19 Schematic diagram of the Hi–Bi elliptical microfiber Sagnac interferometer-based RH sensor. (Reprinted from *Sens. Actuators B*, 231, Sun, L.-P. et al., High-birefringence microfiber Sagnac interferometer based humidity sensor, 696–700, Copyright 2016, with permission from Elsevier.)

sides of a standard single-mode fiber. A pulsed CO_2 laser (SYNRAD 48–5) was used for the fabrication. The output laser beam, after being transmitted through a ZnSe lens had a diameter of ~50 μm and power of 17.5 W. When the high-power laser beam was irradiated onto a fiber surface, a dramatic temperature promotion occurred due to the photon absorption effect, resulting in the sputtering of the fiber material and the deformation of the fiber shape. To obtain the elliptical fiber, they placed a standard single-mode fiber perpendicularly to the laser-irradiating direction, as shown in Figure 16.20a. The whole fabrication procedure was monitored by using a CCD camera. After one-cycle laser milling process, an elliptical fiber could be produced.

Figure 16.21a illustrates the relationship between the wavelength shift and the humidity, varying from 30% to 95% RH. The obtained sensitivity is ~0.2 nm/% RH for the dip with a wavelength of 1560 nm. Such sensitivity is good enough in comparison with sensitivity of many other sensors previously discussed. Measurement also showed that the structure can have a response time of better than 60 ms, suggesting a fast response speed compared to the previous devices.

Chen et al. (2012b) also developed SIs-based humidity sensors. But the configuration proposed by Chen et al. (2012b), even when using coatings, sensitive to water vapors, was not as sensitive as the configuration designed by Sun et al. (2016). The main result of these studies is a confirmation that the chemical etching, performed to reduce the cladding thickness of the PM fiber, promotes to increase the sensitivity of the sensors. According to Chen et al. (2012b), after chemical etching the chitosan-coated PM fiber probe experiences more tension during swelling of the chitosan film. In addition, the evanescent wave in the core has higher penetration depth into the chitosan coating, which results in an increasing sensing signal.

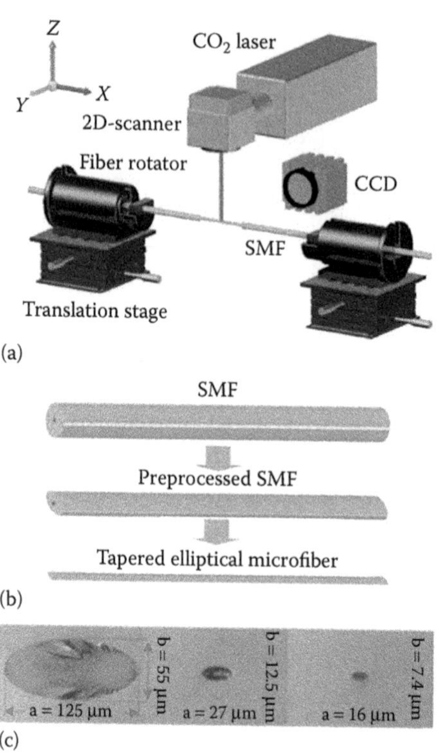

FIGURE 16.20 (a) Schematic of the CO_2-laser-machining system for the elliptic fiber fabrication. (b) Described the fabrication process of the elliptical microfiber. (c) Cross-sectional microscope images of elliptical fibers. (Reprinted from *Sens. Actuators B*, 231, Sun, L.-P. et al., High-birefringence microfiber Sagnac interferometer based humidity sensor, 696–700, Copyright 2016, with permission from Elsevier.)

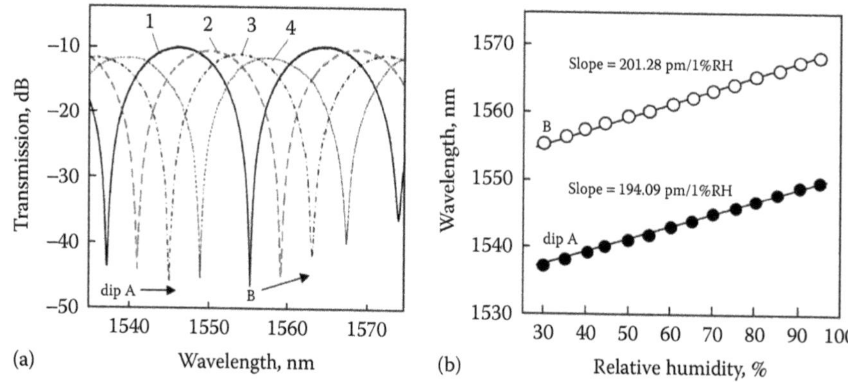

FIGURE 16.21 (a) Recorded transmission spectrum of dips at different RH values (1, 30% RH; 2, 50% RH; 3, 70% RH; 4, 90% RH). (b) Measured dip wavelength shift as functions of relative humidity. (Reprinted from *Sens. Actuators B*, 231, Sun, L.-P. et al., High-birefringence microfiber Sagnac interferometer based humidity sensor, 696–700, Copyright 2016, with permission from Elsevier.)

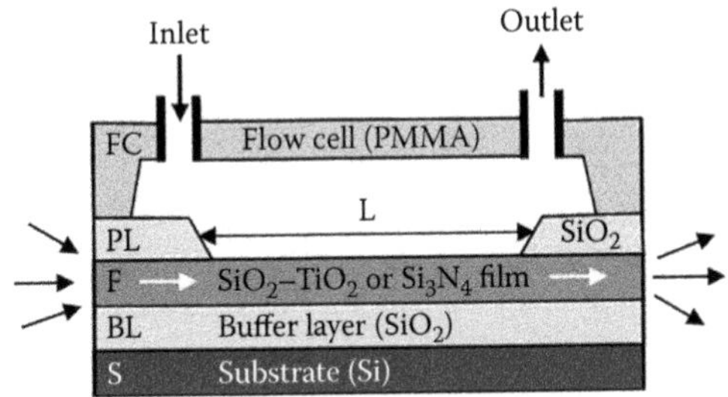

FIGURE 16.22 The cross-sectional view of the waveguide F also shows the flow cell with the sample and the protection layer PL on the waveguide, which defines an interaction window of the length L (L = 20 mm). (Reprinted from *Sens. Actuators B*, 38–39, Lukosz, W. et al., Difference interferometer with new phase-measurement method as integrated-optical refractometer, humidity sensor and biosensor, 316–323, Copyright 1997, With permission from Elsevier.)

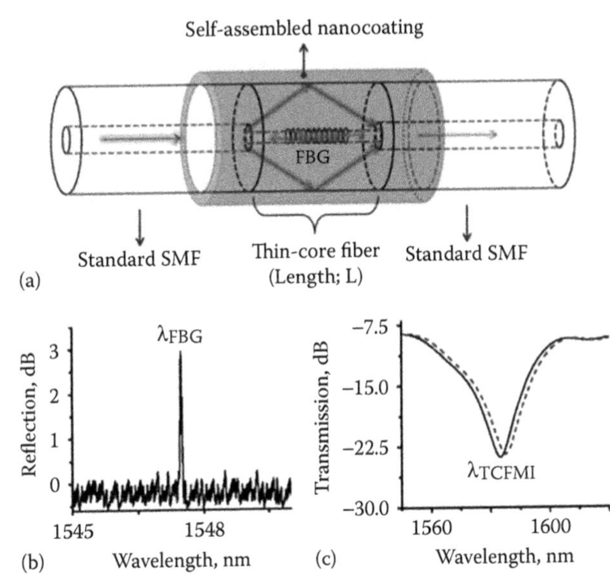

FIGURE 16.23 Schematic configuration of the fiber-optic RH sensor based on an FBG-incorporated TCFMI (a) and its reflection spectrum (b) and transmission spectra (c) before (solid curve) and after (dashed curve) self-assembling of sensing nanocoating. (From Gu, B. et al., *Opt. Exp.*, 19, 4140–4146, 2011. With permission of Optical Society of America.)

16.2.5 Other Variants of Interferometric Humidity Sensors

Lukosz et al. (1997) and Lukosz and Stamm (1991) have developed difference interferometers for the analysis of various parameters including RH. In a difference interferometer, the TE_0 and the TM_0 modes in a planar waveguide are coherently excited, propagate in a common path, and interact with the sample. The sensor response is measured as the time-dependent phase difference $\Delta\Phi(t)$ between the TE_0 and the TM_0 modes induced by the polarization-dependent interaction. For humidity monitoring, the microporous SiO_2–TiO_2 waveguides (n_F = 1.75, d_F = 198 nm) were used (Figure 16.22), as they showed a large effective RI change in the presence of water vapors, attributed mainly to the sorption of molecules in the micropores of the waveguiding film.

16.3 HYBRID SENSORS

The development of hybrid sensors, which involve a combination of both fiber grating and interferometric configurations, is connected with attempts to achieve more effective RH sensing than through the use of either approach separately. As discussed in the Chapter 14 related to fiber grating, both FBGs and LPGs written into conventional fibers are inherently sensitive to the temperature. Therefore, in most cases where hybrid sensors are considered, the involvement of the grating is for the purpose of eliminating the temperature-induced measurement error from the actual RH/moisture sensing results. One of such examples is the work of Gu et al. (2011).

Gu et al. (2011) proposed a fiber-optic RH sensor based on the thin-core fiber modal interferometer (TCFMI) and the FBG hybrid structure in which an electrostatic self-assembled polymer nanocoating was coated on the side-surface of the sensor for humidity sensing (Figure 16.23). A 20 mm long thin-core fiber was inserted into standard single-mode fiber to form an in-fiber modal interferometer. It was demonstrated that the spectrum dip of TCFMI was an ideal sensing signal for RI measurement. The TCFMI was used to measure the humidity-induced RI change of the nanocoating, whereas the FBG inscribed in the middle of TCFMI was used for the compensation of temperature sensitivity. Experiment has shown that the RH sensor had a fast, linear, and reversible response with high resolution in a large RH measurement range. The sensitivities of the transmission dip to RH were 84.3, 87.9, and 97.2 pm/% RH at 20°C, 40°C, and 60°C, respectively.

Another benefit of the hybrid design for interferometric-grating sensing is to improve the measuring technique, that is, to create a probe, and thus to achieve a better resolution in the detection system (Alwis et al. 2013). A typical configuration involving a single grating-based LPG sensor system frequently has the disadvantage of the probe being used in transmission mode. Further, the broad bandwidth of the attenuation bands formed by the propagation mode coupling between the core and the cladding modes constitutes a difficulty when the device is used as a conventional sensor probe. To overcome these

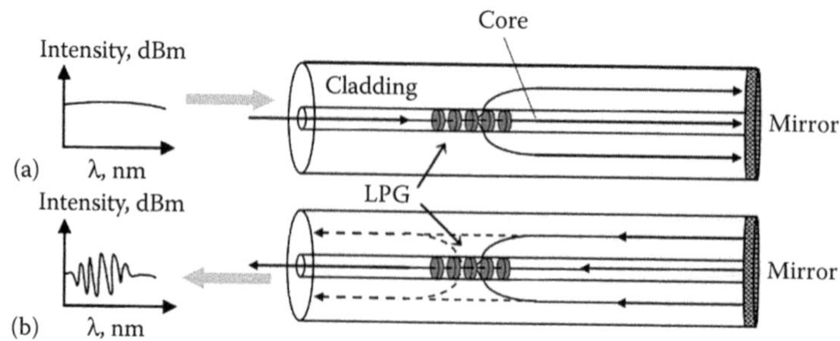

FIGURE 16.24 Light propagation in the SILPG (a) forward propagation path and (b) propagation path of the reflection. (Reprinted from *Proc. Eng.*, 47, Elwis, L. et al., Optimization of a long period grating distal probe for temperature and refractive index measurement, 718–721, Copyright 2012, with permission from Elsevier.)

limitations, an MI-type sensor configuration has been proposed by Lam et al. (2009). They have been using a LPG grating pair formed by coating a mirror at the distal end of the LPG, that is, termed as Self-interfering LPG (SILPG), as can be seen from Figure 16.24, to create a refractometer. This sensor configuration is more convenient to use and is able to overcome the limitations of the single LPG sensor due to the shifts in the attenuation bands being more easily detectable.

16.4 SUMMARY

Conducted analysis showed that interferometric systems should be considered as a promising tool for humidity monitoring. Compared to grating coupler sensors, interferometric systems are more sensitive and exhibit a very high resolution due to the long interaction length (typically in the range of 1–2 cm) of the guided mode with the sample; and the high-accuracy phase shift can be measured. We also believe that interferometric techniques are the most exact and fastest ones from all the existing optical methods of measurement (Kolpakov et al. 2014). The advantages of the interferometric humidity sensing may also include the ability to measure the humidity without the use of hydrophilic material that can be easily damaged when the sensor works in a harsh environment.

But at the same time one should understand that in the way of practical application of interferometric-based sensors having high resolution, the problems might appear, which are associated with the fact that the sensor respond simultaneously to ambient or sample temperature fluctuations. It has been observed that changes in the temperature of the waveguide, at a constant humidity, also induce a phase difference, complicating the calibration of these sensors for RH monitoring. The contributions of the interference gases, refractometric and temperature effects to the phase shift are superimposed and cannot easily be distinguished (Schmitt and Hoffmann 2010).

It is important to note that FPIs have minimal dependence of the parameters on the temperature in comparison with other devices. Moreover, some authors (Sui et al. 2014; Huang et al. 2015a) even noticed the almost complete absence of such a relationship in the range from 20°C to 100°C. However, other authors believe that even for FPIs the temperature compensation is needed (Pevec and Donlagic 2015; Lee et al. 2016; Wang et al. 2016a). To solve this problem, it is proposed to control simultaneously the humidity and the temperature (Gu et al. 2011; Yang et al. 2014).

Low selectivity is another important shortcoming of interferometric-based humidity sensors. The sensors of this type measure the changes in the RI of the atmosphere next to the fiber; consequently, they are sensitive to a variety of contaminants. The need for fairly complicated and expensive measuring equipment can also serve as a deterrent when using interferometric systems for measuring humidity.

REFERENCES

Alwis L., Sun T., Grattan K.T.V. (2013) Optical fibre-based sensor technology for humidity and moisture measurement: Review of recent progress. *Measurement* 46, 4052–4074.

Anderson R., Bilger H.R., Stedman G.E. (1994) Sagnac effect: A century of Earth-rotated interferometers. *Am. J. Phys.* 62(11), 975–985.

Arregui F.J., Cooper K.L., Liu Y., Matias I.R., Claus R.O. (2000) Optical fiber humidity sensor with a fast response time using the ionic self-assembly method. *IEIEC Trans. Electron.* E83-C(3), 360–365.

Arregui F.J., Liu Y., Matias I.R., Claus R.O. (1999) Optical fiber humidity sensor using a nano Fabry–Perot cavity formed by the ionic self-assembly method. *Sens. Actuators B* 59, 54–59.

Arregui F.J., Matias I.R., Cooper K.L., Claus R.O. (2002) Simultaneous measurement of humidity and temperature by combining a reflective intensity-based optical fiber sensor and a fiber Bragg grating. *IEEE Sens. J.* 2(5), 482–487.

Campbell D.P. (2010) Planar-waveguide interferometers for chemical sensing. In: Zourob M., Lakhtakia A. (Eds.) *Chemical Sensors and Biosensors: Methods and Applications*. Springer, Berlin, Germany, pp. 55–113.

Chen L.H., Li T., Chan C.C. et al. (2012a), Chitosan based fiber-optic Fabry–Perot humidity sensor. *Sens. Actuators B* 169, 167–172.

Chen L.H., Li T., Chan C.C. et al. (2012b) Chitosan-coated polarization maintaining fiber-based Sagnac interferometer for relative humidity measurement. *IEEE J. Sel. Top. Quantum Electron.* 18(5), 1597–1604.

Chen Q., Claus R.O., Spillman W.B., Arregui F.J., Matias I.R., Cooper K.L. (2002) Optical fiber sensors for breathing diagnostic. In *Proceedings OFS 15*, Portland, OR, Vol. 1, pp. 273–276.

Consales M., Buosciolo A., Cutolo A. et al. (2011) Fiber optic humidity sensors for high-energy physics applications at CERN. *Sens. Actuators B* 159(1), 66–74.

Corres J.M., Matías I.R., Hernáez M., Bravo J., Arregui F.J. (2008) Optical fiber humidity sensors using nanostructured coatings of SiO_2 nanoparticles. *IEEE Sens. J.* 8(3), 281–285.

Elwis L., Sun T., Grattan K.T.V. (2012) Optimization of a long period grating distal probe for temperature and refractive index measurement. *Proc. Eng.* 47, 718–721.

Grattan K.T.V., Sun Dr. T. (2000) Fiber optic sensor technology: An overview. *Sens. Actuators* 82, 40–61.

Gu B., Yin M., Zhang A.P., Qian J., He S. (2011) Optical fiber relative humidity sensor based on FBG incorporated thin-core fiber modal interferometer. *Opt. Exp.* 19(5), 4140–4146.

Heideman R.G., Lambeck P.V. (1999) Remote opto-chemical sensing with extreme sensitivity: Design, fabrication and performance of a pigtailed integrated optical phase-modulated Mach–Zehnder interferometer system. *Sens. Actuators B* 61, 100–127.

Hu P., Dong X., Ni K., Chen L.H., Wong W.C., Chan C.C. (2014) Sensitivity-enhanced Michelson interferometric humidity sensor with waist-enlarged fiber bitaper. *Sens. Actuators B* 194, 180–184.

Huang C., Xie W., Lee D., Qi C., Yang M., Wang M., Tang J. (2015a) Optical fiber humidity sensor with porous TiO_2/SiO_2/TiO_2 coatings on fiber tip. *IEEE Photon. Technol. Lett.* 27(14), 1495–1498.

Huang C., Xie W., Yang M., Dai J., Zhang B. (2015b) Optical fiber Fabry–Perot humidity sensor based on porous Al_2O_3 film. *IEEE Photon. Technol. Lett.* 27(20), 2127–2130.

Kersey A.D., Marrone M., Davis M.A. (1991) Polarization-insensitive fiber optic Michelson interferometer. *Electron. Lett.* 26, 518–520.

Kolpakov S.A., Gordon N.T., Mou C., Zhou K. (2014) Toward a new generation of photonic humidity sensors. *Sensors* 14, 3986–4013.

Kronenberg P., Culshaw B., Pierce G. (1999) Development of a novel fiber optic sensor for humidity monitoring. In: *Proceedings SPIE Conference on Sensory Phenomena and Measurement Instrumentation for Smart Structures and Materials*, pp. 480–485.

Kronenberg P., Inaudi D., Smith I.F.C. (2000) Development of an "optical hair"-hygrometer: A novel way to measure humidity using fibre optics. In: *Proceedings of the International Conference on Trends in Optical Nondestructive Testing*, Lugano, Switzerland, pp. 467–474, May.

Lam C.C.C., Mandamparambil R., Sun T., Grattan K.T.V., Nanukuttan S.V., Taylor S.E., Basheer P.A. (2009) Optical fiber refractive index sensor for chloride ion monitoring. *IEEE Sens. J.* 9(5), 525–532.

Lee B.H., Kim Y.H., Park K.S., Eom J.B., Kim M.J., Rho B.S., Choi H.Y. (2012) Interferometric fiber optic sensors. *Sensors* 12, 2467–2486.

Lee C.-L., You Y.-W., Dai J.-H., Hsu J.-M., Horn J.-S. (2016) Hygroscopic polymer microcavity fiber Fizeau interferometer incorporating a fiber Bragg grating for simultaneously sensing humidity and temperature. *Sens. Actuators B* 222, 339–346.

Liu N., Hu M., Sun H., Gang T., Yang Z., Rong Q., Qiao X. (2016) A fiber-optic refractometerfor humidity measurements using an in-fiber Mach–Zehnder interferometer. *Opt. Commun.* 367, 1–5.

Lokman A., Arof H., Harun S.W., Harith Z., Rafaie H.A., Md Nor R. (2016) Optical fiber relative humidity sensor based on inline Mach–Zehnder interferometer with ZnO nanowires coating. *IEEE Sens. J.* 16(2), 312–316.

Lopez-Aldaba A., Lopez-Torres D., Ascorbe J. et al. (2016) SnO_2-MOF-Fabry-Perot humidity optical sensor system based on fast Fourier transform technique. In: *Proceedings of European Workshop on Optical Fibre Sensors (EWOFS 2016)*, Limerick, Ireland, pp. 9916–9929, May.

Lukosz W., Stamm C. (1991) Integrated optical interferometer as relative humidity sensor and differential refractometer. *Sens. Actuators A* 25–27, 185–188.

Lukosz W., Stamm C., Moser H.R., Ryf R., Diibendorfer J. (1997) Difference interferometer with new phase-measurement method as integrated-optical refractometer, humidity sensor and biosensor. *Sens. Actuators B* 38–39, 316–323.

Ma Q., Ni K., Huang R. (2017) A carboxy-methyl cellulose coated humidity sensor based on Mach-Zehnder interferometer with waist-enlarged bi-tapers. *Opt. Fiber Technol.* 33, 60–63.

Mitschke F. (1989) Fiber optic sensor for humidity. *Opt. Lett.* 14(7), 967–969.

Ni K., Chan C.C., Chen L., Dong X., Huang R., Ma Q. (2017) A chitosan-coated humidity sensor based on Mach-Zehnder interferometer with waist-enlarged fusion bitapers. *Opt. Fiber Technol.* 33, 56–59.

Pevec S., Donlagic D. (2015) Miniature all-silica fiber-optic sensor for simultaneous measurement of relative humidity and temperature. *Opt. Lett.* 40(23), 5646–5649.

Rao Y.J. (2006) Recent progress in fiber-optic extrinsic Fabry–Perot interferometric sensors. *Opt. Fiber Technol.* 12, 227–237.

Rao Y.J., Jackson D.A. (2000) Principles of fibre-optic interferometry. In: Grattan K.T.V., Meggitt B.T. (Eds.), *Optical Fiber Sensor Technology: Fundamentals Vol. 1*. Chapman & Hall, London, UK, pp. 167–191.

Santos J.S., Raimundo Jr. I.M., Cordeiro C.M.B., Biazoli C.R., Gouveia C.A.J., Jorge P.A.S. (2014) Characterisation of a Nafion film by optical fibre Fabry–Perot interferometry for humidity sensing. *Sens. Actuators B* 196, 99–105.

Schmitt K., Hoffmann C. (2010) High-refractive-index waveguide platforms for chemical and biosensing. In: Zourob M., Lakhtakia A. (Eds.), *Chemical Sensors and Biosensors: Methods and Applications*. Springer, Berlin, Germany, pp. 21–54.

Shao M., Qiao X., Fu H., Li H., Zhao J., Li Y. (2014) A Mach–Zehnder interferometric humidity sensor based on waist-enlarged tapers. *Opt. Laser Eng.* 52, 86–90.

Shao M., Qiao X., Fu H., Zhao N., Liu Q., Gao H. (2013) An in-fiber Mach–Zehnder interferometer based on arc-induced tapers for high sensitivity humidity sensing. *IEEE Sens. J.* 13(5), 2026–2031.

Sharma U., Wei X. (2013) Fiber optic interferometric devices. In: Kang J.U. (Ed.), *Fiber Optic Sensing and Imaging*. Springer Science+Business Media, New York, pp. 29–53.

Su D., Qiao X., Rong Q. et al. (2013) A fiber Fabry–Perot interferometer based on a PVA coating for humidity measurement. *Opt. Commun.* 311, 107–110.

Sui Q., Jiang M., Jin Z., Zhang F., Cao Y., Jia L. (2014) Optical fiber relative humidity sensor based on Fabry-Perot interferometer coated with sodium-p-styrenesulfonate/allyamine hydrochloride films. *Sens. Mater.* 26(5), 291–298.

Sun L.-P., Li J., Jin L., Ran Y., Guan B.-O. (2016) High-birefringence microfiber Sagnac interferometer based humidity sensor. *Sens. Actuators B* 231, 696–700.

Vali V., Shorthill R.W. (1976) Fiber ring interferometers. *Appl. Opt.* 15, 1099–1100.

Wang C., Zhou B., Jiang H., He S. (2016a) Agarose filled Fabry–Perot cavity for temperature self-calibration humidity sensing. *IEEE Photon. Technol. Lett.* 28(19), 2027–2030.

Wang Y., Shen C., Lou W., Shentu F. (2016b) Fiber optic humidity sensor based on the graphene oxide/PVA composite film. *Opt. Commun.* 372, 229–234.

Wong W.C., Chan C.C., Chen L.H., Li T., Lee K.X., Leong K.C. (2012) Polyvinyl alcohol coated photonic crystal optical fiber sensor for humidity measurement. *Sens. Actuators B* 174, 563–569.

Yang M., Xie W., Dai Y., Lee D., Dai J., Zhang Y., Zhuang Z. (2014) Dielectric multilayer-based fiber optic sensor enabling simultaneous measurement of humidity and temperature. *Opt. Exp.* 22(10), 11892–11899.

Yao J., Zhu T., Duan D.W., Deng M. (2012) Nanocomposite polyacrylamide based open cavity fiber Fabry–Perot humidity sensor. *Appl. Opt.* 51(31), 7643–7647.

Yu H.H., Yao L., Wang L.X., Hu W.B., Jiang D.S. (2001) Fiber optic humidity sensor based on self-assembled polyelectrolyte multilayers. *J. Wuhan Univ. Technol. Mater. Sci. Ed.* 16(3), 65–69.

17 Microfiber-Based Humidity Sensors

17.1 MICROFIBERS AND ADVANTAGES OF THEIR APPLICATIONS

Experiments have shown that conventional optical fibers allow limited access to the evanescent field and show inefficient light–environment interactions that make it difficult to use such fibers for many micro/nanoscale-sensing applications (Rajan 2015). A standard single-mode optical fiber (SMF) has a step-index core-cladding structure. The typical diameters of the core and cladding are 8 and 125 μm, respectively. The refractive index (RI) of the fiber core is slightly higher than that of the cladding to achieve the weakly-guiding condition (Snyder and Love 1983). Due to the small RI difference between the core and the cladding, the weakly-guided fundamental modes have a mode field diameter, which exceeds that of the core and thus extends into the cladding with the result that a substantial fraction of the light propagates within the cladding as an evanescent field. However, this evanescent field is not accessible for sensing purposes due to the large diameter of the cladding relative to the mode field diameter. One way to get access to the evanescent fields is by removing the part of the fiber cladding through a mechanical polishing or chemical etching. However, such cladding-removed optical fibers suffer from the residual surface roughness and the poor control over the diameter of the decladded fiber section. The polishing and etching procedures are also complex and time-consuming. The development of optical microfibers (MFs) provides a more efficient and promising solution to meet the challenge. Optical MFs are optical fibers with diameters, which are comparable with the wavelength of the transmitted light. This means that the diameters of optical MFs should be in the range of a few micrometers. For comparison, conventional optical fibers have a typical diameter of 125 μm.

The small diameters of optical MFs can be achieved by using various methods such as drawing fibers from the bulk glass material (Tong et al. 2006), chemical etching (Zhang et al. 2010), the self-growth from silica nanomaterial (Naqshbandi et al. 2012), or by heating and stretching conventional optical fibers to form tapered MFs. The latter approach is preferable due to the advantages such as low fabrication induced losses, and a smooth taper surface. In addition, this method allows the retention of the original fiber input and output ends, which facilitate fiberized input and output connections to other standard optical fibers with negligible losses (Brambilla et al. 2004). Therefore, most of the MF-based devices developed to date have been based on tapered optical MFs. In the case of tapered optical MFs, the total fiber diameter is reduced to a few micrometers. The evanescent field extends into the surrounding medium, effectively propagating outside of the fiber. This means that the evanescent field of an optical MF is accessible and can be utilized for sensing. A typical tapered optical MF has a shape and structure as shown in Figure 17.1. The tapered MF consists of a uniform diameter waist region at the center and two tapering transition regions.

In principle, the optical properties of the tapered MF depend on the geometry of both the transition and the waist regions. However, the waist region plays the key role in sensing because the waist region exhibits the largest evanescent field. This is especially pronounced for adiabatic tapers where light can propagate along the tapered region with negligible loss for the fundamental modes considered (Love et al. 1991). This means that consideration of only the uniform waist region in the design and simulation of optical MFs for sensing purposes to avoid unnecessary complexity in modeling is well-founded.

According to numerous studies (Brambilla et al. 2009; Chen et al. 2013; Bo 2015; Gai et al. 2017), the small diameter of optical MFs provides several advantages over conventional optical fibers, important for sensor applications:

- *More effective access to the evanescent field*: For conventional optical fibers, light is transported within the fiber via total internal reflection. From an intuitive perspective, this means that the optical fiber works as a *light pipe*. For optical MFs, a substantial proportion of light can propagate outside the fiber. In this case, the fiber can be viewed working as a *light rail*. Such a light-guidance configuration allows the evanescent fields to interact with the measurands in the surrounding medium in a much more efficient way. As a result, MF-based sensors can be more sensitive (Lou et al. 2014).

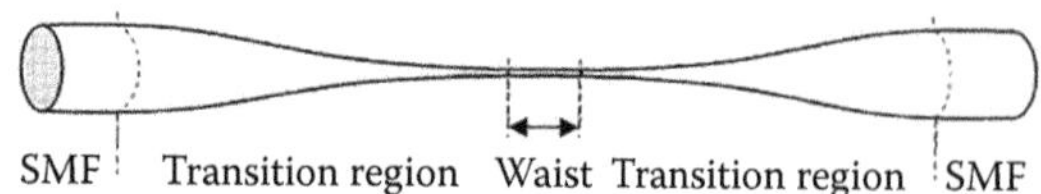

FIGURE 17.1 Schematic diagram of a tapered microfiber with a uniform waist region in the center, which is connected to the original fiber ends via two transition regions.

- *Greater versatility of configuration*: Optical MFs can be bent and manipulated into various structures with low bend loss. Conventional optical fibers suffer from the bend-induced attenuation when the bend radii are of several centimeters (Love and Durniak 2007). Optical MFs can achieve a millimeter- or even micrometer-order bend radii with low bend loss. Thus, the potential for sensor miniaturization can be enhanced with MF-based structures.
- *Low-loss connection*: Low-loss connection to other optical fibers and fiberized components is possible; MFs manufactured by adiabatically stretching optical fibers maintain the original fiber size at their input and output, allowing ready splicing to standard fibers and fiberized components. Insertion losses smaller than 0.1 dB are commonly observed.

17.2 MICROFIBER-BASED STRUCTURES FOR SENSOR APPLICATIONS

As follows from the results of studies, a single tapered optical MF can work as a sensing element in the intensity domain if a strong light–environment interaction was available. According to Tong and Sumetsky (2009), most effectively this condition is satisfied for single MFs with a diameter within the subwavelength-scale. However, MFs with such small diameters (<1 μm) are very fragile, have poor mechanical stability, and are difficult to manipulate. Moreover, overly thin MFs are more likely to exhibit high loss due to a surface degradation and contamination. Therefore, in spite of generating smaller evanescent fields and decreased interaction with environment, it is more practical to develop sensors using MFs, which have relatively large and thus easily manageable diameters (typically over 2 μm) (Bo 2015).

Studies have shown that on the basis of MFs one can implement all types of humidity-sensitive structures previously discussed, including fiber Bragg grading (FBG) structures, Mach–Zehnder interferometers (MZIs), and Michelson interferometers (MI). Due to the strong evanescent fields and the versatility of configuration, MFs can also be configured into a self-coupling knot, loop, or coil structures to form resonators. Such resonators have been attracting great attentions in many sensing fields, due to their advantages of small size, high Q-factor, absolute wavelength measurement, low loss, and so on (Sumetsky et al. 2005; Jiang et al. 2006; Xu et al. 2007; Wu et al. 2009; Zhang et al. 2011; Gai et al. 2017). Examples of such structures are shown in Figures 17.2 and 17.3.

Bo (2015) showed that the optical microfiber couplers (OMCs) and localized surface Plasmon resonance (LSPR) can also be used in the development of optical sensors. The configuration of OMCs is shown in Figure 17.4. The OMC consists of two identical MFs with the entire waist region and the part of the transition regions fused together. Light launched into one input port can be coupled and split between two output ports. Comparative characteristics of MF-based structures discussed above are given in Table 17.1.

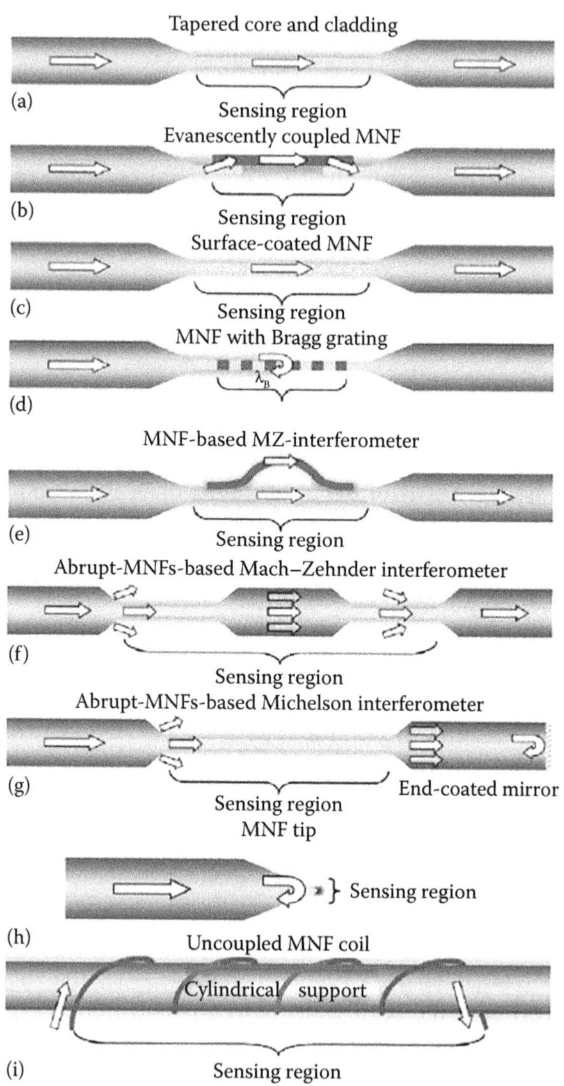

FIGURE 17.2 (a–i) Illustration of common nonresonator-type MNF structures. (From Chen, G.Y. et al., *Open Opt. J.*, 7, 32, 2013. Published by Bentham Open as Open access.)

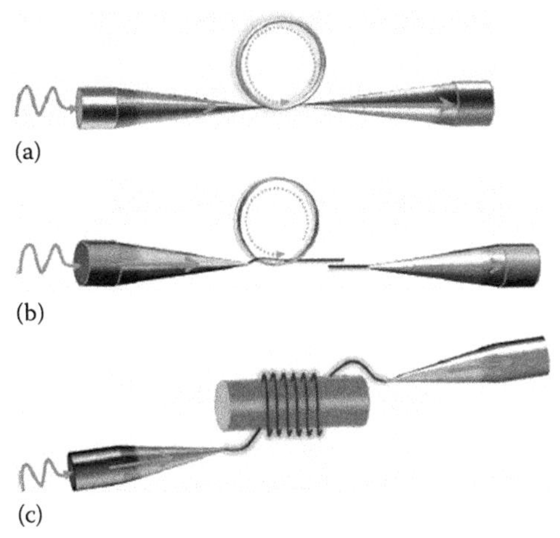

FIGURE 17.3 Schematic of MNF homogeneous (a) loop resonator, (b) knot resonator, and (c) coil resonator. In all structures the evanescent field couples power between different sections of the same MNF. (From Chen, G.Y. et al., *Open Opt. J.*, 7, 32, 2013. Published by Bentham Open as Open access.)

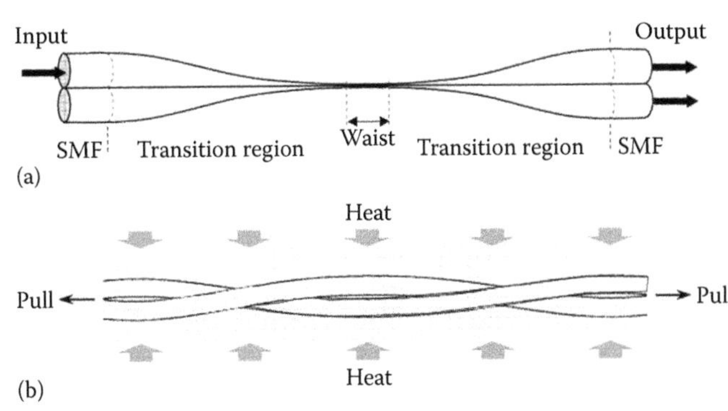

FIGURE 17.4 (a) Schematic diagram of an OMC consisting of two fused microfibers with a uniform waist region and two transition regions. (b) Two fibers are slightly twisted together to ensure close proximity between them during the fusing and tapering process. (Idea from Bo, L., Tapered optical microfibre based structures for sensing applications, PhD Thesis, Dublin Institute of Technology, 2015.)

TABLE 17.1
Comparison of Microfiber-Based Structures for Sensing Applications

Microfiber Structure	Advantage	Limitation
Adiabatic single tapered microfiber	• Simplicity of structure • Easy to fabricate	• Requires subwavelength-diameter to achieve effective light–environment interaction • Unsuitable for wavelength domain interrogation
Nonadiabatic single tapered microfiber	• Relatively large and thus easily manageable fiber diameter • Short transition region for higher compactness	• Requires precise control of taper shape to achieve mode interference • Limited extinction ratio in transmission spectrum
Microfiber-based interferometer	• High RI sensitivity	• Requires two light paths, which can increase set up complexity and limit device miniaturization
Microfiber-based gratings	• High strain and temperature sensitivity • Relatively high mechanical stability of the structure	• Grating inscription is complex and expensive • Limited RI sensitivity
Microfiber-based resonators	• High Q-factor resonance for potential high sensitivity	• Complex manipulation of microfiber to form the structure • Low reproducibility
Optical microfiber coupler	• High RI sensitivity • Easy and inexpensive one-step fabrication	• Straight structure with limited lengthwise compactness
Microfiber combined with WGM	• High Q-factor resonance for high biosensitivity	• Requires precise alignment between the microfiber and the WGM cavity
Microfiber combined with LSPR	• Low-cost fabrication • Suitable for biosensing due to lack of bulk effects of LSPR • Easy biofunctionality of nanoparticles	• Require surface modification of microfiber for immobilizing nanoparticles

Source: Data extracted from Bo, L., Tapered optical microfiber based structures for sensing applications, PhD Thesis, Dublin Institute of Technology, 2015.

LSPR—Localized surface Plasmon resonance; RI—Refractive index; WGM—Whispering gallery mode.

As for the specifics of manufacturing MF-based structures used in sensors, then it is considered in sufficient detail by Gai et al. (2017).

17.3 MICROFIBER-BASED HUMIDITY SENSORS

Currently for realization of humidity sensors on the basis of MFs, various approaches based on the use of MF-based structures (Table 17.2) have been tested. Let us consider some of them in more detail.

17.3.1 Evanescent-Based Humidity Sensors

Evanescent-based sensors have the simplest configuration of humidity sensors based on MFs. However, experiment has shown that these sensors have good performance. In particular, Gu et al. (2008) reported highly versatile nanosensors, using polymer nanofibers or nanowires. For relative humidity (RH) sensing, a polyacrylamide (PAM) nanowire drawn from a PAM aqueous solution was employed. As schematically illustrated in Figure 17.5a, a fiber taper drawn from a single-mode fiber with distal end of about 500 nm in diameter was placed in parallel and close contact with one end of a polymer nanowire. Due to the strong evanescent coupling between the nanowire and the fiber taper, light can be efficiently launched into and picked up from the nanowire within a few micrometers' overlap. Responses to humidity of indicated sensor are shown in Figure 17.6. The estimated response time (baseline to 90% signal saturation) of the sensor was about 24 ms when RH jumps from 10% to 75% and 30 ms when RH falls from 88% to 75%, which are 1 or 2 orders of magnitude faster than those of existing RH sensors.

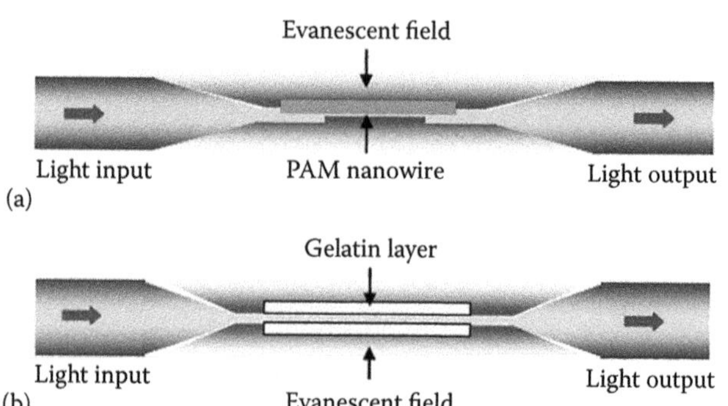

FIGURE 17.5 (a) Schematic of evanescent-coupled PAM nanowire humidity sensor and (b) schematic of a gelatin layer coated subwavelength-diameter tapered optical fiber sensor. (From Zhang, L. et al., *Photon. Sens.*, 1, 31, 2011. Published by Springer as open access.)

Sensor developed by Zhang et al. (2008) also showed a fast response. This RH sensor was based on subwavelength-diameter MF coated with gelatin layer.

TABLE 17.2
Microfiber-Based Humidity Sensors

Measurement Scheme	Sensing Material	Range (% RH)	Response Time	References
Evanescent field	Polyacrylamide nanofiber	35–90	20–30 ms	Gu et al. (2008)
	SiO_2 MF coated with gelatin	9–94	70 ms	Zhang et al. (2008)
Microring resonator	Polyacrylamide MF	5–71	120 ms	Wang et al. (2011)
	SiO_2 sol–gel thin film	30–80	200 ms	Bhola et al. (2009)
	SiO_2 MF	10–90	–	Wu et al. (2011)
	PMMA MF	17–95	~500 ms	
	PMMA MF coated with ZnO	45–80		Irawati et al. (2017)
	SiO_2 MF	50–80	–	Zheng et al. (2013)
	SiO_2 MF coated with GO	30–50	–	Ahmad et al. (2016)
	SiO_2 MF coated with PVA	20–90	~2 s	Shin et al. (2016)
Optical microfiber couplers	Uncoated SiO_2 MF	89–96	–	Bo (2015)
	SiO_2 MF coated with PEO	70–85	–	Bo et al. (2015)
Sagnac interferometer	Uncoated SiO_2 Hi–Bi MFs	30–90	~60 ms	Sun et al. (2015)

GO—graphene oxide; Hi–Bi—high-birefringence; MF—microfiber; PEO—poly(ethylene oxide); PMMA—poly(methyl methacrylate); PVA—polyvinyl alcohol.

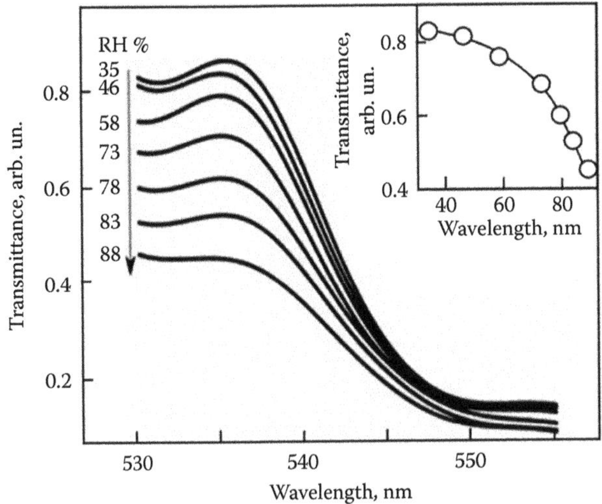

FIGURE 17.6 PAM single-nanowire humidity sensors: Transmittance of an MgF_2-supported 410-nm-diameter PAM nanowire exposed to atmosphere of RH from 35% to 88%. Inset, transmittances of the nanowire at 532 nm wavelength. (Reprinted with permission from Gu, F. et al., *Nano Lett.*, 8, 2757, 2008. Copyright 2008 American Chemical Society.)

The sensing element was composed of a 680-nm-diameter MF coated with an 80-nm-thickness 8-mm-length gelatin layer, and was operated at a wavelength of 1550 nm. As it is known, gelatin has a strong RI dependence on the humidity. The sensor was operated within a wide humidity range (9%–94% RH) with a high sensitivity and good reversibility. Measured response time was about 70 ms. The remarkably fast response of above-mentioned sensors can be attributed to a small diameter of the nanowire and small thickness of gelatin that enable rapid diffusion or evaporation of water molecules.

17.3.2 Optical Microfiber-Coupler-Based Humidity Sensors

Depending on the taper fabrication temperature (Morishita and Yamazaki 2011), various degrees of fusion can be achieved between the MFs within the OMCs (Payne et al. 1985; Morishita and Takashina 1991). For weakly fused OMCs, both fused MFs roughly maintain their original cross section geometry throughout the entire tapering process. Thus such OMCs have a dumbbell-shaped cross section. In this case, the aspect ratio, which is defined as the ratio between the total width and the height of the cross section at the OMC waist region, is usually larger than 1.8 (Morishita and Yamazaki 2011). For strongly fused OMCs, the cross section geometry is altered in a way that the distance between the two MF axes reduces faster than the decrease of the fiber diameters. The resulting OMC can have a cross section with an elliptical or a circular shape (Morishita and Takashina 1991). The aspect ratios for such strongly fused OMCs can vary between 1.0 and 1.8. Compared with weakly fused OMCs, strongly fused OMCs require shorter coupling lengths and less critical diameters to achieve a complete power transfer from one fused fiber to the other. However, strongly fused OMCs also show weaker dependence on surrounding RI changes. In contrast, weakly fused couplers show a stronger dependency on surrounding RIs than the strongly fused ones (Lamont et al. 1985). Therefore, weakly fused OMCs are more suitable for the development of humidity sensors.

Bo (2015) studied the possibility of such structures for the measurement of humidity and found that the transmission spectrum of the uncoated OMC remained relatively unchanged when the RH was lower than 89.5%, and strong changes were observed only at RH exceeding 89.5%. Moreover, the shifts were found to be nonreversible during the experiment. Studied OMC sample had a diameter of approximately 2 μm for each fused MF. According to Bo (2015), observed nonreversible spectral shifts could be a result of the failure of the sensor due to (1) the formation of dew on the OMC surface (Lawrence 2005) and (2) the humidity-accelerated surface degradation (Matthewson and Kurkjian 1988).

To optimize the sensor parameters Bo (2015) used a coating of the fibers by poly(ethylene oxide) (PEO). For these purposes, optical microfiber couplers (OMCs) were immersed in the PEO water solution. It was shown that the surface modification is being accompanied by the improvement of humidity-sensitive characteristics. However, it was found that for the RH range from 25% to 70%, the sensor was relatively insensitive (Figure 17.7). Small linear blue shifts in the dip wavelength were observed with a sensitivity of −0.005 to −0.095 nm/% RH. Only at the RH exceeding 70%, relatively large red shifts in the resonance wavelength occurred. The RH and the resonance wavelength showed linear correlation with a sensitivity of 1.7–2.7 nm/% RH. Bo (2015) believes that the difference in the response between the low RH range (25%–70%) and the high RH range (70%–85%) arises because PEO undergoes a phase change from a semicrystalline to a gel state around 80% RH. The RI of the PEO coating decreases drastically at this phase-transition RH point due to the reduction in the density of the PEO because of an increase in swelling as its water content increases (Acikgoz et al. 2008). As a result, the

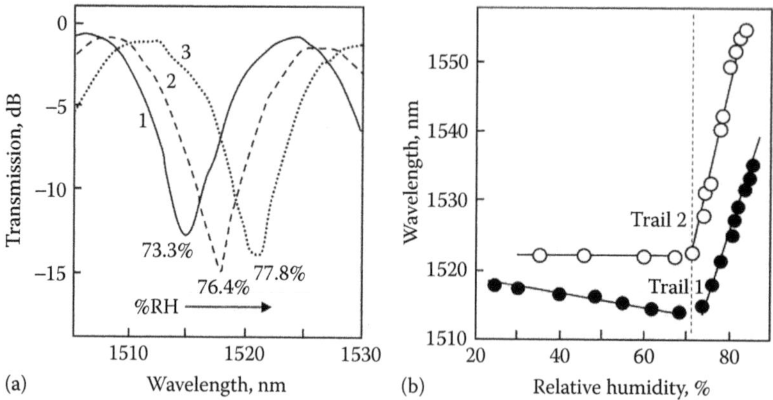

FIGURE 17.7 (a) Examples of spectral responses of the sensor (sample 1) when the environmental RH is 73.3%, 76.4%, and 77.8%. The resonance wavelength shifts to the longer wavelength range as the humidity increases; (b) Resonance wavelength versus RH for the two samples. (From Bo, L. et al.: Optical microfiber coupler based humidity sensor with a polyethylene oxide coating. *Microwave and Optical Technology Letters*, 2015. 57. 457–460. Copyright Wiley-VCH Verlag GmbH & Co. KGaA. Reproduced with permission.)

sensor showed a high sensitivity in the humidity range, which lies above the phase-transition RH point. Finally, Bo et al. (2015) noted that the increase of RH should be limited to 85% because higher RH may cause irreversible changes in the PEO coating and render the sensor unusable. Basing of the obtained results, Bo (2015) concludes that an OMC is a relatively simple and efficient MF structure for developing humidity sensors with high sensitivity. However, a very limited range of RH, where sensors exhibit a high sensitivity makes the use of such sensors in real applications problematic. Meanwhile, there is a confidence that the sensors of this type can be optimized through the use of a more suitable humidity-sensitive material.

17.3.3 Microring Resonator-Based Humidity Sensors

Currently, in the development of humidity sensors there are mainly used knot (Wang et al. 2011; Wu et al. 2001; Yoon et al. 2015; Ahmad et al. 2016; Shin et al. 2016) and loop (Zheng et al. 2013; Irawati et al. 2017) types of microring resonators (MRRs) based on MFs. However, a thin-film approach for fabrication such sensors was also tested (Bhola et al. 2009). The SiO_2-based microring was manufactured using sol–gel technology.

Figure 17.8 shows an example of a single-loop knot resonator fabricated by knotting a MF. Loop resonators can be easily manufactured by coiling a MF (Brambilla et al. 2009). The use of *XYZ* stages to coil the MF allows a great deal of control of the resonator geometry. As a result, MRRs show a strong self-coupling due to the close proximity of the waveguide with itself at the coupling region, with Q-factors of the order of 10^5 (Chen et al. 2013). Still, the geometry stability is based on surface forces; thus, loop resonators are compromised in terms of long-term reliability (Sumetsky et al. 2006). Supporting a MF loop with a certain substrate has been proved to be an effective approach to MF-loop sensors with higher robustness; however, the additional substrate

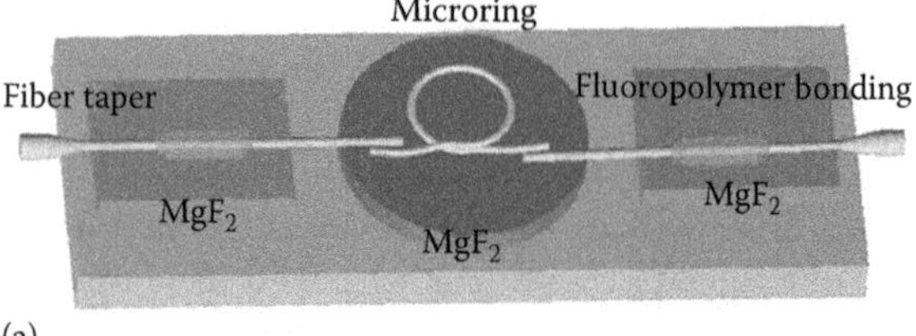

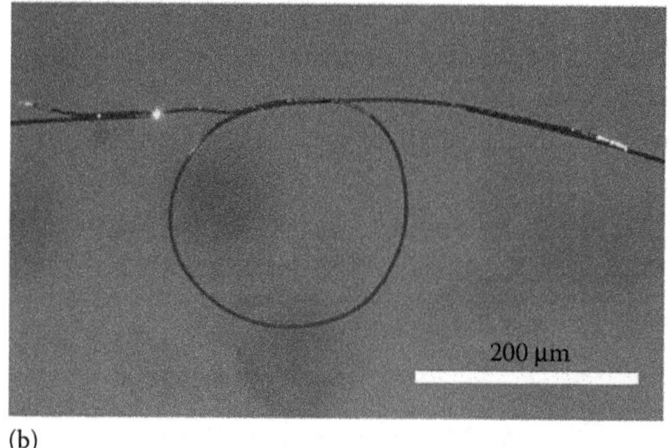

FIGURE 17.8 (a) Schematic diagram of a PAM microring for humidity sensing. (b) Optical microscope image of a 200 μm diameter MgF_2-supported PAM microring assembled with a 2.7 μm diameter PAM microfiber. A 532 nm wavelength light is launched into the microring from the left side. (From Wang, P. et al., *Appl. Opt.*, 50, G7–G10, 2011. With permission of Optical Society of America.)

inevitably increases the complexity and overall size of the sensing element. Embedding the MRR in polymer has been the preferred solution to provide long-term stability, although it can considerably modify the transmission spectrum (Vienne et al. 2007). In contrast, knot resonators exhibit an enhanced temporal stability because of the friction that different sections of the MF exert on one another. However, because the stiffness of the MF is different than that of its pigtails, the bending curvatures needed to manufacture knot resonators induce an enormous stress in the MF, which therefore breaks. The knot resonator can be easily manufactured if the fiber taper is broken; knotting is performed in the region of uniform waist of optical fiber nanowires (OFMs), and no excess tension occurs. As a result, this type of resonator exhibits only one input–output fiberized pigtail.

The optical properties of single-loop resonators can easily be recorded by launching light from a broadband source into one of the MF pigtails and analyzing the transmitted light with an optical spectrum analyzer. Typical transmission spectra of a MRR are shown in Figure 17.9.

Wu et al. (2011) compared SiO_2 (1.2 μm)- and poly(methyl methacrylate) (PMMA) (2.1 μm)-based knot resonators with diameter 2.4–2.8 mm and found that PMMA-based MRR had much better sensitivity (Figure 17.10). The silica MF knot resonator sensor had a humidity sensitivity of ~12 pm/10% RH within a linearity range from 15% to 60% RH, whereas the polymer MF knot resonators sensor had a humidity sensitivity of ~88 pm/10% RH, with a linearity range from 17% to 95% RH. The temporal response of the PMMA-based MRR sensor was <0.5 s. It is important to note that these sensors were fabricated without any surface coating.

The same approach was used by Wang et al. (2011). However, instead of PMMA MF it was used PAM MF, 2–3 μm in diameter, which was also highly sensitive to humidity. To construct a ring resonator, as-fabricated MF was placed on a low-index MgF_2 crystal (RI ~ 1.39) for mechanical support and low-loss operation. The MF was cut and manipulated under an optical microscope by a homemade tungsten probe, which was attached to a manually controlled three-dimensional translation stage for precise movement. By careful micromanipulation, the PAM MF was assembled into a ring shape with an evanescent coupler formed at the overlap region, resulting in a microscale PAM ring. The as-assembled microring was tightly attracted to the substrate via van der Waals force and electrostatic force. For optical characterization, two fiber tapers that were taper drawn from a standard optical fiber were used for evanescently launching light into and collecting signals out of the microring, as schematically shown in Figure 17.8a. For robust humidity sensing, the fiber tapers were firmly bonded on the MgF_2 substrate using a low-index ultraviolet (UV)-cured fluoropolymer (EFIRON PC-373;

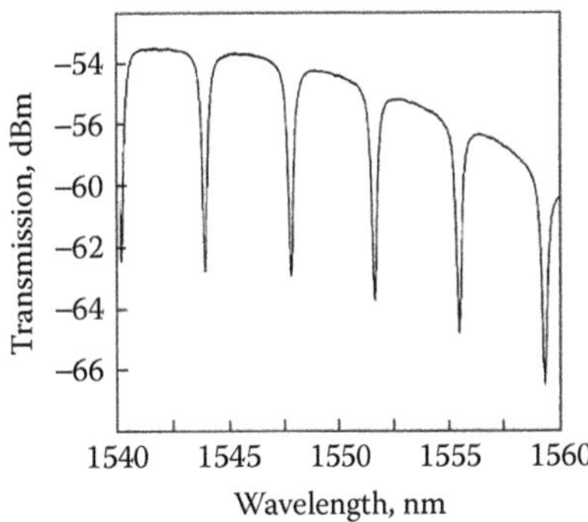

FIGURE 17.9 Transmission spectrum of a loop resonator as recorded from an optical spectrum analyzer (OSA). (From Brambilla, G. et al., *Adv. Opt. Photon.*, 1, 107, 2009. Copyright 2009 With permission of Optical Society of America.)

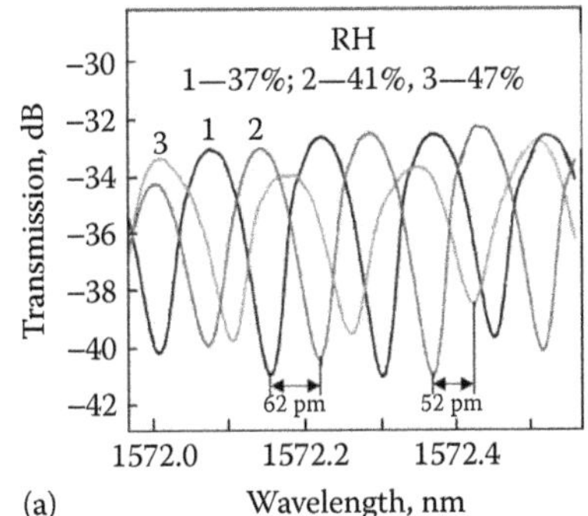

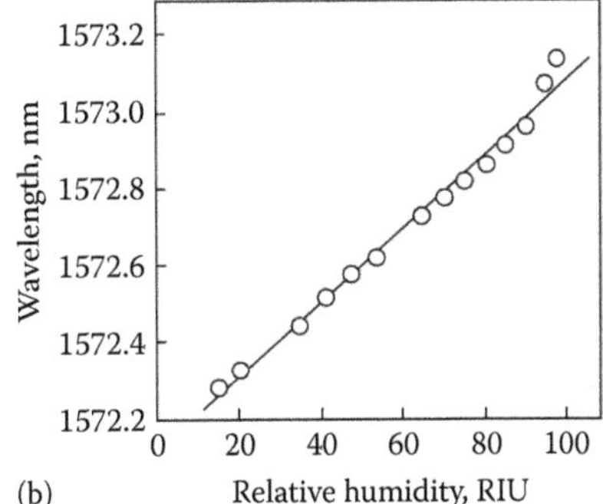

FIGURE 17.10 (a) Spectra of PMMA-based MRR at humidity of 37% RH (1) (black line), 41% RH (2) (dark gray line), and 47% RH (3) (light gray line). (b) Shift of resonance wavelength of the humidity sensor with different levels of humidity. MRR with diameter 2.8 mm was fabricated using 2.1 μm diameter PMMA microfiber. (Reprinted from *Sens. Actuators B*, 155, Wu, Y. et al., Miniature interferometric humidity sensors based on silica/polymer microfiber knot resonators, 258, Copyright 2001, with permission from Elsevier.)

Luvantix Co., Ltd.). It was established that PMA-based MRR sensor was operated within a wide humidity range (5%–71% RH) with measured sensitivity of about 490 pm/% RH and response time of about 120 ms. Thus, Wang et al. (2011) confirmed that because the whole ring is made of RH-sensitive MF with thickness suitable for quick diffusion of water molecules and the spectral shift of the resonance is almost independent on the transmission loss of the whole system, this kind of sensor demonstrates the advantages of very high sensitivity, high stability, and fast response. Similar results have also been demonstrated in other types of MF–ring–resonator sensors.

Despite the fact that the polymer-based MFs provide acceptable sensitivity to humidity, there are studies where researchers try to improve performance through the use of additional surface coatings. Thus, Irawati et al. (2017) on the basis of their research concluded that the coating with ZnO nanorods improved the sensitivity of PMMA-based MRRs to the air humidity. Irawati et al. (2017) believed that the performance of the sensor was enhanced after being coated with ZnO nanorods due to its ability to absorb more water. Ahmad et al. (2016) established that SiO_2 MFs coated with graphene oxide also had better sensitivity to humidity.

It should be noted that at present a maximal sensitivity is achieved by MF-based humidity sensors fabricated on the base of Sagnac interferometer (Sun et al. 2015). At that a SiO_2 Hi–Bi MF with a diameter of 2.17 μm did not have any coating. The high sensitivity was estimated as 200.3 pm/% RH at around $\lambda = 1560$ nm. The sensitivity is higher than the previously reported RH sensors.

17.4 SUMMARY

The above-mentioned results indicate that the MF can really serve as a basic element for optical sensing or for light input/output in miniature photonic humidity sensors. Due to their favorable properties of high-fractional evanescent fields, low dimension, low loss, and high flexibility for optical sensing, MF-based optical resonators and interferometers may offer advantages of high sensitivity, fast response, small footprints, high spatial resolution, and low detection limits (Zhang et al. 2011).

Apart from their excellent performance in laboratory, the MF-based sensors still face a great challenge for commercial application (Gai et al. 2017). First, the existing production process is difficult to precisely control the size and uniformity of MF-based structures, especially microring-based resonators. Second, MFs and especially nanofibers might have a problems connected with stability of their parameters. Research has shown that humidity can accelerate the crack growth in an optical fiber and affect the performance of the fiber device (Muraoka et al. 1993; Armstrong et al. 2000). Tapered optical MFs with small diameters were also tended to degrade in the air because of the formation of cracks on the silica fiber surface due to the moisture absorption (Brambilla et al. 2006). Unfortunately, the study of this effect, which may have an impact on the lifetime of the sensor, is not extensive.

MFs with such small diameters are also mechanically weak and easily contaminated (Bo 2015). Thus, many MF-based devices require a robust fiber packaging technique to ensure the performance of the devices during experiments as well as over time, but must allow for the ingress and egress of a target analyte. Several packaging techniques such as packaging with low-RI polymers (Xu and Brambilla 2007), supporting with substrates (Guo and Tong 2008) and integration with microfluidics devices (Tian et al. 2011) have been investigated. However, these packaging techniques have many disadvantages such as poor reproducibility and long-term instability. Many packaging techniques also sacrifice the MF's sensitivity in detection to achieve the required robustness (e.g., by partially or fully covering the MF and thus weakening the light-environment interactions). The lack of high-performance MF packaging techniques has limited MF sensors to laboratory sensing experiments. Besides, MF is sensitive to many ambient quantities, which can be regarded as an advantage as well as a disadvantage. Many factors interfere with each other and exert a great influence on the accurate measurement. There is no efficient solution for this problem so far.

Nevertheless, the researchers have made great breakthroughs in both production of MF-based structures and their packaging recently. Therefore, it is believed that the study on MF-based sensors will get much wider interests from different fields (Gai et al. 2017).

REFERENCES

Acikgoz S., Bilen B., Demir M.M., Menceloglu Y.Z., Skarlatos Y., Aktas G., Inci M.N. (2008) Use of polyethylene glycol coatings for optical fibre humidity sensing. *Opt. Rev.* 15(2), 84–90.

Ahmad H., Rahman M.T., Sakeh S.N.A., Razak M.Z.A., Zulkifli M.Z. (2016) Humidity sensor based on microfiber resonator with reduced graphene oxide. *Optik* 127, 3158–3161.

Armstrong J.L., Matthewson M.J., Kurkjian C.R. (2000) Humidity dependence of the fatigue of high-strength fused silica optical fibers. *J. Am. Ceram. Soc.* 83(12), 3100–3109.

Bhola B., Nosovitskiy P., Mahalingam H., Steier W.H. (2009) Sol–gel-based integrated optical microring resonator humidity sensor. *IEEE Sens. J.* 9, 740–746.

Bo L. (2015) Tapered optical microfibre based structures for sensing applications. PhD Thesis, Dublin Institute of Technology.

Bo L., Wang P., Semenova Y., Farrell G. (2015) Optical microfiber coupler based humidity sensor with a polyethylene oxide coating. *Microw. Opt. Technol. Lett.* 57(2), 457–460.

Brambilla G., Finazzi V., Richardson D. (2004) Ultra-low-loss optical fiber nanotapers. *Opt. Exp.* 12(10), 2258–2263.

Brambilla G., Xu F., Feng X. (2006) Fabrication of optical fibre nanowires and their optical and mechanical characterisation. *Electron. Lett.* 42(9), 517–519.

Brambilla G., Xu F., Horak P. et al. (2009) Optical fiber nanowires and microwires: fabrication and applications. *Adv. Opt. Photon.* 1(1), 107–161.

Chen G.Y., Ding M., Newson T.P., Brambilla G. (2013) A review of microfiber and nanofiber based optical sensors. *Open Opt. J.* 7(Suppl-1, M3), 32–57.

Gai L., Li J., Zhao Y. (2017) Preparation and application of microfiber resonant ring sensors: A review. *Opt. Laser Technol.* 89, 126–136.

Gu F., Zhang L., Yin X., Tong L. (2008) Polymer single-nanowire optical sensors. *Nano Lett.* 8(9), 2757–2761.

Guo X., Tong L. (2008) Supported microfiber loops for optical sensing. *Opt. Exp.* 16(19), 14429–14434.

Irawati N., Rahman H.A., Ahmad H., Harun S.W. (2017) A PMMA microfiber loop resonator based humidity sensor with ZnO nanorods coating. *Measurement* 99, 128–133.

Jiang X.S., Tong L.M., Vienne G., Guo X., Tsao A., Yang Q., Yang D.R. (2006) Demonstration of optical microfiber knot resonators. *Appl. Phys. Lett.* 88, 223501.

Lamont R.G., Johnson D.C., Hill K.O. (1985) Power transfer in fused biconical-taper single-mode fiber couplers: dependence on external refractive index. *Appl. Opt.* 24(3), 327–332.

Lawrence M.G. (2005) The relationship between relative humidity and the dewpoint temperature in moist air: A simple conversion and applications. *Bull. Am. Meteorol. Soc.* 86, 225–233.

Lou J., Wang Y., Tong L. (2014) Microfiber optical sensors: A review. *Sensors* 14(4), 5823–5844.

Love J.D., Durniak C. (2007) Bend Loss, tapering, and cladding-Mode coupling in single-mode fibers. *IEEE Photonics Technol. Lett.* 19(16), 1257–1259.

Love J.D., Henry W.M., Stewart W.J., Black R.J., Lacroix S., Gonthier F. (1991) Tapered single-mode fibres and devices. I. Adiabaticity criteria. *Optoelectron. IEEE Proc. J.* 138(5), 343–354.

Matthewson M.J., Kurkjian C.R. (1988) Environmental effects on the static fatigue of silica optical fiber. *J. Am. Ceram. Soc.* 71(3), 177–183.

Morishita K., Takashina K. (1991) Polarization properties of fused fiber couplers and polarizing beamsplitters. *J. Light. Technol.* 9(11), 1503–1507.

Morishita K., Yamazaki K. (2011) Wavelength and polarization dependences of fused fiber couplers. *J. Light. Technol.* 29(3), 330–334.

Muraoka M., Ebata K., Abe H. (1993) Effect of humidity on small-crack growth in silica optical fibers. *J. Am. Ceram. Soc.* 76(6), 1545–1550.

Naqshbandi M., Canning J., Nash M., Crossley M.J. (2012) Controlling the fabrication of self-assembled microwires from silica nanoparticles. *Proc. SPIE* 8421, 842182.

Payne F.P., Hussey C.D., Yataki M.S. (1985) Polarisation analysis of strongly fused and weakly fused tapered couplers. *Electron. Lett.* 21(13), 561–563.

Rajan G. (Ed.) (2015) *Optical Fiber Sensors: Advanced Techniques and Applications.* CRC Press, Boca Raton, FL.

Shin J.C., Yoon M.-S., Han Y.-G. (2016) Relative humidity sensor based on an optical microfiber knot resonator with a polyvinyl alcohol overlay. *J. Lightwave Technol.* 34(9), 4511–4515.

Snyder A.W., Love J. (1983) *Optical Waveguide Theory.* Springer Science & Business Media, New York.

Sumetsky M., Dulashko Y., Fini J.M., Hale A. (2005) Optical microfiber loop resonator. *Appl. Phys. Lett.* 86, 161108.

Sumetsky M., Dulashko Y., Fini J.M., Hale A., DiGiovanni D.J. (2006) The microfiber loop resonator: Theory, experiment, and application. *J. Lightwave Technol.* 24, 242–250.

Sun L.-P., Li J., Jin L., Ran Y., Guan B.-O. (2015) High-sensitivity humidity sensor based on microfiber Sagnac interferometer. In: *Proceedings of Asia Communications and Photonics Conference (ACP)*, Hong Kong, OSA, AM3H.4, November 19–23.

Tian Y., Wang W., Wu N., Zou X., Wang X. (2011) Tapered optical fiber sensor for label-free detection of biomolecules. *Sensors* 11(4), 3780–3790.

Tong L., Hu L., Zhang J., Qiu J., Yang Q., Lou J., Shen Y., He J., Ye Z. (2006) Photonic nanowires directly drawn from bulk glasses. *Opt. Exp.* 14(1), 82–87.

Tong L., Sumetsky M. (2009) *Subwavelength and Nanometer Diameter Optical Fibers.* Zhejiang University Press and Springer Press, Berlin, Germany.

Vienne G., Li Y., Tong L. (2007) Effect of host polymer on microfiber resonator. *IEEE Photonics Technol. Lett.* 19, 1386–1388.

Wang P., Gu F., Zhang L., Tong L. (2011) Polymer microfiber rings for high sensitivity optical humidity sensing. *Appl. Opt.* 50(31), G7–G10.

Wu Y., Rao Y.-J., Chen Y.-H., Gong Y. (2009) Miniature fiber-optic temperature sensors based on silica/polymer microfiber knot resonators. *Opt. Exp.* 17, 18142–18147.

Wu Y., Zhang T., Rao Y., Gong Y. (2001) Miniature interferometric humidity sensors based on silica/polymer microfiber knot resonators. *Sens. Actuators B* 155, 258–263.

Xu F., Brambilla G. (2007) Embedding optical microfiber coil resonators in Teflon. *Opt. Lett.* 32(15), 2164–2166.

Xu F., Horak P., Brambilla G. (2007) Optical microfiber coil resonator refractometric sensor. *Opt. Exp.* 15, 7888–7893.

Yoon M.-S., Yoo K.W., Han Y.-G. (2015) Relative humidity sensor based on an optical microfiber knot resonator with a polyvinyl alcohol overlay. *Proc. SPIE* 9634, 96346X.

Zhang E.J., Sacher W.D., Poon J.K.S. (2010) Hydrofluoric acid flow etching of low-loss subwavelength-diameter biconical fiber tapers. *Opt. Exp.* 18(21), 22593–22598.

Zhang L., Gu F.X., Lou J.Y., Yin X., Tong L. (2008) Fast detection of humidity with a subwavelength-diameter fiber taper coated with gelatin film. *Opt. Exp.* 16(17), 13349–13353.

Zhang L., Lou J., Tong L. (2011) Micro/nanofiber optical sensors. *Photon. Sens.* 1(1), 31–42.

Zheng Y., Dong X., Zhao C., Li Y., Shao L., Jin S. (2013) Relative humidity sensor based on microfiber loop resonator. *Adv. Mater. Sci. Eng.* 2013, 815930.

18 Humidity Sensors Based on Special Fibers

18.1 SPECIFIC TYPES OF FIBERS FOR SENSOR APPLICATIONS

Interesting possibilities for increasing the amount of optical power coupled in the evanescent field give using new types of fibers that appeared a few years ago (Bjarklev et al. 2003; Elosua et al. 2006; Sorokin 2012; Jin et al. 2013; Schmidt et al. 2016). Among them, one can mention hollow core fibers (Figure 18.1). The simpler type of hollow core fibers, consist basically of tubular fibers where the optical signal is guided mainly by the cladding, and hence, it is easy to reach the evanescent fields (Peng et al. 2004). If a hollow core fiber is connected between two silica core ones, the optical power is coupled from the core to the cladding in the first transition, and then, from the cladding to the core in the second transition (Matias et al. 2006).

Holey fibers (HFs), other type of fibers, also known as *microstructured optical fibers* (MOFs) and *photonic crystal fibers* (PCFs), are a new type of fibers in which the air holes are incorporated within the silica-cladding region along its entire length; these cavities enhance the interaction between the evanescent field and the gas to be analyzed (Sorokin 2012). A MOF is used simultaneously as a light guide and as a fluidic channel. Parameters such as the size, shape, and relative position of the air holes provide an extra degree of freedom in controlling the light propagation and also the way the optical evanescent signal is affected by the surrounding (Hoo et al. 2002). Optic sensors designed on the base of above-mentioned fibers were intensively studied during the past years (Lee et al. 2009; Skorobogatiy 2009). For example, this kind of microstructured fibers have been employed to measure the acetylene vapors (Hoo et al. 2003), to detect the presence of CO and CO_2 (Charlton et al. 2005), and other gases (Hoo et al. 2005; Konorov et al. 2005).

According to the cross-sectional distribution of the dielectric function, PCFs can be categorized as follows: photonic-bandgap (PBG) fibers (PCF that utilizes the PBG and the defect mode), holey fibers (PCF with air holes along the axis of light propagation), hole-assisted fibers (PCF consisting of a conventional higher-index core with air holes), and Bragg fibers (PBG fiber with concentric rings of different refractive indices [RIs]). PBG fibers can contain periodic sequence of micron-sized layers of different materials (Figure 18.2a), periodically arranged micron-sized air voids (Figure 18.2b), or rings of holes separated by nano-supports (Figure 18.2c). PBG fibers are currently available in silica glass, polymer, and specialty soft glass implementations. An example of the PCF sensor with air holes is also shown in Figure 18.2.

The PCF has a periodic dielectric structure whose periodicity is on the order of a wavelength, giving rise to the PBG. The incident light whose wavelength lies within the PBG cannot propagate through the photonic crystal region and the transmission spectrum exhibits a wide bandgap. By locally breaking the period at the core, one can introduce a photonic defect mode within the bandgap and, as a result, the transmission spectrum has a relatively sharp transmission peak. The spectral position of the center of the transmission peak is highly sensitive to changes in the local environmental conditions. If liquid or gas molecules become bound to the defect, the local environmental condition, such as the RI, changes. Hence, it can be used as the sensing transduction signal. As the light confinement provided by the PBG is very strong and because it is possible to easily adjust the defect mode wavelength across the PBG by finely tuning the structural parameters, PCF sensors have received huge attention since the first demonstration of the detection of a RI change in waveguide configurations. Besides, research has shown that MOFs, and PBG fibers, which are a subset of MOFs, promise a viable technology for the mass production of highly integrated and intelligent sensors in a single manufacturing step (Skorobogatiy 2009). At present, two major application fields of these devices are (1) as gas sensors and (2) as biosensors.

18.2 EXAMPLES OF IMPLEMENTATION OF HUMIDITY SENSORS

At present, there are many humidity sensors designed on the base of photonic crystals (Table 18.1). One example of the humidity sensor implemented based on photonic crystal is presented in Mathew (2013). This

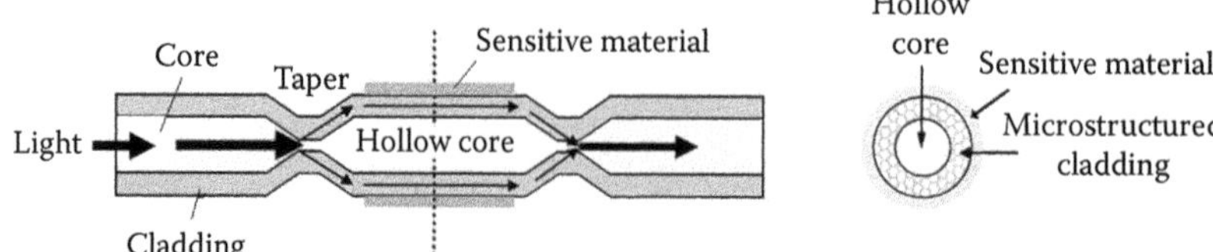

FIGURE 18.1 The hollow core fiber with microstructurated cladding, connected between two standard multimode optical fiber sections. A sensitive material can be also be fixed onto the cladding. (From Elosua, C. et al., *Sensors*, 6, 1440, 2006. Published by MDPI as open access.)

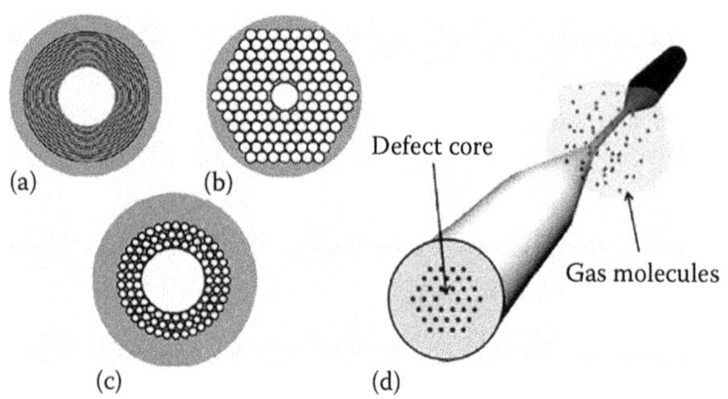

FIGURE 18.2 Various types of hollow core photonic bandgap fibers (HC–PBF) are the following: (a) Bragg fiber featuring a large hollow core surrounded by a periodic sequence of high and low refractive index layers; (b) Photonic crystal fiber featuring a small hollow core surrounded by a periodic array of large air holes; (c) Microstructured fiber featuring a medium size hollow core surrounded by several rings of small air holes separated by nanosize bridges. (From Skorobogatiy, M., *J. Sens.*, 2009, 524237, 2009. Published by Hindawi as open access.); and (d) Photonic crystal fiber microtaper structure for use in the gas sensing. (Reprinted from *Optic. Fiber Technol.*, 15, Lee, B. et al., Current status of micro- and nano-structured optical fiber sensors, 209–221, Copyright 2009, with permission from Elsevier.)

sensor of relative humidity (RH) is based on a reflection type PCF interferometer (PCF-I). A bare silica PCF is used for the fabrication of the PCF-I. PCF-Is based on microhole collapse was selected for sensor design due to the simple fabrication process involved and excellent sensing performance (Choi et al. 2007; Jha et al. 2008; Villatoro et al. 2009a). A reflection-type PCF-I consists of a stub of PCF fusion spliced at the distal end of a single-mode fiber. The key element of the device is the hole collapsed region in the vicinity of the splice point. In addition to the simplicity of fabrication, because such PCF-Is are fabricated using only a splice, they have the advantage of high mechanical stability at high temperatures and over an extended period. A microscopic image of the PCF-I and a schematic of the excitation and recombination of modes in the PCF-I are shown in Figure 18.3. The fundamental SMF mode (single-mode optical fiber) begins to diffract when it enters the collapsed section of the PCF. As a result, the interference pattern of the PCF-I is formed. At that a regular interference pattern in the reflection spectrum of the PCF-I usually suggests that only two modes are interfering in

TABLE 18.1
Photonic Crystal-Based Humidity Sensors

Sensing Method	Sensing Material	Range (% RH)	Response Time	References
Monitoring of reflection in photonic crystal fiber interferometer	A bare silica PCF with microhole collapse	40–100	–	Mathew et al. (2010, 2012a, b)
Photonic crystal fiber interferometer	PCF coated with agarose	20–90		Mathew et al. (2012c, 2013)
Photonic crystal fiber interferometer	PCF coated with PAH/PAA nanolayer	20–95	<1 s	Lopez-Torres et al. (2017)
Monitoring of reflection spectra in PCF-based Michelson interferometer	PVA-coated PCF	30–90	<1 s	Wong et al. (2012)
Monitoring water vapor line absorption	Air-guided hollow core-photonic crystal fiber	0–90	~2 min	Noor et al. (2012)
Fabry–Perot interferometer	SnO_2-suspended-core microstructured optical fiber	20–90	–	Lopez-Aldaba et al. (2016)
Multimode interference	Square no-core fiber coated with SiO_2 nanoparticles	40–98		Miao et al. (2016)

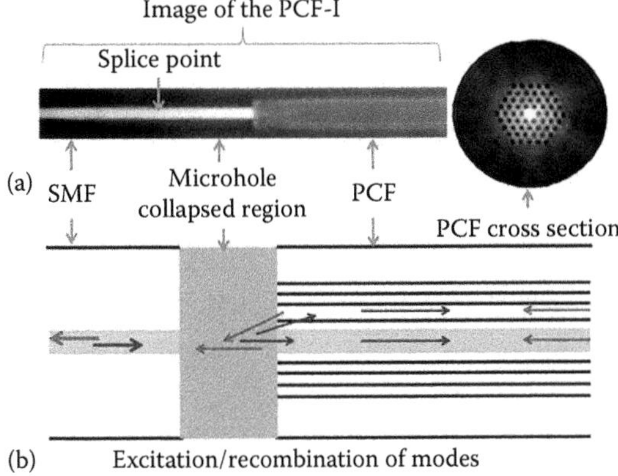

FIGURE 18.3 Microscope image of the PCF-I (a) and a schematic of the excitation/recombination of modes in the hole-collapsed region (b). (Idea from Mathew, J., Development of novel fiber optic humidity sensors and their derived applications, PhD Thesis, Dublin Institute of Technology, 2013.)

the device. More detailed description of the processes taking place in PCF-I, one can find in Choi et al. (2007), Jha et al. (2008), Villatoro et al. (2009a, b), Cárdenas-Sevilla et al. (2011), and Mathew (2013). Simulation shows that the power reflection spectrum of this interferometer will be dependent on $\cos(4\pi\Delta nL/\lambda)$ (Villatoro et al. 2009a). The wavelengths at which the reflection spectrum shows maxima are those that satisfy the condition $4\pi\Delta nL/\lambda = 2m\pi$, with m being an integer. This means that periodic constructive interference occurs when $\lambda_m = (2\Delta nL/m)$. If some external stimulus changes Δn (while L is fixed), the position of each interference peak will change, an effect which allows the device to be used for humidity sensing.

As it was shown before, as a rule the existing fiber-optic humidity sensors are either polymer-based or require the use of a hygroscopic material to detect humidity. The sensor head in the device proposed by Mathew et al. (2010, 2012a, 2012b, 2013) was made of single material (silica). Such construction gives the unique advantage; it does not require any special coatings to measure humidity. Silica surface is hydrophilic and therefore the adsorption of water vapor on the surface occurs when it is exposed to humid air. Tiefenthaler and Lukosz (1985) have shown that adsorption and desorption of water vapor by the surface of a SiO_2/TiO_2 waveguide changed the effective RI of the guided modes. The evolution of an adsorbed water layer structure on silicon oxide at room temperature was considered by David and Seong (2005). They determined the molecular configuration of water adsorbed on a hydrophilic silicon oxide surface at room temperature as a function of RH using attenuated total reflection-infrared spectroscopy. It was found that a completely hydrogen-bonded network of water, which is ice-like, grows up as the RH increases from 0% to 30%. In the RH range of 30%–60%, the liquid water structure starts appearing while the ice-like structure continues growing to saturation. Above 60% RH, the liquid water configuration grows on the top of the ice-like layer. This structural evolution indicates that the outermost layer of the adsorbed water molecules undergoes transitions between different equilibrium states as humidity varies. Also the adsorption isotherm given in David and Seong (2005) shows that the thickness of the adsorbed layer at room temperature increases in an exponential-like manner above 60% RH.

In the case of a PCF-I, a similar adsorption of water vapor changes the effective RI (n_{cl}) of the interfering cladding mode propagating in the PCF. Since this adsorption/physisorption is a reversible process, a modulation of the n_{cl} occurs with respect to the ambient humidity values, which in turn change the position of the interference pattern accordingly. An increase in humidity increases the effective index of the cladding mode, which causes the shift of the interference pattern of a PCF-I toward longer wavelengths. The value of this interference peak shift is exponential-like with respect to RH that means it is identical to the adsorption isotherm of water vapor on silica given in Awakuni and Calderwood (1972) and David and Seong (2005). The shift of the interference pattern is mainly due to the adsorption and desorption of H_2O molecules along the surface of holes within the PCF, at the interface between the air and silica glass. Since the whole device is exposed to humidity, the adsorption and desorption of water vapor on the PCF outer surface and on the end face also contribute to the shift of the interference pattern. But considering the field distribution of the interfering cladding mode shown in Uranus (2010) and Cárdenas-Sevilla et al. (2011) and below the dew point temperature, the main contribution to the interference shift is considered to be due to the adsorption of water molecules within the voids of the PCF. The adsorption on the end face mainly causes a shift in the overall power level of the interference pattern.

Figure 18.4 shows the response to humidity of the PCF-I-based sensors designed by Mathew et al. (2010). It was found that the position of the interference peaks shifted with humidity variations. The interference peak shifts to higher wavelengths with an increase in humidity. Moreover, the shift of the interference pattern is most significant above 70% RH in a manner similar to the water vapor adsorption isotherm. Observed response of the PCF-I-based sensors is much better compared to the reported highest sensitivity for a FBG-based RH sensor. However, in comparison with photometric,

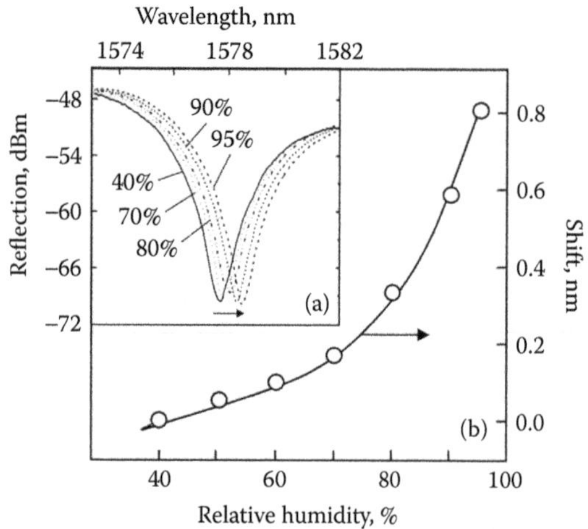

FIGURE 18.4 (a) Reflected spectrum of the photonic crystal fiber interferometer for different ambient relative humidity values; (b) Interference peak shift of the photonic crystal fiber interferometer with respect to relative humidity. (Data extracted from Mathew, J. et al., *Electron. Lett.*, 46, 1341, 2010.)

refractometric, and fluorescence-based humidity sensors discussed earlier in Chapters 12, 14, and 15, this sensitivity is very small, the response is nonlinear and it is observed in a narrow range of humidity, only from 40% to 95% RH.

Then, PCF-I-based sensors have been optimized through the use of additional humidity-sensitive coatings such as agarose (Mathew et al. 2012c, 2013) and poly(allylamine hydrochloride) (PAH) / poly(acrylic acid) (PAA) (Lopez-Torres et al. 2017). As a result, it was managed to significantly increase the sensitivity of the sensor. Comparative characteristics of these sensors are given in Table 18.2. Thus, it was shown that the total thickness of the sensing coating had to be below the penetration depth of the evanescent field to ensure an optimal response. This small thickness of coatings facilitated fast response: response and recovery times below 1 s were observed. In addition, Lopez-Torres et al. (2017) have found that in addition to the traditional approach, based on the analysis of the shift interferometric spectra, which is characterized by nonlinear dependence of RH above 80%, one can use an alternative method based on the fast Fourier transform (FFT) analysis; the interferometric spectrum shift can be monitored by the phase component of its FFT. This approach enhances the performance of PCF-I-based sensors because the resulting characterization is more linear and less noisy than the ones based on the wavelength monitoring (Figure 18.5).

Approximately the same sensitivity was demonstrated by the sensor developed by Wong et al. (2012). A schematic diagram to explain the mechanism of the sensor is shown in Figure 18.6. The sensor consists of a collapsed region between a single-mode fiber (SMF) and a short piece of pure silica PCF (LMA10), which has its end melted into a round tip, as shown in Figure 18.6. The sensor behaves predominately as a Michelson interferometer with the formation of interference fringes due to the phase difference between two paths of light after recombination. To accommodate the interference fringes in the 100 nm spectrum's window, substantial phase difference is needed. This phase difference is caused by the excitation of the cladding mode due to the round tip rather than the commonly used method of substantially increasing the physical length of the sensor. The round tip is used to excite the cladding mode to higher order after reflection, whereas the second collapsed region allows the excited mode to propagate for a distance, accumulating a large phase delay before propagating back in the cladding. Finally, both modes recombine at the first collapsed region to form a Michelson interferometer. Thus, high spatial resolution can be realized with a small sensor.

TABLE 18.2
Comparison of Resolutions of PCF-I Used (PAH/PAA) and Agarose Coatings (Optical Spectrum Analyzer [OSA] Resolution: 0.01 nm)

PAH/PAA PCF-I		Agarose PCF-I	
Total wavelength for a RH change of 75% RH	61 nm	Total wavelength for a RH change of 60% RH	56 nm
Resolution in the lineal range for a RH change of 55% RH	0.035% RH	RH Resolution in the lineal range for a RH change of 40% RH	0.017% RH
Resolution in the lineal range for a RH change of 20% RH	0.004% RH	Resolution in the lineal range for a RH change of 20% RH	0.007% RH
Total resolution for a RH change of 75% RH	0.01% RH	Total resolution for a RH change of 60% RH	0.01% RH

Source: Data extracted from Lopez-Torres D. et al., *Sens. Actuators B*, 242, 1065–1072, 2017.

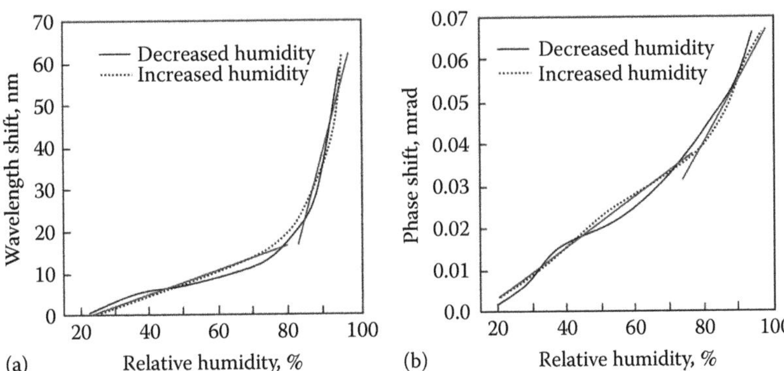

FIGURE 18.5 Characterization of the sensor response as a function of the relative humidity expressed in (a) wavelength shift and (b) FFT phase shift. (Reprinted from *Sens. Actuators B*, 242, Lopez-Torres, D. et al., Photonic crystal fiber interferometer coated with a PAH/PAAnanolayer as humidity sensor, 1065–1072, Copyright 2017, with permission from Elsevier.)

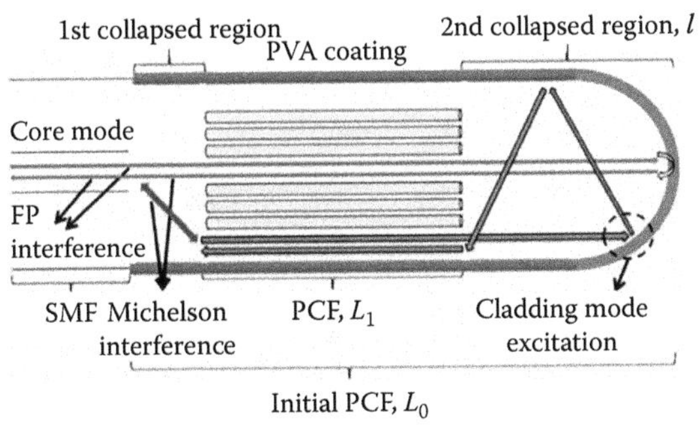

FIGURE 18.6 Schematic diagram showing the operating mechanism of the sensor. (Reprinted from *Sens. Actuators B*, 174, Wong, W.C. et al., Polyvinyl alcohol coated photonic crystal optical fiber sensor for humidity measurement, 563–569, Copyright 2012, with permission from Elsevier.)

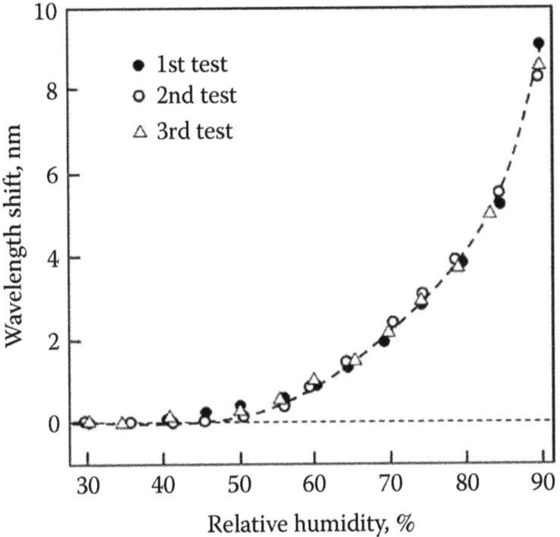

FIGURE 18.7 Wavelength shift with increasing RH for the 9% (w/w) PVA coated PBF-based sensor, three tests over seven days were performed. (Reprinted from *Sens. Actuators B*, 174, Wong, W.C. et al., Polyvinyl alcohol coated photonic crystal optical fiber sensor for humidity measurement, 563–569, Copyright 2012, with permission from Elsevier.)

Polyvinyl alcohol (PVA), used in sensor, is a water-soluble polymer, which has a good swelling ratio. The polymer is able to adhere well to silica and exhibit excellent film formation property. It is easily processed, biocompatible, and is resistant to chemical agents (Shao et al. 2003). As the PVA coating absorbs water molecules from the surrounding, it swells and its RI changes. Hence, variation of RH would cause the physical properties of the PVA coating, such as its swelling degree and RI to change (Gaston et al. 2004). With the PVA coating on PCF cladding, as light is guided to the interface of the PVA coating and PCF cladding, as shown in Figure 18.6, variations in the RI of PVA would affect the propagation of light in the PCF. This changes the effective RI of the cladding mode, n_2, which in turn leads to a phase change in the light. This phase change would be significant due to the presence of higher-order cladding mode, thus high sensitivity can be achieved.

Experiment has shown that a relatively sensitive region was found from 50% to 90% RH (Figure 18.7). The sensitivity for 9% (w/w) PVA sensor was found to be 0.04, 0.15, and 0.60 nm/% RH at 50%, 70%, and 90% RH, respectively. In addition, sensors showed fast response with little hysteresis and high repeatability. This was faster than many similar types of RH optical fiber sensors. The sensor also exhibited little cross-sensitivity to the temperature change and ammonia gas. The temperature sensitivity of the sensor was 0.029 nm/°C, giving a cross-sensitivity of 0.05% RH at 90% RH. Wong et al. (2012) believe that with such performance, the proposed sensor has shown a potential for environmental humidity measurement application. In principle, increasing the thickness, PVA coating should contribute to a larger wavelength shift of interference fringes. However, this should be accompanied by an increase in the response time.

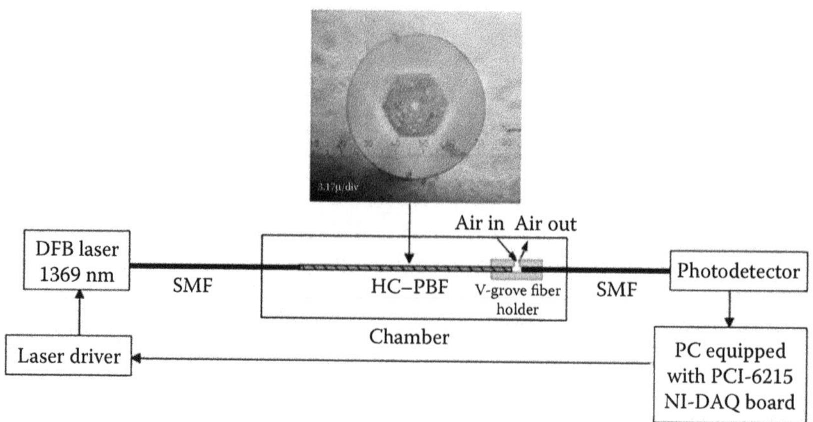

FIGURE 18.8 Experimental setup for humidity detection and HC–PBF (sensor head) cross section. (From Noor, M.M. et al., *Meas. Sci. Technol.*, 23, 085103, 2012, with permission of IOP Publishing Ltd.)

Another version of using photonics crystals for the development of humidity sensors is shown in Figure 18.8. Noor et al. (2012) have demonstrated a novel RH optical fiber sensor based on the hollow core-photonic bandgap fiber (HC–PBF) as the sensor head, using a wavelength scan and a reference scheme. The voltage difference (V) between the reference signal and absorption signal at the peak of the water vapor absorption around 1369 nm band was used to determine the RH level and also to reduce the outside affection such as fiber loss, the fluctuation of a distributed feedback (DFB) laser power, and attenuation by dust in the probed region. A 5 cm HC–PBF with a small gap as an air diffusion hole demonstrated a high humidity sensitivity up to 6.43 mV/1% RH in a wide range of detection from 0% to 90% RH with good linearity, low noise, and good reversibility without requiring any hygroscopic coating on the fiber sensor head. It was assumed that the length interaction between the light and the humid air can be increased with the length of the HC–PBG to further increase the sensitivity of the sensor without addition of light diffraction effect.

Miao et al. (2016) have also used a no-core fiber for manufacturing interference-based humidity sensors. However, for this purpose they used the tapered square no-core fiber (TSNCF) coated with SiO_2 nanoparticles. The schematic diagram of the RH sensor is illustrated in Figure 18.9. The sensor was fabricated by splicing a section of TSNCF with two standard SMF, which was later immersed into the SiO_2 dispersion solution to enhance the SiO_2 nanoparticles deposition on the surface of the TSNCF. As a result of testing of the sensor it has been found that the wavelength shifts reached up to 10.2 nm at 1410 nm and 11.5 nm at 1610 nm for a RH range of 43.6%–98.6%, and the corresponding sensitivities reached 456.21 and 584.2 pm/% RH for a RH range of 83%–96.6%, respectively (Figure 18.10). The temperature response of the proposed sensor has also been experimentally investigated, and the experimental results indicated that it had low temperature sensitivity, about 6 pm/°C for an environmental temperature of 20.9°C–80°C, which is a desirable merit to resolve the temperature cross sensitivity. Miao et al. (2016) believed that the transmission spectrum showed a low temperature sensitivity due to the fact that the sensing head was made of a pure silica rod with a low thermal expansion coefficient and low thermo-optic coefficient. It is important to note that the suggested RH sensor really has several merits such as high humidity sensitivity, low temperature sensitivity, ease of fabrication, and simple and compact configuration.

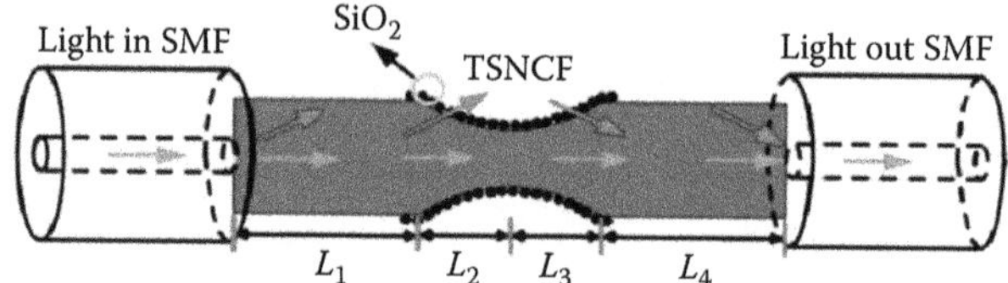

FIGURE 18.9 Schematic diagram of light propagation in the SMF–TSNCF–SMF structure. (Reprinted from *Opt. Fiber Technol.*, 29, Miao, Y. et al., Low-temperature-sensitive relative humidity sensor based on tapered square no-core fiber coated with SiO_2 nanoparticles, 59–64, Copyright 2016, with permission from Elsevier.)

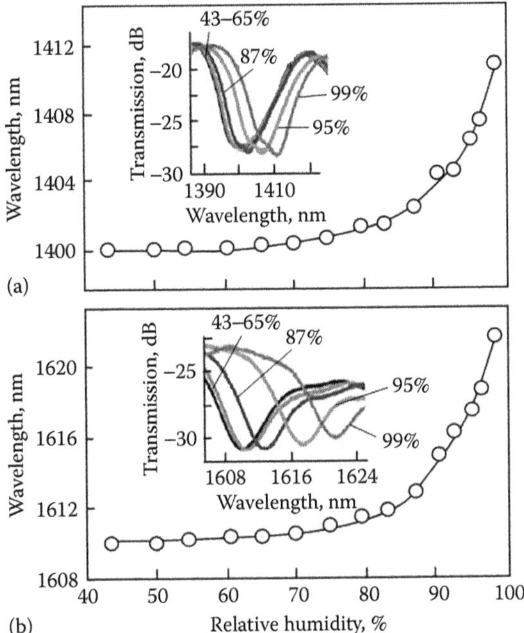

FIGURE 18.10 Transmission spectral responses of different dips in transmission spectra under different environmental RH levels. (a) λ ~ 1400 nm and (b) λ ~ 1620 nm. (Reprinted from *Opt. Fiber Technol.*, 29, Miao, Y. et al., Low-temperature-sensitive relative humidity sensor based on tapered square no-core fiber coated with SiO_2 nanoparticles, 59–64, Copyright 2016, with permission from Elsevier.)

18.3 SUMMARY

The use of specific fibers for humidity measurements is an interesting trend in the development of optical humidity sensors, as it extends possibilities of the fiber-optic technique for sensor design. However, it must also be recognized that the existing versions of the humidity sensors developed on the basis of these fibers did not give any increase in the sensitivity of sensors and did not improve their exploitation parameters. Moreover, there are doubts in the high reproducibility of the parameters of such sensors. Considering the above and the high cost of specific fibers compared with conventional optical fibers, the prospects of entering the market for sensors based on specific fibers seem to be problematic.

REFERENCES

Awakuni Y., Calderwood J.H. (1972) Water vapor adsorption and surface conductivity in solids. *J. Phys. D* 5(5), 1038–1045.

Bjarklev A., Bjarklev A.S., Broeng J. (2003) *Photonic Crystal Fibers*. Springer, New York.

Cárdenas-Sevilla G.A., Finazzi V., Villatoro J., Pruneri V. (2011) Photonic crystal fiber sensor array based on modes overlapping. *Opt. Exp.* 19(8), 7596–7602.

Charlton C., Temelkuran B., Dellemann G. (2005) Midinfrared sensors meet nanotechnology: Trace gas sensing with quantum cascade laser inside photonic band-gap hollow waveguides. *Appl. Phys. Lett.* 86, 194102.

Choi H.Y., Kim M.J., Lee B.H. (2007) All-fiber Mach-Zehnder type interferometers formed in photonic crystal fiber. *Opt. Exp.* 15(9), 5711–5720.

David B.A., Seong H.K. (2005) Evolution of the adsorbed water layer structure on silicon oxide at room temperature. *J. Phys. Chem. B* 109(35), 16760–16763.

Elosua C., Matias I.R., Bariain C., and Arregui F.J. (2006) Volatile organic compound optical fiber sensors: A review. *Sensors* 6, 1440–1465.

Gaston A., Perez F., Sevilla J. (2004) Optical fiber relative-humidity sensor with polyvinyl alcohol film. *Appl. Opt.* 43, 4127–4132.

Hoo Y.L., Jin W., Ho H.L., Ju J., Wang D.N. (2005) Gas diffusion measurement using hollow-core photonic bandgap fiber. *Sens. Actuators B* 105, 183–186.

Hoo Y.L., Jin W., Ho H.L., Wang D.N. (2003) Measurement of gas diffusion coefficient using photonic crystal fiber. *IEEE Photon. Tech. Lett.* 15, 1434–1436.

Hoo Y.L., Jin W., Wang D.N. (2002) Evanescent-wave gas sensing microstructure fiber. *Opt. Eng.* 41, 8–9.

Jha R., Villatoro J., Badenes G. (2008) Ultrastable in reflection photonic crystal fiber modal interferometer for accurate refractive index sensing. *Appl. Phys. Lett.* 93(19), 191106.

Jin W., Ho H.L., Cao Y.C., Ju J., Qi L.F. (2013) Gas detection with micro- and nano-engineered optical fibers. *Opt. Fiber Technol.* 19, 741–759.

Konorov S.O., Fedetov A.B., Zheltikov A.M. (2005) Phase-matched four-wave mixing and sensing of water molecules by coherent anti-stokes Raman scattering in large-core-area hollow photonic-crystal fibers. *Opt. Soc. Am.* 22, 2049–2053.

Lee B., Roh S., Park J. (2009) Current status of micro- and nano-structured optical fiber sensors. *Optic. Fiber Technol.* 15, 209–221.

Lopez-Aldaba A., Lopez-Torres D., Ascorbe J. et al. (2016) SnO_2-MOF-Fabry-Perot humidity optical sensor system based on fast Fourier transform technique. In: *Proceedings of European Workshop on Optical Fibre Sensors (EWOFS 2016)*, Limerick, Ireland, pp. 9916–9929, May.

Lopez-Torres D., Elosua C., Villatoro J., Zubia J., Rothhardt M., Schuster K., Francisco J., Arregui J. (2017) Photonic crystal fiber interferometer coated with a PAH/PAAnanolayer as humidity sensor. *Sens. Actuators B* 242, 1065–1072.

Matias I.R., Bravo J., Arregui F.J., Corres J.M. (2006) Nanofilms onto a hollow core fiber. *Opt. Eng. Lett.* 45, 050503-1–3.

Mathew J. (2013) Development of novel fiber optic humidity sensors and their derived applications. PhD Thesis, Dublin Institute of Technology.

Mathew J., Semenova Y., Farrell G. (2012a) Relative humidity sensor based on an Agarose infiltrated photonic crystal fiber interferometer. *IEEE J. Select. Topics Quantum El.* 18, 1553–1559.

Mathew J., Semenova Y., Farrell G. (2012b) Photonic crystal fiber interferometer for dew detection. *J. Lightwave Technol.* 30(8), 1150–1155.

Mathew J., Semenova Y., Farrell G. (2012c) A high sensitivity humidity sensor based on an agarose coated photonic crystal fiber interferometer. *Int. Soc. Opt. Eng.* 8421, 842177.

Mathew J., Semenova Y., Farrell G. (2013) Experimental demonstration of a high-sensitivity humidity sensor based on an agarose-coated transmission-type photonic crystal fiber interferometer. *Appl. Opt.* 52, 3884–3890.

Mathew J., Semenova Y., Rajan G., Farrell G. (2010) Humidity sensor based on a photonic crystal fiber interferometer. *Electron. Lett.* 46(19), 1341–1343.

Miao Y., Ma X., He Y., Zhang H., Zhang H., Song B., Liu B., Yao J. (2016) Low-temperature-sensitive relative humidity sensor based on tapered square no-core fiber coated with SiO_2 nanoparticles. *Opt. Fiber Technol.* 29, 59–64.

Noor M.M., Khalili N., Skinner I., Peng G.D. (2012) Optical relative humidity sensor based on a hollow core-photonic bandgap fiber. *Meas. Sci. Technol.* 23(8), 085103.

Peng W., Pickrell G.R., Shen F., Wang A. (2004) Experimental investigation of optical waveguide-based multigas sensing. *IEEE Photon. Tech. Lett.* 16, 2317–2319.

Schmidt M.A., Argyros A., Sorin F. (2016) Hybrid optical fibers–An innovative platform for in-fiber photonic devices. *Adv. Opt. Mater.* 4, 13–36.

Shao C., Kim H.-Y., Gong J., Ding B., Lee D.-R., Park S.-J. (2003) Fiber mats of poly(vinyl alcohol)/silica composite via electrospinning. *Mater. Lett.* 57, 1579–1584.

Skorobogatiy M. (2009) Microstructured and photonic bandgap fibers for applications in the resonant bio- and chemical sensors. *J. Sens.* 2009, 524237.

Sorokin Y.V. (2012) Optic fiber on the basis of photonic crystal. In: Yasin M. (Ed.), *Recent Progress in Optical Fiber Research.* InTech, Rijeka, Croatia, pp. 247–270.

Tiefenthaler K., Lukosz W. (1985) Grating couplers as integrated optical humidity and gas sensors. *Thin Solid Films.* 126, 205–211.

Uranus H.P. (2010) Theoretical study on the multimodeness of a commercial endlessly single-mode PCF. *Opt. Commun.* 283(23), 4649–4654.

Villatoro J., Finazzi V., Badenes G., Pruneri V. (2009b) Highly sensitive sensors based on photonic crystal fiber modal interferometers. *J. Sens.* 2009, 747803.

Villatoro J., Kreuzer M.P., Jha R., Minkovich V.P., Finazzi V., Badenes G., Pruneri V. (2009a) Photonic crystal fiber interferometer for chemical vapor detection with high sensitivity. *Opt. Exp.* 17(3), 1447–1453.

Wong W.C., Chan C.C., Chen L.H., Li T., Lee K.X., Leong K.C. (2012) Polyvinyl alcohol coated photonic crystal optical fiber sensor for humidity measurement. *Sens. Actuators B.* 174, 563–569.

19 Surface Plasmon Resonance-Based Humidity Sensors

19.1 PRINCIPLES OF OPERATION

Sensors which use the phenomenon called the surface Plasmon resonance (SPR) have become very popular in recent years. Over the past two decades, the SPR technique has been developed for detection of chemical and biological species (Homola et al. 1999; Homola 2006). At present, surface Plasmon-based sensors are used in a number of commercial instruments produced by companies such as Biacore, Biosensing Instruments Inc, Sensata, and ICx Nomadics. Usually these devices guarantee the measurements of refractive index (RI) with detection limits between 10^{-5} RI and $5{\cdot}10^{-6}$ RI units and better. Experiment has shown that humidity also can be measured using a SPR effect.

In principle, SPR sensors are thin-film refractometers that measure changes in the RI occurring at the surface of a metal film supporting a surface Plasmon (SP). Surface Plasmons, also known as surface Plasmon polaritons, are the surface electromagnetic waves that propagate parallel along the metal–dielectric interface. The utilization of metal is conditioned by the need of electrons in the conduction band that can resonate with the incoming light at a proper wavelength. The charge density wave is associated with electromagnetic waves, the field vectors of which reach their maxima at the interface and decay evanescently into both media. This surface plasma wave (SPW) is a TM-polarized wave (magnetic vector is perpendicular to the direction of propagation of the SPW) and parallel to the plane of interface. The propagation constant (β) of the SPW propagating at the interface between a semi-infinite dielectric and metal is given by the following expression (Homola et al. 1999):

$$\beta = k\sqrt{\frac{\varepsilon_m n_S^2}{\varepsilon_m + n_S^2}} \tag{19.1}$$

where:

k denotes the free space wave number
ε_m is the dielectric constant of the metal
n_s is the RI of the dielectric

A surface Plasmon excited by a light wave propagates along the metal film, and its evanescent field probes the medium (sample) in contact with the metal film. A change in the RI of the dielectric gives rise to a change in the propagation constant of the surface Plasmon, which through the coupling conditions alters the characteristics of the light wave coupled to the surface Plasmon (e.g., coupling angle, coupling wavelength, intensity, and phase). Changes in the polarization can be also detected (Piliarik and Homola 2009).

A resonant interaction between the waves and the mobile electrons at the surface of the metal, that is, the SPR, occurs under certain conditions; and the maximal energy transfer between the incident light energy photons and the electrons at the metal film surface takes place. It is important to note that all non-*p*-polarized light will not contribute to the SPR and will increase the background intensity of the reflected light. The resonance condition depends on the wavelength, the angle at which the light strikes the substrate–metal interface, the dielectric constants of all the materials involved, the RI of the substrate on which the metal film is deposited; and the RI of the sample at the metal surface (Karlsen et al. 1995; Bardin et al. 2002; Kashyap and Nemova 2009). This means that (1) even marginal changes on the metal surface, such as the adsorption of molecules, disturb this oscillation and the change in the intensity of the reflected light dependent on the wavelength or the angle of incidence can be quantified; and (2) the specificity of SPR sensors can be achieved by modifying the metal surface with a functional layer to allow specific binding between the ligand molecules and the measurand (Pockrand et al. 1979; Jorgenson and Yee 1993).

SPR-based devices are well described in the literature. A considerable amount of the literature is available on the principle of operation of the SPR sensing and various configurations (Homola et al. 1999; Homola 2006; Maier 2007; Gupta and Verma 2009; Kashyap and Nemova 2009).

19.2 FEATURES OF SURFACE PLASMON RESONANCE SENSOR DESIGN

The SPR-based sensors are usually classified as label-free devices. In other words, these devices do not need additional tags, dyes, or reagents for performing the optical measurement in opposition to what it happens in fluorometry or colorimetry. Basically, these SPR devices work as highly sensitive refractometers. Therefore, these devices measure changes in the RI that experiences an auxiliary material, previously deposited on the SPR device, which targets the parameter, compound, or substance to be measured.

Analysis of the SPR physics leads to four salient features that govern the geometry and application of SPR sensors (Wilson et al. 2001):

- The SPW cannot be excited by direct illumination. Thus, the exciting photon and the sample must be on opposite sides of the metal layer.
- The evanescent field of the incident photon excites SPW. Thus, the RI of the substrate supporting the metal layer ultimately limits the dynamic range of any SPR sensor.
- The field strength of the SPW decays exponentially into the sample media. Thus, SPR is primarily a surface technique; however, SPR sensors respond to bulk RI changes.
- The real component of the RI (complex permittivity) of the sample has a significantly greater influence on the SPW than does the imaginary component of the RI (complex permittivity) of the sample. Thus, while SPR can be employed in highly absorbing media, bulk absorbance changes of a sample will induce minor changes in SPR reflectance profiles.

Thus, during the design of SPR-based sensor there are three key components, which need to be selected correctly: an optical system, a transducing medium that interrelates the optical and (bio)chemical domains, and an electronic system supporting the optoelectronic components of the sensor and allowing data processing (Wilson et al. 2001). The transducing medium transforms the changes in the quantity of interest into changes in the RI, which may be determined by optically interrogating the SPR. The optical part of the SPR sensor contains a source of optical radiation and an optical structure in which SPW is excited and interrogated. Usually the optical train of an SPR sensor is designed around one of the two sensing philosophies. Constant wavelength sensors relate the SPR angle of incidence of maximum attenuation to analyte concentration. Constant angle sensors relate the wavelength of maximum attenuation to analyte concentration. In the process of interrogating the SPR, an electronic signal is generated and processed by the electronic system. As a rule, the intensity of reflected or diffracted light or its wavelength or angular spectrum is detected by charge-coupled device (CCD) arrays or photodiode arrays. Sensitivity and detection accuracy or signal-to-noise ratio (SNR) are the two parameters that are used to analyze the performance of an SPR sensor. For the best performance, both the parameters should be as high as possible. The detection limit of SPR sensors is often declared in refractive index units (RIU), that is, the lowest relative change in the index of refraction Δn, which can be measured (Homola 1995; Kabashin et al. 2009).

Major properties of an SPR sensor, the sensor sensitivity, selectivity, stability, resolution, and response time, are determined by properties of the optical system and the transducing medium, including analyte-specific receptor (Homola et al. 1999; Homola 2006). In particular, the sensitivity of an SPR sensor utilizing angular interrogation method depends on the amount of shift of the resonance angle with a change in the RI of the sensing layer. The increase in the shift of resonance angle means an increase in the sensitivity of the sensor.

The enhancement and subsequently the coupling between light and a surface Plasmon are performed in a coupling device (coupler). The most common couplers used in SPR sensors include a prism coupler and a grating coupler (Figure 19.1). All these schemes can be used in sensor application. Prism couplers represent the most frequently used method for optical excitation of surface Plasmon's. Since the SPR excitation with a prism is based on total internal reflection, the coupling prism can be replaced by a waveguide layer of the planar structure to get a compact device easily integrated in

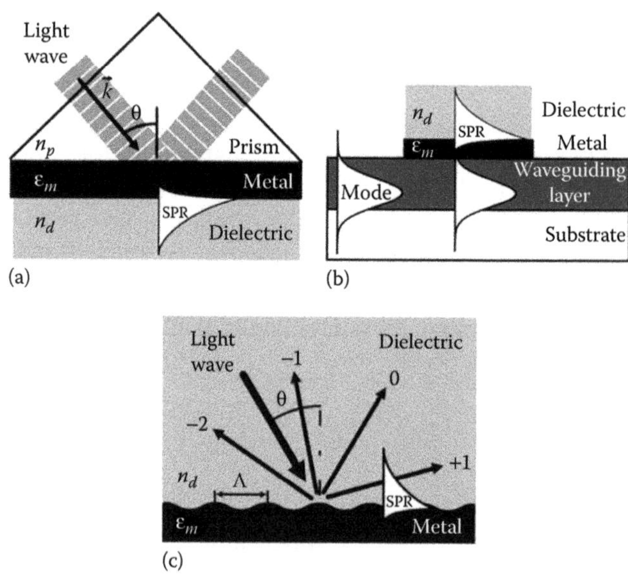

FIGURE 19.1 Most widely used configurations for SPR sensors: (a) prism coupler-based SPR system (ATR method), (b) waveguide coupler, and (c) grating coupler-based SPR system. (Reprinted with permission from Homola, J., *Chem. Rev.*, 108, 462, 2008. Copyright 2008 American Chemical Society.)

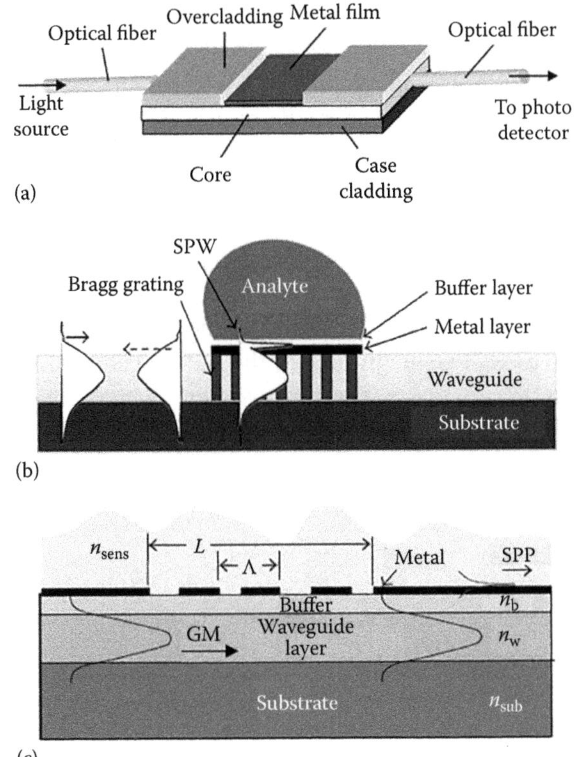

FIGURE 19.2 Diagram of an optical waveguide surface Plasmon resonance sensor: (a) Conventional type of planar SPR sensors; (b) SPR sensing structure with Bragg grating imprinted into the waveguide layer; and (c) LPG engraved on the top of the metal layer. (From Kashyap, R., and Nemova, G., *J. Sens.*, 2009, 645162, 2009. Published by Hindawi as open access.)

any planar schemes. Schematic diagrams of such sensor platform are shown in Figure 19.2. However, it is necessary to take into account that from theoretical analysis and comparison of the sensitivity (Homola et al. 1996) it has been concluded that grating-based SPR sensors using wavelength interrogation are much less sensitive then their prism coupler-based counterparts. When using the angular interrogation mode, instead, the sensitivity of SPR sensors using diffraction gratings depends on the diffraction order and it is almost similar to that of SPR sensors based on prism couplers.

19.3 SURFACE PLASMON RESONANCE PROBE DESIGN

As mentioned earlier, SPR sensors include a metal layer, playing a significant role in sensing effects, and therefore correct selection of the metal used is an important step in the SPR-based sensor design. The thickness of the metal layer usually is in the range from 30 to 50 nm. In principle, the film thickness can be up to 100–200 nm (Rhodes et al. 2008). For thickness values higher than this, the evanescent wave is not able to reach the thin-film surface, so the Plasmons cannot be excited and the whole system behaves as a new waveguide, making the resonance to disappear. At the same time, a decrease in the thickness less than 30–50 nm is also undesirable because at smaller thicknesses the dip in reflective light intensity becomes broader. The most in-depth analysis of metals acceptable for the SPR sensors design was carried out by Stewart et al. (2008), and Kashyap and Nemova (2009). According to Kashyap and Nemova (2009), for metallic coating mainly either silver (Ag) or gold (Au) is used.

One should note that Ag displays the narrowest width of the SPR curve and yields the sharpest and most defined attenuation among of all metals. The sharpness of the resonance curve depends on the imaginary part of the dielectric constant of the metal. Ag, having the larger value of the imaginary part of the dielectric constant shows narrower width of the SPR curve causing a higher SNR or detection accuracy. However, the chemical stability of Ag is poor due to its oxidation. The oxidation of Ag occurs as soon as it is exposed to air and especially to water, which makes it difficult to obtain a reproducible result, and hence the sensor remains unreliable for practical applications. Therefore, Ag films need to be coated with a thin and dense cover such as a dielectric layer or a lesser reactive metal such as Au. But this leads to a decrease in the RI sensitivity. At the same time, Au is a chemically stable metal and therefore, at present, Au is more often employed for SPR sensor applications. In addition, Au demonstrates a higher shift of resonance parameter to the change in the RI of sensing layer. The latter is very important, because SPR is ultimately sensitive to only RI changes at the metal–sample surface, and all selectivity is derived from the inclusion of analyte receptors on the sensor. Au is also applied due to the ability to functionalize the surface with specific receptors. Thus, the Au nanoparticles (NPs) coating on waveguides allows using efficiently the Plasmon scattering for fabrication different fiber-optic sensors based on SPR (Eah et al. 2005). The particles are stabilized by capping materials, which can interact with the analyte. This interaction causes the aggregation of the Au NPs and a change in its absorbance.

Other metals such as copper (Cu), aluminum (Al), and resonant metal film based on bimetallic layers also have the ability to be used for an SPR sensor (Zynio et al. 2002; Kashyap and Nemova 2009; Yang et al. 2010). However, the Cu has some limitations such as Ag. It is chemically vulnerable against oxidation and corrosion; therefore, its protection is required for a stable sensing application. The SPR sensing capabilities of different bimetallic combinations made out of Ag, Au, Al, and Cu were theoretically investigated for the design of SPR-based fiber-optic sensors by Sharma and Gupta (2007). It was established that the sensor with single Au layer was the most sensitive, whereas the sensor with single Al layer was the least. Further, the Cu–Al combination provides the minimum sensitivity for any ratio of their corresponding thickness values. In all the combinations with Au, Ag–Au, and Cu–Au combinations it provides good sensitivity for the small thickness of the inner layer, whereas the Au–Al combination provides larger sensitivity than all other combinations for the larger thickness of the inner Au layer. It implies that a thick Au layer with very thin cover of Al layer (around 2–4 nm) provides quite a large sensitivity. As far as variation of SNR with inner layer fraction is concerned, Cu–Al is better among all the bimetallic combinations, whereas the Ag–Au combination provides the minimum values of SNR. To achieve highest SNR, one should choose a thin Cu layer and a much thicker covering of Al layer. This study implies that there is no single combination of metals that provides high values of both SNR and sensitivity simultaneously. However, the ability to create new structures that can show a large shift of resonance angle as Au film, and narrow resonance curve as Ag film along with the protection of Ag film against oxidation, still exists (Zynio et al. 2002).

A detailed discussion of synthesis and fabrication of Plasmonic nanostructures and their application in optical chemical sensors was carried out by Stewart et al. (2008). The discussion of other aspects of the SPR probe design and fabrication has been published in the literature (Kashyap and Nemova 2009).

19.4 LOCALIZED SURFACE PLASMON RESONANCE

One should note that when developing optical sensors, in addition to the classical SPR effect in recent years the use of so-called localized surface Plasmon resonance (LSPR) effect has received a widespread use (Willets and Van Duyne 2007; Petryayeva and Krull 2011). LSPR is generated by plasmonic nanostructures (metal NPs and nanoholes), in contrast to a continuous metal film used in traditional SPR. At that, NPs should have a size comparable to or smaller than the wavelength of light used to excite the Plasmon. These resonances create sharp spectral absorption and scattering peaks as well as strong electromagnetic near-field enhancements. The phenomenon is a result of the interactions between the incident light and surface electrons in a conduction band. This interaction produces coherent localized Plasmon oscillations with a resonant frequency that strongly depends on the composition, size, geometry, dielectric environment, and particle–particle separation distance of NPs. This means that the position of a strong resonance absorbance peak is highly sensitive to the local RI surrounding the particle. Therefore, LSPR measures small changes in the wavelength of the absorbance position, rather than the angle as in traditional SPR. The magnitude of the spectral shift of LSPR extinction, or the scattering wavelength maximum for small NPs is described by the following relationship (Willets and Van Duyne 2007):

$$\Delta\lambda_{\max} = m\Delta n\left[1-\exp\left(-\frac{2d}{l_d}\right)\right] \tag{19.2}$$

where:

m is the bulk RI response of the NPs, also known as the sensitivity factor (in nm per RIU)
Δn is the change in RI (in RIU)
d is the effective thickness of the adsorbed layer (in nm)
l_d is the characteristic electromagnetic field decay length (in nm)

Using LSPR instead of SPR has a number of significant advantages (https://www.nicoyalife.com):

- The optical hardware needed for LSPR is much less complex because no prism is needed to couple the light, so the instrument can be made smaller and more affordable.
- Since the angle is not important, the instrument is much more robust against vibration and mechanical noise.
- LSPR is not as sensitive to bulk RI changes, which causes errors in experimental data, because it has a much shorter electromagnetic field decay length.
- No strict temperature control is needed, simplifying the instrument.
- The sensor chips can be manufactured at a much more affordable price.
- Easier to use and maintain.

Metals such as Ag and Au, as in the case of SPR, are preferred for use in sensors based on the LSPR effect. Due to the energy levels of *d–d* transitions these metals exhibit LSPR in the visible range of the spectrum (Petryayeva and Krull 2011). It was established that the shape and the size of metallic NPs contribute to spectral properties due to the changes in the surface polarization. Various shapes such as spheres, triangles, cubes, prisms, bipyramids, octahedrons, nanorods, nanoshells, nanostars, and structured array films have been synthesized to tailor LSPR absorption from visible to infrared regions. An increase in edges or sharpness of a NP results in a red shift of extinction spectra due to an increase in the charge separation, whereas increased symmetry results in increases of LSPR intensity (Lu et al. 2009). The number of resonance absorption peaks is determined by the number of modes in which a given NP can be polarized (Liz-Marzan 2006; Lu et al. 2009). Thus, nonspherical NPs tend to exhibit multiple, red shifted peaks in comparison to spherical particles. The size of the NPs influences the relative magnitude of the absorption cross section and the cross section of scatter. For NPs smaller than 20 nm, the predominant process is absorption. An increase in physical dimensions or effective size of NPs increases the cross section of the scatter. This relationship is usually expressed as a ratio of scattering to absorption, and while dependent on size this relationship tends to be independent of the aspect ratio of the NPs (Jain et al. 2006). The LSPR response also depends on the separation distance between NPs (Petryayeva and Krull 2011).

19.5 SURFACE PLASMON RESONANCE FIBER-BASED SENSORS

Currently, optical fiber SPR probes present the highest level of miniaturization of SPR devices, allowing chemical and biological sensing in inaccessible locations where the mechanical flexibility and ability to transmit optical signals over a long distance make the use of optical fibers very attractive (Homola et al. 1999). The development of fiber-optic *SPR* fiber sensors began in the early 1990s of the last century. Real SPR sensors that use fiber devices as sensor heads were first proposed in 1993 (De Maria et al. 1993; Jorgenson and Yee 1993). To fabricate an SPR-based fiber-optic sensor, the silicon cladding from a small portion of the fiber, preferably from the middle, is removed and the unclad core is coated with a metal layer. The metal layer is further surrounded by a dielectric sensing layer. It was established that due to its simplicity and sensitivity, the SPR optical fiber sensor is a very promising tool for vapor sensing and bioapplications (Skorobogatiy 2009). The example of the SPR fiber sensor setup is shown in Figure 19.4. The sensor shown in Figure 19.3, was made of an all silica/silicone fiber with 600 μm diameter. The sensing element was

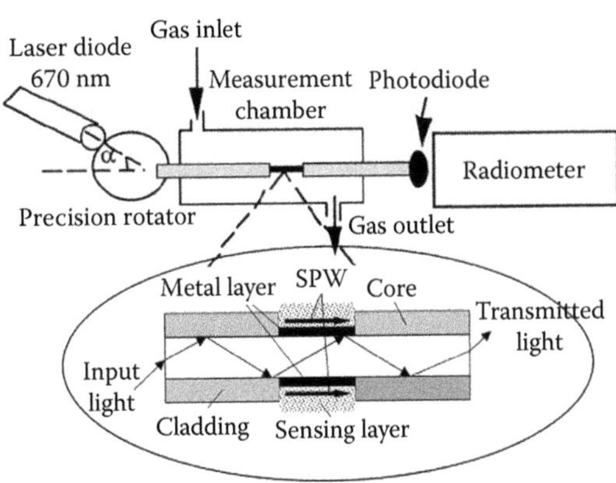

FIGURE 19.3 Experimental setup and typical probe of the SPR fiber-optic sensor. (Reprinted from *Sens. Actuators B*, 74, Abdelghani, A., and Jaffrezic-Renault, N., SPR fibre sensor sensitized by fluorosiloxane polymers, 117, Copyright 2001, with permission from Elsevier.)

prepared by removing the silicone cladding and coating this section with a 50 nm Ag film by thermal evaporation and covering this with thiol and actuator polymer (Abdelghani and Jaffrezic-Renault 2001).

In an SPR-based fiber-optic sensor, all the guided rays are launched and hence, instead of angular interrogation, spectral interrogation method is used. The light from a polychromatic source is launched into one of the ends of the optical fiber. The evanescent field produced by the guided rays excites the surface Plasmon's at the metal–dielectric sensing layer interface. As mentioned earlier, the SPR is a phenomenon, which occurs when light is reflected by thin metal films. The coupling of evanescent field with surface Plasmon's strongly depends on wavelength, fiber parameters, probe geometry, and the metal layer properties. As described earlier, experimentally Plasmon resonance is detected by observing the presence of a minimum in the light reflected in the variation of the angle of incidence on the metal/optical guide interface. This value also depends on the refraction index of the external medium, so that with this method, variations in this index can be detected. Hence, the presence of a chemical species can be detected following a variation in the RI. Using optical waveguides and fibers instead of bulk prism configuration in plasmonic sensors offers miniaturization, high degree of integration, and remote-sensing capabilities. In fiber and waveguide-based sensors, one launches the light into a waveguide core and then uses coupling of a guided mode with a Plasmonic mode to probe for the changes in the ambient environment.

Various fundamental SPR fiber sensor configurations are shown in Figure 19.4. These include symmetrical structures, such as a simple metal coated optical fiber

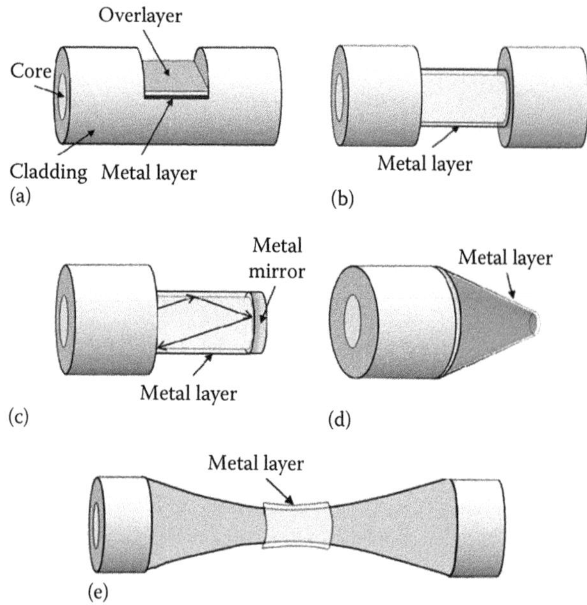

FIGURE 19.4 General SPR fiber-optic sensor schematics with (a) D-shape fiber, (b) cladding-off fiber, (c) end-reflection mirror, (d) angled fiber tip, and (e) tapered fiber. (Reprinted from *Optic. Fiber Technol.*, 15, Lee, B. et al., Current status of micro- and nano-structured optical fiber sensors, 209, Copyright 2009, with permission from Elsevier.)

with and without remaining cladding, tapered fiber, structures without any symmetry, such as side-polished fibers (or D-shape fiber) and one-side metal coated fibers with and without remaining cladding, and structures with modified fiber tips with flat or angled structures. Each of them may contain either overlayer or multilayer structures. The overlayer, which is a layer deposited on the metal layer, can be useful in tuning the measurable range for a fiber-optic SPR sensor. Many SPR fiber sensor systems have also been developed with structures that have been further modified, via the use of several types of gratings, heterocore fibers, nanoholes, and so on. Some approaches employ various gratings to couple light from the core mode to various cladding modes and to provide the phase matching needed to excite SPR on the surface of an optical fiber. A more detailed description of SPR-based fiber-optic sensors can be found in review papers (Wilson et al. 2001; Gupta and Verma 2009; Lee et al. 2009; Skorobogatiy 2009).

Analysis has shown that SPR sensors can be designed on the base of both single-mode and multimode optical fibers (Wilson et al. 2001). Single-mode optical fibers are employed because the single-mode fibers support sharp SPR transitions as only one angle of light (mode) propagates down the fiber at a given wavelength and the fibers preserve the polarization of the incident light. However, there are two disadvantages associated with single-mode fibers for optical sensors. The difficulty of affixing a metal mirror to the distal end of a single-mode fiber prohibits constructing the sensor in a *dip probe* configuration. Hence, the excitation and collection optics must be on opposite ends of the fiber, which leads to a large probe head at the sample interface. Also, the dynamic range, in RIU, of the sensors is very limited. The sensors only respond over a 0.01 RIU range.

SPR sensors using multimode optical fibers can have much smaller probe head. Besides this configuration offers the simplest optical train. Moreover, the interchangeable sensor tips are attached with a standard fiber-optic connector. The dynamic range of multimode sensors can be extended as well. However, as it was noted by Wilson et al. (2001), multimode fiber-based SPR spectral dips are neither as sharp nor as deep as prism-based SPR spectra. Multimode fibers are not polarization preserving so the maximum attenuation is 50% due to the constant reflected background from parallel polarized light that does not excite an SPW. As the fibers support a wide range of angles of total internally reflected light, photons of the same wavelength will have a wide range of electric field wave vectors. Wide range serves to broaden features on the observed SPR reflection spectra. The broader features degrade the precision of commonly employed calibration methods that are based on the location of the spectral minimum. Consequently, when these calibration methods are employed, multimode fibers will demonstrate detection limits on an order of magnitude (or more) worse than prism-based SPR sensors. Accurate and precise calibration of multimode SPR sensors is facilitated by employing multivariate calibration methods such as those based on principal components regression.

19.6 SURFACE PLASMON RESONANCE-BASED HUMIDITY SENSORS

Examples of SPR-based humidity sensors are listed in Table 19.1. It is seen that SPR-based humidity sensors were realized using different approaches based on using planar waveguide, prism cuplers, and optical fibers. Typical platforms used for humidity sensor design are shown in Figure 19.5.

Planar waveguide, shown in Figure 19.5a, was formed in borosilicate glass (BK7) substrates by the K^+–Na^+ ion-exchange process (Weiss et al. 1996a). The substrates were cleaned and placed in a bath of pure KNO_3 at 375°C for 3.5 h, forming a 3 μm-deep waveguide. Next, thin films of SiO_2, TiO_2, Au, and SiO_2 were sequentially deposited by electron-beam evaporation. In this device, the lower SiO_2 layer served as a buffer to control the magnitude

TABLE 19.1
SPR-Based Humidity Sensors

Sensing Platform	Sensing Material	Range (% RH)	Response Time	References
SPR (planar waveguide)	Polyimide waveguide	10–90	–	Reuter and Franke (1988)
	Au/SiO_2/Nafion fluoropolyrner film	20–50	–	Weiss et al. (1996a)
	Au/porous silica layer	10–100	–	Sharma and Gupta (2013)
	Ag/PAA and Ag/PVA	25–65		Komaii et al. (2014)
SPR (prism coupler)	Au/Polyethylene glycol film	13–95	–	Bilen et al. (2008)
	Au/ hydrophobic coating layer	–	–	Li et al. (2014)
	Au film	–	–	Li et al. (2014)
SPR (D-shaped fiber)	PVA embedded in the Au grating	0–70		Yan et al. (2017)
LSPR (special cell)	AuNPs linked with myoglobin	0–90	~5 s	Qi et al. (2006)
LSPR (planar)	AuNPs-ceramic composite	–	–	Iwami et al. (2013)
	AgNPs/PNIPAM-C18	44–100		Li et al. (2014)
	AgNPs/PVP–Nafion	10–90		Powell et al. (2016)
LSPR (fiber-optic)	CuNPs with carbon coating	50–85	–	Luechinger et al. (2007)
	AgNPs/PAH–PAA	20–75	~0.5 s	Rivero et al. (2012, 2013)
	AuNPs	20–90	–	Aneesh et al. (2013)

NPs—nanoparticles; PAA—poly(acrylic acid); PAH—poly(allylamine hydrochloride); PNIPAM—poly (*N*-isopropylacrylamide); PVA—polyvinyl alcohol; PVP—polyvinyl pyrrolidine.

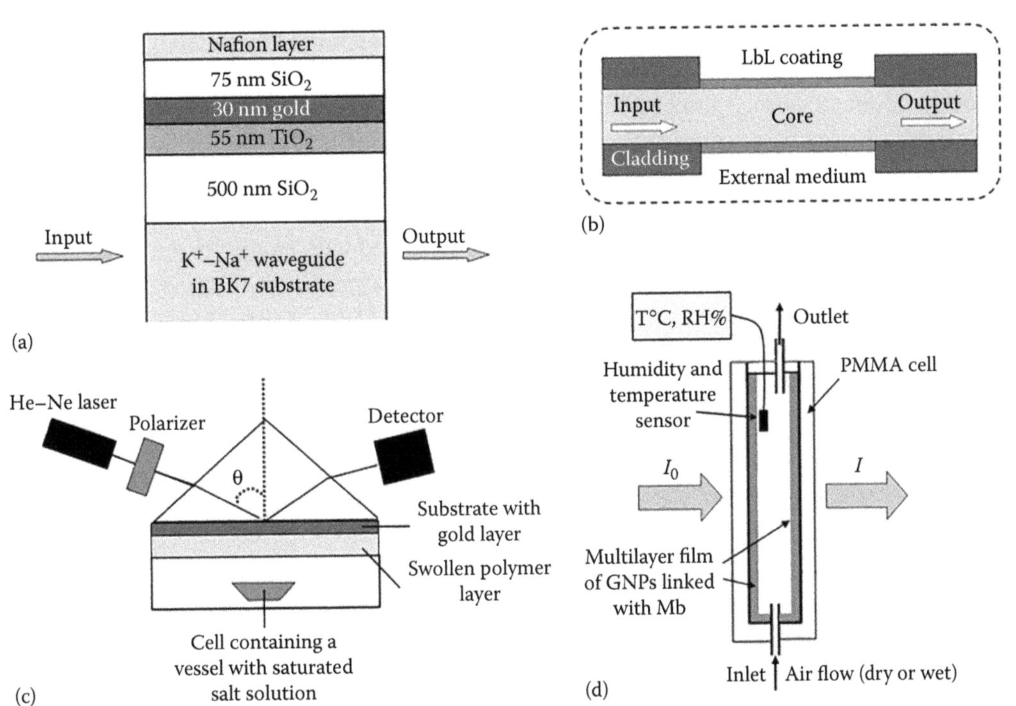

FIGURE 19.5 Variants of platform used for measurements using surface Plasmon resonance: (a) waveguide structure used by Weiss, M.N. and coworkers. (Adapted from *Sens. Actuators A*, 51, Weiss, M.N. et al., A theoretical investigation of environmental monitoring using surface Plasmon resonance waveguide sensors, 211, Copyright 1996, with permission from Elsevier.); (b) Detail of the sensitive region of structure used by Rivero et al. (2012); (c) Typical optical setup used for SPR measurements. (Idea from Bilen, B. et al., *Proc. SPIE*, 7138, 71381G, 2008.); and (d) Detail of experimental setup for measuring the spectral response of the multilayer thin film to RH used by Qi et al. (2006). (Idea from Qi, Z.-M. et al., *Opt. Lett.*, 31, 1854–1856, 2006.)

of the Plasmon-induced losses. The thicknesses of the TiO_2 and upper SiO_2 layers controlled the resonance wavelength of the device. In addition, since the fragile Au layer was not in direct contact with the surrounding environment, this configuration offered a degree of ruggedness seldom found in SPR sensors. For humidity sensing, the device surface was coated by a transducing layer, Nafion fluoropolymer, whose RI is humidity-dependent; the Nafion fluoropolymer swells in the presence of moisture (Grot 1994). A few drops of Nafion fluoropolymer (Aldrich, 1% polymer solution in alcohol) were deposited onto the surface of the SPR waveguide. The solution dried to form a film on the order of 5 μm thick. Variations in the thickness of the Nafion layer did not significantly affect a device performance, because the film thickness was always greater than the penetration depth of the evanescent tail of the Plasmon into the polymer (Weiss et al. 1996b). One should note that BK7 substrates are the most common material used in the manufacture of planar waveguides (Komaii et al. 2014).

When using a prism coupler (Figure 19.5c), the Au layer with thickness 40–50 nm can be applied directly to the surface of the prism or BK7 slides, as it was made by Bilen et al. (2008). The glass prism in this case is made of the same BK7 material with a RI of 1.5151. As the humidity-sensitive material, Bilen et al. (2008) used polyethylene glycol (PEG) with a thickness of ~1 μm. Use of PEG, due to swelling effect, makes it possible to measure the humidity content in the range of 13%–95% RH. However, it should be borne in mind that at about 80% RH in the PEG film the phase changes occur, from a semicrystalline to a physical gel state, accompanied by a strong change in the RI and thickness (Figure 19.6).

Certainly, other materials may be used as substrates. For example, Sharma and Gupta (2013) believe that chalcogenide glass based on the Group VI chalcogen elements (sulfur, selenium, and tellurium) such as 2S2G substrate, is better suited for the development of SPR-based sensors because it has sufficiently higher RI to fulfill the necessary condition $n_c > n_s$ (Bureau et al. 2004). n_c and n_s are RIs of substrate and sensing (dielectric) medium, correspondingly. Moreover, chalcogenide glass is better than other glasses owing to higher thermal and chemical stability (Paivasaari et al. 2007). The presence of buffer layer may reduce the evanescent decay of the SPW, so the thickness of this layer should be minimal. Sharma and Gupta (2013) recommend to use the buffer layer with a thickness of 2–10 nm. But when using substrates with high RI, an efficient coupling of incoming light with the Plasmon-active area can be achieved even at large thicknesses of the buffer layer.

When using a LSPR effect, usually Au and Ag NPs are being incorporated into the polymer matrix, sensitive to humidity. In this case, polymers such as Nafion, polyvinyl alcohol (PVA), poly(acrylic acid) (PAA), poly(allylamine hydrochloride) (PAH), and PEG, are generally used. The interaction with water vapor is accompanied by a homogeneous swelling of the film and subsequent change in Plasmon resonance effects as a result of either increasing spacing between NPs and the change in dielectric constant and index of refraction. This means that the optical resonance properties are sensitive to the change in the ambient humidity. At that, of course, the sensor response will be determined by the rate of diffusion of water vapor into the polymer matrix, that is, it will depend on the thickness and size of the pores in this matrix.

Yan et al. (2017) suggested a different approach to the development of SPR-based humidity sensors. They designed relative humidity sensor based on a D-shaped fiber coated with a polyvinyl alcohol (PVA) embedded in the Au grating (DFPAG) (Figure 19.7). Measurements in a controlled environment showed that the RH sensor can achieve a sensitivity of 0.54 nm/RHU in the RH range from 0% to 70% RH. Moreover, the SPR can be realized

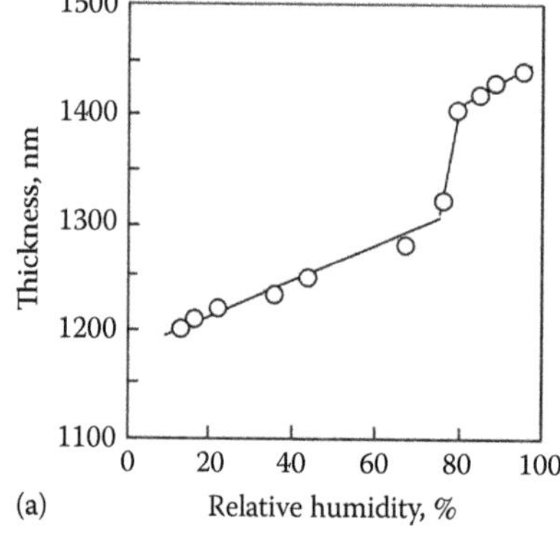

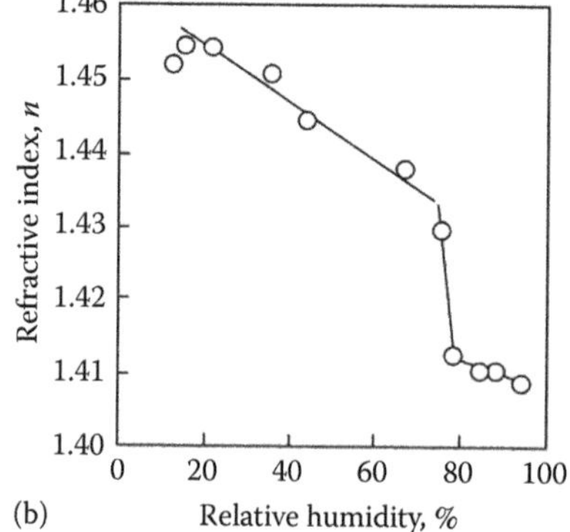

FIGURE 19.6 (a) The thickness and (b) RI of the PEG film as a function of the relative humidity from 13% to 95%. (Data extracted from Bilen, B. et al., *Proc. SPIE*, 7138, 71381G, 2008.)

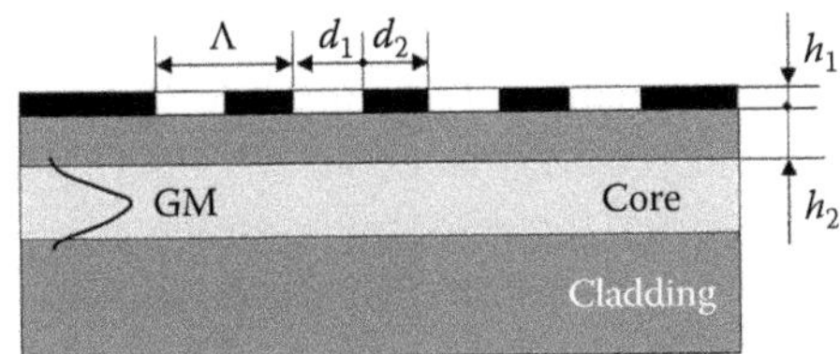

FIGURE 19.7 Schematic diagram of the PVA-embedded metallic grating on a D-shaped fiber surface. (Idea from Yan, H. et al., *J. Nanophoton.*, 11, 016008, 2017.)

and used for RH sensing at the C band of optical fiber communication instead of the visible light band due to the metallic grating microstructure on the D-shaped fiber. The temperature sensitivity of the DFPAG sensor is around 10–40 pm/°C.

The fabrication process for the RH sensor was quite simple and can be described as follows (Yan et al. 2017). A D-shaped fiber was obtained by side grinding and polishing a standard single-mode silica optical fiber (SMF-28, 125 μm in diameter) until we get a polished surface, which is ~10 μm away from the core. The fiber was then coated with a 20-nm thick Au layer on the polished surface by nonmagnetic sputtering. A metallic grating was fabricated by focused ion beam (FIB) milling. Using the phase matching condition, the designed grating period Λ ($\sim\lambda_R/\Delta n_{eff}$) should be around 500 nm, when the exciting SPP wavelength λ_R was in the 1550-nm telecommunication band. The total grating length was only 30 μm. The PVA film was created by the direct dip-coating of PVA onto the Au layer. The response wavelength was influenced by the duty cycle of the grating, but the effect was small compared with the period parameter.

Qi et al. (2006) have shown that as a humidity-sensitive material one can also use a protein such as myoglobin (Mb) applied as a linking agent for AuNPs. Mb is a small molecule with a size of 2.5 × 3.5 × 4.5 nm, making a NP lie very close to its neighbors in the multilayer thin film of AuNPs linked with Mb. Therefore, the interparticle interaction occurs in the multilayer thin film and affects the overall optical properties of the film. With the alpha helical secondary structure, Mb molecules can change their size with changes in the ambient humidity, despite their native or denatured form. A change of the molecular size of Mb would induce a modification of the interparticle spacing that changes the interparticle interaction in the multilayer thin film. AuNPs–ceramic composites can also be used for LSPR humidity sensors design (Iwami et al. 2013). AuNPs were embedded in the pores of ceramics. The average sizes of the NPs and pores were 37.3 and 8.9 nm in diameter, respectively.

Luechinger et al. (2007) have established that LSPR humidity sensors based on the swelling effect could be also realized using polymer free films. Luechinger et al. (2007) found that swelling effect in humid atmosphere was observed for CuNPs covered by carbon. Exposure to controlled atmospheres containing increasing amounts of water vapor demonstrated highly sensitive coloration with spectral shifts of over 50 nm per% RH around 70%–80% RH (Figure 19.8).

Luechinger et al. (2007) and Iwami et al. (2013) correctly pointed out that the most dramatic changes in absorption spectra occur during capillary condensation of water vapor in the micropores of humidity-sensitive material. If this did not happen, the shift of the absorption peak is very slight. This means that the LSPR humidity sensors are best suited to determine the dew point condensation. This option of the use of LSPR-based humidity sensors was proposed by Umeda and coworkers (Numata et al. 2006; Nagasaki et al. 2012; Iwami et al. 2013) and Esmaeilzadeh et al. (2015). Figure 19.9 shows the

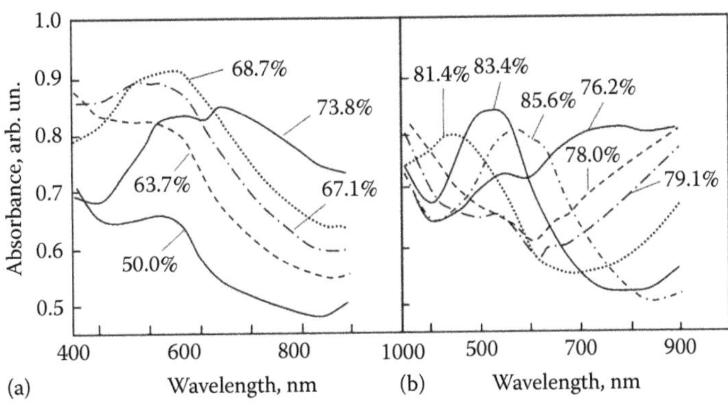

FIGURE 19.8 UV–vis spectra of nanoporous C/Cu films exposed to different relative humidity: (a) starting from 50% to 73.8% RH, and (b) starting from 76.2 to 85.6% RH. Spectra are given without any correction or scaling and illustrate the intense change in coloration for small increases in humidity. (Reprinted with permission from Luechinger, N.A. et al., *Langmuir*, 23, 3473, 2007. Copyright 2007 American Chemical Society.)

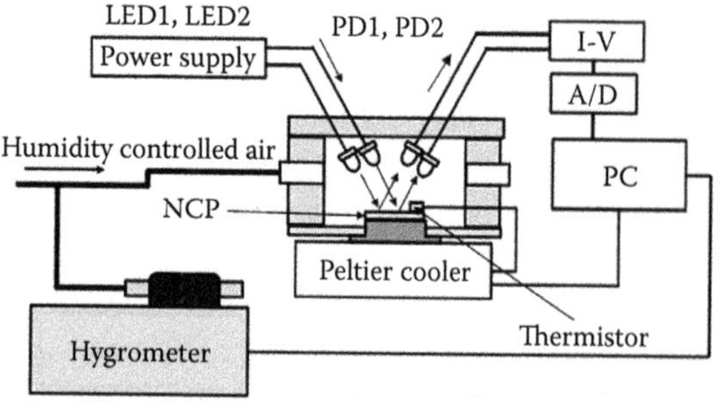

FIGURE 19.9 Schematic of the experimental setup with two-wavelengths readout. (Reprinted from *Sens. Actuators B*, 184, Iwami, K. et al., Plasmon-resonance dew condensation sensor made of gold-ceramic nanocomposite and its application in condensation prevention, 301, Copyright 2013, with permission from Elsevier.)

experimental setup used by Umeda's group for these purposes. The AuNPs–ceramic film was formed on a glass substrate. The thickness of the glass substrate and the AuNPs–ceramic film were 0.7 mm and 1.8 μm, respectively. An Ag/Ni/Cr layer was created on the back side of the glass substrate as a reflection layer. Figure 19.10 shows the peak wavelength as a function of the sensor temperature. In all measurements, the chamber temperature was maintained at 25°C, and the dew point of supplied air was varied at 5.0°C, 7.3°C, 11.5°C, 14.7°C, 17.5°C, and 21.9°C. The solid lines indicate absorption peak wavelengths determined by pseudo-Voigt fitting to

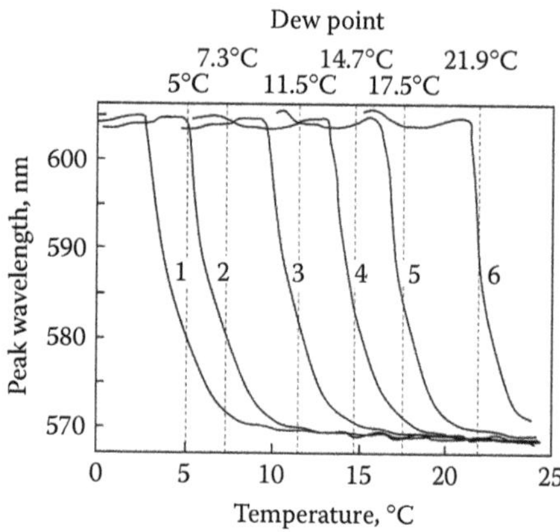

FIGURE 19.10 Peak wavelength as a function of the sensor temperature. In all measurements, the chamber temperature was maintained at 25°C, and the dew point of supplied air was varied at 1°C–5.0°C, 2°C–7.3°C, 3°C–11.5°C, 4°C–14.7°C, 5°C–17.5°C, and 6°C–21.9°C. The solid and dashed lines indicate absorption peak wavelengths and each dew point, respectively. (Reprinted from *Sens. Actuators B*, 184, Iwami, K. et al., Plasmon-resonance dew condensation sensor made of gold-ceramic nanocomposite and its application in condensation prevention, 301, Copyright 2013, with permission from Elsevier.)

the absorption spectra. The dashed lines indicate each dew point. As shown in this graph, the AuNPs–ceramic sensor starts to respond at about 4°C–5°C higher than each dew point, and the response saturates when the peak wavelength reaches 605 nm. The plots crossed their dew points at wavelengths ranging from 580 to 590 nm. Of course, such sensors require calibration because the dew point of the microporous ceramics depends on the pore size. When the pore size is smaller, the capillary condensation of water vapor in these pores occurs with less air humidity. This means that sensors based on microporous ceramics respond to dew condensation at temperatures higher than the real dew point.

19.7 SUMMARY

Comparison of the considered sensors indicates that the simplest variant of SPR-based humidity sensor is sensor, using the LSPR effect. This version is most suitable for practical applications. The use of a prism sometimes prevents the miniaturization of the sample and is not always desirable. Grating coupling does not require any prisms but the preparation of a grating structure on the desired substrate surface is sometimes difficult. The waveguide-based SP excitation is a quite simple and reasonable method. However, the measurements for the waveguide array need complicated setup.

As a rule, in the LSPR-based humidity sensors it is used a measuring setup shown in Figure 19.11. This configuration basically consists of conventional transmission optical setup with a white halogen lamp, connected to one end of the optical fiber, and a CCD-based

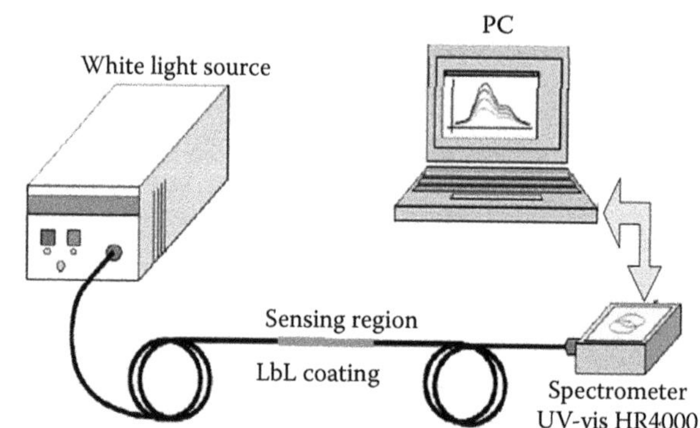

FIGURE 19.11 Schematic representation of the measuring setup Rivero, P.J. and coworkers. (Reprinted from *Sens. Actuators B*, 173, Rivero, P.J. et al., Optical fiber humidity sensors based on localized surface Plasmon resonance (LSPR) and Lossy-mode resonance (LMR) in overlays loaded with silver nanoparticles, 244, Copyright 2012, with permission from Elsevier.)

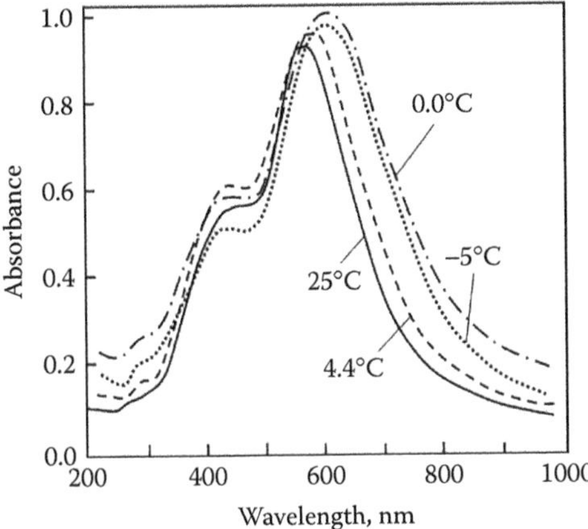

FIGURE 19.12 Absorbance spectra of the AuNCP sensor at the temperatures of 25°C, 4.4°C, 0°C, and −5°C in the chamber with the temperature maintained at 25°C and dew point of 4.47°C. (Reprinted from *Sens. Actuators B*, 184, Iwami, K. et al., Plasmon-resonance dew condensation sensor made of gold-ceramic nanocomposite and its application in condensation prevention, 301, Copyright 2013, with permission from Elsevier.)

ultraviolet-visible spectroscopy (UV–vis) spectrometer used for absorbance spectra measurements, connected to the other end of the fiber.

However, despite the simplicity of the LSPR-based sensor configuration, it must be recognized that given sensors have a significant shortcoming. As previously noted, the parameters of the sensors strongly depend on the size and shape of the NPs, as well as on the NP dispersion in the polymer matrix (Deeb et al. 2013; Powell et al. 2016).

This means that it is very difficult to achieve the desired reproducibility of the sensor parameters. The need to measure the spectral characteristics and to analyze them by PC is also introducing some limitations when using these sensors for applications. We must not forget that the temperature stabilization is also necessary for the correct operation of LSPR-based humidity sensors (Iwami et al. 2013; Li et al. 2014).

The effect of temperature can be very strong, which is shown in Figure 19.12. Just as for the other types of sensors, the readings of LSPR-based humidity sensors may depend on interference effects (Komaii et al. 2014). VOCs can exert the strongest influence, because they, as well as water vapors, can condense in the pores.

REFERENCES

Abdelghani A., Jaffrezic-Renault N. (2001) SPR fibre sensor sensitized by fluorosiloxane polymers. *Sens. Actuators B* 74, 117–123.

Aneesh R., Khijwania S. (2013) Comprehensive experimental investigation of sensitivity enhancement for optical fiber humidity sensor employing localized surface Plasmon resonance spectroscopy. *Proc. SPIE* 8794, 87941M-1.

Bardin F., Kašik I., Trouillet A., Matějec A., Gagnaire H., Chomát M. (2002) Surface plasmon resonance sensor using an optical fiber with an inverted graded-index profile. *Appl. Opt.* 41, 2514–2520.

Bilen B., Skarlatos Y., Aktas G., Inci M.N., Dispinar T., Kose M.M., Sanyal A. (2008) In situ measurement of humidity induced changes in the refractive index and thickness of polyethylene glycol thin films. *Proc. SPIE* 7138, 71381G.

Bureau B., Zhang X.H., Smektala F. et al. (2004) Recent advances in chalcogenide glasses. *J. Non-Cryst. Solids* 345–346, 276–283.

Deeb C., Zhou X., Plain J., Wiederrecht G.P., Bachelot R., Russell M., Jain P.K. (2013) Size dependence of the Plasmonic near-field measured via single-nanoparticle photoimaging. *J. Phys. Chem. C* 117, 10669–10676.

De Maria L., Martinelli M., Vegetti G. (1993) Fiber-optic sensor based on surface Plasmon interrogation. *Sens. Actuators B* 12, 221–223.

Eah S.-K., Jaeger H.M., Scherer N.F., Wiederrecht G.P., Lin X.-M. (2005) Plasmon scattering from a single gold nanoparticle collected through an optical fiber. *Appl. Phys. Lett.* 86(3), 031902.

Esmaeilzadeh H., Rivard M., Arzi E., Légaré F., Hassani A. (2015) Smart textile plasmonic fiber dew sensors. *Opt. Exp.* 23(11), 14981–14992.

Grot W.G. (1994) Perfluorinated ion exchange polymers and their use in research and industry. *Macrornol. Symp.* 82, 175–184.

Gupta B.D., Verma R.K. (2009) Surface plasmon resonance-based fiber optic sensors: Principle, probe designs, and some applications. *J. Sens.* 2009, 979761.

Homola J. (1995) Optical-fiber sensor-based on surface-Plasmon excitation. *Sens. Actuators B* 29, 401–405.

Homola J. (Ed.) (2006) *Surface Plasmon Resonance Based Sensors*. Springer-Verlag, Berlin, Germany.

Homola J. (2008) Surface plasmon resonance sensors for detection of chemical and biological species. *Chem. Rev.* 108(2), 462–493.

Homola J., Koudela I., Yee S.S. (1996) Surface plasmon resonance sensors based on diffraction gratings and prism couplers: sensitivity comparison. *Sen. Actuator B* 54, 16–24.

Homola J., Yee S.S., Gauglitz G. (1999) Surface plasmon resonance sensors: Review. *Sens. Actuators B* 54, 3–15.

Iwami K., Kaneko S., Shinta R., Fujihara J., Nagasaki H., Matsumura Y., Umeda N. (2013) Plasmon-resonance dew condensation sensor made of gold-ceramic nanocomposite and its application in condensation prevention. *Sens. Actuators B* 184, 301–305.

Jain P.K., Lee K.S., El-Sayed I.H., El-Sayed M.A. (2006) Calculated absorption and scattering properties of gold nanoparticles of different size, shape, and composition: Applications in biological imaging and biomedicine. *J. Phys. Chem. B* 110, 7238–7248.

Jorgenson R.C., Yee S.S. (1993) A fiber-optic chemical sensor based on surface Plasmon resonance. *Sens. Actuators B* 12, 213–220.

Kabashin A.V., Patskovsky S., Grigorenko A.N. (2009) Phase and amplitude sensitivities in surface plasmon resonance bio and chemical sensing. *Opt. Exp.* 17(23), 21191–21204.

Karlsen S.R., Johnston K.S., Jorgenson R.C., Yee S.S. (1995) Simultaneous determination of refractive indices and absorbance spectra of chemical samples using surface plasmon resonance. *Sens. Actuators B* 24–25, 747–749.

Kashyap R., Nemova G. (2009) Surface plasmon resonance-based fiber and planar waveguide sensors. *J. Sens.* 2009, 645162.

Komaii R., Honda H., Baba A., Shinbo K., Kato K., Kaneko F. (2014) Simultaneous detection of ammonia and water vapors using surface Plasmon resonance waveguide sensor. In: *Proceedings of International Symposium on Electrical Insulating Materials (ISEIM 2014)*, Niigata, Japan, pp. 284–286, June 1–5.

Lee B., Roh S., Park J. (2009) Current status of micro- and nano-structured optical fiber sensors. *Optic. Fiber Technol.* 15, 209–221.

Li X., Li X., Wang C. (2014) A new method for measuring wetness of flowing steam based on surface plasmon resonance. *Nanoscale Res. Lett.* 9, 18.

Liz-Marzan L.M. (2006) Tailoring surface Plasmons through the morphology and assembly of metal nanoparticles. *Langmuir* 22, 32–41.

Lu X.M., Rycenga M., Skrabalak S.E., Wiley B., Xia Y.N. (2009) Chemical synthesis of novel plasmonic nanoparticles. *Annu. Rev. Phys. Chem.* 60, 167–192.

Luechinger N.A., Loher S., Athanassiou E.K., Grass R.N., Stark W.J. (2007) Highly sensitive optical detection of humidity on polymer/metal nanoparticle hybrid films. *Langmuir* 23, 3473–3477.

Maier S.A. (2007) *Plasmonics: Fundamentals and Applications*. Springer, New York.

Nagasaki H., Kaneko S., Iwami K., Umeda N. (2012) Localized surface plasmon resonance dew sensor for use under low humidity conditions. *Jpn. J. Appl. Phys.* 51, 1–4.

Numata T., Otani Y., Umeda N. (2006) Optical dew sensor using surface plasmon resonance of periodic Ag nanostructure. *Jpn. J. Appl. Phys.* 45, L810–L813.

Qi Z.-M., Honma I., Zhou H. (2006) Humidity sensor based on localized surface plasmon resonance of multilayer thin films of gold nanoparticles linked with myoglobin. *Opt. Lett.* 31(12), 1854–1856.

Paivasaari K., Tikhomirov V.K., Turunen J. (2007) High refractive index chalco-genide glasses for photonic crystal application. *Opt. Exp.* 15, 2336–2340.

Petryayeva E., Krull U.J. (2011) Localized surface plasmon resonance: Nanostructures, bioassays and biosensing—A review. *Anal. Chim. Acta* 706, 8–24.

Piliarik M., Homola J. (2009) Surface plasmon resonance (SPR) sensors: Approaching their limits? *Opt. Exp.* 17(19), 16505–16517.

Pockrand I., Swalen J.D., Gordon II JG., Philpott M.R. (1979) Exciton-surface plasmon interactions. *J. Chem. Phys.* 70, 3401.

Powell A.W., Coles D.M., Taylor R.A., Watt A.A.R., Assender H.E., Smith J.M. (2016) Plasmonic gas sensing using nanocube patch antennas. *Adv. Opt. Mater.* 4, 634–642.

Reuter R., Franke H. (1988) Monitoring humidity by polyimide lightguides. *Appl. Phys. Lett.* 52, 778–779.

Rhodes C., Cerruti M., Efremenko A., Losego M., Aspnes D.E., Maria J., Franzen S. (2008) Dependence of plasmon polaritons on the thickness of indium tin oxide thin films. *J. Appl. Phys.* 103, 093108.

Rivero P.J., Urrutia A., Goicoechea J., Arregui F.J. (2012) Optical fiber humidity sensors based on localized surface plasmon resonance (LSPR) and Lossy-mode resonance (LMR) in overlays loaded with silver nanoparticles. *Sens. Actuators B* 173, 244–249.

Rivero P.J., Urrutia A., Goicoechea J., Matias I.R., Arregui F.J. (2013) A Lossy Mode Resonance optical sensor using silver nanoparticles-loaded films for monitoring human breathing. *Sens. Actuators B* 187, 40–44.

Sharma A.K., Gupta B.D. (2007) On the performance of different bimetallic combinations in surface plasmon resonance based fiber optic sensors. *J. Appl. Phys.* 101(9), 093111.

Sharma A.K., Gupta A. (2013) Design of a plasmonic optical sensor probe for humidity-monitoring. *Sens. Actuators B* 188, 867–871.

Skorobogatiy M. (2009) Microstructured and photonic bandgap fibers for applications in the resonant bio- and chemical sensors. *J. Sens.* 2009, 524237.

Stewart M.E., Anderton C.R., Thompson L.B., Maria J., Gray S.K., Rogers J.A., Nuzzo R.G. (2008) Nanostructured plasmonic sensors. *Chem. Rev.* 108, 494–521.

Weiss M.N., Srivastava R., Groger H. (1996a) Experimental investigation of a surface plasmon-based integrated-optic humidity sensor. *Electron. Lett.* 32(9), 842–843.

Weiss M.N., Srivastava R., Groger H., Lo P., Luo S. (1996b) A theoretical investigation of environmental monitoring using surface plasmon resonance waveguide sensors. *Sens. Actuators A* 51, 211–217.

Willets K.A., Van Duyne R.P. (2007) Localized surface plasmon resonance spectroscopy and sensing. *Ann. Rev. Phys. Chem.* 58, 267–297.

Wilson D.M., Hoyt S., Janata J., Booksh K., Obando L. (2001) Chemical sensors for portable, handheld field instruments. *IEEE Sens. J.* 1(4), 256–274.

Yan H., Han D., Li M., Lin B. (2017) Relative humidity sensor based on surface plasmon resonance of D-shaped fiber with polyvinyl alcohol embedding Au grating. *J. Nanophoton.* 11(1), 016008.

Yang D.F., Lu H.H., Chen B., Lin C.W. (2010) Surface Plasmon resonance of SnO_2/Au bi-layer films for gas sensing applications. *Sens. Actuators B* 145, 832–838.

Zynio S.A., Samoylov A.V., Surovtseva E.R., Mirsky V.M., Shirshov Y.M. (2002) Bimetallic layers increase sensitivity of affinity sensors based on surface Plasmon resonance. *Sensors* 2(2), 62–70.

20 Lossy-Mode Resonance-Based Humidity Sensors

20.1 LOSSY-MODE RESONANCE

As it was shown before, standard surface Plasmon resonance (SPR)-based sensors used a complex optical setup, including a prism, to excite the surface Plasmon wave. This set up presents some important drawbacks, such as its big size and the presence of fragile mechanical parts. The optical fiber configuration, discussed in Chapter 19, overcomes these disadvantages and allows the miniaturization of these devices, adding these features to the typical advantages of optical fiber sensors. However, even fiber-type SPR-based devices have some limitation. First, there are just a few stable metals that allow the generation of SPR in the visible and near-infrared spectral regions, such as gold or silver. In addition, the SPR is only visible with TM polarization of light. The reason for this is that the SPR can be understood as conductivity fluctuations of collective surface density charge oscillations. These charge densities have to be excited by an external electric field. TM-polarized light is the only polarization that contains an electric field component perpendicular to the silica—thin-film interface. Therefore, the TM-polarizing is the only way to obtain these charge densities and, consequently, to excite the SPRs (Schasfoort and Tudos 2008). Thus, the light must be previously polarized, leading to an increase in the final device cost. In addition, the conditions for a SPR generation are very specific and for this reason, the SPR is produced in a limited spectral region that depends on the deposition metal.

Recently, it was established that the using of metal oxides such as Indium Tin oxide (ITO), Titanium dioxide (TiO_2), or Indium oxide (In_2O_3) instead of noble metals, allows realization of devices using resonance of a different nature, lossy-mode resonance (LMR). The description of the nature of this effect one can find in Del Villar et al. (2012, 2017), Arregui et al. (2014), Kaur et al. (2014), and Corres et al. (2015). This LMR is a type of resonance that overcomes some of the limitations of SPRs and never has been observed before in an optical fiber configuration. An appearance of attenuation maxima of the light propagating through the optical waveguide can be obtained for specific thickness values and at certain wavelengths or incidence angles (Del Villar et al. 2012). This is due to a coupling between waveguide modes and a particular lossy mode of the semiconductor thin-film (Marciniak et al. 1993). Since the phenomenon occurs when the lossy mode of the thin-film is near cutoff, there are thin-film thickness values that lead to transmission attenuation maxima (Batchman and McWright 1982). The same phenomenon can be observed if the variable was the wavelength, but not the thickness. If the thin-film thickness is fixed, a resonance will be visible in the electromagnetic spectrum for those incident wavelength values, where there is a mode near the cutoff in the overlay. At that, one should note that setups for LMR and SPR generation are the same (Figure 20.1). Moreover, LMRs and SPRs have similarity in the shapes of optical spectrum. Typical LMR spectrum for TiO_2 film is shown in Figure 20.2.

Conditions of the LMR appearance are shown in Figure 20.3 and Table 20.1. It is seen that the conditions are very different from the SPR generation (Del Villar et al. 2017). SPRs are obtained when the real part of the thin-film permittivity is negative and higher in magnitude than both its own imaginary part and the permittivity of the material surrounding the thin-film, whereas LMRs occur when the real part of the thin-film permittivity is positive and higher in magnitude than both its own imaginary part and the material, surrounding the thin-film. According to this, the expressions for the SPR and LMR generation as function of the permittivity ($\varepsilon = \varepsilon' + i\varepsilon'$) and the refractive index (RI) (n is the real part and k the imaginary part) are presented in Figure 20.3 under the assumption of a substrate RI of 1.45 (with small variations, this is the value of silica in a broad range of the optical spectrum [Malitson 1965]) with a surrounding medium refractive index (SRMI) of 1 (air). Also, in Figure 20.3 a map is plotted containing the regions where LMRs and SPRs can be obtained. As long as the real part of the thin-film permittivity is positive and the absorption coefficient (k) is low, many materials except pure metals (typical of SPRs) can induce LMRs. For instance, several contributions using metal oxides or polymer coatings (Del Villar et al. 2012), two-layer-coated structures combining both materials (Zamarreno et al. 2010), or even immune-sensors consisting of multilayers of polymers

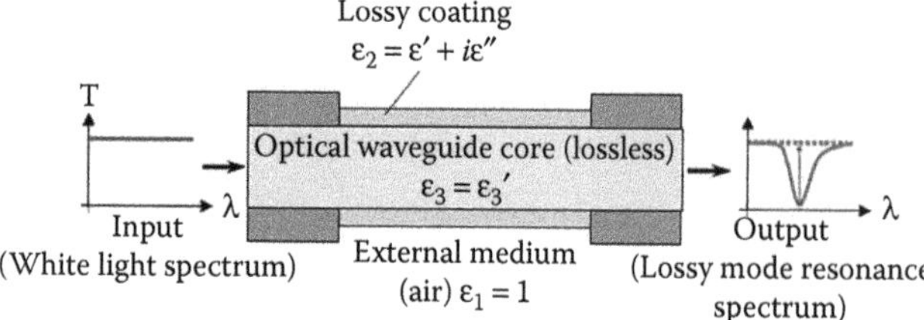

FIGURE 20.1 Schematic of the waveguide coated with a thin-film of an optical absorbing material that fulfills the conditions to generate LMRs. (From Arregui, F.J. et al., *Proc. Eng.*, 87, 3, 2014. Published by Elsevier as open access.)

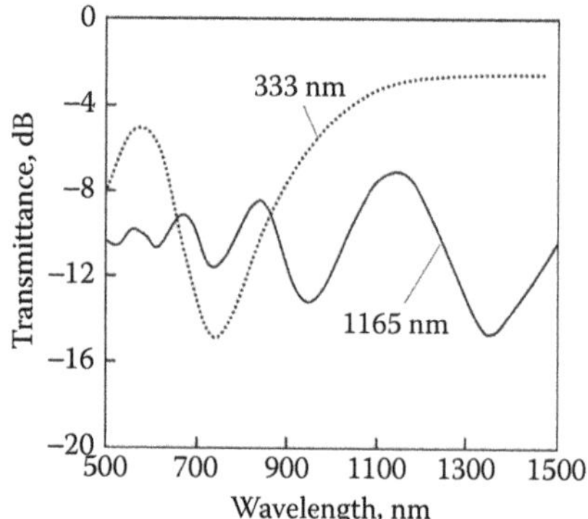

FIGURE 20.2 Experimental transmission spectra for two different thickness values (333 and 1165 nm) of TiO_2/PSS-coated CRMOF (the surrounding medium refractive index is 1.321). (Reprinted from *Sens. Actuators B*, 240, Del Villar, I. et al., Optical sensors based on lossy-mode resonances, 174–185, Copyright 2017, with permission from Elsevier.)

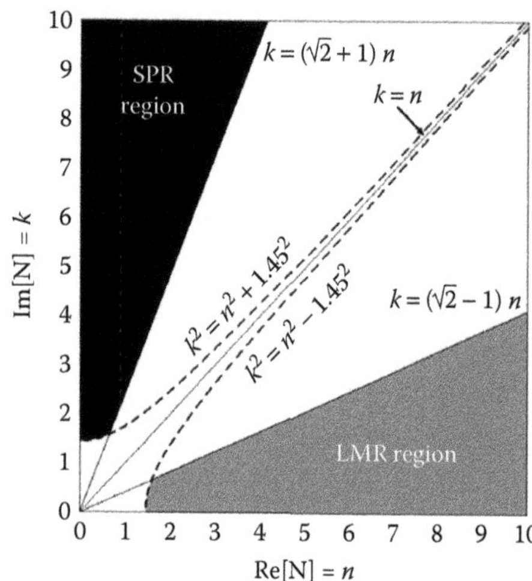

FIGURE 20.3 Conditions for LMR and SPR generation. (Reprinted from *Sens. Actuators B*, 240, Del Villar, I. et al., Optical sensors based on lossy-mode resonances, 174–185, Copyright 2017, with permission from Elsevier.)

TABLE 20.1
Conditions for LMR and SPR Generation

Types of Resonance	ε Conditions	*n*, *k* Conditions ($n > 0, k > 0$)
SPR (Surface Plasmon Resonance)	$\varepsilon_2' < 0$ $\lvert\varepsilon_2'\rvert > \lvert\varepsilon_2''\rvert$ $\lvert\varepsilon_2'\rvert > \lvert\varepsilon_3'\rvert$	$k_2 > (\sqrt{2}+1)n_2$ $k_2^2 - n_2^2 > 1.45^2$
LMR (Lossy-Mode Resonance)	$\varepsilon_2' > 0$ $\lvert\varepsilon_2'\rvert > \lvert\varepsilon_2''\rvert$ $\lvert\varepsilon_2'\rvert > \lvert\varepsilon_3'\rvert$	$k_2 < (\sqrt{2}-1)n_2$ $n_2^2 - k_2^2 > 1.45^2$

Source: Data extracted from Del Villar, I. et al., *Sens. Actuators B*, 240, 174–185, 2017.

and antibodies (Socorro et al. 2014) have successfully produced LMRs so far.

As the main advantages of LMR-based sensors in comparison with the SPR-based devices one needs to allocate the followings (Arregui et al. 2014):

- To produce LMR-based sensors, noble metals are not needed. The possibility to use transparent metal oxides with high chemical and thermal stability offers new technological and measurement opportunities in sensing applications.
- When the material is properly selected at the LMR measurement, it is possible to refuse the usage of an optical polarizer. This simplifies enormously the fabrication of the sensor or even the experimental setup of the optical devices, where only an incandescent light source and an optical spectrometer are necessary to make the experiments.
- The spectral position of LMRs can be fine-tuned just by changing the thickness of the lossy coating. Even more, instead of having a unique optical resonance, several resonances can appear when the thickness or the lossy coating is increased and all these peaks can be used for sensing or other applications.

20.2 FEATURES OF LOSSY-MODE RESONANCE-BASED SENSOR DESIGN

The first step in order to design an LMR-based sensor is to choose an adequate material for the thin-film (Del Villar et al. 2017). Metallic oxides and polymers are among the materials that permit the generation of LMRs. However, the metal oxide group is more adequate than the polymer one because the material RI is typically higher. This means that a higher sensitivity can be achieved (Del Villar et al. 2012). In addition, metallic oxides present a much flatter surface, which permits the development of many LMR refractometers based on ITO (Del Villar et al. 2010a, 2010c), In_2O_3 (Del Villar et al. 2010c), TiO_2/PSS (Del Villar et al. 2012), and SnO_2 (Sanchez et al. 2014) coatings. Among these, ITO is the only material that enables the generation of LMRs and SPRs at shorter and longer wavelengths, respectively.

With regard to the optimization of the LMR-based sensor performance, then besides the imaginary part of the thin-film RI, three other parameters must also be considered: (1) the real part of the RI, (2) the thin-film thickness, and (3) the SRMI (Del Villar et al. 2012). In particular, an increase in the film thickness and in the external medium RI leads to an increase in sensitivity and the red-shift of the LMR wavelength (Figures 20.4 through 20.6). It is important to note that all studied metal oxides satisfy the conditions for LMR generation in the entire optical spectrum (typically 400–1500 nm) and follow the rules indicated by Del Villar et al. (2012): the thin-film RI and SRMI must be increased to obtain a higher sensitivity (Figures 20.5 and 20.6). For example, TiO_2, with a RI of 1.97–1.74 in the wavelength range 400–1700 nm, permits one to obtain a sensitivity of 2872 nm/RIU in the RI range of 1.321 (water) to 1.43 (close to silica index). In_2O_3, with a RI of 2.09–1.84 in the wavelength range 400–1700 nm, permits one to obtain a sensitivity of 4068 nm/RIU (Del Villar et al. 2010c).

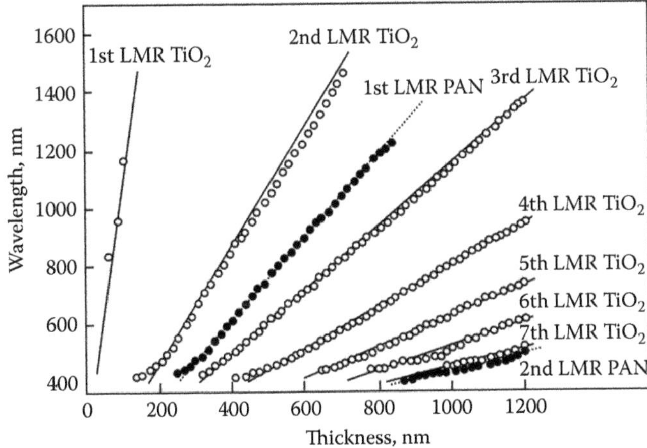

FIGURE 20.4 LMR wavelength as a function of coating thickness for two different materials: PAH/PAA and TiO_2/PSS. PAH—poly(allylamine hydrochloride), PAA—and poly(acrylic acid), PSS—poly(sodium 4-styrenesulfonate). The SMRI is 1.321 (water). Simulation data, continuous line; experimental data, circles. (Reprinted from *Sens. Actuators B*, 240, Del Villar, I. et al., Optical sensors based on lossy-mode resonances, 174–185, Copyright 2017, with permission from Elsevier.)

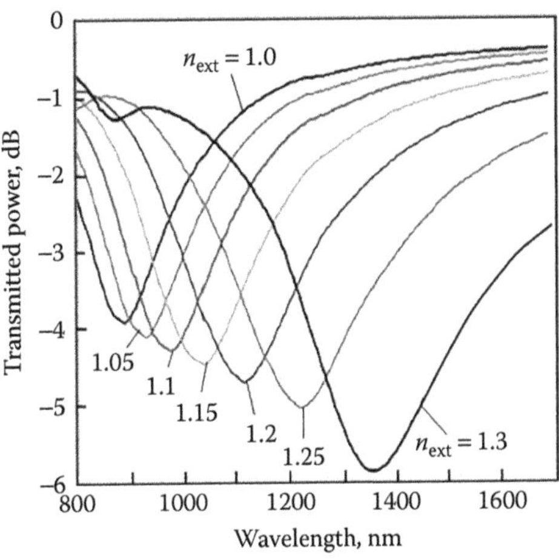

FIGURE 20.5 Response of the LMR In_2O_3-based fiber-optic sensor to changes in the external medium refractive index between 1 and 1.3. In_2O_3 was deposited onto the uncladded core of a plastic cladding optical fiber (FT200EMT, Thorlabs, Inc.). (From Corres, J.M. et al., *Opt. Lett.*, 40, 4867, 2015. With permission of Optical Society of America.)

SnO_2, with higher RI than TiO_2 and In_2O_3, permits one to obtain a sensitivity of 5390 nm/RIU (Sanchez et al. 2014). Analysis showed that with Al_2O_3 and an adiabatic tapered single-mode fiber (ATSMF) structure it was possible to attain 6008 nm/RIU (Zhu et al. 2015).

Earlier it was noted that an increase in the film thickness is necessary to achieve higher sensitivity. However, this approach to increasing the sensor sensitivity has limitations. The detection method is based on tracking the central wavelength of the LMR. Since the LMR peak is progressively red-shifted as a function of the RI and the thickness of the thin-film, there is a moment when the LMR can no longer be monitored due to the limited spectral range of the spectrum analyzer. For example, due to ITO's special characteristics, the general rule of increasing the thickness for a higher sensitivity is not applicable when the LMR band approaches the SPR region (Del Villar et al. 2017).

It was also established that the limit wavelength, which depends on the RI dispersion, can be controlled with the thin-film fabrication conditions (Del Villar et al. 2015). For instance, with DC sputtering and a post-annealing it was possible to obtain ITO-coated LMR-based sensors with a limit wavelength of 800 nm (Lopez et al. 2012), whereas with dip-coating technique this value was shifted to 1400 nm (Del Villar et al. 2010a).

The selection of optimum angle for exciting LMRs is also based on the influence of two parameters: (1) the RI and (2) the thickness. Analysis of the results obtained has shown that the best angle range for SPR excitation was 40°–75° (Homola 2006; Rhodes et al. 2006), whereas for

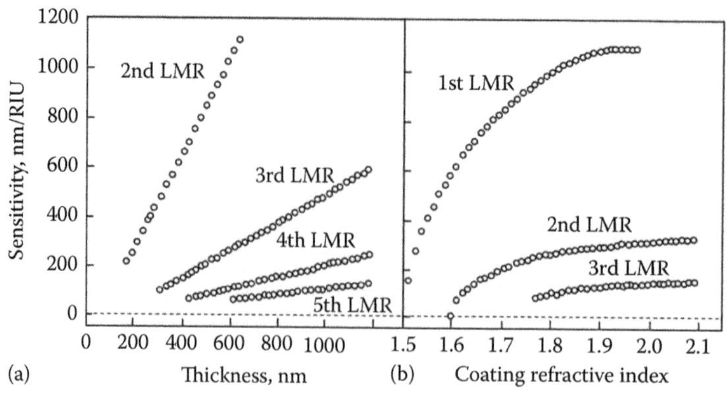

FIGURE 20.6 Results of simulation of SSMRI (sensitivity to variations of SMRI expressed as wavelength shift versus refractive index variation nm/RIU) as a function of (a) coating thickness (the coating refractive index is that of TiO_2/PSS film) and (b) coating refractive index (the coating thickness is 600 nm). (From Del Villar, I. et al., *Appl. Opt.*, 51, 4298, 2012. With permission of Optical Society of America.)

LMRs the optimum angles approached 90° (Del Villar et al. 2015). Even though it is possible to move away from 90° by modifying the RI (increasing the imaginary part) or changing the thickness, it is easier to obtain the LMR for angles approaching 90° (Del Villar et al. 2010b). That is why the simplest and the most effective structure for the LMR generation consists of the optical fiber configuration, where the incident angles approach 90°.

Large enough spectral width of LMRs is a shortcoming of LMR-based sensors (Del Villar et al. 2017). It was shown that to solve this issue, optimized optical fiber waveguides can be used such as ATSMF (Socorro et al. 2014) and D-shaped fiber (Zubiate et al. 2015). The application of D-shaped fiber has an additional advantage, which is related to the guidance of modes in the thin-film in pairs.

20.3 LOSSY-MODE RESONANCE-BASED HUMIDITY SENSORS

LMR-based sensors used for measuring the humidity are listed in Table 20.2. As expected, the humidity sensors have been developed mainly on the basis of conductive oxides such as ITO, In_2O_3, SnO_2, and TiO_2. The coating

TABLE 20.2
LMR-Based Humidity Sensors

Sensing Material	Range (% RH)	Spectral Range, nm	Maximum Sensitivity, nm/% RH	References
In_2O_3 uncoated	30–80	1250–1330	0.2	Ascorbe et al. (2015)
SnO_2 uncoated	20–90	1520–1680	1.9	Sanchez et al. (2014), Ascorbe et al. (2016)
rGO-coated hollow core fiber	30–90	1495–1500	0.11	Gao et al. (2016)
PAH–PAA	20–80	650–750	0.56	Sanchez et al. (2013)
ITO/PAH–PAA	20–80	1000–1100	1.08	Hernáez et al. (2009),
		1400–1500	5.4	Zamarreno et al. (2010)
In_2O_3/PAH–PAA	20–80	1450–1510	0.93	Sanchez et al. (2012)
ITO/Agarose	20–80	1350–1410	0.75	Hernáez et al. (2010)
TiO_2/PSS	20–90	1050–1250	1.43	Zamarreno et al. (2011),
		900–1150	3.54	Socorro et al. (2015, 2016)
AgNPs/PAH–PAA	23–75	740–810	0.94	Rivero et al. (2012, 2013)
		1140–1180	0.455	

ITO—Indium Tin Oxide; PAH—poly(allylamine hydrochloride); PAA—poly(acrylic acid); PSS—poly(sodium 4-styrenesulfonate); rGO—reduced graphene oxide.

thickness varied in the range of 100–200 nm, which ensured the work of sensors in the spectral range of 700–1600 nm. For these metal oxides deposition on the surface of the fibers, it was usually used a sol–gel dip-coating method, previously described by Ota et al. (2002). Just Ascorbe et al. (2015) have used a pulsed DC-sputtering for deposition a layer of In_2O_3.

As optical fibers, in most sensors they were used the polymer-clad silica fiber (PCS) optical fibers (Thorlabs Inc.) with 200 μm core diameter. The preparation of the optical fibers before metal oxide deposition consisted of removing the buffer, the plastic cladding, and cleaning the fibers in an ultrasonic bath with detergent, ultrapure deionized (DI) water, and acetone, consecutively. The deposition process was repeated up to 10 layers with a 0.5–1 h annealing process at 500°C between each layer and a postannealing process at 300°C for 3 h under nitrogen atmosphere. From the tested metal oxides, maximum sensitivity demonstrated sensors, using SnO_2.

In addition to indicated oxides, in the humidity sensors were used reduced graphene oxides (Gao et al. 2016) and polymers PAH–PAA (Sanchez et al. 2013). According to the sensitivity, these sensors were not inferior to sensors developed on the base of uncoated metal oxides. This indicates that besides metal oxides other conductive materials may also be used.

However, the sensitivity of these sensors, regardless of the material used, is significantly inferior to the sensitivity of the sensor developed on the basis of double coating. In these sensors, the first layer of metal oxides was used to generate the LMR, whereas the second layer was a highly hydrophilic material (a hydrogel), which under changing humidity conditions experienced a variation in its properties. This caused a wavelength shift of the LMR. As the second layer were tested agarose (Hernáez et al. 2010), poly(sodium 4-styrenesulfonate) (PSS) (Zamarreno et al. 2011; Socorro et al. 2015, 2016), and PAH–PAA (Hernáez et al. 2009; Zamarreno et al. 2010; Sanchez et al. 2012). The polymers used here, agarose, poly(allylamine hydrochloride) (PAH), poly(acrylic acid) (PAA), and PSS, are the well-known polymers whose optical properties (thickness) vary with the relative humidity of the medium. An additional PAH–PAA polymeric coating was deposited by using the traditional layer-by-layer (LbL) electrostatic self-assembly method (Goicoechea et al. 2009). An agarose layer was created onto the ITO coating using the boiling water method as reported by Arregui et al. (2003). The coating thickness was varied in the range of 100–500 nm. Hernáez et al. (2010) have established that the use of agarose allows increasing the sensitivity of ITO-based sensors up to 0.75 nm/% RH. Zamarreno et al. (2010) used the same structure, but the second thin-film was a polymer (PAH/PAA) deposited with the LbL technique. The best samples had a sensitivity of 1.2–5.4 nm/% RH (Figure 20.7). Approximately the same sensitivity was performed by sensors on the base of TiO_2/PSS (Zamarreno et al. 2011; Socorro et al. 2015, 2016). Zamarreno et al. (2010) have also shown that the sensitivity of the sensor depended on the thickness of the second layer; the ITO/PAH-PAA sensor with the higher number (100) of bilayers (higher thickness) showed a higher sensitivity of 5.4 nm/% RH compared to the 1.2 nm/RH% obtained for the sensor with a lower number of bilayers (20). It was also found that the increase in thickness, as one would expect, considerably slowed down the sensor response. Sensors with a smaller thickness had a reduced response time, which without a doubt is the advantage of these sensors.

In this respect, the results reported by Rivero et al. (2012, 2013) deserve interest. In their research, they

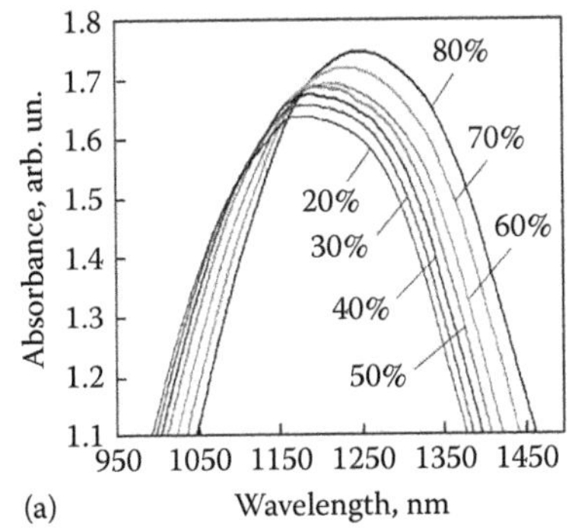

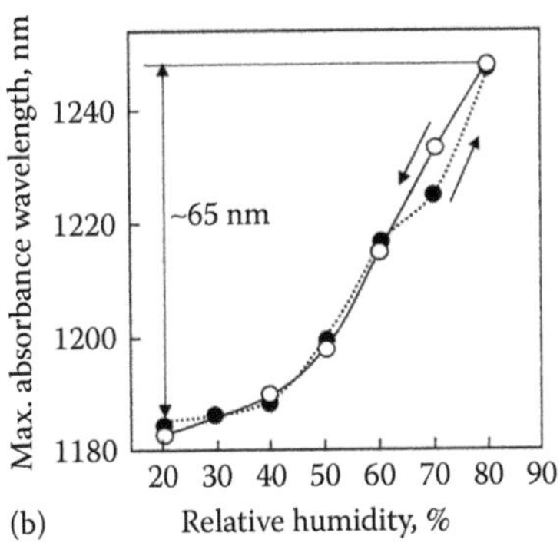

FIGURE 20.7 (a) Absorption spectra when the ITO/PAH–PAA sensor is subjected to different RH values and (b) maximum attenuation wavelength versus RH values from 20% to 80% and down again. (Reprinted from *Sens. Actuators B*, 146, Zamarreno, C.R. et al., Tunable humidity sensor based on ITO-coated optical fiber, 414–417, Copyright 2010, with permission from Elsevier.)

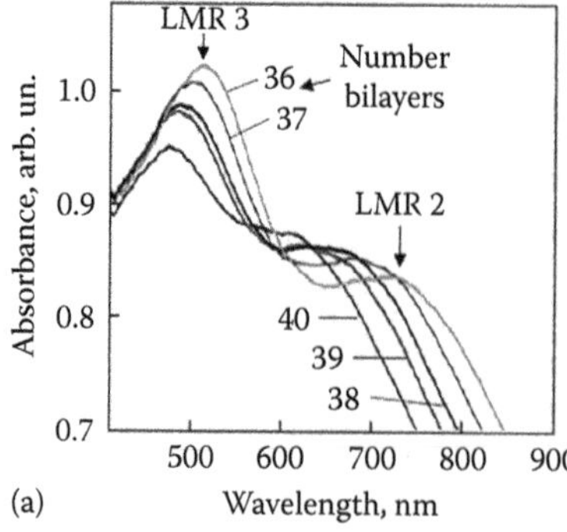

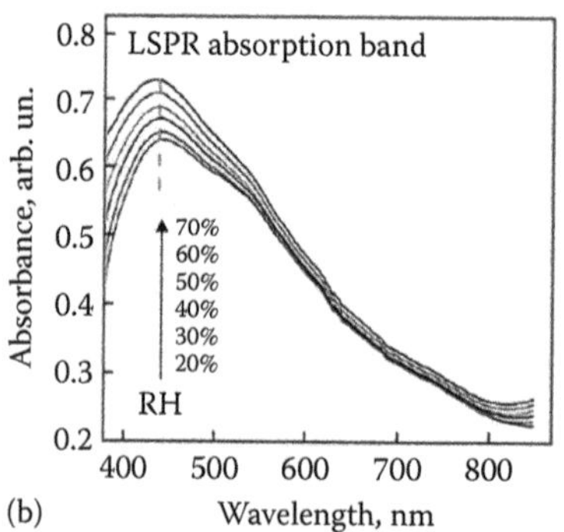

FIGURE 20.8 (a) UV–vis spectroscopy of the sensor as a function of the number of bilayers (from 36 to 40 bilayers). (b) Spectral response of 15 bilayers device (LSPR) to RH changes from 20% to 70% RH at 25°C. (Reprinted from *Sens. Actuators B*, 173, Rivero, P.J. et al., Optical fiber humidity sensorsbased on Localized Surface Plasmon Resonance (LSPR) and Lossy-mode resonance (LMR) in overlays loaded with silver nanoparticles, 244–249, Copyright 2012, with permission from Elsevier.)

studied the sensors manufactured on the basis of silver nanoparticles embedded in a PAH/PAA matrix. A technology of the AgNPs embedding into a polymer matrix was described in Rivero et al. (2013). It was established that in such sensors one can observe both LMRs and the localized surface Plasmon resonance (LSPR) (Figure 20.8a). At small thicknesses LSPR dominates, whereas for large thicknesses of the coating (>400 nm) LMR begins to dominate. The highest sensitivity attained in such sensors was 1 nm/% RH (Rivero et al. 2012). This sensitivity corresponded to LMR. The shift of the LSPR absorption band at the same change in the humidity (20%–70% RH) did not exceed 1 nm (Figure 20.8b). At that, which is very important, the sensor response was very fast, ~0.5 s, that is, fast enough for *in situ* monitoring (Figure 20.9).

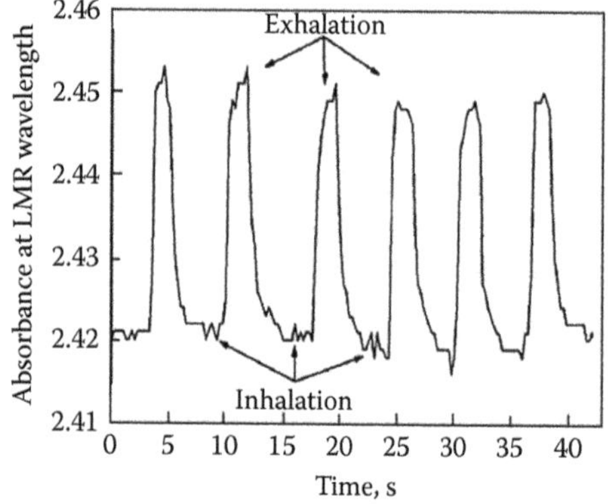

FIGURE 20.9 Response of the AgNPs/PAH–PAAH sensor to several consecutive human breathing cycles. (Reprinted from *Sens. Actuators B*, 187, Rivero, P.J. et al., A Lossy Mode Resonance optical sensor using silver nanoparticles-loaded films for monitoring human breathing, 40, Copyright 2013, with permission from Elsevier.)

As mentioned earlier, several LMRs can be observed in the spectrum. Sanchez et al. (2013) compared the performance of two relative humidity sensors working at the first and the second LMR and found that as expected, the most sensitive device was the sensor where the first LMR was tracked.

20.4 SUMMARY

In spite of the fact that LMR-based sensors is still a young research field, one can conclude that LMR could become a competitor of other traditional optical sensing platforms such as SPRs, long-period fiber gratings, or microring resonators for designing humidity sensors.

LMR-based sensors are fairly simple to manufacture and use, and moreover, have a high sensitivity. However, like most sensors, previously discussed, the LMR is characterized by the lack of selectivity for molecular recognition. Certainly, one can offer different approaches to solving this problem (Massie et al. 2006). In particular, it was shown that the selectivity problem can be overcome with multivariate data analysis. However, this complicates the measurement process and increases the cost of the devices and prevents their widespread use.

REFERENCES

Arregui F.J., Ciaurriz Z., Oneca M., Matías I.R. (2003) An experimental study about hydrogels for the fabrication of optical fiber humidity sensors. *Sens. Actuators B* 96, 165–172.

Arregui F.J., Del Villar I., Corres J.M. et al. (2014) Fiber-optic lossy mode resonance sensors. *Proc. Eng.* 87, 3–8.

Ascorbe J., Corres J.M., Matías I.R., Arregui F.J. (2015) Humidity sensor based on Lossy Mode Resonances on an etched single mode fiber. In: *Proceedings of 9th International Conference on Sensing Technology, IEEE*, Auckland, New Zealand, pp. 365–368, December 8–10.

Ascorbe J., Corres J.M., Matias I.R., Arregui F.J. (2016) High sensitivity humidity sensor based on cladding-etched optical fiber and lossy mode resonances. *Sens. Actuators B* 233, 7–16.

Batchman T.E., McWright G.M. (1982) Mode coupling between dielectric and semiconductor planar waveguides. *IEEE J. Quantum Electron.* 18, 782–788.

Corres J.M., Del Villar I., Arregui F.J., Matias I.R. (2015) Analysis of lossy mode resonances on thin-film coated cladding removed plastic fiber. *Opt. Lett.* 40(21), 4867–4870.

Del Villar I., Arregui F.J., Zamarreno C.R. et al. (2017) Optical sensors based on lossy-mode resonances. *Sens. Actuators B* 240, 174–185.

Del Villar I., Hernáez M., Zamarreno C.R., Sánchez P., Fernández-Valdivielso C., Arregui F.J., Matias I.R. (2012) Design rules for lossy mode resonance based sensors. *Appl. Opt.* 51(1), 4298–4307.

Del Villar I., Zamarreno C.R., Hernáez M., Arregui F.J., Matias I.R. (2010a) Lossy mode resonance generation with indium-tin-oxide-coated optical fibers for sensing applications. *J. Lightwave Technol.* 28, 111–117.

Del Villar I., Zamarreno C.R., Hernáez M., Arregui F.J., Matias I.R. (2010b) Generation of lossy mode resonances with absorbing thin-films. *J. Lightwave Technol.* 28, 3351–3357.

Del Villar I., Zamarreno C.R., Sanchez P., Hernáez M., Valdivielso C.F., Arregui F.J., Matias I.R. (2010c) Generation of lossy mode resonances by deposition of high-refractive-index coatings on uncladded multimode optical fibers. *J. Opt.* 12, 095503.

Del Villar I., Zamarreno C.R., Hernáez M., Sanchez P., Arregui F.J., Matias I.R. (2015) Generation of surface plasmon resonance and lossy mode resonance by thermal treatment of ITO thin-films. *Opt. Laser Technol.* 69, 1–7.

Gao R., Lu D.-F., Cheng J., Jiang Y., Jiang L., Qi Z.-M. (2016) Humidity sensor based on power leakage at resonance wavelengths of a hollow core fiber coated with reduced graphene oxide. *Sens. Actuators B* 222, 618–624.

Goicoechea J., Zamarreno C.R., Matias I.R., Arregui F.J. (2009) Utilization of white light interferometry in pH sensing applications by mean of the fabrication of nanostructured cavities. *Sens. Actuators B* 138, 613–618.

Hernáez M., Zamarreno C.R., Fernandez-Valdivielso C., Del Villar I., Arregui F.J., Matias I.R. (2010) Agarose optical fibre humidity sensor based on electromagnetic resonance in the infra-red region. *Phys. Status Solidi* (c) 7, 2767–2769.

Hernáez M., Zamarreno C.R., Matías I.R., Arregui F.J. (2009) Optical fiber humidity sensor based on surface plasmon resonance in the infra-red region. *J. Phys. Conf. Series* 178, 012019.

Homola J. (2006) *Surface Plasmon Resonance Based Sensors*. Springer, New York.

Kaur D., Sharma V.K., Kapoor A. (2014) High sensitivity lossy mode resonance sensors. *Sens. Actuators B* 198, 366–376.

Lopez S., Del Villar I., Zamarreno C.R., Hernáez M., Arregui F.J., Matias I.R. (2012) Optical fiber refractometers based on indium tin oxide coatings fabricated by sputtering. *Opt. Lett.* 37(2012), 28–30.

Malitson I.H. (1965) Interspecimen comparison of the refractive index of fused silica. *J. Opt. Soc. Am.* 55, 1205–1209.

Marciniak M., Grzegorzewski J., Szustakowski M. (1993) Analysis of lossy mode cut-off conditions in planar waveguides with semiconductor guiding layer. *IEEE Proc. J. Optoelectron.* 140, 247–252.

Massie C., Stewart G., McGregor G., Gilchrist J.R. (2006) Design of a portable optical sensor for methane gas detection. *Sens. Actuators B* 113, 830–836.

Ota R., Seki S., Ogawa M., Nishide T., Shida A., Ide M., Sawada Y. (2002) Fabrication of indium-tin-oxide films by dip coating process using ethanol solution of chlorides and surfactants. *Thin Solid Films* 411, 42–45.

Rhodes C., Franzen S., Maria J.P., Losego M., Leonard D.N., Laughlin B., Duscher G., Weibel S. (2006) Surface plasmon resonance in conducting metal oxides. *J. Appl. Phys.* 100, 54905.

Rivero P.J., Urrutia A., Goicoechea J., Arregui F.J. (2012) Optical fiber humidity sensorsbased on Localized Surface Plasmon Resonance (LSPR) and Lossy-mode resonance (LMR) in overlays loaded with silver nanoparticles. *Sens. Actuators B* 173, 244–249.

Rivero P.J., Urrutia A., Goicoechea J., Matias I.R., Arregui F.J. (2013) A Lossy Mode Resonance optical sensor using silver nanoparticles-loaded films for monitoring human breathing. *Sens. Actuators B* 187, 40–44.

Sanchez P., Zamarreno C.R., Hernáez M., Del Villar I., Fernandez-Valdivielso C., Matias I.R., Arregui F.J. (2012) A Lossy mode resonances toward the fabrication of optical fiber humidity sensors. *Meas. Sci. Technol.* 23, 014002.

Sanchez P., Zamarreno C.R., Hernáez M., Del Villar I., Matias I.R., Arregui F.J. (2013) Considerations for lossy-mode resonance-based optical fiber sensor. *IEEE Sens. J.* 13, 1167–1171.

Sanchez P., Zamarreno C.R., Hernáez M., Matias I.R., Arregui F.J. (2014) Optical fiber refractometers based on Lossy Mode Resonances by means of SnO_2 sputtered coatings. *Sens. Actuators B* 202, 154–159.

Schasfoort R.B.M., Tudos A.J. (Eds.) (2008) *Handbook of Surface Plasmon Resonance*. The Royal Society of Chemistry, Cambridge, UK.

Socorro A.B., Del Villar I., Corres J.M., Arregui F.J., Matias I.R. (2014) Spectral width reduction in lossy mode resonance-based sensors by means of tapered optical fibre structures. *Sens. Actuators B* 200, 53–60.

Socorro A.B., Hernáez M., Del Villar I., Corres J.M., Arregui F.J., Matias I.R. (2015) A comparative study between SMS interferometers and lossy mode resonace optical fiber devices for sensing applications. *Proc. SPIE* 9517, 95171U-1.

Socorro A.B., Hernáez M., Del Villar I., Corres J.M., Arregui F.J., Matias I.R. (2016) Single-mode—multimode—single-mode and lossy mode resonance-based devices: A comparative study for sensing applications. *Microsyst. Technol.* 22, 1633–1638.

Zamarreno C.R., Hernáez M., Del Villar I., Matias I.R., Arregui F.J. (2010) Tunable humidity sensor based on ITO-coated optical fiber. *Sens. Actuators B* 146, 414–417.

Zamarreno C.R., Hernáez M., Sanchez P., Del Villar I., Matias I.R., Arregui F.J. (2011) Optical fiber humidity sensor based on lossy mode resonances supported by TiO_2/PSS coatings. *Proc. Eng.* 25, 1385–1388.

Zhu S., Pang F., Huang S., Zou F., Dong Y., Wang T. (2015) High sensitivity refractive index sensor based on adiabatic tapered optical fiber deposited with nanofilm by ALD. *Opt. Exp.* 23, 13880–13888.

Zubiate P., Zamarreno C.R., Del Villar I., Matias I.R., Arregui F.J. (2015) High sensitive refractometers based on lossy mode resonances (LMRs) supported by ITO-coated D-shaped optical fibers. *Opt. Exp.* 23, 8045–8050.

21 Ellipsometry-Based Humidity Sensors

Ellipsometry (Azzam and Bashara 1977; Schubert 2005; Tompkins and Irene 2005; Fujiwara 2007; Losurdo and Hingerl 2013) is a noninvasive optical measurement technique. It is known for its ability to determine thicknesses and dielectric functions of thin layers with very high accuracy, thereby providing access to fundamental physical parameters of the sample.

Ellipsometry is based on the physical fact that linearly polarized light, incident onto a dielectric surface at some angle Ψ, will be reflected as elliptically polarized light (Tompkins and Irene 2005; Haus 2010). The name ellipsometry is derived from the fact that the final state is usually elliptical polarization. Ellipsometry thus requires the measurement of the state of polarization of the reflected light. This is usually done by rotating an analyzer in the reflected light path and recording the light intensity that reaches the detector as a function of the rotating angle (rotating analyzer ellipsometry [RAE]). Ellipsometers used for these purposes usually consist of a light source (a laser beam with a specific wavelength or a lamp to vary the wavelength) radiating on the sample and a detector receiving the reflected light (Figure 21.1). As the optical parameters of the sample are wavelength dependent, the simplest solution is to employ a laser light source. The angle of incidence (AOI) can be set. Also, ellipsometers may contain an additional polarization—optical components, for example, retardation plates as compensators.

The optical devices of an ellipsometer and the sample can be arranged in a different order. Depending on this arrangement, three or four Stokes parameters can be measured. A polarizer is placed behind the light source, and an analyzer is placed in front of the detector. A compensator can be optionally placed on the light source or the detector side. Null ellipsometry has to be mentioned at this point as the first exercised type of ellipsometry. Null ellipsometry can be carried out using the human eye as a detector and was first used at the end of the nineteenth century. The principle is to rotate polarizer and analyzer so that the minimal light intensity is detected and the ellipsometric angles are determined from the rotation angles. There are also sophisticated ellipsometers with the ability to make an analysis at different wavelengths, that is, so-called spectroscopic ellipsometers. These devices employ continuum light sources, and the reflected spectrum is analyzed with a monochromator.

The results of an ellipsometric measurement are the angles Ψ and Δ dependent on the wavelength and/or the AOI. By applying an adequate model (Arwin et al. 2004), the optical constants (refractive index [RI] n and extinction coefficient k) and the thickness d of multilayer systems can be determined using above-mentioned data (Fujiwara 2007). Thus, ellipsometry is a nondestructive and sensitive optical measuring method for the analysis of thin films. However, it is necessary to know that these parameters for the sample may not be obtained very accurately if both are unknown. Some of assumptions used during calculations are invalid and therefore limit the reliability of the absolute values of the thickness and the RI which are obtained.

The ability to determine average monolayer properties gives great opportunities for studying the processes of absorption on the surface of sensing materials, and thus, to judge about the humidity change of the environment (Krull et al. 1987; Chen et al. 1999). For example, Figure 21.2 shows a humidity influence on the Ψ, and Δ measured for P(MTGA-*r*-AA) films. The same results obtained for TiO_2 samples are shown in Figure 21.3.

However, it must be recognized that this method is not simple, and requires careful measurements and calculations, and therefore is more suitable for the study of the properties of surfaces and thin films (Archer 1963; Alvarez-Herrero et al. 1999, 2001, 2004; Chen et al. 1999; Sergeev et al. 2012; Acikgoz et al. 2015; Ogieglo et al. 2015), than for practical use as a tool for determining the humidity of the air. For example, it was shown that this method, Atmospheric Ellipsometric Porosimetry (EPA), can be used for pore characterization of studied films (Baklanov et al. 2000). This is especially important for humidity sensors, because their parameters are largely controlled by the pore size in the humidity-sensitive materials. According to Defranoux et al. (2007), the pore size and the change of the optical properties and thickness can be measured by EPA even using adsorption and desorption of wet air at atmospheric pressure (Figure 21.4). This noncontact and nondestructive technique is an effective and unique

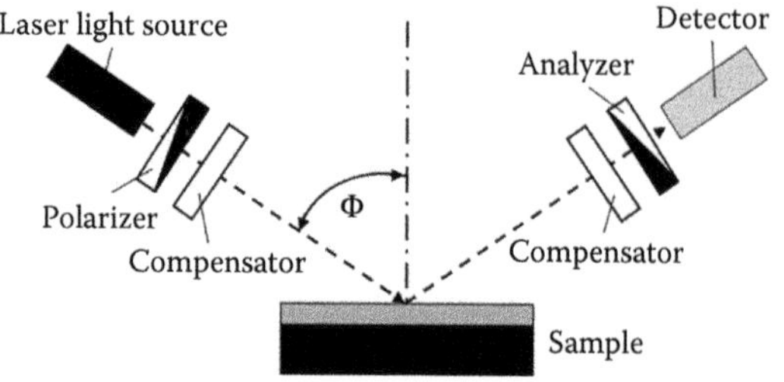

FIGURE 21.1 Schematic setup of an ellipsometer. (From Korotcenkov, G. et al., Optical and fiber optic chemical sensors, in *Chemical Sensors*, Vol. 5: Electrochemical and Optical Sensors, Korotcenkov, G. (Ed.), Momentum Press, New York, 2011. With permission.)

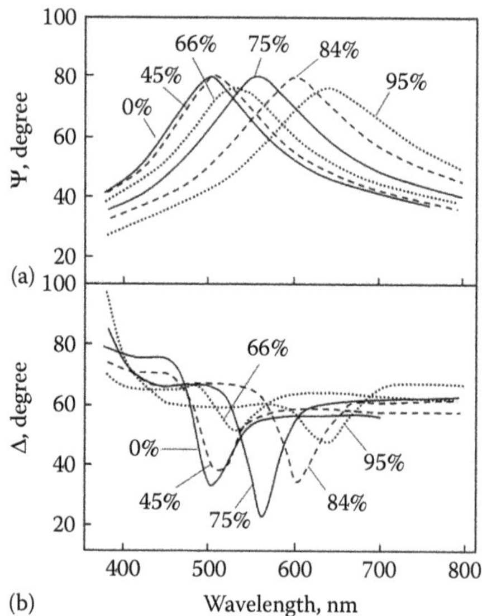

FIGURE 21.2 Ellipsometric angles Ψ (a) and Δ (b) as a function of wavelength for a series of humidities for poly(methoxytriethylene glycol acrylate) (P(MTGA-*r*-AA)). (Reprinted with permission from Chen, W.-L. et al., *Macromolecules*, 32, 136, 1999. Copyright 1999 American Chemical Society.)

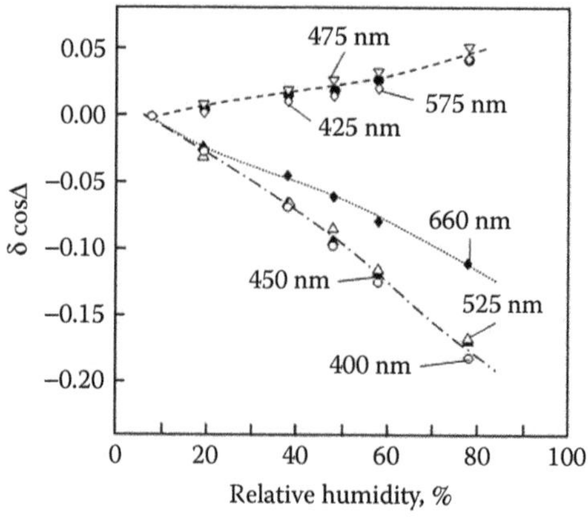

FIGURE 21.3 Relative humidity influence on the phase difference cosΔ, which quantify the polarization change during spectroscopic ellipsometry, measured for different wavelengths. Measurements were carried out for TiO_2 samples. (Data extracted from Alvarez-Herrero, A. et al., *Thin Solid Films*, 349, 212–219, 1999.)

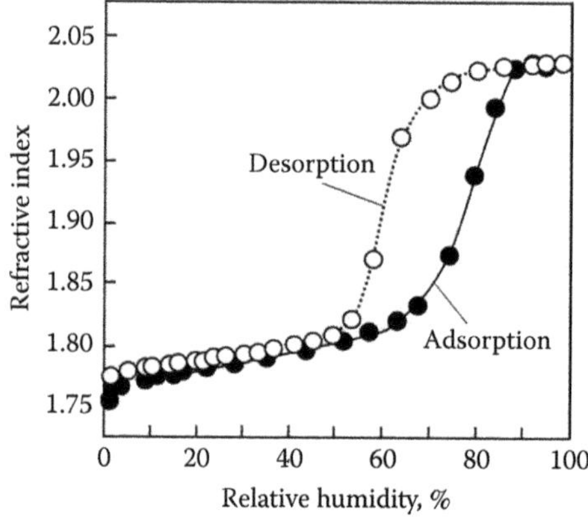

FIGURE 21.4 The change of refractive index of mesoporous TiO_2 thin films versus the relative humidity. (Data extracted from Defranoux, C. et al., 2007, Ellipsometric porosimetry: Fast and nondestructive method of porosity characterization of mesoporous thin films: Example on cubic TiO_2. In: *Proceedings of 211th ECS Meeting*, May 6–10, Chicago, IL, 2007.)

method to characterize porosity, pore size distribution (PSD), and Young modulus of thin porous films. It does not require to scratch the film, does not need low temperature or low pressure. Advantages of this method in comparison with conventional methods are given in Table 21.1.

However, one should recognize that, despite the above-listed advantages of this method, the use of ellipsometry measurements to control the properties of humidity-sensitive materials is very limited. Archer (1963) have shown that adsorption of water vapor strongly depends on the chemical nature of the substrate surface and prehistory of the samples (Figure 21.5), which means that the parameters of humidity sensors depend on the technology chosen for the preparation of humidity-sensing material.

Granberg et al. (2012) studied polyethyleneimine/nanofibrillated cellulose (PEI/NFC) films prepared by layer-by-layer (LbL) technique using ellipsometry measurements and found that the maximum swelling ratio of the LbL films was almost independent of the number of layers within the film. But, the swelling ratio can be affected by the charge density of the NFC (higher charge—larger swelling) and can be reduced considerably by cross-linking via heat treatment. Cross-linking

TABLE 21.1
Comparison of Realized Possibilities of Different Types of Adsorption Porosimeters

	Method of PSD Analysis (Type of Porosimeter)		
Parameter/Properties	**Classic Microbalance**	**Surface Sensor**	**Ellipsometric**
Temperature	Liquid nitrogen	Liquid nitrogen and room temperature	Room temperature
Absorptive	Nitrogen	Nitrogen	Organic vapors, water vapors
Pore size (nm)	2–25	2–25	2–25
Ability to analyze a film on top of Si wafer	Very limited (depends on the tool sensitivity)	No	Yes
Surface area required	Very large	Area of surface sensor	Less than 1 mm
Measurement errors related to film swelling	Can be significant	Can be significant	Film swelling can be measured and taken into account
Compatibility with microelectronic technology	No	No	Yes

Source: Baklanov, M.R. et al., *J. Vacuum Sci. Technol. B*, 18(3), 1385, 2000.

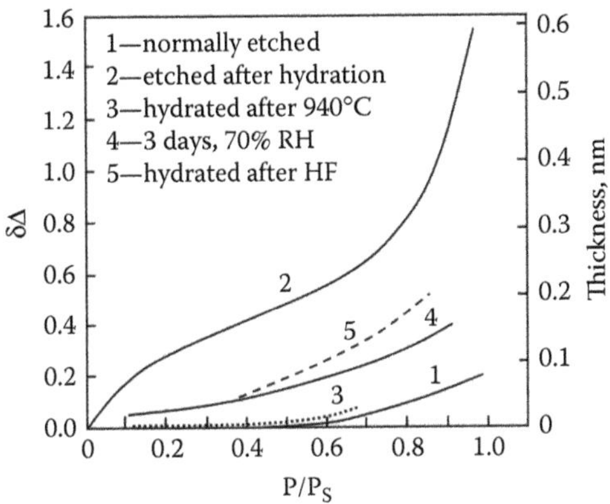

FIGURE 21.5 Water adsorption on silicon surface at 25°C: 1—normally etched (HNO_3:HF=3:1) surface; 2—etched surface after hydrating treatment; 3—hydrated surface after heating to 940°C for 30 min; 4—three days in water vapor at 70% RH (25°C) after heat treatment; 5—hydrated surface after immersion in HF. (Data extracted from Archer, R.J., Measurement of the physical adsorption of vapors and the chemisorption of oxygen and silicon by the method of ellipsometry. In: *Proceedings of Symposium on Ellipsometry in the Measurement of Surfaces and Thin Films*, September 5–6, Washington, D.C., 255–280, 1963.)

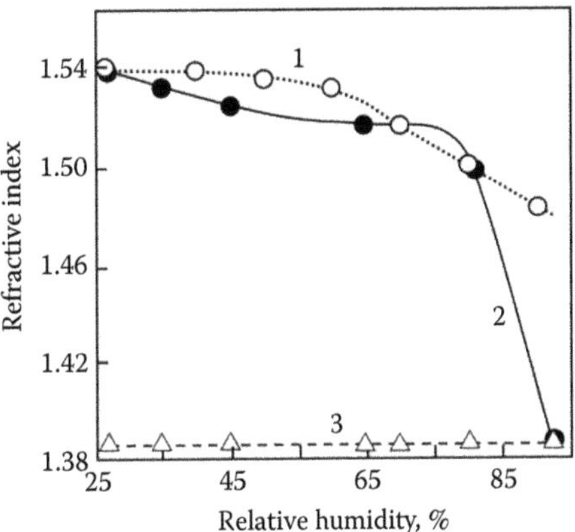

FIGURE 21.6 The influence of the environment relative humidity on the chitosan films and substrates refractive indices determined by ellipsometry measurements: 1—Chitosan acetate; 2—Chitosan citrate; 3—Magnesium fluoride (substrate). (Adapted from Sergeev, A.A. et al., *Phys. Procedia*, 23, 115, 2012. Published by Elsevier as open access.)

reduced the moisture induced film swelling by 10%–15%. It was also established that the dynamic swelling and deswelling curves were almost exponential, showing a fast initial response within the order of seconds, followed by a slower response reaching its limiting value ~20 min later. It is likely that the slow response shows a relaxation process within the LbL films as it is adjusting its interactions to include more water molecules.

Sergeev et al. (2012) using ellipsometry measurements established that eventually observed thresholds of relative humidity (RH) at which it considerably affects optical properties of chitosan films are almost equal to those at which the rate of change of chitosan films RI accelerates (Figure 21.6). In the case of chitosan acetate, this threshold equals to 60%–75%, for the case of chitosan citrate it equals to 55%–75%. It seems that these values of the RH are the lower limits at which hydrophilic properties of the chitosan commence to emerge substantially. The further increase of

RH causes change of chitosan films structure and, as a result, a change of their optical properties. Based on the results obtained, Sergeev et al. (2012) concluded that ionic forms of chitosan is a prospective material for humidity sensors.

REFERENCES

Acikgoz S., Yungevis H., Sanyal A., Incic M.N. (2015) Humidity sensing mechanism based on the distance dependent interactions between BODIPY dye molecules and gold thin films. *Sens. Actuators A* 227, 21–30.

Alvarez-Herrero A., Fort A.J., Guerrero H., Bernabeu E. (1999) Ellipsometric characterization and influence of relative humidity on TiO_2 layers optical properties. *Thin Solid Films* 349, 212–219.

Alvarez-Herrero A., Heredero R.L., Bernabeu E., Levy D. (2001) Adsorption of water on porous Vycor glass studied by ellipsometry. *Appl. Opt.* 40(4), 527–532.

Alvarez-Herrero A., Ramos G., del Monte F., Bernabeu E., Levy D. (2004) Water adsorption in porous TiO_2–SiO_2 sol–gel films analyzed by spectroscopic ellipsometry. *Thin Solid Films* 455–456, 356–360.

Archer R.J. (1963) Measurement of the physical adsorption of vapors and the chemisorption of oxygen and silicon by the method of ellipsometry. In: *Proceedings of Symposium on Ellipsometry in the Measurement of Surfaces and Thin Films*, Washington, DC, pp. 255–280, September 5–6.

Arwin H., Poksinski M., Johansen K. (2004) Total internal reflection ellipsometry: Principles and applications. *Appl. Opt.* 43(15), 3028–3036.

Azzam R.M.A., Bashara N.M. (1977) *Ellipsometry and Polarized Light.* North-Holland, Amsterdam, the Netherlands.

Baklanov M.R., Mogilnikov K.P., Polovinkin V.G., Dultsev F.N. (2000) Determination of pore size distribution in thin films by ellipsometric porosimetry. *J. Vacuum Sci. Technol. B* 18(3), 1385–1391.

Chen W.-L., Shull K.R., Papatheodorou T., Styrkas D.A., Keddie J.L. (1999) Equilibrium swelling of hydrophilic polyacrylates in humid environments. *Macromolecules* 32, 136–144.

Defranoux C., Bondaz A., Kitzinger L., Walsh C. (2007) Ellipsometric porosimetry: Fast and non destructive method of porosity characterization of mesoporous thin films: Example on cubic TiO_2. In: *Proceedings of 211th ECS Meeting*, Chicago, IL, May 6–10, Abstract #1320.

Fujiwara H. (2007) *Spectroscopic Ellipsometry-Principles and Applications.* John Wiley & Sons, Chichester, UK.

Granberg H., Coppel L.G., Eita M., de Mayolo E.A., Arwin H., Wågberg L. (2012) Dynamics of moisture interaction with polyelectrolyte multilayers containing nanofibrillated cellulose. *Nordic Pulp Paper Res. J.* 27(2), 496–499.

Haus J. (2010) *Optical Sensors: Basics and Applications.* Wiley-VCH Verlag, Weinheim, Germany.

Korotcenkov G., Cho B.K., Sevilla III F., Narayanaswamy R. (2011) Optical and fiber optic chemical sensors. In: Korotcenkov, G. (Ed.), *Chemical Sensors: Comprehensive Sensor Technologies.* Vol. 5: Electrochemical and Optical Sensors. Momentum Press, New York, pp. 311–476.

Krull U., Hum A., Vanderberg E.T. (1987) Surfactant characterization at an air/water interface by direct reflection ellipsometry. *Anal. Chim. Acta* 202, 215–221.

Losurdo M., Hingerl K. (2013) *Ellipsometry at the Nanoscale.* Springer, Berlin, Germany.

Ogieglo W., Wormeester H., Eichhorn K.-J., Wessling M., Benes N.E. (2015) In situ ellipsometry studies on swelling of thin polymer films: A review. *Prog. Polym. Sci.* 42, 42–78.

Schubert M. (2005) *Infrared Ellipsometry on Semiconductor Layer Structures.* Springer, Berlin, Germany.

Sergeev A.A., Voznesenskiy S.S., Bratskaya S.Y., Mironenko A.Y., Lagutkin R.V. (2012) Investigation of humidity influence upon waveguide features of chitosan thin films. *Phys. Procedia* 23, 115–118.

Tompkins H.G., Irene E.A. (Eds.) (2005) *Handbook of Ellipsometry.* William Andrew Publishing, Norwich, NY.

22 Design and Fabrication of Optical and Fiber-Optic Humidity Sensors

22.1 GENERAL COMMENTS

In general case, an optical fiber sensor system and a planar waveguide-based humidity sensor include an optical source (Laser, light emitted diode [LED], Laser diode [LD], etc.), optical fiber, sensing, or modulator element (which transduces the measurand to an optical signal), an optical detector (photodiode [PD]) and processing electronics (oscilloscope, optical spectrum analyzer, etc.). So, an optical fiber-based humidity sensor is a complex instrument, development of which requires a consideration of many factors (Dakin and Culshaw 1988; Brenci and Baldini 1991; Modlin and Milanovitch 1991; Burgess 1995; Grattan and Meggitt 1995; Vurek 1996; Allen 1998; Bansal 2004; Korotcenkov et al. 2011; Pérez et al. 2013), which are as follows:

- There is not a universal solution for critical devices in the topologies of optical fiber sensors, because each type of measurement strategy forces the specifications and the requirements for those devices. All components of optical sensors must be selected for matching the wavelength spectra of involved phenomena, and according to the measurement strategies. Excitation source must cover the excitation band of the optical sensor; the optical fibers should introduce low attenuation in the involved wavelengths; and the photodetector device has to process all the light emitted by the optical sensor.
- The successful development of an optical humidity sensor is intrinsically linked to the nature of the *physical platform* on which it is based. When careful consideration of the transduction mechanism dictates the design of the sensor platform, this leads to development of highly efficient integrated optical sensors that combine the delivery/collection of light with the intrinsic sensing functionality (McDonagh et al. 2008).
- The choice of the *transducer* in the case of humidity sensing is based on the optical property change in the material due to the water vapor presence. In particular, a recognition element (or indicator phase) should absorb light in a spectral region within the fiber transmission window. In many cases, the optical fiber itself plays an active role as a transducer, for example, microbend sensor. It is clear that when designing a recognition element (or indicator phase) we need to answer the questions such as what humidity reagents are supplied within it, how they selectively and sensitively produce an observable optical response when placed into contact with the water vapor-containing gas phase, on what timescale and how stably and reversibly is the response obtained, and in what manner the significant humidity reagent is immobilized within the indicator phase on the surface of the optical fiber, or the planar optical waveguide? The response timescale factor depends on the factors such as the rate of water vapor penetration into the indicator phase (a molecular recognition element), and the rate of the chemical reactions, producing an optical response (Murray 1989).
- Usually optical detection is based on the type of the *light modulation* technique used. The choice of the modulator is based on the kind of sensing desired form the optical fiber sensor. For phase modulation, interferometric detection is used, whereas for polarization modulation the polarization analyzer is used. However, in a wavelength modulation the spectrum analyzer is used. In the case of intensity modulation, which is common for most humidity sensors, the detectors used are the conventional PDs depending on the wavelength of the light source and the wavelength region of the sensor response. Sensor setups, where the response is very weak, usually use a high sensitivity APD (avalange photodiodes), which have very high gains.
- Spectrophotometric-type humidity sensors are the simplest, and at most require the realization of an optimized photometric cell to be connected to the fiber. In the case of transducer detectors, the humidity reagent can be immobilized directly on the fiber or on a solid external support, which will constitute the probe, referred to as the optode. It is necessary, however, to

distinguish the case in which the chromophore is immobilized at the fiber's extreme tip or along it on the core or on the clad. In the first case, the signal levels obtainable are generally weak, since the modulation of the optic signal comes from a thin layer of reagent connected to the fiber: this makes it necessary to use sophisticated and costly electronic and optical components (laser sources, lock-in, photomultipliers, etc.). By specially treating the tip-end surface of the fiber, sufficiently to increase the sites available for attaching the chromophore, it is possible to obtain a partial improvement in the signal-to-noise ratio (SNR). Instead, in the case of the immobilized chromophore along the fiber, modulation of the luminous intensity, even if due to a thin layer of reagent, it occurs on a section of fiber that is long enough to guarantee good signal levels. In general, better results are obtained in the extrinsic-optode sensors, since a larger surface is available for attaching the chromophore, even if the realization of a special *envelop* makes it necessary to attach the support to the tip-end of the fiber. Thus, special attention should be paid to the search for the most appropriate *envelop*, because this can weigh heavily on the performance of the probe, and in particular on the response time. In fact, it must be kept in mind that a free exchange for the water vapor being investigated must be guaranteed between the inside of the optode, where the chromophore is located, and the external environment (Brenci and Baldini 1991).

- When designing the fiber-optic humidity sensors, we need to take into account that their *operating parameters*, in particular sensor response, depend on many factors. For example, it was established that operating parameters of fiber-optic humidity sensors mainly depend on the humidity-sensitive dye, primarily in terms of its physical structure and thickness. For example, calculations for luminescence-based sensors show that the increase in the sensor thickness by more than about one fiber diameter leads to the diminishing of the incremental signal increase (Modlin and Milanovitch 1991). On account of that it is convenient to get a thin and uniform film. The metal layer thickness in surface Plasmon resonance (SPR) sensors is also an important parameter, which determines sensitivity, resonance coupling, handling, and robustness of the sensors. Ideally, the metal layers could be printed directly onto plastic substrates and integrated into polymer waveguides or optical fibers (Kashyap and Nemova 2009).

  Factors affecting the formation of absorbing species in the sensor include also the reagent composition, reaction kinetics, temperature, secondary reaction pathways, and the reagent and reaction-product stability. Similar calculations can be performed for single-fiber absorbance sensors. In the case of absorbance sensors, light has to pass through the sensor and some fraction of that light must be returned to the detector. Dye layer thickness variations affect not only the amount of incident light that is absorbed but also the amount of light that is returned which, in turn, affects the signal level and sensitivity. If the dye matrix swells or shrinks as the ionic environment changes, the sensor geometry and its response changes as well. The sensor may be sensitive to drying-wetting cycles, which also can change its dimensions and put substantial strains on the bond between the sensor material and fiber. It is necessary to take into account that some dyes are photochemically unstable. When a molecule absorbs a photon, the molecule enters an excited state. It may lose that energy by converting it to heat, re-emitting it as luminescence, or causing a chemical reaction. The reaction could disrupt the bond between the dye and polymer and thus allow the dye to be lost from the sensor. The reaction could also convert the dye to an insensitive form. It was established that these inactivating reactions may be facilitated by the presence of oxygen (Vurek 1996).
- *Fiber connection* also requires special attention (Mignani and Baldini 1996), because the optical fiber connection carries the light intensity from the source to the probe and returns the intensity modulated by the measuring parameter to the detector. Generally, fibers having diameter larger than 100 μm are used to maximize the source's coupling efficiency. The connection can be either *single fiber*, in which case the fiber serves for both lighting and detection, or *two fibers*, in which case one fiber serves for lighting and one for detection. The single-fiber connection, which is evidently more compact and thus reduces probe dimensions, requires a device upstream of the fiber that separates the lighting and detection channels. However, the beam divider should always be characterized by intrinsic low losses and crosstalk to avoid covering

up the measuring signal and also impairing the SNR. Often bi- or trifurcated fiber bundles are used, with random distribution of the fibers inside the bundle or with special distributions (linear, semicircles, concentric circles, etc.). In such case, the source(s) and detector(s) are connected to the branches of the bundle and the common termination is connected to the measuring probe.

- The *kinetics of sensor response* is also strongly dependent on many parameters. For example, the rate of signal formation in the humidity sensor is a function of both the permeation rate through the membrane and the rate of reaction with the humidity reagent. Each variable is affected by several physical and chemical parameters. In particular, permeation of a water vapor through the membrane is affected by the membrane composition, thickness, matrix temperature, and porosity. As analytes exchange with the sensor chemistry by diffusion so that the response time will vary with sensor dimensions. It is necessary to note that requirements to response and recovery times depend on the field of application. In online monitoring applications, the response and recovery time should be as low as possible, meanwhile in other typical applications, for example, in the determination of the humidity during fruit storage, this is not so critical. The response time of sensors can be reduced by making the support membrane more permeable for water vapor from surrounding environment. Another contributor to the overall system response time is the time required by the instrument to obtain enough data for a satisfactorily small signal variation (Vurek 1996). Sensor temperature coefficients are also variables to be understood. The temperature may change the dye properties and change the cell thickness, and will certainly affect the sensor response. For these reasons, optic humidity sensors should include a local temperature sensor. At that the humidity sensor temperature coefficient must be well characterized for correct measurements. The conditions of storage also affect the operating parameters of sensors.
- As we shown before, various technologies can be used for *sensor fabrication*. The choice of this technology is based on different economic, physical, and performance requirements of the device and the approach that is taken in sensor design (Burgess 1995). For example, some humidity sensors must be inexpensive to fabricate and only need to function reliably for a relatively short period of time. In contrast, a variety of industrial and environmental applications require sensors, which can operate in situ and continually for months or even years with occasional servicing. Thus, the size and per point cost of the sensor are less of an issue. In these systems, the quality of the long-term analytical data that the device can provide is the paramount feature. Therefore, specific attention during humidity sensor design requires aspects such as sensor reliability, accuracy, and robustness. We need to say that in many specific applications it is very difficult to satisfy all requirements.

22.2 REASONS OF UNCONTROLLED INTENSITY MODULATION IN OPTICAL SENSORS

As it was indicated earlier, the light intensity is the simplest solution for most of optical fiber sensors. However, the use of light intensity introduces some problems in measurement processes because the light intensity is also sensible to other variables. Mignani and Baldini (1996) have noted that uncontrolled intensity modulation is one of the main problems of such optical sensors, including humidity sensors. Since the information is contained in the intensity of the guided light, any modulation not correlated to the state of the measuring parameter upsets the measurement. They selected the following reasons of intensity modulation:

- *Sensitivity to light propagation*: Perturbation of light intensity has a lot of causes. For example, an incorrect interpretation can be caused by fluctuations in the light source, optical fiber couplings (source-to-fiber, fiber-to-sensor, and fiber-to-photo detector), and changes in the attenuation of fiber due to curvature, optical fiber length, and so on. To solve this problem, a reference system must be used to compensate for the undesired intensity fluctuations, despite the increase in system complexity and thus the cost it entails. Source intensity fluctuations can be offset by normalizing the measurement signal with a reference signal proportional to the source intensity. In practice, this implies that a single-source and single-detector-based configurations are used, and both signals travel the same optical path at every point in the system except at the point of sensing. Since they both travel on the same fiber, their ratio provides a measurement devoid of propagation accidents. However, it is difficult to meet these requirements at the same time. Currently, there are several techniques for separation and identification of sensing and the referencing signals (Adamovsky 1987; Murtaza and Senior 1995) (Figure 22.1): (1) wavelength domain multiplexing (WDM), (2) spatial separation, (3) mode multiplexing, (4) polarization multiplexing, (5) time-domain and time-division multiplexing (TDM), and (6) frequency-domain and frequency-division multiplexing (FDM).

Spatial separation is the simplest approach, which does not require complex and expensive electronics, and works with almost all types of intensity sensors. Signals are identified by separation within different transmit and receive components, so that they can remain physically isolated (Figure 22.1a). At that two fibers made of the same material and of the same length are used. One fiber connects the source, sensor, and photodetector and forms a signal channel. Another fiber that is placed alongside the first one but bypasses the sensor–transducer makes up the reference channel. As both fibers are subjected to the same external conditions and experience the same variations in the light source intensity, these effects can be easily factored out.

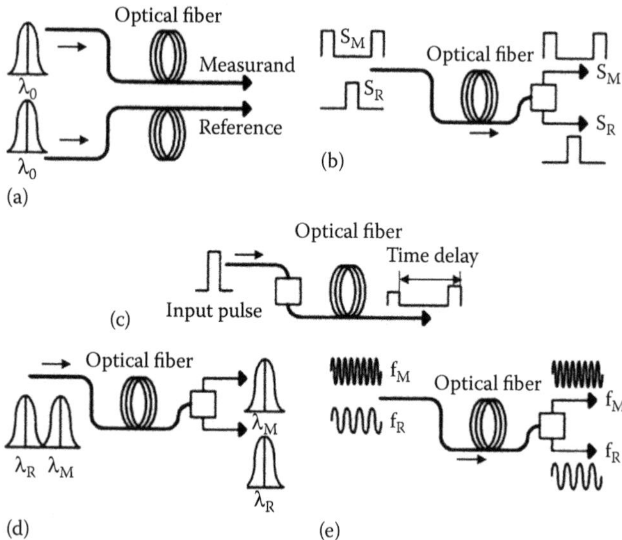

FIGURE 22.1 Major techniques used for the measurand and the reference signal separation: (a) spatial separation, (b) temporal separation (or TDM), (c) temporal separation through optical pulse division, (d) wavelength separation (or WDM), and (e) frequency separation (or FDM). (Idea from Murtaza, G., and Senior, J.M., Schemes for referencing of intensity-modulated optical sensor systems, in *Optical Fiber Sensor Technology*, Grattan, K.T.V., and Meggitt, B.T. (Eds.), Chapman and Hall, London, UK, 383–407, 1995.)

In the case when each signal is contained within a separate spectral band, both signals can be transmitted on the same optical fiber link (Figure 22.1d). One of the wavelengths is sensitive to disturbances that occur at the sensor–transducer. This wavelength forms the signal channel. The second wavelength makes up the reference channel and lies outside the bandwidth of spectral sensitivity of the sensor, and the signal at this wavelength is not affected by the measurand. The presence of spectral spacing between the signal channels enables a wavelength separation by the use of simple wavelength demultiplexing elements. However, multiplexing components may also be necessary when the optical signals are launched from separate light sources.

Temporal separation can be employed when the signals are generated from separate optical sources as illustrated in Figure 22.1b. It is achieved by time division multiplexing (TDM) of the two optical signals. The optical signals are then transmitted within separate time slots and therefore they can be separated and identified from their time of arrival at the point of detection. As illustrated in Figure 22.1c, temporal separation can also be produced between pulses of light obtained by dividing the optical power in a single parent pulse into two pulses followed by the introduction of a time delay between them.

An alternative signal separation method called the frequency-division multiplexing (FDM), shown in Figure 22.1e, is sometimes employed instead of the TDM method. The FDM operation involves electrical modulation of each optical source at a different carrier frequency. The optical signals are then separated by the use of electrical filtering of the photocurrent into the respective frequency components. An advantage with both TDM and FDM signals is that they can be allowed to propagate into the same optical fiber link as well as enabling detection with a common photodetector (Murtaza and Senior 1995). Although most of these schemes can be adapted for fiber-optic and integrated optical-sensing, but

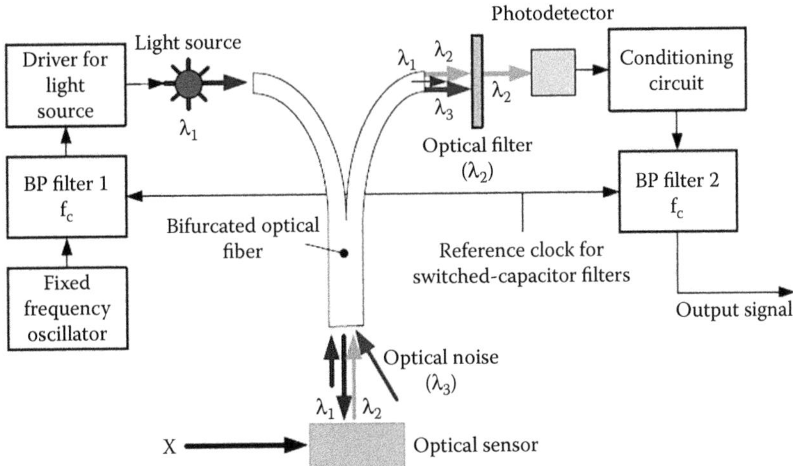

FIGURE 22.2 General idea of a light intensity measurement system based upon an optical fiber extrinsic sensor with bifurcated fibers. In case of sensors without modification in wavelength, the emission of sensor has the same wavelength that exciting light ($\lambda_1 = \lambda_2$). (From Pérez, M.A. et al., Optical fiber sensors for chemical and biological measurements, in *Current Developments in Optical Fiber Technology*, Harun S.W., and Arof H. (Eds.), InTech, Rijeka, Croatia, 265–291, 2013. Published by InTech as open access.)

the use of integrated optics gives us some extra possibilities for sensor referencing.

- *Insufficient lighting power*: The main problem determined by active and passive optical components is insufficient lighting power in the optical sensors. This depends somewhat on the power emitted by the source, and also on the quantity and quality of the required passive components. The requisites of the source, that is, compactness, adequate power, and suitable wavelength, are not always fulfilled, especially when a source that is visible or beyond the range of typical telecommunication wavelengths is required. The requisites of the optical components are compactness and the capacity of each to perform several functions. The number of passive components should be minimized by accurate assessment of system power requirements.
- *Design and manufacturing specifications*: The probes require accurate workmanship and hand assembly and must be designed to produce a high back-transmitted signal simultaneously with a fast response time. Major requisites are optimized housing, miniaturization, seal, and transducer stability.

During the measurement, it is also necessary to take into account the appearance of an additional noise that reduces the accuracy of measurement. The effect of noise could be very important in extrinsic sensors because the light must leave the optical fiber to reach the sensor, and return into the fiber. During the external path of light, optical noise could be added to the signal, reducing the SNR. Optical filters between the fiber end and the photodetector can increase this ratio by reducing the presence of external sources of light. In addition, the use of a DC source for exciting must be substituted by a fixed frequency source and a narrow band-pass filter after the photodetection to reduce the bandwidth and to increase de SNR. The use of synchronic switched-capacitor filters for both, excitation and received signals, improves the operation of the system because it provides large stability of central frequency (Grillo et al. 2006). All these solutions are shown in the block diagram of Figure 22.2. However, it is necessary to take into account that in the case of optical sensors with wide-spectrum sensitivity, too narrow optical filters allow us a heavy reduction of optical noise, but the use of them implies the decrease of the total light power, resulting in a poor SNR.

22.3 LIGHT SOURCES AND DETECTORS

22.3.1 Light Sources

As it was shown earlier, optical and fiber-optic humidity sensors need light sources. Of course, for optimal gas sensor operation these light sources should correspond to the following requirements (Zosel et al. 2011):

- Rugged construction
- High emissivity

- Long lifetime
- Low cost
- Small size
- Low power consumption
- High pulse rates and, hence, a low thermal time constant for light modulation to offset thermal background signals

At present, there are many types of light sources, which operate using different principles and can be used for these purposes. Therefore, the choice of the *light source* should depend first on the wavelength region in which the transducer shows the maximum response. For example, for some applications a He–Ne laser (632.8 nm), Argon laser (364 and 488 nm), or mercury lamp (184.45, 253.7, 365.4, 404.7, 435.8, 546.1, and 578.2 nm) can be used. The most popular mercury lamp for optical microscopy is the HBO 100 (a 100-watt high-pressure mercury plasma arc-discharge lamp), which has the highest radiance and mean luminance, due to its very small source size, of the commonly used lamps of any wattage would be an ideal source. One should note that the intensity and the position of the peaks depend on the pressure in the lamp (Figure 22.3). However, analysis shows that in fiber-optic sensor (FOS) and different portable humidity sensors semiconductor-based light sources, LEDs and LDs offer the best advantages in terms of size, cost, power consumption, and reliability (Grattan 1995; Selvarajan and Asoundi 1995). LEDs and LDs are ideal substitutes for replacing bulky, high-power consuming light sources such as gas lasers and tungsten–halogen or deuterium lamps, which are commonly used in conjunction with PD array detectors in spectroscopic instrumentation. Features of LED include very low coherence length, broad spectral width, low sensitivity to back-reflected light, and high reliability. LEDs and other broad band sources may be used with multimode fibers and in intensity type of sensors. LED sources can be easily integrated with PD-arrays detectors through fiber optics, waveguides, and microscope objectives. Simplicity of use and low power consumption make LEDs a much more attractive option than tungsten lamps in microfluidic devices (O'Toole and Diamond 2008). LDs on the other hand exhibit high coherence, narrow line width, and high optical output power, all of which are essential in interferometric sensors. Single-mode fibers may require coherent LD sources as well. High-performance Mach–Zehnder and Fabry–Perot type sensors also need single-mode lasers. However, LDs in general are susceptible to reflected (feedback) light and temperature changes. They are also less reliable and more expensive. Information about commercially possible diode lasers one can find in Werle et al. (2002).

The choice of source for exciting light depends also on measurement type (intensity, and time or frequency domain) (Pérez et al. 2013). For intensity and frequency domain measurement, the source must produce a DC+AC light signal. The operation frequency of AC component does not have important restrictions in the case of intensity, but must be properly selected for frequency domain operation, according to the expected time delay produced by the sensor response. LEDs and LDs are excellent solutions for these applications. Pulsating sources are the right selection for time-domain measurement; in this case, the total energy of pulse and its duration are the most important parameters that must be taking into account in the design process. Pulse lasers are the best choice for this kind of measurement, because it is possible to obtain extremely short pulses. Other solutions, such as short-arc pulse lamps (Xe, H_2, etc.) could be used in a design (Campo et al. 1997), but they have some inconvenient: they cannot concentrate the light into the fiber tip and, consequently, need additional—and expensive—optical systems (parabolic mirrors, lenses…)

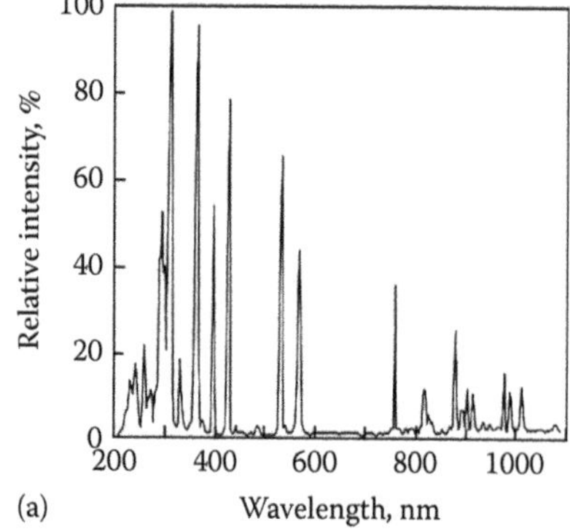

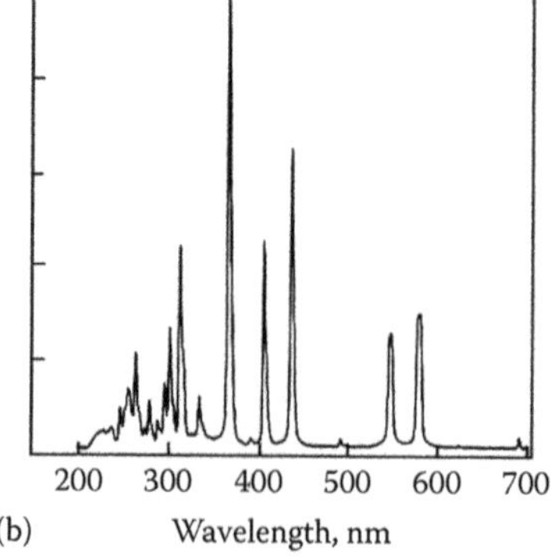

FIGURE 22.3 Emission spectrum of the (a) high-pressure and (b) medium-pressure mercury vapor lamp. (From http://www.zeiss-campus.magnet.fsu.edu.)

to do it. Moreover, pulse lamps are used to produce wide-emission spectrum, forcing the addition of optical filters to reduce the complete spectrum, and to adequate it to the wavelength band.

The details of the laser packaging can be critically important for many applications as well (Allen 1998). Fiber-coupled systems are often preferred for their simplicity and the ease with which the laser light can be transported over distances of hundreds of meters. For stable transmission, single-mode fibers, where a single transverse mode structure is supported in the waveguide, are superior. Larger diameter, multimode fibers tend to exhibit uncontrolled bending losses and interference effects associated with multiple transverse modes. Depending on the guided wavelength, the core diameter for single-mode fibers ranges between 5 and 10 μm, introducing extreme alignment tolerances at the coupler. Careful attention by the manufacturer to alignment and thermal stabilization of this alignment is critical for a well-behaved device (Allen 1998).

22.3.2 Photodetectors

The photodetector is the device that provides an electrical signal in function on received light signal; its choice is quite similar to the selection of excitation source, because it must have a spectral response, including the emission spectrum. Too wide spectral response would include undesirable optical noise, and narrow spectral response reduces the total power of desirable signal; in both cases, the effect becomes negative for SNR. A common solution for photodetector is the PD, a low-volume and low-cost device (Jones 1995; Selvarajan and Asoundi 1995; Pérez et al. 2013; Sood et al. 2014). PDs are extremely versatile and have been employed in various configurations. PDs are popular because of their rapid response and wide linear range, which are typically three and four orders of magnitude better than photo transistors (PTs). However, PDs have high noise generation, large dark current, poor sensitivity, and parameter dependence on temperature. Solutions such as APDs increase the sensitivity (Barragán et al. 2000). The APD can sense low light levels due to the inherent gain because of avalanche multiplication, but need large supply voltage, typically about 100 V. In addition, signal includes additional noise (avalanche noise) and the sensitivity strongly depends on temperature. Sometimes, PDs and APDs should be refrigerated to keep a constant temperature by means of Peltier cells and control closed loops for temperature (Prieto et al. 2004).

Photodiode arrays (PDA) are also the most suitable detectors in optical humidity sensors (Jones 1995; Selvarajan and Asoundi 1995). The use of a PDA detector in combination with a single LED could be considered superfluous to the general requirements of small, simple, and low cost, optical devices. However, O'Toole and Diamond (2008) believe that there are many additional advantages that a multiwavelength detector can offer, even when coupled with a single LED source. For example, simultaneous reference measurements can be performed at regions of the spectrum where there is no change in absorbance to reduce effects such as drift, turbidity, baseline noise, changes in refractive index (RI), and so on. In addition, the signal can be measured over a range of wavelengths where changes in absorbance are occurring. For example, in cases where two absorbing species are at equilibrium, with the equilibrium position affected by the analyte, this often manifests as two regions, one increases, whereas the other decreases, pivoted around an isosbestic point. For systems like this, measurements at the isosbestic point allow correction for drift, whereas measurements at either side of the isosbestic point enable true analytical events to be validated and distinguished from artifacts. The incorporation of these additional features must always be balanced with the fact that a more complex multiwavelength detector is required along with more sophisticated software. In practice, the solution employed must be fit for purpose, but there is no doubt that the price-performance index is continuously improving, as technologies for producing fluidic manifolds with integrated computational and communications capabilities become more ubiquitous.

Regarding materials preferable for PD fabrication, we can say that Si PDs are good for visible and near-infrared (IR) wavelengths, Ge, InGaAs/InP, InGaAsP/InP, InGaAsSb, AlGaAsSb/InGaAsSb, and HgCdTe PDs are good for 0.9–2.2 μm range (Refaat et al. 2010), whereas wide bang gap semiconductors such as GaN and AlGaP are acceptable for ultraviolet (UV) spectral range (Table 22.1) (Monroy et al. 2003). Generally, there is no bandwidth limitation due to the detector as such, although the associated electronic circuits can pose some limitation.

When the emission level is low (power signal is similar to noise equivalent power [NEP], PDs do not have enough sensitivity or introduce intolerable noise level. In this case in optical sensors photomultiplier tubes (PMTs), light dependent resistors (LDRs), and phototransistors (PTs) could be used (O'Toole and Diamond 2008). However, PTs in comparison with PDs and APDs have slow response time. Most PTs will have response times measured in tens of microseconds, which is approximately 100 times slower than PDs. They also have the disadvantage of having small active areas and

TABLE 22.1
Performance of Some Commercial UV Photodetectors

Material	Spectral Range (nm)	λ_{max} (nm)	Responsivity (A/W)
Si	200–1100	850	0.14 at 254 nm
GaP	200–520	440	0.15 at λ_{max}
Diamond	130–225	200	0.15 at λ_{max}
SiC	219–380	275	0.16 at λ_{max}
GaN	250–360	360	0.1 at λ_{max}
GaN	200–365	360	0.1 at 325 nm
AlGaN	200–320		~0.8 at 310 nm
AlGaN	200–280		0.03 at 275 nm

Source: Data extracted from Monroy, E. et al., *Semicond. Sci. Technol.*, 18, R33–R51, 2003.

high noise levels. We need to note that the various noise mechanisms associated with the detector and electronic circuit limit the ultimate detection capability. Both thermal and shot noises are two main noise sources and need to be minimized for good sensor performance. LDRs while inexpensive and small in size are also not commonly employed in optical sensors as they possess disadvantages of slow response times in comparison to PDs and they are nonlinear devices. LDRs are more ideally employed for light presence/absence detection than for accurate measurements of light intensity. Therefore, in these cases, a photomultiplier tube (PMT) must be used, to guarantee a good behavior of light to electrical signal conversion (O'Toole and Diamond 2008). In the past, PMTs are complex, expensive; they have a large volume and need high voltage power sources. But, in the present, they are compact solutions, with low voltage supply (5 or 12 V), and reasonable cost. PMTs provide low dark current, produce low noise, and have high sensitivity, being an excellent solution for most of optical fiber sensor based on luminescence phenomenon.

22.3.3 CCD and CMOS Matrixes

As the photodetectors, charge-coupled device (CCD) and complementary metal-oxide semiconductor (CMOS) matrixes, which in many cases are called digital cameras, are widespread in recent years in the development of optical sensors of various types, where the analysis of the spectral characteristics is required. Figure 22.4 shows CMOS sensors. CCD sensors have the similar view. Digital cameras have become extremely common as the prices have come down. At present such sensors are being

FIGURE 22.4 CMOS linear image sensor fabricated by Hamamatsu (Japan). (From http://www.hamamatsu.com.)

manufactured by various companies, including Sony, Toshiba, Nicon, BaySpec, Hamamatsu, and many others (www.teledynedalsa.com; www.hamamatsu.com). Both CCD and CMOS image sensors start at the same point—they have to convert light into electrons. In principle, CCD and CMOS cameras are a 2D array of thousands or millions of tiny photoreceivers, each of which transform the light from one small portion of the image into electron but performs this task using a variety of technologies. CCDs use a special manufacturing process to create the ability to transport charge across the chip without distortion. In a CCD device, the charge is read at one corner of the array. An analog-to-digital converter turns each pixel's value into a digital value. In a CCD image sensor, pixels are represented by p-doped MOS capacitors. These capacitors are biased above the threshold for inversion when image acquisition begins, allowing the conversion of incoming photons into electron charges at the semiconductor-oxide interface. This process leads to very high-quality sensors in terms of fidelity and light sensitivity. CMOS chips, on the other hand, use traditional manufacturing processes to create the chip—the same processes used to make most microprocessors. In most CMOS devices, there are several transistors at each pixel that amplify and move the charge using more traditional wires. The CMOS approach is more flexible because each pixel can be read individually. The principle of the analysis of the emission spectrum when using a CCD or CMOS sensors can be understood from the illustration shown in Figure 22.5.

On account of the manufacturing distinctions, there have been some noticeable differences between CCD and CMOS sensors:

- CCD sensors create high-quality, low-noise images. CMOS sensors, traditionally, are more susceptible to noise.
- As each pixel on a CMOS sensor has several transistors located next to it, the light sensitivity of a CMOS chip tends to be lower. Many of the photons hitting the chip hit the transistors instead of the PD.

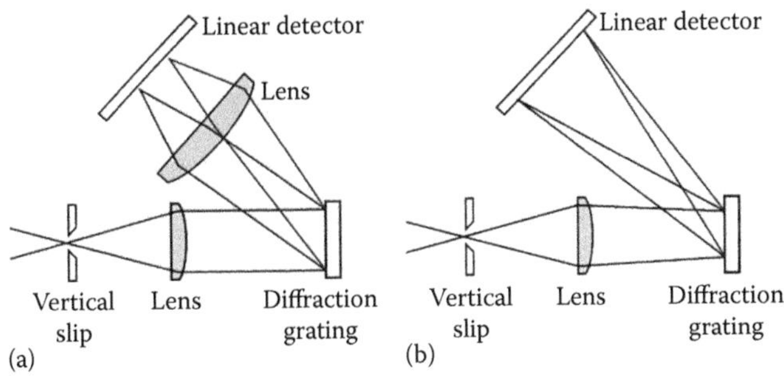

FIGURE 22.5 Horizontal schematics and light paths in two variants (a,b) of spectrometers using CCD or CMOS detectors.

- CCDs use a process that consumes a lot of power. CCDs consume as much as 100 times more power than an equivalent CMOS sensor. This means that CMOS sensors should be considered as low-power devices, which have great battery life.
- CMOS chips can be fabricated on just about any standard silicon production line, so they tend to be extremely inexpensive compared to CCD sensors.
- CCD sensors have been mass produced for a longer period of time, so they are more mature. They tend to have higher quality and more pixels.

Based on these differences, CCDs tend to be used in devices where the high resolution and excellent light sensitivity are necessary. If we do not have such strict restrictions, the use of lower-cost CMOS sensors becomes more preferable. However, it should be noted that recent advances in the CMOS sensor technology have allowed significantly improve the parameters of these sensors to the point where they reach the parity with CCD devices in many industrial, scientific, and commercial applications previously dominated by CCD sensors. These advances include lower noise, thinner and more compact packages, and more on-chip functions. New CMOS sensors now have characteristics comparable to CCDs, and they offer certain advantages such as simpler operation and simpler external circuit design. At present CMOS linear image sensors suitable for UV–vis, VIS, and VIS-NIR spectrometers are optimized for spectroscopy applications (http://www.hamamatsu.com). For example, the pixel shape is typically tall and narrow to better match the light distribution from a spectrometer's grating. Some of these sensors also have UV sensitivity without the need for additional UV coatings, and they exhibit excellent linearity, an important benefit in spectroscopy measurements. CMOS sensors for spectroscopy also may include an electronic shutter function and a built-in timing generator, and are operated from a single power supply.

22.4 OPTICAL MATERIALS USED IN OPTICAL AND FIBER-OPTIC SENSORS

For optimal sensing *appropriate optical materials* are required as well (Allen 1998; Smith 2000). To be useful as an optical material, a substance must meet certain basic requirements. It should be able to accept a smooth polish, be mechanically and chemically stable, have a homogeneous index of refraction, be free of undesirable artifacts, and of course transmit (or reflect) radiant energy in the wavelength region in which it is to be used (Smith 2000). For example, the material for the widows has been chosen to provide good transmission properties at the wavelength of interest (Table 22.2 and Figure 22.6). Fused-silica windows are satisfactory for most visible and near-IR applications of moderate temperature. However, special IR-grade quartz should be used for applications near 1.4 μm due to a material absorption band in normal quartz that can affect the properties of

TABLE 22.2
Characteristics of Some Optical Materials

Material	Transmission Range, μm	Application
Crystal quartz (SiO_2)	0.12–4.5	Birefringent
Fused-silica (SiO_2)	0.12–4.5	Windows, substrates
Calcite ($CaCO_3$)	0.2–5.5	Birefringent
Rutile (TiO_2)	0.43–6.2	Birefringent
Sapphire (Al_2O_3)	0.14–6.5	Substrate, windows
Magnesium fluoride (MgF_2)	0.11–7.5	IR optics, low reflection coatings
Lithium fluoride (LiF)	0.12–9.0	Prisms, windows, lenses
Calcium fluoride (CaF_2)	0.13–12	Prisms, windows, lenses
Barium fluoride (BaF_2)	0.25–15	Windows
Sodium chloride (NaCl)	0.2–26	Prisms, windows
Potassium bromide (KI)	0.25–40	Prisms, windows
Silicon (Si)	1.2–15	IR optics
Germanium (Ge)	1.8–23	IR optics
Zinc Selenide (ZnSe)	0.5–22	Windows, optics
AMTIR (Ga/As/Se)	0.7–14	Optical elements
Gallium arsenide (GaAs)	1–15	Platform for IO, electro-optics
Magnesium oxide (MgO)	0.25–9	Substrate

Source: Data extracted from Smith, W.J., *Modern Optical Engineering*, 3rd ed., McGraw-Hill, New York, 2000.

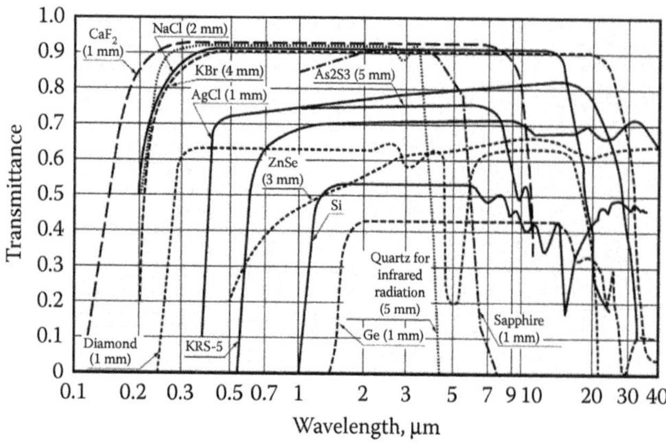

FIGURE 22.6 Transmittance of various optical materials. (From Hamamatsu Photonics K.K., www.hamamatsu.com.)

TABLE 22.3
Common Infrared Coating Materials

Material	Wavelength (μm)	Refractive Index @ (μm)
Sapphire	0.1–5	1.76
BaF_2	0.2–11	1.47
CaF_2	0.2–8	1.41
Silicon	1.1–10	3.42
ZnSe	0.6–20	2.49
KBr	0.5–25	1.54

Source: Data extracted from Smith, W.J., *Modern Optical Engineering*, 3rd ed., McGraw-Hill, New York, 2000.

antireflection coatings and lead to severe etalons. At the same time crystal quartz and calcite are infrequently used for windows because of their birefringence, which limits their usefulness almost entirely to polarize prisms and the like. Sapphire, which is extremely hard, is also used for windows, interference filter substrates, and occasionally for lens elements. It is slightly birefringent, which limits the angular field over which it can be used.

The halogen salts have good transmission and refraction characteristics. Worthy of special mention is calcium fluoride, or fluorite. This material has excellent transmission characteristics in both UV and IR, which make it valuable for instrumentation purposes. In addition, its partial dispersion characteristics are such that it can be combined with optical glass to form a lens system, which is free of secondary spectrum. However, its physical properties, as well as the properties of other halogen salts, are not outstanding since it is soft, fragile, resists weathering poorly, and has a crystal structure, which sometimes makes polishing difficult. In exposed applications, the fluorite element can sometimes be sandwiched between glass elements to protect its surfaces. In the near-IR, sapphire optics can be useful as well. In the mid-IR, common materials are chalcogenide glasses, ZnSe, CaF_2, and KBr (Kraft 2006). Indicated materials have been used to create thin-film coating such as an antireflection coating and filter material (Table 22.3).

Germanium and especially silicon are widely used for refracting elements in IR devices. Their extremely high index of refraction is a joy to the lens designer since the weak curvatures, which result from the high index tend to produce designs of a quality, which cannot be duplicated in comparable glass systems. But, special low-reflection coatings are necessary in this case since the surface reflection is very high, for example, 36% per uncoated germanium surface. Zinc sulfide, zinc selenide, and amorphous material transmitting infrared radiation (AMTIRs) such as Ge–As–Se, As–Se, and As–Se–Te (http://www.amorphousmaterials.com), are also widely used in IR systems.

Common to refractive elements is the problem of chromatic aberration. Therefore, *mirror optics* is usually applied in particular in the mid-IR. Reflecting coatings applied to a surface for operation in the visible and near-IR ranges can be silver, aluminum, chromium, and rhodium. However, the mirrors used are usually aluminum-coated ones. Gold is preferable for the far-IR spectral range devices. By selecting an appropriate coating, the reflectance may be achieved of any desired value from 0 to 1. The best mirrors for broadband use have pure metallic layers, vacuum-deposited, or electrolytically deposited on glass, fused silica, or metal substrates. Before the reflective layer deposition, to achieve a leveling effect a mirror may be given an undercoat of copper, zirconium–copper, or molybdenum. While near-IR mirrors are usually protected by thin SiO_2 layers, unprotected mirrors have to be used in the mid-IR. Disadvantages of mirror optics are the elevated space consumption and the higher prices in comparison to refractive optics, especially comparing nonstandard mirrors against nonstandard lenses. In total, mirror optics are so preferable to fibers and refractive optics, at least in the mid-IR, that in some technical applications they are used to replace waveguides to transport IR radiation among source, sensor head, and spectrometer. More detailed descriptions and parameters of components used during IR and UV–vis measurements can be found in the literature (Kraft 2006; Haus 2010).

22.5 REFLECTORS

Although polished bulk metals are occasionally used for mirror surfaces, most optical reflectors are fabricated by evaporating one or more thin-films on a polished surface, which is usually glass (Smith 2000). The spectral

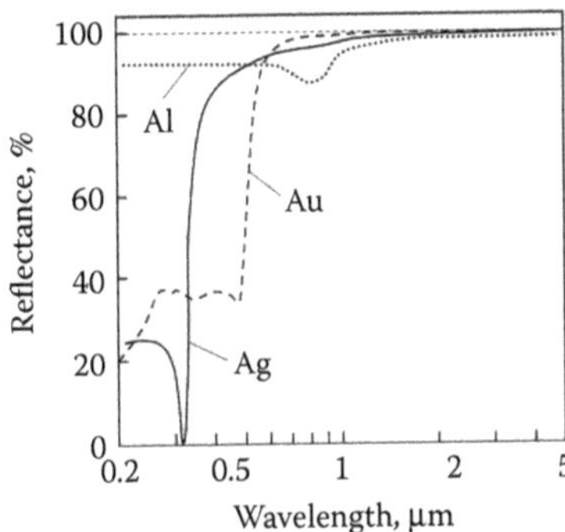

FIGURE 22.7 Spectral reflectance for evaporated metal films on glass. Data represent new coatings, under ideal conditions. (Data extracted from Smith, W.J., *Modern Optical Engineering*, 3rd ed., McGraw-Hill, New York, 2000.)

reflectance characteristics of several evaporated metal films are shown in Figure 22.7. With the exception of the curve for rhodium, the reflectivities given here can seldom be attained for practical purposes. The workhorse reflector material for the great majority of applications is an aluminum film deposited on a substrate by evaporation in vacuum. Aluminum has a broad spectral band of quite high reflectivity and is reasonably durable when properly applied. However, it is necessary to take into account that the silver coating will tarnish and the aluminum film will oxidize, so that the reflectances tend to decrease with age, especially at shorter wavelengths. Therefore, the high reflectivity of silver is only useful when the coating can be properly protected. Almost all aluminum mirrors are also *overcoated* with a thin protective layer of either silicon monoxide or magnesium fluoride. This combination produces a first-surface mirror, which is rugged enough to withstand ordinary handling and cleaning without undue scratching or other signs of wear. A run-of-the-mill protected aluminum mirror can be expected to have an average visual reflectance of about 88%. One should note that in addition two, four, or more interference films may be added to improve the reflectance where the additional cost can be accepted. This enhanced reflectivity within the bandpass of the mirror is obtained usually at the expense of a lowered reflectivity on either side.

Dichroics and semireflecting mirrors constitute another class of reflector (Smith 2000). Both are used to split a beam of light into two parts. A dichroic reflector splits the light beam spectrally, in that it transmits certain wavelengths and reflects others. A dichroic reflector is often used for heat control in projectors and other illuminating devices. A *hot mirror* is a dichroic, which transmits the visible region of the spectrum and reflects the near-IR. A *cold mirror* does just the reverse, in that it transmits the IR and reflects the visible. For example, a cold mirror introduced into the optical path will allow undesired heat in the form of IR radiation to be removed from the beam by transmitting it to a heat dump. These mirrors have the advantage over heat-absorbing filter glass in that they do not themselves get hot and thus do not require a fan for cooling. A semireflecting mirror is, nominally at least, spectrally neutral; its function is to divide a beam into two portions, each with similar spectral characteristics.

22.6 THE SELECTION OF FIBERS

22.6.1 Features of Fiber Used

Due to telecom applications, the classical step index fibers in silica are very cheap (~0.10 $/m). Usually these optical fibers consist of four components: a core, a cladding, coating, and a light-impermeable jacket (Figure 22.8).

However, beyond these specific applications, a large number of new fibers is nowadays available with different materials and geometries. Some of them can be customer made. Among the available optical fibers, one can distinguish (Lecler and Meyrueis 2012):

- The single-mode and multimodes optical fibers, with core diameter from 5 μm to the millimeter.
- The polarization maintaining fibers (generally with an elliptical core).
- The step index fibers and the graded index ones with several index profiles.
- The fibers whose core and cladding are in silicate or polymer (plastic). A lot of improvements have been achieved concerning plastic fiber (Peters 2011): loss decrease, temperature resistance, homogeneity, and so on.
- The silica fibers with several possible doping (germanium, fluorine, alumina, erbium, etc.).
- The microstructured fiber in the radial direction such as the photonic crystal fibers.
- The Bragg grating fiber (with a periodic modulation of the RI in the longitudinal direction).

Such a large set of fibers on the market gives great scope for designing various optical sensors.

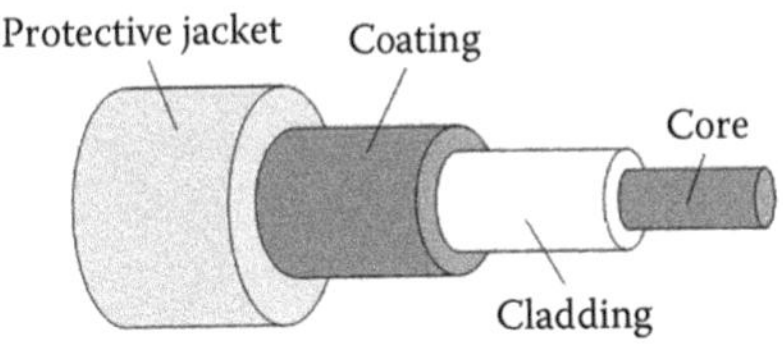

FIGURE 22.8 Schematic diagram of fiber.

22.6.2 The Selection of Fiber

It is evident that, in order to be able to perform a detection using fiber optics, the *working wavelengths* must be included within the transmission windows that are characteristic of the fibers (Brenci and Baldini 1991). In principle, for the development of sensors one can use either glass or polymer (plastic) optical fibers (POF) (Peters 2011). Comparison of various fibers is shown in Figure 22.9.

At present most of the humidity sensors are developed basing on glass fibers. Glass optical fiber is most often (though not always) fabricated from very pure silica or SiO_2 (Syms and Cozens 1993). This has a RI of $n = 1.458$ at $\lambda_0 = 850$ nm. Dopants useful for fabrication of silica-based fibers include germania (GeO_2) and phosphorus pentoxide (P_2O_5), both of which increase the RI of silica, and boric oxide (B_2O_3) and fluorine (F_2), which reduce it. Thus, a typical fiber might consist of a GeO_2:SiO_2 core with a SiO_2 cladding. Alternatively, a pure SiO_2 core could be used, with a B_2O_3:SiO_2 cladding. The boundary between the core and cladding may either be abrupt (step-index fiber) or gradual (graded-index fiber). Glass and especially silica-based fibers have better thermal stability than plastic fibers. In addition, glass fibers exhibit higher light transmission characteristics, which range from the visible end of the electromagnetic spectrum to the near-IR region (Figure 22.10). In particular, silica fibers with a low OH content guarantee low attenuations for wavelength values, which are between 500 and 1750 nm. Both UV and mid-IR absorption bands result in a fundamental limit to the attenuation, which one can achieve in the silica system. This occurs despite the fact that the Rayleigh scattering contribution decreases as λ^{-4}, and the UV Urbach absorption edge decreases even faster with increasing λ. Therefore, for $\lambda < 500$ nm, as the wavelength decreases, the attenuation increases up

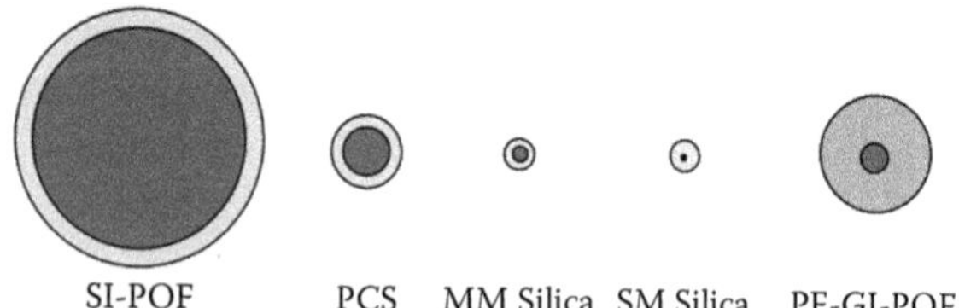

FIGURE 22.9 Relative comparison of diameters in different kinds of fibers. SI-POF = step-index polymeric optical fiber; PCS = plastic cladding silica; MM silica = multimode silica; SM silica = single mode silica; PF-GI-POF = perflorinated graded-index POF. The light color represents the cladding and dark color the core. (Reprinted from Werneck, M.M., and Allil, R.C.S.B., Optical fiber sensors, in *Modern Telemetry*, Krejcar, O. (Ed.), InTech, Rijeka, Croatia, 3–40, 2011. Published by InTech as open access.)

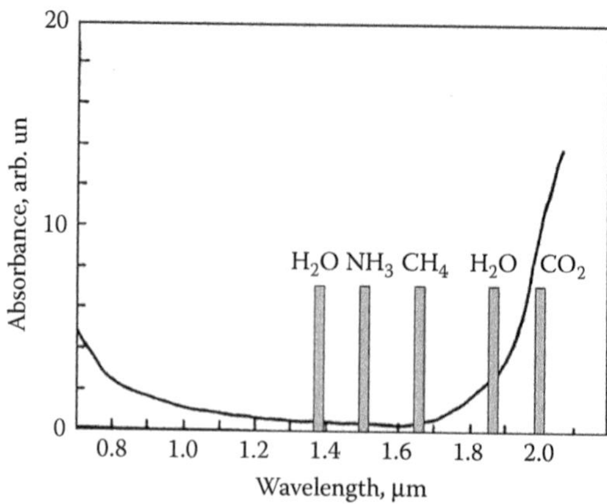

FIGURE 22.10 Absorption bands of some common gases within the transmission window for silica fiber. (Reprinted from *Sens. Actuators B*, 29, Dakin, J.P. et al., Progress with optical gas sensors using correlation spectroscopy, 87–93, Copyright 1995, with permission from Elsevier.)

to values on the order of 3 dB/m for $\lambda \sim 200$ nm, making the use of very short fiber lengths necessary. The IR absorption increases with long wavelengths, and becomes dominant beyond wavelengths of about 1.6 μm, resulting in a fundamental loss minimum near 1.55 μm (Bass and Van Stryland 2002).

However, plastic (polymer)-based fibers are also used for the development of humidity sensors (Muto et al. 2003; Tay et al. 2004; Zhang et al. 2010a; Peters 2011). Comparing POF and silica fibers by the attenuation, silica fibers are much better. However, when constructing a fiber sensor using POF instead of silica, we have some additional advantages such as (1) more resistance to strain (larger modulus of elasticity), which means more reliable networks; (2) cheaper peripherical components; (3) easy handling; and (4) no need of special skill for splicing and connectorization. In addition, plastic fibers are cheaper than their counterpart, less brittle and less expensive to use. Moreover, due to their larger diameter, it is simpler to work with open optics and easy handling. The additional flexibility of these materials makes them attractive for large-core, short-length applications in which one wishes to maximize the light insertion. Therefore, the low-cost, low-temperature processes by which polymers can be fabricated has led to continued research into the applications of plastic fiber to technologies, which require low cost, easy connectivity, and that are not loss-limited. Poly(methyl methacrylate) (PMMA) is the most commonly used material in polymer optical fiber fabrication.

Another advantage of polymer-based optical fibers (POF) is the ability to manufacture humidity sensors that use the moisture sensitivity of the fibers themselves.

In particular, PMMA based fibers have such properties (Chen et al. 2015). PMMA strongly absorbs water (Zhang et al. 2010a; Woyessa et al. 2016b). Other polymer optical-fiber such as TOPAS and ZEONEX, have been reported to be insensitive to humidity (Yuan et al. 2011; Markos et al. 2013; Woyessa et al. 2016a, 2017b). The affinity for water of PMMA leads to a swelling of the POF and an increase of RI. Woyessa et al. (2017a) have shown that the humidity-sensitivities of PMMA POF Bragg gratings (POFBGs) and TOPAS step index POFBGs at 850 nm, fabricated in their group, were 45 and 0.45 pm/% RH, respectively. Thus the humidity-sensitivity of PMMA POFBG is 100 times larger than TOPAS POFBGs.

Of course POFs have disadvantages as well. POF only transmits visible and near-IR light; there are several strong absorption regions in the near-IR, which tend to limit the application of plastic fibers in this region. So, we cannot use the available technology of telecommunications such as 1300 and 1500 nm telecom windows. Additionally, POF has a very high attenuation in the visible spectrum. The other issue is the temperature because plastic materials cannot withstand high temperatures as much as glasses. POFs can operate only up to 70°C–85°C. However, some specials POFs have been developed mainly for harsh environment. In particular, Woyessa et al. (2017a) developed and characterized a polycarbonate (PC) based POFBG humidity sensor that can operate beyond 100°C in the relative humidity range 10%–90%. The PC-based sensor gave a RH sensitivity of 7.31 ± 0.13 pm/% RH in the range 10%–90% RH at 100°C and a temperature sensitivity of 25.86 ± 0.63 pm/°C in the range 20°C–100°C at 90% RH. The humidity-sensitivity of PC POFBGs was 6 times smaller than PMMA POFBG. However, at ambient relative humidity PC POFBGs can operate up to 125°C, whereas PMMA POFBG can only be operated up to 90°C. Thus, PC POFBGs humidity sensors can be used in several different applications areas where humidity measurement at high temperature is required.

Slow response of POF-based humidity sensors is also disadvantage of these sensors (Chen et al. 2015). The process of water absorption or desorption in PMMA can be described by the diffusion theory of mass transfer. Therefore, the wavelength change of POFBG-based sensors induced by environmental humidity consists of two parts: (1) humidity-dependent RI change and (2) humidity-induced volumetric change in the fiber core. This means that sensors made of different PMMA fibers may exhibit different response times and sensitivity. For the same type of PMMA fiber the humidity response time of these sensors is determined by the fiber geometry. At present, there are two ways to improve the response time of a POFBG humidity sensor. First, the decrease of the thickness of the cladding; in this case the moisture can reach the fiber core in a shorter time. Second, reduction of fiber diameter, which will reduce the time of the moisture diffusion process. Zhang et al. (2012a) have shown that the response time of a POF Bragg grating sensors really can be improved by reducing the fiber diameter using chemical etching. It was found that the greater the volume of cladding removed, the faster the response of the POFBG on the humidity change. Alternatively, a laterally accessible microstructured polymer fiber with a side-slotted cladding was suggested for further improvement in humidity time response (Cox et al. 2007; Zhang et al. 2012a). However, such slotted specialty fiber is not commercially available and also has the drawbacks of high loss and butt-coupling difficulty (Cox et al. 2007). As for the attempts of Zhang et al. (2012b) reduce the response time via the using micromachining technology and D-shape fibers, then even in this case, the time of response of the POF-based sensor with cladding diameter of 130 μm on the 10% step RH change was more than 30 min.

In the mid-IR region, the materials, which exhibit better IR transparency, are usually used for fiber fabrication. IR optical fibers may be defined as fiber optics transmitting radiation with wavelengths greater than approximately 1.8–2 μm (Harrington 2002). IR fiber optics may logically be divided into three broad categories: glass, crystalline, and hollow waveguides. These categories may be further subdivided based on the fiber material, structure, or both, as shown in Table 22.4. A key feature of current IR fibers is their ability to transmit longer wavelengths than most oxide glass fibers can. In some cases, the transmittance of the fiber can extend well beyond 20 μm, but most applications do not require the delivery of radiation longer than about 12 μm. Two of the most important representative materials suitable for application in IR fibers are the heavy-metal fluoride glasses and the chalcogenide glasses. While both classes exhibit better IR transparency, neither has yet improved to the point of serious competition with silica materials.

The materials, indicated in Table 22.4 provide a wide variation in the range of transmission for different fibers (Table 22.5). For example, fibers from heavy-metal fluoride glasses and the chalcogenide glasses have attenuations on the order of tenths of dB/Km. However, research has shown that there is a significant extrinsic absorption that degrades the overall optical response of above-mentioned fibers (Harrington 2002). Most of these extrinsic bands can be attributed to various impurities, but, in the case of the hollow waveguides, they are due to interference effects resulting from the thin-film coatings used to make the

TABLE 22.4
Categories of IR Fibers with a Common Example to Illustrate Each Subcategory

Main	Subcategory	Examples
Glass	Heavy metal fluoride (HMFG)	ZrF_4–BaF_2–LaF_3–AlF_3–NaF (ZBLAN)
	Germanate	GeO_2–PbO
	Chalcogenide	As_2S_3 and AsGeTeSe
Crystal	Polycrystalline (PC)	AgBrCl
	Single crystal (SC)	Sapphire
Hollow waveguide	Metal/dielectric film	Hollow glass waveguide
	Refractive index < 1	Hollow sapphire at 10.6 μm

Source: Data extracted from Harrington, A., Infrared fibers, in *Fiber Optic Handbook: Fiber, Devices, and Systems for Optical Communications*, Bass M., and Van Stryland, E.W. (Eds.), McGraw-Hill, New York, 2002.

TABLE 22.5
Selected Properties of Optical Fibers Used in Sensors

Fiber/Parameter	Wavelength (μm)	Attenuation	Refractive Index, Core	Maximum Use Temperature (°C)
Silica	0.2–4	0.5 dBm^{-1} (1.5 μm)	1.458	800
Chalcogenide	3–10	0.5 dBm^{-1} (6 μm)	2.9	300
Fluoride	0.2–4.3	0.02 dBm^{-1} (2.6 μm)	1.51	250
Sapphire	0.2–4	20 dBm^{-1} (3 μm)	1.7	>1500
Single-mode photonic crystal	0.4–3	<1 dBm^{-1} (16 μm)	1.0 (core) 1.46 (cladding)	800
AgBr/Cl	3.3–15	0.7 dBm^{-1} (10.6 μm)	2.0	400
PMMA	0.4–0.8	0.1 dBm^{-1} (0.6 μm) 30 dBm^{-1} (0.8 μm)	1.492	80

Source: Data extracted from suppliers and Fernando, G.F., and Degamber, B., *Int. Mater. Rev.*, 51, 65–106, 2006.

waveguides. Besides, Harrington (2002) also believes that, compared to silica, the IR fibers usually have higher losses, larger RIs, and *dn/dT* values.

As it was shown before, the spectral transmission of an optical fiber is determined by the material used in its core and cladding. In particular, Table 22.6 shows a list of typical acceptance angles, transmission ranges, and numerical apertures. The numerical aperture (NA) is an important parameter for the familiar light-beam approach. It provides a connection between the maximum angle of incidence θ_{max}, at which the core of the fiber will take in light that will be contained within the core, and the RIs of corresponding materials:

$$\mathrm{NA} = n_0 \cdot \sin(\theta_{max}) = \frac{\sqrt{n_1^2 - n_2^2}}{n_0} \tag{22.1}$$

The RI n_0 belongs to the medium surrounding the fiber. For air, $n_0 = 1$.

As it was indicated before, depending on the wavelength of the light input, waveguide geometry, and distribution of its RIs, several modes can propagate through the fiber, resulting in the so-called single- and multimode optical fibers. Table 22.7 provides a summary of the typical properties of various optical silica-based fiber types (e.g., single- and multimode). Both fiber types are used in the construction of fiber-optic humidity sensors.

While making a choice between single-mode and multimode fibers we need to take into account that a large core diameter and a higher numerical aperture give to multimode fibers several advantageous in comparison with single-mode fibers (SMFs), because they make it easier for fiber connection to be made and light

TABLE 22.6
Acceptance Angles (θ_c) and Numerical Apertures (NA) of Different Types of Optical Fibers

Fiber Material	Transmission Range (nm)	Θ_{max} (deg)	NA (λ < 587 nm, $n_0 = 1$)
Glass	400–1600	26.5	0.45
		32.5	0.54
		35.0	0.57
		40.0	0.64
		60.0	0.86
Quartz	250–1000	12.5	0.22
		15.0	0.26
PMMA	400–800	27.5	0.46
		30.0	0.50

Source: Data extracted from Haus, J., *Optical Sensors: Basics and Applications*, Wiley-VCH, Weinheim, Germany, 2010.

TABLE 22.7
Typical Characteristics of Single-Mode and Multimode Silica-Based Fibers

Element	Single-Mode Fiber, μm	Multimode Fiber, μm	Material
Core	2–10	50–150	Silica based
Cladding	80–120	100–250	Doped silica
Coating	250	250	Polymer
Protective jacket	900	900	Plastic

Source: Data extracted from Krohn, D.A., *Fiber Optic Sensors: Fundamentals and Applications*, Vol. 138, Instrument Society of America, Research Triangle Park, NC, 1988.

is launched into a multimode fiber with ease (Grattan and Meggitt 1995). In addition, multimode fibers permit the use of LEDs rather than laser as the radiation source, which saves costs. LEDs are cheaper and have longer lifetime. On the other hand, as the number of modes increases, the effect of modal dispersion in multimode fibers increases as well. Differences among the group velocities of the modes result in a spread of travel times and are accompanied by the broadening of a light pulse as it travels through the fiber. This effect limits how often adjacent the pulses can be launched without resulting in pulse overlap at the far end of the fiber. Modal dispersion therefore limits the speed at which multimode optical fiber communication systems can operate. Therefore, for intensity modulation-based sensors multimode optical fibers are generally used, while SMFs are used in interferometric fiber sensors where the phase changes are measured. It is necessary to note that developed during last decades graded-index multimode fibers, which use variations in the composition of the glass in the core to compensate different path lengths of the modes, have much better parameters. They offer hundreds of times more bandwidth than conventional step index fibers.

22.7 FIBER PREPARATION

There are different practical implementations of FOS using standard optical fibers (Elosua et al. 2006) that require fibers in different versions, including side-polished fibers, tapered fibers, and D-fibers (Figure 22.11). In many cases, it is also necessary to remove a cladding.

As a rule silica-based fibers have a cladding made of silica, which is difficult to remove or modify. A possible solution is to polish the fiber, eliminating the cladding or at least a part of it (Senosiain et al. 2001; Gastón et al. 2003, 2004), or to use a chemical attack known as etching. This last solution is probably the most employed, and consists of soaking the optical fiber in a hydrofluoric acid solution (Segawa et al. 2003; Khalil et al. 2004; Sumdia et al. 2005). The etching solution consists of hydrogen fluoride (HF), deionized water, and

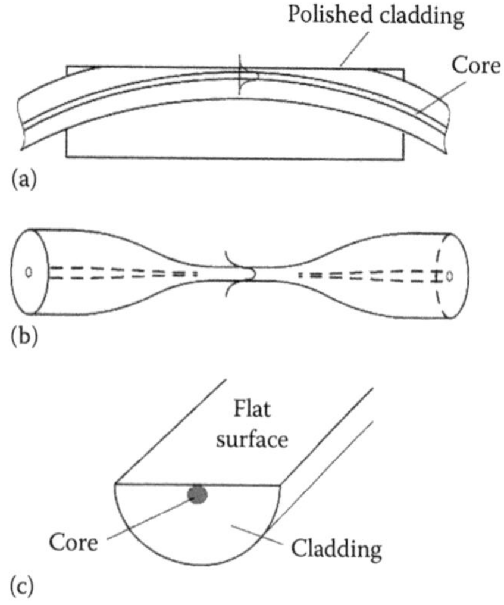

FIGURE 22.11 (a) Side-polished fiber, (b) Tapered fiber; and (c) D-fiber. (Reprinted from *Opt. Fiber Technol.*, 19, Jin, W. et al., Gas detection with micro- and nano-engineered optical fibers, 741–759, Copyright 2013, with permission from Elsevier.).

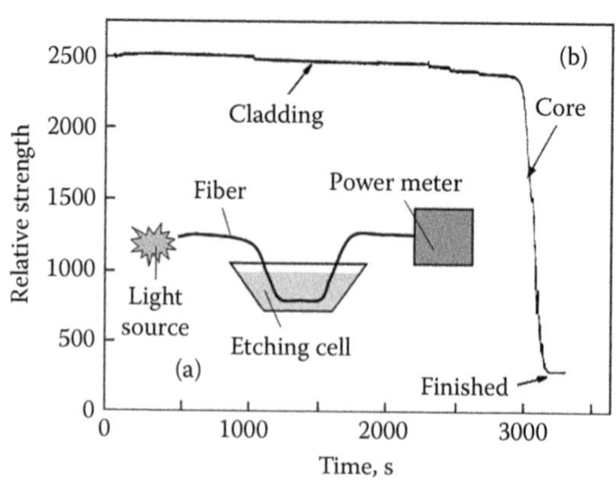

FIGURE 22.12 (a) Experimental setup of the etching process. (b) The relationship between the output power and the etching time. The volume ration among HF, deionized water, CH_3COOH in the corrosion solution used here is 1.3:1:1. (Adapted from Zhuang, X. et al., A novel approach to evaluate the sensitivities of the optical fiber evanescent field sensors, in *Fiber Optic Sensors*, Yasin, M. (Ed.), InTech, Rijeka, Croatia, 165–184, 2012. Published by InTech as open access.)

CH_3COOH. The function of HF is to etch silica with the well-known reactions:

$$4HF + SiO_2 \rightarrow SiF_4 + 2H_2O \quad (22.2)$$

$$6HF + SiO_2 \rightarrow H_2SiF_6 + 2H_2O \quad (22.3)$$

By adding CH_3COOH as a buffer, the etching process becomes much more gently and the quality of the core surface would be improved greatly. The velocity of etching is mainly influenced by the parameters such as the concentration of HF, the microstructure of silica, and the temperature. The higher the concentration of the HF in the corrosion solution, the more F^- reacts with silica, the higher the speed of the etching is. And when the temperature is high, the reaction between HF and the silica is exquisite and the etching speed is high.

One should note that handling this acid is very dangerous, and polished fibers are fragile to handle and difficult to replicate. As the reaction goes on, the concentration of F^- in the corrosion solution decreases gradually, and the etching speed slows down. However, the radius of the fiber becomes small, which indicates the area–volume ratio of the fiber becomes bigger. In this case, the speed of the etching will increase. So it's really difficult to control the speed of the etching speed carefully. To solve this problem, it is proposed to monitor the output power of the fibers (Zhuang et al. 2012). As shown in Figure 22.12a, a power meter is used to monitor the transmitted power, which is a function of the core diameter. Figure 22.12b shows the relationship between the output power and the etching time. The output power decreases slightly as the cladding is etched by the HF. When the cladding is eaten up and the core is etched by the etching solution, the strength of the output power decreases dramatically. When the core is eaten up, nothing is left in the sample cell, the relative strength of the output power keeps constant, which is the strength of the background noise. Zhuang et al. (2012) believe that according to the relative strength of the output power the radii of the silica-based fibers can be estimated accurately.

A possible alternative to etching in HF solution is plastic cladding fibers (PCS) (Malis et al. 1998; Potyrailo and Hieftje 1998; Cherif et al. 2003; Okazaki et al. 2003). The cladding of these fibers can be removed easily either mechanically or with nonhazardous solvents such as acetone, allowing more reproducible sensors.

Side-polished fiber can be fabricated by polishing a curved fiber embedded in, for example, a quartz block (Figure 22.13a). The cladding region on one side of the fiber can then be removed. Hussey and Minelly (1988) developed a wheel-polishing method, which allows the fabrication side-polished fibers with a length of several centimeters. It is important that such approach allows producing planar fiber-optic chip (FOC), which allows the use of advantages of both fiber-optic and planar technologies (Beam et al. 2007, 2012). According to Beam et al. (2012), fabrication of the FOC begins with stripping the jacket off of a small central section, 2–4 cm, of an optical fiber to expose the cladding. The stripped section of the optical fiber is then mounted in a V-groove

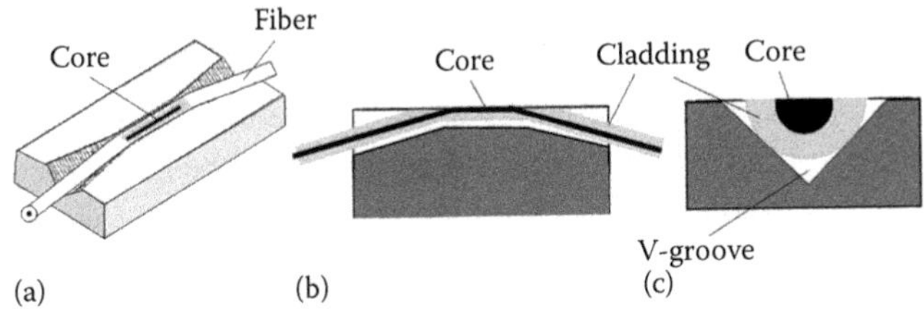

FIGURE 22.13 Fiber-optic chip (FOC) schematic of a side polished fiber mounted in a V-groove where red represents the exposed fiber core sensing platform. (a) Top down view; (b) and (c) are cross sections of the sensing platform along and across the fiber. (Adapted from Beam, B.M. et al., *Appl. Spectrosc.*, 61, 585, 2012. With permission of Optical Society of America.)

substrate using a two-part epoxy; the V-groove acts as a platform for spectroscopic investigation as well as supports the fragile fiber during the polishing procedure. Prior to mounting the fiber, the edges of the V-groove block must be polished to a 2° taper. Using a custom built assembly jig to keep tension on the fiber during hotplate curing, the fiber is laid into the V-groove and optical grade epoxy (Epotek 301) is applied liberally to ensure permanent immobilization of the fiber. The first generation FOC platforms were produced by side-polishing an optical fiber in a glass V-groove mount (Beam et al. 2007), but subsequent improvements on the FOC manufacturing process include replacing the glass V-groove with a customized etched silicon wafer support (Kendall 1979), improved polishing processing, and finally generating arrays of side-polished fibers. A planar interface would be advantageous for using standard planar deposition technologies including Langmuir–Blodgett (LB)-deposition (Doherty et al. 2005; Flora et al. 2005). In addition, due to its more robust supported platform, a planar design would be amenable for integration into microfluidic systems and sensor arrays. The disadvantages of this sensor are its time-consuming fabrication process and difficulty to fabricate long length of side-polished fibers.

Tapered optical micro/nanofibers (Figure 22.13b) are typically fabricated from a standard fiber. Tapering a SMF involves reducing the cladding diameter (along with the core) by heating the waist of the fiber while pulling the fiber's ends. When the optical fiber is tapered, the core-cladding interface is redefined in such a way that the light propagation inside the core penetrates to the cladding, which plays the role of the new core, and the external medium is the new cladding. The four heat sources most widely exploited in achieving micron-sized tapers are a flame (Kieu and Mansuripur 2006), a focused CO_2 laser beam (Kakarantzas et al. 2001), a microfurnace and an electronic arc formed between

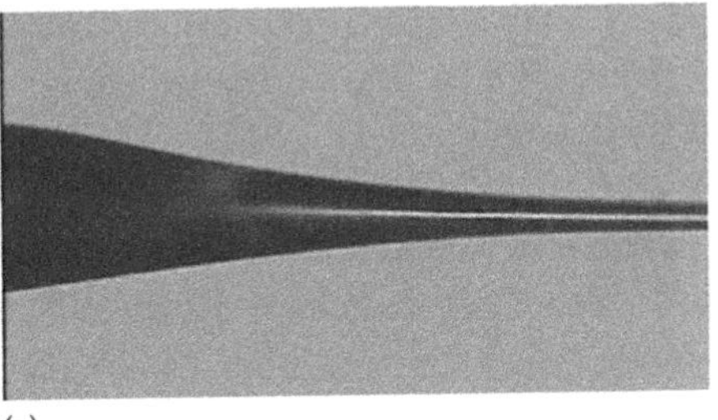

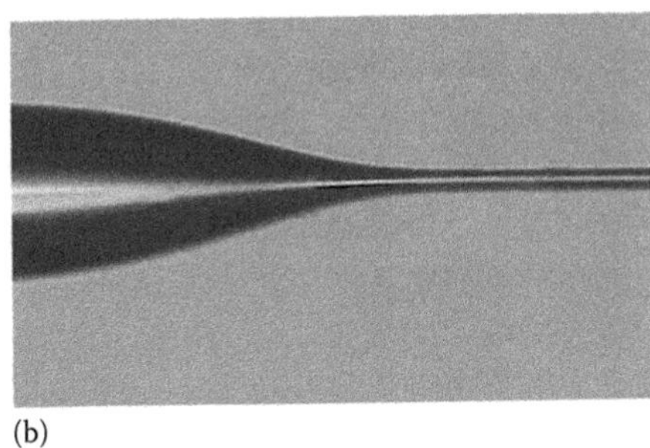

FIGURE 22.14 (a) The adiabatic tapered and (b) the non-adiabatic tapered. (Reprinted from Zibaii, M.I. et al., *Meas. Sci. Technol.*, 21, 105801, 2010. Published by IOP Publishing Ltd. as open access.)

a pair of electrodes, such as a fusion splicer (Corres and Arregui 2006). Also, chemical etching using hydrofluoric acid has been reported (Haddock et al. 2003). The tapered fiber consists of a conical segment, where the diameter of the fiber gradually decreases, a relatively long taper waist section, where the diameter of the fiber is small and uniform, and a second conical segment with its waist merging into the SMF again. Depending on the pulling conditions (pulling speed, length of the heated zone, pulling temperature, etc.), one can fabricate biconical tapers with different shapes and properties (Zibaii et al. 2010). Fiber tapers may be divided into two distinct categories: adiabatic and nonadiabatic (Figure 22.14).

A tapered fiber can be considered adiabatic if the main portion of the power remains in the fundamental mode (LP_{01}) and does not couple to higher order modes as it propagates along the taper. To avoid coupling between the fundamental mode and higher order modes, the taper local length scale must be much larger than the coupling length between these two modes. In other words, the relative local change in the taper radius has to be very small (small taper angle), as shown in Figure 22.14a (Snyder and Love 1983; Lacroix et al. 1988). It has been shown that nonadiabatic fiber tapers (abrupt taper angle, as shown in Figure 22.14b can be made so that coupling occurs primarily between the fundamental mode of the unpulled fiber and the first two modes of the taper waveguide (LP_{01}, LP_{02}), where, due to the large difference of the RIs of air and glass, the taper normally supports more than one mode (Zibaii et al. 2010).

Tapered optical silica-based micro/nanofibers are typically fabricated by the flame-brushing technique described in Brambilla et al. (2004), Jarzebinska et al. (2009), and Brambilla (2010) in which the fiber is stretched when it is heated with a stationary or an oscillating flame torch. For example, in Korposh et al. (2013), a single-mode silica optical fiber was tapered using the heat and pull technique. First, the polymer buffer coating was removed from a 50 mm long section in the middle of a ~1 m length of the single-mode optical fiber using a mechanical stripper. The stripped section of the optical fiber was then fixed on a three-axis flexure stage (NanoMaxTM, Thorlabs) and exposed to the flame produced by a gas burner (max temperature 1800°C) for approximately 60 s while the ends of the fiber were pulled in opposite directions using translation stages. The waist diameter of fiber tapers can be reduced down to less than one micrometer. For example, low loss biconical tapers with a submicron-diameter taper waist and a waist length of up to ~10 cm have been fabricated with the flame-brushing technique (Brambilla et al. 2004). However, the thin tapered region is not easy to handle, making it difficult to be used as practical sensors.

Typical configuration of the setup used for the manufacture of plastic nonadiabatic tapered optical fiber (NATOF) is shown in Figure 22.15. The CO_2 laser beam (SYNARD, 48-1SAL, 30 W full power) was focused via a Zn–Se lens of 2.5 cm focal length (Zibaii et al. 2010). The resultant spot size was about 120 μm on the fiber length. After collision with the SMF, the laser beam arrived at the Zn–Se lens 2 and a gold-coated mirror, which was embedded under lens 2 in the focal length. The reflected beam was then focused on the SMF. Therefore by using this setup, both sides of the SMF were heated. One end of the fiber was fixed on a translation stage and the other end was attached to a 5.12 g weight, which was on a slope so that the fiber was subjected to constant tension during the laser irradiation. The value of the utilized weight has been optimized to fabricate tapers with desired structures. The geometric characteristic of the fabricated taper depends on the power, number of pulses, beam shape, and duty cycle of the CO_2 laser. In our setup, the frequency, duty cycle, and number of pulses of CO_2 laser were 430–450 Hz, 40%–50%, and 90–100, respectively. A microscope was employed to monitor the tapering process.

D-fibers (Figure 22.11c) are drawn from a conventional preform but with half of the cladding region removed. This fiber allows continuous access to the evanescent field, thus very long interaction length. The use of D-fiber for gas detection has been demonstrated (Stewart et al. 1997) but the sensitivity is very low, on the order of 0.01% that of the open path sensor of the same length. Calculation shows that the use of a high index overlay on top of the flat surface enhances the sensitivity but the improvement factor is modest, typically 6–10 (Muhammad et al. 1993; Stewart et al. 1997).

With regard to the optical fibers prepared for SPR-based sensors, the most common variant is shown in Figure 22.16. The fabrication of such sensors Cennamo and Zeni (2014) have described as follows. The plastic optical fiber had a PMMA core of 980 μm and a fluorinated cladding of 20 μm. The taper ratio (r_i/r_o) was about 1.5 and the sensing region (L) was about 10 mm in length. The thickness of gold layer was about 60 nm. The sensor was realized starting from a plastic optical fiber, without protective jacket, heated (at 150°C) and stretched with a motorized linear positioning stage until the taper ratio reached 1.5. After this step, the POF was embedded in a resin block, and polished with a 5 μm polishing paper in order to remove the cladding and part of the core. After 20 complete strokes following a *8-shaped* pattern to

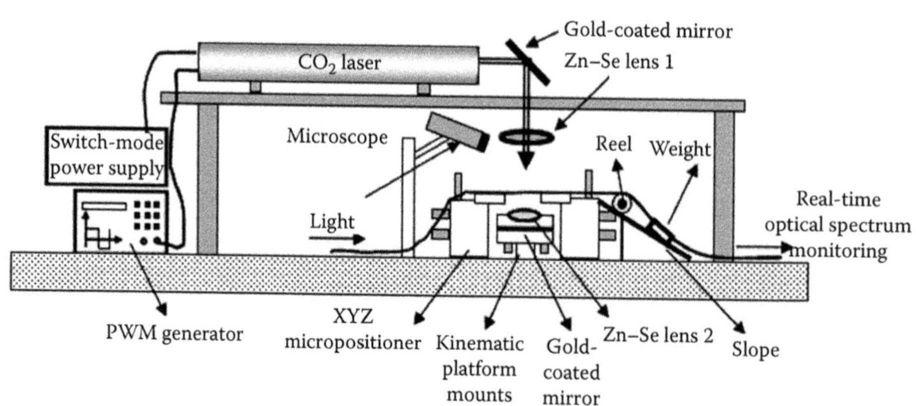

FIGURE 22.15 The fabrication system for a tapered fiber by the heat-pulling method. (Reprinted from Zibaii, M.I. et al., *Meas. Sci. Technol.*, 21, 105801, 2010. Published by IOP Publishing Ltd. as open access.)

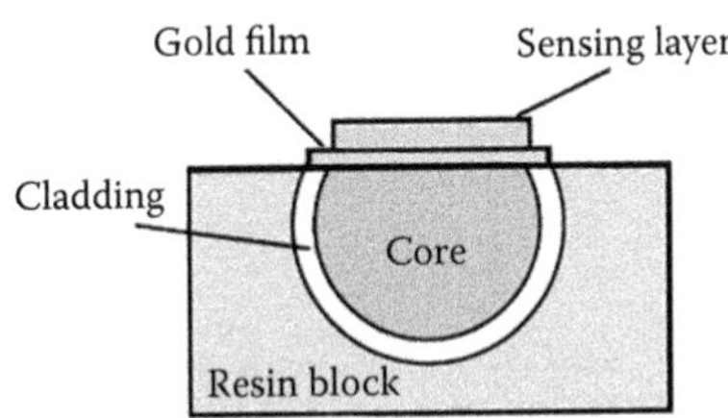

FIGURE 22.16 Section view of sensor system based on SPR in tapered POF. (Reprinted from Cennamo, N., and Zeni, L., Bio and chemical sensors based on surface plasmon resonance in a plastic optical fiber, in *Optical Sensors—New Developments and Practical Applications*, Yasin, M. et al. (Eds.), InTech, Rijeka, Croatia, 119–140, 2014. Published by InTech as open access.)

completely expose the core, a 1 μm polishing paper was used for another 20 complete strokes with a *8-shaped* pattern. The thin gold film was sputtered by using a sputtering machine. On the top of planar gold film it is possible to apply a layer of sensing material for the selective detection of analytes.

22.8 FABRICATION OF FIBER GRATING

Normal optical fibers are uniform along their lengths. In a simple fiber grating, the RI of the fiber core varies periodically along the length of the fiber, as shown in Figure 22.17.

As it was shown in Chapter 14, fiber gratings such as Bragg gratings and long-period gratings (LPG), being the main part of refractometric sensors, are intrinsic devices that allow controlling over the properties of the light propagating within the fiber; they are used as spectral filters, as dispersion compensating components and in wavelength division multiplexing systems (James and Tatam 2003; Dewra et al. 2015). As it was indicated before, fiber gratings consist of a periodic perturbation of the properties of the optical fiber, generally of the RI of the core. Thus, the fabrication of fiber gratings relies upon the introduction of a periodic modulation of the optical properties of the fiber (Kashyap 1999; James and Tatam 2003; Dewra et al. 2015). This may be achieved by permanent modification of the RI of the core of the optical fiber or by physical deformation of the fiber.

According to Hill (2002), an optical method is the most used for the fiber grating fabrication, including the fiber Bragg grating (FBG). Writing a fiber grating optically in the core of optical fiber requires irradiating the core with a periodic interference pattern. Historically, this was first achieved by interfering light that propagated in a forward direction along an optical fiber with light that was reflected from the fiber end and propagated in a backward direction (Hill et al. 1978). This method for forming fiber gratings is known as the *internal writing technique*, and the gratings were referred to as *Hill gratings*. The Bragg gratings, formed by internal writing, suffer from the limitation that the wavelength of the reflected light is close to the wavelength at which they were written (i.e., at a wavelength in the blue–green spectral region). However, in 1989, Meltz et al. (1989) have shown that the transverse holographic inscription technique where the laser illumination came from the side of the fiber is much more flexible. Exactly these methods are described in subsequent Sections 22.8.1–22.8.3. These fiber grating fabrication techniques may be classified to three main categories: (1) two-beam holographic, (2) phase mask, and (3) point-by-point techniques (Chen 2013; Dewra et al. 2015). Each technique has its merits and limitations and should be employed according to the specification requirement of the gratings to be fabricated. One should note that these techniques can be used in the manufacture of planar Bragg grating humidity sensors as well (Sparrow et al. 2009).

22.8.1 Two-Beam Holographic Technique

Figure 22.18 shows the two-beam holographic UV-inscription system (Hill 2002). The light from an UV source is split into two beams that are brought together so that they intersect at an angle, θ. The interferometric method usually uses an UV writing beam at 244 or 248 nm. The intersecting light beams form an interference pattern that is focused using cylindrical lenses on the core of the optical fiber. Unlike the internal writing technique, the fiber core is irradiated from the side, thus

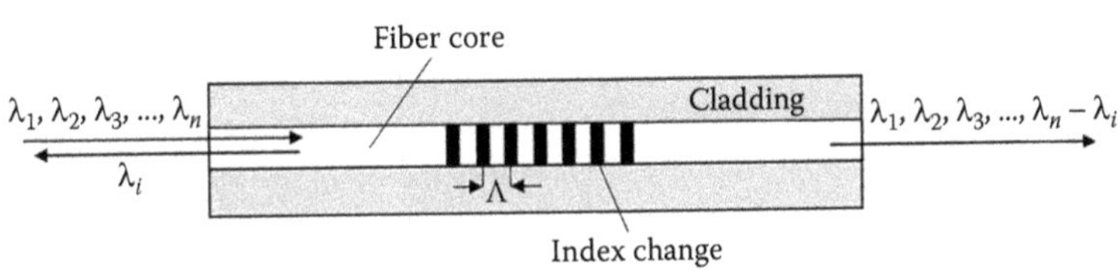

FIGURE 22.17 Schematic diagram of a fiber Bragg grating.

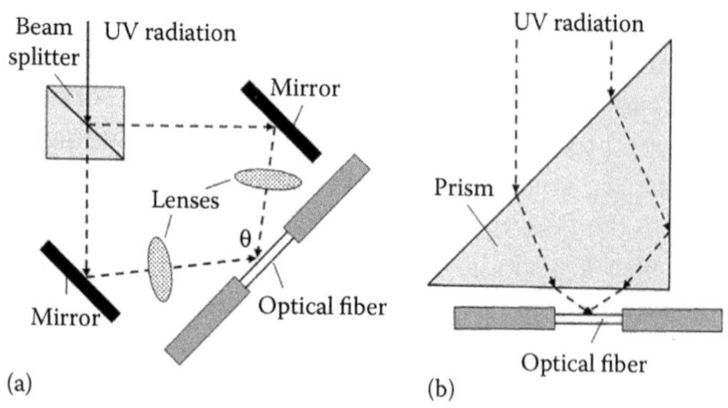

FIGURE 22.18 Fabrication of Bragg gratings using (a) interferometric scheme and (b) using prism method. (Idea from Claus, R.O. et al., Optical fiber sensors, in *Fiber Optic Handbook: Fiber, Devices, and Systems for Optical Communications*, Bass, M., and Van Stryland, E.W. (Eds.), McGraw-Hill, New York, 2002.)

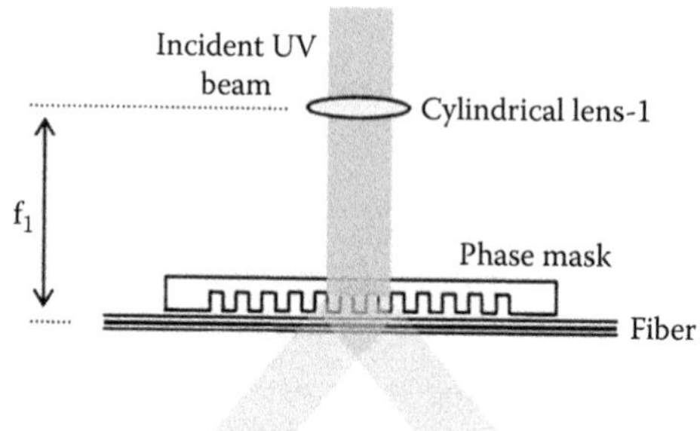

FIGURE 22.19 Fiber grating inscription by UV-beam scanning across a phase mask. (Reprinted from Chen, X., Optical fibre gratings for chemical and bio-sensing, in *Current Developments in Optical Fiber Technology*, Harun, S.W., and Arof, H. (Eds.), InTech, Rijeka, Croatia, 205–235, 2013. Published by InTech as open access.)

giving rise to its name *transverse holographic technique*. The technique works because the fiber cladding is transparent to the UV light, whereas the core absorbs the light strongly. Since the period, Λ, of the grating depends on the angle, θ, between the two interfering coherent beams through the relationship $\Lambda = \lambda_{UV}/2\ \sin(\theta/2)$, the Bragg gratings can be made so, that reflect light at much longer wavelengths than the UV light that is used in the fabrication of the grating. The holographic technique for grating fabrication has two principal advantages. Bragg grating could be photoimprinted in the core of fiber without removing the glass cladding. Moreover, the period of the photoinduced grating depends on the angle between the two interfacing coherent UV light beams. This means that the two-beam holographic method has ability to write gratings with arbitrarily selected wavelengths simply by adjusting the angle (θ) between the two beams (Meltz et al. 1989). Thus even through UV light is used to fabricate the grating, Bragg grating could be made to function at much longer wavelengths in particular in the range of 750–2000 nm, that is, in the spectral region that are of interest for the fiber-optic communication and optical sensing.

22.8.2 Phase-Mask Technique

The phase-mask technique, based on near-contact UV-beam scanning a phase mask, is one of the most effective techniques for reproducible FBG inscription (Claus et al. 2002) (Figure 22.19). The grating written by phase mask technique has a period of

$$\Lambda = \frac{\lambda_{UV}}{2\sin\theta_m} = \frac{\Lambda_{PM}}{2} \tag{22.4}$$

where Λ_{PM} is the period of the phase mask. The Bragg wavelength is then given by Erdogan (1997)

$$\lambda_B = 2n_{eff}\Lambda = n_{eff}\Lambda_{PM} \tag{22.5}$$

The phase mask is made from a flat slab of silica glass, which is transparent to UV light. On one of the flat surfaces, a one-dimensional periodic surface relief structure is etched using photolithographic techniques. The shape of the periodic pattern approximates a square wave in the profile. The optical fiber is placed almost in contact with and at right angles to the corrugations of the phase mask. UV light, which is incident normal to the phase mask, passes through and is diffracted by the periodic corrugations of the phase mask. The phase mask technique has the advantage of greatly simplifying the manufacturing process for Bragg gratings, while yielding high-performance gratings. In comparison with the holographic technique, the phase mask technique offers easier alignment of the fiber for photoimprinting, reduced stability requirements on the photoimprinting apparatus, and lower coherence requirements on the UV laser beam, thereby permitting the use of a cheaper UV excimer laser source (Matsuhara and Hill 1974; Hill 2002). Furthermore, there is the possibility of manufacturing several gratings at once in a single exposure by irradiating parallel fibers through the phase mask. The capability to manufacture high-performance gratings at a low per-unit grating cost is critical for the economic viability of using gratings in some applications. A drawback of the phase mask technique is that a separate phase mask is required for each different Bragg wavelength (Hill 2002).

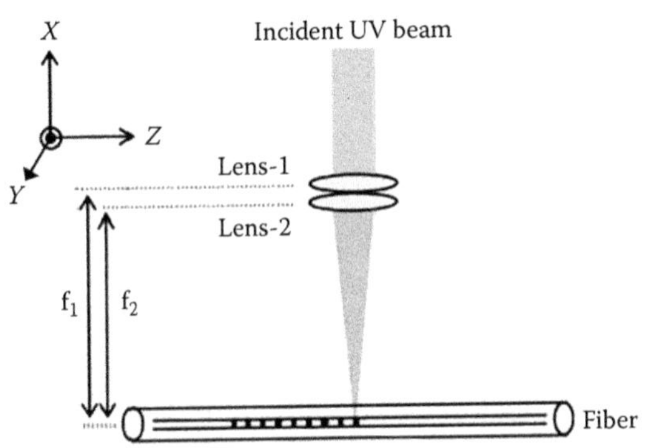

FIGURE 22.20 Schematic of LPG fabrication using point-by-point technique. (Reprinted from Chen, X., in *Current Developments in Optical Fiber Technology*, Harun, S.W., and Arof, H. (Eds.), InTech, Rijeka, Croatia, 205–235, 2013. Published by InTech as open access.)

22.8.3 Point-by-Point Technique

The third main grating fabrication technique is the point-by-point technique. In this method each index perturbation of the grating are written point-by-point (Shenk Strasser et al. 1996). As the grating is written a point at a time, it is a flexible method to alter the grating parameters, such as length, periodicity, and strength. Limited by the focused spot size of UV-beam, it is difficult to control translation stage movement accurately enough to write FBG structures, which in general have typical periods of ~0.5 μm at 1550 nm. Thus, the point-by-point technique is mainly used to fabricate LPG with periods ranging from 10 to 600 μm. As shown in Figure 22.20, in the point-by-point inscription system, two cylindrical lenses are added to focus the writing beam on the fiber to an approximate spot size of 20 × 20 μm in *z*- and *y*-dimension and a shutter is computer-programed to switch on/off with a 50:50 duty cycle to realize period-by-period print. The system has a great flexibility in fabricating LPGs with different periods, lengths, and strengths (Chen 2013).

22.8.4 Features of Long-Period Grating

22.8.4.1 UV Long-Period Fiber Gratings

As it was indicated before, the period of an FBG is approximately half a micrometer whereas the period of an LPG is typically several hundred micrometers (Bhatia and Vengsarkar 1996). Therefore, in comparison with Bragg gratings, the fabrication procedure of LPG is simpler (Claus et al. 2002; Martinez-Rios et al. 2012). Experiment has shown that for these purposes one can use methods such as the UV photosensitivity, the residual thermal stress, the photoelasticity, and the geometrical

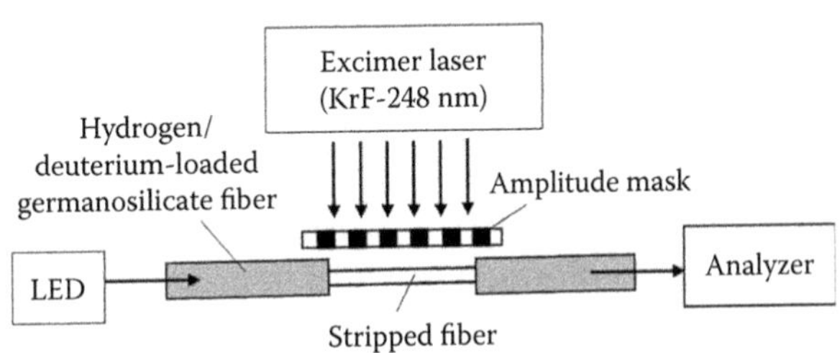

FIGURE 22.21 Setup to fabricate long-period gratings using an amplitude mask. (Idea from Claus, R.O. et al., Optical fiber sensors, in *Fiber Optic Handbook: Fiber, Devices, and Systems for Optical Communications*, Bass, M., and Van Stryland, E.W. (Eds.), McGraw-Hill, New York, 2002.)

modulation of the cladding structure (Hill et al. 1990; Dianov et al. 1997b; Hwang et al. 1999; Rao et al. 2004). In the first case, an UV light source is used to write the core index modulation in Ge-doped fibers by the photosensitive mechanism. For example, to fabricate LPGs, hydrogen-loaded (3.4 mol%), germanosilicate fibers may be exposed to 248 nm UV radiation from a KrF excimer laser through a chrome-plated amplitude mask possessing a periodic rectangular transmittance function. This is the most widely utilized method for the fabrication of LPGs (James and Tatam 2003). This method was also used in Bhatia and Vengsarkar 1996, Blows and Tang 2000, Chen et al. 2000, Guan et al. 2000. The RI change is associated with the formation of Ge-related glass defects. Figure 22.21 shows a typical setup used to fabricate such gratings. The laser is pulsed at approximately 20 Hz with pulse duration of several nanoseconds. The typical writing times for energy of 100 mJ/cm^2/pulse and a 2.5 cm exposed length vary between 6 and 15 min for different fibers. The fibers with high photosensitivity have been developed by co-doping the core with boron and germanium (Williams et al. 1993) and by hydrogen loading (Lemaire et al. 1993). The point-to-point technique that uses the same mechanism was described in previous Section 22.8.3. Both techniques are relatively simple to implement; however, the first one is more attractive to mass production, although the point-to-point technique is more flexible and cheaper than the amplitude mask technique. The main limitation of these techniques is that they only work in UV photosensitivity fiber.

The postfabrication tuning of the characteristics of a fiber gratings transmission spectrum is possible by reducing the fiber diameter by etching. The reduction in the cladding diameter changes the effective index of the cladding modes, resulting in an increase in the central wavelengths of the attenuation bands. The etching also changes the electric field profile of the cladding modes, resulting in a change in the overlap integral and a concomitant change in the coupling efficiency, and minimum transmission of the attenuation bands (Vasiliev et al. 1996; Zhou et al. 2001).

22.8.4.2 Residual Thermal Stress Long-Period Fiber Gratings

The heating and fast cool down of the glass alter the viscosity, it in turns, slightly modified the RI. In this case, the heating process and follow by a fast cool, of the glass can be used to freeze a periodical index modulation across the optical fiber structure (core and cladding). There are two used methods to generate the periodical index modulation by CO_2 radiation (heating absorption) and electrical arc (James and Tatam 2003; Martinez-Rios et al. 2012) (Figure 22.22). Both methods can be used in any kind of fiber. These methods induce a RI profile in the transversal plane, allowing the coupling to symmetric and asymmetric cladding modes. In the CO_2 radiation method, the optical fiber under tension is radiated through a germanium lenses with an IR signal at 10.6 μm point-to-point (Davis et al. 1998a, 1998b). At this wavelength, the silica is not transparent and absorbs the energy from the CO_2 laser, so that the fiber is heated and cooled at room temperature. The required power in optical fibers is 10 W in continuous wave CO_2 lasers. Davis et al. (1998a, 1998b) using this technique produced LPGs with high temperature stability and polarization insensitivity and with spectral characteristics that are unchanged even after annealing at 1200°C. The CO_2 laser exposure of the fiber was originally thought to result in the densification of the glass, and/or the relaxation of tensile stresses built into the cladding of fibers such as Corning SMF28 during fabrication (Kim et al. 2000). However, there is evidence to suggest that the change in the RI is a result of breakage of Si–O–Ge chains (Drozin et al. 2000). Structural LPGs in photonic crystal fibers can also be created using a CO_2 irradiation (Kakarantzas et al. 2002).

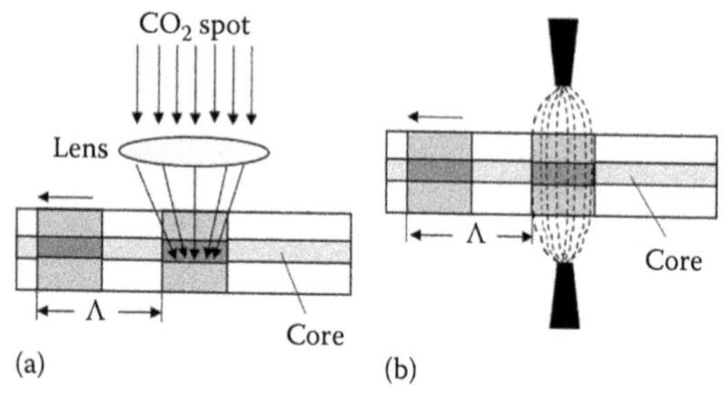

FIGURE 22.22 LPFG inscription by thermal stress techniques; (a) CO_2 irradiation, (b) electric arc discharge. (Reprinted from Martinez-Rios, A. et al., Long period fibre gratings, in *Fiber Optic Sensors*, Yasin, M. (Ed.), InTech, Rijeka, Croatia, 275–294, 2012. Published by InTech as open access.)

On the other hand, in the electric arc discharge method, the optical fiber is heated by an electric arc generated usually by a fusion splicing machine approach (Bjarklev 1986; Palai et al. 2001; Rego et al. 2001). For standard optical fibers, the parameters of electric arc discharge are: 200 ms, current 50 mA at tension of few grams. This technique relies upon a combination of up to four effects to generate the periodic modulation of the fiber properties. The mechanisms exploited include the induction of microbends into the fiber (Hwang et al. 1999), the periodic tapering of the fiber (Kakarantzas et al. 2001), the diffusion of dopants (Dianov et al. 1997a, 1998), and the relaxation of internal stresses (Rego et al. 2001). Such LPGs have been shown to operate at temperatures of up to 800°C without permanent modification of their properties (Humbert and Malki 2002), and, if annealed appropriately, they may operate at temperatures up to 1190°C (Rego et al. 2001). Typically, the electrodes of a fusion-splicing machine are used, exposing a region of fiber with a length of the order of 100 μm to the arc, limiting the minimum period of LPG that may be fabricated. Chemical etching of the cladding of a fiber to produce a corrugated structure has been shown to allow the generation of an LPG (Lin et al. 2001). According with the price arc discharge method is cheaper than CO_2 laser.

As it was shown the spectral characteristics of thermal induced long-period fiber gratings (LPFGs) are high stability in wide range of temperature, low loss insertion loss. The disadvantages of thermal induced LPFGs are: the loss bands shift and the depth decrease of the bands with the time. James and Tatam (2003) noted that other methods can also be used for the LPGs fabrication. For example, LPGs have been fabricated using irradiation with femtosecond (fs) pulses in the near-IR (800 nm) (Kondo et al. 1999). The irradiation is believed to cause a densification of the glass, and the RI change, and LPG spectrum is stable at temperatures up to 500°C.

22.8.4.3 Mechanical Stress-Induced Long-Period Fiber Gratings

The optical RI of glass can also be modulated when the glass is exposed to mechanical stress (Poole et al. 1994; Narayanan et al. 1997; Savin et al. 2000; Lin et al. 2001; Jiang et al. 2002; Martinez-Rios et al. 2012). This change in optical RI is due to the photo-elastic response of glass. In this case, the photo-elasticity can be used to induce a temporal periodical modulation of the RI in the core

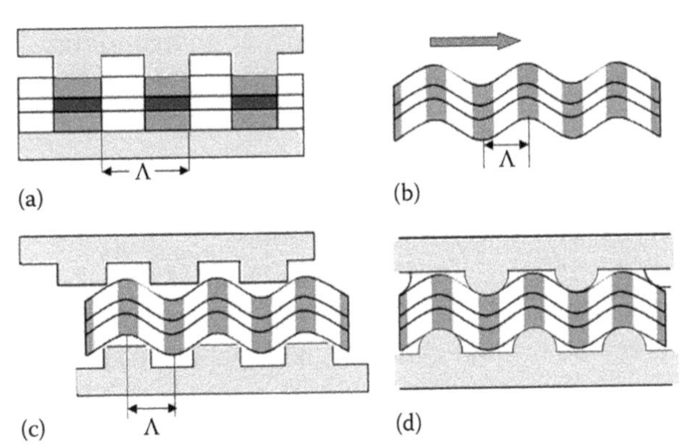

FIGURE 22.23 Stress-induced LPFGs, pressure points (a,b) microbending (c,d). (Reprinted from Martinez-Rios, A. et al., Long period fibre gratings, in *Fiber Optic Sensors*, Yasin, M. (Ed.), InTech, Rijeka, Croatia, 275–294, 2012. Published by InTech as open access.)

and the cladding to generate LPFGs. According to this, microbending and periodical pressure points in the fiber can be used to induce LPFGs. Different techniques have been reported to generate microbending such as: metallic grooved plates, strings, and flexure acoustic waves. In the case of periodical pressure points; metallic grooved plates, coils springs, and torsion. Figure 22.23 illustrates a typical example of periodical microbending and pressure points. As, its flexibility and cost relation plates can be used for microbending or pressure is one of the most used techniques. This technique presents many variants in the corrugated design.

The photoelasticity of the glass allows the inscription of LPFGs practically in any kind of fiber, this include, standard telecommunication fibers, photonic crystal fibers, between others. The amplitude of the RI periodical perturbation is the order of 10^{-4}–10^{-5}, so the pressure requite is less than 500 gr/cm, in the microbending case, the amplitude of the perturbation needed is in order of 200–500 μm. The stress induced LPFGs present loss bands with attractive and flexible spectral transmission optical properties. The loss bands are erasable, simple control of the depth (0–20 dB), bandwidth control (10–50 nm) and a large tuning range >250 nm. The drawbacks are high sensitivity to ambient temperature, and insertion losses are in the order of 0.2–0.3 dB (Martinez-Rios et al. 2012).

22.8.4.4 Etching-Induced Long-Period Fiber Gratings

The effective index in the core can be modulated by the geometric modulation of the optical fiber diameter by etching chemical method (Martinez-Rios et al. 2012). The etching induced LPFGs are also known as

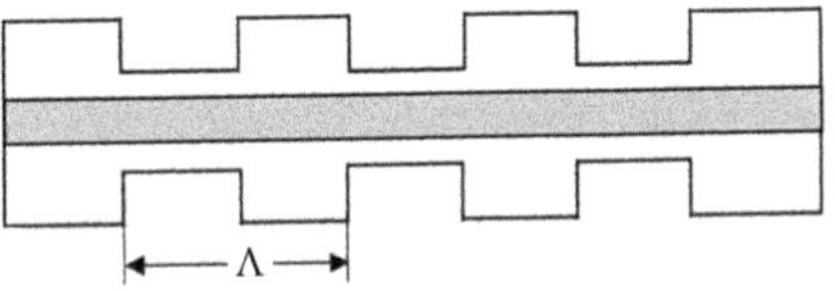

FIGURE 22.24 Corrugated LPFG. (Reprinted from Martinez-Rios, A. et al., Long period fibre gratings, in *Fiber Optic Sensors*, Yasin, M. (Ed.), InTech, Rijeka, Croatia, 275–294, 2012. Published by InTech as open access.)

corrugated LPFGs. This method, consist in locate a photomask over the fiber, and then the optical fiber is immersed in a hydrofluoric acid solution, in this way alternated regions of the fiber are protected of the hydrofluoric solution meanwhile other regions are divested hydrofluoric attack on the silica glass. The diameter in the desvasted regions is controlled by the exposition time, as illustrated in Figure 22.24. This method is not as popular as UV, thermal or mechanical stress because, their low repeatability and the difficult control of corrugated structured of the fiber. Physically the corrugated LPFGs are fragile to tension or bending stress. The loss bands are high sensitivity to twist because the torsion stress focus in the core diameter.

It was shown that ion implantation (Fujumaki et al. 2000), diffusion of dopants into the core (Dianov et al. 1997a, 1998) and relaxation of mechanical stress (Kim et al. 2000) can also be used for the fiber grating fabrication. A fiber grating has also been achieved by tapering the fiber (Kakarantzas et al. 2001).

22.8.5 Microfiber Gratings

In principle, a microfiber Bragg gratings (MFBGs) is a miniaturized copy of a standard FBGs. However, due to its much smaller diameter and much more compact overall size, to obtain an evident grating effect, the index-contrast of the microfiber grating should be higher (Lou et al. 2014). Examples of the fabrication of gratings on the base of microfibers can be found in Liang et al. (2005), Zhang et al. (2010b), and Lou et al. (2014). Liang et al. (2005) and Zhang et al. (2010b) fabricated a FBG on the silica microfiber with diameter of 6 μm. Fang et al. (2010) reported FBGs fabricated using a 2 μm diameter silica microfiber. Fiber gratings with even smaller diameter (1.8 μm) were fabricated by Liu et al. (2011). A microfiber LPG has also been reported for optical sensing. Using a fs laser to periodically modify the surface of wavelength-diameter (1.5–3 μm) microfiber, Xuan et al. (2009) reported a microfiber LPG with periods of 10–20 μm, which exhibited strong resonant dip as high as 22 dB around 1330 nm with only 10 periods.

22.8.6 Optic Fiber Sensors Fabrication by Laser Micromachining

Earlier we discussed method of fiber grating using UV radiation. However, it was established that fiber-optic grating sensors written by UV laser exposures cannot be used for high-temperature measurement due to its poor long-term stability under high temperature of more than 300°C. Moreover, fabrication of fiber-optic interferometric sensors is time consuming and labor intensive, and ease to be contaminated and damaged (Kersey et al. 1983).

Studies carried out in recent years have shown that many problems in manufacturing fiber grating (FG)-based sensors can be solved by using fs lasers, in particular 157 nm excimer lasers. These lasers are good candidates due to their high instantaneous power and high single-photon energy, offering the ability to ablate optical fiber sensing structures with high surface quality, and/or to change RI of normal silica fiber without special pretreating and doping. In particular, in contrast to traditional technology this technique does not require Ge-doped fibers with photorefractive effect.

When fs-laser is focused onto optical fiber, the energy absorption takes place through nonlinear phenomena such as multiphoton absorption (A number of photons could be absorbed simultaneously, when material is irradiated by high-power laser light beam.), tunneling and avalanche ionization. As long as the absorbed energy is high enough, the catastrophic material damage occurs, which leads to the formation of special structures (Glezer et al. 1996). Intensive index modulation of up to ~10^{-3} could be induced in fiber by fs-laser. If the absorbed energy increases further, a microstructure in optical fiber can be formed with designed shape and size.

Examples of manufacturing FG-based and LPG-based sensors can be found in Kondo et al. (1999), Mihailov et al. (2003), Hindle et al. (2004), and Martinez et al. (2004). The experimental setup of the fabrication technique is shown in Figure 22.25.

It was found that the gratings fabricated using fs-lasers are stable and do not erase after two weeks at 300°C (Mihailov et al. 2003). Other studies have shown that the FBGs obtained were stable at temperature of up to 900°C (Martinez et al. 2004). More detailed discussion of this technology one can find in Rao and Ran (2013). Rao and Ran (2013) believe that 157 nm excimer laser and fs-laser micromachining systems are powerful tools to fabricate novel in-line all-fiber FOS. These sensors have many advantages over conventional FOS, such as direct formation, easy mass-production with low-cost,

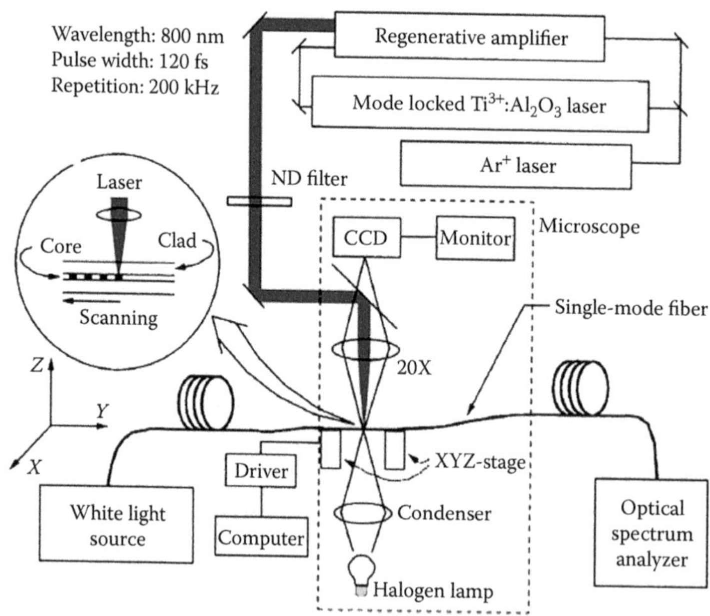

FIGURE 22.25 Experimental setup of fabricating LPFGs with femtosecond laser. The laser pulses used to induce refractive index changes were obtained from a regeneratively amplified Ti^{3+}:Al_2O_3 laser pumped by an Ar^+ laser. The pulse width was 120 fs, the wavelength was 800 nm, and the repetition rate was 200 kHz. The fiber, which was mounted on a computer-controlled XYZ stage, was irradiated point-by-point. An optical spectrum analyzer (OSA) simultaneously monitored the transmission spectra of the LPFGs that were being fabricated. (Reprinted from *Opt. Fiber Technol.*, 19, Rao Y.-J., and Ran Z,-L., Optic fiber sensors fabricated by laser-micromachining, 808–821, Copyright 2013, with permission from Elsevier.)

good reproducibility, excellent optical performance, and high temperature stability, making it ideal to form a new generation of in-line all-fiber-optic sensors for many applications.

22.8.7 Fiber-Optic Couplers

Distribution of optical signals into two or more output fibers is a task that constantly arises in the manufacture of optical sensors. This problem is usually solved using fiber couplers (Figure 22.26). An optical coupler is a passive optical component in optical fiber systems that can combine or split transmission data (optical power) from optical fibers. A coupler basically brings together two or more fibers cores to interact with each other via evanescent field (Section 14.1.1). The necessary interaction between fibers and its invariance during sensor operation is achieved by fusing fibers together (Figure 22.27). Several techniques described in Davis et al. (1986) have been developed for this operation. One of such technologies was developed by Sheem (1980). According to this technology, the fibers are cleaned carefully after removing the jacket. Then they are twisted together and while remaining twisted they are etched to remove most of the cladding, leaving approximately 1 or 2 μm of cladding around the core. The diameters of the fibers after etching are less than 10% of their initial diameters, therefore the resulting sections of fibers are quite fragile. The joint is held fixed (ruggedized) by either potting in an index matching material or by fusing under an axial tension that prevents the fiber from sagging due to gravity and at the same time stretches the fiber slightly, thus forming a biconical taper. The biconical taper arrangement mentioned above is shown in Figure 22.26.

When two or more fiber-optic cables are fused together, light entering an input fiber can appear at one or more output fibers and its power distribution potentially depending on the wavelength and polarization. This coupler performs the function of an optical splitter (Figure 22.28a). A fiber-optic coupler can also combine the optical signal from two or more fibers into a single fiber. In this case, we have optical combiner (Figure 22.28). Based on the type of inputs and outputs, couplers are categorized as star couples (multiple input and output) and tees (single input, two outputs).

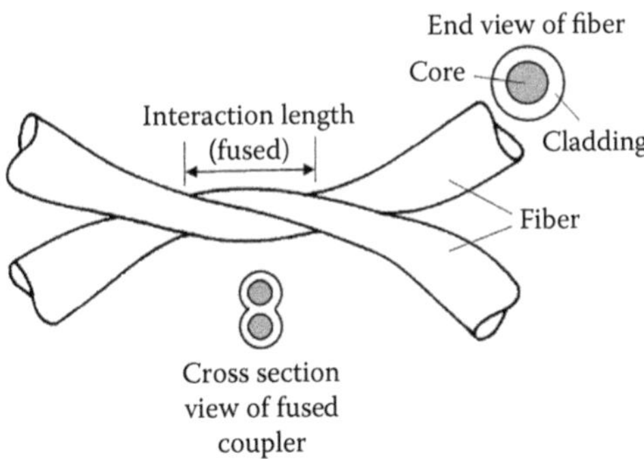

FIGURE 22.26 A biconical (tapered) fiber-optic coupler.

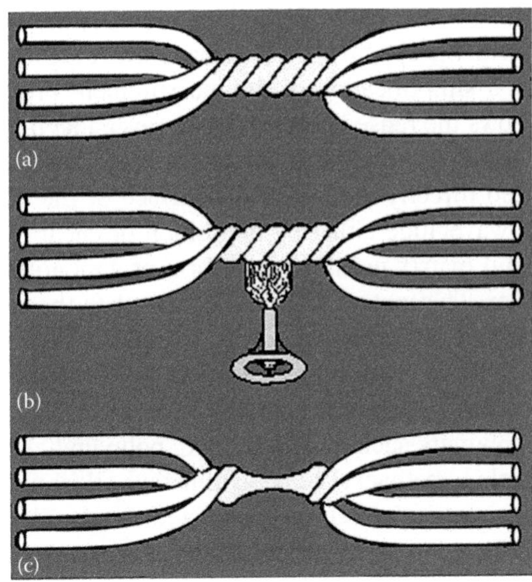

FIGURE 22.27 Fabrication of a fused biconical taper coupler (star coupler): (a) fiber twisting, (b) fiber fusing or melting, and (c) fiber tapering.

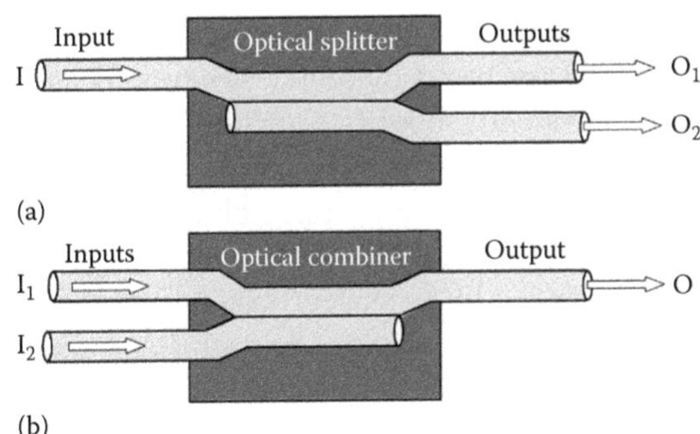

FIGURE 22.28 (a) Optical splitter and (b) optical combiner.

22.9 COATING OF FIBERS

22.9.1 Techniques of Fiber's Immobilization

In optical humidity sensors, the humidity recognition element is usually immobilized on an inert and stable solid matrix. Therefore, it is evident that, in the case of optode sensors, operating parameters depend strongly on the recognition element and technique utilized for the immobilization of the humidity-sensitive reagent.

Immobilization facilitates the integration of the reagent phase with the optical fiber or planar waveguide, contributes to a simplification of the measurement procedure and allows the reuse of recyclable or reversible reagents. The immobilized reagent system can be conveniently localized directly on one of the surface of the fiber or planar waveguide either by deposition or encapsulation in a solid matrix usually in the form of a monolith or a thin-film. The matrix serves to encapsulate the reagent such that it is accessible to the humidity while being impervious to leaching effects. The porous membrane on which the reagent is immobilized can be fabricated from polymers such as polystyrene, polyvinyl chloride, polymethyl methacrylate, polydimethyl siloxanes, and cellulose derivatives (Davletbaev et al. 2014). The main requirements to the material of the reagent carrier are optical transparency, high sorption rates, ease of synthesis, processability, inertness to the reactants, stability in acidic and alkaline media, and high sensitivity to analytes (Zolotov et al. 2002). At present, the natural cellulose polymer is the most widely used as a matrix. It is believed to be an ideal support for indicators since it has a high permeability for water and ions, and it can be used in both acidic and basic solutions. Various indicators can be either covalently bound to cellulose, or physically entrapped in it (where the hydrophobic interactions may also be involved). In the covalent binding method, the cellulose is first hydrolyzed and then activated. Finally, cellulose is soaked in the reagent solution and then dried. The process can be one-stage or multistage and some reagents can be added to improve the performance of the sensing membranes. The same process is used for other solid carriers. Controlled pore glass, silica gel, carbon nanotubes, and alumina have also been employed as the solid support for the immobilized reagent. Silica gel usually is synthetically produced from solutions of sodium silicate or silicon tetrachloride, or substituted chlorosilanes/orthosilicates. The active surface of silica gel with a large surface area is of great importance in the adsorption of tested agents. The modification of the silica gel surface for the analytical reactions is carried out in two separate ways as follows: the organic functionalization, where the modifier is an organic group; and inorganic functionalization where a group fixed on the surface can be an organometallic composite or metal oxide (Collinson 2002). Recognition element can also be situated on a measurement cell, which is coupled to an optical fiber system.

For the immobilization of the reagents in the molecular recognition element of a humidity sensor, a variety of techniques that vary in complexity and applicability can be used (Chrisey and Hubler 1994; Gardner and Bartlett 1995; Harsanyi 1995, 2000; Kumar and Sharma 1998; Matsumoto et al. 1998; Pique et al. 2003; Kern and Gadow 2004). For example, for incorporating polymers into humidity sensors, polymers can be presynthesized, stretched into sheet forms separately from the sensor structures, and only then can be attached (typically by gluing) to inorganic sensor surfaces. Polymer films can be also formed directly on the sensor surfaces. In this case, the synthesis (polymerization) and shaping process occur on the sensor surface. Certainly, the second approach is more progressive and is therefore the more commonly used technique for the fabrication of thin polymer layers used in the majority of humidity sensors. These chemical methods are described in detail in various comprehensive reviews (Malkin and Siling 1991; Kumar and Sharma 1998; Malinauskas 2001; Reisinger and Hillmyer 2002). A part of polymers formed by chemical polymerization can be deposited spontaneously on the surface of various materials immersed in the polymerization solution. The distribution of the resulting polymers between the precipitated and deposited forms depends on many variables and varies within a broad range. To coat materials with a polymer layer, it is desirable to shift this distribution toward the surface-deposited form, whereas a bulk polymerization should be diminished as much as possible. This can usually be achieved by choosing appropriate reaction conditions, such as the concentration of the solution components, the concentration ratio of oxidant to monomer, the reaction temperature, and an appropriate treatment of the surface of the material to be coated by conducting polymers. Although a bulk polymerization cannot be completely suppressed, a reasonably high yield of surface-deposited polymers can be achieved by adjusting the reaction conditions (Malinauskas 2001). Some fibers can be coated with polymer layers by immersing them into an electrolyte, where electrochemical polymerization takes place, or by a vapor-phase treatment of oxidant-containing carriers with the monomers (Malinauskas 2001).

When developing humidity sensors, besides chemical methods, physical methods can also be used. Physical immobilization could be achieved through adsorption, entrapment, or electrostatic attraction. The physical methods for immobilization involve simple procedures and mild conditions; however, the binding of the reagent molecule is weak. Therefore, due to the weak adhesion, the detachment from the support or from the fiber is observed sometimes, which makes the use of probes realized in this manner unsuitable for many applications. However, the choice of deposition technique, physical or chemical, depends on the physicochemical properties of the material, the film quality requirements, and the substrate being coated.

One should note that prior to the deposition of humidity-sensitive materials, fibers, and substrates are generally subjected to chemical treatment aimed at cleaning the surface and improving adhesion of the coating. For example, Korposh et al. (2013) have cleaned the previously stripped section of the optical fiber with concentrated sulfuric acid (96%), then rinsed several times with deionized water, and finally treated with 1 wt% ethanolic KOH (ethanol/water = 3:2, v/v) for about 10 min with sonication to functionalize the surface of the silica core with OH groups.

According to Brenci and Baldini (1991), Elosua et al. (2006), and McDonagh et al. (2008), generally accepted immobilization techniques have the following advantages and disadvantages:

LB films: Until now, the realization of LB films with an incorporated chromophore has been a very promising technique for the realization of fiber-optic detectors, also because these films are, in principle, applicable to the outside surface of the fiber, thus realizing evanescent-wave sensors that are extremely compact and miniaturized (Acharya et al. 2009). The LB technique is based on the deposition of layers with hydrophobic and hydrophilic behavior. Special equipment is required to accomplish this deposition technique, in which the bilayer to deposited onto the sensor is spread onto ultrapure water, forming a nanometric surface; when the substrate, in this case the optical fiber, is introduced in the solution, a new layer get deposited onto the surface (Gutierrez et al. 2001). The possibility of depositing ordered films with known and controlled thickness (in the range ±2.5 nm) is the main advantage of the LB technique. In principle, the LB technique can be used to prepare mono- and multimolecular layers and architectures with high perfection, different layer symmetries, and molecular orientations. This method, however, has very low technological effectiveness, making its application in a real chemical sensor fabrication processes unlikely. The number of polymers, which can be used for preparing polymer films with the LB method, is also very limited.

Adsorption: Adsorption involves the reagent molecules being held on the surface of the solid substrate through physical forces such as hydrophobic forces and hydrogen bonding. The simplest method consists of dissolving the compound of interest in a solvent that does not alter its sensing properties and then, dipping the fiber in this mixture. This technique is known as a dip coating. One should note that the dip-coating method continues to be one of the most widely methods used in the technology of FOS fabrication (Kaneko and Nittono 1997; Bariain et al. 2000). This method is a more commercially attractive alternative to costly vapor deposition technologies (Kern and Gadow 2002; Herbig and Libmann 2004). Depending on the number of times the fiber planar or waveguide are dipped and other parameters such as immersion speed, or the type of humidity dye, the new claddings will have different thicknesses and morphology (Arregui et al. 2002). An alternative option consists of leaving the fiber dipped in the solution for a certain time (Scorsone et al. 2004). This procedure is employed when preparing sensors in transmission configuration. The fiber is dipped perpendicularly into the mixture, so a thin layer is fixed along it. In the case of reflection sensors, the layers deposited onto the end of the pigtail do not conform a regular shape but one similar to a match head. Although simple, when preparing reflection sensors, this method is limited by its poor reproducibility.

The process of dip-coating offers several unique properties, including the ability to use a wide range of compositions and the ability to apply coatings to complex substrates, both of which make this process advantageous for coating fibers with sensing materials. For example, Herbig and Libmann (2004) found that by using a dip-coating procedure, the sol–gel technique could be used to apply a coating to individual fibers.

The sol–gel process provides a relatively benign support matrix for the immobilization of analyte-sensitive reagents and dyes. The basic process involves the hydrolysis and polycondensation of the appropriate metal alkoxide solution to produce a porous glass matrix in which the reagent is encapsulated in a nanometer-scale cage-like structure and into which the analyte molecules can diffuse (Brinker et al. 1992; Podbielska et al. 2006). The sensing material is added to the sol–gel solution while it is still in the liquid phase, and then the fiber is dipped into the mixture. After drying the deposition, an optically uniform porous matrix doped with the analyte fixed onto the fiber is achieved. These mixtures are usually made of silica, the same material as standard optical fiber, so then the transmission losses are minimized. The main advantage of this technique is the possibility it offers for producing these glass structures, with the chromophore incorporated, and shaped to the form of the fiber. In this way, a section of fiber is realized and it is sensitive to a certain chemical species, and the problem of the optode simply remains the connection between two fibers, with all the ensuing advantages of simplicity, compactness, and miniaturization. Sol–gel solutions deposited with dip-coating technique are a typical combination used in recent years to implement evanescent wave sensors. Standard sol–gel materials have a nonordered, amorphous structure where diffusion of analytes can be limited by the random microporosity of the structure. Mesostructured porous films, on the other hand, have a large open porosity, which can offer enhanced diffusion and accessibility for analytes. For example, mesoporous sol–gel sensor films can be realized via an evaporation-induced self-assembly (EISA) approach (Grosso et al. 2004; Nicole et al. 2004). These films have a highly structured mesoporosity and excellent optical quality.

Screen printing, spin-coating, and dip-coating technologies are also suitable when depositing polymers in the paste forms. A screen printing and spin-coating are intended usually for flat semiconductor substrates and cannot deposit easily uniform films on complex geometries, like in optical fibers, whereas a dip-coating technique can be successfully used for the deposition of uniform coatings onto optical fibers. The dipping method technique depends on the solubility of the conductive polymers, previously synthesized by the chemical polymerization technique. In this case, the surface to be coated is enriched either with a monomer or an oxidizing agent, and is then treated with a solution of either oxidizer or monomer, respectively. A major advantage of this process is that the polymerization occurs almost exclusively at the surface; no bulk polymerization takes place in the solution. For some polymer materials, the surface can be enriched with a monomer by its sorption from the solution. Enrichment of the surface by an oxidizer can be achieved either by using an ion-exchange mechanism or by deposition of an insoluble layer of oxidizer. The disadvantage of this process is that it is limited by materials that can be covered or enriched with a layer of either monomer or oxidizer in a separate stage, preceding the surface polymerization. In addition, this method is not useful for controlling the thickness of the coatings on the nanometer scale.

In *electrostatic attraction*, the reagent molecule is immobilized through the charged sites occurring in its structure and held by strong coulombic forces on ionic sites in the solid support, which is usually an ion exchange resin. This deposition technique is also known as a molecular self-assembly (Liu et al. 1997), electrostatic self-assembly (ESA) (Arregui et al. 1999; Song 2014), or layer-by-layer (LbL) assembly (Renneckar 2009). During last two decades the number of works on this topic has increased exponentially and some reviews permitted to understand the current state of the art (Schreiber 2000; Caruso 2001; Adams et al. 2003; Schonhoff 2003; Hammond 2004; Thunemann et al. 2004; Richardson et al. 2016). The versatility of LbL method for the synthesis of materials permits the application of this technique to design or fabrication of different structures on the tip or the cladding of the optical fiber.

This method takes advantage of a property of the matter, which is called a self-assembly. Certain type of molecules is able to orient and assemble spontaneously, forming complex structures. The LbL-assembly is produced due to electrostatic attraction forces between molecules. These forces can be stronger or weaker depending on whether these binds are covalent, ionic, hydrogen or van der Waals bonding. As a consequence, the self-assembly is possible with substances, which may present some kind of charge when dissolved, leading to forming positive or negative charged compounds. Thus, several layers of positive and negative materials can be assembled, forming a thin-film in a nanometer scale onto the desired substrate (Figure 22.29). This characteristic is very interesting in terms of fabrication, since an accurate control of the process can be done, stopping the deposition when desired. The final results of the process will strongly depend on how the deposition is designed.

Among the advantages this technique gives, it is interesting to remark the versatility, not only in the materials to be deposited (polymers, biological substances, and metal oxides), but also

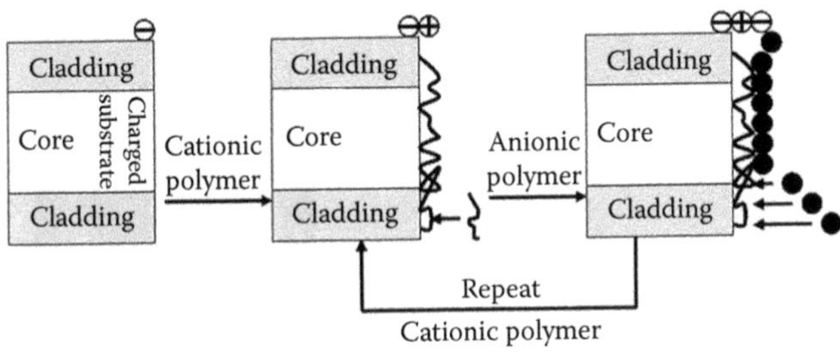

FIGURE 22.29 ESA method schematic construction process. (Reprinted from Elosua, C. et al., *Sensors*, 6, 1440, 2006. Published by MDPI as open access.)

in the substrates used and their size, since the deposition can be carried out in a high quality manner as long as the dipping recipient is large enough. In this sense, a LbL-assembly can be performed onto a wide variety of substrates, such as ceramics, metals, and polymers of different shapes and forms, including planar substrates, prisms, lenses, and fibers (Del Villar et al. 2004). However, it is important to know which compounds are susceptible to be deposited by this method, because not all the substances can be charged to be deposited. In particular, the main group of substances that can be used for LbL-assembling is that formed by polyelectrolytes such as polycationic and polyanionic solutions, which have opposite charged polymers chains.

The LbL-assembly technique involves different steps, which can be scaled to an industrial level due to their easiness to be carried out. This process can be described as follows. First of all, the substrate is cleaned and treated to create a charged surface. After this step, the substrate is alternately dipped into solutions of cationic and anionic substances to create a multilayer thin-film. If the initial charge of the substrate is positive and the materials to deposit are polyelectrolytes, the first monolayer will be a polyanion, whereas if it is negative, the first deposited monolayer will be a polycation. The result of a positive monolayer and a negative monolayer can be called *bilayer*. In this way, a multilayer structure is formed by electrostatic attraction between each bilayer and the bilayer previously deposited. After each polyelectrolyte immersion, the substrate is rinsed in ultrapure/deionized water to remove the excess of material. At the end of the process, the molecular species of the anionic and the cationic components and the long-range physical order of the layers determine the resulting coating properties. It is important to note that the polyanions and polycations overlap each other at molecular level, and this produces a homogeneous optical material. The composition and thickness of an individual bilayer can be controlled by adjusting the deposition parameters (concentration of solutions, pH, temperature, immersion times, etc.) (Choi and Rubner 2005). Typical individual layer thickness values ranged between 0.5 and 15 nm. However, it has been studied with other materials that the range can be widened to 60 nm (Del Villar et al. 2005). The humidity-sensing material can also be added into one or both of the polymeric solutions.

To perform a classic LbL-assembly, the first approach is to prepare the positive and negative solutions and alternately immerse the substrate in them. However, there is an increasing interest in depositing using LbL-assembly but varying not only the deposition parameters but the way of deposition. In this sense, instead of just dipping, the polyelectrolytes can be sprayed, like in *spray-coating* (Schaaf 2012) or stirred while depositing, like in agitating LbL-assembly (Fu 2011). The goal of spray-coating is to increase the smoothness of the thin-film by depositing little drops of material, what is an enhancement in comparison to just dipping the substrate in the solution.

Tolstoi (1993) and Tolstoy (2006) has shown that the principles of LbL-assembly can also be used in the deposition of metal oxides. Description of this technology and its possibilities can be found in Tolstoy et al. (2010), and Korotcenkov et al. (2015).

Covalent bond: At the moment, coupling of the optically-sensitive reagent by means of a covalent bond seems to be most promising, because it guarantees compactness for the probe and, at the same time, completely avoids losses of the chromophore. Chemical immobilization yields a more stable reagent phase, since the reagent molecules are strongly bound to the solid support. A much-followed method for the coupling of the chromophore exploits a silylation process, by means of which the reagent is coupled either to the fiber or to a fixed support. This method often involves several reaction steps in the synthesis or modification of the reagent and/or the support material in order to realize stable chemical bond(s) between them. The solid substrate usually is activated by functionalizing it through the introduction of an aldehyde or amino group.

Interaction of reagent and support matrix: Analysis has shown that parameters of optical sensors are strongly dependent on the matrix. It was found that in order to be efficiently encapsulated in the support matrix, the reagent must be soluble and homogeneously entrapped in the support material (McDonagh et al. 2008). In the case of sol–gel matrices, for example, it was established that the counterion and sol–gel solvent should be chosen to facilitate homogeneous distribution of the dye in the matrix. Research has also established that the optical properties of some reagents are very sensitive to the environment; thus, often these reagents will experience spectroscopic shifts when encapsulated in a solid matrix. Many optical sensor reagents photodegrade under conditions of high-intensity illumination, and in many cases the matrix has an influence on the degree of photodegradation. In general, reagents entrapped in rigid supports exhibit increased photostability due to reduced ligand photodegradation compared to a solution environment (Orellana et al. 2005). It has been shown that reagents encapsulated in sol–gel matrices had higher photostability than in polymer matrices. It has been suggested that the more organic polymer environment provided less stability than the more inorganic sol–gel matrix.

22.9.2 Features of Fiber Coating by Physical Methods—Evaporation and Sputtering

For deposition of metals onto fibers, conventional methods such as vacuum deposition, electron beam evaporation or sputtering can be used.

Molecular beam epitaxy (MBE) and chemical/physical vapor deposition (PVD) (Asatekin et al. 2010; Mahan 2010) can be also used for forming various coatings. In principle, the formation of coatings on the fibers using these methods takes place in the customary manner. It is only necessary to take into account the following factors:

- Deposition takes place on a small area and, as a result, the process has very low effectiveness of use of the deposited material.
- All sides of the fiber need to be covered uniformly. A scanning electron microscope (SEM) image of a fiber with a covering, deposited using the PVD method without rotation is shown in Figure 22.30. When coating the fiber end, this problem is absent.

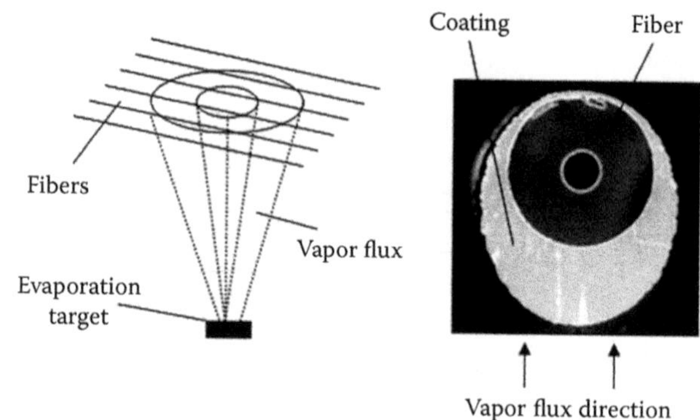

FIGURE 22.30 Schematic diagrams and SEM image for fiber coating by thermal evaporation.

- The fiber may be too long and have poor thermal contact with the substrate's holder.
- The fiber may be polymeric, which may impose a series of limitations on the thermal parameters of deposition.
- The fiber has lowered mechanical durability, whereas the covering needs to have high adhesion because the fiber may bend.

This means that coating methods for fibers do not require specific changes in the deposition process. Therefore, to deposit sensing material on the fiber, the same modes may be used as for deposition on substrates with large particle sizes. The details of the deposition on the fiber become apparent only when specific constructions of the substrate's holders and heaters are used, which fix the fiber's position, rotation, uniform heating, and displacement (Kashima et al. 1991; Kaneko and Nittono 1997). However, at present there are no fixed rules for elaborating these units. Every designer resolves this problem individually, taking into account the deposition method being used and available equipment. For example, an effective way of improving the uniformity of a fiber's PVD covering using the PDV method is through simultaneous use of several deposition sources located around the fiber (Kashima et al. 1991), as shown in Figure 22.31.

The tool, which is used to hold the fibers in the coating chamber, is another area that is critical to the successful production of fiber coatings. As most fibers must be coiled and packed into the tool, it is important that the fiber remain unstressed. This requirement may be a necessitate additional tooling in a standard coating chamber. For example, tooling in the form of a drum can be used for this purpose. Even in such cases, however, the number of fibers that can be coated is limited.

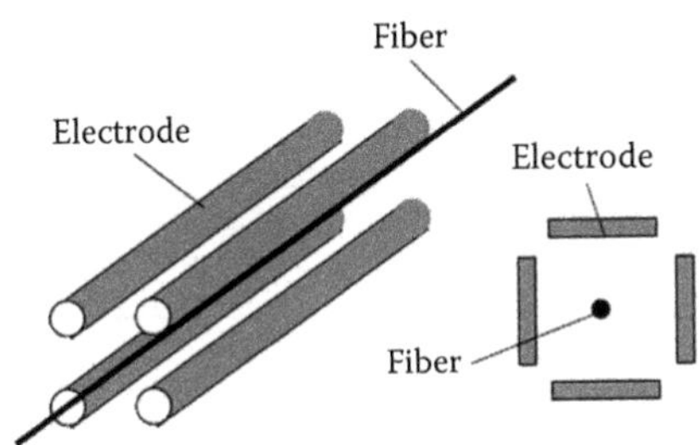

FIGURE 22.31 Apparatus with multielectrode configuration. (Reprinted from *Mater. Sci. Eng. A*, 139, Kashima, T. et al., Development of the quadrupole plasma chemical vapour deposition method for low temperature, high speed coating on an optical fibre, 79, Copyright 1991, with permission from Elsevier.)

Another problem can be connected with the application of epoxy for fiber fixing. Outgassing of the epoxy in a vacuum chamber affects the adhesion of the coating to the fiber, the packing density of the coating, and the RI of the deposited films.

During deposition of the sensing material on polymer fibers using magnetron sputtering, it is necessary to take into account that, depending on the evaporation conditions, PVD-coated fibers can present quite different surface properties (Dietzel et al. 2000). Deposition of reactive metals induces both textural and chemical changes on the fiber surface. Nitrogen and carbon react with the polymer surface and form new structures as a result of an etching process. A higher concentration of reactive gas leads to a more drastic recombination of the polymer surface and results in a fibroid fiber structure. Dietzel et al. (2000) have also established that different noble and reactive gases produce different layer adhesion. Thus, nitrogen and acetylene plasmas lead to better layer adhesion than argon plasma does.

Temperature control of the fiber during coating is another parameter that must be monitored. Research has shown that the RIs of a *cold* coated film are usually less than that of a hot film, even with the use of ion-assisted deposition. In many cases, these RIs are also nonuniform. If the fibers have epoxy or plastic jackets, as is usually the case, they cannot be exposed to high temperatures. The temperature controller of the coating system is usually programed to a temperature between 35°C and 90°C; however, the heat from the electron-beam gun or ion gun will raise the temperature of the fiber during coating. If the temperature of the fiber exceeds a safe limit, the ion or electron gun must be shut off to allow the fiber to cool down.

In addition to densifying the coating, an ion gun may also be used to clean the surface prior to coating, thus improving adhesion of the coating to the fiber tips. For this to be possible, the ion gun must be capable of handling oxygen gases, because an oxide coating is the most-robust that can be applied to a fiber.

Of course, there are many different methods that can be used to form coatings on the surface of optical fibers. For example, Bialiayeu et al. (2011) have developed the method to prepare optimized gold coatings for IR SPR sensors (Chapter 19) by electroless plating. This method is based on the reduction of metallic ions from a solution without applying an electrical potential (Hrapovic et al. 2003). Moreover, they have shown that tilted FBG sensors could be used to monitor in real-time the growth of gold nanofilms up to 70 nm in thickness and to stop the deposition of the gold at a thickness that maximizes the SPR (near 55 nm for sensors operating in the near-IR at wavelengths around 1550 nm). For coatings that are too thin, the SPR resonances broaden and weaken. On the other hand, when the thickness is too large, the light cannot tunnel across the metal to excite the Plasmon polariton on the outer surface. Bialiayeu et al. (2011) have shown that films, deposited using developed method, were highly uniform around the fiber circumference. The authors believe that plating is a conformal process, whereas all exposed surfaces of the same material in the plating bath are coated simultaneously and uniformly. This means that plating is a batch process in which a large number of devices can be coated simultaneously. Therefore, plating is an ideal method to coat optical fibers with nanoscale layers of metals.

It is clear that every method discussed earlier has its own advantages and disadvantages, making it necessary to choose the appropriate method of film deposition of forming for each specific engineering application, taking into account the properties of the deposited material and the possible consequences for the sensors parameters during the application of this method. The ideal method in any application should meet the following criteria:

- Compatibility with the process of manufacture of the humidity sensor
- No impairment of, or effect on, the properties of the fibers or waveguides used in the device
- Ability to deposit the required type of material with the required thickness and structure
- Improvement in the quality of the designed sensor
- Ability to coat the fiber or waveguide uniformly with respect to both size and shape
- Cost-effectiveness in terms of the cost of the substrate, depositing material, and coating technique
- Ecologically clean and safe for attending personnel

22.10 PROTECTION AND PACKAGING OF FIBER-OPTIC SENSORS

Due to the fact that the fiber-optic humidity sensors are mainly used in the laboratory, the problem of packaging of humidity sensors has been neglected. Although this problem, including the development of appropriate packaging for fiber sensors, without a doubt is an important one on the way to promote fiber-optic humidity sensors on the market, because sensors in the appearance that is used in most of the studies, cannot emerge the market. Therefore, the packaging FOS is next big task after a sensor design.

In the humidity sensors, the main factor affecting the sensor readings should be humidity and therefore any other factors that may affect the operation of sensors should be minimized (Yoshida et al. 1993). This means that the signal loss caused by the fiber bending needs to be considered for the requirements of sensor packaging design.

To eliminate the influence of this factor, the packaging must ensure the absence of any displacements of the fibers during operation. In other words, a strong bonding between the optical glass fiber and the packaging is important for sensor application. In addition, the material employed for bonding fibers during the packing should be highly durable and should not affect the operation of the sensors during the monitoring period. The installation of FOS, especially Bragg gratings on the substrate of the sensor packaging, also requires a longitudinal uniform bonding. Otherwise, the nonuniform bonding of the gratings will cause several reflected Bragg wavelength peaks when the gratings are strained. In this regard, planar humidity sensors have a significant advantage (Figure 22.32); their packaging does not cause as many problems as the packaging of FOS. As a result, the reproducibility of parameters and operational stability of such sensors should be better.

To resolve this problem in relation to the FOS, Zhuang et al. (2012) developed a special cell, which consisted of a cap and a bottom. The characters of the cap and the bottom are illustrated in Figure 22.33. The sizes of the cap and the bottom are both 1 mm × 20 mm. There are two holes fabricated in the cap, which are the channels for the sample flow in and the waster drain out. The fiber slit in the bottom is used to fix the sensing fiber, which will facilitate the process of encapsulation greatly. The slit is 2.5 mm in length, 250 μm in width and 150 μm in depth. The work of the cone that manufactured in the bottom is to guide the sample into or out of the sample pool. The island stand in the sample pool is used to support the sensing fiber. The function of the slit decorated in the island is to fix the sensing fiber, which can avoid the sensing fiber from shaking. The slit is 140 μm in width and 100 μm in depth. The slit is 50 μm shallower than the fiber slit. The fiber with cladding is thicker than the fiber whose cladding is removed. If these slits are made with the same depth, the sensing fiber will warp, the repeatability of the sensor will be poor and the results will be distorted. So, using slits with different depth to support different parts of the fiber can help the sensing fiber keep straight in the sample pool. Sample pool is the place where the analyte interacts with the evanescent field. The width of the sample pool is 200 and 150 μm in depth. The length of the pool can be chosen discretionary according to experimental requirements. Both the cap and the bottom are fabricated based on Micro-Electro-Mechanical Systems or MEMS technologies. In order to obtain high sensitivities, the radius of the sensing fiber is usually very small, such as 4 μm. So the strength of the sensing fiber is weak, and the sensing fiber is fragile and easy to break. In order to solve this problem, the fiber is fixed to the bottom first. Then, the cap is bound to the bottom to form the sensor's cell using glue.

The same situation with sensor signal instability may occur if we do not take into account the thermal expansion coefficients of the bonding compound and of the

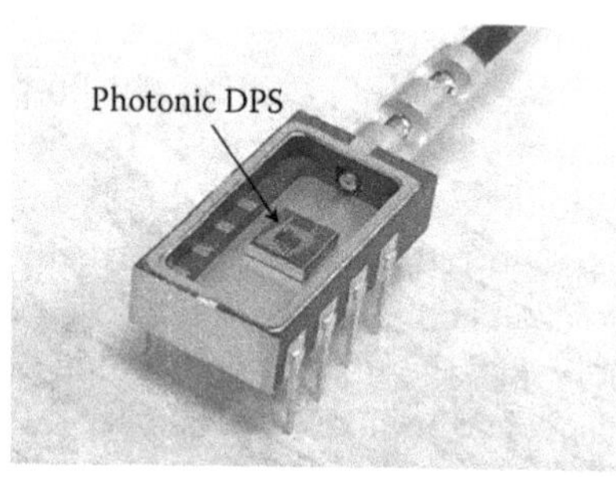

FIGURE 22.32 Packaged planar dew point sensor (DPS). (Reprinted by permission from Macmillan Publishers Ltd. *Sci. Rep.*, Tao, J. et al., 2016, copyright 2016.)

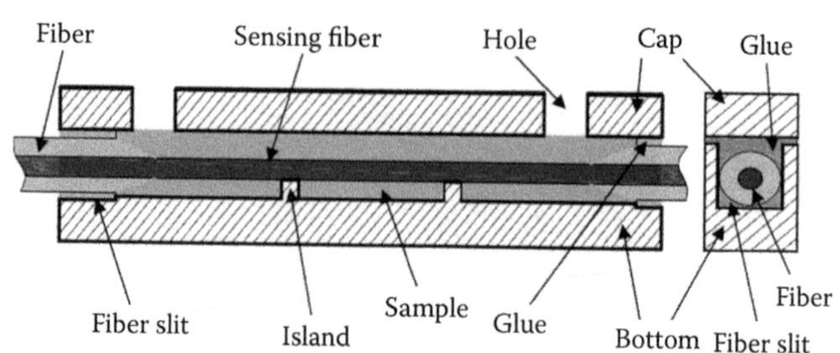

FIGURE 22.33 The schematic structure of the sample cell. (Reprinted from Zhuang, X. et al., A novel approach to evaluate the sensitivities of the optical fiber evanescent field sensors, in *Fiber Optic Sensors*, Yasin, M. (Ed.), InTech, Rijeka, Croatia, 165–184, 2012. Published by InTech as open access.)

platform used for sensor packaging; a mismatch of the expansion coefficients of the packaging, bonding compound and fibers when the temperature changes may be accompanied by a change in the fiber bending with a corresponding change in the propagation conditions of light in the fiber. It is possible that to solve this problem, for certain types of sensors the temperature stabilization is required. Mechanical and thermal fatigue as well as chemical aggression of environment can also decrease the life of FOS (Murata and Nakamura 1991; Yoshida et al. 1993). Therefore, special attention and various measures have to be taken to relieve these adverse effects on optical fiber health monitoring.

REFERENCES

Acharya S., Hill J.P., Ariga K. (2009) Soft Langmuir–Blodgett technique for hard nanomaterials. *Adv. Mater.* 21, 2959–2981.

Adamovsky G. (1987) Referencing in fiber optic sensing systems. NASA Technical Memorandum 89822. Lewis Research Center, Cleveland, Ohio.

Adams D.M., Brus L., Chidsey C.E.D. et al. (2003) Charge transfer on the nanoscale: Current status. *J. Phys. Chem. B* 107, 6668–6697.

Allen M.G. (1998) Diode laser absorption sensors for gas-dynamic and combustion flows. *Meas. Sci. Technol.* 9, 545–562.

Arregui F.J., Matias I.R., Liu Y., Lenahan M., Claus R.O. (1999) Optical fiber nanometer-scale Fabry-Perot interferometer formed by the ionic self-assembly monolayer process. *Opt. Lett.* 24, 596–598.

Arregui F.J., Otano M., Fernandez-Valdivielso C., Matias I.R. (2002). An experimental study about the utilization of liquicoat solutions for the fabrication of pH optical fiber sensors. *Sens. Actuators B* 87, 289–295.

Asatekin A., Barr M.C., Baxamusa S.H., Lau K.K.S., Tenhaeff W., Xu J., Gleason K.K. (2010) Designing polymer surfaces via vapor deposition. *Mater. Today* 13, 26–33.

Bansal L. (2004) *Development of a Fiber Optic Chemical Sensor for Detection of Toxic Vapors.* PhD Thesis, Drexel University.

Bariain C., Matıas I.R., Arregui F.J., Lopez-Amo M. (2000) Optical fiber humidity sensor based on a tapered fiber coated with agarose gel. *Sens. Actuators B* 69, 127–131.

Barragán N.A., Pérez M.A., Campo J.C., Alvarez J.C. (2000) Photo detection of low level phosphorescence produced by low level oxygen concentrations using optical system based on fiber optical and avalanche photodiode. In: *Proceedings of the IEEE International Symposium on Industrial Electronics*, Puebla, Mexico, pp. 596–601, December 4–8.

Bass M. and Van Stryland E.W. (Eds.) (2002) *Fiber Optic Handbook: Fiber, Devices, and Systems for Optical Communications.* McGraw-Hill, New York.

Beam B.M., Burnett J.L., Webster N.A., Mendes S.B. (2012) Applications of the planar fiber optic chip. In: Yasin M. (Ed.), *Recent Progress in Optical Fiber Research.* InTech, Rijeka, Croatia, pp. 387–412.

Beam B.M., Shallcross R.C., Jang J., Armstrong N.R., Mendes S.B. (2007) Planar fiber-optic chips for broadband spectroscopic interrogation of thin-films. *Appl. Spectrosc.* 61(6), 585–592.

Bhatia V., Vengsarkar A.M. (1996) Optical fibre long-period grating sensors. *Opt. Lett.* 21, 692–694.

Bialiayeu A., Caucheteur C., Ahamad N., Ianoul A., Albert J. (2011) Self-optimized metal coatings for fiber plasmonics by electroless deposition. *Opt. Exp.* 19(20), 18742–18753.

Bjarklev A. (1986). Microdeformation losses of single-mode fibres with step-index profiles. *J. Lightwave Technol.* 4(3), 341–346.

Blows J., Tang D.Y. (2000) Gratings written with tripled output of Q-switched Nd:YAG laser. *Electron. Lett.* 36, 1837–1839.

Brambilla G. (2010) Optical fibre nanotaper sensors. *Opt. Fiber Technol.* 16, 331–342.

Brambilla G., Finazzi V., Richardson D. (2004) Ultra-low-loss optical fiber nanotapers. *Opt. Exp.* 12, 2258–2263.

Brenci M. and Baldini F. (1991) Fiber optic optrodes for chemical sensing. In: *Proceedings of the 8th IEEE Optical Fiber Sensors Conference*, New York, pp. 313–319, January 29–31.

Brinker C.J., Hurd A.J., Schunk P.R., Frye G.C., Ashley C.S. (1992) Review of sol-gel thin film formation. *J. Non Cryst. Solids* 147–148, 424–436.

Burgess L.W. (1995) Absorption-based sensors. *Sens. Actuators B* 29, 10–15.

Campo J.C., Pérez M.A., Mezquita J.M., Sebastián J. (1997) Circuit-design criteria for improvement of xenon flash-lamp performance (lamp life, light-pulse, narrowness, uniformity of light intensity in a series flashes). In: *Proceedings of IEEE Applied Power Electronics Conference and Exposition, APEC '97*, Atlanta, GA, pp. 1057–1061, February 27.

Caruso F. (2001) Nanoengineering of particle surfaces. *Adv. Mater.* 122, 11–22.

Cennamo N., Zeni L. (2014) Bio and chemical sensors based on surface plasmon resonance in a plastic optical fiber. In: Yasin M., Harun S.W., Arof H. (Eds.), *Optical Sensors—New Developments and Practical Applications*, InTech, Rijeka, Croatia, pp. 119–140.

Chen K.P., Herman P.R., Tam R., Zhang J. (2000) Rapid long-period grating formation in hydrogen loaded fibre with 157 nm F_2 laser radiation. *Electron. Lett.* 36, 2000–2001.

Chen X. (2013) Optical fibre gratings for chemical and bio-sensing. In: Harun S.W., Arof H. (Eds.), *Current Developments in Optical Fiber Technology.* InTech, Rijeka, Croatia, pp. 205–235.

Chen X., Zhang W., Liu C., Hong Y., Webb D.J. (2015) Enhancing the humidity response time of polymer optical fiber Bragg grating by using laser micromachining. *Opt. Exp.* 23(20), 25942.

Cherif K., Mrazek J., Hleli S., Matejec V., Abdelghani A., Chomat M., Jaffrezic-Renault N., Kasik I. (2003) Detection of aromatic hydrocarbons in air and water by using xerogel layers coated on PCS fibers excited by an inclined collimated beam. *Sens. Actuators B* 95, 97–106.

Choi J., Rubner M.F. (2005) Influence of the degree of ionization on weak polyelectrolyte multilayer assembly. *Macromolecules* 38, 124–166.

Chrisey D., Hubler G. (Eds.) (1994) *Pulsed Laser Deposition of Thin Films*. Wiley, New York.

Claus R.O., Matias I., Arregui F. (2002) Optical fiber sensors. In: Bass M., Van Stryland E.W. (Eds.), *Fiber Optic Handbook: Fiber, Devices, and Systems for Optical Communications*. McGraw-Hill, New York, Chapter 14.

Collinson M.M. (2002) Recent trends in analytical applications of organically modified silicate materials. *Trends Anal. Chem.* 1(21), 30–38.

Corres J.M., Arregui F.J. (2006) Design of humidity sensors based on tapered optical fibers. *J. Lightwave Technol.* 24(11), 4329–4336.

Cox F.M., Lwin R., Large C.J., Cordeiro C.M.B. (2007) Opening up optical fibres. *Opt. Exp.* 15(19), 11843–11848.

Dakin J., Culshaw B. (Eds.) (1988) *Optical Fiber Sensors-Principles and Components*, Vol. I. Artech House, Boston, MA.

Dakin J.P., Edwards H.O., Weigl B.H. (1995) Progress with optical gas sensors using correlation spectroscopy. *Sens. Actuators B* 29, 87–93.

Davis C.M., Carome E.F., Weik M.H., Ezekiel S., Einzig R.E. (1986) *Fiber Optic Sensors Technology Handbook*. Optical Technologies—A Division of Dynamic System INC, Herndon, VA.

Davis D.D., Gaylord T.K., Glytsis E.N., Kosinski S.G., Mettler S.C., Vengsarkar A.M. (1998a) Long-period fibre grating fabrication with focused CO_2 laser beams. *Electron. Lett.* 34, 302–303.

Davis D.D., Gaylord T.K., Glytsis E.N., Mettler S.C. (1998b) CO_2 laser induced long-period fibre gratings: Spectral characteristic, cladding modes and polarization independence. *Electron. Lett.* 34, 1416–1417.

Davletbaev R., Akhmetshina A., Gumerov A., Davletbaeva I. (2014) Optical sensors based on mesoporous polymers. In: Yasin M., Harun S.W., Arof H. (Eds.), *Optical Sensors—New Developments and Practical Applications*. InTech, Rijeka, Croatia, pp. 47–66.

Del Villar I., Matias I., Arregui F.J., Claus R.O. (2004) Fiber-optic nanorefractometer based on one-dimensional photonic-bandgap structures with two defects. *IEEE Trans. Nanotechnol.* 3, 293–299.

Del Villar I., Matias I.R., Arregui F.J. (2005) LBL-based in-fibre nanocavity for hydrogen-peroxide detection. *IEEE Trans. Nanotech.* 4, 187–193.

Dewra S., Vikas, Grover A. (2015) Fabrication and applications of fiber Bragg grating—A review. *Adv. Eng. Tec. Appl.* 4(2), 15–25.

Dianov E.M., Karpov V.I., Grekov M.V., Golant K.M., Vasiliev S.A., Medvekov O.I., Khrapko R.R. (1997a) Thermo-induced long period fibre grating. In: *Proceedings of 11th International Conference on Integrated Optics and Optical Fibre Communications and 23rd European Conference on Optical Communications, IOOC-ECOC 1997*, Edinburgh, UK, Vol. 2, pp. 53–56, September 22–25.

Dianov E.M., Karpov V.I., Kurkov A.S., Grekov M.V. (1998) Long period fibre gratings and mode-field converters fabricated by thermodiffusion in phosphosilicate fibres. In: *Proceedings of 24th European Conference on Optical Communication* (*ECOC*), Madrid, Spain, Vols. 1–3, pp. 395–396, September 20–24.

Dianov E.M., Stardubov D.S., Vasiliev S.A., Frolov A.A., Medvedkov O.I. (1997b) Refractive-index gratings written by near-ultraviolet radiation. *Opt. Lett.* 22(4), 221–223.

Dietzel Y., Przyborowski W., Nocke G., Offermann P., Hollstein F., Meinhardt J. (2000) Investigation of PVD arc coatings on polyamide fabrics. *Surf. Coat. Technol.* 135, 75–81.

Doherty W.J., Simmonds A.G., Mendes S.B., Armstrong N.R., Saavedra S.S. (2005). Molecular ordering in monolayers of an alkyl-substituted perylene-bisimide dye by attenuated total reflectance ultraviolet-visible spectroscopy. *Appl. Spectrosc.* 59(10), 1248–1256.

Drozin L., Fonjallaz P.-Y., Stensland L.K. (2000) Long-period fibre gratings written by CO_2 exposure of H_2-loaded standard fibres. *Electron. Lett.* 36, 742–743.

Elosua C., Matias I.R., Bariain C., Arregui F.J. (2006) Volatile organic compound optical fiber sensors: A review. *Sensors* 6, 1440–1465.

Erdogan T. (1997) Fiber grating spectra. *J. Lightwave Technol.* 15, 1277–1294.

Fang X., Liao C., Wang D. (2010) Femtosecond laser fabricated fiber Bragg grating in microfiber for refractive index sensing. *Opt. Lett.* 35, 1007–1009.

Fernando G.F., Degamber B. (2006) Process monitoring of fibre reinforced composites using optical fibre sensors. *Int. Mater. Rev.* 51(2), 65–106.

Flora W.H., Mendes S.B., Doherty W.J. III, Saavedra S.S., Armstrong N. (2005) Determination of molecular anisotropy in thin films of discotic assemblies using attenuated total reflectance UV–visible spectroscopy. *Langmuir* 21(1), 360–368.

Fu Y. (2011) Facile and efficient approach to speed up layer-by-layer assembly: Dipping in agitated solutions. *Langmuir* 27, 672–677.

Fujumaki M., Ohki Y., Brebner J.L., Roorda S. (2000) Fabrication of long-period optical fibre gratings by use of ion implantation. *Opt. Lett.* 25, 88–90.

Gardner J.W., Bartlett P.N. (1995) Application of conducting polymer technology in microsystems. *Sens. Actuators A* 51, 57–66.

Gastón A., Lozano I., Pérez F., Auza F., Sevilla J. (2003) Evanescent wave optical-fiber sensing (temperature, relative humidity and pH sensors). *IEEE Sens. J.* 3, 806–811.

Gastón A., Pérez F., Sevilla J. (2004) Optical fiber relative-humidity sensor with polyvinyl alcohol film. *Appl. Opt.* 43, 4127–4132.

Glezer E.N., Milosavljevic M., Huang L., Finlay R.J., Her T.H., Callan J.P., Mazur E. (1996) Three-dimensional optical storage inside transparent materials. *Opt. Lett.* 21(24), 2023–2025.

Grattan K.T.V. (1995) Sources for optical fiber sensors. In: Grattan K.T.V., Meggitt B.T. (Eds.), *Optical Fiber Sensor Technology*. Chapman and Hall, London, UK, pp. 45–74.

Grattan K.T.V., Meggitt B.T. (Eds.) (1995) *Optical Fiber Sensor Technology*. Chapman and Hall, London, UK.

Grillo G.J., Pérez M.A., Florencias A.E. (2006) Synchronic filter based on switched capacitor filters for high stability phase-detectors systems. In: *Proceedings of the Instrumentation and Measurement Technology Conference, IEEE-IMTC 2006*, Sorrento, Italy, pp. 1997–1981, April 24–27.

Grosso D., Cagnol F., Soler-Illia G.J., Crepaldi E.L., Amenitsch H., Brunet-Bruneau A., Bourgeois A., Sanchez C. (2004) Fundamentals of mesostructuring through evaporation-induced self-assembly. *Adv. Funct. Mater.* 14, 309–322.

Guan B.-O., Tam H.-Y., Ho S.-L., Liu S.-Y., Dong X.-Y. (2000) Growth of long-period gratings in H_2-loaded fibre after 193 nm UV inscription. *IEEE Photon. Technol. Lett.* 12, 642–644.

Gutierrez N., Rodriguez-Mendez M.L., De Saja J.A. (2001) Array of sensors based on lanthanide bisphthalocyanine Langmuir–Blodgett films for the detection of olive oil aroma. *Sens. Actuators B* 77, 437–442.

Haddock H.S., Shankar P.M., Mutharasan R. (2003) Fabrication of biconical tapered optical fibers using hydrofluoric acid. *Mater. Sci. Eng. B* 97, 87–93.

Hammond P.T. (2004) Form and function in multilayer assembly: New applications at the nanoscale. *Adv. Mater.* 16, 1271–1293.

Harrington A. (2002) Infrared fibers. In: Bass M., Van Stryland E.W. (Eds.), *Fiber Optic Handbook: Fiber, Devices, and Systems for Optical Communications*. McGraw-Hill, New York, Chapter 14.

Harsanyi G. (1995) Polymeric sensing films: New horizons in sensorics? *Sens. Actuators A* 46–47, 85–88.

Harsanyi G. (2000) *Sensors in Biomedical Applications: Fundamentals, Technology and Applications*. Technomic, Basel, Switzerland.

Haus J. (2010) *Optical Sensors: Basics and Applications*. Wiley-VCH, Weinheim, Germany.

Herbig B., Loebmann P. (2004) TiO_2 photocatalysts deposited on fiber substrates by liquid phase deposition. *J. Photochem. Photobiol. A* 163, 359–365.

Hill K.O. (2002) Fiber Bragg gratings. In: Bass M., Van Stryland E.W. (Eds.), *Fiber Optic Handbook: Fiber, Devices, and Systems for Optical Communications*. McGraw-Hill, New York, Chapter 9.

Hill K.O., Fujii Y., Johnson D.C., Kawasaki B.S. (1978) Photosensitivity in optical fiber waveguides: application to reflection fiber fabrication. *Appl. Phys. Lett.* 32(10), 647.

Hill K.O., Malo B., Vineberg K.A., Bilodeau F., Johnson D.C., Skinner I. (1990). Efficient mode conversion in telecommunication fibre using externally written gratings. *Electron. Lett.* 26(16), 1270–1272.

Hindle F., Fertein E., Przygodzki C. et al. (2004) Inscription of long-period gratings in pure silica and germano-silicate fiber cores by femtosecond laser irradiation. *IEEE Photon. Technol. Lett.* 16(2004), 1861–1863.

Hrapovic S., Liu Y., Enright G., Bensebaa F., Luong J.H.T. (2003) New strategy for preparing thin gold films on modified glass surfaces by electroless deposition. *Langmuir* 19(9), 3958–3965.

Humbert G., Malki A. (2002) Electric-arc-induced gratings in non-hydrogenated fibres: Fabrication and high-temperature characterizations. *J. Opt. A: Pure Appl. Opt.* 4, 194–108.

Hussey C.D., Minelly J.D. (1988) Optical fibre polishing with a motor-driven polishing wheel. *Electron. Lett.* 24, 805–807.

Hwang I.K., Yun S.H., Kim B.Y. (1999) Long period fibre grating based upon periodic microbends. *Opt. Lett.* 24, 1263–1265.

James S.W., Tatam R.P. (2003) Optical fibre long-period grating sensors: characteristics and application. *Meas. Sci. Technol.* 14, R49–R61.

Jarzebinska R., Cheung C.S., James S.W., Tatam R.P. (2009) Response of the transmission spectrum of tapered optical fibres to the deposition of a nanostructured coating. *Meas. Sci. Technol.* 20, 034001.

Jiang Y., Li Q., Lin C.H., Lyons E., Tomov I., Lee H.P. (2002) A novel strain-induced thermally tuned long-period fibre grating fabricated on a periodic corrugated silicon fixture. *IEEE Photon. Technol. Lett.* 14, 941–943.

Jones J.D.C. (1995) Optical detectors and receivers. In: Grattan K.T.V., Meggitt B.T. (Eds.), *Optical Fiber Sensor Technology*. Chapman and Hall, London, UK, pp. 75–104.

Jin W., Ho H.L., Cao Y.C., Ju J., Qi L.F. (2013) Gas detection with micro- and nano-engineered optical fibers. *Opt. Fiber Technol.* 19, 741–759.

Kakarantzas G., Birks T.A., Russell P.S. (2002) Structural long-period gratings in photonic crystal fibers. *Opt. Lett.* 27, 1013–1015.

Kakarantzas G., Dimmick T.E., Birks T.A., Le Roux R., Russell P.St.J. (2001) Miniature all-fibre devices based on CO_2 microstructuring of tapered fibres. *Opt. Lett.* 26, 1137–1379.

Kaneko T., Nittono O. (1997) Improved design of inverted magnetrons used for deposition of thin films on wires. *Surf. Coat. Technol.* 90, 268–274.

Kashima T., Matsuda Y., Fujiyama H. (1991) Development of the quadrupole plasma chemical vapour deposition method for low temperature, high speed coating on an optical fibre. *Mater. Sci. Eng. A* 139, 79–84.

Kashyap R. (1999) *Fiber Bragg Grating*. Academic Press, San Diego, CA.

Kashyap R., Nemova G. (2009) Surface plasmon resonance-based fiber and planar waveguide sensors. *J. Sens.* 2009, 645162.

Kendall D.L. (1979) Vertical etching of silicon at very high aspect ratios, *Annu. Rev. Mater. Sci.* 9, 373–403.

Kern F., Gadow R. (2002) Liquid phase coating process for protective ceramic layers on carbon fibers. *Surf. Coat. Technol.* 151–152, 418–423.

Kern F., Gadow R. (2004) Deposition of ceramic layers on carbon fibers by continuous liquid phase coating. *Surf. Coat. Technol.* 180–181, 533–537.

Kersey D., Jackson D.A., Corke M. (1983) A simple fibre Fabry–Perot sensor. *Opt. Comm.* 45, 71.

Khalil S., Bansal L., El-Sherif M. (2004) Intrinsic fiber optic chemical sensor for the detection of dimethyl methylphosphonate. *Opt. Eng.* 43, 2683–2688.

Kieu K.Q., Mansuripur M. (2006) Biconical fiber taper sensor. *IEEE Photon. Technol. Lett.* 18(21), 2239–2241.

Kim C.S., Han Y., Lee B.H., Han W.-T., Paek U.-C., Chung Y. (2000) Induction of the refractive index change in B-doped optical fibers through relaxation of the mechanical stress. *Opt. Commun.* 185, 337–342.

Kondo Y., Nouchi K., Mitsuyu T., Watanabe M., Kazansky P., Hirao K. (1999) Fabrication of long-period fibre gratings by focused irradiation of infra-red femtosecond laser pulses. *Opt. Lett.* 24, 646–648.

Korotcenkov G., Cho B.K., Gulina L.B., Tolstoy V.P. (2015) Synthesis of metal oxide-based nanocomposites and multicomponent compounds using Layer-by-Layer method and prospects for their application. *J. Teknologi (Sci. Eng.)* 75(7), 16–26.

Korotcenkov G., Cho B.K., Sevilla III F., Narayanaswamy R. (2011) Optical and fiber optic chemical sensors. In: Korotcenkov G. (Ed.), *Chemical Sensors: Comprehensive Sensor Technologies.* Vol. 5: Electrochemical and Optical Sensors. Momentum Press, New York, pp. 311–476.

Korposh S., James S., Tatam R., Lee S.-W. (2013) Fibre-optic chemical sensor approaches based on nanoassembled thin films: A challenge to future sensor technology. In: Harun S.W., Arof H. (Eds.), *Current Developments in Optical Fiber Technology.* InTech, Rijeka, Croatia, pp. 237–264.

Kraft M. (2006) Vibration spectroscopic sensors. In: Baldini F., Chester A.N., Homola J., Martellucci S. (Eds.), *Optical Chemical Sensors.* Springer, Dordrecht, the Netherlands, pp. 117–155.

Krohn D.A. (1988) *Fiber Optic Sensors: Fundamentals and Applications*, Vol. 138. Instrument Society of America, Research Triangle Park, NC.

Kumar D., Sharma R.C. (1998) Advances in conductive polymers. *Eur. Polym. J.* 34(8), 1053–1060.

Lacroix S., Gonthier F., Black R.J., Bures J. (1988) Tapered-fiber interferometric wavelength response: The achromatic fringe. *Opt. Lett.* 13(5), 395–397.

Lecler S., Meyrueis P. (2012) Intrinsic optical fiber sensor. In: Yasin M. (Ed.), *Fiber Optic Sensors.* InTech, Rijeka, Croatia, pp. 53–76.

Lemaire P.J., Atkins R.M., Mizrahiu V., Reed W.A. (1993) High pressure H_2-loading as a technique for achieving ultrahigh UV photosensitivity and thermal sensitivity in GeO_2 doped optical fibers. *Electron. Lett.* 29, 1191–1193.

Liang W., Huang Y., Xu Y., Lee R.K., Yariv A. (2005) Highly sensitive fiber Bragg grating refractive index sensors. *Appl. Phys. Lett.* 86, 151122.

Lin C.-Y., Chern G.-W., Wang L.A. (2001) Periodical corrugated structure for forming sampled fibre Bragg grating and long-period fibre grating with tunable coupling strength. *J. Lightwave Technol.* 19, 1212–1220.

Liu Y., Meng C., Zhang A., Xiao Y., Yu H., Tong L. (2011) Compact microfiber Bragg gratings with high-index contrast. *Opt. Lett.* 36, 3115–3117.

Liu Y., Wang A., Claus R. (1997) Molecular self-assembly of TiO_2/polymer nanocomposite films. *J. Phys. Chem. B* 101, 1385–1388.

Lou J., Wang Y., Tong L. (2014) Microfiber optical sensors: A review. *Sensors*, 14, 5823–5844.

Mahan J.E. (2000) *Physical Vapor Deposition of Thin Films.* Wiley, New York.

Malinauskas A. (2001) Chemical deposition of conducting polymers. *Polymer* 42, 3957–3972.

Malis C., Landl M., Simon P., and MacCraith B.D. (1998) Fiber optic ammonia sensing employing novel near infrared dyes. *Sens. Actuators* B 51, 359–367.

Malkin A.Y., Siling M.I. (1991) Scientific principles of present-day and future technologies of synthesis and processing polycondensation polymers. *Rev. Polym. Sci.* 33, 2135–2160.

Markos C., Stefani A., Nielsen K., Rasmussen H.K., Yuan W., Bang O. (2013) High-Tg TOPAS microstructured polymer optical fiber for fiber Bragg grating strain sensing at 110 degrees. *Opt. Exp.* 21(4), 4758–4765.

Martinez A., Dubov M., Khrushchev I., Bennion I. (2004) Direct writing of fiber Bragg gratings by femtosecond laser. *Electron. Lett.* 40(19), 1170–1172.

Martinez-Rios A., Monzon-Hernandez D., Torres-Gomez I., Salceda-Delgado G. (2012) Long period fibre gratings. In: Yasin M. (Ed.), *Fiber Optic Sensors.* InTech, Rijeka, Croatia, pp. 275–294.

Matsuhara M., Hill K. (1974) Optical-waveguide band-rejection filters: Design. *Appl. Opt.* 13, 2886–2888.

Matsumoto Y., Yoshida K., Ishida M. (1998) A novel deposition technique for fluorocarbon films and its applications for bulk- and surface-micromachined devices. *Sens. Actuators A* 66, 308–314.

McDonagh C., Burke C.S., MacCraith B.D. (2008) Optical chemical sensors. *Chem. Rev.* 108, 400–422.

Meltz G., Morey W.W., Glenn W.H. (1989) Formation of Bragg gratings in optical fibers by a transverse holographic method. *Opt. Lett.* 14(15), 823–825.

Mignani A.G., Baldini F. (1996) Biomedical sensors using optical fibres. *Rep. Prog. Phys.* 59(1), 1–28.

Mihailov S.J., Smelser C.W., Lu P., Walker R.B., Grobnic D., Ding H., Henderson G., Unruh J. (2003) Fiber Bragg gratings made with a phase mask and 800-nm femtosecond radiation. *Opt. Lett.* 28(12), 995–997.

Modlin D.N., Milanovitch F.P. (1991) Instrumentation for fiber optic chemical sensors. In: Wolfbeis O.S. (Ed.), *Fiber Optic Chemical Sensors and Biosensors*, Vol. 1. Chemical Rubber Company, Boca Raton, FL, pp. 237–302.

Monroy E., Omnes F., Calle F. (2003) Wide-bandgap semiconductor ultraviolet photodetectors. *Semicond. Sci. Technol.* 18, R33–R51.

Muhammad F.A., Stewart G., Jin W. (1993) Sensitivity enhancement of D-fiber methane gas sensor using high-index overlay. *IEE Proc. J. Optoelectron.* 140(2), 115–118.

Murata N., Nakamura K. (1991) UV-curable adhesives for optical communications. *J. Adhesion* 35, 251–267.

Murray R.W. (1989) Chapter 1 Chemical sensors and microinstrumentation: An overview. In: Murray R., Dessy R.E., Heineman W.R., Janata J., Seitz W.R. (Eds.), *Chemical Sensors and Microinstrumentation*; ACS Symposium Series. American Chemical Society, Washington, DC, pp. 1–19.

Murtaza G., Senior J.M. (1995) Schemes for referencing of intensity-modulated optical sensor systems. In: Grattan K.T.V., Meggitt B.T. (Eds.), *Optical Fiber Sensor Technology*. Chapman and Hall, London, UK, pp. 383–407.

Muto S., Suzuki O., Amano T., Morisawa M. (2003) A plastic optical fibre sensor for real-time humidity monitoring. *Meas. Sci. Technol.* 14(6), 746–750.

Narayanan C., Presby H.M., Vengsarkar A.M. (1997) Band-rejection fibre filter using periodic core deformation. *Electron. Lett.* 33(4), 280–281.

Nicole L., Boissiere C., Grosso D., Hesemann P., Moreau J., Sanchez C. (2004) Advanced selective optical sensors based on periodically organized mesoporous hybrid silica thin films. *Chem. Commun.* 20, 2312–2313.

Okazaki S., Nakagawa H., Asakura S., Tomiuchi Y., Tsuji N., Murayama H., Washiya M. (2003) Sensing characteristics of an optical fiber sensor for hydrogen leak. *Sens. Actuators B* 93, 142–147.

Orellana G., Moreno-Bondi M.C., Garcia-Fresnadillo D., Marazuela M.D. (2005) The interplay of indicator support and analyte in optical sensor layers. In: Orellano G., Moreno-Bondi M.C. (Eds.), *Frontiers in Chemical Sensors: Novel Principles and Techniques*, Springer Series on Chemical Sensors and Biosensors, Vol. 3. Springer-Verlag, Berlin, Germany, pp. 189–225.

O'Toole M., Diamond D. (2008) Absorbance based light emitting diode optical sensors and sensing devices. *Sensors* 8, 2453–2479.

Palai P., Satyanarayan M.N., Das M., Thyagarajan K., Pal B.P. (2001) Characterisation and simulation of long period gratings fabricated using electric discharge. *Opt. Commun.* 193, 181–185.

Pérez M.A., González O., Arias J.R. (2013) Optical fiber sensors for chemical and biological measurements. In: Harun S.W., Arof H. (Eds.), *Current Developments in Optical Fiber Technology*. InTech, Rijeka, Croatia, pp. 265–291.

Peters K. (2011) Polymer optical fiber sensors—A review. *Smart Mater. Struct.* 20, 013002.

Pique A., Auyeung R.C.Y., Stepnowsk J.L., Weir D.W., Arnold C.B., McGill R.A., Chrisey D.B. (2003) Laser processing of polymer thin films for chemical sensor applications. *Surf. Coat. Technol.* 163–164, 293–299.

Podbielska H., Ulatowska-Jarza A., Muller G., Eichler H.J. (2006) Sol-gel for optical sensors. In: Baldini F., Chester A.N., Homola J., Martellucci S. (Eds.), *Optical Chemical Sensors*. Springer-Verlag, Dordrecht, the Netherlands, pp. 353–385.

Poole C.D., Presby H.M., Meester J.P. (1994) Two mode fibre spatial-mode converter using periodic core deformation. *Electron. Lett.* 30, 1437–1438.

Potyrailo R.A., Hieftje G.M. (1998) Oxygen detection by fluorescence quenching of tetraphenylporphyrin immobilized in the original cladding of an optical fiber. *Anal. Chim. Acta* 370, 1–8.

Prieto M., Braña E., Campo J.C., Pérez M.A. (2004) Thermal performance of a controlled cooling system for low-level optical signals. *App. Ther. Eng.* 24, 2041–2054.

Rao Y.J., Zhu T., Ran Z.L., Wang Y.P., Jiang J., Hu A.Z. (2004). Novel long-period fibre gratings written by high-frequency CO_2 laser pulses and applications in optical fibre communication. *Opt. Commun.* 229(1–6), 209–221.

Rao Y.-J., Ran Z.-L. (2013) Optic fiber sensors fabricated by laser-micromachining. *Opt. Fiber Technol.* 19, 808–821.

Refaat T.F., Ismail S., Koch G.J. et al. (2010) Backscatter 2-μm Lidar validation for atmospheric CO_2 differential absorption Lidar applications. *IEEE Trans. Geosci. Remote Sens.* PP(99), 2055874 (1-9).

Rego G., Okhotnikov O., Dianov E., Sulimov V. (2001) High-temperature stability of long-period fibre gratings using an electric arc. *J. Lightwave Technol.* 19, 1574–1579.

Reisinger J.J., Hillmyer M.A. (2002) Synthesis of fluorinated polymers by chemical modification. *Prog. Polym. Sci.* 27, 971–1005.

Renneckar S. (2009) Nanoscale coatings on wood: Polyelectrolyte adsorption and layer-by-layer assembled film formation. *ACS Appl. Mater. Interface* 1, 559–566.

Richardson J.J., Cui J., Björnmalm M., Braunger J.A., Ejima H., Caruso F. (2016) Innovation in Layer-by-Layer assembly. *Chem. Rev.* 116(23), 14828–14867.

Savin S., Digonnet M.J.F., Kino G.S., Shaw H.J. (2000) Tunable mechanically induced long-period fibre gratings. *Opt. Lett.* 25, 710–712.

Schaaf P. (2012) Spray-assisted polyelectrolyte multilayer buildup: From step-by-step to single-step polyelectrolyte film constructions. *Adv. Mater.* 24, 1001–1016.

Schonhoff M. (2003), Self-assembled polyelectrolyte multilayers. *Curr. Opin. Coll. Interf. Sci.* 8, 86–95.

Schreiber F. (2000) Structure and growth of self-assembling monolayers. *Prog. Surf. Sci.* 62, 151–256.

Scorsone E., Christie S., Persaud K.C., Kvasnik F. (2004). Evanescent sensing of alkaline and acidic vapours using a plastic clad silica fibre doped with poly(omethoxyaniline). *Sens. Actuators B* 97, 174–181.

Segawa H., Ohnishi E., Arai Y., Yoshida K. (2003) Sensitivity of fiber-optic carbon dioxide sensors utilizing indicator dye. *Sens. Actuators B* 94, 276–281.

Selvarajan A., Asoundi A. (1995) Photonics, fiber optic sensors and their applications in smart structures. *J. Non-Destruct. Eval.* 15(2), 41–56.

Senosiain J., Díaz I., Gastón A., Sevilla J. (2001) High sensitivity temperature sensor based on side-polished optical fiber. *IEEE Trans. Instrum. Meas.* 50, 1656–1660.

Sheem S. (1980) Fiber-optic gyroscope with [3 × 3] directional coupler. *Appl. Phys. Lett.* 37, 869.

Shenk Strasser T.A., Chandonnet P.J., DeMarco J., Soccolich C.E., Pedrazzani J.R., DiGiovanni D.J., Andrejco M.J., Shenk D.S. (1996) UV induced fiber grating oadm devices for efficient bandwidth utilization. In: *Proceedings of Optic Fiber Communication, OFC96*, San Jose, CA, pp. 360–363, February 25.

Smith W.J. (2000) *Modern Optical Engineering*, 3rd ed. McGraw-Hill, New York.

Snyder A.W., Love J. (1983) *Optical Waveguide Theory.* Chapman & Hall, London, UK.

Song J. (2014) Antimicrobial polymer nanostructures: Synthetic route, mechanism of action and perspective. *Adv. Colloid Interface Sci.* 203, 37–50.

Sood A.K., Dhar N.K., Polla D.L., Dubey M., Wijewarnasuriya P. (2014) Nanostructured detector technology for optical sensing applications. In: Yasin M., Harun S.W., Arof H. (Eds.), *Optical Sensors—New Developments and Practical Applications.* InTech, Rijeka, Croatia, pp. 165–208.

Sparrow I.J.G., Smith P.G.R., Emmerson G.D., Watts S.P., Riziotis C. (2009) Planar Bragg grating sensors—fabrication and applications: A review. *J. Sens.* 2009, 607647.

Stewart G., Jin W., Culshaw B. (1997) Prospects for fibre-optic evanescent-field gas sensors using absorption in the near-infrared. *Sens. Actuators B* 38, 42–47.

Sumdia S., Okazaki S., Asakura S., Nakagawa H., Murayama H., Hasegawa T. (2005) Distributed hydrogen determination with fiber-optic sensor. *Sens. Actuators B* 108, 508–514.

Syms R.R.A., Cozens J.R. (1993) *Optical Guided Waves and Devices.* Academic Press, London, UK.

Tao J., Luo Y., Wang L., Cai H., Sun T., Song J., Liu H., Gu Y. (2016) An ultrahigh-accuracy miniature dew point sensor based on an integrated photonics platform. *Sci. Rep.* 6, 29672.

Tay C.M., Tan K.M., Tjin S.C., Chan C.C., Rahardjo H. (2004) Humidity sensing using plastic optical fibers. *Microwave Opt. Technol. Lett.* 43, 387–390.

Thunemann A.F., Muller M., Dautzenberg H., Joanny J.-F., Lowen H. (2004) Polyelectrolyte complexes. *Adv. Polymer Sci.* 166, 113–171.

Tolstoi V.P. (1993) Synthesis of thin-layer structures by the ionic layer deposition method. *Russ. Chem. Rev.* 62(3), 237–242.

Tolstoy V.P. (2006) Successive ionic layer deposition. The use in nanotechnology. *Russ. Chem. Rev.* 75(2), 161–175.

Tolstoy V., Han S.D., Korotcenkov G. (2010) Successive ionic layer deposition (SILD): Advanced method for deposition and modification of functional nanostructured metal oxides aimed for gas sensor applications. In: Umar A., Hahn Y.B. (Eds.), *Metal Oxide Nanostructures and Their Applications, Vol. 3, Applications (Part 1).* American Scientific Publishers, Stevenson Ranch, CA, pp. 384–436.

Vasiliev S.A., Dianov E.M., Varelas D., Limberger H.G., Salathe R.P. (1996) Post-fabrication resonance peak positioning of long period cladding-mode-coupled gratings. *Opt. Lett.* 21, 1830–1832.

Vurek G.G. (1996) Optical sensors for biomedical applications. In: Schultz J.S., Taylor R.F. (Eds.) *Handbook of Chemical and Biological Sensors.* IOP, Bristol, UK, Chapter 15.

Werle P., Slemr F., Maurer K., Kormann R., Mücke R., Janker B. (2002) Near- and mid-infrared laser-optical sensors for gas analysis. *Opt. Lasers Eng.* 37, 101–114.

Werneck M. M., Allil R.C.S.B. (2011) Optical fiber sensors. In: Krejcar O. (Ed.), *Modern Telemetry.* InTech, Rijeka, Croatia, pp. 3–40.

Williams D.L., Ainslie B.J., Armitage J.R., Kashyap R., Campbell R. (1993) Enhanced UV photosensitivity in boron codoped germanosilicate fibers. *Electron. Lett.* 29, 45–47.

Woyessa G., Fasano A., Stefani A., Markos C., Nielsen K., Rasmussen H.K., Bang O. (2016a) Single mode step-index polymer optical fiber for humidity insensitive high temperature fiber Bragg grating sensors. *Opt. Exp.* 24(2), 1253–1260.

Woyessa G., Fasano A., Markos C., Rasmussen H.K., Bang O. (2017a) Low loss polycarbonate polymer optical fiber for high temperature FBG humidity sensing. *IEEE Photon. Technol. Lett.* 29(7), 575–578.

Woyessa G., Fasano A., Markos C., Stefani A., Rasmussen H.K., Bang O. (2017b) Zeonex microstructured polymer optical fiber: Fabrication friendly fibers for high temperature and humidity insensitive Bragg grating sensing. *Opt. Mater. Exp.* 7(1), 286–295.

Woyessa G., Nielsen K., Stefani A., Markos C., Bang O. (2016b) Temperature insensitive hysteresis free highly sensitive polymer optical fiber Bragg grating humidity sensor. *Opt. Exp.* 24(2), 1206–1213.

Xuan H., Jin W., Zhang M. (2009) CO_2 laser induced long period gratings in optical microfibers. *Opt. Exp.* 17, 21882–21890.

Yoshida J., Yamada M., Terui H. (1993) Packaging and reliability of photonic components for subscriber network systems. *IEEE Trans. Comp. Hybrids Manuf. Technol.* 16(8), 778–782.

Yuan W., Khan L., Webb D.J., Kalli K., Rasmussen H.K., Stefani A., Bang O. (2011) Humidity insensitive TOPAS polymer fiber Bragg grating sensor. *Opt. Exp.* 19(20), 19731–19739.

Zhang C., Zhang W., Webb D.J., Peng G.-D. (2010a) Optical fibre temperature and humidity sensor. *Electron. Lett.* 46, 643–644.

Zhang W., Webb D., Peng G. (2012a) Polymer optical fiber Bragg grating acting as an intrinsic biochemical concentration sensor. *Opt. Lett.* 37(8), 1370–1372.

Zhang W., Webb D.J., Peng G.-D. (2012b) Investigation into time response of polymer fiber Bragg grating based humidity sensors. *J. Lightwave Technol.* 30(8), 1090–1096.

Zhang Y., Lin B., Tjin S.C., Zhang H., Wang G., Shum P., Zhang X. (2010b) Refractive index sensing based on higher-order mode reflection of a microfiber Bragg grating. *Opt. Exp.* 18, 26345–26350.

Zhou K., Liu H., Hu X. (2001) Tuning the resonant wavelength of long period fibre gratings by etching the fibre's cladding. *Opt. Commun.* 197, 295–299.

Zhuang X., Li P., Yao J. (2012) A novel approach to evaluate the sensitivities of the optical fiber evanescent field sensors. In: Yasin M. (Ed.), *Fiber Optic Sensors.* InTech, Rijeka, Croatia, pp. 165–184.

Zibaii M.I., Latifi H., Karami M., Gholami M., Hosseini S.M., Ghezelayagh M.H. (2010) Non-adiabatic tapered optical fiber sensor for measuring the interaction between alpha-amino acids in aqueous carbohydrate solution. *Meas. Sci. Technol.* 21, 105801.

Zolotov Y.A., Ivanov V.M., Amelin V.G. (Eds.) (2002) *Chemical Test Methods of Analysis.* Elsevier, New York.

Zosel J., Oelßner W., Decker M., Gerlach G., Guth U. (2011) The measurement of dissolved and gaseous carbon dioxide concentration. *Meas. Sci. Technol.* 22, 072001.

23 Integrated Humidity Sensors

23.1 INTRODUCTION

As it was shown in Chapters 11, 12, 15, and 16, a planar waveguide is a promising platform for development of gas sensors, including humidity sensors (Ballantine and Wohltjen 1986; Franke et al. 1993; Ansari et al. 1997; Skrdla et al. 2002; Fuke et al. 2009). More recently, researchers have started to exploit the advantages of planar integration as a way to allow enhanced functionality devices to be made, in which microfluidics and multiple sensor elements can be incorporated into a single device, integrated optical sensor (IOS). In addition, integrated optics (IO) allows for on-chip thermal compensation. Therefore, the integration of the light source and detector onto the same chip platform with the sensing element to complete the standalone *sensor-on-a-chip* platform has been a field of intensive investigation. It is believed that this approach provides a superior degree of compactness, stability, and reliability.

In general, *IOSs* are a branch of microoptics and IO (Hunsperger 2002; Palais 2002; Pollock and Lipson 2003; Iga and Kokubun 2005), which emerged from the successful techniques of microelectronics together with micromechanics and other microtechnologies. These devices have a common physical operating principle, in that they all operate by having a dielectric waveguide in which the propagating mode is allowed to partially interact with the measurand, and where the optical path change associated with that interaction is measured (Sparrow et al. 2009).

Schematic structure of an integrated optical sensing system is shown in Figure 23.1. As it is seen, the optical read-out system forms the heart of a much more extended sensing system, which generally encompasses functions to connect the optical to the electrical domain (light sources and optoelectrical detectors) and electronics (for processing the electrical signals obtained from the detectors and sometimes for controlling the state of some optical functions) (Lambeck 2006). Also a fluid delivery system may come into play if a well-controlled transport of small sample volumes to the sensing region or if some pretreatment of the samples is required. It often contains a microfluidic system, which is applied on the top of the sensing region. In principle all functions, such as chemical, optical, electronic, and the microfluidic ones, can be integrated into one monolithic chip. However, we need to note that for economic reasons (yield and universality) a hybrid system is mostly preferable.

Comparing IO sensors with microelectronic devices, it is necessary to say that whereas microelectronics chiefly focuses on silicon technology, IO still makes use a variety of materials like glass, silicon, metal oxides ($LiNbO_3$, ZnO, TiO_2, Ta_2O_5, SiO_2, etc.), polymers, liquid crystals, and III–V compound semiconductors, each with different waveguide fabrication technologies. IO combines thin-film and microfabrication techniques to make optical wave guides and devices in planar form and offers the advantages of compactness, power efficiency, and multifunction in single chip capabilities. Currently in microelectronics very small device dimensions and very large package densities are commonly realized. A similar advance is essential in photonics if optical systems are to be competitive and compact in comparison to electronic systems. In other words, miniaturized optical components integrated on a chip, due to better protection from thermal drift, moisture and vibration, lower power requirements, and lower cost due to the possibility of batch fabrication, will greatly enhance the application potential of IOSs. A big part of all results of the R&D done for IO systems for the strong market of optical telecommunication is also beneficial to the IO sensor field. Not only because a lot of useful expertise (principles, simulation software, and technologies) has been or is being developed in this area, but also because this strong market allows for an economically acceptable running of the facilities (e.g., foundries), which are also needed for producing the sensors (Lambeck 2006). Lambeck (2006) believes that in contrast to fiber optics, IO has greater flexibility in choosing the materials and structures of the optical systems giving additional degrees of freedom for an effective optimization. In addition, sensor arrays can be built on one single optical chip allowing one single optical chip to be the heart of a multipurpose sensing system, for example, for measuring simultaneously the concentrations of all tested analytes. Systems can be developed for mobile *in situ* field analysis with instant results or on-site data logging and communication with a base site.

At the same time, according to Lambeck (2006), IO sensors also show some weak points, the most relevant being the disturbing effects of contamination by chemical entities, scattering due to inhomogeneities (causing parasitic optical signals), and the possibility of degradation of

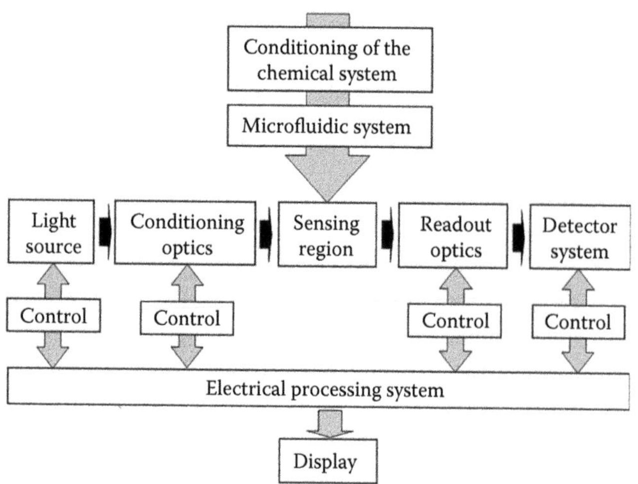

FIGURE 23.1 Schematic structure of an integrated optical sensing system. (Idea from Lambeck, P.V., *Meas. Sci. Technol.*, 17, R93–R116, 2006.)

the properties of sensitive materials due to photobleaching. The latter excludes, for example, the application of some chemically sensitive materials, which can be incorporated harmlessly in nonoptical sensors. Relatively high cost of IO sensors can also be considered as disadvantage of such devices. The microtechnological facilities needed for the production of the IO sensors (clean rooms, equipment, and skilled labor force) are expensive.

23.2 BASIC PRINCIPLES

Materials and processes usually used in IOS fabrication are listed in Table 23.1. IO in glass has advantages because of the same material basis as fibers, so that efficient light coupling is possible. However, IO in solid-state materials such as silicon or on semiconducting substrates like GaAs or InP allows simple integration of optical and electronic functions. Absorption is relatively high but acceptable for short path lengths in IOSs. Semiconductor waveguides are often made as ridges, because of the ease with which a mesa structure may be made by etching.

There are two basic semiconductor waveguide types: *heterostructure guides*, which operate by the refractive index (RI) differences obtained between different materials, and *homostructure guides*, which use the index reduction following from an increase in the majority carrier density (Syms and Cozens 1993). At present the most often semiconductor waveguides are fabricated on the base of GaAs/$Ga_{1-x}Al_xAs$ and InP/$Ga_{1-x}In_xAs_{1-y}P_y$ heterostructure systems. The variation of the dopant content between the layers in homostructure also provides the RI difference necessary for a waveguide. However, the RI changes possible in homostructures are much lower than in heterostructures, giving reduced confinement in any guide structure. A further problem arising in homostructures is that the increased carrier concentration in the substrate causes a rise in absorption, which can result in unacceptable propagation loss. Examples of semiconductor waveguides are shown in Figure 23.2.

Ridge guides can also be made in amorphous material on semiconducting substrates (Syms and Cozens 1993). One common system involves silicon oxynitride guides, constructed on a silicon dioxide buffer layer, which in turn lies on a silicon substrate. Silicon oxynitride has a relatively high RI, but one which is lower than that of the silicon, for which $n \sim 3.5$. The two are therefore separated by a thick layer of SiO_2, which acts as a low-index

TABLE 23.1
Materials and Processes for Integrated Optics

Substrate	Guiding Layer	Fabrication Process
Glass or fused quartz	Various glasses;	Sputtering or e-beam
	Ta_2O_5, Nb_2O_5;	evaporation;
	Polymers	Solution deposition;
Glass	Mixed metal oxide layers	Ion migration and ion exchange;
$LiNbO_3$ or $LiTaO_3$	$LiNbO_3$ and other metal	Chemical etching;
	oxide layers	LPE, VPE, MBE, and MOCVD;
GaAs or InP	$Ga_{1-x}Al_xAs$;	Diffusion; Ion implantation;
	$Ga_{1-x}In_xAs_{1-y}P_y$	E-beam and laser writing;
Quartz crystal	Metal oxides	Micromachining
Silicon	SiO_2	

Source: Data extracted from Selvarajan, A. and Asoundi, A., *J. Non-Destruct. Eval.*, 15, 41–56, 1995.

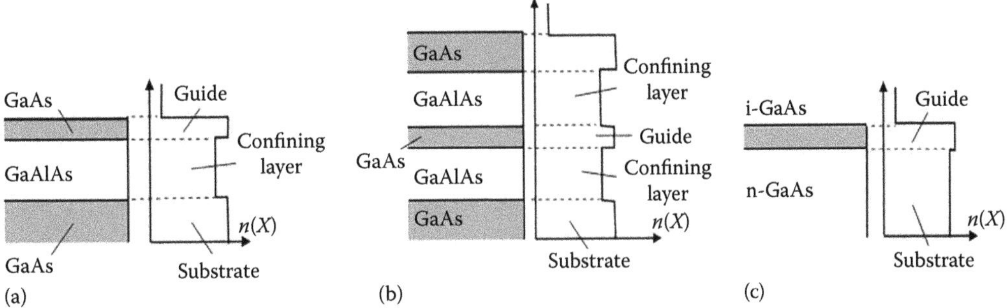

FIGURE 23.2 (a) A GaAs/GaAlAs planar waveguide; (b) A double heterostructure GaAs/GaAlAs planar waveguide; and (c) A planar waveguide in GaAs formed by changes in free carrier concentration. (Idea from Syms, R.R.A. and Cozens, J.R., *Optical Guided Waves and Devices*, Academic Press, London, UK, 1993.)

($n \sim 1.47$) spacer. Other systems use doped silica (e.g., a SiO_2/TiO_2 mixture) on the top of the silica buffer layer. These are important, since they allow the fabrication of waveguides in a form compatible with very large scale integration (VLSI) electronics. ZnO can also be used as the waveguiding material. Its high RI (–2.0) enables a high surface sensitivity. Furthermore, ZnO shows the electro-optic effect: by applying an electric field across a layer of this material, the optical RI is changed, which can be used for linearizing the sensor signal Heideman et al. (1996).

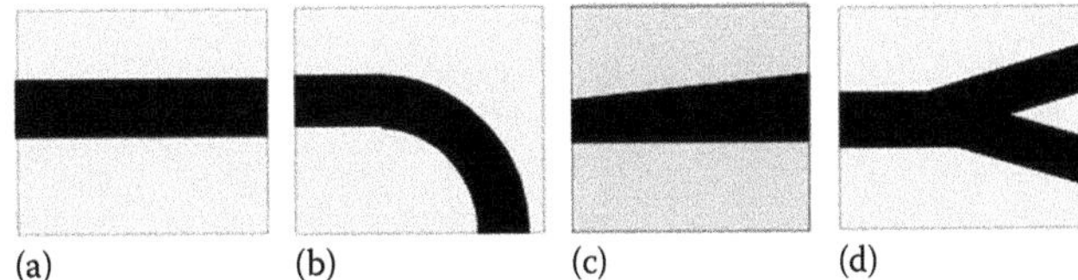

FIGURE 23.3 Top view of several IO basic functions: (a) straight segment, (b) bend, (c) taper, and (d) *Y*-junction. (From Lambeck, P.V., *Meas. Sci. Technol.*, 17, R93, 2006, with permission of IOP Publishing Ltd.)

We need to note that integrated optical structures besides planar waveguides discussed before in Section 11.5 may consist of various basic passive and active device components such as *Y*-junctions, directional couplers (Section 22.8.7), lasers, photoreceivers (Section 22.3), different interferometer types (Chapter 16), electro-optical switches, and phase and amplitude modulators (Syms and Cozens 1993; Teichmann 1995). Many of these active and passive elements, which were designed for telecommunication, can be fabricated in planar form acceptable for IO. For example, besides long straight waveguides there are other configurations of waveguides with other functions. By introducing *z*-dependence of the RI distribution, a quartet of other basic functions of waveguides can be defined: the straight segment, the bend, the taper (continuous increase or decrease of a core dimension), and the *Y*-junction (Lambeck 2006). Top views of these basic functions are depicted in Figure 23.3. All purely optical higher level IO functions and circuits can be considered as being configured from these four basic functions. Although at first sight these basic functions seem to be very simple, research has shown that above-mentioned configurations exhibited very special properties. For example, if the *Y*-junction configuration is symmetric it obviously will function as a 3 dB power splitter and both output channels will support the half of the input power. In certain asymmetrical configurations, however, the *Y*-junction behaves like a mode splitter, implying, for example, that the power of a TE_0-mode (transverse electric) propagating through a bimodal central input branch is completely transferred to one branch and the power of the TE_1 mode to the other branch.

Mirror, which offers a simple method of changing the direction of a guided beam, is also the element of IO (Syms and Cozens 1993). It can occupy considerably less space than a bend, and can theoretically be more efficient. Such a device can operate by the reflection of guided modes from a flat, highly reflective interface, arranged exactly orthogonal to the guide. A suitable interface can be made by cutting a trench across the guide, as shown in Figure 23.4a. This can be done using a highly anisotropic etch process—reactive ion etching (RIE) and focused ion-beam micromachining are both suitable. The interface must then be metalized to increase its reflectivity. Although the integrated mirror is conceptually a simple device, it is hard to make in practice, because any reduction in surface flatness, orthogonality or reflectivity will lower the overall reflection efficiency. Mirrors can also be used to turn a beam through 90°, for example, by making an interface at 45° at an intersection between two orthogonal ridge guides, as shown in Figure 23.4b. The device is similar to the direct reflector, but there is no need for coating to increase reflectivity—the interface itself is

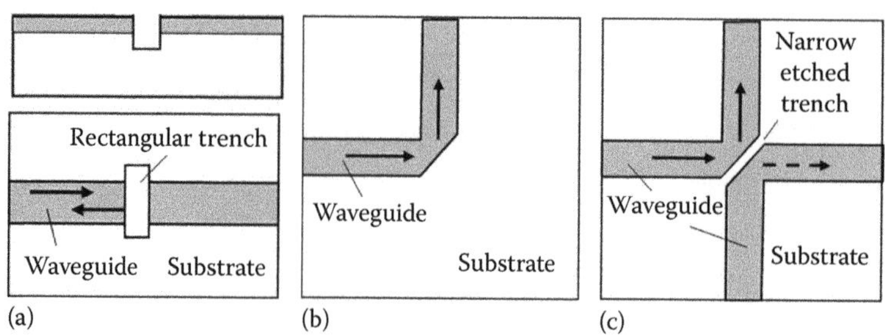

FIGURE 23.4 (a–c) Schematic diagrams of IO devices with mirror. Explanations are in the text. (From Korotcenkov, G. et al., Optical and fiber optic chemical sensors, in *Chemical Sensors*, Vol. 5: Electrochemical and Optical Sensors, Korotcenkov, G. (Ed.), Momentum Press, New York, 2011. With permission.)

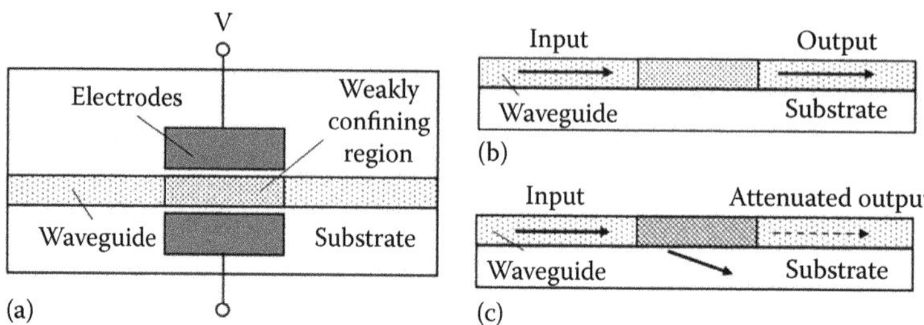

FIGURE 23.5 A cutoff modulator: (a) schematic, (b) operation with no applied voltage, and (c) operation with a voltage sufficient to induce cutoff. (Idea from Syms, R.R.A. and Cozens, J.R., *Optical Guided Waves and Devices*, Academic Press, London, UK, 1993.)

normally totally reflecting, due to the large index difference between the ridge guide material and its surround (air). This particular technique is important, because it allows the construction of bendless integrated optical circuits (Syms and Cozens 1993). Figure 23.4c shows a related device, formed from two 90° reflectors separated by a small air-gap. This is made using a narrow etched trench, so that the light can cross the air-gap by optical tunneling. The device then acts as a frustrated-total-internal-reflection beam splitter, and the splitting ratio is controlled by adjusting the gap width.

Planar couplers, polarizer, and phase modulator required for IO, can easily be realized as a single chip on lithium niobate ($LiNbO_3$). A planar $LiNbO_3$ phase modulator consists of a titanium diffused monomode channel waveguide and a pair of oriented electrodes. Through application of an electrical voltage to the electrodes, the RI of the waveguide, and hence the phase of the optical signal passing through it can be altered.

Intensity modulation can also be organized in planar waveguide. Figure 23.5a shows a schematic of the device, which consists of a pair of phase-modulator electrodes operating in a region of weak optical confinement. With no applied voltage, the guided beam is transmitted through the active section with little attenuation (Figure 23.5b). If a voltage of the correct sign is applied, this section approaches cutoff, and the guided beam suffers heavy attenuation through conversion to radiation modes (Figure 23.5c).

It was shown that the phase modulation can also be converted into amplitude modulation, by inserting a phase shifter in one arm of an interferometer. The simplest is the Mach–Zehnder interferometer (MZI), shown in Figure 23.6. The structure consists of two back-to-back *Y*-junctions, separated by a region containing a pair of identical electro-optic phase modulators.

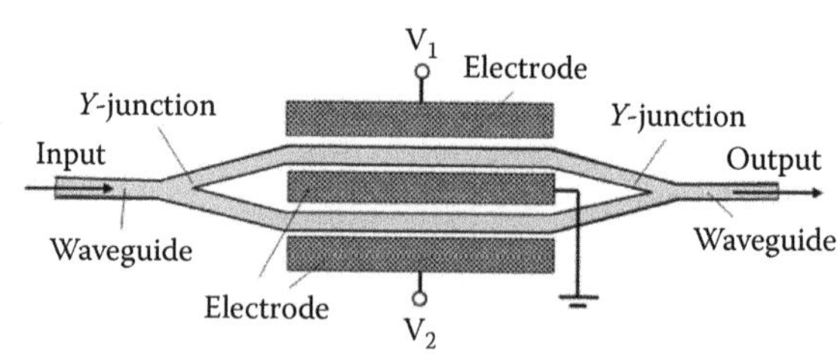

FIGURE 23.6 An electro-optic channel waveguide Mach–Zehnder interferometer, constructed using *Y*-junctions. (From Korotcenkov, G. (Ed.), *Chemical Sensors*, Vol. 5: Electrochemical and Optical Sensors, Momentum Press, New York, 2011. With permission.)

23.3 TECHNOLOGIES APPLIED IN INTEGRATED OPTICAL SENSORS

There are several ways by which waveguides and devices in the IO form can be realized (Selvarajan and Asoundi 1995; Burck et al. 1996). The type of material chosen more or less decides the process technology to be employed. In the case of glass, wet and dry ion-exchange techniques are commonly used for fabricating mostly passive IO components such as splitter/combiners. Polymer waveguides on glass and other substrates, on the other hand, are formed by spin or dip coating. Although this process is simple, precise thickness and uniformity control are difficult. Plasma polymerization and Langmuir Blodgett method of formation are other techniques used in the case of polymers. The most popular material for IO, $LiNbO_3$, as we wrote earlier, can be processed either using metal in-diffusion (usually titanium) or by proton exchange in weak acids. Epitaxial methods such as Liquid phase epitaxy (LPE), Molecular Beam Epitaxy (MBE), and Metal-Organic Chemical Vapor Deposition (MOCVD) are appropriate for the growth of crystalline layers and quantum well structures in semiconductors. Amorphous layers on silicon are also useful in certain passive component development such as grating and channel waveguides in sensing and communication applications.

Microstructuring technologies designed for microelectronics can be successfully applied for fabrication the optical structures such as gratings, microlenses, and synthetic optical elements as well. For example, the wave-guiding and grating structures in Bragg-based optical sensors can be fabricated with a number of approaches such as ultraviolet (UV) written waveguides and gratings, corrugated/etched Bragg gratings, or even Bragg gratings through a selective precipitation of nanoparticles. Moreover, a direct UV writing technology was commercialized by Stratophase Ltd. This technology was developed in the University of Southampton, United Kingdom (Sparrow et al. 2005). The method allows the inscription of waveguides and gratings onto planar substrates. This process utilizes the photosensitivity of Germanium doped silica. Exposure of the silica to UV light at 244 nm causes an increase in the RI. The three layer samples are fabricated by flame hydrolysis deposition such that the silica is deposited onto a silicon substrate. The middle layer is doped with germanium to provide a photosensitive layer that may be exposed to UV to raise the RI. High-pressure hydrogen loading is used as a method of enhancing the photosensitive response immediately prior to UV exposure. This process is illustrated in Figure 23.7. Controlled, localized irradiation of the silica with a sufficiently high fluence after hydrogen loading allows well-defined waveguide structures to be created. For the work reported here, a frequency doubled argon-ion laser is used to produce the UV beam for irradiation. The beam is conditioned and focused down to a 5 μm diameter spot in a fixed location. Samples for UV writing are mounted on the air-bearing translation stage, which allows three dimensional movement of the photosensitive substrate to a spatial resolution better than 10 nm. A more detailed description of this technology can be found in the review (Sparrow et al. 2009).

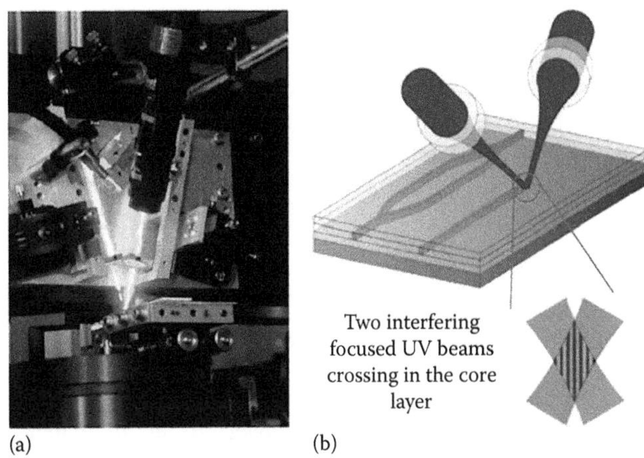

FIGURE 23.7 (a) Diagrammatic representation of the UV writing process with crossing UV beams focused onto the photosensitive substrate; (b) Diagrammatic representation of the UV writing process with crossing UV beams focused onto the photosensitive substrate. (From Sparrow, I.J.G. et al., *J. Sens.*, 2009, 607647, 2009. Published by Hindawi as open access.)

In contrast to telecommunication applications, most sensor applications do not require an optimized coupling efficiency between optical fiber and integrated optical channel waveguide. More important for achieving commercially attractive IOSs, is a low cost, reproducible connection of optical fibers to the channel waveguides on the chip. Therefore, a self-aligning fiber-to-chip connection unit shown in Figure 23.8, which requires no polishing steps and no external fiber-to-chip alignment equipment, is good approach to fabrication of integral chemical sensors (Heideman et al. 1996). However, Burck et al. (1996) have noted that due to strong light losses in such

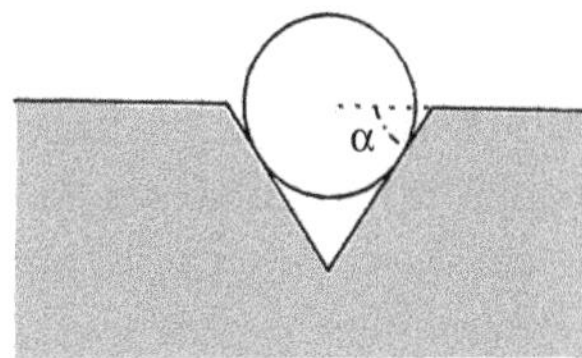

FIGURE 23.8 Schematic representation (cross section) of a fiber in a V-groove in silicon.

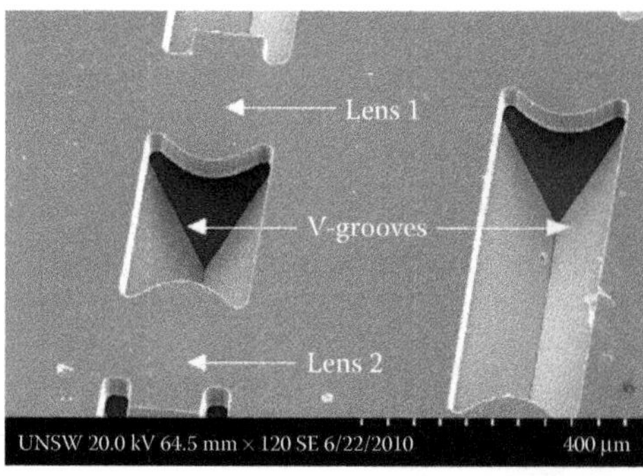

FIGURE 23.9 The SEM image of fabricated microlens pairs with V-grooves. (From Michael, A. et al., *Current Developments in Optical Fiber Technology*, InTech, Rijeka, Croatia, 541–585, 2013. Published by InTech as open access.)

configuration a long interaction length of the waveguide structure is necessary to get an acceptable sensitivity. V-shaped structures are also a basis for optical microplatforms, where different elements are mechanically positioned and fixed with high precision, and so on (Bartelt 1995). V-shaped channels (Figure 23.8) usually formed by chemical etching Michael et al. (2013). One should note that silicon is a typical substrate for integration of fiber-optic elements with other components of hybrid devices. For example, Figure 23.9 shows the SEM image of the planar lens pairs with V-grooves for optical fiber insertion and alignment. Technologies acceptable for the IO fabrication were discussed in details by Heideman et al. (1996), Spannhake et al. (2009), and Michael et al. (2013). On the basis of conducted analysis one can conclude that though a great potential of monolithic IOS, hybrid technology till now remains basic technology for IOSs fabrication. Klainer et al. (1997) believe that using miniature waveguides on a hybrid, semiconductor integrated circuit (IC) at present is an ideal solution for designing a handheld unit that is rugged, reliable, and small-sized. The hybrid IC solves the problems associated with existing optical waveguide configurations, where the entire sensor is not integrated. Therefore, the hybrid IC eliminates the concerns such as optical alignment and the need for an external light source and detector. The commercial IC package is encapsulated so that only the waveguides are exposed to the analyte. Thus, this approach is acceptable for mass production. For example, the hybrid configuration, including a dual in-line IC package with sense and reference waveguides mounted above the top of the IC, is already available commercially (Saini and Coulter 1996). In such IOSs a temperature sensor and a processor with memory are combined to form the electronics module, which converts the output of the sensor module to concentration. The serial memory provides the performance and calibration information about the specific sensing chemistry or biochemistry. This information can be provided for each interchangeable plug-in sensing module.

23.4 INTEGRATED HUMIDITY SENSORS

Unfortunately, at present there were only a few attempts to develop integrated humidity sensors. Skrdla et al. (1999) have presented a planar waveguide water vapor sensor that utilized an organic dye encapsulated in a sol–gel thin film. The device demonstrated a sensitivity range of 1%–70% RH with a response time of <1 min. Bhola et al. (2009) have performed a planar optical microring resonator hygrometer with a sensitivity of 16 pm/% RH in the range 0%–72% RH and a response time of ~200 ms. Prosposito et al. (2011) have described a potential hygrometer consisting of a sol–gel waveguide containing a Bragg grating on a planar substrate. Rosenberger et al. (2012) have reported about a planar-integrated Bragg grating device fabricated from the bulk poly(methyl methacrylate) (PMMA) and this device was demonstrated as a relative humidity (RH) sensor. The planar polymer Bragg grating sensor showed a sensitivity toward a RH of about 42 ± 2 pm/% RH over a range from 35% to 85% RH. Voznesenskiy et al. (2013) fabricated integrated-optical sensors for a RH detection based on waveguide films of chitosan, which is humidity-sensitive material. Sensitivity varied in the range of 0.005–0.2 dB/% RH. The main advantages of these sensor systems included a wide dynamic range (up to 20 dB) and low (<3%) hysteresis of the signal. Eryürek et al. (2017) have presented integrated optical humidity sensors based on the chips containing SU-8 polymer microdisks and waveguides fabricated by a single-step UV photolithography. The microdisk RH sensor consisted of an integrated pair of microdisks and a waveguide, both made out of SU-8 on a 4 inch commercial (100) Si wafer (wafer thickness is 500–550 μm, resistivity: <0.05 Ωcm) (Figure 23.10). The wafer incorporated a 5-μm-thick thermal-oxide layer with the index of refraction lower than that of SU-8 to ensure the light guiding inside the SU-8 layer. Fabrication of the sensor device was carried out using standard UV photolithography. The sensing was achieved by recording spectral shifts of the whispering gallery modes (WGMs) of the microdisk microresonators. Between 0% and 1% RH, an average spectral shift sensitivity of 108 pm/% RH was observed during sensor testing.

However, one should recognize that listed above works represent a demonstration of opportunities rather than real integrated humidity sensors. For example, Skrdla et al. (1999) and Voznesenskiy et al. (2013) in their studies have used a prism coupling to launch light

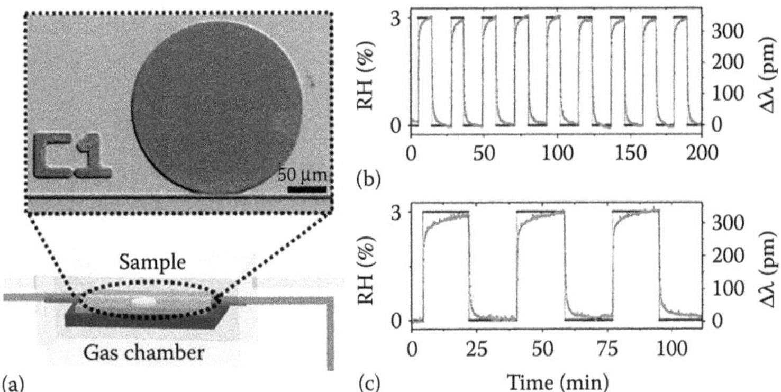

FIGURE 23.10 (a) The SEM image of the microdisk in the humidity sensor developed by Eryürek et al. (2017); (b) The sensor response on multiple cycles of humidity during on–off experiments; and (c) The second set of on–off cycles is 45 days after the first set of cycles. Both long-term and short-term experiments show good repeatability. (Reprinted from *Sens. Actuators B*, 242, Eryürek, M. et al., Integrated humidity sensor based on SU-8 polymer microdisk microresonator, 1115–1120, Copyright 2017, with permission from Elsevier.)

into the sensor. But this method is not practical for the waveguide-based chemical sensing due to inherent reproducibility problems. In addition, such approach poorly performs a miniaturization and integration. Versions of humidity sensors, which can be regarded as truly integrated sensors, were developed by Lukosz et al. (1997), Wörhoff et al. (2004), and Wales et al. (2013) (Table 23.2).

Wörhoff et al. (2004), using standard packaging technology in combination with IO, developed MZI-based sensor allowing for multispecies sensing. The design resulted in 40 mm long and 4 mm wide chips each containing 5 MZI channels. Configuration of this device is shown in Figure 23.11. One should note that the elaboration of this device was initiated by Heideman and Lambeck (1999). To realize the passive IO circuits, the SiON technology was used. By varying the O/N ratio in the gas phase during CVD or PECVD, the RI can be set between $n = 1.46$ (SiO_2) and 2.01 (Si_3N_4). ZnO was used as an electro-optic modulator. The main requirement for an efficient fiber-to-chip coupling is a good match of the field profiles of the single-mode fiber mode and the waveguide channel mode. To obtain a large overlap between the modal field of the fiber and the channel waveguide, either the fiber can be adapted to the channel waveguide (lensing and/or tapering the fiber) or vice versa. For lateral alignment in the core positioning the well-known wet-etched V-shaped groove, obtainable in silicon (100) wafers, and a double mask technique were applied (Heideman and Lambeck 1999). Testing conducted by Wörhoff et al. (2004) showed that elaborated sensors can detect the humidity in the range of 15%–90% RH.

Wales et al. (2013) have used a different approach to elaboration of integrated humidity sensors. Configuration of their device is shown in Figure 23.12. The flame hydrolysis deposition was used to deposit a fully dense, glassy germanium-doped silica layer onto a thick thermal oxide (~17 μm) coated a silicon wafer (~1 mm). Into this doped core layer, the channel waveguides, containing Bragg gratings were written using a pair of interfering UV-laser beams ($\lambda = 244$ nm). A process known as direct UV grating writing allows for the simultaneous

TABLE 23.2
Examples of Integrated Humidity Sensors

Waveguide	Substrate	Configuration	Sensing Material	References
$LiNbO_3$	$LiNbO_3$	Michelson interferometer	Polyimide	Izutsu et al. (1986)
Porous SiO_2–TiO_2	Si	Difference interferometer	Uncoated	Lukosz and Stamm (1991), Lukosz et al. (1997)
SiO_xN_y/SiO_2	Pyrex	Mach–Zehnder interferometer	Uncoated	Wörhoff et al. (2004)
Doped silica	Undoped silica/Si	Bragg grating	Al_2O_3	Wales et al. (2013)
Si_3N_4	Si	Microring resonantor	Uncoated	Tao et al. (2016)
SU-8	Si	Microdisk microresonantor	Uncoated	Eryürek et al. (2017)

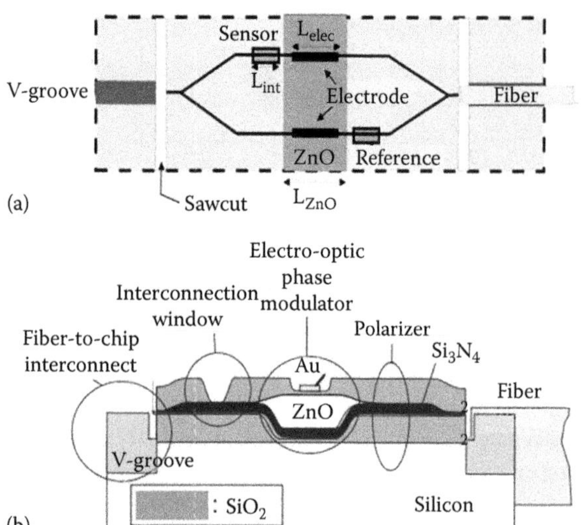

FIGURE 23.11 Top view (a) and cross section (b) of the integrated optical MZI configuration. (Reprinted from *Sens. Actuators B*, 61, Heideman R.G. and Lambeck P.V., Remote opto-chemical sensing with extreme sensitivity: Design, fabrication and performance of a pigtailed integrated optical phase-modulated Mach–Zehnder interferometer system, 100, Copyright 1999, with permission from Elsevier.)

formation of both waveguides and, through modulation of the laser intensity, Bragg gratings. The UV-written waveguide dimensions were defined by the thickness of the planar core layer and the width of the UV beam, both approximately 5 μm. Multiple Bragg gratings were written into a single waveguiding channel, each Bragg grating being 2 mm long and with a period between 525 and 542 nm, satisfying the Bragg condition to reflect at wavelengths between 1520 and 1570 nm. Grafting the thin-film with aluminium oxide was used to prevent hydrolytic degradation of silica and to improve sensitivity and long-term stability.

A commercial optical fiber *pigtail* was aligned and robustly attached to the integrated Bragg grating device via a commercially available *v-groove* optical fiber assembly and UV-cured adhesive (Wales et al. 2013). Solubilization of the UV-cured adhesive by water vapor was not expected; the UV-cured adhesive is very insoluble in water. Regardless, the *pigtail* joint was routinely protected by a polymer sheath, which did not interfere with operation of the sensor, and acted to prevent/lessen the degradation of the UV-cured adhesive due to ingression of solvent vapor during the solvent washing/steps to regenerate the sol–gel thin film. The reflectance spectrum of the device was interrogated through the optical fiber pigtail by an infrared erbium fiber amplified stimulated emission (ASE) source (1520–1570 nm) and was analyzed by an optical signal analyzer (OSA) connected via an optical circulator. The results obtained by Wales et al. (2013) demonstrated that the RI change on physisorption of water into the pores of a silica sol–gel and silica modified with Al_2O_3 was sufficient to be detected by the high-precision Bragg gratings. The sensitivity of the aluminum oxide modified thin film device, in the range 0%–65% RH, was 0.69% ± 0.05% RH/pm.

With regard to the problems associated with the use of integrated sensors to measure humidity, they are common to all integrated chemical sensors. At present, there is no material well suited for all function purposes, and therefore none of them has so far emerged as the key material for IOS fabrication. Moreover, in spite of developments in the field of humidity-sensitive optical materials, it remains difficult to find for sensing material a combination of all subfunctions such as sensitivity, specificity, stability, and reproducibility, and the range response time, which matches completely the requirements of an application and which can be integrated into a complete

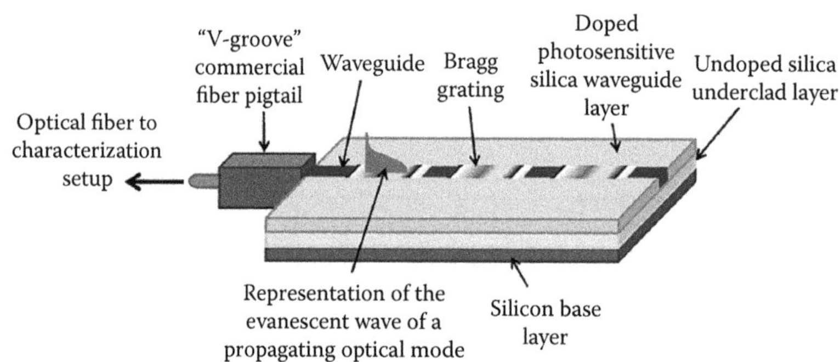

FIGURE 23.12 An illustration of the Bragg grating refractometer devices designed by Wales, D.J. and coworkers. (Reprinted from *Sens. Actuators B*, 188, Wales, D.J. et al., An investigation into relative humidity measurement using an aluminosilicate sol–gel thin film as the active layer in an integratedoptical Bragg grating refractometer, 857, Copyright 2013, with permission from Elsevier.)

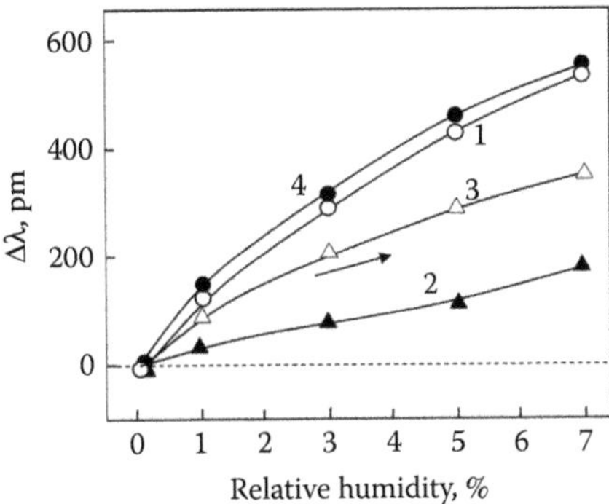

FIGURE 23.13 Responses to humidity obtained for four microdisk sensors. (Reprinted from *Sens. Actuators B*, 242, Eryürek, M. et al., Integrated humidity sensor based on SU-8 polymer microdisk microresonator, 1115–1120, Copyright 2017, with permission from Elsevier.)

sensing system, which can be offered for a price affordable by the market (Lambeck 2006). Thus, depending on application requirements, IOS technology faces a broad variety of materials, processing, and hybridization techniques such as electron beam writing and laser beam writing, silicon bulk and surface micromachining, microelectro-mechanical systems (MEMS), optoelectronics integration on a semiconductor substrate, and so on. Certainly, the lack of any generally accepted solution for IO in terms of the material and fabrication technology creates limitations in the design and manufacture of IOS with reproducible parameters (Figure 23.13). It is also necessary to take into account the specificity of the effect of humidity on the components of IOS; the stability of the packaging and UV-curable epoxy adhesives used to manufacture photonic integrated circuits (PICs) in optical network systems can be severely impacted in high humidity environments (Yoshida 1993). However, this problem is also common for all IOSs, used for environment monitoring.

In conclusion, we would like to note that the operation of optical humidity sensors and numerous gas sensors is based on the same principles and therefore integrated devices designed to detect gases can be also used to measure humidity. It is only necessary to replace the sensing material. Therefore, the readers can refer to further information on this topic of sensor technology, contained in comprehensive reviews of integrated optical chemical sensors, published by Kunz (1999), Lambeck (2006), Kashyap and Nemova (2009), Sparrow et al. (2009), Zourob and Lakhtakia (2010), and Passaro et al. (2012), and in the books, related to IO (Hunsperger 2002; Iizuka 2002; Pollock and Lipson 2003; Iga and Kokubun 2005).

REFERENCES

Ansari Z.A., Karekar R.N., Aiyer R.C. (1997) Humidity sensor using planar optical waveguides with claddings of various oxide materials. *Thin Solid Films* 305, 330–335.

Ballantine D.S., Wohltjen H. (1986) Optical waveguide humidity detector. *Anal. Chem.* 58, 2883–2885.

Bartelt H. (1995) Optical microsensors. In: Gopel W., Hesse J., Zemel J.N. (Eds.), *Sensors: A Comprehensive Surve.* Wiley-VCH Verlag, Berlin, Germany, pp. 259–274.

Bhola B., Nosovitskiy P., Mahalingam H., Steier W.H. (2009) Sol-gel-based integrated optical microring resonator humidity sensor. *IEEE Sens. J.* 9, 740–747.

Burck J., Zimmermann B., Mayer J., Ache H.-J. (1996) Integrated optical NIR-evanescent wave absorbance sensor for chemical analysis. *Fresenius J. Anal. Chem.* 354, 284–290.

Eryürek M., Tasdemir Z., Karadag Y., Anand S., Kilinc N., Alaca B.E., Kiraz A. (2017) Integrated humidity sensor based on SU-8 polymer microdisk microresonator. *Sens. Actuators B* 242, 1115–1120.

Franke H., Wagner D., Kleckers T., Reuter R., Rohitkumar H.V., Blech B.A. (1993) Measuring humidity with planar polyimide light guides. *Appl. Opt.* 32(16), 2927–2935.

Fuke M.V., Vijayan A., Kanitkar P., Kulkarni M., Kale B.B., Aiyer R.C. (2009) Ag-polyaniline nanocomposite cladded planar optical waveguide based humidity sensor. *J. Mater. Sci.: Mater. Electron.* 20, 695–703.

Heideman R.G., Lambeck P.V. (1999) Remote opto-chemical sensing with extreme sensitivity: Design, fabrication and performance of a pigtailed integrated optical phase-modulated Mach–Zehnder interferometer system. *Sens. Actuators B* 61, 100–127.

Heideman R.G., Veldhuis G.J., Jager E.W.H., Lambeck P.V. (1996) Fabrication and packaging of integrated chemo-optical sensors. *Sens. Actuators B* 35–36, 234–240.

Hunsperger R.G. (2002) *Integrated Optics: Theory and Practice*, 5th edn. Springer, Berlin, Germany.

Iga K., Kokubun Y. (Eds.) (2005) *Encyclopedic Handbook of Integrated Optics*. CRC Press, Boca Raton, FL.

Iizuka K. (2002) *Elements of Photonics*, Vol. 1. Wiley, New York.

Izutsu M., Enokihara A., Suets T. (1986) Integrated optic temperature and humidity sensors. *J. Lightwave Technol.* LT-4(7), 833–836.

Kashyap R., Nemova G. (2009) Surface plasmon resonance-based fiber and planar waveguide sensors. *J. Sens.* 2009, 645162.

Klainer S.M., Coulter S.J., Pollina R.J., Saini D. (1997) Advances in miniature optical waveguide sensors. *Sens. Actuators B* 176, 38–39.

Korotcenkov G., Cho B.K., Sevilla III F., Narayanaswamy R. (2011) Optical and fiber optic chemical sensors. In: Korotcenkov G. (Ed.), *Chemical Sensors: Comprehensive Sensor Technologies. Vol. 5: Electrochemical and Optical Sensors*, Momentum Press, New York, pp. 311–476.

Kunz R.E. (1999) Integrated optics in sensors: Advances toward miniaturized systems for chemical and biochemical engineering. In: Murphy E.J. (Ed.), *Integrated Optical Circuits and Components, Design and Applications*. Dekker, New York, Chapter 10, pp. 335–381.

Lambeck P.V. (2006) Integrated optical sensors for the chemical domain. *Meas. Sci. Technol.* 17, R93–R116.

Lukosz W., Stamm C. (1991) Integrated optical interferometer as relative humidity sensor and differential refractometer. *Sens. Actuators A* 25–27, 185–188.

Lukosz W., Stamm C., Moser H.R., Ryf R., Diibendorfer J. (1997) Difference interferometer with new phase-measurement method as integrated-optical refractometer, humidity sensor and biosensor. *Sens. Actuators B* 38–39, 316–323.

Michael A., Kwok C.Y., Al Hafiz Md., Xu Y.W. (2013) Optical fibre on a silicon chip. In: Harun S.W., Arof H. (Eds.), *Current Developments in Optical Fiber Technology*. InTech, Rijeka, Croatia, pp. 541–585.

Palais J.C. (2002) Micro-optics-based components for networking. In: Bass M., Van Stryland E.W. (Eds.), *Fiber Optic Handbook: Fiber, Devices, and Systems for Optical Communications*. McGraw-Hill, New York, Chapter 10.

Passaro V.M.N., de Tullio C., Troia B., La Notte M., Giannoccaro G., De Leonardis F. (2012) Recent advances in integrated photonic sensors. *Sensors* 12, 15558–15598.

Pollock C.P., Lipson M. (2003) *Integrated Photonics*. Kluwer, Boston, MA.

Prosposito P., Palazzesi C., Michelotti F., Fogiletti V., Casalboni M. (2011) Hybrid sol-gel Bragg grating loaded waveguide by soft-lithography and its potential use as optical sensor. *J. Sol-Gel Sci. Technol.* 60, 395–399.

Rosenberger M., Koller G., Belle S., Schmauss B., Hellmann R. (2012) Planar Bragg grating in bulk Polymethylmethacrylate. *Opt. Exp.* 20, 27288–27296.

Saini D.P., Coulter S.L. (1996) Fiber sensors sniff out environmental pollutants. *Photonics Spectra* 5, 91–94.

Selvarajan A., Asoundi A. (1995) Photonics, fiber optic sensors, their applications in smart structures. *J. Non-Destruct. Eval.* 15(2), 41–56.

Skrdla P.J., Armstrong N.R., Saavedra S.S. (2002) Starch-iodine films respond to water vapor. *Anal. Chim. Acta* 455, 49–52.

Skrdla P.J., Saavedra S.S., Armstrong N.R., Mendes S.B., Peyghambarian N. (1999) Sol–gel-based, planar waveguide sensor for water vapor. *Anal. Chem.* 71, 1332–1337.

Spannhake J., Helwig A., Schulz O., Muller G. (2009) Micro-fabrication of gas sensors. In: Comini E., Faglia G., Sberveglieri G. (Eds.), *Solid State Gas Sensing*. Springer, New York, pp. 1–46.

Sparrow I.J.G., Emmerson G.D., Gawith B.E., Smith G.R. (2005) Planar waveguide hygrometer and state sensor demonstrating supercooled water recognition. *Sens. Actuators B* 107, 856–860.

Sparrow I.J.G., Smith P.G.R., Emmerson G.D., Watts S.P., Riziotis C. (2009) Planar Bragg grating sensors—fabrication and applications: A review. *J. Sens.* 2009, 607647.

Syms R.R.A., Cozens J.R. (1993) *Optical Guided Waves and Devices*. Academic Press, London, UK.

Tao J., Wang L., Cai H., Sun T., Song J., Gu Y. (2016) A novel photonic dew-point hygrometer with ultra-high accuracy. In: *Proceedings of International IEEE Conference MEMS 2016*, Shanghai, China, pp. 893–896, January 24–28.

Teichmann H. (1995) Integrated optical sensors: New developments. In: Gopel W., Hesse J., Zemel J.N. (Eds.), *Sensors: A Comprehensive Surve*. Wiley-VCH Verlag, Weinheim, Germany, pp. 222–258.

Voznesenskiy S.S., Sergeev A.A., Mironenko A.Y., Bratskaya S.Y., Kulchin Y.N. (2013) Integrated-optical sensors based on chitosan waveguide films for relative humidity measurements. *Sens. Actuators B* 188, 482–487.

Wales D.J., Parker R.M., Gates J.C., Grossel M.C., Smith P.G.R. (2013) An investigation into relative humidity measurement using analuminosilicate sol–gel thin film as the active layer in an integratedoptical Bragg grating refractometer. *Sens. Actuators B* 188, 857–866.

Wörhoff K., Heideman R.G., Gilde M.J., Blidegn K., Heschel M., van den Vlekkert H. (2004) Flip-chip assembly of an integrated optical sensor. In: *Proceedings of the Symposium IEEE/LEOS Benelux Chapter*, Ghent, Belgium, pp. 25–28, December 2–3.

Yoshida J. (1993) Packaging and reliability of photonic components for subscriber network systems. *IEEE Trans. Compon. Hybrids Manuf. Technol.* 16, 778–782.

Zourob M., Lakhtakia A. (Eds.) (2010) *Chemical Sensors and Biosensors: Methods and Applications*. Springer, Berlin, Germany.

24 Outlook—State of the Art and Future Prospects of Optical and Fiber-Optic Sensors

24.1 PROSPECTS OF OPTICAL AND FIBER-OPTIC SENSORS APPLICATION

Summarizing the consideration of optical and fiber-optic sensors (FOSs) one can state that the optical and fiber-optic technologies represent great opportunities for the development of a variety of devices capable of controlling the humidity. These devices can have a sufficiently high sensitivity and good operation speed necessary for *in situ* monitoring (Muto et al. 2003); and the sensing element may have a very small size. But at the same time, we must recognize that the fabrication of such sensors does not come to agreement with mass production. In addition, special sources and radiation detectors, and personal computer (PC) with special programs for signal processing are required for their functioning. This means that the detection systems and measuring systems may be complex, and their operation requires precise installation procedures and qualified professionals. This naturally substantially restricts the use of such facilities, since such sensors and devices, using them can be expensive. Therefore, it is difficult to expect that the optical and FOSs displace from the market of electronic and electrical humidity sensors, which, as well as optical and FOSs, do not possess the required selectivity, but considerably cheaper in production and in operation.

At the same time, optical and FOSs, which have a number of significant advantages, without a doubt can be used in the security systems and devices for special purposes, which are developed on different principles (Wolfbeis 1991, 1992; Burgess 1995; Cámara et al. 1995; Kersey 1996; Gansert et al. 2006; Consales et al. 2008; Korotcenkov et al. 2011). The main advantages of FOSs, including fiber-optic humidity sensors in comparison to their conventional electronic counterparts can be summarized as follows:

1. They allow *in situ* determination and real-time analyte monitoring.
2. They are excellent for applications where the measurement is to be made over a long period, with the major instrumental problems being window fouling and calibration.
3. They are easy to miniaturize because optical fibers have very small diameters. Therefore, FOSs can be used in the fields, where a measurement of humidity is needed but the accessibility is limited in space, and where an electronic sensor could be more difficult to locate.
4. They are fairly flexible: optical fibers can be bent within certain limits with no damage.
5. They can be used in hazardous places and locations of difficult access because of the ability of optical fibers to transmit optical signals over long distances (between 10 and 10,000 m). In addition, sensors have excellent corrosion resistance, and they are potentially resistant to ionizing radiation. Optical fiber humidity sensors have immune to radio frequency interference and electromagnetic interference. They are potentially resistant to ionizing radiation. In addition, sensors are nonmetallic and MRI compatible. The immunity to electric and magnetic fields suggests its applications in high-voltage installations, where it can be important to monitor humidity. For example, optic sensors do not use electricity and consequently they can be used for monitoring of inflammable liquids or gases because of the absence of sparks.
6. Fiber-optic chemical sensors are much safer in explosive environments compared with sensors involving electrical signals, where a spark may trigger a gas explosion.
7. Multielement analysis is possible when using various fibers and a single central unit.
8. They normally permit nondestructive analysis.
9. They perform a high-voltage insulation and absence of ground loops and hence one can avoid any necessity of isolation devices such as optocouplers.
10. Optical fibers can carry more information than electrical cables. Fiber-optic sensing is very versatile, because the intensity, wavelength, phase, and polarization of light can all be exploited as measurement parameters, and several wavelengths

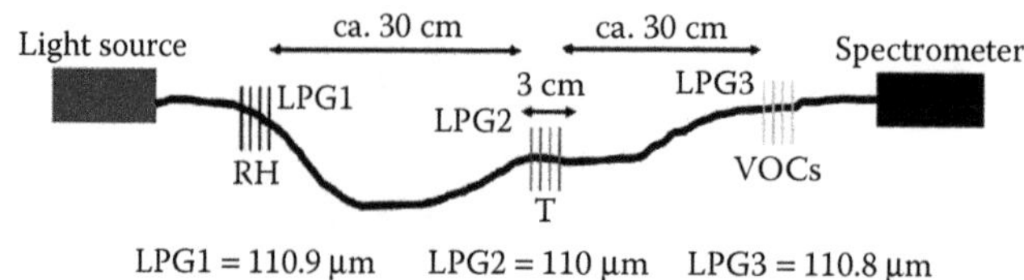

FIGURE 24.1 Schematic illustration of the LPG sensor array for multiparameter measurements; the individual LPGs were used to measure: LPG1 relative humidity (RH); LPG2 temperature; and LPG3 volatile organic compounds (VOCs). Each LPG sensor was designed with optimized response to a particular measurand. The first sensor was with no surface modification. The second one was modified by a mesoporous coating of silica nanoparticles (SiO_2 NPs), and the third one was modified with a coating of SiO_2 NPs infused with a functional material, p-sulphanatocalix[8]arene (CA[8]). The LPGs were fabricated with periods such that they operated at or near the phase matching turning point. (Reprinted from *Sens. Actuators B*, 244, Hromadka, J. et al., Multi-parameter measurements using optical fibre long period gratings for indoor air quality monitoring, 217–225, Copyright 2017, with permission from Elsevier.)

launched in the same fiber in either direction form independent signals. This gives the possibility of monitoring several chemicals with the same fiber sensor or even simultaneously monitoring unwanted environment parameters variations, which could drastically affect the chemical concentration measurements, such as the temperature or disturbance of the fiber. For example, FOSs can also provide simultaneous monitoring of parameters such as temperature and humidity inside microwave ovens. Hromadka et al. (2017) have shown that an array of three long-period gratings (LPGs) fabricated in a single optical fiber and multiplexed in the wavelength domain can be used to measure simultaneously temperature, relative humidity (RH), and volatile organic compounds (VOCs), which are key indoor air quality (IAQ) indicators (Figure 24.1).

11. Probes are often easy and inexpensive to build.
12. The sensor element can be a long length of fiber providing an extended sensing element: this can be used to enhance sensitivity by wrapping the fiber in a compact form to create the transducer head, or it can be used to provide spatial averaging of the measurand of interest.
13. Another option is the ability to spatially discriminate the measurand at different locations along a fiber length (Figure 24.2). This leads to the capability to perform distributed sensing in addition to local measurement. Such ability is a powerful sensing tool, which is not generally possible in using conventional sensor technologies. If the fiber is not sensitive along its entire length, but is locally sensitized at various points, the system becomes a *quasi-distributed* sensor system (Kersey 1996). In addition, the optical

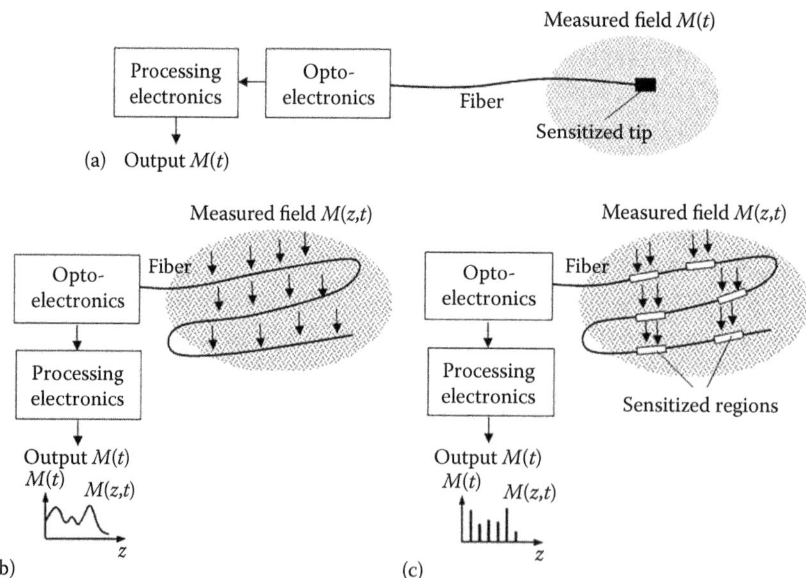

FIGURE 24.2 (a) Point, (b) distributed, and (c) quasi-distributed sensing.

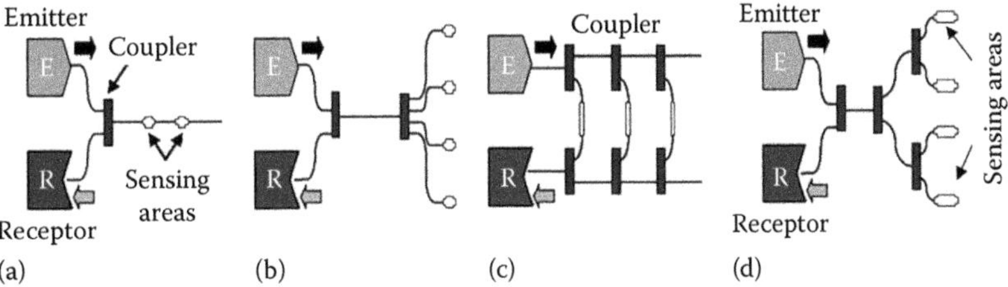

FIGURE 24.3 Examples of optical fiber sensor networks: (a) line, (b) star, (c) ladder, and (d) tree. (From Elosua, C. et al., *Sensors*, 6, 1440, 2006. Published by MDPI as open access.)

fiber sensors permit coupling of the interrogating and the response signal in the same fiber, which simplifies the sensor system and also makes possible the implementation of a multipoint sensor network (Figure 24.3). These properties are all very attractive when is important to control gas concentrations in large spaces or in pollutant control applications (Elosua et al. 2006).

However, the optical and fiber-optic humidity sensors have found their area on the sensor market; further research is needed, aimed at optimizing the manufacturing technology, increasing the life time of sensors, and increasing the reproducibility of the parameters.

24.2 WHAT CONSTRAINS THE USE OF OPTICAL AND FIBER-OPTIC SENSORS?

As it was noted by Cámara et al. (1995), fiber-optic sensors, including humidity sensors, have the following disadvantages:

1. In a similar way to their electronic counterparts, optical and fiber-optic humidity sensors are secondary devices.
2. The number of reversible reactions is very limited, so in many cases probes have to be regenerated after use.
3. The area of interaction between the chemical and the material is very small, that is, 8–10 μm in diameter, in the case of single-mode fiber and 50–200 μm in diameter, which directly affects the sensitivity achieved from this type of sensors.
4. Ambient light may distort the measurement and some of the reagents immobilized at the sensor surface, mainly dyes, are unstable and can be bleached by ultraviolet (UV) radiation or washed out by solvents.
5. The properties of the indicator may vary. Besides they usually have lower dynamic ranges than electrodes. There are other shortcomings of the fiber-optic-based humidity sensors, which should be resolved, such as the presence of hysteresis, long response time and relatively low accuracy, and so on.
6. In some cases, the concentration of the immobilized indicator is unknown and two optodes prepared similarly can have different analytical characteristics.
7. Some limitations arise from the somewhat restrictive spectral windows that can be effectively accessed using available low-cost fibers. Frequently, an analyte will not have an intrinsic spectral feature within a usable window or will simply be too weak an absorber in any optical path length that could reasonably be accommodated in the remote cell.
8. Stability and durability are also an important issue for polymer-based humidity sensors. Many papers and reviews dealing with the development of FOSs have been published in the literature but only a few have reported applications to environmental monitoring. It was also established that some polymers undergo degradation or an irreversible swelling process when exposed to certain environments (Pejcic et al. 2007). Polymer degradation is a change in the properties—tensile strength, color, shape, conductivity, and so on—of a polymer or polymer-based product under the influence of one or more environmental factors such as heat, light, and chemicals (Carraher 2008). For example, Jiang et al. (2005) reported that the sensitivity of polypyrrole (PPy)/polyvinyl alcohol (PVA) composite sensor was only maintained for two weeks, while the sensitivity of pure PPy sensor was maintained more than one month. Of course, this lifetime is too short. Irreversible changes in polymers under influence of UV radiation and oxidizers such as ozone and NO_2 also

limit appreciably of the application of polymers for designing sensors aimed at environment control. Experiment has shown that the reaction of ozone with the polymer is irreversible and the average useful lifetime is found to range from 20 to 3000 ppb hours only. It has been reported that ozone and other oxidizing components in the polluted atmospheres of industrial centers could either initiate or accelerate the photochemical destruction of polymers (Razumovskii and Zaikov 1982). As a result, in the case of an ozone-containing atmosphere, the lifetime of polymer-based sensors was considerably shorter (Muller et al. 2011). Polymer sensors used for environmental control also have a significant disadvantage in terms of their sensitivity to UV radiation. Moreover, it was found that polymer degradation was almost always faster in the presence of oxygen (air) and moisture. However, that longtime instability, which is accompanied by the temporal drift and degradation of sensor performance, is a big drawback of all types of sensors-based on polymers (Kumar and Sharma 1998; Kondratowicz et al. 2001; Bai and Shi 2007). It is clear that environment stability of sensors depends on the type of polymer material and therefore, this feature of the polymers should be taking into account during selection of polymers for humidity sensor design (Carraher 2008). On account of the low stability, even ultrasensitive sensors are unsuitable for real application in industry and environment monitoring.

9. The big problem of modern developments in the field of optical and FOSs is also poor reproducibility of parameters. Unfortunately, the methods used for the deposition of humidity indicators on the surface of optical fibers mostly have problems in controlling the thickness and uniformity of the coated layer. The variety of humidity-sensitive materials, used in developed optical and FOSs, with the almost complete absence of studies aimed at establishing the correlation between the parameters of humidity-sensitive materials, their synthesis technologies, and sensor performance, also creates certain difficulties in choosing the humidity indicators with optimal properties. As a rule, each team uses a different set of materials, which will not be repeated in the studies carried out by other teams. Some researchers believe that PVA is a very promising material for humidity detection (Kolpakov et al. 2014). According to Wong et al. (2012), this material allows fabrication of very fast and selective humidity sensors with response times of <1 s. However, this statement requires verification and correct comparison with other materials.
10. Disadvantages also include the poor elastic property that makes optical fiber extremely brittle. This is not particularly desired for sensing applications. Consequently, the handling, treatment, and operation of FOS require extreme care. In addition, each process requires technical skills. With the current technology, *in situ* assembly of the optical system is a cumbersome and exhaustive process, since the measurement setup of the optical fiber sensor system comprises different modules (e.g., light source, coupler, and receiver), difficult splicing process and the constant need for careful treatment. Although the method of manufacturing fiber gratings has improved considerably, current technology only manages to produce one at one time and most processes still require manual handling. In addition, the cost of running fiber grating writing facilities is high due to expensive laser, masks, (both phase mask and amplitude masks) and the requirement for a highly skilled operator.

Certainly, in the past decade, it has been done a lot to improve the parameters of optical and fiber-optic humidity sensors, to optimize their production technology and the development of different measuring systems (Alwis et al. 2013; Kolpakov et al. 2014). Optical fiber humidity sensors have been already used to facilitate the remote sensing and continuous monitoring of humidity in diverse applications such as the baking and drying of food, cigar storage, civil engineering to detect water ingress in soils or in the concrete in civil structures, medical applications, and many other fields (Yeo et al. 2006, 2008; Alwis et al. 2013). For example, Caponero et al. (2013) illustrated the flexibility of the fiber-optic humidity sensors to monitor humidity condition in resistive plate counters (RPCs) packaged in a steel box. This is important, as the humidity influences the working point of the RPC; these devices are nuclear and subnuclear physics detectors based on the ionization of a gas medium. Due to the RPC operating in a high electromagnetic field environment, the FOS is the best candidate for installation on an RPC due to the fact that FOSs are immune to electromagnetic interference. Other examples of fiber-optic humidity sensors applications are listed in Table 24.1. The integration of

TABLE 24.1
Humidity/Moisture Application-Specific Sensors

Sensor Application	References
Biomedical Measurements	
Breathing sensor	Akita et al. (2011), Favero et al. (2012)
Recognition of devoiced vowels	Morisawa et al. (2010)
Breathing air-flow monitor	Kang et al. (2006)
Climate/Agricultural Monitoring	
Turbidity sensor	Bilro et al. (2011)
Flood monitoring	Kuang et al. (2008)
Canopy water content sensor	Clevers et al. (2008)
Water stress detection of Poplar plantation	Eitel et al. (2006)
Water content sensor for vegetation	Sims and Gamon (2003)
Simultaneous measurement of several parameters	Arregui et al. (2002), Gaston et al. (2003)
Structural Health Monitoring (SHM)	
Water absorption and detection in concrete	Yeo et al. (2006), Kaya et al. (2013)
Building stone condition monitoring	Sun et al. (2012)
Dew detection	Mathew et al. (2012)
Quality Control Applications	
Water content measurement in ethanol	Xiong and Sisler (2010), Srivastava et al. (2011)
Water detection in jet fuel	Puckett and Pacey (2009), Zhang et al. (2009)
Humidity detection in oil-paper insulation of electrical apparatus	Rodriguez-Rodriguez et al. (2008)
Other Applications	
Water leak detection	Cho et al. (2012)
Water detection in optical fiber splice enclosures	Hsu et al. (2011)
Dew detection inside organ pipes	Baldini et al. (2008)

Source: Alwis, L. et al., *Measurement*, 46, 4052–4074, 2013.

optical fibers in reduced and more robust structures has led to a new generation of FOSs with lower detection limits, increased sensitivity, higher selectivity, reduced response time, and in some cases, reversibility and long-term stability. But these efforts are insufficient, as most of the researches carried out in this area are still aimed at demonstrating the feasibility of the humidity measurement. This means that the individual samples were made, which were not intended for sensor market. Currently, there are only a few commercially available fiber-optic humidity sensors manufactured by O/E Land Inc. (Table 24.2). Therefore, there is a constant need to develop a compact optical fiber humidity sensing system with high reliability, high sensitivity, and low cost.

TABLE 24.2
Performance of Commercial Optic Humidity Sensors

				Range, % RH	
Sensor	**Sensitivity pm/% RH**	**Operation Temperature**	**Size, mm**	**Minimum**	**Maximum**
O/E Land FBG-Based Sensors					
OEFHS-100A, OEFHS-100B	4.5	0°C–80°C	6 × 40 or 3 × 40 6 × 60 or 3 × 60	10	100

Source: http://www.o-eland.com/SensorProducts/OEFHS-100.php.

24.3 OPTICAL AND FIBER-OPTIC SENSORS—WHAT DETERMINES OUR CHOICES?

It is clear that when developing the humidity sensors and humidity control systems, one can use a variety of approaches. Comparative characteristics of some of them are given in Tables 24.3 and 24.4. However, a more detailed comparison of these methods must be sought in the relevant chapters of this book. It is seen that each of the optical and FOS classes has intrinsic benefits and limitations for application in measurement systems.

For example, adsorption-based humidity sensors use diode lasers and light-emitting diodes, which are reliable, high speed, miniature, and consume little energy, and photoreceivers and spectrographs with charge-coupled devices (CCD) or complementary metal-oxide semiconductor (CMOS) detectors that connected with PC, can be portable, relatively cheap, and due to incorporation of microprocessor many external and temporal factors affecting on measurement results can be eliminated (Wilson et al. 2001). However, as with nonoptical humidity sensors, issues of integration, packaging, and calibration must be carefully considered and streamlined in the design process to meet stringent size, weight, and power constraints. In addition, humidity sensors based on measuring the change in the output intensity of the device require light sources with high radiation stability. Otherwise, it is difficult to separate the response of the sensor from humidity with changes in laser power and also to vary losses in optical splices and connectors.

TABLE 24.3
Advantages and Disadvantages of Different Fiber-Optic Sensors

Advantages	Disadvantages
Grating	
Increased sensitivity	Expensive signal processing and interrogation technology; problem of temperature and strain signal division
Interferometer	
Extremely flexible geometry; high sensitivity; wide area distribution	Problem of multisensing signal division; sensitivity stabilization problem; tuning control problem
Amplitude	
Low cost; simplicity; easy installation; possibility of coverage of wide area	In many cases low sensitivity and presence of mobile mechanical elements; high requirements to stability of light source intensity
Nonlinear	
Possibility of distributed sensor fabrication	Low sensitivity; high cost

TABLE 24.4
Summary of the Advantages and Limitation of the Studied Technologies

Technology	Advantages	Limitations
Etched FBG	Well-developed technology Multiplexing capability	Fragility Low sensitivity High cost
Tilted FBG	Well-developed technology	Low sensitivity High cost
LPG	High sensitivity	Fabrication Temperature cross-sensitivity
LPG-based interferometer	High sensitivity	Fabrication Device length
Abrupt taper	High sensitivity	Fragility
CDM-based interferometers	Low-cost	Reproducibility
Multimodal interferometer	Low-cost Low-temperature cross-sensitivity	Reproducibility Broader resonance

Source: Gouveia, C.A.J. et al., Refractometric optical fiber platforms for label free sensing, in *Current Developments in Optical Fiber Technology*, Harun, W.S. and Arof, H. (Eds.), InTech, Rijeka, Croatia, pp. 345–373, 2013.

FBG—fiber Bragg grating; CDM—code division multiplexing; LPG—long period grating.

Luminescence sensors are extremely sensitive to small quantities of water vapor in the air, but the broadband character of luminescence spectra and the ubiquitous presence of naturally fluorescent compounds lead luminescence sensors to suffer from a critical lack of selectivity. Also, luminescence methods are sensitive to temperature fluctuations and other environmental factors that quench fluorescence. In addition, the lifetime of many of these sensors is short.

Some scientists (Kolpakov et al. 2014) believe that the most prospective direction is the development of interferometric-based humidity sensors based on photonic crystal. It was demonstrated experimentally (Mathew et al. 2011) that a simple design, using a laser diode as an interrogator is possible to use with this kind of sensors. Interferometers based on Michelson or Mach–Zehnder layouts or even Fabry–Perot intracavity were also demonstrated showing high sensitivity and great potential for the various applications. Meanwhile, most of the sensors that are based on optical fibers require the use of a spectrum analyzer as the interrogator. The technology, which allows the use of low-cost laser diodes as the interrogators, is expected to be attractive for industrial exploitation. However, these sensors do not solve the problems inherent for interferometric-based sensors such as low selectivity and the effect of temperature.

Refractometric-based sensors also have significant problems associated with low selectivity. Refractometric sensors, which use tapered fibers, due to its highly reduced cladding diameter have an enhanced evanescent interaction and have long been explored for refractive index (RI) measurements by monitoring the transmitted optical power. In spite of high sensitivity and very compact size (few millimeters), however, these structures are very fragile and special packaging is needed. It should also be borne in mind that in the fiber Bragg grating (FBG)- and LPG-based sensors the coating with humidity-sensitive polymer, which swells on absorbing moisture, in addition to changes in the RI, exerts a tensile force on the FBG and LPG. Thus, the swelling of the polymer can cause bending stresses in the fiber, which causes the reflection spectra to widen because of the chirp in the grating (Bhola et al. 2009). In addition, fiber grating sensors are also highly sensitive to temperature, therefore they need an extra mechanism to compensate temperature changes. FBG-based configurations are more attractive for the purpose of multipoint sensing due to their very narrow spectral response, while LPG-based sensors provide a powerful platform for advanced optical sensing in various other applications. Also, modern versions of refractometric humidity sensors such as FBG and LPG-based and interferometric-based sensors, due to the nature of their manufacture, are complex to manufacture. Multimode interference-based refractometers are also interesting solutions that rely on the concept of reimaging effects of multimodal interferometer (MMI) patterns present in multimode waveguides (Gouveia et al. 2013). In these devices, the transmitted spectral power distribution is highly sensitive to the optical path length of the multimode fiber and its surrounding refractive index (SRI). Usually based on singlemode-multimode-singlemode structures, they can be easily fabricated and applied in different situations. However, these configurations are also difficult to reproduce and present very broad spectral resonance making for instance multiplexing a very difficult task. The comparison of the most relevant evanescent field-based fiber refractometers is presented in Table 24.5.

Thus the use of fiber-optic techniques imply the use of read-out circuitry that requires on-chip phase modulation, a tunable wavelength or multiple wavelengths, where retrieving the phase information with the help of this additional hardware requires rather complicated signal processing electronics. In addition, the required an extremely high sensitivity over a large measurand range, requires highly coherent (single-line) laser sources. All together this makes these sensors quite expensive. No doubts, the features of these instruments cannot restrict their use in areas where very small concentrations have to be detected, for example, in security

TABLE 24.5
Summary of the Performance Parameters of the Most Relevant Works on Fiber-Based Refractometers

Configuration	Measurement Method	Sensitivity, nm/RIU	Resolution, RIU	References
Microfiber FBG	Spectral shift	100	–	Zhang et al. (2010)
TFBG	Spectral shift	10	10^{-4}	Chan et al. (2007)
Bare LPG	Spectral shift	1481	–	Shu et al. (2002)
HRI-coated LPG	Spectral shift	>9000	–	Pilla et al. (2012)
Mach–Zehnder LPG	Phase	–	1.8×10^{-6}	Allsop et al. (2002)
Fabry–Perot LPG	Spectral shift	–	2.1×10^{-5}	Mosquera et al. (2010)
LPG/FBG	Normalized optical power	–	2×10^{-5}	Jesus et al. (2009)
Abrupt taper	Spectral shift	1150	8.2×10^{-6}	Zibaii et al. (2010)
CDM-based Mach–Zehnder	Spectral shift	188	–	Ma et al. (2012)
MMI	Spectral shift	148	–	Biazoli et al. (2012)

Source: Gouveia, C.A.J. et al., Refractometric optical fiber platforms for label free sensing, in *Current Developments in Optical Fiber Technology*, Harun, W.S. and Arof, H. (Eds.), InTech, Rijeka, Croatia, pp. 345–373, 2013.

CDM—code division multiplexing; HRI—high refractive index; MMI—multimodal interferometer; RIU—refractive index unit; TFBG—tilted fiber Bragg grating.

systems or environmental monitoring. However, it should be understood that in many other applications much simpler and cheaper sensors are preferred. Another problem with optical humidity sensors is that the response times of most developed devices are relatively slow due to the thickness of the moisture absorbing layer in which water molecules have to diffuse through to interact with the guided optical wave (Bhola et al. 2009). Therefore, when developing the systems for practical application one must take into account this distinction, because as mentioned earlier more or less relates to sensors that use different principles.

With regard to the comparison of fiber-optic and integrated sensors, the fiber-based sensors compared to integrated optics are very cheap and have the strong advantage that they can be easily applied for distributed sensing. The sensor can be incorporated in the fiber that transports the light from the source to the detector and the different sensors can be read out in, for example, the time domain (OTDR) or wavelength domain (in case Bragg reflectors are used). Other advantages of FOSs, such as flexibility and small size of detection probes, have been shown previously at the beginning of this chapter. Nevertheless, fiber sensors cannot compete with integrated optics with respect to (1) robustness; (2) compact optical circuitry, enabling a higher complexity (e.g., multiple sensors on one chip); (3) design flexibility with respect to the geometry as well as the choice and combination of materials (e.g., active and passive materials); (4) ease of access to the optical path in evanescent-field sensing; (5) potential of integration with microelectronics, micromechanics, and micro total analysis systems; and (6) potential of *cheap* batchwise mass production.

We hope that presented in this book detailed information on various optical and fiber-optic-based RH sensing schemes proposed by different teams will allow basing on cross-comparison to make selection of suitable sensing method for specific applications. Many believe (Kolpakov et al. 2014) that the application of fiber-based humidity sensors will grow exponentially throughout the next decade. Initially, optical humidity sensors will satisfy specific unfulfilled applications in the chemical industry, medicine, and security systems, where the humidity control would be beneficial and present electronic sensors cannot satisfy the need. Another attractive feature of the fiber-based technology is the relative ease of integrating centralized remote monitoring and control over a number of separate facilities related to the low cost and low weight of the optical fiber cable in comparison with a copper cable. An optical interrogation module can be designed to allow simultaneous interrogation of tens or even hundreds of sensors. This can be installed in a remote office allowing the operator to monitor a set of sensors covering an area of up to a few miles in radius. In addition, recent scientific advances should allow lower cost dedicated systems by avoiding the relatively high price of interrogation modules, which are presently a significant disadvantage of fiber-based sensors.

REFERENCES

Akita S., Seki A., Watanabe K. (2011) A monitoring of breathing using a hetero-core optical fiber sensor. *Proc. SPIE* 7981, 79812W.

Allsop T., Reeves R., Webb D.J., Bennion I., Neal R. (2002) A high sensitivity refractometer based upon a long period grating Mach-Zehnder interferometer. *Rev. Sci. Instrum.* 73(4), 1702–1705.

Alwis L., Sun T., Grattan K.T.V. (2013) Optical fibre-based sensor technology for humidity and moisture measurement: Review of recent progress. *Measurement* 46, 4052–4074.

Arregui F.J., Matias I.R., Cooper K.L., Claus R.O. (2002) Simultaneous measurement of humidity and temperature by combining a reflective intensity based optical fiber sensor and a fiber Bragg grating, *IEEE Sens. J.* 2, 482–487.

Bai H., Shi G. (2007) Gas sensors based on conducting polymers. *Sensors* 7, 267–307.

Baldini F., Falciai R., Mencaglia A.A., Senesi F., Camuffo D., Valle A., Bergsten C.J. (2008) Miniaturised optical fibre sensor for dew detection inside organ pipes. *J. Sens.* 2008, 321065.

Bhola B., Nosovitskiy P., Mahalingam H., Steier W.H. (2009) Sol-gel-based integrated optical microring resonator humidity sensor. *IEEE Sens. J.* 9(7), 740–747.

Biazoli C.R., Silva S., Franco M.A., Frazao O., Cordeiro C.M. (2012) Multimode interference tapered fiber refractive index sensors. *Appl. Opt.* 51(24), 5941–5945.

Bilro L., Prats S., Pinto J.L., Keizer J.J., Nogueira R.N. (2011) Turbidity sensor for determination of concentration, ash presence and particle diameter of sediment suspensions. *Proc. SPIE* 7753, 775356.

Burgess L.W. (1995) Absorption-based sensors. *Sens. Actuators B* 29, 10–15.

Cámara C., Perez-Conde C., Moreno-Bondi M.C., Rivas C. (1995) Fiber optical sensors applied to field measurements. *Tech. Instrum. Anal. Chem.* 17, 165–193.

Caponero M.A., Polimadei A., Benussi L. et al. (2013) Monitoring relative humidity in RPC detectors by use of fiber optic sensors. *J. Instrum.* 8(3), T03004.

Carraher C.E. Jr. (2008) *Polymer Chemistry.* CRC Press, Boca Raton, FL.

Chan C.-F., Chen C., Jafari A., Laronche A., Thomson D.J., Albert J. (2007) Optical fiber refractometer using narrowband cladding-mode resonance shifts. *Appl. Opt.* 46(7), 1142–1149.

Cho T., Choi K., Seo D., Kwon I., Lee J. (2012) Novel fiber optic sensor probe with a pair of highly reflected connectors and a vessel of water absorption material for water leak detection. *Sensors* 12, 10906–10919.

Clevers J.G.P.W., Kooistra L., Schaepman M.E. (2008) Using spectral information from the NIR water absorption features for the retrieval of canopy water content. *Int. J. Appl. Earth Observ. Geoinform.* 10, 388–397.

Consales M., Cutolo A., Penza M., Aversa P., Giordano M., and Cusano A. (2008) Fiber optic chemical nanosensors based on engineered single-walled carbon nanotubes. *J. Sens.* 2008, 936074.

Eitel J.U.H., Gessler P.E., Smith A.M.S., Robberecht R. (2006) Suitability of existing and novel spectral indices to remotely detect water stress in *Populus* spp. *Forest Ecol. Manag.* 229, 170–182.

Elosua C., Matias I.R., Bariain C., Arregui F.J. (2006) Volatile organic compound optical fiber sensors: A review. *Sensors* 6, 1440–1465.

Favero F.C., Villatoro J., Pruneri V. (2012) Microstructured optical fiber interferometric breathing sensor. *J. Biomed. Opt.* 17, 037006.

Gansert D., Arnold M., Borisov S., Krause C., Muller A. (2006) Hybrid optodes (HYBOP). In: Popp J. and Strehle M. (Eds.), *Biophotonics: Visions for Better Health Care.* Wiley-VCH Verlag, Weinheim, Germany, pp. 477–518.

Gaston A., Lozano I., Perez F., Auza F., Sevilla J. (2003) Evanescent wave optical fiber sensing (Temperature, relative humidity, and pH sensors). *IEEE Sens. J.* 3, 806–811.

Gouveia C.A.J., Baptista J.M., Jorge P.A.S. (2013) Refractometric optical fiber platforms for label free sensing. In: Harun W.S. and Arof H. (Eds.), *Current Developments in Optical Fiber Technology.* InTech, Rijeka, Croatia, pp. 345–373.

Hromadka J., Korposh S., Partridge M.C., James S.W., Davis F., Crump D., Tatam R.P. (2017) Multi-parameter measurements using optical fibre long period gratings for indoor air quality monitoring. *Sens. Actuators B* 244, 217–225.

Hsu Y., Wang C., Wang Z. (2011) A remote water sensing system with optical fiber networks. In: *2011 7th International Conference on Network and Service Management (CNSM)*, Paris, France, October 24–28.

Jesus C., Caldas P., Frazao O., Santos J.L., Jorge P.A.S., Baptista J.M. (2009) Simultaneous measurement of refractive index and temperature using a hybrid fiber Bragg grating/long-period fiber grating configuration. *Fiber Integr. Opt.* 28(6), 440–449.

Jiang L., Jun H.-K., Hoh Y.-S., Lim J.-O., Lee D.-D., Huh J.-S. (2005) Sensing characteristics of polypyrrole–poly (vinyl alcohol) methanol sensors prepared by in situ vapor state polymerization. *Sens. Actuators B* 105, 132–137.

Kang Y., Ruan H., Wang Y., Arregui F.J., Matias I.R., Claus R.O. (2006) Nanostructured optical fibre sensors for breathing airflow monitoring. *Meas. Sci. Technol.* 17, 1207–1210.

Kaya M., Sahay P., Wang C. (2013) Reproducibly reversible fiber loop ringdown water sensor embedded in concrete and grout for water monitoring. *Sens. Actuators B* 176, 803–810.

Kersey A.D. (1996) A review of recent developments in fiber optic sensor technology. *Opt. Fiber Technol.* 2, 291–317.

Kolpakov S.A., Gordon N.T., Mou C., Zhou K. (2014) Toward a new generation of photonic humidity sensors. *Sensors* 14, 3986–4013.

Kondratowicz B., Narayanaswamy R., Persaud K.C. (2001) An investigation into the use of electrochromic polymers in optical fibre gas sensors. *Sens. Actuators B* 74, 138–144.

Korotcenkov G., Cho B.K., Sevilla III F., Narayanaswamy R. (2011) Optical and fiber optic chemical sensors. In: Korotcenkov G. (Ed.), *Chemical Sensors: Comprehensive Sensor Technologies.* Vol. 5: *Electrochemical and Optical Sensors.* Momentum Press, New York, pp. 311–476.

Kuang K.S., Quek S.T., Maalej M. (2008) Remote flood monitoring system based on plastic optical fibres and wireless motes. *Sens. Actuators A* 147, 449–455.

Kumar D., Sharma R.C. (1998) Advances in conductive polymers. *Eur. Polym. J.* 34, 53–1060.

Ma Y., Qiao X., Guo T., Wang R., Zhang J., Weng Y., Rong Q., Hu M., Feng Z. (2012) Mach-Zehnder interferometer based on a sandwich fiber structure for refractive index measurement. *IEEE Sens. J.* 12(6), 2081–2085.

Mathew J., Semenova Y., Farrell G. (2011) A miniature optical humidity sensor. In: *Proceedings of the 2011 IEEE Sensors Proceedings*, Limerick, Ireland, pp. 2030–2033, October 28–31.

Mathew J., Semenova Y., Farrell G. (2012) Photonic crystal fiber interferometer for dew detection. *J. Lightwave Technol.* 30, 1150–1155.

Morisawa M., Natori Y., Taki T., Muto S. (2010) Recognition of devoiced vowels using optical microphone made of multiple POF moisture sensors. *Electron. Commun. Jpn.* 93(9), 12–18.

Mosquera L., Saez-Rodriguez D., Cruz J.L., Andres M.V. (2010) In-fiber Fabry-Perot refractometer assisted by a long-period grating. *Opt. Lett.* 35(4), 613–615.

Muller J.B.A., Smith C.E., Newton M.I., Percival C.J. (2011) Evaluation of coated QCM for the detection of atmospheric ozone. *Analyst* 136, 2963–22968.

Muto S., Suzuki O., Amano T., Morisawa M. (2003) A plastic optical fiber sensor for real-time humidity monitoring. *Meas. Sci. Technol.* 14, 746–750.

Pejcic B., Eadington P., Ross A. (2007) Environmental monitoring of hydrocarbons: a chemical sensor perspective. *Environ. Sci. Technol.* 41(18), 6333–6342.

Pilla P., Trono C., Baldini F., Chiavaioli F., Giordano M., Cusano A. (2012) Giant sensitivity of long period gratings in transition mode near the dispersion turning point: an integrated design approach. *Opt. Lett.* 37(19), 4152–4154.

Puckett S.D., Pacey G.E. (2009) Detection of water in jet fuel using layerby-layer thin film coated long period grating sensor. *Talanta* 78, 300–304.

Razumovskii S.D., Zaikov G.Y. (1982) Effect of ozone on saturated polymers. *Polym. Sci. USSR* 24(10), 2305–2325.

Rodriguez-Rodriguez J.H., Martinez-Pinon F., Alvarez-Chavez J.A., Jaramillo-Vigueras D. (2008) Polymer optical fiber moisture sensor based on evanescent-wave scattering to measure humidity in oil-paper insulation in electrical apparatus. In: *Proceedings of IEEE Sensors Conference*, Lecce, Italy, pp. 1052–1055, October 26–29.

Sims D.A., Gamon J.A. (2003) Estimation of vegetation water content and photosynthetic tissue area from spectral reflectance: A comparison of indices based on liquid water and chlorophyll absorption features. *Remote Sens. Environ.* 84, 526–537.

Shu X., Zhang L., Bennion I. (2002) Sensitivity characteristics of long-period fiber gratings. *Lightwave Technol. J.* 20(2), 255–266.

Srivastava S.K., Verma R., Gupta B.D. (2011) Surface plasmon resonance based fiber optic sensor for the detection of low water content in ethanol. *Sens. Actuators B* 153, 194–198.

Sun T., Grattan K.T.V., Srinivasan S., Basheer P.A.M., Smith B.J., Viles H.A. (2012) Building stone condition monitoring using specially designed compensated optical fiber humidity sensors. *IEEE Sens. J.* 12, 1011–1017.

Wilson D.M., Hoyt S., Janata J., Booksh K., Obando L. (2001) Chemical sensors for portable, handheld field instruments. *IEEE Sens. J.* 1(4), 256–274.

Wolfbeis O.S. (1991) *Fiber Optic Chemical Sensors and Biosensors*, Vol. 1. CRC Press, Boca Raton, FL.

Wolfbeis O.S. (1992) *Fiber Optic Chemical Sensors and Biosensors*, Vol. 2. CRC Press, Boca Raton, FL.

Wong W.C., Chan C.C., Chen L.H., Li T., Lee K.X., Leong K.C. (2012) Polyvinyl alcohol coated photonic crystal optical fiber sensor for humidity measurement. *Sens. Actuators B* 174, 563–569.

Xiong F.B., Sisler D. (2010) Determination of low-level water content in ethanol by fiber-optic evanescent absorption sensor. *Opt. Commun.* 283, 1326–1330.

Yeo T.L., Eckstein D., McKinley B., Boswell L.F., Sun T., Grattan K.T.V. (2006) Demonstration of a fiber-optic sensing technique for the measurement of moisture absorption in concrete. *Smart Mater. Struct.* 15, N40–N45.

Yeo T.L., Sun T., Grattan K.T.V. (2008) Fiber-optic sensor technologies for humidity and moisture measurement. *Sens. Actuators A* 144, 280–295.

Zhang C., Chen X., Webb D.J., Peng G.-D. (2009) Water detection in jet fuel using a polymer optical fibre Bragg grating. *Proc. SPIE* 7503, 750380.

Zhang Y., Lin B., Tjin S.C., Zhang H., Wang G., Shum P., Zhang X. (2010) Refractive index sensing based on higher-order mode reflection of a microfiber Bragg grating. *Opt. Express* 18(25), 26345–26350.

Zibaii M.I., Latifi H., Karami M., Gholami M., Hosseini S.M., Ghezelayagh M.H. (2010) Non-adiabatic tapered optical fiber sensor for measuring the interaction between alpha-amino acids in aqueous carbohydrate solution. *Meas. Sci. Technol.* 21, 105801.

Index

Note: Page numbers followed by f and t refer to figures and tables respectively.

G

H

I

J

K

L

M

Q

R

S

T

U

V

W

Y

Z

Milton Keynes UK
Ingram Content Group UK Ltd.
UKHW050131071024
449327UK00029B/2541